THE X-MAS WAR

SCOTT MALENSEK

"The X-MAS War." ISBN 1-58939-281-7 (softcover); 1-58939-282-5 (hardcover).

Published 2002 by Virtualbookworm.com Publishing Inc., P.O. Box 9949, College Station, TX , 77842, US. ☐ Scott Malensek. All rights reserved.

Manufactured in the United States of America.

AUTHOR'S NOTE

This book is an anthology. It is comprised of four other books: *Black Rain For
 Christmas*, *The Secret War In South Asia, Sixth Fleet Under*, and *The Sugar-
Sweet Smell of Fear*. While each book was written as a unique tale, all of their stories take place within a single, overlapping time frame. That time frame is:
 The XMAS War.

OTHER BOOKS BY SCOTT MALENSEK

- Black Rain for Christmas, 2001,
- The Secret War in South Asia, 2002
- The Sugar-Sweet Smell of Fear,2002,
- Sixth Fleet Under, 2002,
- Black Rain for Christmas-Illustrated edition, 2002,
- The X-Mas War, 2002,
- The Ignored War, 2004
- How Did It Come To This? , 2004,
- America's War with Saddam 1990-2003, 2004,
- Saddam's Ties to Al Queda, 2005,.com
- The Weekend Warriors, 2002,
- 50+ Ways to Play with Your Paintballs, 2003,
- Reparations And America's 2nd Civil War, 2023
- Money For Mayhem, 2018,
- Tom Clancy: Between The Lines, 2023
- Dancing With Dragons: Close Calls With Nuclear Weapons, 2024,
- INFRARED: Civil War, 2024

History shows us that human conflict has far from passed. In places like Afghanistan, Lebanon, Bosnia, Rwanda, Somalia, Chechnya, and Kosovo, the 20th Century ended with incredibly brutal parting shots. Given the historical trends of conflict, and the social hatreds that remain, tomorrow's wars will undoubtedly be far more horrific than any of those in the past. The terrorist attacks and subsequent war in Afghanistan are reinforcing this idea.

Even in the face of such dangers, there are still people out there trying to put their fingers in the dike and hold back the flow of history. They go through innumerable daily hardships not for pay or glory, but in the veiled hope that they may buy the rest of us a few more years, months, days, or just hours of peace.

This book is dedicated to the men and women of
The United States Armed Forces.
May their sacrifices be remembered and
honored by us all.

CONTENTS

SECTION 1 1

SECTION 2 176

SECTION 3 315

SECTION 4 463

SECTION 5 511

SECTION 6 562

SECTION 7 708

SECTION 8 734

SECTION 1

"The origins of every war can be found in the settlement of earlier conflicts."

Saturday, October 28, 1939, 1315 hours

Washington, D.C., War Department,

U.S. Army Signal Corp

Special Intelligence Services Division (SISDiv)

Major William Meijer walked down the empty halls of an overcrowded and occupant-free government building. Except for the mountains of papers and filing cabinets on either side of the hall, he was alone. Everyone else was enjoying the fall weekend. It was their last chance to rake leaves, clean gutters, cut grass, and otherwise prepare themselves for the winter. Many of his colleagues were also using their free time to prepare for a string of Halloween parties that had accidentally been scheduled on the same night.

William had drifted from the after-hours social scene ever since his wife had divorced him two years earlier. Many of his friends had never even heard of divorce until his wife's lover, an attorney, had served him the papers. They all remained congenial at work, but the Major felt whispers as he walked past the water cooler. They weren't harsh of him. It was worse than that. He felt pitied. A fighting man at heart, the sentiment disgusted him. At least he had his work.

America wasn't at war, and there was no place for a fighting man in the peacetime army, but as far as desk jobs went his was about as interesting as possible. Besides, it was because of that job that he was certain war was coming. There was going to be another world war, and America was on the verge of desperately needing fighting men, like him, once again. William convinced himself that he wouldn't be in Washington more than another year. He was wrong. In the military, American or otherwise, the standard had always been, if you did something too well, you'll get stuck doing it forever. He knew this maxim, but duty, honor, and country compelled the West Point graduate to persevere and excel at his assignment-code-breaking.

As he drew closer to his office, William heard the snapping of a typewriter from behind the closed door. He put down his briefcase and tried the door without the keys. It was open. Inside was a young buck sergeantmaybe 17 years old at best. Sergeant Joe Barber was a Japanese-American from Hawaii.

He'd enlisted after his parents' florist was burned by a group of drunken sailors on leave. They were fueled by bourbon and anti-Japanese propaganda. All were arrested, but no one was seriously punished. Joe's family was left without income. His sister still lived in Japan so, after a few weeks of fruitless searching for work, Joe's father signed a waiver and let him enlist. Within a year this 17-year-old was supporting his parents, a new wife, and his first child, Walter.

William was surprised and caught between pride and contempt. What was this Jap doing? He'd been through all of the usual security checks, but this office contained some of the most important secrets in the world. Scrutiny and suspicion weren't just commonplace. They were a requirement; part of the job.

Joe spotted the Major, stopped typing immediately and snapped to attention. Eyes squinting and smile half-heartedly opening, William told him to carry on.

"What's going on, Joe? I thought everyone was going to at least one party tonight. Don't you have something better to do than be here? Let me guess, you're up for promotion again?"

Joe was professional and sharp in his answer.

"Sir, it's the end of the month, and I want to make sure we get all our payables taken care of. We didn't get our budget allotted until last Wednesday, so we're a little compressed for time. I hope no one minds, but I had the building maintenance supervisor let me in."

William's suspicions had been sufficiently satisfied. He knew that the War Department hadn't allocated them their department funds until last week, so there was no need to research his story. The Jap's alibi checked out.

"Okay, Joe. I'm not gonna make an issue out of it. You really shouldn't be in here alone, though, even if it is on the up and up. People could talk, and I

think you know how delicate your being here is. Okay?"

"Yes sir. I'll make sure I'm not alone anymore. Will you be staying long?"

"Nah. I think I might actually check out one or two of those parties tonight. How 'bout you?"

Joe's face had been soldier-like until this point, but William detected a hint of frustration when he answered a simple "No sir". The Major tried to remain as open minded as possible when it came to racial matters. He knew that he had often misjudged people just because of the color of their skin, and finally, he realized that he was doing it again. William knew Joe to be a solid American and a good man despite his Japanese parents.

"Well, I'll tell you what. Let's both cut outta here around dinner. We'll stop by Colonel Freidman's little get-together to be social, then we can call it a weekend. Okay?"

Joe relaxed and finally made direct eye contact.

"It's a deal, sir. Can I get you anything? Coffee?-"

"No thanks, sergeant. I'll be in my office. I wanna go over some paperwork too."

William opened the door to his office. It was a mass of scrolled blueprints, stacked documents, books, and small unusual looking pieces of machinery. There were no windows, just his desk, a chair and the obligatory empty filing cabinet buried under the mess. He put down his briefcase, took a deep breath, loosened his tie, and began filing. After all, he really didn't have any pressing work to do. He just had to get out of the house so he could feel like he was doing something, anything even remotely productive.

William's official duty was to serve as a liaison officer between various other intelligence operations-specifically foreign, code-breaking exercises. Technically he was part of Operation Magic-a joint U.S. Army and Navy program to decipher Japanese diplomatic and military codes. He also routinely worked with a combined U.S. Army, Navy, and British Intelligence operation called Ultra, which was working on German codes.

The afternoon slipped by slowly at first, but once his filing was in a rhythm, it passed a bit more quickly. Just before 1500, he heard the phone ring. Joe's sporadically smooth typing ceased as he answered it, but William didn't pay too much attention. If it was a call for him, Joe knew exactly where he was. William heard Joe's conversation end, then came the knock. It was three simple taps that would unwittingly lead to the opening of untold doors in the years to come.

"Major, I'm sorry to disturb you, but it's a Lord Blankenship calling all the way from England-some place called Bletchley Park?"

William recognized Bletchley Park as the headquarters for the most secret of British intelligence projects-including deciphering of foreign codes. Lord Blankenship was their liaison officer-his counterpart. Like William, Lord Blankenship wasn't a career code-breaker. Both men left that to people who were decidedly...unique. Before Joe could close the door or acknowledge William's nod, the phone was to his ear.

"Lord Blankenship, what a surprise. How are things over there in jolly 'ole?"

"Everything is simply fine over here, Major. Is there something we should know about?"

"No, sir. I'm sorry. I was just trying to-"

"Ah yes, American greetings. Forgive me, major. I forgot myself for a moment there. Actually, I must say things are very interesting right now. Very interesting."

With the awkward "hello and how do you do's" out of the way William's curiosity climbed. The British were notorious with their lack of enthusiasm. Surely something was going on.

"Well, you've certainly gotten my attention, sir. How can we help you? It must be something big for you to give us a call in the middle of the night."

"Yes, that's right. It is only afternoon on the other side isn't it? Well, we've run into a bit of terrific luck over here. I'd like to discuss it in detail with you as soon as possible. In fact, there's a bomber fueling up right now to bring me across

tonight. The plane should be ready in about an hour. It seems we've got our hands on one of their little gadgets over here."

William knew that by "gadget", Lord Blankenship was referring to a long-sought-after deciphering machine that the Germans were using to send and receive coded messages. Still, it was too incredible to believe.

"Did I hear you correctly, sir? Did you say you've actually got one?"

His voice was slightly confused with a tinge of disbelief at the apparent good fortune-incredibly good fortune. Lord Blankenship, however, remained British, restrained and nonchalant.

"Yes, Major. It's actually quite the coup. It seems our Polish counterparts made it out of Warsaw a few weeks ago, but they got stuck at the port of Hela on the Baltic. When the city finally fell, they made it out one of the last boats. It's really only part of their little odyssey. In fact, they ran into all kinds of trouble just getting here."

William was shocked. The Polish military started the war with 800,000 troops. In only a few weeks they lost over 100,000 killed, tens of thousands more wounded, and the remainder captured. In fact, the Germans had cut through them like a hot knife through butter. Most everyone in had shrugged it off as Polish ineptitude, but if what Lord Blankenship was claiming was true, the Polish Intelligence service had actually captured one of the mysterious German deciphering machines. The British and Americans had been trying for years!

There was a moment of silence while everything settled into William's mind. Then Lord Blankenship continued.

"You see, Major, our Polish friends here have had a great disdain for the Germans ever since they leveled Warsaw. I can't say I'd blame them though. Can you?"

"No, sir. I don't think anyone expected that aircraft alone could do that kind of damage. And who would give such an order against civilians? Imagine, airplanes destroying a city in only a few days. I saw the newsreels, and it looked pretty bad. Now, you say they actually grabbed one of the gadgets, sir? I mean, did I hear you correctly?"

"Well, actually, major, they claim to have overrun one of the Hun's division headquarters outside of Krakow. Gerry was able to destroy his machine, but they did capture a great deal of documents, including blueprints and something else that we think you'll find extremely interesting. Apparently Gerry has passed on one of his machines to the Japanese. Our Polish friends were only able to get out with some of the plans that they captured, but there are a few blueprints that show modifications to the German machine, and those drawings are detailed in Japanese. It looks like the Emperor is planning to make a similar machine for his messages. We're making copies as we speak, and my little middle of the evening journey across the pond will begin as soon as they're ready."

"Quite right, your Lordship, but you did say you've actually got a gadget, sir?"

"Oh yes. That's what we've all found to be one of the most incredible parts of this little tale. When our Polish friends were hemmed in back at Hela, they found themselves a machine shop and had one made. We were, of course, concerned that the Hun might be able to find out from those machinists, but not to worry. They were all killed and the shop was destroyed by Gerry's artillery in the city's last days. Awful shame for those workers, but top-notch luck for us, I must say. More to the point, Major, the Poles seem to have had an easier time of getting that blasted machine across the Baltic than they did those documents and plans. We don't have a complete set of prints for the Japanese gadget, but everyone ever here thinks that you'll simply be thrilled when you see it tomorrow. I'm scheduled to land in Washington just before noon. Could you be a good chap and pass on the word to some of your people for me? I'd like to get back here as soon as possible. As you can imagine, we've got a lot of work ahead of us."

"Lord Blankenship, you can rest assured that there's going to be twenty very excited people waiting for you when you arrive. Would you pass on our thanks to everyone who was involved in getting this information? I'm sure we'll make very good use of it and thank you sir! Please, take care, and know that we'll be waiting for you hours before you even get to this side of the pond."

"Right, Major. I will see you tomorrow. Goodnight."

In one indiscernible motion, William hung up his black iron phone and opened the door. Joe had restarted typing, but when the door flew open, he sensed it was important and snapped to attention at the mention of his name.

"Joe, I want you to personally call each and every officer involved with Operations Magic, Ultra, and especially Purple. Tell them Lord Blankenship called. I spoke with him, and I'm the one who ordered you to call them on the weekend. Tell them Blankenship is coming in tomorrow with a gadget. I'm calling a meeting for tonight at...1900 hours. Is that clear? Call all of the officers. Don't call any of the enlisted personnel even if another officer tells you to. If they've got any questions, I'll tell them everything myself. And one other thing...no one, I repeat NO ONE other than those officers on the phone list that I've mentioned is to ever hear about this conversation or my call from Lord Blankenship. Consider both of those to be of the utmost top secret classification. Is that clear?"

"Sir, yes sir. I'll get to work on it immediately. 1900 Hours. This office, sir? Wouldn't you prefer to use Colonel Freidman's office?"

Colonel Freidman was in charge of Operation Magic: a tireless, completely devoted, some would say obsessed officer.

"You better call him first, Joe. I'd like to use his office. It's got the most chairs. In fact, I think I'll call him myself. Just go ahead, call everyone else, and tell them to meet here. I'm sure Colonel Freidman will let us use his place, but he might have a better idea. I'll let everyone know where the meeting will be held when they get here."

William closed the door and began digging through his papers trying to find Colonel Freidman's home phone number. Joe knocked and opened the door at the same time.

"Here's the Colonel's number, sir."

He knew William would never find it his piles of papers, but he didn't want to embarrass his boss.

"I just happened to have it in front of me, sir-thought you could use it."

Joe smiled, closed the door and began his phone calls. The British and Americans had been trying to break the German and Japanese codes since 1926, and in one fell swoop, a handful of Polish refugees made all their work useless. It would take years to fully crack either of the Axis nation's codes, but by December 7, 1941, American and British code-breakers would be able to forecast the imminent war with Japan. Years later, some would even allege that at least the British would know of the Pearl Harbor attack before it happened. By late 1942, after countless commando operations and lucky chances to grab more of the German Enigma machines, the allies would be able to secretly monitor all German and Japanese coded radio traffic almost without exception.

Tuesday, June 25, 1940 Stow Herald FRANCE FALLS TO NAZIS!

A second armistice between German and France was signed yesterday. France has officially fallen after losing almost 100,000 troops while only charging the invading Germans with 40,000 casualties. The cost for both sides is still a far cry from the millions that were killed, wounded, or left homeless in the attrition-style trench warfare of the First World War. After a little more than twenty-odd years Germany has finally defeated France.

Monday, July 21, 1941

Chicago Times

Japs *Move into Indochina*

Forces from the Empire of Japan will be formally moving into the one-time French colony of Indochina. Prior to the fall of France a year ago, the Yunnan railway running from Haiphong harbor to Kunming in southern China served as a major supply route for the Chinese in their fight against invading Japanese forces. Since the creation of the German-controlled Vichy French government, Japan has encouraged her Axis partner to break China's supply chain from Britain and the United States. Today, after months of pseudo-negotiations and legal leapfrogging Japanese forces were "invited" to use the cities of Cam Rahn Bay and Saigon as naval bases. This effort comes with little surprise since Japanese aircraft have been operating from airfields along the northeastern and southeastern portions of the former French Colony for several months.

Friday, September 26, 1941

Miami Daily

<u>**U.S. Puts Pressure on Japan**</u>

As a result of the Japanese entry into Indochina, the United States has officially threatened to cease and block all exports to Japan. Japanese aggression has been met with increasingly harsh diplomatic response ever since July 1937 when Japanese military forces that were allegedly on maneuvers clashed with nearby Chinese troops.

An embargo of this nature-including vitally needed oil, tin, rubber, food, and other basic resources-might well lead to a war between the Empire of Japan and the United States of America. The Island Empire in the east currently draws at least 80% of its resources from the United States. Sources in the U.S. State Department denounce talk of war commenting, only that the embargo is simply the line on the map that the Japanese have been looking for. Most seem to believe once the Emperor knows just how long he can ignore the United States, it is expected that the Japanese will withdraw from Indochina and seek other supply sources.

Economically speaking, the loss of markets in Indochina to the Japanese will do little to hinder

President Roosevelt's New Deal which continues to pull the U.S. from the grips of its 12-year old economic depression. However, the embargo against Japan will significantly affect west coast exporters until new markets can be developed. One San Francisco shipping company manager explained that the embargo is typical of overseas shipping both on the west coast with Asia, and even more so on the east coast with Europe and the Nazi submarine-infested waters of the Atlantic.

Sunday, December 7, 1941

Cleveland Press EXTRA <u>JAPS BOMB HAWAII!!!</u>

Months of back-and-forth political pressure over the Japanese occupation of China and French Indochina finally came to a crescendo yesterday. Faced with the threat of an imminent embargo of raw materials-including precious oil Japanese naval Forces conducted several air raids on the U.S. Navy and Army facilities around Pearl Harbor.

Witnesses claim that the sky was filled with hundreds of bombers. The U.S. fleet was caught completely by surprise and thousands of sailors died in the early morning sneak attack. U.S. Army Air Corps planes were also caught on the ground, and at least a hundred aircraft are reportedly destroyed. The Navy Department has declined to list specific losses, but at least three battleships have been sunk in the shallow waters of Pearl Harbor. Rumors persist that almost every ship was damaged, burning, or sunk.

President Roosevelt went directly to Congress and a few short hours later, it was declared that a State of war now exists between the Empire of Japan and the United States of America. Many believe that it is only a matter of time before Japan's Tripartite Treaty partners-Germany and Italy-come to the Emperor's aid and declare war on the United States. Most everyone acknowledges that the spreading World War has finally pulled in the United

States.

Saturday, August 8, 1942, 1153 hours

Savo Sound

(Between Florida, Tulagi, and Guadalcanal Islands)

U.S.S. Quincy

Tom Chamberlin watched the dark tropical horizon silently, waiting for the inevitable. He wasn't on lookout duty, but the word around the ship was that a Japanese Task Force was coming down the Solomon Islands; "The Slot." Odds were that the Japanese were looking for a fight. His ship, the Northampton class cruiser U.S.S. Quincy, was one of the ships supporting America's first amphibious operations since the Spanish-American War. Yesterday morning U.S. Marines had landed on Tulagi, Florida, and Guadalcanal Islands.

For the most part, the landings went unopposed. It was a good thing too. The one and only practice landing had been cancelled because there were too many accidents and personnel injuries. 19,000 Marines went on the combat stage with no rehearsal, and, thankfully, only a small crowd of Japanese attended. Where were they? Tulagi only had squad and platoon level defenses, backed up by a few heavy weapons. The landings around Florida Island were stalled by a handful of well-armed, well-entrenched, and well-surrounded Japanese, but that was only for a few hours. Guadalcanal, the big island and the centerpiece of the entire operation, was disturbingly quiet.

The attack's main objective was to secure an airfield that the Japanese were about to finish building on Guadalcanal. By the end of the first day, the Marines had secured their position on the beach, moved inland, secured the airfield, and setup a defensive perimeter. Only snipers and small units had opposed them.

Last year, Tom's ship had been in the icy, winter-waters of the North Atlantic escorting "neutral" American ships to Great Britain during its Blitz from the Germans. After Pearl Harbor, she'd been ordered to the Pacific to help bolster the shattered fleet.

Tom felt the soft Pacific breeze wash over him and watched the typical glory of yet another sunset fading into night. The stars were coming out of their deep sleep, and he could smell the jungle flowers on Guadalcanal a few miles to the west. Small roaming rainsqualls meandered through the bay.

Even with the excitement and long hours over the past few days, he couldn't sleep. Many of his shipmates didn't have that problem. There was a feeling of security and safety in these waters. The Marines on the islands meant that the Japanese were, at least for now, inactive. Tulagi and Florida islands would serve as protective barriers from the east. Savo Island, a rocky mound in the middle of the ocean, split the northeastern side, and the U.S. Navy's three aircraft carriers were guarding the south. There were also four other allied cruisers in the bay, and several destroyers to protect the transports supplying the Marines. If the Japanese were coming down the slot, they were moving into a geographical bear trap.

The doors leading below decks had been left open to let the cooler evening air inside, and, around 1:40 in the morning, Tom heard some commotion on deck. He went with some of his friends to the gun deck to see what was going happening. Two small seaplanes, like the ones that ships launch for reconnaissance, were flying at medium altitude over the bay. Everybody commented on what idiots they were flying around with their identification lights on. If the Japanese sent down a night fighter, they'd be sitting ducks. A few minutes later, one of the planes dropped a green-white illumination flare to the south, near the cruisers, HMS Canberra and U.S.S. Chicago. A friend of Tom's on the bridge called down to him, and saw the looks on growing crowd's faces.

"Relax! They must've seen a Jap sub or something!"

No one knew for sure what was going on because a rainsquall was drifting between the two groups of ships, and only the green-white flickering suggested any trouble. Three minutes later, they could see flashes of gunfire through the rain followed by echoing thunder. Lightning, muzzle flashes, explosions, flares, and searchlight illuminated a small raincloud on the horizon.

Captain Moore sounded the General Quarters alarm on the Quincy, and men ran for their battle stations. Before most could get below deck, the small Japanese Task Force slipped through the rain, fog, and smoke closer to Quincy. More searchlights came from the Japanese ships, this time illuminating Tom and his ship.

The cruiser U.S.S. Vincennes had been moored in front of the Quincy and it was already moving straight toward the Japanese. Quincy and the cruiser U.S.S. Astoria followed. Great towers of water splashed all around as Japanese shells came closer. After only a few misses, there was a huge explosion on the fantail of the Quincy. Another followed directly on the bridge. More explosions and near misses continued. Tom was making his way aft, but a series of explosions amidships stopped him in his tracks. Each felt like it was blowing the ship apart. The sound was deafening. He could hear the screams of his friends above and below decks. Several fires started, and the Japanese, by now at point blank range, had a well lit, burning target.

Tom could hardly breathe through the smoke and concussion. He tried to run to the starboard side, to head aft from there, but when he got to the other side of the ship, a torpedo hit the starboard aft section. The shockwave threw him high in the air, and he slammed into one of the starboard secondary gun mounts. With only the wind knocked out of him, Tom stumbled to his feet, uninjured. A quick check for all his body parts reassured him until another shell impacted on the port side. Pieces of bodies, and slippery black blood that flashed red in each explosion seemed to be everywhere. Tom fell flat on his back again. The explosions faded, and the numbness from the blasts comforted him as he drifted into unconsciousness.

He awoke coughing and treading water. Not even a hundred feet away, the Quincy started to roll over. He swam to get away from any suction that the ship might create when it went down. There was one more huge explosion, and then the bow went under. The stern rose high and turned upside down in the air. With propellers still turning, the Quincy dove for the bottom.

To the south, the cruisers Canberra and Chicago were burning. Vincennes sank about fifteen minutes after Quincy. Astoria was nowhere to be seen. The U.S. Navy had suffered the worst blue water defeat in its history.

Tom found himself alone in the battle. He could hear explosions and men calling out in the distance, but wherever he swam, he found no one. All of his friends and shipmates were gone. He was alone. He wondered if that was how he would die, and how long would it take?

Hours after losing his ship without ever having reached his battle station, the destroyer U.S.S. Elle rescued Tom. In the morning, the Elle would have to fire on the Canberra to scuttle it, but until then, she crowded her decks with survivors, and picked up more while crewmen stood watch with rifles to keep away sharks.

Tom found a shipmate of his by the name of Harris. He was so happy to have found Tom that he demanded Tom follow him below decks to give him a present. They made their way down to the engineering section. There, Tom met an engineer who offered him a cup of coffee from one of the pots brewing on a steam line. Cold, wet, and exhausted, Tom tested its temperature, and took a healthy sip only to realize that it was filled halfway with bourbon. The entire watch laughed until an officer from one of the upper decks came down to see what was so funny. Everyone kept it a secret, but Harris was right. It was the best goddamned cup of coffee he had ever had.

He had finally come to experience the war. Most of his friends were gone. Some, he remembered, died right in front of his eyes on the *Quincy,* but most were simply never seen again. There was a vacuum of loneliness about him, but it was nothing like the feeling of separation that he had felt while treading water for hours. At least he was on a ship with other sailors. He was alive, and he was not alone.

Tuesday, January 15, 1991, 5:47 PM

Copley, Ohio

Tom fluttered his wrinkled 68-year-old eyes and realized it was yet another memory of the war. He could taste the bourbon and smell the scents of battle: cordite, saltwater and Guadalcanal's sweet honeysuckle. Even though the sensations and memories were as clear as when they had actually happened, a quick glance around reminded him that he was not floundering in Iron Bottom Bay, but rather at home watching television with his grandson Johnny.

Johnny was 16, and he really didn't need a babysitter, but Tom's daughter and son-in-law both needed to be away on separate business trips at the same time. To make matters worse, Johnny had recently fallen in with the wrong crowd. His father had caught him smoking one of those marijuana cigarettes.

Johnny needed an adult who could actually be involved in his life. Although Tom could put a good deal of blame on his son-in-law, he had to recognize that his own daughter was also working too many hours. Women's liberation was great for women, he often wondered what it did to those who depended on women? Behind every great man is a great woman keeping him up. There was a time when a woman could take pride in being a housewife. Tom was living in 1991 and 1942 at the same time.

Johnny Chamberlin was coming off a terrible week at school. Monday, he'd failed his Trigonometry exam. Tuesday, his girlfriend of nearly three full months told him that she wanted to see other people, but she also wanted to remain friends. Wednesday, he was reminded that his Physics midterm, due on less than two weeks, was to be a five-page report, typed! Thursday, his dad noticed a door ding in "the family's" new Dodge Daytona, and his driving privileges were revoked indefinitely. Today, he failed a pop quiz in British Literature, and was reminded about both a test in his U.S. History class next Monday, and his Photography class assignment due next Tuesday. His world was closing in on him from all directions, and it was time for the weekend!

Normally he would be getting ready for another night of watching videos at Michelle's. She wanted to see other people now, and all the rest of his friends already had plans for the night. He was too young to go to a bar, and since he wasn't allowed to drive anywhere but to school, Johnny was destined for a night in front of the TV. At least the war was on. In fact, you couldn't get away from it. Every channel on TV and radio was talking about the U.S.

forces bombing Iraq.

Wednesday, January 16, 1991, 4:17 PM

Undisclosed village, People's Republic of China

While the United States, and its coalition allies, were attacking Iraqis in Kuwait, a small convoy of unmarked limousines and military trucks made its way down a distant dirt path that a map alleged to be a road. Walt Barber, A middle aged businessman, road in the lead limousine to the ancient village. Sipping his tea, he gazed out a tinted window at the muddy streets and cinder block buildings of his destination. A military technological revolution was coming to fruition in the Gulf, and no one would be able to escape its future presence. Walt warmed his cup in the car's microwave, and realized that even this birthplace of civilization would be set on a collision course with destiny today.

The car slowed to a stop behind the three escort trucks: two Chinese, and one North Korean. Escorting troops jumped through the canvas at the rear of their trucks and assumed positions inside and outside a small church. The meeting was about to begin. Walt was early, as usual. The Chinese didn't like outsiders inspecting a meeting that they had prepared, but, since he was the man responsible for the occasion, a good deal of tolerance was given. It was his special relationship with so many powerful men around the world that had enabled him to set up the meeting.

Inside a church, where bread and wine had been consecrated into the body and blood of Christ, a buffet of roast pig, fish, rice, vegetables, and tea was carefully being put together by some of the Chinese troops who had come from the second truck. Walt inspected the food carefully. He sampled the vegetables, and, true to his vegetarian lifestyle, ignored the meats. Even in this dusty old church, a presidential reception was quickly being prepared in front of his eyes. As the final arrangements were being made, more and more troops took their posts. Walt inspected them himself, as though it was his place to tell Chinese soldiers how to appear or present themselves in front of their ruler.

Outside, he could hear helicopters approaching from the north. A few seconds later, more helicopters thundered through the valley from the west. The meeting was about to begin. Seven Chinese Mi-8 helicopters appeared from the north and landed single file in the street to the left of the small church. While the dust swirled, their troops, over a hundred, ran from the helicopter doors into defensive positions throughout the village. Before the last troops were unloaded, the three North Korean helicopters from the west began circling above. They too landed single file in the street, but this time to the right of the small church, facing the Chinese helicopters. Again, troops ran from helicopters and into the town. These were North Korean troops. A handful of Chinese and North Korean officers casually approached the church, and Walt leaned on the doorway waiting for them to climb the few steps toward the portal.

Walt spoke several languages, including various Chinese and North Korean dialects. Everyone knew this, but they chose to address him in his native English. Each formally introduced himself with as much military decor as possible. Each wanted to show as much national pride as possible. It was a matter of honor. Although the introductions and resume reciting was brief, it demanded enough attention for everyone to have forgotten the slight that Walt had done by waiting for them to introduce themselves. He was, after all, only an advisor, and held no substantive office or title anywhere in the world-save confidant, advisor, assistant, and investor. In truth, he was somewhat of a kingmaker.

The Chinese officers opened the introductions.

"I am General Sadis of the People's Army. This is General Ahr Phuhn of the People's Air Force, and Colonel Naht Synn, Special Security Commander for the President. We are pleased to finally meet you, sir. Your reputation is well known and respected."

With military precision in timing, the North Korean officers followed.

"Mr. Barber, I am General Soo Pi of the North Korean Air Force and this is Commander Tong Yoo of our Special Naval Operations Branch. This meeting is long overdue. We are very pleased that you have been involved from the beginning. It has surely been a logistical feat to have-" Walt interrupted.

"Pardon me, please. I'm sorry to interrupt your thanks, sir, but, as you've pointed out so well, this has been a difficult effort, and it is far from over.

Rather, it is only beginning. If you would all follow me, I will show you the arrangements inside. We have much to do."

With the pleasantries ended, the men were shown the interior of the church where the meeting was about to begin. They all raised the noses, nodded their heads, and dressed their faces grimly to show reluctant satisfaction. An officer who has trained for hardship at the front might appear soft if pleasure was shown towards comfort that was being created.

Outside, more limousines began arriving. More security men, this time all wearing suits, low level government officials, businessmen, and finally, with great family like fanfare, the patriarchs arrived: Deng Xiaoping, leader of the People's Republic of China, from a limousine to the left of the church, and Kim Il Sung from a matching limousine to the right. According to the protocols that were agreed upon beforehand, the head security men from both nations met on the steps of the church and, as one, motioned for their leaders to exit. The two old men left their respective vehicles at the same time, and with the help of their personal secretaries, were walked to the steps of the church. Decades of responsibility forced them to move slowly and take a good deal of time. In order to begin the dialogue as soon as possible, Walt walked down to greet them. The crowd of officers trailed, while the lower level officers and businessmen waited at the top of the steps. There was a lot to cover, and everyone had very busy schedules.

"My friends, it has been a long and difficult journey, and in the heat of the day. May I offer you some of the refreshments prepared for you? There's a great feast arranged in the building. Even if you're as hungry as you must be thirsty, there will still be plenty."

"Walt," Deng responded, "The sight of everyone gathered together has filled this old spirit with energy, but some tea would be welcomed. Would you like to join me, Kim Il Sung?"

Through Walt, acting as interpreter, Deng and Kim had conversed many times on the phone and in confidential written correspondence. Eccentric as either man was, they could still feel comfortable skipping the titles and using proper names.

"Comrades...let's relax for a while before we begin. Tea sounds wonderful."

They climbed the few steps that the church offered, and thanked Walt as they moved. Both sat down at their prepared locations, drank their tea, and caught their breath. They spoke of their families, mutual friends, the weather at home, and finally current events. Walt was artistic in his control of the conversation's topics. It was a skill that hadn't existed until his days at Georgetown University's law school.

The war in the Persian Gulf wasn't a direct concern to anyone present, but Walt had convinced both men to meet for that very reason. When current events entered the conversation, Walt really went to work. He explained how, even though neither nation was anything but remotely involved, the economics of the situation were about to create a waterfall effect on the communist world.

Walt moved the topic to his primary purpose for this meeting.

"The Soviets are on the verge of collapse, my friends. They've tried their best to stop this war, and everyone has ignored them. Their international respect is evaporating like a summer rain in the city. If it weren't for their ICBMs, no one would even pay them the attention that they're groveling for right now. Afghanistan cracked the Soviet steel shield of security, Azerbaijan has punched a hole, and I expect that their fruitless efforts to negotiate a solution in Kuwait will bring them down. I tell you both this, in the greatest confidence. I have spoken with members of the Soviet military who are whispering of another revolution. This time, it will be a revolution to preserve Socialism. If they fail and the Soviet Union collapses, your two nations will suffer. There will be nothing to stop the west from putting your people out to the fields in the next century. International respect will fade, and your people, along with most of Asia's people, will begin to feel the lure of the west's decadence."

Kim was confused. "Why are you telling us about a revolution in the Soviet Union? Mikhail's not a strong leader from the revolution, but even if he were to let his guard down, the Politburo would never allow the west to overpower the U.S.S.R. Seriously old friend, don't you think you're overestimating the effects of Glasnost? This is not like you to be so reckless and..."

"Forgive me Kim Il Sung, but I must ask Walt Barber, have there been whispers of revolution in our countries? Are we to be abandoned by our own military too? What shall we do?"

"My friends, I've heard no talk of revolution in my travels throughout your people's homes. I have tried to warn the Soviets of their weakness, but my cries have fallen of deaf ears. Even the Politburo is blinded by the past. They still see the might of Stalin and Khrushchev. The Kremlin walls are blocking their view of the rest of the world. Mikhail's been kissed by the West, and whether or not he realizes it, he's falling in love with the trappings of capitalism. To answer your question of 'what shall we do?' Well, I had hoped we could meet and develop some possible plans. There is a great deal of turmoil ahead in the west. It may weaken either, or even both of you, but it may also be an opportunity."

Several of the businessmen who had arrived and conglomerated near the church's front door began to drift towards the table where two kings and a kingmaker were sipping tea. Walt began another round of introductions.

"I would like you to meet some people who have some ideas on how we may prepare for the coming storm. These men are prepared to use their influence and positions to try and balance the world's markets if the Soviets do crumble and the coup should fail. Obviously, investment capital, expenses, fees, and commissions would need to be addressed. Until your nations develop further in terms of financially liquid assets, these men would be willing to invest their own money in your people with the understanding that any profits from their investments would serve only as interest on a debt. Direct principal payments would be made over a period of 100 years, at a rate of 1% a year. The financial and market backing of these men, coupled with the willpower of your nations, ensures that any future economic vacuum left by the Soviets will easily be filled in less than 10 years. In addition, the structure that would be created to fill that vacuum will bring economic resources, wealth, industry, technology, and world influence to your nations."

The two old men trusted Walt as much, if not more than anyone else, but faith in mankind had weathered both over the decades. They started to look curious, confused, and finally skeptical. Walt could feel it as well as see it. He paused to let their minds wander then catch up.

Deng Xioping had some questions.

"I believe you are suggesting additional capitalist investment in our countries, but I've tried for years to make my people's lives better through the same method. Why would these men succeed, and why are they so eager to undertake such a difficult and even dangerous journey?"

Walt expected questions, and this was one he'd rehearsed. "It's true that these men, and there are others who we will speak with later, are already very wealthy, and it's true that they have no interest in the people's revolution. Even with all that they have, they remain capitalists at heart. It's that self-preservation that's driving them to confront their competitors on all fronts. There are hundreds of competing banks, investment companies, trading companies, and manufacturers who built post-World War II Japan into an economic powerhouse. Those same people are spreading their interests into other nations like Indonesia, Thailand, Southern

Korea, Nationalist China, and the list goes on. These men behind me don't want to help the revolution at all, but they do want to prevent their competitors from controlling the world. Their competitors are the ones we must watch. These men want a long-term investment which will secure their futures, and the futures of their families."

The old men understood how western capitalism was taking over the eastern half of the world. They also understood the danger of capitalism in a strict communist society. Kim Il Sung, the Great Leader of North Korea (many would say dictator), had more questions.

"If the Soviets are about to collapse because of capitalism's influence on their people, and the Chinese have already tried to open their markets to the western capitalists already, how can we hope to maintain our internal and external security? How can we survive and succeed where Mikhail and Deng have already tried?"

"Comrade Kim," Walt answered, "A centrally-governed capitalist society can indeed exist in a socialist country if it is very carefully monitored and steered. Unlike the other capitalists who have worked with the Soviets or the Chinese, these men will work with the central government to make sure that things do not spill out of control."

Walt saw that Kim Il Sung was wavering so he tried to explain further. "Other investors would do anything to get their goods sold, but these men want to work with you to protect their investments while using the finished goods to attack their competitors' economic base. They understand that as long as a strong socialist government exists, their investments are protected. Your central governments will still monitor and control prices internally. Exported goods will be controlled by market prices, and even those can be raised or lowered by taxes, tariffs, fees, and subsidies from the central government. This kind of arrangement has never been attempted before because the capitalists were never able to work out an investment agreement like the one that we are discussing right now. In the corrupt nations of the west, everyone in the government takes a piece of their profits; here, they will spread some of their wealth, but the central government will control how those funds are distributed."

Kim and Deng had become cautious men with time, and they seemed convinced, but Walt knew each would have to discuss his idea with other advisors and again with each other. It was a complex and secretive plan, so Walt had Dan, his personal assistant, prepare written proposals for both men's advisors. While they talked, Dan passed out the proposals and held a private question and answer session so that everyone understood how the great worker's revolution was going to be saved by cutting a deal with some shady, international capitalists.

Walt continued. "Now, although I'm nearly certain that it's already too late, the Soviets have not fallen. If these men and I are wrong, then there is nothing to think about or plan for. If we are right, the time to begin preparing may be right now."

He explained his plan in further detail, then deftly shifted the conversation over to the war in Kuwait again, and finally to life at home in the two old men's capital cities. Politics, even for the most weathered and savvy of politicians, brings a headache to anyone if spoken of too seriously. Fueled with complex accounting and talk of international financing, it was a guarantee for internal cranial crushing.

It was more interesting and relaxing to reminisce about the Great March with Mao, or the war for reunification on the Korean Peninsula. Victory and defeat both held fond moments in the old minds of the two communist dictators. The meeting continued after darkness fell. Their political conversations had only taken a few minutes. The rest of their talks were concentrated on old memories. Walt started to recognize that the two communist legends were subconsciously saying their final good-byes.

By 7:30 p.m. they were slowly escorted to their limousines, and then home. Dan Greene, Walt's personal assistant, had remained in the shadows, socializing with other businessmen. He made sure cups were kept warm and filled while his boss served as interpreter, advisor, lobbyist, confidant, and friend. Without Walt nearby, Dan was somewhat empty. Walt was more than a friend or employer. In his mind, he was the better half of a team in which he was proud to be a part.

A cup of tea in hand, Dan walked down the steps towards the dust cloud where the limousines had been. Walt's silhouette was a thin shadow in the swirling beige.

"Sir, if I may?"

"Thank you, Dan. I think things went as well as could be expected. Don't you?"

"Yes sir, but do you really think all this is going to be necessary? Is the future already so well planned that it can't be changed?"

"It better be, Dan. We've invested more than just fortunes on those Soviet Generals. If we're playing cards, we've just asked the dealer for four new cards, and someone's gonna call us real soon if they're not good enough to beat everyone. They better come through. This economic plan with Deng and Kim is a good one, but it might be too optimistic to fully succeed."

Dan knew about the special arrangement that Walt had created with the Soviet generals, but something told him that Walt hadn't prepared many contingencies should they fail. Walt was normally systematic, even methodical about having several backup plans if things didn't go the way he wanted. Aside from the negotiation with the Chinese and North Korean, there were no more alternative plans in the works.

"Dan, get in touch with Tony Kerns over at GTI (Global Technology Inc.). I'm sure he's a busy man right about now, but I want to meet with him as soon as this Kuwait thing's over. There's gonna be a lag between the Soviet and the Chinese events. Tony may be able to help us put something together for the interim. Also, schedule a meeting with our friends at the U.S. Department of

Agriculture. Kim's gonna try and fight this thing as much as possible, and he may need to be nudged a bit later on. Besides, I think we'll at least need to get a little more proactive in how Kim gets his rice during the next few years."

"Yes sir. Should we get tech guys to work on that NYSE system-hacking issue?"

"No. Have them run some tests. We probably won't need them to do anything for at least another year, but some smaller transfers may help our operational costs, and also serve as good tests. I want them to focus more on that Strategic Protocol software for the Pentagon. It might wind up being worth more than the entire NYSE if it works as advertised."

His instructions passed on, Walt and Dan stood silently as hundreds of communist troops appeared in small groups and boarded their helicopters to leave. In a few minutes, everyone was gone, and only the men in suits remained. As the dust settled, and the staccato of helicopters faded in the mountain valley, they made their way inside the church for their own meeting, followed by a buffet dinner; compliments of the Chinese People's Liberation Army.

Thursday, January 17, 1991, 2:19 PM

Moscow, Union of Soviet Socialist Republics

In an unobtrusive building along the Moscow River, Soviet defense Minister Dmitry Yazov climbed the stairs to an empty hallway on the third floor. His aides and security team waited outside the doors to the building. A few steps down the corridor to the right, and he found his destination, room 316. Defense Ministers rarely needed to knock on a door in Russia, but the man adorned in medals, gold trim, and a perfectly pressed gray uniform waited patiently for an answer to his four taps. The door opened, and Dan Greene invited him inside, graciously. They passed through a lobby where a young woman toggled through constantly ringing phone lines. After the lobby, they walked through a small corridor, past the kitchenette, and to the open office doorway of Walt Barber.

Dmitry was used to dealing with powerful men in ostentatious 18th century rooms adorned with gold leaf. Government offices were sometimes mistaken for the palaces that they once had been under the Czars. Walt sat behind a typical bureaucrat's desk with nothing on any of the walls expect a map of the Earth, a few awards, degrees from various universities, and a small painting of Thomas Jefferson. While Dmitry glanced at the unusually austere surroundings, Walt rose from behind his desk and welcomed him in perfect Russian with the vice-like handshake that all military men require.

"Welcome, Defense Minister! Things are well I trust? Especially now that the Americans are so preoccupied, eh?"

Dmitry's stern face, stereotypical of a nearly 70-year-old general, cracked a wide-open grin. With the war in the Persian Gulf the Soviet Military could breathe a little easier. There was no possible way that even the Americans could conduct military operations while they had such a concentration of forces in Saudi Arabia. Certainly they could defend themselves, but offensive operations in the post-World War II era would require incredible amounts of logistical resources. It was almost inconceivable that two or more offensives could be conducted simultaneously.

"Comrade Barber, the socialist revolution is always on guard against the treachery of the west, but....you're correct. This is a time for us to turn our attentions away from our enemies outside the state, and cleanse ourselves from within. To that end, I have come as you asked. The committee we spoke of earlier is still in its infancy, so, what is it you desire of me?"

Walt's attempt at breaking the ice with the stoic general had come and passed. His charm wasn't wasted on the old man, but Dmitry was a busy man, carefully watched by the secretive forces abounding in Soviet Moscow. Time was never a luxury at their meetings.

"Dmitry, I have spoken with the most powerful members of the revolution in the east. While I was able to maintain the anonymity required by the committee, they were obviously very curious, and asked many questions."

Dmitry sat down in one of Walt's guest chairs while Dan closed the door, and Walt leaned on his desk. His conversations with Kim Il Sung, and Deng Xioping never demanded details, but they were worried about the coming coup. Walt had his own concerns and used the meeting to conceal them as his own. "Dmitry," he continued, "They asked me about the Dead Hand system. Everyone is worried that two of three awful events may occur. While you and I expect that the launch codes for the state's nuclear deterrent will be secure and safe in your hands, there is still the fear that the soft forces of Gorbachev's regime may either change the codes, or that the codes may be lost somehow. If they are lost, I am told that a failsafe system called "Dead Hand" will enable the automatic launch of all nuclear weapons. How real is this possibility, and how well are we prepared for it?"

Dmitry had no idea that anyone outside the inner Soviet military circles even knew of the Dead Hand system. His shock was well hidden, but still evident through his wrinkled eyelids. There must have been a security leak. His surprise was replaced by a faint chuckle at the irony. After all, he had become a security leak himself. Since Walt was one of his co-conspirators, there was no reason to hide the system from Walt.

"What do you know specifically of Dead Hand, and who has told you?"

Walt hadn't really learned of the Dead Hand system from either the North Koreans, or the Chinese. In fact, one of the software support companies that he had recently acquired once assisted in the creation of the communications software used in the Dead Hand system. The company specifically worked on the secure communications link between the Dead Hand missile, and all of the other nuclear launch sites.

"From what little I've learned, through several different people, the Dead Hand system is the automatic launch of a single missile upon the incorrect authentication of launch code orders from the central command. The idea, correct me if I'm wrong, is that if the central command is wiped out in a sneak attack by the west, then this missile will launch and broadcast a signal to all other launch sites. That signal is translated as...'It took God six days to create the Earth, and we will destroy it in one.' My list of confidants is long, and must remain personal in matters of this nature."

Dimitry understood confidentiality with Walt, and even his iron heart skipped a beat at the thought of losing his own secure relationship with the man.

"You are very correct about the Dead Hand, but you need not worry about it. As Defense Minister, I have placed a former aide of mine in command of the Dead Hand launch site. Colonel Alexov Kryuchov. His uncle is the head of the KGB, and a member of our committee. Alexov has been passed over for promotion because his views were considered overzealous by the current regime. We have complete confidence in his abilities, and his decision making. If the moment should ever come, Alexov is the man we will want to have in a position such as this."

Dmitry was convinced in Alexov's loyalty to the revolution, and to him personally. During is assignment to the Defense Minister, Alexov had found out that his wife was having an affair. True to his nature, Alexov waited until he found his wife with the other man, then murdered them both. Dmitry, a firm believer of the 'eye for an eye' part of the bible, arranged for Alexov's promotion to the Dead Hand command site instead of prison or execution. It was fairly easy, after all, Alexov's wife was from a common family, and the young Corporal that was servicing her was too. Neither would be missed. No one had anything to gain by the loss of such a promising officer, and the position was open at the new Dead Hand site.

Walt nodded and smiled with the reassurance. Dmitry could be trusted more than most. His conviction, and his political position, made it easier to stay with the truth, as a habit, rather than get caught up in lies. With this in mind, Walt also knew that Dmitry was the principal member of their committee: the committee that would soon take over the government he was currently working for.

"Is there anything else, Walt? If not, I have something to ask of you."

Walt shook his head and casually motioned with his hands that he had no other questions, but wanted Dmitry to continue. He expected the Defense Minister to ask about the Chinese and North Korean response to the planning of the coup, but what else could Dmitry want of him?

"Of course I would like to know what our friends to the east think of the committee, but some of the other committee members have expressed concerns about the west, and of course our nation's economy. Interior Minister Pugo would like to know how we will bring new life to our economy. Foreign Minister Bessmertnkh is concerned that, because of the war in the Gulf and our nation's history with international banks the actions of the committee might further isolate the socialist revolution rather than enhance its growth. Myself, I have concerns that west will see the actions of the committee as a chance to strike. The last thing that anyone wants to see is American troops in Vladivostok again."

Walt could feel Dmitry's concern, and even that of the committee members. By now the committee had taken on a life of its own, and, even though he had a lot to do with its creation, Walt no longer felt as though he needed to nurse it as a newborn entity. The child was already becoming a man.

"Dmitry, once the Soviet Union has re-established itself as a true world power, its history with international banks, and its relationship with other nations will change. You can assure Foreign Minister Bessmertnkh that he will no longer have to look towards the floor when he meets with other leaders. They will shrug their shoulders with fear, and he will be making demands upon them instead of asking them for handouts. Interior Minister Pugo is correct in assuming that international banks will never be eager to loan money to the Soviet Union, but, I have tentatively arranged for several investment groups to begin working with our comrades in the east. This arrangement will allow those investors to produce goods at a lower cost, and the state, while monitoring the investments, will be able to provide the people with a much higher standard of living. These investment groups are most eager to develop their relationships with the Soviet Union because of the advanced technology already in place here. The weak government and precarious economy has kept them out. They fear losing their investment contacts or even the investments themselves."

Dmitry looked unmoved. Walt could feel his reassurance was sinking in. It was, but there was still the question of the west in relation to the activities of the committee.

"I too have concerns about the Americans and the committee's actions. They very well may see it as a time for opportunity, and, in fact, it will be one since the Soviet Union will be in a state of change. Change means discord. We must make sure of two things to offset their opportunity. First, the committee will have to act completely to limit the amount of discord. Second, the committee's actions will have to take place swiftly, and at a time when the Americans will be least likely to react rashly. One of their elections, during one of their countless scandals, or possibly during one of their imperialistic actions like this gulf war-"

"No," Dmitry interrupted, "We are not yet ready. We will need at least another year before all of the commanders have been convinced or replaced."

"Fine, Dmitry. I didn't mean to imply that now was the time, only sometime in the future when conditions are similar to now militarily. It might be simple enough to act when a passive President is in power, or one who is too indecisive to make a quick decision. There will be plenty of chances. When the committee is ready, it should wait for a chance like one of those I've mentioned. Only with those things in mind can we come close to being certain that the Americans will not react illogically."

Dmitry stood to leave, apparently satisfied with Walt's ideas.

"I must go. My aides have no idea why I am here. I think my guards believe I have a young lady friend. If I stay too long, people might actually get the right idea about the wrong things." Walt rose laughing.

"Dmitry, I'm sure my receptionist would be honored to go to lunch with you if you want to maintain a facade. Let's ask her..." Dmitry laughed.

"Thank you, and your receptionist is very pretty, but I need to return to the Ministry. There are duties that still need attending. Have you anything else for me?"

By now they had reached the front lobby, and, as Dan returned, he interrupted their conversation.

"Sir, Mr. Bush called about the U.S. food shipments that we discussed the other day."

The issue hadn't slipped his mind entirely, but he had forgotten to bring it up with Dmitry.

"Dmitry, I need the committee to help me out with some arrangements." His laughter faded, and the solemnness of Dmitry's face returned.

"Certainly, what can we do for you?"

"I need Interior Minister Pugo and Foreign Minister Bessmertnkh to make sure that grain supplements to the Democratic People's Republic of North Korea are curtailed indefinitely. This seems like a terrible thing to do, but my reports indicate that the Soviet people have little to spare as it is. Besides, I think their leader may need a taste of what we are trying to prevent before he will throw his weight behind us. Only after he sees the danger of depending on the west will he realize the righteousness of our proposal."

Dmitry grinned with approval. He looked like a car salesman that had finally gotten a customer to sign.

"Walt, I'm sure they'll make the arrangements you need. If a little nudge is needed, then our friends to the east will get a little nudge. Call me with any more news, but....not too often, eh."

They all laughed, even the receptionist who had no idea what was going on. Professional colleagues were about to bring down an empire, starve a nation, and send the world on a sinuous path to Armageddon.

While Dan closed the door to the hallway, the receptionist took another call.

"Mr. Barber? Tony Kerns is on line three returning your call." Walt thanked her and told her he would take the call in his office.

After a short walk, he sat down, pulled a file from his briefcase, and picked up the phone.

"Tony, how's it going? Is it nice and sunny in L.A., or is it as cold as it is here?"

Tony laughed. Outside in L.A., the thermometer on his window read 78 degrees. It had to be 20 below in Moscow.

"C'mon Walt, it's not the heat, it's the humidity!"

They both laughed and remembered the dry desert environment where they had first met south of Palm Springs only a few years ago.

"Seriously Walt, what can I do for you?"

"Tony, I've got some friends who are looking to invest."

Tony put on his investment broker salesman hat, and, even though he still sounded reminiscent, there was a tone of used car salesmanship added to his voice.

"Well that's understandable Walt, this is a good time to invest in almost anything, except oil. In fact-"

"Funny you should mention oil. That's the area they're looking to move into."

Tony was confused. The oil market was always volatile, but since the invasion of Kuwait, oil prices were bouncing all over. No one knew how the market would effect the region's supply and distribution, either in the present or the future.

"I dunno, Walt. That's a speculative area at best right now. No one knows what's going to happen with this Kuwait thing. Now, you know me, I'm not a gambling man, especially with my clients. Have you suggested more conservative methods of investment? Well, first, what kind of turn-around are they looking for?"

"The people I'm referring to have plenty of other investments already. They're looking for something with a little more potential. As for they're turn around, I'd say at least 20 years, maybe longer."

"Well, I can come up with some numbers and suggestions for you to review, but I'm still not comfortable with the oil industry right now. How soon do you need figures?"

Walt didn't want to arise Tony's suspicions by sounding to eager. If he sensed the ambition in his voice, Tony would know that a big deal was in the works. Although he was a friend, Tony was a businessman first in all respects, and he was very familiar with Walt's ability to create networked situations. Insider trading was a whispered word since the 1980's, but when anyone got a call from Walt Barber looking for a favor, something was bound to be in the works. Odds were that Tony wouldn't spill the beans, but he might use any information Walt gave him to fill his own pockets, and those of his clients.

"Ah Tony, these people know what's going on over there, and they're not idiots. They're not in a particular rush. Just send some numbers to my office in

New York. I'll be there in a few days, and then we'll get together again. You got any good prospects in the meantime?"

As careful as Walt had been not to sound overly enthusiastic, Tony had caught on, and knew something was in the works. He could sense the urgency even if it wasn't spoken.

"Actually Walt, I think you may be right. This Kuwait thing could work out to be something if everything goes okay. I'm looking at my monitor now, and-"

Walt interrupted again. "Let's just wait and see, Tony. A couple of days, and we'll get a feel for it. Until then, we don't want too many people to spot the potential, right?"

Tony sounded like a kid caught eyeing the cookie jar. "Oh, yeah. Absolutely. I'm sure there's others out there glancing at it, but I'm not gonna suggest anything. Can I handle anything for you right now, or are you comfortable with your current portfolio? Maybe should I call it a book!?"

"No Tony, thanks though. If you can just get me those numbers, I'm sure we'll have some things to talk about in a few days. Listen, I don't want to keep you on the phone too long. I know how you are, so I'm gonna get going. Besides, this satellite delay is driving me insane. I'll never get used to it."

"Okay Walt. Good to hear from you again. When you're in L.A., stop by, alright?"

"Sure Tony," Walt replied, "same for you if you're in New York."

Both men hung up their phones and considered what the other was planning. Tony was curious, but confused. Walt rose from his desk and looked out his window. Across the smooth gray waters of the Moscow River, he stared at the white Soviet Parliament building. With an overcast sky, and the freshly fallen snow, the building seemed like nothing more than simple lines in a black and white photograph. In recent years, the activities inside that building, and the presence of the structure itself, had weakened the Soviet Union to this point; this point in history when revolution simmered beneath a frozen goulash of nationalities. The only way to release their combined flavor from bureaucratic stagnancy was to bring the entire mess to a slow boil, add in new spices to control the taste, and skim off the fat from the west.

In a few years, Dmitri and his committee would regain control of the Soviet Union. Using capital from the Soviets, the Chinese, and the North Koreans, Walt would organize the purchase of countless holdings in oil companies around the world. Through front companies, and international personal accounts, he could force oil prices to remain increasingly low. Low oil prices meant that his manufacturing and industrial investments in the communist countries would have a better chance at surviving their infancy. Once those investments were firmly established, the capital raised by liquidating some of the oil holdings would allow them to economically bring down the west's share of international markets. First the weaker areas in Southeast Asia and Indonesia would falter then Europe, South America, and finally North America.

Every time he thought of his career's objectives, Walt laughed. It all seemed like a pipe dream, a fantasy, or just some idea to strive towards at the time. Someone once said, "Be careful what you wish for. You might just get it." Walt's face turned to ice, and he stopped laughing when he remembered that it was his boy scout of a father who had told him.

Saturday, February 2, 1991, 6:05 am

Iraq/Kuwait Border

From 40,000 feet, Captain Al Meyers and his Weapons Officer Phil Sidney crossed the border from Turkey into northern Iraq. Above them, the black evening sky was pin-holed with countless stars and wisps of high altitude clouds. Thousands of feet below, a thick layer of overcast hid any signs of the Earth.

Having just finished refueling, the lone F-111 bomber, callsign Wolfman one-eight, cruised peacefully towards its high-priority target. Somewhere under the overcast, their escort package made sure that the approach to the target was uneventful. Made up of four F-15C Eagle fighter planes, an EF-111 Raven radar and electronics jamming plane, and a pair of F-16C Falcon fighter-bombers loaded for air-to-ground attack. The strike package was ready for almost any type of enemy action.

Al Meyers glanced at the unbalanced load on his bomber. Under the port side wing of their F-111 was a 2000-liter fuel tank. Under the starboard wing was a single, experimental, laser-guided bomb, the GBU-28C. Nicknamed "Deep Throat," the bomb had been designed, built, and shipped to an American airbase in Saudi Arabia, all within 48 hours. The guidance system and control fins were the same as any other strap-on conversion kit. The nose and tailfin assemblies converted normal, 5000-pound bombs into precision guided munitions. Between the tailfins, at the rear of the bomb, and the laser seeking/guiding unit at the front of the bomb, the unique design of the "Deep Throat" bomb hung from the wing. The "Deep Throat's" body was actually made from a section of 8-inch Howitzer barrel that one of the designers had seen lying around a U.S. Army base. After the barrel's ends were cut off and machined to fit the guidance package, the inside was filled with a special molten high explosive.

Like other bombs to follow, it was built to penetrate the Iraqi command's super bunkers. Those super bunkers were built to handle near-misses from anything-even nuclear weapons. There were four super bunkers in southern Iraq between Baghdad, Kuwait, Saudi Arabia, and Iran, and two more bunkers just outside Baghdad itself. All of them were built of special steel-reinforced concrete as much as 150 feet below the surface.

No one had any real way of determining what was going on inside the bunkers, but it was unusually easy to figure out which one was being used by the

Iraqi high command at any given time. No one could possibly get an agent or Special Forces team discreetly inside, but a CIA sponsored sale of modified fax machines back in the 1980's had ensured that at least some of the communications would be detected and monitored.

Years before the Gulf War, an American company wanted to sell special secure communication fax machines to the Iraqis. The Iran/Iraq war, and the international sanctions placed on both countries at the time was interfering with the sale. Somehow, the CIA found out about the possible sale, and contacted the manufacturer. As long as the manufacturer added a small circuit board, manufactured by a CIA front company, the world community would give special consideration to the deal. The manufacturer agreed, the deal was brokered, and the secret was maintained-until Al, Phil, and the rest of the Wolfman flight came within 30 miles of the first bunker.

Phil Sidney sat in the right side of the cockpit. While he scanned his sensor displays, and listened to half a dozen radio frequencies at a single time, he heard the Wolfman flight leader's voice.

"Wolfman one-eight, this is Wolfman one-one. Squawk the target."

Phil switched one of his newly installed multifunction displays (MFD'S) to a special display, then he pressed a series of buttons around it. Finally, with the press of one more button, three times, he waited. A custom electronics addition to their plane sent a low voltage, low frequency signal to the CIA built circuit board in one of the special fax machines located underground ahead. The low frequency allowed the signal to go through the ground, but the power wasn't enough. Phil made a few modifications to the avionics by editing the computer program on his MFD. Again he pressed buttons, then one of them three times. A light blinked on his MFD three times in response. The fax machine had been pinpointed.

With the exact location appearing on his screen, Phil gave Al the course correction and time to target. A short while later he gave Al the two minute warning and another course correction. Then, like a scene from a bad movie, he heard an unknown voice on one of his many radio frequencies.

"Bandits bearing 043, angels 1 and climbing. Looks they might actually try us this time."

Another voice responded.

"Wolfman flight, bandits confirmed. Wolfman one-two and one-three break low and engage. Everybody else keep goin'."

"Two has a tally on low lead. Three take the trailers."

There was a pair of clicks as Wolfman one-three, one of Al and Phil's fighter escort pilots, acknowledged his orders by clicking his radio mike twice. Below Al and Phil, somewhere in the overcast soup, a dogfight was beginning. Al could tell something was happening by his right seater's focused attention, but there was no time to ask what it was. If it were important, Phil would tell him. They continued on toward their target.

There was information coming from over a hundred different sources, but Phil was still able to discern that the target was in sight. He swung the infrared camera into position and watched the black and white image as if there were no clouds obscuring it. He recognized the target from the reconnaissance films of the area that he had watched during the pre-mission briefing. Another toggle of the fax machine's circuit board, and his MFD highlighted the exact location on the screen. Phil told Al to keep it steady, and, after a few seconds of aim, he dropped their massive bomb. They both felt the plane lighten, and, since it was almost empty anyway, Al dropped their large external fuel tank on the other wing to help compensate for the changed aerodynamics of the situation.

Their plane continued on. The clouds didn't obscure the infrared camera, but they did block the laser that Phil was using to guide the bomb. He tried in vain to make the bomb stay on target. Falling from almost 7 miles in the sky, at 300+ miles an hour, didn't allow much room for error. It did allow plenty of room for crosswinds to blow it a few feet off target. And so it happened. Phil watched the bomb miss the indicated area on his screen.

Their strike completed, Al notified the rest of the flight. "Wolfman one eight to Wolfman flight; let's get outta here."

Phil had shown no signs of exuberance. His lack of words was clear. Al expected the mission's results without even having to watch the video. He made sure that he was only on the intercom then he asked Phil about the strike.

"How close did we come?"

Phil shook his head in frustration. "It's tough to tell distances from this angle, but it hit just on the edge of our aim point. It could have been a foot, or a football field. I just can't tell perspective. The BDA (Bomb Damage Assessment) guys'll be able to tell us. We'll know about an hour or so after we get back I suspect. One thing's for sure, they're gonna see one hell of a hole outside if they do come out of that bunker. I'm sure they'll at least have a silent moment of reflection. I mean, only an idiot would get back inside that place after seeing a hole like that. I doubt we'll be coming back here. I shouldn't have even tried it through these clouds. I dunno why I did."

Al tried to console his bombardier. "We had to take the shot. There was no bringing the thing home this time, not on a rush order like this one. Remember when they loaded that bomb on our wing? It was still warm. The explosives were still cooling inside! I don't think they'd be too appreciative when we brought it back because of cloud cover. The only way we could have gotten a better shot would have been to get beneath the clouds. That wouldn't have worked with all the flak and those fighters. No way. This was our best chance. We took it, and you got damn close. It's like you said. Only an idiot would get back in that bunker with a hole like that outside! Don't sweat it. Let's go home."

The F-111 Aardvark banked to the left and headed for home, in Saudi Arabia. Other air strikes with the similar bombs were also being conducted against fax machine signals. Few met with success. Even years after the war, the U.S.A.F. would only acknowledge the use of a handful. The specific number would never be known to anyone except their builders, and a few Pentagon accountants.

On the ground below, the bomb had, in fact, hit the target area, but just on the edge as Phil had suspected. It plunged over 170 feet into the sand before detonating a few feet from the bunker. Inside the bunker, the walls of sub level 4 collapsed. The chemical weapons production level, level three, also collapsed and crushed both scientists and storage drums alike. When the bunker was evacuated, the small crater was filled with cement, and the bunker hidden from the world for the remainder of the millennium.

Phil may have missed his target, but the accuracy and effect of the makeshift guided munitions now meant that there was no place safe to hide. Even places that were relatively safe in a nuclear war would never be sanctuary again. No one in Wolfman flight realized the strategic effect that their mission would have on the rest of the world, but even Phil's near miss was good enough to render all but the biggest nuclear weapons obsolete.

Since the advent of strategic bombing in World War I, military planners had dreamt of a time when they could decisively strike far behind the lines and take away their enemy's ability to wage war. World War II was the first time that airpower could actually be used in such a way, but the inaccuracy of bombs dropped from the planes of the period meant that either thousands of bombs had to be dropped to take out a single small target, or one huge bomb like at Hiroshima and Nagasaki. Either way, taking out a target meant that the entire area had to be destroyed. Those areas were then of little use to the allies when they advanced and seized them.

Deep Throat (later re-named "Bunker Buster" by a politically correct media) proved that airpower had finally come of age. One bomb from one plane could take out any target and leave hardly any collateral damage. Governments could now project military power without having to worry about accidental genocide. Most importantly, one highly accurate bomb on one sophisticated plane was far cheaper than a thousand World War II B-17 Flying Fortresses with tens of thousands of inaccurate bombs. Ultimately, government accountants would determine what weapons a military could use, and the choices for the future were clear. Nukes were never cheap.

Sunday, February 24, 1991, 4:01 am

Iraq/Kuwait Border

19-Year-old Private First Class (PFC) Robert Mamush drove his HUMMV through the sticky, black, desert sand. In the passenger seat, his platoon leader, a green Lieutenant fresh out of ROTC continued reading maps. In front and behind them, a long column of AAV-7 amphibious armored personnel carriers transported the rest of their platoon. Artillery, mortars, and bombs thundered from all over the endless horizon. Oil fires appeared as orange flickering lights between the soot-covered sand and the featureless black sky above. In a single column they moved forward through the Saddam Line; a barrier of alternating minefields, anti-tank ditches, and oil-filled pits of flame. Bob could see the white talcum powder sand that was being turned up from the treads of the AAV-7's. Except for their column of vehicles, the vehicle tracks in the sand, and the oil fires, there were no features anywhere to be seen.

In the back of the HUMMV, the platoon radioman was listening to a transmission. He passed the headset to the lieutenant. Bob paid no attention. After months of sitting in the sand, he was desperate to get into a fight. Typical of most young Marines, Bob was filled to the brim with *esprit de corps*. Added with the relative innocence youth still provided, and his own aggressive "hair-on-fire" attitude, he was almost blind with rage towards the Iraqi troops ahead.

The noise of the HUMMV's motor, the loud drone of the AAV-7's engines, the constant distant thunder of explosions, and the radio's indiscernible conversation was all interrupted by a huge explosion somewhere up ahead. Several more followed, and Bob could even see a mushroom cloud that looked like a nuclear explosion. Eyes wide open with disbelief, he looked at the lieutenant about to ask for an explanation. They had to yell to be heard over the noises all around them.

"FAB's! The Air Force is dropping Fuel Air Bombs to detonate the minefields. The bombs hit, then spread an explosive aerosol vapor. A smaller charge sets off the cloud after a while. Not very good against armor or bunkers, but the concussion sets off mines for hundreds of feet around! See that big one over there!"

The lieutenant pointed to the mushroom cloud ahead on the right of the column.

"That must be one of those daisy-cutter bombs we got briefed about a couple of days ago!"

Bob didn't remember the briefing, and the lieutenant could see it in his expressionless face.

"You remember. It's that 10,000-pound bomb that's so big they can only drop it out of a C-130 cargo plane! Son of a bitch! It's gotta suck to be on the receiving end of that, huh!?"

"Damn straight L-T!" Replied Bob. "I just hope it leaves some of them bastards for us. You know!? I wanna get to Baghdad and personally let Saddam know how much I appreciate our little vacation at the mother of all beaches! OOO-YAHH!"

"You and me both, Bob!"

The radioman handed the headset back to the lieutenant again, and he began to converse with the company commander who was already 3 kilometers ahead. The lead elements of the Marine units attacking the Saddam Line were breaking through, and only a handful of Iraqis were offering resistance. He handed the headset back to the radioman and turned to Bob again.

"Well Private, it looks like we'll be home in no time. So far, they've just been emptying their clips until somebody returns fire, then it's white flag city! You might not get your chance for payback! At least you'll get home soon, as long as it keeps up like this! I know my wife's gonna be happy as hell! How

'bout you!? You got anybody back in the states!?"

Bob's immediate frustration at the lack of vengeance was immediately replaced by the thoughts of his girlfriend Meagan, and the thought of being with her again.

"Yeah, I gotta girlfriend! She's great, super nice, and damn good lookin' too! We're sort of serious! I think we'll probably get married someday, but we're not ready for that yet! I dunno. We'll see! How about you, you got anybody besides your wife?!"

"Oh yeah! I been keepin' in touch with my parents, my aunts, uncles, everybody in my entire family's been writing and sending me stuff! Check it out!"

The lieutenant reached in his right thigh pocket and pulled out a thick wad of letters.

"I've got more in my pack! Like I said. They've been sending me stuff too! My one uncle sent me a flask with some Jack in it since he knew we couldn't get anything over here! I got cookies and all kind of food too! When we get settled in tonight, I'll show you guys what's left in my pack!"

Bob was impressed. It was the first time that the lieutenant had really seemed like a regular Marine, and not a fresh green officer. The months in the sand, the countless drills, the endless digging, and all the rest of the innumerable suffrages had finally brought them together, as Marines.

Random red and white tracer rounds zipped into the sky from ricochets and loose ground fire ahead of the column. Everyone's attention was again focused on the task ahead. They were still in Kuwait, and there were still Iraqi troops who very much wanted to kill U.S. Marines.

After a few minutes, the AAV-7 in front of them rose up and over the sand berm that was the outer edge of the Saddam Line. When it drove down the opposite side, the 10-foot tall machine disappeared from view. Bob drove through the breach next. At the high point, he could see the fighting all along the horizon. Muzzle flashes from artillery and mortars, automatic weapons fire, illumination flares, heavy weapons fire from tanks, AAV-7's, HUMMV's, and countless other vehicles. Added to the oil fires, hell was boiling from the left to the right as far as the eye could see.

Undaunted, they pressed forward. A few minutes later, they passed between two triangular Iraqi fortifications. Each fort, 2 kilometers long on all sides, contained three smaller triangular forts, one for each company in the battalion. Each company's fort, about 700 meters long, contained three even smaller forts, one for each platoon. Surrounding the two large forts, there were anti-tank ditches, minefields, and trenches filled with oil.

The Air Force had worked the forts over well. Anti-tank ditches had been turned into a moonscape. Oil-filled trenches were hit with napalm and were almost burned out. Minefields were covered with huge black smudges surrounded by small craters where mines had been set off by FAB's. The forts themselves looked like Swiss cheese from all the breaches and craters. Every once and a while, Bob saw streams of red tracers race toward an invisible location. A few moments later, the firing would stop as the target had been either destroyed or captured.

For a while, Bob was glad to have something to watch besides the empty horizon. Then he saw rows of captured Iraqi troops. These men who were starving and shaken from bombing, were now relieved to be captured. Some were wounded and blood-stained bandages did little to hide the pain on their faces, or the missing limbs. Some were so shaken, they seemed like zombies. Others were happy and blew kisses at them as they drove by. A few wore the faces of hardened soldiers whom, even as prisoners, looked as if they wanted revenge far worse than he did. When he saw the first bodies, he didn't wonder why.

They passed a few Iraqi trucks that had been destroyed by either artillery or air attack. The vehicles were shredded wreckage and could barely be recognized as trucks. Like everything else he had seen in the area, they were black with soot, oil, and their own scorching. In the ashes around the trucks, Bob noticed some of the piles were shaped like stick figures made of burnt logs. One had a head that had not burned, and the man's expression was that of extreme pain cast for eternity. A stray dog tugged at what might have once been an arm. When the HUMMV was closest, it startled the dog. Not bound by any sort of morals, it tugged at the arm until it came off, and then ran away into the darkness.

Until then, he had begun to regard the bodies with the same respect as an animal roadkill. The dog scene brought home a new respect for the dead. Those men probably had wives, families, girlfriends, just like he and the lieutenant. This was reality up close, and personal.

The war continued for 99 more hours, and then the allies called a ceasefire. The surviving Iraqi forces were driven from Kuwait, and the U.S. stigma of Vietnam had faded. Bob returned home to a victory parade, veteran's benefits, and a hero's welcome. A few days later, after his girlfriend decided they should "see other people," Bob decided to remain in the Marine Corps indefinitely. He had seen both the real world, and the world that so many Americans lived in, like Meagan. Hers was too different from the one he had experienced. Bob changed that first night in Kuwait, and he expected the rest of the world to understand his new perspective on life, but only other combat veterans could understand; other Marines.

Monday, September 2, 1991, 6:00pm

Channel 4 News

The images and music on the television played with a fast moving tempo indicating that however fast the world's events occur; Channel 4 News would be the place to find out the latest, up to the minute information.

"Good evening everyone, I'm Tom Waushinski, and this is Channel 4 News. Tonight, after several landmark days of protests, the Soviet Communist party is in ruins. The coup attempt to remove President Gorbachev and his administration has utterly failed. In fact, the very reforms that the members of the coup had been trying to suppress, have now gained even more momentum. Parliament leader, Boris Yeltsin, rallied crowds all over Moscow to protect vital centers of power from seizure by the military. For the most part, the military backed down when faced with killing unarmed citizens. Some are saying that the nearly bloodless revolution created in response to the coup attempt is already over."

"Last Thursday, President Gorbachev, whom the coup leaders had described as 'ill', returned to Moscow in what appears to be perfect health. The conspirators, as they are now being called, are not in the same condition. While Gorbachev remains the official head of the Soviet Communist Party, the President of the Soviet Union, Boris Yeltsin, the leader of Parliament, is clearly the man in control. 72 hours after he had stood on a tank in front of the Parliament building and declared the coup illegal, the entire affair is over, and the conspirators are either in custody or otherwise removed."

"Let's go down the list. Among the arrested were: Vice President Gennady Yanayev, Chairman of the KGB Vladimir Kryuchkov, and Defense Minister Dmitry Yazov. Foreign

Minister Bessmertnykh, Military Chief of Staff Mikail Moiseyev, and Parliament Speaker Lukyanov were all fired, while Prime Minister Valentin Pavlov and Interior Minister Boris Pugo were both hospitalized for 'heart attacks'. By Wednesday the 28th, at 3:00pm local time, a three-mile long column of tanks was leaving Moscow. It would appear that, for the first time in their nation's history, a

secure democratic government is in charge."

"White House officials have been in constant contact with Moscow throughout the crisis. Information coming from the White House, through unofficial channels, is that President Bush was deeply worried about the Soviet nuclear launch codes. During the three days of his captivity, Gorbachev knew that the coup leaders could launch the missiles if they wanted. Two people, each with a separate code, would have to send their codes to the Soviet Generals Staff Main Operational Directorate in Moscow in order for the authorization to be processed. Those two people, up until now, have been the President of the Soviet Union, and the Defense Minister, Dmitry Yazov-one of the coup conspirators. When the conspirators broke in through his door and arrested him, they seized his half of the launch codes. For three days, Mikhail Gorbachev, a prisoner in his own vacation retreat, painfully wondered not only about his life and the life of the Soviet Union, but also about the future of the entire world. President Bush's intelligence advisors had informed him about the Soviet launch code system. The conspirators assured him that the nuclear arsenal was secure. Finally, a direct line phone conversation between Presidents Bush and

Gorbachev quelled his fears, at least for now."

"In other news today, Wall Street, while still shuddering from the news in the Soviet Union, saw major movement in technology stocks. One company, Global Technologies Incorporated, surged from $11 a share to $23 1/2 fueled by rumors that it has landed a major communications hardware contract with the Pentagon. In local news, a west side woman's 9 month old baby was attacked by the family's pet python, and we'll be seeing more traffic delays in the near future as the city plans to close the 7th St. bridge for the next 8 to 12 months. Plus, we'll have sports and weather from John and Chris. We'll also have a local University professor's point of view on what the Russian revolution's effects will be right here at home. Please stay tuned. We'll be right back."

Friday, July 8, 1994, 6:00pm

Channel 4 News

"Good evening everyone. I'm Tom Waushinski, and this is Channel 4 News. There's uneasy news from the far east tonight. Kim Il Sung, North Korea's "Great Leader" has died of what is being called a heart attack. After weeks of being out of the public eye, and amid speculation about his condition, communist officials made the announcement early this morning. The few westerners that have met with him noted that he appeared to be in normal health weeks earlier. Sources at the State Department declined any official comment as yet, since they are waiting to see who succeeds Kim as that countries new leader. For years many have believed that his son, Kim Il Jong, would take over his father's position, but Il Jong has yet to be seen. There are many in the west who feel it is really a handful of military leaders who have been managing the nation. Over the past several years, North Korea has endured several poor harvests and the World Health Organization has annually claimed that the people are starving while military personnel continue to appear fit and healthy at every parade."

"In other news tonight..."

Monday, February 17, 1997, 6:00pm

Channel 4 News

"Good evening everyone. I'm Tom Waushinski, and this is Channel 4 News. They're calling it the end of an era in China today. The last communist leader to march with Mao during the Chinese communist revolution has passed away. Official statements from the Chinese government indicate that Premier Deng Xioping fell ill several days ago and expired due to a heart failure. Deng Xioping initiated many of China's market and human rights reforms in recent years, and broke bonds with the more traditional elements of the government. It was only his past ties to Mao Tse Tung that shielded him from ousting. While largely responsible for loosening Chinese society, he was also responsible for the recent Tiananmen Square massacre where thousands of protesters had gathered to call for even more government reforms. Hundreds were killed, possibly thousands injured, and an unknown number remain in prisons today. Leaders all over the world are caught between praising Deng Xioping and chastising him. However, they are unanimous in the agreement when it comes to one thing. Everyone seems to readily acknowledge that China is a far less closed nation because of him, and that Deng's mark will forever be left on the Chinese people."

Wednesday, February 19, 1997, 6:00pm

Chicago, Illinois

Downtown Hilton Hotel

Under a wintry Illinois overcast, with a windy city offshore breeze, seven oriental men, wearing three-piece suits, custom-made shoes, and no jewelry of any kind, stepped out of their airport limousine/van. One by one they stepped into a six-inch mesh of road salt, ice water, and colorless gray slush. Not a man flinched or took the time to wipe off his unusually unprotected shoes. Only the luggage attendant noticed their odd behavior, which was accented by the lack of luggage from the van. Each man carried only a briefcase. Frustrated with the inability to collect gratuity, the luggage attendant sought other prey.

Inside, flanked by rose-colored marble columns, wall paintings, and gold leaf whenever practicable, the seven men walked in a single file. They walked in step. Dan Greene, international businessman and Personal Assistant to Walt Barber, immediately recognized the gentlemen. He had been waiting for them. Dan rose from one of the lounge chairs in the lobby and introduced himself. As he made eye contact with the lead man, they all stopped and completely focused all attention on every miniscule movement he made. As he reached out for a hand to shake, six of them turned and looked around the lobby trying to find something, someone. The last of the seven men stepped forward and addressed Dan in perfect English.

"Mr. Dan Greene. It is a pleasure to see you again. When was it last, 1990 or 91?"

"Mr. Naht Synn, I believe it was 1991, but I could be wrong. I trust your trip to the hotel wasn't too difficult?"

Dan had learned the art of doublespeak from Walt. He still had a long way to go before reaching the plateau of his master's skill level, but he didn't sound like a telemarketer anymore. Even the simplest, most benign sentence could hold a double meaning. In this case, Dan wanted to find out if Colonel Naht Synn suspected that his covert entry into the United States from China had been detected.

"The weather was fine, and we had no trouble at the airport."

Outside, on the east side of Chicago snow was still blowing from an afternoon squall, and another approached from the west; due to arrive within the hour. Dan concluded that the Chinese had, indeed, successfully infiltrated the United States. Since the end of the Cold War, and the presence of a completely carefree Democratic Presidential administration, entry controls and monitoring had been loosened in favor of "...better, more worthwhile, government funding programs." Their entry was not a particularly difficult feat.

Dan kept the conversation fluid.

"I imagine that you're eager to begin the meeting. I have taken the liberty of reserving the entire floor. Here are your rooms' key cards, and your receipts. You'll notice that they are all in the names of different, privately owned American businesses, and no individual names are present. Please feel free to choose any room on the floor that you feel is satisfactory for your needs. If you would be so kind, I'll lead you to Mr. Barber, he's in a private meeting room on the top floor. Since your trip was uneventful, I believe we can dispense with the stairwell and use the elevator."

Dan pointed to the hotel Concierge across the lobby who casually and swiftly rose and pointed to one of the hotel's attendants waiting by an elevator. Dan had arranged that the elevator be held for their private use in anticipation of the Colonel's uneventful journey. Colonel Naht Synn and four others followed Dan, the other two men located the stairwell and made their way up in a fruitless attempt to check each floor as the elevator passed. When the elevator finally reached the top floor, there were two very tired Chinese agents waiting and panting while the doors opened. The men inside followed Dan out of the elevator, and the two men rejoined the security formation around the Colonel.

As they turned from the floor's elevator lobby down the hall to the right, Walt was waiting in the hall outside the meeting room.

"Gentlemen, it's good to see you again, but I hate to have it be under these circumstances. I was just on my way to see if you'd arrived yet.

Obviously you have. Was there a problem?" Colonel Naht Synn shook his head.

Dan answered. "Their trip sounds like it was uneventful, sir. None the less, I would think everyone would like some refreshment. Shall I have some brought up?"

Walt looked at the Colonel who was again shaking his head, then replied.

"I think we'll be all right for a while, Dan. Thanks for everything. Would you make sure that the rest of the gentlemen have all their needs taken care of. I think we'd like to get this meeting started. Why don't we all meet back here in about an hour or so?"

The Colonel motioned for his escort to go with Dan, and the two parties separated. Dan took the Chinese agents on a tour of the rooms on the floor so that they might decide upon their security needs, while Walt and Colonel Naht Synn went into the meeting room.

Inside the meeting room, the Colonel removed his overcoat, and neatly draped it over a lounge chair in the corner; rubbing out the unwanted folds before leaving it. His shoes were still wet and cold. Walt, meanwhile, poured a cup of coffee for himself, two creams, no sugar. Both men finished at the same time, but, when they were both seated and comfortable, it was Walt who opened the conversation.

"Colonel, again, I offer my condolences to you for the loss of Comrade Deng Xioping. He will be remembered around the world for the good that he's done."

Naht Synn's blank expression changed slightly and hinted at a latent frustration.

"Mr. Barber, Den Xioping was a weak man from the moment he came to power. It was his ties to the great Mao Tse Tung that kept him safe, not his personal prowess. The only good he ever did was to keep China from completely falling into the hands of western capitalists like you and your associates. By the way, it's General Naht Synn now. Yesterday I was promoted to the Military Commission as Special General for Internal Security and Counter Espionage. I have not yet had the time to review your meetings with Deng Xioping, or with others in China, but I must inform you that because of your unusual influence I will make it a priority task. The decision to make this trip and to have this meeting was not taken lightly at all; especially with all that is going on in China now. I will ask you directly, why have you taken such a great personal interest in China."

Walt's "glad to see you" expression filtered away during the General's dagger-like speech. He had known about the General's appointment to the military commission months ago. It was only a matter of time before Deng Xioping would perish and more conservative elements in the Chinese government could take control. One of the methods they were bound to take was the appointment of Colonel Naht Synn to the Military Commission, a commission that arguably controlled all true power in China. Walt took a sip of coffee and slowly put the cup back on its saucer.

"General, I understand your tone, and I understand your mistrust. It is, after all, your job to be skeptical of things that seem too good for China. I'll admit to you, I am not a humanitarian. I'm not a socialist, a communist, a fascist, or even a capitalist-despite my surroundings. I'm also not a philosopher or a power monger. What I am is simple."

Walt continued. "You see General, everyone has a gift. They may be born into aristocracy, they may be talented, or they may have been born with the ability to develop certain special skills that will make them unique. I can't quite describe what my individual gift is. I do know that I, like every single person in this world, have to work hard to make something of myself, to make a difference. There are billions of religious zealots in the world, and for every rabbi, priest, or mullah, there is an atheist, a realist, a fallen believer, or someone who hasn't taken the time to decide their fate after death. I am the latter."

General Naht Synn crossed his arms. Subconsciously he was indicating that either his attention had been grasped, or he was becoming defensive, but Walt had rehearsed a speech for such an occasion, and he went on unabated.

"I see the good in most religions and political orientations. I also see the terrible conflicts that have been created by them. One thing about the hereafter is certain: what we do here is what's really important. If a person who goes to church, temple, mosque or shrines everyday then does nothing else with their life, few believe that they will have any more of a chance at a wonderful afterlife than someone who picked his nose all day. If there is an afterlife, its quality will be based on how we lived in this life. If one falls in with the atheists and the undecided of the world, then our only value is the legacy that we leave with the life that we live here, now."

Sensing that his answer was becoming too complex and lengthy for consumption, Walt tried to wrap it up.

"My intentions toward China, toward any nation or Corporation that I deal with, are not harmful. I only want to leave a legacy. God willing, years from now, when I pass on as Deng Xioping has, I will remain in people's minds as someone who did something with his life. If that is true, than any afterlife I encounter should be grand and rewarding. In the absence of a world beyond this one, I will still have made a difference."

The general rose and walked over to pour himself a cup of coffee. His trip had been non-stop, and the only sleep he was able to take was a few hours on each plane he rode. The coffee was long overdue. Despite it's steaming hot warmth, he took down an entire cup, black, with one healthy drag. Then he poured himself another and turned to Walt.

"Mr. Barber, contrary to your own thoughts and words, it sounds to me as though you in fact are a socialist, religious, philosopher with a gift for capitalism and diplomacy. The only values you carefully did not embrace would be considered as vices to someone in my position. You should be congratulated on not being a communist, a fascist, a religious fanatic, or a capitalist pig." Walt laughed, but Naht Synn continued.

My investigation will inevitably reveal your true nature possibly to you as well as to me. In the interim, we should discuss your future plans."

Walt was surprised, and didn't bother to hide it, though he did down play the effect.

"Walt, I'm talking about your recent acquisitions in the electronics market, particularly those that are working with the United States Military."

Walt was always very careful to only buy and sell companies through other corporations and other people. Over the past few years, since the beginning of the Gulf War, Walt had bought large quantities of stock in a multitude of companies that made electronics for the U.S. military. He was getting money, albeit fractions of a cent at a time, for every wristwatch, night vision goggle, computer, radio, and radar that the Pentagon bought. Since the U.S. military had to buy every type of item imaginable, the list was almost endless.

How had General Naht Synn found out about his purchases? Even Dan didn't know how deeply he was involved with the U.S. military's electronics market. The general would have had to work to find out about the stock purchases. Why? Was this why Naht Synn called the meeting? What did he want from Walt? Walt remained silent, but there were more questions for him.

"Mr. Barber, of the many things that you said you are and are not, I wonder if you intentionally left out criminal. I'm by no means accusing you of a crime or something immoral, but I wonder where your allegiances are? You've had unprecedented access to world leaders, including government and private leaders in China. At the same time, you have been actively reinforcing America's greatest assets, her electronics that multiple the strength of her military. In which bed do you lay, sir?"

Coffee break was over. Walt had to be very careful. He leaned over to the general in a private manner and spoke directly.

"My business is my own business, general. If you feel the need investigate it, then there's nothing I can do to stop you. I'm sure you've realized that in this global economy, finances and diplomacy often cross political lines. In Middle-Eastern politics, they have a saying: 'My enemy's enemy is my friend.' As you know, I do posses American citizenship. I spent a good portion of my childhood here in the states, including a substantial portion of my education. These days, I believe more in the principles of freedom that this country was based on rather than its bureaucratic functions. It will always be illegal to murder or rape. This should be true the world over, but I'm afraid that the United States' bureaucracy may be strangling its original fundamentals. Too often anymore, lands are taken, rights are ignored, and people are incarcerated solely for the benefits of serving a bureaucrat, a lawyer, or a corporate mogul. Since the Nuremberg trials after World War II, the U.S. military has had a policy that allows its members to disobey orders that they find immoral or unjust. The Constitution allows for a certain degree of civil disobedience too."

Walt found himself preaching again, and attempted to summarize rather than bore.

"My allegiance is with the rights of every man to the pursuit of life, liberty, and happiness. I believe in these and all of the others that the founding fathers of the United States believed in. I am an American, but I am more over a citizen of the world, with basic American ideals. I am, however, not bound completely to the current bureaucratic laws of the land. I would certainly not blindly follow the orders of our current leaders or probably not those of any of our leaders for the past sixty years. My allegiance is to world and to myself."

The two men stopped for a while and searched for answers in each other's eyes, then it was the general's turn again.

"Mr. Barber, would you agree that the balance of power between the United States and the Soviet Union offered a certain degree of tense peace for the entire world?"

Walt was attentive as he considered the provocative query.

"Yes, to a certain degree. I think that the arrangement was too dangerous, but it did allow for a great deal of growth and enlightenment for the entire world- perhaps the greatest in peacetime history."

"Too dangerous, Mr. Barber? Have the past six years been more or less secure? There are 50,000 nuclear warheads scattered around the globe with no central control and, in many cases, no control at all. Many would say that the situation is far more dangerous."

Walt nodded in focused agreement. He could see that the general was finally getting to the point of the meeting. He wanted something. He wanted something that Walt could give him from the U.S. military electronics market assets that he controlled. What?

"When we first met years ago, I listened to you discuss your investment plans for China and Korea. I heard that you tried to bolster the Soviet Union by assisting in that coup that failed. You were right in bringing investors into China and Korea back then. You were right in supporting the coup even though it clearly backfired and became a catalyst for the Soviet Union's collapse."

Walt could feel the general's need. He didn't know what it was, but Naht Synn was coming as close as he possibly could to pleading with him.

"Mr. Barber,... Walt, If there was a way to ensure a world balance of power, similar to that which existed during the Soviet Union's reign, without the threat of global Armageddon, how interested would you be? How interested would someone who believes in civil disobedience, the socialist ideals of America's founding fathers, and the corrupt, possibly illegal government in control today? Walt, I think I know of a way that you can bring that balance to the world. I think I know how you can leave a legacy, not just a legacy of Walt Barber, but a legacy of peace."

The cat was out of the bag, and they both knew it. Walt acted nonchalant about the offer.

"Oh general, I'd love to have the opportunity, but I'm afraid I don't have the influence to do something like that. In reality, that's all I'm trying to do, but I see my purpose as driving down that road rather than getting to that destination. Hopefully, I'll make it possible for someone else to get there. I guess that's my hope for a legacy. I'm not out to save the world, but I'd love to be part of helping someone else do it. Of course, I think you have something much more specific and much more difficult in mind."

The unofficial interview was over for Walt. Now he had to find out if he passed the general's tests.

"Now that you've traveled half way around the globe to come here and check me out, what is it you really want, general?"

General Naht Synn smiled and they both laughed. As coy as each had tried to be, sometime or another, the veil of darkness had to be lifted. Someone had to be trusted-to a degree, and this was the time.

"Walt, I think you're familiar with a company called Global Technologies Incorporated, GTI."

"Yes, I have some friends there, and I might have invested a little in it. What about it?"

"It turns out that I've got some friends there too. In fact, it's come to my attention that my friends don't have all the access they'd like. I was wondering if I could introduce you to them. Maybe you could introduce my friends to your friends. I would like to see their careers advance at GTI."

Walt sat back and thought for a while. Why GTI? Obviously GTI had people who were either serving as informants to the Chinese, or were working directly for them. The U.S. defense department probably would have detected any Chinese agents at GTI during their annual defense supplier review investigation. It must be informants. What could the informants tell the Chinese that would allow the Chinese to offset American global domination in the same way that the Soviets had?

Walt quickly thought about GTI for a while. A publicly owned company based in Cheyenne, Wyoming, they got their start when someone from the U.S. Air Force out at the NORAD Cheyenne Mountain facility decided that personnel stationed there needed some computer training. GTI was nothing more than a local computer shop at the time. Computer training lead to assistance in maintaining the lesser-used computers at NORAD. When computer-networking software became reliable enough, GTI was called in to install the local terminal network. That led to the modification and the eventual creation of computer network hardware that could connect standard PC's to the various unique military hardware configurations.

Since the Gulf War, the Pentagon had been trying to put a system in place that would enable all electronically gathered and entered data to be shared between units and services. The idea was that every AWAC's plane radar contact, every satellite and NORAD detection, every tank, every ship, every field commander, would all have complete access to any intelligence data regarding enemy disposition. Eventually the Pentagon wanted soldiers to be able to call in cruise missile strikes from 1500 miles away instead of just local artillery. There were other desires too. This included the near-priceless ability for the National Command Authority back in Washington to be able to watch battles on the other side of the globe develop in real-time rather than hear about their outcome minutes, hours, or days later. Walt understood, now.

"General, I'll have to think about that for a while. Is there any particular rush, or can I take a few days?"

"Walt, I'm sure my friends, our friends, will take some time to make things come together, but, if you need a few days, please, take them. When should I check back with you? I can't be away from Beijing for too long."

Walt rose and walked over to the window. Looking out at the city, he was already thinking the proposal over. When he bought the shares of GTI, he never thought that the small computer shop would become such a world power asset. Even after he heard that GTI was leading the bidding for the military network, Walt didn't dream of this situation. He had no contingency plan! What to do?

Obviously the general wanted his informants at GTI promoted so that they could leak information about the U.S. military's data communications network. That information would allow the Chinese-or anyone with access to it-to monitor and maybe even ambush U.S. military deployments. Lives hung in the balance.

"General, I think for security purposes, it might be better that my friends become yours, and your friends disappear. Quite honestly, an arrangement like that might be more financially expensive, but it would be a much more secure scenario. It would be a shame if too many people got involved. I like the keep it-simple approach when it comes to these kinds of matters; the fewer, more reliable people, the better. Wouldn't you agree?"

"Absolutely, Walt. I'll see to it that my friends aren't the weak link as soon as I speak with your friends. How financially expensive do you think that might be? I'll need some time to put things in order if your friends want an arrangement that's too lucrative."

Walt knew that if he gave the general a number, he could choose his own commission on the treasonous deal, and still have plenty of money to pay off the people he could trust at GTI. As for the informants, as soon as he introduced General Naht Synn to his friends at GTI, the general's informants would probably have to be killed. At least the general could be counted on for that. One doesn't rise to the position he has without getting his fingers get sticky.

"General, Dan should be dining with your associates right now. Why don't I take you to join them, and I'll make some calls. I expect the numbers to be in the twelve figures, possibly thirteen. How much time would that take?"

General Naht Synn hid his shock well. Some expression slipped through, but Walt couldn't tell if he had thrown out a high or a low price. It bothered him to have been caught so unprepared. If he had a pet peeve or fetish, it was always being prepared. As a sort of social irony, Walt actually tended to operate best on the fly. He was strong at planning, but off the cuff deals were where he always made the most money. Despite what he had told the general, that's all that really mattered to him: not religion, philosophy, politics, etc; just money. His true allegiance was the almighty U.S., not the government, just the dollar.

"I can have the funds tomorrow night, Walt, providing we come to an arrangement before noon tomorrow."

Walt had just made a verbal agreement that would make him a trillionaire. Not a cautious man, the number did have a unique effect.

"You know general, considering the values that we're talking about here, maybe I should come up with a bit of a ruse for the people over at GTI. That would limit the sources on my end to just Dan and myself, and it would cut down on the overhead. It's a lot safer and cheaper if this arrangement is as private and singular as possible. I'd also like to limit the number of people on your end. I understand that the numbers will make that difficult, but if we put a little time in efforting the problem tonight, maybe we can come up with a subtle way for both of us to make this even more discreet."

Naht Synn admired Walt's attention to detail, and he had done things like this before. The Chinese Special General for Internal Security and Counter Espionage was not a trusting man, but the risk in such an expensive venture would be better calculated by Walt, so he thought about it and agreed. After some food and refreshments, they would get together and work out the covert details of the biggest intelligence coup since the Allies cracked the German and Japanese codes in World War II. Ironically, that was an operation in which Walt's father had participated. The two men rose normally from the table where they had just agreed to change the balance of power in the world. The conversation switched to the weather, the city, and pleasantries about each other's reputation. Walt escorted the general to a suite down the hall, past the elevator lobby, where Dan and the Chinese agents were eating from an unattended buffet that had been setup. Walt casually told Dan that he had some phone calls to make and would need about an hour. Dan understood and already planned ways to keep the Chinese occupied. A tour of the entire hotel would probably do it. They all agreed to meet at the Kitty O'Sheas, the hotel bar, at about 8:00pm. By that time, the die would already be well cast.

Saturday, February 7, 2010

Islamabad, Pakistan

Our Holy Lady of Divinity Church

3:48pm local

43

The afternoon sun was unusually warm for a late winter day. Inside the small neighborhood church, everyone saw it as God's wedding gift to the lucky couple being married. The sky was laced with decorative high level strato-cumulous clouds, and a brisk breeze broke the gentility of a 67-degree day.

At the front of the altar, the Christians proclaimed their vows to each other while family, friends, and co-workers watched with innocent awe and remembrance. For almost an hour, the 119 people present listened and prayed to their God. They sang with hearts that were filled with the peace and hope of the moment. It was a moment when they could forget the rest of their world and be together as a community. In this building, they were all Christians, and no one was an outcast because of their faith.

Beyond the church's small portal in the impoverished streets of the city, the world was different. Only 3% of the people in the country had the same religious beliefs, and fewer had accepted the Western tradition of marriages that had not been arranged by the families. A Christian congregation with Western traditions in a fundamentally Islamic country was not something readily accepted by most of the neighborhood. Still, throughout the centuries, most people were content to let the Christians live their lives as long as they kept to themselves. The Christians had a similar perspective, and everyone was able to live together, not as friends, but as tolerant neighbors.

The pair of 19-year old lovers sealed their bond with a delicate kiss to the applause of all those present. They walked down the aisle to the lobby in the front of the church, and there they waited to greet and thank all those who had come to share the moment with them. Small flowers fell from the bouquets of the bridal party, and petals from their delicate blossoms danced in the wind as people opened the doors to the outside. There they waited for the traditional exit. It was a poor neighborhood, so instead of rice or birdseed, the crowd prepared to toss wildflowers to the happy couple as they left for their honeymoon. Laughter and more blossoms rode the brisk breeze that chased away the sun's warmth.

The bride's uncle, husband to her oldest sister, was an American. He was the odd one in the crowd. Though everyone tried to make him feel welcome, his differences in dress, stature, and skin made him feel awkward. To hide it, he hid behind his video camera and prepared to catch the couple's exit on film. The only thing more unusual than his presence was the very camera that he was using to hide himself.

Finally, the couple emerged and stood on the steps of the little church.

Everyone cheered as they do at all weddings for every religion in every nation. They kissed for the crowd, who cheered and laughed louder while throwing more of wildflowers into the sun only to be carried away by the wind. The couple made their way down the three steps to the street and thanked individuals once more. Everyone smiled, laughed, and forgot where they were.

Around them, their Islamic neighbors had mixed reaction. Some went about their normal business. Others seemed to flee from the unholy crowd. A few stopped to watch the happy scene. Regardless of their religion, these were happy and peaceful people in a land that was fraught with tension, poverty, and often violence.

While everyone's attention was on the smiles, the hugs, and the wildflower petals in the breeze, that violence suddenly appeared. To the American, the shots sounded like firecrackers at first. Everyone else immediately recognized them as a matter of sad familiarity. The crowd tried to disperse and hide. Those who couldn't find cover dove for the ground. The young groom embraced his bride in an effort to shield her. The American who had only the day before been resting quietly in his suburban home stood alone. He had no experience with airborne lead, and as such he had little respect for it. Instead of seeking cover or protecting his wife of 7 years, he froze in place and filmed.

His camera caught the images for the world to see later that night on the evening news and in the next day's newspapers. A rusty and dented Toyota pickup truck that had once been some shade of red was stopped in the street not 5 feet from the closest person in the crowd. From the passenger side of the truck, a man with a red cloth wrapped around his face was seen firing an old AK-47 assault rifle. In the back, 4 more men, also with the red cloth draped faces were firing similar weapons. One man fired his rifle in a long fully automatic burst. While he reloaded, another man fired small bursts. A third gunman picked his targets more carefully and fired one shot at a time. The last man fired off an entire magazine of ammunition in one long burst.

Instead of reloading, the last gunman dropped his rifle and tossed white phosphorus grenades into the church. The grenades thumped and echoed from inside as they exploded. White smoke flooded out of the doors and glowing orange fragments went everywhere. Streaks of white, smoky trails chased the shrapnel. Peaceful banners hanging from the ceiling caught fire immediately.

So did the dress that the bride's mother had made for her.

The hit squad's attack lasted just under a full minute. In that time, they had randomly fired all over the church, pockmarked its façade, and rained stained glass from the small rose window everywhere. On the ground, the wounded were executed one shot at a time. Frozen in place, first by ignorance, then awe, then terror, the American was left unscathed, and everything was caught on video.

Saturday, February 7, 2010 Washington D.C.

Office of Homeland Security

International Terrorist Movement Monitoring Division

6:08pm local

The small office was nearly empty. Out of the 46 people who normally worked there, only 3 remained. Everyone else was off for the weekend. Even the three people in the office didn't need to be there. They were just trying to get some paperwork done.

The small division's responsibilities were more exciting in name than in function. At first glance, anyone walking through the generic government building might look at the plate on the door and be impressed. It implied a sense of importance in America's defense against the world's worst criminals. Instead, the forty-six employees were little more than a liaison and information-filing group.

As a matter of habit more than anything else, the office's television set was turned on to monitor one of many 24-hour cable TV news channels. The division was typically being sent electronic information and news from dozens of different government agencies; including the Central Intelligence Agency, and the National Security Agency, the Federal Bureau of Investigation. Still, as every American learned over the years, there's no beating a 24-hour news channel for finding out about breaking events. It was the exception and not the rule when any of their information-feeding agencies notified the division about anything more than a developing event. Breaking news events needed cable.

Even with the television turned on, everyone usually went about their normal office routine. Form after government form always had to be filled out. In an era of the paperless office, the biggest change in document control was that forms that were once filled out in triplicate only needed to be filled out once and saved on computer. That left more time for more paper forms to be required. Dealing with all the different agencies also meant dealing with all of their required documentation. It was a war against bureaucracy, and this office was the frontline.

Among the people in the office was a 24-year old man named Manuel Luka. Having graduated from Georgetown University with degrees in criminal investigation and Eastern European studies, he was one of the office's bright stars. Manny came to the office fresh out of college, when the Office of Homeland Security was first being created back in 2001. It was a chaotic time, and through the confusion, he was hired into the Terrorist Movement and Monitoring Division.

At the time, most of the government's attention centered on the infamous Al Quaeda terrorist network. Despite his Eastern European studies background, Manny found himself documenting the movement of Central Asian terrorists. It was a boring job that was really nothing more than summarizing information from multiple government agencies, and then inputting the summaries into the computer network database. A high-school educated, administrative temporary employee could have done his job. The only real reason he was there was because his supervisor liked his resume, and he was sure that someday he would be more far more useful in some other capacity.

Through the years, Manny did his job well. Actually, he did it too well. In time he had come to memorize all of the known terrorists in the database. He knew their names, their hometowns, the names of their family members, and he knew their individual histories. It wasn't something that Manny tried to memorize, but re-writing reports, constantly entering the same information over and over again resulted in a sort of passive memorization. By memorizing so much specific data, Manny had slowly become one of the foremost authorities on Central and East Asia's known terrorists.

Attrition in government agencies, as in 21st Century business, was fast and furious. When administrators came and went, so did pieces of their administrations. Some people left for other careers. Others moved on because of financial cutbacks or disagreements with supervisors. Manny was a quiet and generally average person who kept a low profile. He did his job, and he got his paycheck. He always felt a better than average sense of patriotism, and money was his primary motivation. Even at only 24-years of age it was almost as strong as his red, white, and blue. While the names on the reports that came to his desk for summary were changing, Manny remained. One by one, their replacements began calling Manny for information instead of sending it to him.

The result of becoming a walking database was that he had far more work to do, and he was still a salary employee with no overtime benefits. The only perk from working so much overtime was that it made him more valuable and less likely to be let go in some sort of financial cut or administrator scuffle. Manny valued his job security above and beyond all. Besides, a new political appointee had been brought into his division. That appointee had been trying to make a name for himself, and to prove that he was qualified for the job. Already two of Manny's friends had been let go by Manny's new boss. The money was nice. The patriotism of his job was nice, and his tasks were interesting, but the security is what kept him from quitting and working in a different agency or going to the private sector.

On this snowy and quiet Saturday afternoon, Manny had come to work to finish his summary of an interrogation report. It was a CIA summary of a

Pakistani ISI (Inter-Service Intelligence; i.e. secret police) interrogation report. The subject of the interrogation was a Pakistani named Munthmar Hussein a.k.a. "Papa." Papa had been arrested by the Pakistani ISI on a wimpy charge of carrying a weapon-specifically, an outlawed tribal knife traditional worn by most Pakistani tribesman. It was just an excuse to interrogate the man after a NSA (National Security Agency) operated electronic spy satellite had monitored one of Papa's phone calls.

The satellite, one of many spinning around the planet, was a signal intelligence receiver. The satellites received the electrical signals between almost every phone on Earth. Those signals were compared to smaller signals, and when there was a match, the origin of all the signals was triangulated to determine their location. Whenever possible, the telephone accounts were determined. Digital transcripts of phone calls were stored in a database. Electronic signatures from different voices and languages were constantly being updated-billions of times a second.

There was so much information that a computer program was written to automatically sort through it all. That program depended on keywords. Words and phrases like "...shoot the President..." were digitally recorded in multiple languages and dialects. All of the phone calls monitored were then digitally compared to the digital recordings of the keywords. Whenever a keyword was located, the recording and all of its information was saved in the database and compared with other digital recordings of people on the various CIA watch lists.

Manny's subsequent summary report went as follows:

On Thursday, February 5, 2010 at 10:37am local time, Papa used his home telephone in Islamabad, Pakistan to call a man using a cellular phone in the disputed Kashmir province near the town of Shengus. The conversation was recorded based on Munthmar Hussein's previous record of involvement in the Pakistani group known as "The Protectors of the Koran." During previous investigations into that group, telephone conversations were recorded with his voice prior to an assassination attempt on Pakistani General Omar Hunandri. The call was initially triggered when electronic signatures, common to low level keywords, were detected in a high level order. Those key words were: mission, guns, holy, Christian, American, and example. Individually, those words are all benign, but when the NSA satellite recognized them in that order, it searched the watch list database to determine the caller. The caller's voice was definitely Munthmar Hussein's, and the conversation was recorded.

Details of the conversation were scattered due to interference. Typically the NSA's term 'interference' usually refers to some sort of sophisticated and deliberate scrambling usually from military sources to protect their conversations. In this case, the person that Munthmar called was a man using a cellular phone with an unknown battery condition. The call could have been broken up by the high altitude or the presence of the nearby mountains-specifically Mt. Heramosh a few miles to the northwest. The NSA has supplied a fragmentary transcript for management in the Office of Homeland Security database.

During that conversation, Munthmar seems to be referring to a scheduled attack on Americans. The NSA immediately notified the CIA and the Pakistani ISI. They in turn arrested Munthmar and interrogated him using local procedures.

According to the ISI, no further information was yielded from the interrogation. Munthmar was treated for some minor injuries that were allegedly incurred during his arrest. He was charged with illegally carrying a weapon and resisting arrest. To limit the apparent threat to Americans, he was tried, convicted, and fined $10,000. It was hoped that the huge fine would have imprisoned him indefinitely, but another Pakistani paid the fine, and Munthmar was immediately released.

The man that paid Munthmar's fine, Sheik Mohammed Hanjour, is also on the ISI and the CIA's watch lists. Sheik Hanjour is a selfproclaimed prophet of Islam. He has a small following estimated at less than 100 people. The average follower has very little education and comes from the destitute poor in the Islamabad, Pakistan area. While the Sheik has a near empty record of criminal arrest, most of his followers are on the watch lists. Sheik Hanjour has had many private visits with high level members of the Al Quaeda network. It is assumed that the Sheik's teachings are little more than a recruitment tool used to fill the ranks of the Protectors of the Koran terrorist group.

Given that the Sheik has no recorded income and survives on the benevolence of others, there are questions about his involvement. Where did he get the $10,000? Who supplied it to him? $10,000 is currently a huge sum of money in Pakistan. Why did someone pay so much money to release Munthmar Hussein from the ISI? If there is a terrorist attack against American's in the works, who is funding it, and what is Munthmar Hussein's involvement?

The CIA is researching the issue of the $10,000, but at the time and date of this summary memorandum there is no more information about the apparently impending attack. The ISI is still keeping a close watch on Munthmar Hussein, Sheik Mohammed Hanjour, and a few other members of the Protectors of the Koran group. It is recommended that the State Department issue a traveler's advisory to all Americans currently in Pakistan or traveling there in the near future.

Manny finished his summary report, saved it in his computer, emailed it to all the usual recipients, and finally headed off to the office fax machine. After dialing the number and waiting for the familiar screeching sound of two fax machines communicating, he pressed "SEND" and his day was over. All that was left was to file a hard copy of the fax in the main filing cabinet near the receptionist.

He walked to the long row of horizontal cabinets and began to flip through them. If everything weren't where it was supposed to be, then no one else would be able to find it except him. It took a minute or so, and while he flipped pages, the television next to the receptionist desk behind him spewed out the evening news. Manny was oblivious to the anchorman until the man on the box reported that there had been a drive-by shooting at a Christian church in Islamabad, Pakistan.

Instantly, Manny flipped around to see the report. Having just written the report about a potential attack involving Americans in Pakistan, he was curious. The he saw the footage that the American wedding guest had filmed. The news anchorman had warned that the footage was graphic, but it was more than Manny had expected. Bodies lay as they had fallen. The street was covered with blood, wildflowers, and empty casings from the assassins' assault rifles. Each of the bodies had small holes in their heads, and large exit wounds. The carnage was terrible.

The report also showed interviews with some of the survivors. The American said that he "…looked right into the eyes of the terrorists." The badly burned bride cried through her bandages as she pleaded to know the cause of the attack. Her new husband who had shielded her from the bullets was hanging on to life after being shot twice in the back. Even if he survived, doctors told the reporters that he would be paralyzed from the waist down. The report convinced Manny. This was the attack that he had just written about, and it wasn't against Americans, it was orchestrated for the Americans to see.

Manny knew that terrorists, at their core, seek to accomplish some sort of gain through terror first, damage second. The best way to create terror is to illustrate it, and the best way to illustrate something was to make sure that it was caught on camera. That's why they were intent on NOT shooting the American: so he could film it. Everyone else was too poor to own a video camera, and that fact sealed their fate. The couple's survival was probably a fluke on the terrorists part, but it played in their favor. Now, the media had more than bodies, blood, and bullet holes. They had witnesses to cry in front of the cameras.

Manny stopped trying to file his report and headed back to his computer. A few miles away someone else was working late, and when Manny sat down at his computer, an email had already arrived from his NSA liaison. The email did little more than state Manny's own thoughts. Was this the attack involving Papa?

It was late on a Saturday, and instead of finally being able to leave, Manny was compelled to do more research. He pulled out all the information that he could on Papa, the Sheik, and then searched the internet news dot-coms for the latest information on the church attack. Finally, he went to make calls to see what kind of "less public" information he could scrounge up. The only problem was that most everyone was at home on a Saturday night.

Everyone was off for the weekend except his CIA contact. Over at CIA, his contact had the same problem as Manny: yet another new political appointee was trying to show that they were qualified for their job, and heads were rolling. Instead of sending emails back and forth, Manny called his contact to see if there was any more information from the Pakistanis about the church. His contact was evasive and reluctantly yielded no information. It was obvious that he was in some sort of awkward position. He did, however, give Manny the name and number of a contact in Pakistan's ISI, Captain Monsoor Asaad.

Manny thanked him, wished him well, and immediately tried the ISI. The satellite communication delay was awkward, as was the inherent language difficulty, but he was able to get through. Captain Asaad told Manny that they had already arrested a suspect, and Manny was correct in guessing that it was Munthmar Hussein-Papa. After delicately sharing information for a few minutes, Asaad told Manny that he was surprised that Manny was so well informed. Eventually, the two finished comparing notes, and they thanked each other. Manny told Asaad that if he was ever in D.C. to look him up, and they'd have lunch. Asaad did the same.

By 9:30 that night, Manny was finally done rewriting his report. Once more, he emailed, faxed, and filed copies. It was time to go home. The other two people in the office had long since left, and there were only a few lights left to turn off. He shut off the television, locked the doors, turned, and took one step.

In the office, the phone started to ring. *"SHIT!"* screamed in his mind. Like a juggling rodeo clown with a loose bull, Manny scrambled for his keys, found them, picked out the right one, unlocked the door, and ran inside to get the phone. When he picked it up, there was no one there. Completely frustrated, Manny finally made his way home-again.

After meandering his way through Washington's endless security checkpoints, he finally got in his car and headed for home. Manny fought his way through 6 lanes of slush-filled Beltway traffic all the way to his apartment in Maryland. At the same time he was hanging up his keys, Manny started kicking off his shoes.

Somehow he had gotten a cold and annoying piece of snow inside his shoes. His feet were like ice, and he peeled off his socks to help get them dry and warm. Barefoot, he made his way to the kitchen to give his cat some overdue dinner. As soon as the shoes and socks were off, he started feeling better. The grateful orange cat, Bob, was affectionate, and Manny was feeling relaxed after the long and frustrating unpaid Saturday at work. He flopped down on his couch and reached for the television remote control. It was time to turn off his brain.

The TV came on, and the phone rang in his kitchenette. Once more, Manny dropped what he was doing and ran to answer the phone. This time there was someone there. It was his new boss, and he was calling from the office.

Manny's boss wanted him to come in to talk about the Pakistan attack. When he informed his new boss about how he had just left, the appointee huffed and decided to tell him the news over the phone. Captain Asaad had called Manny's CIA contact to get Manny's phone number. No one answered when he called the office, so Asaad made a few more calls and got the home phone number for Manny's new boss. He in turn, called Manny to express his great displeasure at receiving phone calls in the middle of a Saturday night from Pakistani secret police on the other side of the globe!

Asaad had been impressed with Manny's knowledge of church attack, Munthmar Hussein, the Sheik, The Protectors of the Koran, and the network as a whole. He was so impressed that he wanted to take Manny up on his offer to have lunch, but he couldn't make it to Washington with the investigation underway, so he wanted Manny to fly to Pakistan! His boss didn't want to upset anyone in the least, understood the offer was an opportunity to show how closely his division would work with intelligence services around the globe, and he had already made the arrangements.

Manny had no choice. He'd never been outside the U.S. except to Canada, but that was a private matter, and he wasn't too proud of it. He wasn't afraid to leave the U.S., but Pakistan?! The only thing he could do was quit his job and search for another. Before he could think, he agreed to his boss' "suggestion," and he started packing his suitcase.

His boss told him to go to Andrews Air Force Base at 7:00am Sunday morning. Arrangements had already been made for him to ride an Air Force C141 cargo plane that was on its way to Karachi. It would stop in Islamabad on its way. At 12:15 in the morning, Manny slipped a note and a key under his landlord's door. Manny couldn't believe that in 24 hours he would be in Pakistan!

He walked back to the couch and tried to surf the channels for something interesting. Everywhere he turned there seemed to be a commercial. The only channel that didn't seem to be showing anything but ads was the same 24news channel he started with.

They were still discussing the problems in Pakistan. One of the experts pointed out that a Multi-Denominational Religious Conference was going to be held in Washington, and that provided some hope toward quieting the religious issues in the region-particularly the Christian/Muslim relationship in Pakistan. They talked about the international difficulties around the country, the threat of nuclear war between India and Pakistan, and about America's continuing refusal to support Pakistan as an ally.

As he drifted off to sleep, he tried to suppress the idea. His final thought for the day was of Bob. At least the cat would get fed.

While his blinds were still open, and the sun was still hidden, Manny's alarm went off. It was far too early for him, so he hit the snooze button. Seven minutes later, it went off again, and he hit snooze again. This time, he was awake and already frustrated. He knew the alarm would be going off again soon, and he had to get on the plane to go to Pakistan. It was still unbelievable.

As he wrestled with the concept, Bob jumped on his chest in the hopes of convincing Manny to feed him. He turned off the alarm and began getting ready. The suitcase was already packed, and he expected traffic at 6:00am on a Sunday would to be light even on the beltway. Still, he left in a relative hurry. Bob the cat wasn't pleased, but his landlord would hopefully remember to feed him. Besides, inconceivable as it was, Manny had to catch a plane to the other side of the planet!

He was right about the traffic. The Beltway was barren at 6:45am. It was also free of snow and slush. His frustration from the alarm and the mandatory flight faded as he realized that he had discovered the perfect commute time for when he returned. Then he wondered, when would he return? The trip was open ended, and seemed hastily-organized by a political appointee whose only objective was to get some on-scene face time for his division. Manny wondered if perhaps he would be able to come back after only a day or two?

At Andrews Air Force Base, security was tight. In the event that a passenger flight was hijacked and turned into a missile by a terrorist, fighter interceptors were on alert and waiting at the end of the runway. Armed guards circled the perimeter in armored pickup trucks with machine guns on the back. At the gate, Manny pulled up and was sternly asked to state his business. Awkwardly, he told the guard his name and that he was supposed to catch a flight. Oddly enough, the guard immediately recognized his name and directed him to the parking area, the

THE X-MAS WAR

flight line, and even his specific plane! Manny had to wonder if perhaps his new administrator was more efficient than he had expected.

He parked his car in a parking lot that the gate guard had directed him toward, then he walked up to the plane. A young man, possibly 18 or 19-years old stood at the long and fat plane's door near the front left side of the plane. In his hand there was a clipboard, so Manny asked him if this was the plane to Islamabad? The kid put the clipboard behind his back and looked at Manny square in the eye.

"You must be Mr. Luka? We were told to expect you. Yeah, this is the plane to Pakistan. You can go on in and find a place to sit. This isn't an airline, so there's no real seat assignment. We're not first class either, but our squadron commander, Colonel Earley, will be taking the reins for part of the trip, so I'm sure we'll have a smooth run."

Manny knew a bit about the military, and he recognized the Sergeant stripes on the kid's shoulder. He thanked the young man and climbed into the plane. Inside, he found a seat behind the flight cabin, and tossed his suitcase in an empty bunk nearby. A minute or so later, the Sergeant came in and moved it to the back of the plane with some other flight bags for the crew. The plane was filling with other members of the flight crew, and on his way to move the bag the Sergeant stopped to introduce Colonel Earley to Manny. The Colonel was happy to see Manny's suitcase removed from his bunk, but he wasn't too impressed with Manny.

"Ah yes, Mr. Luka. I've known your boss for years. He's a good man. We served together during the Gulf, but of course he never calls me unless he needs something. This time it's your ride on one of my planes. Last time he made my wife and I sit at some sort of D.C. "Look who's here" $1000-a-plate dinner! I hear he's hangin' his hat over at Homeland Security now. So, are you some sort of emergency diplomat, or are you a spook, a spy?"

Manny was taken back. This was all news to him. Manny barely knew anything about his boss. He did know that military people, retired or active, need to be shown polite respect, so he unfastened his seatbelt and stood up to talk with the Colonel and offer a strong handshake.

"Sir, it's a pleasure to meet you. I'm afraid I really don't know my new boss too well yet. I can tell you this, I'm not a diplomat in the least, and I am definitely not some sort of spook. Over at Homeland Security, our mandate is to develop emergency procedures and assist the coordination between different agencies. As far as I know, I'm just going over to get some face time for our division."

The Colonel laughed and headed to the cockpit. On his way he stopped and turned toward Manny.

"Mr. Luka, you really might not know it, but you're a spook."

He laughed and sat down in the pilot seat. Manny was confused and didn't know what to say. The Colonel knew it too, and he told him what to do next.

53

"Have a seat, Mr. Luka. Your adventure is about to begin. We're taking off as soon as our Sergeant decides to stop sitting around eavesdropping and finishes getting us ready to taxi out."

In the background, the rear cargo ramp started to close. Then the Sergeant appeared from the cargo bay and ran over to the plane's door to close it. Manny sat in his seat and buckled up. The Colonel was already taxiing the plane out to the runway and communicating with the tower. They were on their way.

Once airborne, the plane began its journey. Most people take flight for granted, but Manny had only been in a plane a few times, and never for more than three hours. Day by day, Manny poured over reports about terrorist movements from all over the globe, and he had come to take it as a matter of fact that for someone to go from Europe to Indonesia was no big deal.

As the C-141 flew from Washington D.C., to Denver, to Seattle, to Anchorage, to Tokyo, then Bangkok, and finally Islamabad, Manny finally realized that traveling to the other side of the globe was still a matter of great endurance. He had left Sunday morning just before 0700 local time, and after all of the refueling stops, he landed Tuesday afternoon at 1400 local time. Because of the time zones and International Date Line, he had never even seen Monday. It was no wonder that his boss made the arrangements so hastily. If he had taken a commercial flight, it could have taken him a week!

None of the refueling stops had been long enough to let anyone off the plane. Their longest stop was in Tokyo, where they had to open the rear cargo doors, and he had to help the Sergeant push out a wheeled pallet with boxes on it. They were wrapped up in plastic so there was no way to tell what was in them, but the Sergeant told him that it was food for the Japanese to drop as humanitarian aid over Afghanistan. Manny thought it was odd that their C-141 didn't just fly over Afghanistan and drop it on their own, but he also understood the nature of a bureaucracy. Paperwork and government organization were things he knew a great deal about. Even in Tokyo, the stop wasn't even an hour. He had been able to walk around the plane, but it was great to finally leave it!

The plane continued on to Karachi, then Islamabad. Manny thanked the Colonel and the crew. Meanwhile, the Sergeant opened the front left cabin door to the plane, and waved goodbye as he walked away from the plane. Manny stopped in the middle of the open taxiway and watched as refueling trucks were already connecting to the huge fuel tanks in the wings. When he turned around to head for the airport concourse, an old jeep pulled up next to him. Out of the passenger side, a brightly decorated Pakistani Army officer stepped out before it even came to a complete stop.

"Mr. Luka! I'm Captain Asaad. It's a great pleasure to finally meet you." The two men shook hands while Asaad's driver came around and threw Manny's suitcase in the back of the jeep.

"We should keep moving, Mr. Luka. This airport seems like any other, but unfortunately we have an insurgency problem in this country, and things are not as safe as they seem."

To Manny, the place looked almost the same as Tokyo or any other airport that he had just visited, but from reading the reports, he knew that Asaad was speaking the honest truth. Everyone climbed back into the jeep, and the driver began a speed-free run that would have turned any NASCAR driver green with envy. The last time Manny had been in a car with a driver that courageous was back in college, and there had been a lot of beer involved.

Asaad saw Manny's white knuckles gripping the open side of the jeep and tried to calm him by talking shop.

"Mr. Luka, may I call you Mirko?"

Mirko was Manny's birth name, but he had since changed it to Manuel. Even after that, everyone still preferred to call him Manny. It was obvious that Asaad had done his homework, but it still surprised Manny to know that the Pakistani secret police had a detailed file on him!

"My friends call me Manny. I changed it from Mirko years ago."

"Ah yes…Manny. I'm sorry. Manny, I thought you might find it interesting to see the latest on the Church attack investigation. The people higher-up in the chain of command have decided that the potato was too politically hot for them to handle, so I'm leading it myself now."

Manny hadn't understood why someone so low on the chain of command was assigned to handle an incident that was so politically delicate. The question was clear in his face.

"You see Manny, this is what you would call a fundamentally Islamic country. It's only because of a matter of timing that there is a democratic government in power right now. It wasn't always that way in the past, and if most people have it their way, it may not always be in the future. There are only a few Christians in this country, but to a great many people they are symbols of Western Imperialist governments. Those governments are not very well liked by many people, and many of those same people see our government as a mere puppet of the West. The upset people cannot strike back at our government, because we would imprison them forever. They cannot strike back at the governments of the West; we are all very poor over here, and it takes a lot of money to travel. Instead, they attack the symbols of the Westlike the church. Everyone knows that the West, including you and your government, will never be scared off by such acts, but it does relieve some of the political stress."

Manny understood the problem and told Asaad. "Captain, or may I call you Monsoor?" Asaad laughed and nodded in agreement.

"Monsoor, this is an age old problem. We all know it, and we all understand it, but very few people can do anything about it. There are people frustrated with

Western-backed governments and what is sometimes perceived as harsh rule. There are those that want every Islamic country to be another Iran. There are those as well that just hate Christians. It's a very bad situation. On the other side of it, there are billions of people that see such frustration venting attacks as acts of barbarism. This church attack is a good example. Given all of its symbolism and its high profile, why have your seniors passed it down to you?"

Asaad turned to watch the road while he thought about his reply. Their jeep raced through streets congested with pedestrians, animals, makeshift shops, and even cars. All the while, Asaad thought of how to answer.

"Manny, this is turning out to be a messy incident, and many people do not want to get their hands dirtied. As soon as it was tossed on one desk, it was passed to another and another. Unlike the others, I will not pass it on for the protection of my career. It's not that I so desperately want to see justice done, but I do not like to hide from things.'"

The driver interrupted and turned to Manny. "Captain Asaad is a relentless man, Mr. Luka. He's always the first to go through a door and arrest a suspect. Nothing deters him: not doors, and not even bullets. That's why they call him 'Bulletproof.'" He's been shot on three separate occasions, and he has a hundred scars to prove it!"

Asaad laughed and turned to Manny. "You had better just call me Asaad. Your pronunciation is a bit awkward, and it's a little easier. There are not that many scars, but he's right about the doors. I will not let a closed door block my path. If I want to get into a room, I will get inside. If I want to solve a mystery, I will solve it, and I will solve this one."

Manny was tired, a bit edgy, and not as quiet as usual. He had to ask the question even though he might normally have been so inquisitive.

"So, Asaad, I don't understand. Why are you so determined to solve this mystery?"

Asaad turned around and looked Manny square in the eyes. He saw a man who hadn't slept in days, but he also saw a man that was honest and innocent. Those characteristics might have just been elements of Manny's inexperience, but Asaad also hoped that they were part of his true nature. He told the driver to pull over so that they could get some coffee at the next café. As the jeep stopped, Asaad motioned for Manny to come out of the jeep and join him. When the driver pulled the jeep across the street to park, Asaad explained why he was so determined to bring the perpetrators of the church attack to justice.

"Manny, as I said earlier, there are few Christians in this country, and they are not all viewed very favorably. Many are hated just because of what they passively represent. I was born a Christian. Today I am a Muslim, but I feel a kind of relationship with them. I know what goes on in a church, and it is not some sort of secret imperialist meeting. Christians are peaceful people, and very few will stand up for them here. It's political suicide. I see it as a chance to show everyone in Pakistan and in the rest of the world that justice crosses religious boundaries, even in an Islamic nation. I think that if I do not do this, then no one will."

They sat and had two cups of coffee each. The driver waited with the jeep across the busy street. Everything seemed normal. The conversation was as familiar as Manny's usual phone calls except that the setting was far from his clean and orderly cubicle, and the people around them were the complete opposites of his normal crowd. Asaad didn't offer to pay, but as soon as Manny pulled out his wallet, he motioned for him to put it away.

"We do not have to pay. There is an understanding that we have with the merchants in this country, but thank you for the gesture. We should go; I think there may be trouble soon."

He nodded toward a small crowd of 5-10 young men that were arguing on the sidewalk about 100 yards away. The driver saw the nod and brought the jeep over to pick them both up. While they were pulling away, Manny looked back and saw that the crowd had already quadrupled in size. Asaad saw him looking back.

"Trouble can spring up very quickly in our country, Manny. That is another problem that we will face, but we will face it tomorrow. I'm sure that you could use some rest. We will take you to your hotel. I have made all of the arrangements already."

"Is it the same arrangement as with the café manager, or will my people be able to foot the bill for this?"

"No no no. It is neither. The hotel's manager understands why you are here, and he was more than happy to help out. I'm sure you will find everything surprisingly comfortable. You may wish to report your safe arrival before you retire. My driver will remain here in case you wish to go anywhere. I suggest you follow his advice if you do. As you see, problems can develop very quickly in our country, and even before Afghanistan and the attacks on their Taliban, there are many many people here who do not like westerners."

They pulled into the hotel. Manny thanked Asaad, and the Bulletproof Captain of the Pakistani ISI secret police disappeared into the crowded sidewalk. While Manny watched Asaad, the driver parked the jeep across the street and waved. When he turned around, a bellboy was already holding his suitcase and waiting to escort Manny to his room.

The hotel was luxurious, even by American standards. Manny tipped the bellboy and flopped on the bed for some long overdue sleep. On its way to the pillow, Manny's head glanced at the room before the heavy eyelids crashed shut. His last thought of the remarkable day was how closely the room resembled an American hotel room. Then he slept.

Wednesday, February 11, 2010 Islamabad, Pakistan

Our Holy Lady of Divinity Church

7:30am local

Manny's sleep deprivation was finally relieved by the long night's sleep. It wasn't until the concierge personally called him that he finally wrested himself from slumber. Asaad had arranged the call, and he arrived exactly thirty minutes afterward. Manny's late arrival the previous night had meant that he was unable to do more than leave a voice mail message for his boss, and his purpose in Pakistan was still vague. For all Manny knew, he was just there to get some face time for the division. Asaad assumed that he was there for a more in-depth collection of information, so when he arrived they immediately went to the scene of the church massacre.

The neighborhood was poor by western standards. People wore rags. Shops were little more than tents and wagons parked at the side of the street. Beggars were plentiful, and there was the unique smell of a third-world slum or ghetto. Manny was a typical American suburbanite with no experience in such neighborhoods. To him a poor neighborhood was a place with lots of graffiti, some bullet holes, blaring rap music with lots of bass, drug dealers, and a population of people oppressed by their minority status. It was his suburbanite view of the world, and now he was getting an idea of what poverty truly was. In the U.S., Manny believed that there was at least some chance of leaving a ghetto, but in the third-world streets of Islamabad, Pakistan, there was no chance to change one's station. It was startling to him, and although he tried, he could not hide it.

Asaad didn't seem to notice Manny's epiphany. He nonchalantly watched the streets while the driver raced through to the church. People had to jump out of their way as the jeep drove straight into the crowd. Everyone knew that if they were hit there would be no ambulance, there would be no lawsuit, and there would be no justice. In this country, the government controlled all, and the ISI had absolute power. All that anyone could do was stay out of their way, and so they did.

At the church Manny finally saw what the pictures on TV and on the internet could not convey. Days after the incident, the streets were still maroon with blood and glittering with shards of stained glass. Before the jeep stopped, Asaad stepped out and headed for the doors of the church. Manny followed after the brakes stopped squeaking and the jeep stopped moving. On his way up the steps, his feet kept sticking to the blood-stained areas. Crunching glass and shards of stone seemed to be louder than the surrounding hustle and bustle of the crowded neighborhood's daily routine. He made additional note of the gaping hole where the rose window had been. As he drew closer, he noticed the pockmarks in the walls from the bullets. Peering inside, he saw nothing but blackness. Sunlight couldn't pass through the soot in the inside of the windows, and it couldn't reflect off the scorched wooden interior.

While Manny was surveying the scene, Asaad was already inside and talking into his cell phone. He wasn't speaking English, and Manny didn't understand a word. It also struck him as odd that the church had lost its echo. The soot and charring had built up so heavily that it acted as an insulator. There was near total silence. Even if he could have understood Asaad, he wouldn't have been able to hear him fully.

Manny just took in the scene and imagined what had happened that day. Eventually, Asaad finished his conversation and came over to tell him what they knew. Asaad explained that the ISI had rounded up some suspects while Manny was asleep. Among those arrested was Munthmar Hussein-'Papa.' This time, the ISI hadn't filed any charges. They had put agents in the streets to see who was passing the word of his arrest. They were also in the process of setting up a bank account-monitoring program to alert them in case anyone began taking out large sums of money in preparation for another bond.

From all the suspects arrested, they had determined that The Protectors of the Koran had conducted the massacre. Their intention was to gain media attention from the West, and maybe even to help scare Christians away from Pakistan or other Muslim nations. Sheik Mohammed Hanjour hadn't come up with the idea. He was asked to approve it, but that wasn't enough to put him away-even in Pakistan.

Manny was curious. He had been reading reports for years, but he never understood why supposedly holy people-devout believers-advocated or even promoted violence, even though their religion was supposed to be so peaceful.

He asked Asaad.

"Why is it that whenever we chase the tail of these kinds of things, there's always some religious fanatic near the top? I thought Islam was supposed to be real peaceful, but it seems like the holiest of Muslims are the ones who incite the violence. I mean, how can anyone claim to be a proponent of peace, then turn right around and tell his followers to go and machinegun some people at a wedding?"

As a leader in the ISI, Asaad had to first come to grips with the motivation of the people that he had to face before he could combat them. He knew the question well, since he had asked it himself many times. He also knew that every Westerner had the same misconception that Manny had, and he was happy to have the opportunity to try and explain it to at least one Westerner.

"Islam is a very peaceful religion. It is more peaceful than Christianity, Buddhism, or Hinduism. There's a long list of reasons why Muslims, or anyone for that matter, conduct these attacks. I could go on for days listing them. Ultimately, there is no one reason that a peace-loving Muslim decides to go out and slaughter people. You must know this: there is no greater power than that of love. Someone who loves a person or a people completely is far more dangerous than someone who does nothing but act out of hate, greed, or thirst for power. Whether it's an Islamic terrorist acting on behalf of others, or a soldier defending others, it comes down to a matter of who has the greater love for the greater number of people. That love is easily changed to conviction, and the man with the most conviction will win any contest in sports, in business, or especially in battle."

Manny was still a bit confused. Asaad had made some sense. All animals defended those that they cared for-usually their offspring or their mates. Human beings have the ability and the need to defend entire extended families, neighborhoods, and even nations. When they see others being threatened, harmed, or oppressed in some way, they react with all their love, all their conviction.

"So, you're saying that the terrorists murdered all these people because they were such loving people? That they were just acting in defense of their families and their religion? This church has been here for a long time, and I assume that so have the Christians attending it. Were the Christians threatening the Muslims around here?" Asaad shook his head.

"No. They weren't a direct threat, but they were a symbol of the threat that the rest of the neighborhood perceived. You see, these people have nothing-no money and very little freedom. They know that if they had more money, they would be able to do more with their lives. They could even get away with more. Let's face it, even in the United States of America, if you can afford the most expensive lawyers, you can get away with murder. If you can afford to campaign, you can be President. If you can afford to own a company, you can control other people. As long as you've got enough money, you can buy whatever you want. That is a form of freedom, and that these people will never have. They hear about the United States with all of its wealth and freedom. They know that they have no money, and very little freedom. They know that they are at least ignored by the West, and some assume that their government is little more than a puppet of the West. You and I know that is not the case, but to a man in this neighborhood that can't even read, it's difficult to explain the order of things. They turn to their religion for answers, and there are some, like Sheik Mohammed Hanjour, who use their religious status as vocal points against government polices. Your Martin Luther King did the same, I believe."

Manny interjected. "Yeah, but he advocated non-violent protests to bring about change. That's what he was all about."

He had thought about all of this before, and he agreed with Asaad. Money is both power and freedom, and he knew that all three were powerful motivators. Still, Manny thought that someone in the chain of events had to convince the apparently peace-loving Muslims that the best way to express their love for the families, other Muslims, or their God was to go out and kill other people. Someone had to convince them that the Koran's doctrines were to be followed explicitly except in matters of murder and sometimes suicide for the alleged sake of Islam. He also believed that the person or people making that theological jump, and that emotional leap, were the true terrorists. They were more dangerous than the people who had pulled the triggers were.

Like most Westerners, Manny believed that the true terrorists, the instigators, had probably acted out of either selfish or delusional motivation. In any case, it was not worth getting into a deep and culturally-divided conversation with Asaad. Manny had seen enough of the devastation, and he was already feeling ready to go home.

Asaad could tell that Manny still had questions. He also knew that their conversation was slipping into the cultural divide, and it was best to move on. There was nothing more to see at the church anyway. Outside a curious crowd had surrounded the jeep. The driver had armed himself with an assault rifle from the back of the small vehicle. When Asaad stepped out onto the steps, the crowd opened and cleared a path to the jeep and down the street. Either he or his ISI uniform had clearly been recognized.

As soon as the crowd cleared, the driver tossed the rifle in the back seat and started the engine. Manny climbed in back, and the driver stepped on the gas while Asaad still had one foot out the door. Again they zipped down the streets, through the crowd of dashing pedestrians and across the town. An hour later, they pulled up to a modern building, an ISI headquarters station.

Asaad led the way up the steps that were more familiar to him than any home he'd ever had. Manny and the driver followed, but the jeep and assault rifle remained unattended in the street. While he was just about to remind the driver about the weapon, Asaad turned and told him that they were in the safest corner of the city. No one would dare even protest within blocks of the ISI building. It was a good indication of the reputation that the ISI had with the locals.

Inside, Manny couldn't help but notice the familiarity of the place. The ringing phones, the cubicles and computers, the coffee in the air, and even the indiscernible drum of hushed conversation all reminded him of his Washington D.C. office. It was a stark contrast to the neighborhood around church, but Islamabad was proving to be a place of contrasts. His hotel and the building seemed normal to him, but the neighborhoods and people seemed as unusual and foreign as he had expected. This was not a city of mud huts and illiterate people. There were decrepit old buildings and a level of poverty that he'd never experienced before, but there were dashes of Western-style modernism. He saw that as hope. The terrorists and many of the indigenous people saw it as an encroaching threat. Now he was beginning to understand.

Asaad lead them to his cubicle where he asked them to wait for a moment while he checked his voicemail and email. One of his co-workers came up and asked him a question, but Manny didn't understand the language. He wondered if perhaps he should have taken some sort of Arabic language courses in college instead of Serbian. The two men seemed to argue for a moment, then Asaad's co-worker walked away in a huff.

The ISI Captain finished listening to voicemail and looking at email, so he turned his attention to a long-ignored pile of snail-mail and hand-written messages from his departmental receptionist. After reading one of them, he handed it to Manny.

"It looks like you'll be leaving tomorrow. Your superiors seem to feel you're needed back home. I had hoped to show you some of our more peaceful attractions in the city. Perhaps we'll have time later today or tonight?"

Manny read the note. It was hard to read the handwriting, but he managed. It told him that his return flight was to be on the same U.S.A.F. plane that had dropped him off, and it was scheduled to arrive at 1330 hrs. local time, Thursday, February 12, 2010. Manny's face-time PR mission for the division was complete. What he had accomplished was still a big mystery, but at least he had gotten out and had a little adventure. He's also been given a rare opportunity to see the coordinated secret war that intelligence services around the globe were had been fighting everyday since September 11, 2001.

Asaad pulled out his cell phone and told Manny that he would contact the hotel to make sure everything was in order. The man on the other end must not have like something that Asaad had said because he had to raise his voice. It was the first time that Manny had seen any breach of professionalism in the man. Their conversation went on for another minute or so. Eventually, Asaad regained his composure and hung up. He smiled, apologized, and said there was some misunderstanding about the duration of Manny's stay, but it was all taken care of.

With his messages checked, Asaad asked Manny if he would like to actually see the terrorists that they had rounded up before heading home. Manny was unsure, but he had to reply in the positive. If he had refused, it would have been a slight against the man's work. Through his tone, demeanor, and charisma, it was obvious that Captain Asaad was a man whose work was his life.

Besides, a tiny part of him was curious. He knew that Pakistani interrogations were notoriously rough. In the worst cases, torture had been reported. The ISI was simply not known for messing around in security matters. He had the same curiosity that brought little boys to poke at road kills or made drivers go slowly past a car accident. "How bad could it really be?" he wondered.

They walked down a hall to what looked like a fire stairwell. Through the doors, there were stairs leading up to the floors above, and to what Manny assumed was the basement. They went down two floors, then into the subbasement where the interrogation rooms were kept. There were three more floors to go down, and Manny could hear people crying for help through the doors below him. Their cries were muffled and dim, and neither Asaad nor his driver flinched or batted an eye.

Manny had always imagined a torture cell to be a dirty, dank, and smelly dungeon, but this place was as clean and white as a hospital. If he didn't know that he was in an ISI headquarters, that's where he would have assumed he was. Even the stairwell was white and clean.

Asaad explained the building's layout.

"As I'm sure you know we employ a variety of interrogation techniques. We usually begin with primary interrogations, like the one that yielded our initial information about the possible attack on Americans from Munthmar Hussein, and those are held either on scene or in one of the second floor interrogation rooms. They are little more than your basic empty gray room with a desk. Secondary interrogations are held here on the level 2 basement holding rooms."

They walked through the stairwell door and into a room where four ISI guards snapped to attention. Banks of security monitors surrounded them. Asaad greeted them nicely, and they relaxed while he signed himself, his driver, and Manny on the interrogator list. One of the guards opened the door, and another joined them as they walked through to a well-lit hallway with wide doors on both sides. At a desk at the far end sat two more guards who remained seated, but were attentive to the visitors.

Asaad continued.

"On this floor, the visitors as we like to call them are kept in solitary confinement, but that's usually for their own safety. Rather than risk being a source of information, some of them would be killed on the spot if we left them in a general prison population. In these rooms, we monitor the interrogations through hidden cameras. Other hidden cameras here in the corridor are watching us right now. The rooms are empty except for a chair, a blanket, a bucket, and a small window near the ceiling. We hide our camera in the bottom of the light switch where the fastening screw would normally fit. No one has ever noticed that one of the screws was missing on the switch cover, and they remain secure."

Manny looked through a window in a door and into an empty room that Asaad pointed out. It was dim, greenish-gray in color, and the window was too thin and high for anyone too see out of. Besides, they were in the basement.

There was nothing to see.

Asaad described otherwise.

"In some cases, sleep deprivation techniques are required, and the small window assists in imitating the rise and set of the sun. By slowly varying the time between the imitation sunlight coming through the window, within a few days a visitor can become convinced that weeks, even months have passed. In a reverse action, we can also use it to lengthen days during interrogation periods. This is sometimes more effective than most techniques because the visitor may become convinced that an impending event has come and passed while in reality it would be days or weeks away. After learning that there might be an attack on Americans in the near future, this is where Munthmar Hussein was brought. There was not enough time to convince through light manipulation him that the event had already passed, so we had to move him downstairs. Other techniques used on this floor are as passive as the light manipulation. We can use a myriad of other methods here, most of which would be familiar to any of the law enforcement communities in the United States."

They walked on, and the guards at the far desk watched warily. This time, it was Asaad's driver that spoke with them. They handed him a clipboard to sign, and when he handed it back, they presented him with a pass for each of the three men. The guard that had accompanied them led them back to the original gate where the driver signed them out, and then they headed back into the stairwell.

On their way down another flight, Asaad turned and warned Manny.

"Now on this floor, there are some interrogation methods that are less popular in your country, but I imagine you know that there are certain American agencies that employ them regularly in the name of your 'National Security.' We are less secretive about them because they serve as a deterrent to many could-be visitors. Some of us take a great deal of pride in what we do here, despite how horrible things might seem. Each of us feels that we are saving lives through the things that we must do, and that gives us the strength to carry out the hardest orders with a sense of pride. These men are not sadists. They are professionals doing what are sometimes terrible tasks, and they are doing them extremely well. They are not doing this for fun or for money, but for the safety of their families and the security of our nation."

Manny was starting to wonder if his decision to see Munthmar and the other terrorist suspects was such a good idea. He wondered what he might see, and if he might be able to stomach it. There wasn't a lot of time to think about turning back. He couldn't if he wanted to. Besides, it didn't take them long to walk down the two flights of stairs to the next level of hell.

Once more they passed through the stairwell doors to a security checkpoint. Asaad's driver presented the pass he had obtained on the higher floor, and then he signed all three men in on the interrogator list. Then they passed through another gate to another sterile hall. Once more a guard accompanied them.

Asaad directed Manny to another door's window, and then he continued his tour.

"On this floor, we use rooms with a little more biological suggestion. The rooms are similar in layout to the ones above. They all have the same lights and windows and hidden cameras, but they are also equipped with medical hookups on the side. Do you see those small fittings on the far wall? We can bring in a medical team and use anesthesia to calm or relax visitors who have become either too energetic, too resistant, or are having some sort of medical difficulty. Very often, this is one of our most effective forms of intelligence gathering."

They walked down the hall, and Manny stopped occasionally to look through the small windows in the doors. Most of the rooms had people in them, but they all seemed like patients in an intensive care ward, and not Asaad's "visitors." When he peered into one of the rooms, the guard accompanying them pointed out that the visitor inside was Munthmar Hussein. Manny couldn't believe his eyes. He looked just like an average Pakistani in an intensive care wing.

Manny turned and asked Asaad about Hussein.

"The report I read indicated that he had resisted arrest and needed to be treated for his wounds. What happened? How was he wounded?"

64

"The arresting agents needed to restrain him, and his wrists were badly bruised in the scuffle. When he was finally restrained, we went through some of our usual procedures, and now he is resting comfortably as you can see. The guards tell me that since his subsequent arrest after the church massacre there has been no more significant information from him."

Asaad motioned for them to continue to the guard desk at the end of the hall. This time, he talked with the guards, signed the clipboard, and received the pass for the next level downstairs. They walked back through the gate, signed out on the interrogator list, and into the stairwell once more. As soon as he stepped out of the door, Manny heard the screams from below again. This was going to be it. He began to imagine the horrors he might see, but it was difficult to envision the dungeon like atmosphere given the hospital setting.

Asaad opened the door and presented their pass to the guards. They in turn presented the usual clipboard to sign, and the driver took care of the matter. Through the gate, the screams grew louder. More than one individual was being interrogated/tortured.

Manny started to feel the heavy uneasiness that was associated with screams of terror from other human beings. He'd never heard it before. The guards, Asaad, the driver, they were old hands to the shrill sounds, shrieking, pleading, and they didn't even seem to notice the same sound that was tingling Manny's spine. The door opened to the now familiar faceless hall of doors, and Manny followed the crowd. Right away Asaad stopped and proudly pointed out the room with Sheik Monsoor Hanjour inside.

Sheik Hanjour was a Pakistani man of unusually dark complexion. His robes had been removed, and he was wearing nothing more than a cloth around his privates. The light in the room was very dim, but Manny could see two men taking turns with the 50+year old man. One interrogator forced him to stand up, and the next made him sit down. He could hear a third man asking questions, but he couldn't see him through the small window. It was hardly the kind of thing that yielded the kinds of screams coming from at least two rooms farther down the hall. He had read about this kind of torture in the debriefing reports of pilots who had been shot down over Iraq in Gulf War I.

"This is a much more physically challenging interrogation method, Manny. Two teams of interrogators have been working in shifts with Sheik Hanjour since his arrest yesterday. He has not been given any food, and only water and Ipecac syrup. I do not think it would be a good idea to open the door. These kinds of things typically have a very very bad odor. That-more than anything-is the cause for rotating the interrogators."

Asaad stopped his tour and read a clipboard that was hanging in a tray next to the door.

"It seems our friend the Sheik here has already produced a few leadsthough not at the same time. After the interrogators switch, they report their findings in this report, and if you put some of this information together, it appears that he was supposed to meet with someone wearing a red sash at a small café near the church today. Our agents already have an operation underway there. Perhaps after we finish this little tour, we can go and watch. I think it would be good for you to see how quickly our intelligence can be collected. Would you like to see that?"

Manny thought for a moment, and a blood-chilling scream pierced the silence from down the hall.

"Captain, I think I've already seen enough here. It's obvious to me that your interrogation techniques are far more professional than has been reputed, but of course even that kind of a reputation is not a bad thing. After all, I imagine that the fear from the ISI's reputation probably assists in the interrogations."

Manny had fluffed the hard man, and it appeared that to have gone over well. Now all he had to do was bring it home, and he could get out of that torture hell. The screams had already gotten to him, and he knew that he might not ever sleep the same way again. He continued.

"I think that perhaps we should get over to that café as soon as we can. I wouldn't want to miss that opportunity. There's already so much that I'd like to see more of in Islamabad. I guess I'll just have to finish the tour of this facility on the next trip along with all of the cultural sites that you had mentioned earlier."

It worked. Asaad was sold on the idea. There were few things that he enjoyed as much as fieldwork, and he quickly agreed with Manny. They immediately walked back to the guard station, signed out, and headed back up the stairs. Manny glanced back down the stairwell. There were three more floors below the scream-filled one that he had just been on. No sounds came from those floors, and he wondered what went on there. He also knew better than to ask Asaad because he might have told him-or worse, shown him.

Back upstairs on the main floor, Asaad told Manny and his driver to wait while he found out the café operation's details. He left for a minute, and the two men just looked around in the awkward silence. Finally Asaad came back and told the driver where to take them. He hustled ahead and brought the jeep around while Manny and Asaad strolled through the office and down the steps.

Just as they reached the last step, the jeep wheeled around a bend and squealed to a stop. Manny had learned by now. He hopped in and grabbed on. As soon as Asaad's body was halfway off the pavement, the jeep was rolling and pedestrians began dashing out of its way.

The café was only a few minutes away, but Asaad had the driver go past the site. The street was much wider than the ones around the church or most of the neighborhoods that Manny had seen. After surveying the scene, Asaad had the driver park the jeep behind a building across the street. Once more the typically crowded city streets of Islamabad were packed with carts, street vendors, pedestrians, scooters, and occasionally cars. The driver reached in the back and grabbed his assault rifle again. Asaad led Manny and the driver into the building through a small rotten wooden door at the end of the alley.

Inside, there was a Pakistani family standing in a corner of a small and poorly lit room. Two other ISI agents and two Army commandos waited with them. Asaad continued through the room and into a small gift shop at the front of the building. Two more commandos and another agent were already there. The commandos were waiting behind the counter, and the agent was waiting in a shadow-filled corner near the front of the room-opposite the door.

Asaad spoke with the agent for a moment then came back to talk with Manny and his driver.

"The operation's scene commander is across the street in the back of the café. I just spoke with agent-in-charge on this side, and he tells me that Sheik Hanjour was supposed to meet with a man named Ali Azmaath. He's sometimes called 'The Ox' because of his stature. Azmaath is wanted in several countries, and I believe he is on your watch list for having ties to Osama Bin Laden. Our records indicate that Azmaath apparently runs a spiritual retreat somewhere in Kashmir."

Manny was impressed. This was just like the reports that he normally put together, but here the Pakistanis were doing it on the fly, and they were doing it very well even as information was still coming in.

Asaad continued.

"As I'm sure you know, Kashmir and Jammu are very troubled regions. There have been many terrorist groups infiltrating into the region to try and throw off the balance that we share with India. Of course, India blames the ISI, and we can't convince them otherwise even though they've captured most of the terrorist troublemakers. This Ox man's spiritual retreat is more than likely a front for a terrorist training camp. We have no evidence, since we don't even know specifically where it is, but I'm convinced that is probably the case. If we can speak with him for a bit back at headquarters, maybe he can tell us a little more."

Manny was eager to see things come together. Maybe this would be one of those lynchpin intelligence events where one man's information leads to two people who might lead them to an entire organization.

"So how will this all happen?"

"The man is supposed to meet with the Sheik around 5:30, during the evening meal. He'll be wearing a red sash around his waist. We hope that he'll think the Sheik is simply late, and he will take a table for himself. Either way, when he is spotted, the commandos will secure him and the scene while our agents arrest him and bring him back for questioning."

It was exciting to Manny, and he had to try hard to keep from squirming in the chair he'd found at the back of the shop. Asaad stood in the shadows with the agent near the front of the store, and his driver waited behind the counter with the two commandos. They arrived in mid-afternoon, and had to wait for hours until the sunset, and the evening meal crowd started to arrive.

Across the street at the café, there were plenty of people picking up bags of food, and Manny found it noteworthy that even in this supposedly backward country, people bought take-out just like Americans. While he was watching the crowd and thinking of take-out, the commandos and agents were becoming more alert. Everyone's weapon was loaded, cocked, and safety switches turned off. It had been like that for hours, and trigger fingers were becoming tired. So were nerves. It took everything each man had to stay focused on the mission.

As the street started to fill with dim orange light from the buildings instead of the dust and smog-choked sun, another crowd started to gather.

Manny recognized them as the same type of troublemakers that had driven he and Asaad from a similar café in the morning. This time, Asaad didn't want to leave, and they all watched as the crowd grew from a handful of people, then a few dozen, and finally hundreds; all within less than 30 minutes.

They were supposedly protesting the American involvement in Afghanistan, Palestine, and the very existence of Israel. More than anything, they seemed to protest everything that they could. Manny thought if someone had told them smoking was morally wrong then perhaps they would protest cigarette makers. The crowd reminded him of those groups of people that he had seen in college who would protest anything and everything just to protest.

The ISI and commandos had other thoughts. They were wondering how the crowd would change the flow and initiative of the situation. Would the crowd interfere, or would they run from them?

As the protesting crowd drew closer to the people picking up their food or sitting at the café, it started. A man near the center of the outside sitting section stood up and unfurled a red cloth. In front of everyone, he wrapped it around his waist and stood looking around.

Over the loud voices of the protesters, shots rang out. There were several individual pops across the street, then a burst of gunfire. People from the café ran in panic, and many of the protesters did too. Some were confused and didn't know where to run so they sought cover next to the buildings on either side of the street. While Manny looked around trying to make sense of the confusion there was an explosion behind the building. The ISI agent ran out of the building and across the street in an attempt to grab Ali Azmaath, The Ox. The commandos were up and on their way to help him when Asaad's driver raised his assault rifle and shot them both. Then he turned his weapon on Asaad and Manny. Both were sitting targets.

"Sit down and wait, Captain. This will only take another minute or so."

Asaad was quiet so Manny kept his mouth shut. Having seen men killed for the first time, there was more than a bit of fear in his veins. It didn't paralyze him, but it didn't make him heroic either.

As the popping sound of gunfire faded, and screams from the terrified and the wounded faded, Ali Azmaath walked through the shop's front door with a dozen other men. He sat down in front of Asaad and the two men stared at each other's eyes for what seemed to Manny as forever.

"Captain Asaad, I trust you will accept my invitation as a guest at my retreat house?"

Asaad was silent and refused to speak with the man. One of The Ox' cohorts backhanded Asaad to get him to speak. The sounds of the street seemed to fade, and silence filled the room.

The Ox told his men, "Take Asaad, and this one too. We'll deal with them when we've had time to give thanks to Allah and ask for his inspiration."

The men put pillow cases over both of their heads, plastic zip-ties on their hands, and then Manny felt the prick of a needle in his arm. He remembered walking out toward the back of the store and slipping on what he guessed was blood. He assumed that the explosion in the back of the store was probably a grenade, and the slippery floor was most likely the remains of the two commandos and two ISI agents that were in the back with the four-person family that owned the shop. Then he wondered how many people had died in the manic minute's worth of shooting.

Thursday, February 12, 2010

15 Miles NNW of Shengus, Kashmir

9:12am local

The smell of a farm woke Manny. He hadn't remembered passing out, but whatever the needle pinprick was, it must have knocked him out. The pillowcase was gone, so was everyone else. He was alone on the floor of a room with cold stone walls and a straw-covered dirt floor. He could hear a strong wind blowing through the space under the rough wooden door across from him. There were no windows, and very little light came inside through the small half-space under the door.

Manny tried to stand up, but became dizzy and fell almost immediately. He heard footsteps coming, and before he could even shut his eyes to feign sleep, Ali Azmaath, The Ox, stepped in to greet him.

"Mr. Mirko Luka. I believe it is safe to assume that you are an American, and since you are in the company of Captain Asaad, you must be some sort of spy for the CIA-"

"I'm just a data entry guy from Washington on a visit to show the flag and see how things are done. I'm not a spy-"

The Ox shook his head from left to right.

"Enough, Mirko. We're not going to hurt you. This is not the ISI. I'll release you in a little moment, but first I want to have your undivided attention."

The Ox stopped for a moment and peered into Manny's eyes. It was important to him to see a man's soul, and the eyes have always been the truest windows to it.

"Mr. Luka, I regret having to arrest you like this, but an American with Captain Asaad is not someone you leave behind when doing a mission like this one. Captain Asaad has been a very bad man to my friends, and someone had to put an end to his reign of terror. That is exactly what we have done."

Manny fought the urge to shake, and hoped that The Ox didn't see his fear.

"If you're not going to kill me, then what's going to happen to me. There's no one back home that'll pay a ransom, and my government doesn't negotiate with…um, people who take hostages."

The Ox laughed and pulled out a huge knife.

"We're not going to take you hostage, Mr. Luka, and I believe the word you were trying to hard to avoid is 'terrorist.'"

Azmaath cut the zip-ties that bound Manny's hands. "I'm setting you free, Mr. Luka, but I don't think you'll really want to leave."

"People call me Manny, and what makes you think I won't leave?"

The Ox helped him to his feet and through the doorway. Outside, Manny didn't see Islamabad anywhere. How could he when instead of the relatively flat terrain around the city, snow-capped mountain peaks surrounded him and illuminated the bright green terraced valley below?

"Well Manny, I think you'll have a tough time leaving here if you do want to. As you can see, you're not in Islamabad anymore. The tall mountain over there to our north is Rakaposhi Paravat, and the one over there on the left to our south is Heramosh. On the other side of that is the town of Shengus." The Ox laughed again and then continued.

"Will that help you find your way home? Maybe it would help if I told you that to the north are Tajikistan, Afghanistan, and China. To the east is India, and a few hundred miles to the south is Islamabad, Pakistan. All of those nations claim this area as their own. Does that help you at all?"

He laughed again, but Manny didn't find it funny. He was in a terrorist camp in the middle of the most disputed and war-torn area of the world. All the while he kept wondering how he had wound up there, and how he was ever going to make it home? He also wondered if he was alone.

"Where is Captain Asaad?"

The laughter faded from The Ox. They walked across the dirt road and into a large meeting house. Outside of the community kitchen, The Ox helped Manny into a chair on a terrace. He was overlooking the fields that descended on the north side of the little village he was in, and he watched as locals tended to a herd of goats while others carried baskets of food up the road from the valley to the west.

Manny surveyed the peaceful scene with the thoughts of his own predicament overpowering in his own mind. Then he saw why The Ox had helped him into the chair. Four men were pushing another man down the dirt road. They were headed down to the river to the west. The man they were pushing wore a pillowcase on his head, but he also wore an ISI uniform. Manny knew it was Asaad.

Minutes later, when they reached the river at the bottom of the valley they pushed him in. When the shock of the icy water made him stand up and try to walk out, each of the three men fired a burst of bullets into him. It was almost a mile away, but Manny could see the crystal clear water run red, and the whitecaps of the rapids to the south faded into pink with blood. Manny knew he was alone.

The Ox had been behind Manny watching.

"Manny, it looks like Captain Asaad's reign of terror has finally come to an end. As I said, no harm will come of you here. I do ask that you regard yourself as a guest, and not some sort of prisoner. If you do decide to leave, you will be doing so at some great risk, and you will be undertaking that on your own accord."

Manny turned around and looked at the man that had just drugged and kidnapped him. With the corpse in the distant background, it was hard for The Ox to miss Manny's displeasure.

"This is not a bad place to stay, Manny. It is not the terrorist camp that you envision. It is a spiritual retreat of sorts. This area is filled with different religious hideaways. Here, we take time out to worship Allah and thank him for all his mercy. We spend our days discussing politics, but only their relationship to Islam. You have just seen the sad side of things here, I hope that you will try to keep an open mind and observe the peace that we have here. After you've had a while to recover from your trip, I will show you around our little village. Until then, I'll have some water brought to you, please relax and try to enjoy the view."

Azmaath had already forgotten that the focal point of the mountain scenery was now the pink-stained river and the corpse drifting down it. Manny hadn't, but he wasn't about to try and piss off his kidnapper by saying something about the pathetic body of Captain Asaad in the water.

The Ox left him, and Manny had a few minutes to begin thinking about his situation. By now, his boss would know that he was not on the U.S.A.F. plane taking him back to Washington. He might have already talked with the ISI and found out about the kidnapping. What happened next was anyone's guess. Ali Azmaath seemed comfortable enough, and it didn't seem as though he believed some sort of rescue operation was coming after him. He must have felt that he left too few clues, and if that was the case then Manny was really stuck.

Maybe when his strength returned, he could begin walking home, but which way was that, he wondered? Manny considered walking down the road to the next village, but there was no telling how a lost American would be welcomed in this crazy corner of the world. One village might welcome him as a hero; another might try to kill him. He was stuck.

Even walking itself would have been a far-fetched idea, and he knew it. It was still February, and he was high in the mountains. While some fields were already as green as spring, he knew that the cold at night would probably kill an unweathered American desk jockey like himself.

That led to the one question that was surprisingly not in the forefront of his mind, "How did he wind up in such a mess?" Manny wasn't prone to selfpity. Instead, he liked to try and think his way out of situations. It didn't matter if it was something trivial like how to coordinate three reports into one, or if it was something as life threatening as his current situation. It was his nature to use his head first and not dwell on the trivial. As serious as his situation was, the question of how chance had put him there was trivial in comparison to how he was going to get home.

While he was pondering the obvious questions, an old woman from the village (probably in her late 40's, but old by local standards) brought him some milk and sweet herb bread. He thanked her and complimented her efforts by smiling thoughtfully. It was as fake as it could be, but the cultural difference camouflaged his face, and she appreciated his thanks.

As the sun settled behind the huge mountains, The Ox returned.

"Manny, we've just finished our afternoon prayers; perhaps you would like to come inside and join our discussion, or shall I show you to your room. It will be too cold to stay outside soon."

Manny had calmed down over the past few hours. The food had helped a lot with that, and so had a newly-found strength in his legs. He could stand and walk again.

"Thank you. Mr. Azmaath-"

Ali interrupted him with laughter.

"If I may call you Manny, then surely you must call me Ox, as everyone else does. Agreed?"

"Fine, Ox. I'm still a little tired from this entire experience. How about I just turn in for the night?"

"That's perfectly understandable. Rest assured. It was nothing but the finest morphine that calmed you for your journey. You need not worry about your health in any way. Here the air and water are pure. The herds are well tended, and the fields are chemical free. We forbid drinking and other stimulants in the village, so your body will most likely be better than ever if you decide to leave us. Now, if you'll follow me, I'll take you to your little cottage."

The Ox walked up the sloping dirt road, toward the east and the sheer cliff that sheltered the village. At one of the small stone buildings he stopped and opened the thick wooden door. It was a simple one-room building with four walls, four shuttered windows, and the door. The ceiling was the underside of the clay roof, and the floor was a layer of straw over dirt. A small stack of thick, hand-spun wool blankets next to the fireplace were his only comforts.

"I'm sure this is a far cry from the hotel that Captain Asaad arranged for you back in Islamabad, but here we all stay in similar surroundings. Some have more people or more blankets, etc., but there is no presidential suite. We also do not have room service. We will eat twice a day in the study building where we just came from. Prayers are before and after each meal, and at midday. The rest of the time is either spent with chores or discussion. I would also mention that it is not safe to leave at night. There are many animals around here, and while our watchmen are alert, the nights are cold. Besides, sometimes they will shoot first and ask questions later. Goodnight Manny, we will talk more tomorrow."

When the Ox left, Manny walked over to the fireplace. It still had coals in it from the previous occupant the night before. He tossed in some straw and a few logs. The fire grew, and the cold started creeping in. He crammed straw around the base of the door and wrapped himself in wool blankets. Soon he was actually feeling cozy-not comfortable, but cozy. With his eyes getting dry and heavy, Manny threw another log on the fire and went to sleep.

Morning eventually came, and his fire was still smoldering. He put another log on it to keep it going, and after making sure that the sun was up and the "watchmen" were gone, Manny made his way back to the main study building. Inside, the same woman who had brought him food the day before greeted him. She had more of the same milk and sweet herb bread waiting for him. He accepted it and again pretended that it was wonderful. It was good. Almost any food is good when one is hungry, but it wasn't anywhere near as wonderful as he pretended.

While Manny ate, he could hear the others saying their prayers. The sound reminded him of Buddhist monks or isolated Christian chants. He couldn't help but see the irony in their similarities. Without knowing any of the languages, they all sounded alike to him.

When they finished, he heard The Ox begin his discussion. It was exactly as Manny had expected-pure anti-Western propaganda. He couldn't help but shake his head and steal a snicker. In that most tranquil and spiritual setting, Ali Azmaath, The Ox, was preaching-justifying-violence against the "infidels" as he called them.

Manny also realized that The Ox was the type of man who was the real root of terrorists. He was an instigator. He was the kind of man that took relatively illiterate people and sold them a bag of rhetoric that they could not resist.

Most of the terrorists that Manny had read about came from desperately poverty-stricken areas. Many were downtrodden Egyptians, economically oppressed and discriminated Palestinians, Saudi Arabian have-nots, or Afghans-of which there are no wealthy.

Judging by his experience with reviewing intelligence reports, Manny also surmised that there are two types of people in the world: leaders and followers. While some people were only satisfied by taking on the responsibility of leadership roles, others found their satisfaction-their comfortin letting others make the big decisions for them. Manny had seen no sign of a hierarchy beyond The Ox, so he felt that it was safe to assume there were nothing but followers in the company of the instigator.

The perfect terrorist recruits were people who were generally illiterate, dirt-poor, and preferred to be followers. The Ox seemed to know that, and the teachings of Sheik Mohammed in the poor illiterate areas of Islamabad would serve as a perfect recruiting station. Unsuspecting recruits were then sent to The Ox' mountain hideaway for more spiritual discussion. From what little Manny had understood in the background, their discussion was definitely antiWestern. Anyone who was prone to being a follower would sit quietly and buy into the propaganda. Wannabe leaders would question the discussion.

The only question remaining to Manny was "why?" Why did people such as The Ox go to such great lengths and efforts to instigate terrorist attacks on the West? While sipping his tea and admiring the view around the village from the terrace, Manny brainstormed.

Money was his first guess, but there was nothing to indicate that The Ox was profiting from his efforts. He hadn't seen any elaborate Saudi Arabianstyle palace or army of servants. If there was wealth being accrued, it was well hidden.

Revenge was a possibility since anyone could have at one time or another suffered a tragedy and then blamed the West for it. Everyone loses family members at one time or another, and maybe he suffered a loss at the hands of a Westerner or a Western-backed authority. That was common enough in the Islamic World.

Manny didn't think that the man was the religious fanatic that he seemed. The Koran, like all religious texts was based on peace and love. It took a lot of thin rationalization to hide that fact, and Manny didn't think that any truly religious scholar of any holy text could make that rationalization.

Was Asaad right? Were The Ox and his followers so fanatical because they had such a deep-rooted love for their fellow Muslims? With endless religious study, a financially oppressed background, an illiterate group of followers could very well be convinced to do just about anything out of love for their fellow Muslims. Manny was starting to see into the mind of the terrorists. Skills could be trained at any one of thousands of camps around the world, but commitment was another matter. Terrorists needed to be the most committed people in the world, and he had a front row seat for watching the recruiting and selection processes.

While staring at the scenery and contemplating his situation, Manny spotted Asaad's body at the side of the river. It was almost a mile away, but he saw it clearly. The white pillowcase had burst into red as at least one highpowered bullet found its deadly mark. The Ox walked in and saw Manny locked in a stare, and intuitively knew what he was looking at. "Captain Asaad was not a pious man. You do know that, Manny?" Manny continued to stare in silence.

"I could tell you the terrible things that he has done in the name of his country, but I believe you already know. I've been told that you've even been to the interrogation rooms."

Manny had forgotten that it was Asaad's driver who had ambushed them in the store. The same driver had taken him to his hotel and accompanied them on Asaad's torture cell tour. It was apparent that the driver had done more than just ambush them. He was an informant, and since Manny hadn't seen him at the village, he surmised that he was still undercover at the ISI in Islamabad.

"I know Captain Asaad wasn't an angel, Ox. I saw the rooms. I heard the screams, and I felt the fear. I'm not afraid to tell you that even here I'm still afraid. I don't know you, and I can't tell how honorable your word is just yet. At least with Asaad, I knew that he was not trustworthy. He never pretended to be that way. He never tried to give me his word, and he never told me that I was safe. Now he's dead. You had him shot, and you're telling me I'm safe. Until I see otherwise, I have to assume that you might have told Asaad the same thing."

The Ox was surprised. He rightfully had taken Manny as a quiet dataentry guy from the safe and comfortable United States. He didn't expect the man to be so straightforward. It was a welcome surprise to him.

"I never told Asaad he was safe, Manny. From the moment we arrested him, he knew his fate. Now, you know yours. You are safe here. At the very least, you must know that you're safer here than almost anywhere else on the planet right now."

It was an intriguing statement, and Manny turned for clarification. The Ox saw the look in his eyes and since Manny clearly wasn't going anywhere, he explained himself.

"Manny, it was bad enough before the attacks on your country back in 2001, but ever since then, people like Asaad have become very ambitious. They were desperate to pin the blame on Bin Laden, Al Quaeda, the Taliban, and anyone else that they could go after. Since then, the intelligence world has been trying to find countries or organizations that they could attack as a matter of retribution, but also as a matter of public relations. The truth is that any group of people who oppose a Western nation, particularly the United States, finds its members threatened, imprisoned, tortured, or killed. Even non-violent protesters are shot routinely. Look at how terribly Palestinian protesters were treated before that day in September, and ever since. Now, it's seems as though Israel has a license for genocide, and they are but one example. You've been in the torture chambers. You know I'm speaking the truth."

Manny knew it was inevitable that the Palestinian issue would be brought up. It seemed that was the crying card of choice to play for Islamic extremists. He didn't deny their sad situation, and he was saddened by the nightly news' pictures of protesters being fired on, but Manny also knew that the violence was a circular chain that neither side had the courage to break.

While he had those thoughts, he didn't want to get into an argument with his kidnapper, and that was how he still viewed The Ox. He thought of his situation again, and decided that it was a good way to change the subject. The Ox was obviously still on a propaganda roll from his discussion, and he didn't seem to want to stop even after his discussion period with the others.

"I really don't want to debate the issues with you. I know Asaad was not a great guy. I've said that, and I'm not going to try and say he did anything for humanity's sake. I don't know his motivations. I do know that I want to go home, and I can't. You say I can, but it's obvious that I can't even try. Right now, all I can think about is going home. Since I can't realistically think about that, I think about maybe burying Asaad's body. I'm no fan of his, but the sight sickens me. It's hard to find peace in an area littered with familiar corpses."

Their relationship was becoming a poker game. Neither Manny nor The Ox wanted to tip their hand and tell what they really thought of each other. Manny tried to look as though his statement was innocent and benign to protect himself by not insulting his kidnapper. The Ox responded by pretending not to be offended.

"I understand your feelings Manny. I don't think anyone here would like to help in burying Asaad. They all feel a certain amount of justice is seeing him treated poorly. If you wish to see to him, there is a spade in a shed on the eastern side of the village. You can help yourself. As I said, you will not be harmed here. You may walk freely. I'll even show you the way."

They walked off the terrace, out of the building, and down the street toward the edge of town. As they came over a slope, Manny couldn't help but notice that the road continued into a cave, at the base of the sheer cliff, on the eastern side of town. The Ox pointed out the small shack. Then he opened the door and even handed Manny the spade.

"If you would like, I can go with you, Manny."

Manny tried to be polite, though he really didn't want to listen to too much propaganda from The Ox.

"It's up to you. I guess I could use some help getting him out of the river. C'mon, let's go."

They walked down the street toward the river and Asaad's body. Villagers and followers went about their individual routines, but one looked out of place to Manny. Near the far end of the village, Manny spotted another Caucasian man. He wondered if there was another American prisoner at the village. The man didn't seem restrained in anyway, but neither was Manny.

As they walked, Manny and the man made eye contact. Immediately, he walked over to The Ox and began speaking in Serbian! Manny understood everything in their conversation.

The man was a Muslim Serb who had come to the village years ago, after the siege of Sarajevo in 1995. The Serb asked if Manny knew about the

"upcoming mission in the United States" and about some sort of "MultiDenominational Religious Conference."

"No. He knows nothing of the mission. We never should have taken him, but we couldn't leave him behind. We'll get rid of him tomorrow."

The Ox was referring to Manny's release. He planned to have Manny dropped off in front of the nearby Pakistani firebase a few miles away. Both the Serb and The Ox knew that killing an American would bring down a great deal of unwanted attention from the United States. Public attention was a good thing for The Protectors of the Koran, but military attention could be devastating to their plans.

All the while Manny kept walking toward Asaad's body. He didn't want to be seen eavesdropping, and he wasn't eager for introductions to terrorists. He was surprised to find out all the information, but he also knew that it was useless unless he could communicate it with someone in the outside world. Most importantly, he heard The Ox talking about getting rid of him, and that struck a cord of near panic in his heart.

The Ox and the Serb faded from earshot, and Manny never seemed to notice that he was alone. He knew it, but he didn't want to interrupt their conversation. More and more he wanted to get rid of the sight of Asaad, and he wondered if anyone would come along to take care of him when it was his time. There were miles and miles of beautiful picturesque valley to gaze upon, but his eyes always found Asaad's corpse, and in his mind he saw his own.

The river was loud and fast as it passed by. Manny could see the rounded blue, gray, and red rocks under its crystal clear wavy surface. It would have been a nice sight except for the bloated body of an ISI agent caught in a fallen tree at the bank. Asaad still wore his pillowcase, and he was thankful for that. Looking at his face would have been too much for Manny.

He pulled on the tree until the body was close enough for him to grab and drag to shore. It was like trying to drag 200 pounds of limp meat wrapped in soaked cloth, and he wanted to vomit at the absence of life. Manny kept his cool and dragged the body almost 30 yards up the slope toward the village. Then he started digging. It took him most of the day to get just 3 ½ feet down. Then he was too tired to continue. As he went to pull Asaad's body into the shallow grave, he felt something familiar in the lower right front jacket pocket. It was Asaad's cell phone!

Manny looked around to see if anyone had seen him pull it out of the pocket. He couldn't be sure. People were all over the village, so he casually put it in his pocket and immediately went back to burying the body. When he was done, Manny tamped down the soft soil and headed back to the village. It was almost time for the evening meal. During the long walk back up the hill, the suspense and hope of the secret cell phone in his pocket was driving him mad.

From time to time he put his hand in his pocket and felt the dead man's phone. Could it be his ticket home? Manny hadn't taken the time to check to see if it even worked. He realized that the cold water had probably killed the battery, but there was still hope. If it did work, Manny doubted that a cell phone tower was anywhere within 200-300 miles. That meant that he'd have to use a satellite to make a call, and there was no way to tell when one would be available. Besides, whom would he call?

He didn't have many phone numbers memorized. Most of the important ones were kept in his passport, and The Ox had that. If he asked for it back or tried to steal it, then there was no telling what might happen. Manny started to try and remember phone numbers, but the stress of the entire moment was too much, and the only numbers he could think of was his own, his parents, and his work phone number. The number for the Office of Homeland Security International Terrorist Movement Monitoring Division would do fine, he decided.

Manny sat and had dinner with the entire village in the main study building. They had some sort of roasted meat with more sweet bread and green vegetables that tasted like either onions or cabbage but looked like white carrots. It was remarkably good food, and it took his mind off the stares and conversation that clearly revolved around him.

When dinner was over, The Ox immediately began his discussion. It was in some sort of Arabic language, so Manny felt he could slip away. He made his way out of the hall, through the kitchen, down the terrace, and back to his designated cottage.

Inside, he built up his fire, piled straw around the door, and finally pulled out the cell phone. He turned it on, and when he saw the words "NO SIGNAL" come on the liquid crystal display, he immediately turned it off. It worked! Now all he had to do was find a way to get a satellite.

Manny knew that on a starry night, anyone who watched the sky for a few minutes would be able to see at least a few satellites cross from horizon to horizon. There was no way to tell if the cell phone would reach the satellite, or even if any of the moving stars were a communications satellite, but he had to try.

Leaving his cottage was dangerous with the watchmen outside, so he opened one of the windows, wrapped himself in the wool blankets, and watched the stars. Far from the ambient light of any cities, Manny could even see the Milky Way's band across the winter sky.

He sat for hours even though his nose and ears were numb. Whenever he saw a satellite, he checked his watch. After a while he couldn't take the cold anymore, so he closed the window and curled up next to the fire. It was too hard to sleep with the hope of rescue on his mind.

While he pondered the timing of the satellites, he figured out a formula. Manny didn't know specifically what time it was since he didn't know the time zone. Instead he based his assumptions on Islamabad time. Starting at about 8:15pm, Manny had seen a satellite cross over the narrow band of sky between the two mountain peaks every 50 minutes minus 10 minutes for each pass. His formula wasn't entirely accurate, but he figured out that a satellite should fly overhead at about 10:00pm.

At 9:55 he opened the window and watched for it, and it appeared. Frantically, Manny turned on the cell phone and tried to call work. The liquid crystal display read "WEAK SIGNAL" and flashed "ROAMING," but he heard the dial tone. Then he heard it dial, and there was a pause. Finally, he heard ringing, and a feint voice answered!

"Good afternoon, Office of Homeland Security, Division of International Terrorist Movement Monitoring Division, this is Kimberly. How may I direct your call?"

Manny knew that he could lose the satellite at any second, but he had figured out a way to make sure that his call was traced and recorded. Before saying anything, Manny coughed out the same words that had triggered the monitoring of a call from Munthmar Hussein. Those words had triggered the NSA's listening satellite to record and trace the call from Hussein, and Manny figured that they should trigger it for him too.

"Mission, guns, holy, Christian, American. Kimmy, It's Manny. Forget about those things I just said. I can't talk very long. I'm being held prisoner in some sort of mountain village. The guy that's keeping me here is named Ali Azmaath. They call him the Ox. Right now, I'm okay, but they killed the Pakistani ISI Captain that I came over here to meet with. His name was Captain Asaad. Put the word out to get a triangulation on this call, and get me the hell outta here. I've gotta go. I can't tell if anyone's listening, and I don't know how long this battery or the satellite will hold out."

"Manny! My God! We thought you'd been killed. We all saw it on the news and-"

"Kimmy, I'm counting on you. Call NSA and get them to trace this call. They'll know what to do, and tell everyone I'm okay for now. Oh yeah, you better call my parents too. Thanks Kimmy."

Manny turned off the phone right away. He hoped that he had left it on long enough to triangulate the position. At least he was sure that it was recorded. He knew from his reports that anything going through those satellites could be download or bounced to the NSA's satellites. At least he had gotten the word out.

Satisfied that the wheels of a rescue operation might already be turning. Manny allowed himself a quick smile before fear overtook him. What if one of the watchmen were right outside the cottage? What if his cottage was bugged? What if someone found him with the phone tomorrow? Where could he hide it other than in his pocket? There were a lot of questions and a lot of fears that needed to be addressed.

He tried to calm himself. When no one came bursting through his door, he convinced himself that no one had heard him. If they had, he would at least live until tomorrow. There was only one more thing to take care of before going to sleep. Manny turned the ringer off on Asaad's cell phone. Then he put some more wood on the fire and tried to relax. It was a losing fight, but he had to try.

Morning came all to quickly. A few shafts of light poked through the cracks around the windows and door and lit the smoke filled room. Manny wondered if he had been heard making the call, and if this would be his last morning. He shook off the sleep and headed outside. There was no firing squad waiting, so he headed for the main study building. There was something out of place, though. In front of the main building there were three beat-up pickup trucks with big machine guns mounted in the back. Behind them there were two shiny new white SUV's and what looked like a jeep. They were hard to miss.

Manny went down a small path and onto the terrace. With Asaad's body out of sight, it was almost out of mind. That and the belief that, on the other side of the globe, people were working to get him home brought him a new sense of peace. He actually enjoyed the view. Voices from the kitchen area and a discussion from the main study hall brought him back to the reality of his surroundings.

The woman that normally brought him food stepped out from the kitchen and spotted him. Immediately she hurried over to him and shuffled him away. Something was obviously going on that he wasn't supposed to know about. Perhaps someone had overheard him using the cell phone in the middle of the night.

Manny knew that he couldn't run far. The high altitude, unknown territory, and the harsh evenings made that improbable. He thought about trying his luck in the cave, but given that this was a terrorist recruitment center and terrorists had a habit of using underground infrastructure, he chose not to go there. Sitting in his cottage would have been just as bad as walking into the kitchen saying, "Here I am. I called in the commandos last night. Now go ahead and shoot me!" At the very least, he could go for a little sightseeing walk to make himself scarce.

Manny walked a wide arc around the vehicles in the street, and he headed back down toward the river. On his way, he avoided looking back or to his left at the village. Instead he pretended to admire the terraced fields on his right to the north. It amazed him that this primitive culture had found a way to grow food year round in such a harsh environment. He also figured that they had probably been doing things the same way for hundreds or even thousands of years.

Near the river, the dirt road turned right and headed north along an embankment. Manny stopped and did an obvious 360 sightseeing turn to survey the area. On the other side of the river, a small beach of round cobbles and boulders covered the distance between the water and vertical rock face of a mountain reaching hundreds-perhaps thousands-of feet into the sky. To the north, the road continued to twist and turn until it disappeared over the sloping hill layered with terraced fields. To the east the road climbed up to the village and disappeared into the large cave opening. To his south lay the grassy fields with coniferous trees, tangles of driftwood, and Asaad's shallow grave.

Having surveyed his surroundings, Manny thought about how and when he might be rescued. When the brainstorms and visions passed, he reminded himself of his situation and walked into the trees to spend the day away from the village. The last thing he wanted was to see someone, or overhear more and get himself killed before help could arrive.

Friday, February 13, 2010

USAF Base at Doho, Qatar

1:12pm local

United States Marine Corps Major Chris Mosby walked off his CH-46 Sea Knight helicopter and straight toward the airfield's duty station. On his way, a U.S. Air Force officer walked over and greeted him.

"Major, I'm Colonel Earley with the 452nd Tactical Airlift Wing. I'll be taking your people in."

They shook hands and smiled professionally. Major Mosby still had many questions, and he hoped that the Air Force Colonel might be able to help him out.

"Colonel, I was basically told to take three companies and our MEU's (Marine Expeditionary Unit) sniper platoon over here and await orders. They didn't even tell us what to bring, so all we've got is beans and bullets for about a week. Do you know what we're here for?"

"Oh yeah, Major. I do indeed. You're gonna love this. Last week I took some desk jockey from Homeland Security over to Pakistan on what was supposedly a face-time visit. He said they wanted to show that Homeland Security is on the ball, working closely with foreign services. Well, he didn't look like a spook (spy), but he was put on a routine flight at the last minute, and the whole thing had that CIA feel to it. You know what I'm saying?"

Mosby was a veteran of the Afghanistan war of 2001, and he had some experience with intelligence services. None of it really impressed him. The historical dislike of most military field officers remained current.

"Oh, I know the type Colonel. Spooks suck in my book. I don't think I've ever heard of them giving good intelligence reports. Everything they ever gave me was either already in an atlas or on the internet. They're not a very impressive bunch. So what happened with the one you dropped off?"

They walked through a gate, past a pair of armored pickup trucks loaded with Air Force Security personnel. Behind them, thirty U.S.M.C. CH-46 Sea Knight helicopters (affectionately called Bullfrogs by the Marines) unloaded Major Mosby's three companies of U.S. Marines. Almost five hundred people were falling in to follow him through the gate, and wherever the Air Force Colonel led them.

"So I dropped this spook off in Islamabad, Pakistan. From what I hear, he was checking out a terrorist shooting at some church. The next day, he and one of their ISI guys-"

Major Mosby interrupted him.

"What's ISI?"

"The ISI is their secret police. Not a very friendly bunch of guys either. Anyway, our spook and some ISI people are sitting around some café waiting to nab one of the guys involved with this church shooting. The guy shows up at the café, and just as they're about to nab him, a couple hundred antiAmerican protesters show up with guns. The ISI guys are wiped out, and our guy is missing. I guess it was pretty messy, and identification of the bodies was rough. We wrote him off as dead."

Major Mosby imagined the story as it was told. All the while he kept looking behind him to make sure that everyone came off the helicopters safely.

Colonel Earley paused to make sure that he still had the Major's attention.

"Go on. I'm listening."

"Maybe not, Major. Turns out that Homeland Security gets a phone call from their guy yesterday. He says he's alive and being held in some mountain village."

Mosby laughed.

"I know it's not very descriptive, but get this, the guy actually used his head. Right off the bat, he said some things that triggered a trace and triangulation by an NSA electronic emission monitoring satellite. So we actually know his location! Someone sent over a UAV (unmanned aerial vehicle-remote controlled spyplane), and we're getting the feed right now. The geeks running the UAV are set up on a projection screen over there in that hangar."

Mosby stopped while the Colonel pointed to a hangar big enough to park cargo planes inside. Then he turned and with one finger waved for his company commanders to come over. Three U.S.M.C. Captains and a 1st Lt. from the sniper platoon jogged over.

"Have everyone fall in by company over there in that hangar for a briefing."

They ran back to their units and barked out orders. Soon hundreds of Marines were jogging into the hangar and sitting down on foldout chairs that the Air Force had setup.

"Okay Colonel. Why us? Why aren't they sending in Delta, Rangers, or some sort of hostage rescue team? We're just a bunch of mud Marines. We don't have any special training for Hostage rescue. There is a small team back on the ship, but…wait a sec. This isn't just a go in and grab one guy thing is it?"

"You just made the understatement of the year Major. C'mon. Let's go inside. You're gonna love this."

They walked into the hangar through the smaller, regular door. The big aircraft entry doors had been open to let the crowd of Marines get inside. While they were sitting down, loud motors hummed and closed the doors. At the far end of the hangar there was a table with a dozen Air Force technicians manning computers and facing the same direction as the Marines; toward the large projection screen that had been setup like a movie theatre.

Colonel Earley walked over to the computer-packed table and faced the Marines. In no time, he had their undivided attention. There was no screwing around. The Marines knew when they had to pay extra close attention.

"Okay. Listen up! I'm Colonel Earley of the 452nd Tactical Airlift Wing. I am your senior mission commander until you're on the ground. The situation is this: an American intelligence agent is loosely being held in a small village 15-Miles NNW of Shengus, Kashmir. Now Kashmir is a province that both India and Pakistan claim as their own. Tajikistan, China, and Afghanistan also border it. This is a very hot region. No one wants to upset the balance. That's why there are very few actual terrorist training camps, but there are lots of 'spiritual retreats.' Now, if something does go wrong, the powers that be don't want to have more of our guys grabbed. That's one of the reasons that we're sending in all of you. Another reason is that the normal hostage rescue teams, Delta, Rangers, even the FBI and U.S. Marshall's teams are all busy. There's a lot of small unit Special Forces-type action going on all over the globe right now. As you probably know the 82nd Airborne is busy in Afghanistan right now, so they're unavailable, too. You're not the first choice for this mission, but I think you're the best choice."

There was a round of morale-lifting "Oo-Yahs!" interjected by the Marines. Mosby nodded his head for them to stop, and it was like an off switch. Then Colonel Earley continued.

"There's more than just our spook in this village too. After we got the call from him, we dispatched a UAV from the Afghan Theatre of Ops to check out the area. That's what these guys at this table are doing right now."

Col. Earley tapped one of the technicians on the shoulder. The technician gave a quiet, "Yes sir" and the projection screen came to life. It showed the aerial view of the village where Manny was being held in clear color imagery from a few thousand feet above. The Colonel tapped the technician again, and the image zoomed in closer. It centered on the vehicles that Manny had seen and avoided. The Colonel asked the technician to zoom in closer, and as he did, the Marines fell eerily silent at the technological marvel.

"What you are seeing is live footage from our UAV over the village. Am I the only one here who thinks that those nice $60,000 SUV's look a bit out of place in this picture?"

Everyone laughed except Mosby and the company commanders. They were intent on the seriousness of the situation.

The Colonel pointed to a spot on the monitor in front of one of the technicians. The image on the screen panned to the side and centered on the terrace where Manny had spent so much time looking at the scenery and eating local homemade sweetbread. There were people on the terrace, and the camera zoomed in further until they filled the screen that blanketed the huge aircraft hangar's wall.

The Colonel continued.

"This is another reason that you guys are going in and not some small commando team. You all see these people? This fat one here with the white robe is Saudi Prince Abdualla. He's one of Bin Laden's old business buddies. The word is that he's a big bank-roller of anything anti-American. If you worked in a drive-thru anywhere in the middle east and the cooks had instructions to spit on burgers going to Americans, well then he probably owns a piece of the restaurant." No one laughed.

"This guy here, the one in the nice and shiny uniform, We're not sure who he is, but the uniform is one that's worn by Chinese Intelligence officersspecifically, a Colonel. It's a little early for Halloween, so we'd like to find out a little more as to why he's in the same village where one of our people are being held. Now this guy here, the one with the turban and the red rag around his waist, we think he's the one behind that attack on a church the other day and the attack where our guy was nabbed. We don't know his name yet, but the boys and girls back at CIA would like to. They'd also like to know who these two American-looking businessmen types are. I myself find them a little out of place."

The Air Force technicians chuckled, but the rest of the room was silent. The Marines were starting to realize how important the village was, and just how dangerous their mission could become.

The Colonel didn't chuckle, and he wasn't stoic. Instead, he let everyone feel their emotions through, and when quiet returned, he continued.

"My mission is to carry you guys wherever you want to accomplish your objectives. Those objectives are outlined on the detailed briefing reports here on this table. There should be enough copies for all your officers, Major Mosby. It's enough to say that your primary mission objective will be to safely return the American or any Americans held against their will in the village or the immediate surrounding area. The secondary objective is the capture of these men and anyone else that basically seems like a terrorist or terrorist supporter. Major Mosby and the rest of your officers will brief you on their plan for the mission after they've had a chance to review the briefing reports. Until then, sit tight and try to relax.

It'll be a while before we've worked out the specifics of this operation, and even then, the flight over there is gonna be a long one."

Major Mosby, his staff, and his company commanders began reviewing the briefing reports. They had been hastily put together, as had the entire operation, but they were sufficient.

Mosby immediately saw the operation in his mind. They would ride in Colonel Earley's C-141 transport planes out of the Persian Gulf, over the Arabian Sea, through the Pakistan air combat corridor, into the Afghanistan Theatre of Operations, then over to Kashmir. The Marines would have to parachute in at night, into mountainous terrain, and without a large drop zone.

Echo Company would jump first. They could secure the terraced fields on the north side of the village. While the other companies came in, Echo would setup a base-of-fire line, and protect them from any attacks out of the village or from the road on the other side of the farms. This is where Mosby planned to have his headquarters.

Golf Company would act as a blocking force between the terraced farms and the cliff face of the mountain on the east side of the village. After contact had been made, Golf Company could either reinforce, or go into the village along the cliff's face.

Golf Company would make a frontal assault up the hill. They'd get covering fire from Echo Company on their left, and then they could get into the village.

The sniper platoon would have to make their way through the brush to the south, and block anyone from escaping or reinforcing. The foliage was thick in the woods to the south along the river. Putting an entire company in there was just asking for an ambush, but a few two-man sniper teams in the brush would be perfect.

After Golf and perhaps Fox company secured the village, they'd all regroup in the terraced fields, and begin the march out. There was a Pakistani artillery firebase 14 miles away, but if they were under dependable control, then they would have been called upon to do the mission. Mosby would have to make a polite visit, but he wanted to have everyone else march on to the nearby Pakistani supply depot in the town of Gilgit on Highway 35-another 23 miles further to the west. From there, they could catch a ride into Pakistan or Afghanistan, where they could meet up with some other American units.

Colonel Earley told the Major that he was able to scrounge up some air support from the Afghanistan Theatre of Operations. They'd be able to get plenty of mid-air refuelings. An E-3 Sentry AWACs (Airborne early Warning and Airborne Control) airplane was going to leave Bahrain and accompany them through Afghan airspace. It brought along four F-15C Eagle fighter planes to intercept any airborne threats. The UAV that they were watching was going to loiter over the area for another two hours, then it would go home, refuel, and return before their operation was on the ground. Lastly and most importantly to

Major Mosby, three F-15E Strike Eagle fighter-bombers would go in with them to take out any ground problems that needed big stuff dropped on them.

Once the plan was detailed, Mosby told his company commanders to get ready. Each Marine was to bring nothing except their weapon, ammunition, and night-vision equipment. All heavy weapons, food, water, and extra ammunition were to be packed on pallets for separate dropping. He was counting on a swift and silent surprise night operation, and Mosby wanted his men and women to hit the ground ready to fight.

Colonel Earley made the phone calls to the different squadron commanders of the participating air support aircraft. Everything was set. They would leave Doho, Qatar at 2100 hours local time. From there, they would rendezvous with their air support F-15E Strike Eagles, the E-3 Sentry AWACs plane, its four F-15C Eagle fighter plane escort, all about 2200 hours over the Arabian Sea. Then they'd all meet with a mid-air refueling plane and tank-up over Afghanistan at about 2330 local. The plan was to fly in at about 12,000 feet in a single file formation over the drop zone. There, the Marines would jump out and land on the 8000 foot high, terraced farm drop zone code-named appropriately "The Gardens." At that point Colonel Earley's job was over, and Major Christopher Mosby had the ball.

The planning was quick and straightforward. It would take almost 4 hours to load the transport planes and launch the air support aircraft. In the meantime, the Marines were allowed to get some food at the base commissary. Given that they were in a hurry, there was some line jumping, and they were all asked to leave by base security personnel. It didn't matter too much though. They had taken most of the food that was cooked anyway.

When they returned to the hangar, Major Mosby had the companies form up and load the planes. A task that normally took a good deal of time was accomplished in far less, since Colonel Earley had already given instructions to load the Marines' palletized equipment. Since their weapons, ammo and night-vision equipment never left their side, all the Marines had to do was board the planes.

Shortly after, they taxied out to the runway and took flight. Colonel Earley's plane led the way with Echo Company on board. Major Mosby rode in the second plane with Fox Company, and Golf Company trailed in the last plane. According to plan, they linked up with their different air support aircraft over the Persian Gulf, and all the planes were mid-air refueled by a trio of overworked U.S. Air Force KC-135 Stratotankers. After refueling, they headed north through the Pakistani Military Air Corridor to the American controlled airspace over Afghanistan. There, they turned right and headed to Kashmir.

Saturday, February 14, 2010

15 Miles NNW of Shengus, Kashmir

0111 hours local

Snow-capped mountain peaks glowed above the transport planes' dark silhouettes against the night sky. It seemed as though the planes were flying at low altitude, since they were only a few thousand feet off the valley floor, but they were actually at the pre-planned height of 12,000 feet. Four long hours after leaving the hot and dry U.S. airbase at Doho, Qatar, the Marines stood up and prepared to parachute into Kashmir. Everyone hooked up their static ripcords to a long cable running the length of their plane. They checked each other's parachute and gear, then sounded off that they were ready to go. The loadmasters on the three planes opened doors on the port and starboard aft sections while the pilots slowed to near stalling speed. A red indicator light started blinking as they approached the drop zone. Soon it was keeping pace with almost 500 combat-loaded Marines. Finally, it changed to green; and from the red-lit cargo bays of the C-141's, past the silent thumping of the green jump indicator lights, men and women jumped out of their airplanes.

Major Mosby wasn't the first to jump, but he was the first to leave his plane. The near 200 mile an hour wind hit his body like a thousand boxers' punches. His feet were the first to catch the slipstream from the C-141, and they were kicked out from under him as soon as he was out of the plane. Mosby felt himself falling toward the ground back first, but he was used to it, and when the static ripcord connecting his parachute to the plane finally pulled away, he felt the familiar jerking sensation of an open parachute. It felt like it was going to rip his arms off, and all of his equipment tried to stretch his body like a rubber band.

There was no time to make sure that his chute had opened properly. Even if it hadn't, he was too close to the ground to try and cut it away then open his reserve chute. Instead, Mosby glanced toward the horizon and recognized his swaying motion underneath the black nylon canopy. He swung backwards, forwards, and when he started to swing backwards again he hit the ground. He let his body slump like a sandbag, and then he rolled.

In one motion Mosby stood back up, deflated the parachute, rolled it up, and disconnected himself. In his peripheral vision, he saw the hundreds of other Marines parachuting into the muddy terraced farms. He put a heavy rock over his parachute and grabbed his M16A2 assault rifle. As best he could tell, they still had the element of surprise, but how long that would last was a matter of luck.

Mosby's Marines were hard chargers. They had been trained primarily in charging beaches or seizing helicopter landing zones. Most only had two or three parachute jumps under their belts, and as far as he knew, no one had any sort of Special Forces training. Of course, with all the in depth training that Marines were exposed to, many in the defense community classified the entire Marine Corps as a Special Forces unit. Either way, they were not in their most familiar environment, and it remained to be seen how well they could perform on this unusual adhoc mission.

While he pondered his concerns, Mosby looked around for his company commanders and noticed that everyone had taken the extra time to duct tape all their loose gear to their bodies. It was a little thing to silence a buckle on a knife, but when 450+ knife buckles rattled, Mosby knew it would make more than enough noise to wake the terrorists. It reassured him that perhaps his men weren't just the last choice for the mission. Maybe they were the best choice.

His company commanders were nowhere to be seen at first. There was a lot of running around, and in the dark, even with night vision goggles on, Mosby couldn't tell faces and rank in the confusion. Almost 2 minutes passed, and then he saw why. Each company commander had given a detailed brief to his entire company, and everyone knew exactly where they were supposed to be. Echo Company was taking up positions along the edge of The Gardens. Fox Company and the sniper platoon were on their way down to the river, and Golf Company was moving out toward the rocky cliff face to the west. Even individual platoons and squads seemed to know exactly where they were supposed to be.

The end result was a deployment that went swiftly, silently, and perfectly. Even the best Special Forces teams in the world had things go wrong. Mosby looked around and couldn't believe how perfectly things had gone, but he knew the truth. As much as planning and background training had put the men on the ground, it was really still a matter of luck that no one spotted them-or had they?

Always the pragmatic, Mosby wondered if maybe they were spotted and they were walking into an ambush. He knew that there was no way the three, huge transport planes went undetected-not while they were flying only 3000 feet above the ground! One question stuck out above all, "How could anyone have possibly not noticed the planes?"

They hadn't gone unnoticed, and he was about to find out. In the village, the night watchmen spotted the planes before they heard them, and they put out the alert. Right away, the potential terrorists were called to arms. The Ox passed the word to his potential recruits, "The Americans were coming to kill them all." When faced with the do or die prospect, everyone picked up a weapon.

Along the north side of the village, three of the cottages filled were with groups of four, five, and six people respectively. Others brought in ammunition and RPG's (Rocket-Propelled Grenade/i.e. lightweight missile launchers). A heavy machine gun was set up in the kitchen, and pointed out of the terrace toward where the road from the village met the river and turned north. Another heavy machinegun was carried through the south side of the village and set up in the driftwood near Asaad's shallow grave. Some men took up station in the buildings on the west side of the village to block the road. Most everyone else headed into the cave.

Manny had heard the chaos in the village. It didn't take him long to realize that his help was arriving, and the last thing he wanted to do was poke his head out the door to see what was happening. It was more than likely that The Ox or one of the other terrorists might decide to come in and kill him before help could rescue him. With that in mind, Manny grabbed a baseball bat-sized piece of firewood and hid behind the door. He hoped that if someone did come to execute him, it would only be one person.

From three sides, the Marines were surrounding the village. Major Mosby waited in The Gardens to see what was happening. He watched as Golf Company started to bunch up along the stream and the cliff face on the east side of the village. On the other side of Echo Company, Mosby saw two of Fox's platoons make it down across the east/west stream that divided the terraced farms from the village. Fox's third platoon was still filtering across in four-man groups. Then, the terrorists drew first blood.

A long sequence of white illumination flares popped into the clear night sky from the center of the village. More seemed to come from every window. They were fired almost directly at Fox Company down by the river and at Echo Company's line up in The Gardens.

Like all good infantry, the Marines were trained to stop and they became motionless under the light of a flare. It was the motion that would attract fire, not necessarily the sight of the Marines. The flares also blinded the Marines wearing night-vision goggles, and until the flares faded, they couldn't move to remove their goggles without suddenly being a target. Their lack of motion wasn't a natural response; it was an innate, trained response that had been drilled into them so hard that no matter how badly they wanted to remove the goggles, they couldn't make their muscles move.

The ready terrorists opened fire with more than just flares. The heavy machinegun in the woods next to Asaad's grave near the river immediately took a toll on the Marines in Fox Company that were caught in the open. Two were killed instantly, and four more were wounded. Then the heavy machinegun in the kitchen joined in. It killed three and wounded six more. Other terrorists with assault rifles on the west end of the village took their toll by picking off some of the wounded that were caught in the open.

Echo Company tried to provide covering fire, but a heavy machinegun on top of the cliff face at the east side of the village was firing over Golf Company and raining .50 caliber rounds on to The Gardens. In addition to the incoming lead, there was a great deal of fire coming from three of the stone cottages along the north side of the village.

Minutes earlier, the area had been quiet and dark. Since the flares went up, yellow, red, and green tracers criss-crossed the open space between the village and the terraced farms. More flares came up from inside the village and out from the windows of the small stone buildings, and the night flickered with artificial magnesium light that hung from small fluttering parachutes all over the sky.

In the village, most of the would-be terrorists were frantically carrying documents and equipment into the cave. A small group of four people had a case of flares and a few pistols, and they were running all around the buildings firing randomly into the sky. Villagers were running into the thick woods and brush on the south side of the village. A few others were coming out of the cave with more weapons and ammunition. Some setup a small 60mm mortar and started lobbing rounds toward Golf Company on the east side of the village.

A few Protectors of the Koran members had navigated their way through the cave and were now on top of the cliff face with another heavy machinegun. They tossed white-phosphorus (WP) incendiary fragmentation grenades down on Golf Company. One terrorist had found a nice little crop of rock and was sniping at all three companies ruthlessly.

The Serbian that Manny had overheard talking with The Ox wasn't just any Serb. He was one of the most effective snipers of the Yugoslav Wars. He had hundreds of kills. Mothers, daughters, fathers, sons, brothers, and sisters were as much his targets as soldiers were. The Protectors of the Koran had recruited him not for his devout religious beliefs, but for his skills with a rifle. High on the mountain, with his favorite 7.62mm bolt action sniper rifle, the Serb wasn't training, and he wasn't working. He was enjoying his art.

Casualties were mounting. The small arms fire from the north side of the village wounded ten people in Fox Company and four people in Echo Company. The heavy machinegun on the mountain's cliff wounded four more in Fox Company and five in Echo Company. Worst of all, the WP grenades falling on the bunched-up Golf Company killed ten and wounded eighteen. The Serbian sniper killed two people in Echo Company, and wounded three more.

Mosby was standing between the two that were killed. They had each taken a single bullet in the head under their helmets. He glanced at both for a moment, then the familiar bass bark of the three Russian-made DShKM 12.7mm heavy machineguns brought his attention back to the situation at hand. Those guns were designed to take down aircraft or lightly armored vehicles, and the terrorists were using them at point blank range against his Marines. A glance toward Fox Company reminded him of their power when he saw a Marine cut in half at the waist by a single bullet.

In the village, the return fire from the Marines was getting intense. The centuries-old cement that held the stones together in the cottages was crumbling. Manny was forced to go down on the floor when the wooden shutters on his windows started coming apart from Marine bullets. Splinters went everywhere, and a choking dust filled the one-room building as the bullets started pounding their way through the gaps between the wall's stones. Manny knew that he couldn't hide in the crumbling building for much longer.

On the other side of the combat zone, Major Mosby continued to watch what was happening. One-third of Fox Company was still on his side of the stream. The rest was caught in the open, and taking heavy fire from the village in front of them and from the woods to their right. Mosby called the Fox Company commander on his helmet-mounted headset, but there was no answer. He was dead, and Mosby guessed it, so he ran along Echo Company's frontline and down to the Fox Company platoon that still hadn't crossed the stream.

The platoon's lieutenant was badly wounded from the sniper on the mountain. Mosby told the men that Echo Company would provide all the covering fire possible, and that they should all follow him. Then he sprinted through the stream and toward the road where the rest of the company was pinned down. Without looking behind him, Mosby kept running along the road. Some of the Sergeants yelled for everyone to follow him, and everyone was on their feet headed into the village.

The surviving members of the sniper platoon stayed in place. Without orders, they split up and took on both heavy machineguns that were chewing up the company. A few well-placed rifle shots into the machineguns' muzzle flashes, and they were silenced. Only one terrorist in the building at the edge of the village remained in Fox Company's path. Mosby stopped at the edge of the village and picked off the small group of bad guys who had been firing flares all over. The rest of Fox Company charged into the main study building and the small building at the edge of the village. They lost five more men, and another six were wounded.

On the other side of the village, Golf Company was still getting pasted. An endless rain of grenades, the heavy machinegun, and the sniper on top of the mountain's cliff were really taking their toll. A few four-man fire teams were eventually able to take out the terrorists that were dropping the grenades on them, and a full platoon had gotten into the closest building in the village. From there, the fighting became house to house.

Mosby leaned back on the wall of the building at the edge of the village. Three Marines plowed through the ancient wood door and sprayed the interior with bullets. The terrorists inside tried to spin around and return fire, but they were too slow, and even over the deafening sounds of gunfire and explosions, he heard their bodies slump down to the dirt floor. He screamed at the Marines in response.

"What the hell do you think you're doing! We came he to rescue one of our own and to capture some of these guys! If we kill'em all, we'll have lost all these Marines for nothing! Now go clear the rest of this Fucking place! Find our guy, get some prisoners, and be careful, dammit!"

The Marines had been who had just had their company chopped to pieces around them by heavy machineguns showed not one iota of fear until Mosby raised his voice. Then they looked like deer in the headlights. All three were so paralyzed that not a one could cough out a "Yes sir." Instead they nodded and ran off to clear the next building. Mosby didn't normally strike fear in the hearts of his Marines, but those three were near their breaking point, and Mosby was lucky enough to catch them at that moment. Manny was lucky too, since he was in the building that they went to clear next. Thanks to Mosby's timely words, they didn't kick in Manny's door and strafe him.

One four-man fire team at a time, Golf Company continued to claw its way into the village. They cleaned out three cottages and finally broke through to the main road and the cave entrance. The first man to make it down the alley from to the cave took a round in the chest and was thrown backward 6 feet-lightly wounded. The next two Marines eliminated the terrorist that had just shot their buddy.

His body slumped over and fell on a pile of Claymore mine clackers (detonators). The mines were carefully placed all around Golf Company's position, but since the man responsible for setting them off was dead, the company was saved. If he would have been able to set them off, there was no doubt that the company would have been entirely wiped out. It was an all or nothing deal, and the terrorist took all the hits while accomplishing nothing.

A few Marines from Golf Company took up positions outside the cave to make sure that no one went in or out without their permission. The rest of the Company was already tending to the dead, dying, and wounded. A few of the buildings were turned into makeshift aid stations, and medics were working furiously to save lives. Those who just needed patching or stitching refused to leave the combat areas until their comrades were cared for. Fox Company continued to clean out the village, and Echo Company finally came across when they silenced the heavy machinegun on the mountain, and the sniper disappeared.

Major Mosby surveyed the situation. Almost 200 villagers were rounded up and kept under guard. They seemed innocent enough, but how they had come into the care and welfare of The Protectors of the Koran terrorist organization left some doubt to that innocence. He would determine how to deal with them later.

Only a few of the terrorists seemed to have been killed in the melee. There was the one building where the Marines sprayed-and-prayed their way in behind a hail of lead. The snipers had picked off the heavy machinegun teams in the kitchen and the small woods. Golf Company was a little less selective in their fire, and they tallied fourteen dead terrorists. Mosby was disheartened to see that the members of Golf Company had executed many, but he understood why it happened, and he decided not to put that detail in his report.

Instead, he decided to list the movements of the Companies, the enemy's actions, and then the statistics. Twenty-seven members and potential members of The Protectors of the Koran terrorist organization were killed. Three more were wounded. Sixty-four terrorists were captured, and at least twelve disappeared into the cave.

The Marines had a bad time of it. Given the terrain, there was really only one method of attack, and the terrorists had planned well against it. They were also well armed, and at least some were well trained and experienced. Echo Company lost two Marines killed and twelve wounded-all lightly. Fox Company had twenty of its one hundred twenty members killed and another twenty-one wounded- three seriously. Golf Company had bunched up and been caught in the worst part of the terrorists pre-planned ambush zones. Twenty-three Golf Company Marines were dead, and another forty-one were wounded- six seriously. Mosby knew that it was one of the worst days in Marine history, and search as he did, he also knew that there was no other way to have made the attack.

At least his mission had been mostly successful. Fox Company had rescued the American and taken most of the terrorists. Golf Company had stumbled on a filing cabinet with lots of other names, and while Mosby was adding up the stats, one of the Marines from Echo Company found a briefcase with hundreds of thousands of American dollars in it. They also found a laptop computer that would hold an unknown amount of information.

The only manner in which they failed was in grabbing the specific terrorists that they had seen on the big screen back at the air base on Qatar. There was no sign of the man known as The Ox, the mysterious businessmen, the Saudi Prince Abdualla, or the Chinese intelligence officer. Mosby was determined not to let the death of forty-five Marines be diminished by the lack of mission success. He knew that the 'brass-ring' prizes of the mission had probably left long before their planes ever lifted off from Qatar.

Echo Company's commander brought Manny to Major Mosby.

"Major, I can't tell you how good it is to see you. One minute I'm in a shop in Islamabad, and the next I'm here. Where are we? When are we going home? Is there anything I can do to help? I'm just-"

"Look here Mr. Luka. I don't exactly know how you came to be here or why you were in Pakistan, but I do know that this is a fishy situation. You and I will have to have a nice little sit down later. Maybe tomorrow when we pack up and start our little walk outta here. Right now, if you have any medical training, you can help out the medics in that building over there. If not, then stay close, and I don't want you to go anywhere without at least four of Marines with you at all times. I've got some things to take care of right now. You'll have to excuse me, and, oh yeah…give your thanks to the forty-five Marines that died trying to get you outta here. Golf Company's lining them up in that building over there."

Mosby was irritated at Manny, but it wasn't Manny's fault. He was more frustrated that so many of his men and women had to die to rescue someone that he thought was a spy. His solace came from the knowledge that every terrorist dead or captured was at least one American that would live. That's what Mosby believed they had died for…not for Manny.

The Golf Company commander was busy dealing with his dead and wounded, when the Company's Gunnery Sergeant asked for permission to enter the cave. The young Captain was too preoccupied to think about it, and he granted the salty Sergeant permission. Twelve Marines went in and began their hunt.

After Mosby called his superiors to report his units' status, he met with the Golf Company Captain and asked if his Gunnery Sergeant was alive. Only then did the Captain realize the gravity of his snap decision. Mosby had Echo Company secure the village from any possible counter-attacks while Fox Company dealt with the villagers, captured terrorists, and the all of the seized documents etc. Then he had the rest of Golf Company go into the cave to check it out, search for the escaping terrorists, and find out what happened to the Gunnery Sergeant's recon patrol.

By the time everyone was organized, a Corporal from Golf Company came out of the cave. He couldn't find his Company commander or any of the other officers in his Company, so he reported directly to Major Mosby. He was completely out of breath.

"Sir...the 'Gunny' wants some more people to come in and help us check out the rooms-I mean the cave. See, they've got a garage and some offices setup in there, sir. So far we haven't found any tangos (military term for terrorists), but...I dunno, sir. It's really weird in there, sir."

Mosby was patient, but he did have a limit. There were dead Marines lined up in one of the buildings, and his deep-seated blood thirst for vengeance was crying out.

"What's so weird about it? C'mon! Speak up Marine. I need a little more information than 'It's weird in there.' Now, what exactly did the 'Gunny' say?"

The Corporal caught his breath and continued.

"The Gunny just said for me to come back here and tell the old man that he needs reinforcements. He also told me that we could probably put the whole company in there. As far as weird goes, sir...the place looks like a normal cave, but about 200 meters in, there's walls and stuff. It seems like you're in a building. They've got trucks, cars, and lots of crates with weir-I mean, markings that we couldn't read on them."

By this point, the Echo and Fox Company commanders had joined them. Mosby checked with each to make sure that the situation was well in hand, then he told them that he was going into the cave with the rest of Golf Company. Everyone understood, and as soon as Golf Company was assembled, they headed in with the Corporal leading the way.

It was exactly as he had described. The dirt road continued into the cave, curved to the right, and opened into a wider section where vehicles and crates were lined up on the right side. It was as if they had parked in a parking garage except that the walls weren't smooth poured concrete. Fluorescent shop-lights hung from bolts mounted in the rock ceiling. There were three beat up pickup trucks with a DShKM heavy machineguns mounted in the bed of each. Then there was another pickup truck, much newer and in much better condition, but it was unarmed. Next to it was a Russian-built jeep and an old ¾ ton Mercedes cargo truck.

A few pallets with crates lined the back wall. Their markings were rubbed, faded, and generally illegible. Mosby had them opened and found all kinds of things. Most of the crates and boxes had American MRE's (Meals Ready to Eat-field rations for soldiers). There was a lot of miscellaneous stuff too; including blank paper, a few computers, lights, blankets, books, music CD's, and over two hundred DC truck batteries. Next to all the crates were forty-two 55-gallon drums of gasoline.

Mosby had a squad check everything for boobytraps while he and the rest of Golf Company went to find the Gunnery Sergeant. They passed down a long room that was lined with desks and computers. Some were still turned on. At the end of the room, there was a hallway that turned left and right.

The Corporal led them to the snaking hallway on the left. About 50 meters down the hallway, it opened up into a huge room with flimsy bunk beds and religious posters everywhere. At the far end, there was a ladder climbing up into a hole in the ceiling, and another hallway leading to the left. Mosby had four men go up to check out where the ladder went, and he sent a squad of ten Marines with the Corporal to link up with the Gunnery Sergeant.

After a few minutes, the Gunnery Sergeant came into the room with the Corporal and the other Marines that he had taken into the cave.

"Sir, this damn cave goes on forever. We could be here for weeks. Now, there's a lotta places for them tangos to hide, and there's a lotta outstanding places for them to get the jump on us. How long d'you wanna us to check this place out?"

"We'll keep looking for a few more hours, Gunny. I think we outta put our casualties on some of their trucks and drive out tomorrow morning. It's getting a little late to leave right now, and I don't want us to be on that road in the dark-night vision gear or not. Besides, that'll give us time to get a good look around, treat our wounded, get some rest, and then properly blow the hell out of this place. You read me Gunny?"

The Sergeant was normally a simple and sometimes pessimistic man, but his smile said it all. Mosby wasn't unusually familiar with his Marines, but they had been under fire before. They knew that Mosby wouldn't toss their lives away or make a bad decision. He had a reputation for always seeing the best methods clearly and immediately. The Sergeant had his own opinions about searching the potential ambush alley inside the cave complex, and he was happy to see that Mosby had the rest of the big picture in mind also.

The Marines that went up the ladder came down a few minutes after the Gunny. They reported that the ladder led up to another cave. That cave was about 300 meters long and had four rooms attached. There was nothing in the rooms except some blankets and candles, but they did find the cave's exit. It came out on top of the cliff above the village. There, they found the 12.7mm heavy machinegun, lots of ammunition, three dead terrorists, and a spot where the sniper might have been. They brought the gun and ammo inside, and left the bodies where they had fallen. It was assumed that the sniper escaped to somewhere else on the mountain. The Marines didn't want to stay outside of the cave for too long and risk becoming sniper bait, so they came back and reported to Mosby.

Golf Company continued to explore the caves until midnight. Echo Company brought the trucks out of the cave, loaded them with dead Marines, and prepared to move the wounded in the morning. Fox Company maintained a perimeter around the village and watched their prisoners.

An hour before the sun rose, Mosby met with his company commanders. Echo Company was ready to move. They had also taken the time to load as many computers and documents as they could into the back of the unarmed pickup truck. Fox Company had some trouble with a few of the terrorists, but it was nothing that the pissed-off Marines couldn't handle, and the terrorist's resulting injuries were all minor. Golf Company believed that it had mapped out the entire cave complex, but there was some speculation about secret passages, and everyone wondered where the escaped terrorists had gone.

They took some satisfaction in the knowledge that if the terrorists did come out of some secret passages, they would be welcomed by an inferno. Golf Company dumped the drums of gasoline all over the cave. A downhill slope from the garage area to the lowest room made the job easy.

As the sun turned the sky blue and the mountains' crests glimmered gold, the Marines loaded their wounded and headed for home. Echo Company took the lead on foot. Golf Company walked escort alongside the trucks, and Fox Company brought up the rear. Everyone thought that it would have been nice if the terrorists had stored more trucks, but no one complained about the windfall gifts that the terrorists left behind-including the transportation.

As the Marine column turned the bend and crested the terraced farms where they had landed, the village exploded. Out of the wall of flames, a single figure ran for his life. One of the Marines from Golf Company had stayed behind to set off the gasoline with a flare gun. The village was turned to cinders within minutes, but the cave kept thumping out explosions. Oxygen, coming down the air vent with the ladder in it was feeding the flames slower than the gas could burn. As the smoke built-up, it smothered the fire, and a few seconds later, when more air was pulled in from the airshaft, the cave would thump out another roaring explosion. The blasts continued for as long as the Marines were within earshot.

Within a few hours, the column came within sight of the Pakistani firebase, and a few minutes later they were greeted by a patrol. The Pakistani commander offered to guide the Marines back to the supply depot, and Mosby agreed. He also had the word silently passed among his Marines to watch out for a double-cross from the Pakistanis. They were allies, but as far from civilization as they were, the intentions of low level field commanders could not be unquestionably depended upon. His concerns were well based, but not warranted.

Three hours later, they arrived at the supply depot. Mosby greeted yet another questionable Pakistani officer. There was a convoy headed for Islamabad the next morning, but Mosby wanted his wounded evacuated immediately. The commander agreed.

Mosby had been unable to arrange for an American Dustoff (military term for helicopter medical evacuation flight), so the Pakistanis had three helicopters from Islamabad make the flight up to take out the wounded. A few armed Marines from Golf Company accompanied them to ensure their safety.

In the middle of the next morning, the Pakistani convoy was pulled together. Three companies of Pakistani infantry and an artillery battery were being sent back to Islamabad.

Only one company was remaining at the supply depot. Mosby thought it was a bit odd, but he ignored the movement. He knew that the most well informed people did not always order military movements. It was his opinion that too often, pencil or button pushing staff officers and beaurocrats made stupid decisions that grunts in the field had to pay for in blood. He also believed that Manny was one of those button-pushers.

Some of his Marines boarded into empty Pakistani trucks, and they pulled into column with the other Pakistanis. The Marines carrying the wounded, prisoners, and contraband in the liberated terrorists' vehicles also pulled into the convoy. Finally, they headed back to Islamabad.

A few hours later, they boarded some of Colonel Earley's U.S.A.F. C141 Starlifter transport planes, and they headed back to Doho, Qatar. There, the wounded would be cared for, and once there, Mosby could turn over his prisoners and contraband. They would have to be debriefed, but then he'd finally be able to dump off Manny and begin putting the costly mission behind him.

Though the mission was a success, the heavy casualties kept it from feeling like a victory. Mosby hated the terrorists passionately, but he also respected their conviction, courage, and sometimes their skill. In contrast, he had no respect for Manny, and he did greatly dislike Manny's stereotype, but he didn't hate him. More than anything, Mosby just wanted to get back to his ship. He had a lot of letters to write, and phone calls to make.

Sunday, February 15, 2010

Gaithersburg, Maryland

7:28pm local

Manny was home at last-physically. Since his return, he had been debriefed time and again. Now he drifted in and out of sleep in a hopeless effort to fight off the combination of jet lag, days lost, and stress. At least it was the weekend. Most employers would have given him a few weeks off, but Manny's new division administrator wanted him back ASAP. It was probably the best thing anyway.

Mentally, he was still in Kashmir. Even though he was a captive, the isolated mountainous region had burned its beauty and peace into his soul. He couldn't forget the shine that the mountains had in the morning, the clarity of the deep and fast river, the wild purity of the woods, the ancient structure of the terraced farms, or the kindness of the woman that always brought him his sweet bread in the morning. His mind was caught in a whirlpool of emotion.

While thoughts of majesty colored his memories, the horrific images of the experience were tossed in and caused shivers. He remembered Asaad's grisly death. He saw the faces of the terrorists. He heard their voices and wondered what the Serb had meant by the mission in the United States? Was there another September 11 in the works?

Manny also remembered the horrors of the Marines rescue raid. He had hidden during most of it, but the incredibly furious small arms fire and the bellowing booming of all the heavy machineguns still reverberated in his bones. His lungs were clean, but he still felt the choking dust from the stone walls that were shattered by Marine bullets.

The smells stayed with him as well. There was the dry dirt smell of the dust in his prison cottage. He wanted to sneeze at the spent gunpowder smell, and vomit at the sweet stench of corpses and coppery odor of blood on the dirt road. The best smells that he could remember were the gasoline smell from the cave exploding and the sweet bread and tea smell of the kitchen in the main study building.

There were the sights too. The starry night sky had been filled with red, orange, and white flickering flares from the terrorists. Smaller glowing lines of red, yellow, and green tracer bullets had criss-crossed the valley. Muzzle bursts from the heavy machineguns flashed yellow against everything close to the guns. The sound, smells, and sights made the scene surreal in his mind.

Once more he awoke in a cold sweat. He had turned the heat in the apartment up to 80 degrees, but it still wasn't warm enough to burn off the memories. His body was cold with sweat, and his mind still felt the mountain cold. As he made a pot of coffee and fed his cat, he wondered if he would ever warm up.

When his coffee was finally ready, Manny filled his cup and sat down for another day/night in front of the television. The couch was a familiar luxury, and he couldn't believe that he'd been sleeping on the ground for as long as he had been. Some of his memories were perfectly clear, but others seemed so distant. Getting back into his routine clouded the clarity and helped push the worse memories deeper into his mind.

The TV news didn't help the situation. He'd been watching everyday since his return, but there was no report on any of the news channels about the costly U.S. Marines raid that had rescued him. He knew there wouldn't be one, since it was a secret and special operation, but he also wondered about Major Mosby and the letters he had to write to the families that had lost sons and daughters. How could one explain and justify their sudden and mysterious deaths?

There were news reports from Pakistan, and they further twisted Manny's mind. There were three new stories of violent anti-American protests around the country. One in Peshawar was particularly violent, and the footage reminded him of the protesters that had been present when he and Asaad were abducted. Tear gas, smoke, bodies, and chaos filled the street scene on TV and in Manny's memory.

The 24-hour news channel staff seemed as hungry as ever to speculate on the Pakistani problems. Expert guests were interrogated on five channels, while more and more anti-American crowds protested anything and everything dealing with the U.S. Even the 13-year old Operation Desert Storm was protested.

The rise in anti-American sentiment reportedly came from several major sources. Religious differences were a constant thorn in the relationship. America's inability to reward its allies for their support in military operations was another. America itself was the biggest reason for the sentiment.

A large portion of Pakistan's predominately Muslim population was fundamentalist. Since the 1979 Iranian revolution, there had been no love lost between radical Muslim fundamentalists and the United States. As Muslim extremists sought to create a more fundamental religious government, they had to defame their government's relationship with the United States in order to push away the threat of American intervention into any revolution-Islamic or otherwise. That was how it had happened in Iran, Lebanon, Afghanistan, and Indonesia. Before those countries fell, the Communists used the same methods in Cuba and Nicaragua. Only once had rebels driven off American military support for a regime, and that war-the Vietnam War-took 20 years. Muslim extremists in search of an Islamic government would need to protest and drive off U.S. military support before any other attempts to change their government.

Ironically, the U.S. was doing a nice job of ignoring its ally even without the negative media attention of Muslim extremists. Most of the TV experts reported that Pakistan had always been a stalwart ally of the United States, but in order to remain unbiased in the region, and to prevent a nuclear war, the

United States had rarely returned the favor. Even during its failed humanitarian efforts in Somalia, after Pakistani forces provided the vehicles used to rescue trapped U.S. Army troops, the Americans showed almost no tangible gratitude. Most recently, the United States promised to improve relations with the Pakistani government after its support in Operation Enduring Freedom and the American War on Terror. That support never came in an amount or manner signifigant and public enough for the Pakistani population to appreciate.

It was no wonder that people in Pakistan were starting to lose faith in both the United States and the Pakistani government. Pakistan had several problems with the U.S. There were the religious differences and cultural differences. The one-sided alliance proved to be a continuing public relations nightmare; and the rich, free, boisterous, flashy, and comfortable United States itself left poverty-stricken Pakistanis with one question, "Why do the Americans have to hoard all the money and power in the world?" As far as it may or may not have been from the truth, people's hearts and minds were pushed in that direction.

Even though he fully understood the alleged TV experts, Manny had a tough time accepting the rise in Pakistani anti-American sentiment. It seemed like those kinds of protests had gone on for years, and he couldn't find many reasons that they would suddenly grow in their intensity. The only reason one that made sense was that perhaps some of the terrorist instigators-people like The Ox-were up to something.

Monday, February 16, 2010 Washington D.C.

Office of Homeland Security

International Terrorist Movement Monitoring Division

9:50am local

Manny was surviving on coffee and over-the-counter caffeine supplements. His body was still in a shambles from the time change and the mental stress that kept him from a steady sleep. Back at work, everyone seemed happy to see him. Someone brought in a cake and decorated his cubicle. Even his new administrator gave him a pat on the back and a promise to never send him overseas again. There was no apology, but Manny was content with the promise.

There were the phone calls, too. Manny had spent most of his normal days on the phone with his contacts at different intelligence, law enforcement, and safety agencies. It seemed each and every person on his phone list wanted to talk to him. Some were polite and tried to avoid asking any potentially stressful questions. Others were more straightforward.

When his contact over at the CIA called him, his stress level went through the roof. There was the usual pleasantries, "Hey, how'ya doin'? So glad you're back." Then Manny got the surprise of his life.

"Say Manny, you remember that ISI Captain? I thought I read in your debriefing report that he was killed over there?"

Manny was slightly taken back by the matter-of-fact tone, but in certain circles he did have to treat his experiences as intelligence data. He swallowed hard, and then he answered as professionally as he could.

"Yeah. It was pretty bad. I watched'em do it...not a nice thing to see."

"Well, I think there's something weird going on over there. We've got another transmission trace that our system triangulated near that village. It's a call to someone in Beijing. Now, the computer doesn't have a file on the Chinese voice, but it did identify the other one as Asaad's. The margin of error is real slim. There's probably only one chance in millions, but maybe this is the one chance? Do you know if Asaad had an evil twin like in the movies, or could it have been somebody else that was executed?"

Manny was furious, and he didn't hide it. His professionalism evaporated, and the stress of the past few days exploded.

"What? That can't be right. Are you sure about this? I watched them shoot the man. Sure, he had a pillowcase over his-SHIT! GODDAMNIT! How? WHY!

SON OF A BITCH! THAT BASTARD! I CAN'T BELIEVE I DIDN'T EVEN CHECK HIM! I EVEN BURIED THAT SON OF A BITCH! Hold on a minute…"

Everyone in the office was attentive and silent. Even the phones stopped ringing. No one wanted to be obvious in their eavesdropping, but they all had visions of an office worker going on a shooting spree. The division administrator hustled out of his office and stood in the doorway of Manny's cubicle.

"What's going on Manny? Is there a problem here?"

"I'd say there's a huge problem! They baited us and we fell for it. Asaad's alive. It's no wonder they called the guy 'Bulletproof.'"

The room was silent. His new boss had come over because he thought that Manny was having a breakdown, but immediately he understood the depth of the situation.

"Oh my God. Are you sure? How do you know? Who's on the other end?"

His CIA contact could hear their conversation.

"Manny? Are you there?"

"Yeah. I'm still here. Were there any keywords?"

"I was just about to tell you. They spoke in numbers. While you were gone, one of our guys wrote a new subroutine that scans all calls from people on the watch list and listens for numbers. We've always had a problem with tangos using codes, but usually only professional spooks use numbers. I'll email you a copy of the transcript in a few minutes. Right now we're doing a digital-data comparison-sort through the database to see if it matches any previous codes that we've been able to identify. This is a new feature, and I don't expect much. Do you have any ideas why the Captain faked his death, and now he's making secret coded calls to the Chinese?" "Lemme call you back. We're gonna effort this right now." Manny hung up his phone and walked past his boss.

"Let's go in your office and talk about this."

The new man followed and motioned for a few of the others in the office to leave their cubes and join them. Everyone hurried to grab pen and pad. The door was already closing before the last person was seated.

Manny told them his whole story and the new information that he had from the CIA. Only his boss had read his debriefing report, and everyone else sat in awe as he told his tale. Then they began to list questions and match them with brainstormed answers.

Why had Asaad insisted that Manny-of all people-come to Pakistan? Everyone agreed that it wasn't because Manny would have been a valuable hostage. No one had ever asked for a ransom while he was being held. The only thing that Manny really offered was a chance to get information into the American intelligence system.

Manny mentioned that he had heard The Ox talking about getting rid of him, but some of his coworkers pointed out that if they had wanted him dead, they would have killed him with Asaad-or whoever died in Asaad's place. They all started to believe that Asaad knew that Manny spoke Serbian, and that detail could have been passed to The Ox. If that happened, then it was very convenient for The Ox to discuss an upcoming mission in the United States in front of Manny, and the only way to pass that information on any easier would have been to tell him in English.

Manny understood, but he was still convinced that The Ox was going to kill him. Part of him still couldn't accept that The Ox had even killed Asaad. One of his co-workers mentioned how unusual Asaad's death had been. He was betrayed by his driver, who disappeared immediately afterward. They seemed to want to kill him, but they waited until Manny could see it happen. They disfigured his face and then left the body where Manny could do nothing but stare at it all day. When Manny finally wanted to bury the faceless body, conveniently he found a cell phone. It seemed very careless that they had gone to all that effort and then left a cell phone on the ISI agent's pocket.

The line of deductive reasoning fell into place quickly. If Manny could believe that Asaad was alive, then he had to believe that his death was a deliberate hoax, and since Asaad hadn't reported back to the ISI, it was safe to assume that he was part of the scheme. Since the execution was a hoax, it was a good bet that the cell phone was left there on purpose. That purpose would have to be so that Manny could at least report Asaad's death. If Manny was expected to call the intelligence community, then The Protectors of the Koran might very well have expected the rescue team. Judging from the wellorganized defense of the village, they probably expected it.

Manny was convinced. The larger question as to why The Protectors of the Koran had created this elaborate scheme still remained. If it was assumed that Manny was used for discreetly and inadvertently passing information, then what information could only have been moved through such a scheme? If there was an attack being planned against some sort of cross-denominational conference in the United States, then why did they plant that information?

Manny remembered the 24-hour TV news experts that he had been watching since his return. Pakistan was rapidly showing signs of antiAmerican discontent. The government over there had a reputation for being ruthless from time to time, and no one ever really believed that it had the respect and will of the diverse people behind it. Manny remembered all the files and computers found in the caves, the fully fueled trucks, and even the extra gasoline. Whatever information was on those computers and in those files was probably the information that they wanted the Americans to have.

While he was thinking about everything that they had found in the caves he realized that there were a few things missing-most notably ammunition. There was plenty to fight the Marines with, but there wasn't enough to arm a small army of terrorists. There was just enough to use and make it look acceptable. There wasn't any huge ammunition storage. If the terrorists could afford all those new computers, the nice construction and office built into a cave in the high Himalayan Mountains, then why didn't they have a stockpile of ammunition there? He knew the answer: because they only left what they wanted to be found.

Manny looked at the division administrator and they both tried to ask the same question at the same time, "Exactly what kind of information was in those files and computers?" Everyone looked around and three people bailed from the room to make calls and find out. Everyone else sat for a moment, thought about the question again, then quickly left to make calls in the hopes that someone else might be able to answer the question.

Manny and his boss stepped out of the office last. Neither man had taken three steps before they both stopped in front of the TV by the receptionist desk. It was tuned to one of the sensational 24-hour news channels, and the scenes were surprising. The reporter was discussing the deteriorating situation in Pakistan. More pictures of anti-American protesters were being shown, and finally the latest source of the riots followed. It was allegedly a security tape of an American CIA agent directing the torture of Sheik Mohammed Hanjour. In reality, it was well-edited security camera footage of Manny being given a tour of the ISI interrogation floors by Captain Asaad. None of that mattered to the Pakistanis who saw the tape.

The media had told them that it was an American spy torturing a Pakistani cleric in Pakistan with the full cooperation of the ISI. The media had been given the story by their confidential source that had given them the video. Everyone in Manny's office new it was a setup, but the phones immediately started ringing from officials all over Washington D.C. They needed an explanation.

Manny's boss turned and tried to break the increasing tension in the office.

"Well Manny, I guess you're definitely not going back over there now. I think if I were you I'd even try to avoid going near their embassy right now." No one laughed, but the tension did stop increasing.

Manny walked back to his cubicle and his phone lines were lit. The first one he picked up was his contact at CIA.

"Manny? Did you just see that footage on TV? Buddy, if that's not a setup, then I don't know what one would look like. Did you guys figure anything out over there?"

"I can't believe I'm on the international news as a CIA agent. On a normal day that would be pretty cool-but not as some sort of torture expert. I just can't believe it."

"We can't either, Manny. I don't think most people will. At least most people over here won't buy it. Of course, we KNOW we didn't send you over there. Still, I think there's a lotta people over there that're just looking for some excuse to get pissed at us. That's why it screams setup."

103

"Well, I <u>know</u> it was a setup. Listen, we're pretty sure the intel that the Marines brought back with me is planted. We just haven't figured the whole thing out yet. It looks like they wanted us to make security tight at the MultiDenominational Religious Conference that's going on here in D.C. Exactly why the terrorists want us to make it tough there is beyond us. The only thing I can possibly think of is that they might want to sacrifice someone in a bad hit attempt at the conference. That would be another method of passing on more false information. The other option is they've got a plan they're so confident will work that the only way to make it better would be to have it work while cameras see our best security efforts fail. Either way, it doesn't tell me why they went to all this effort to get us all the data in the files and computers. It looks like we've got a multitude of things going on, and the only commonality is that the bad guys went to a lot of work and a lot of risk to plant the information. I dunno. Do you have anything for me?"

"It looks like the bulk of the information collected so far can be piled into three groups: police, military, and political. There's information indicating certain individuals from each group were planning to make attacks on the other two. The types of information are all over the place. We've got actual plans, names, receipts for purchases, pictures, prints-all kinds of stuff. The list goes on and on. We've already passed on a bunch of data to the ISI, and they're rounding up tangos as we speak."

"Maybe that's it, then, don't you think?"

"What do mean by 'it'?"

"If they planted information that leads to arrests from all three groups of power, then the ISI would have to attack all three groups. They're making the ISI dismantle the power structure. Maybe the reason they made it look like each of the groups was going to attack the other two was to gain the support of the other two in rounding people up. That's gotta be it. The Protectors of the Koran are playing on the weak relationship between the different power structures to have them tear each other apart. Then they can move in."

"You're definitely making sense, but how would they move in? They're not that big."

"They're not that big yet. 95% of that population are quiet Muslims. They're like the infamous silent-majority over here. 3% of the people are Christian, and 2% are radical, Muslim fundamentalists or extremists. If the radicals can convince the 95% that the government is screwed up, then the 95% and 2% take power together. That would explain the fake tape that was sent to the networks to make me look like a CIA torturer. It all fits together!"

"How does the tape do that?"

"The tape makes the current government look like puppet of the United States, and I certainly play the part of The Great Satan in their little movie. It defames the Pakistani government, embarrasses the U.S., and moves to rally their silent-majority to find a better, more independent and righteous government than the one they've got now."

"Manny, I think it's time you left Homeland Defense and came over here."

"No way, man. I am never going overseas ever again."

"Seriously Manny, I think you'd better get over here right away so we can talk more about what's going on with their side and with ours. I'm gonna have my boss make a few calls...in fact... he's motioning to me right now that he's already started. You might as well say your good-byes and get in your car now."

In the background, Manny heard the phone ring. The receptionist answered and passed it to his boss. The wheels were in motion to transfer Manny to the CIA. His tour at the Office of Homeland Security was about to end.

"I'll see you in about 30 minutes, Manny."

Not five minutes after Manny was off the phone, his boss came back and gave him the "good" news. Manny wasn't so sure about how good it was, since he was very comfortable at Homeland Security, but the way in which his boss put it to him was that he could either move up or move out.

On the positive side, he was told that there were going to be some budget cuts in the division, and that if he didn't move out, then his boss had been advised by the CIA that Manny had become too expensive for him to keep in the office. The CIA had no idea of how expensive Manny might have been, but they knew that they wanted him, and the Office of Homeland Security was obliged to cooperate and coordinate between all agencies and departments concerning security issues within the United States. Manny was transferred.

Monday, February 16, 2010 Langley, Virginia

Central Intelligence Agency

Terrorist Tracking Center

11:11am local

Manny pulled his car up the security gate. The guard asked for his identification, and when Manny pulled out his driver's license, the guard went back in his little booth to make a call. He came back a moment later and told Manny where to park and where to enter the building. His contact would meet him inside at the security checkpoint.

He parked his car and looked around. Manny had never been to CIA headquarters. It was a strange building with a glass-like façade attached to the outside. It looked like they had tried to build a building around the original structure. In reality, it was an attempt to block feint sounds and conversations from being monitored by outside spies and sensitive satellites hundreds of miles above. Manny knew nothing of the purpose or causes for the building in the glass bubble, but it seemed odd to him. The entire setting was foreign. For a moment he wondered if he had just gone into another form of captivity.

He walked through the parking lot and into the lobby that the guard had directed him towards. Inside, his CIA contact, Carl Crain, waited and greeted him. They had spoken for years over the phone, but Manny had never met Carl. He was just under 6 foot and about 170 pounds. He had a nice balding patch of hair at the back of his head, and his eyes carried overworked bags under them. He looked average.

At 35, Carl was typical of the new generation of CIA analysts. The Greatest Generation created the agency. The Baby Boomers ended the Cold War (surprising themselves when it happened), and then they fumbled into the New World Order and through the millennium. After the Al Quaeda conflict ended, many chose either early retirement or moved on to other fields rather than try to keep up with the monthly leaps in technological capabilities and requirements. As was happening in public service organizations around the U.S., the Boomers left early, and the one-time stereotyped 'slackers' of Generation X were handed the ball. What no one expected was that people like Carl or Manny grabbed the ball and ran with it. Between 2000 and 2011 it was expected that 85% of the Federal government would retire and hand their reins to a new generation. Carl, Manny, and hundreds of thousands of others were stepping up to the plate.

"Manny! It's good to finally meet you. I'm Carl Crain. Welcome to Central Intelligence."

"Hey there, Carl. It's nice to finally meet you. We've been on the phone now for…what, three or four years now?"

They walked past the security checkpoint and down hallway after hallway. On their way, they navigated a maze of cubicle dividers that had been setup in the halls.

"Just over four and a half. Too long without meeting. I'm sorry it took you getting caught up in all of this to finally meet. You'll probably get sick of me really soon, though. For the time being, we're gonna keep you on board as an analyst like myself. Really, you'll be more like an intern though. All you've gotta do is follow me around, watch what I do, see who I talk to, and generally learn how things are done around here. You won't have your own office for a while. You'll have to share mine. I.T.'s putting in a terminal for you right now, and you should have some network rights by the end of the day." Manny was impressed.

"I'll tell ya, you guys sure know how to make things happen. It would take us a few weeks to get all that stuff setup for a new guy over at our place. I can't believe you had me moved as quickly and easily as you did."

"Well, this is your place now, and there's nothing easy about making things happen. We just get the best people, and we expect the best from them. By the way, you're getting a 50% pay raise too, and here's your parking permit, your temporary identification card, and your security keycard. You can get a permanent

ID card sometime when you've got a few extra minutes. Right now, we've gotta go straight to a meeting. I won't even have time for introductions, but it doesn't matter since everybody there won't know each other. It's an inter-departmental thing."

Manny was completely lost in the maze. They had gone down four hallways, through a waiting lounge, through a copy room, and into an office area. Now they were in a hallway again.

"Carl, I am so lost right now. It's incredible."

Carl knew Manny was still tired from his ordeal in Kashmir, and he could see the lingering stress from the kidnapping. He also saw the confusion of being thrown into his new job.

"Don't sweat it. Like I said, no one's going to know you in the meeting. Half the people won't even know me. Just be cool. They'll have coffee there, and sometimes there's even donuts left over from morning meetings. We're just gonna go and listen, but if you do have something to say, just cough it out. Things are really changing around here. I dunno if it's a post-Al Quaeda thing, a new generation thing, a leadership thing, or what, but everyone seems to be much more focused on getting tangos instead of grandstanding or political bickering. You really won't even find any office politics here. It's unusual, unexpected, and very welcome."

They walked into a huge, well-lit auditorium. It looked like a theatre with rows of desks instead of just reclining chairs. Each desk had a nice flat screen monitor below it and a glass window to view it though the desk's surface. Manny thought it was a neat idea. The only other place he had seen it done was on TV news anchor desks. A keyboard and mouse slid out on a tray from beneath the desk at each seat. He and Carl sat down next to each other in the back of the room.

There were at least a hundred other people in the room. Manny and Carl got comfortable. Once they settled, Carl spotted the coffee cart. There were donuts too! Just as Carl started to stand up, a woman took to the podium and began her presentation. She introduced herself, but her name and position never registered in Manny's mind. However, he assumed that she was going to brief them on the situation in Pakistan.

Behind the woman the huge screen came to life. Under Manny's glass desktop, he saw that his monitor had the same image. It was a map showing everything from Iran on the left to Bangladesh on the right. The map was almost unrecognizable with all of the strange symbols on it.

Carl was only half-listening to the briefing, and he could see the confusion on Manny's face.

"Obviously, we're digitally linked into a lot of different systems. It's part of our Strategic Protocol data network. They call it SP 2.0. It's similar to the MilNet-something that the Pentagon's been working on, and we're already trying out. Of

SCOTT MALENSEK

course, they don't know that we're trying it out! You can zoom in on any area with the mouse. Just click and drag, and a box will open up.

You can choose any size or location to zoom in on."

Manny reached for his mouse and tried it. He zoomed in on Pakistan. The map was much clearer, but he still didn't understand the blue, orange, light-blue, yellow, gray, and red symbols.

Even though they were at the back of the auditorium, Manny kept his voice down so he wouldn't disturb anyone.

"What're all these symbols. Is there a key or a legend someplace?" Carl kept his voice down too.

"They're military units. Just double-left mouse click on them, and it'll open a temporary text window that tells you what each symbol is."

Manny tried it. He double clicked on a blue squarely that had the letters RGR in it and a single dot on top. A window appeared next to the symbol. The window described the unit as a U.S. Ranger squad from the 2nd Ranger Battalion. Its operation was named L34, and there were other menu options offered in the window. Manny clicked on the L34 and another window opened up. This time it listed all the text details about their mission-to setup an ambush on highway A74 outside the Pakistani town of Bela and capture a suspected terrorist named Mohamed Hussein. Then he clicked on Hussein's name and another window opened up revealing his entire dossier. Manny couldn't believe the data at his fingertips.

While listened to the briefing, some of the symbols, mostly aircraft and ships, were moving. Some of the planes disappeared as they landed. Others suddenly appeared as planes took off or ships left port. Behind the speaker, on the big screen in the auditorium, Manny saw an orange box appear next to the one he had just clicked on.

He looked down at his monitor and clicked on it. The window described it as an unknown tango unit of twenty-five to fifty people. Manny clicked and dragged to zoom in as tight as he could on the location. He tried three times to get closer and closer. On the third try, the screen changed to a topographic overlay on top of a satellite image. He was watching the firefight live on his desk from the other side of the world! The satellite was almost 150 miles in the sky, but he thought it was as though he were 1000 feet off the ground. "Hey Carl, check this out. Is this live?"

Carl leaned over and saw Manny's screen.

"What're you looking at?"

Manny pointed to the auditorium screen.

"You see that orange box next to the blue one between Karachi and Bela? I think it's a Ranger team that ran into somebody."

108

Carl zoomed in on his monitor and checked the information by double clicking a few times.

"Wow. Looks like they're really in it."

Manny looked back at Carl with an urgent look on his face.

"Shouldn't we tell someone or do something?"

"Don't worry about it. The Navy's seeing the same thing right now. I'm sure they're already directing a plane to the area for support. Pretty impressive Manny. I've tried for a long time to actually catch an operation in action, but I've only seen it happen a few times over the years. You're either real good or real lucky, and I'm betting on the latter."

Manny watched his screen for a few minutes. He could clearly see twelve U.S. Army Rangers. They were moving around in groups of four. Each one seemed to have a huge backpack either on their back, or nearby. They were on a highway overpass, and there were three pickup trucks on one side of the small bridge. Manny couldn't see how many people there were around the trucks, but he counted at least twenty. He couldn't see the tracers, hear the guns, smell the smoke, or feel the fear, but he recognized it. In combat, surrounded by those unique sensory experiences, people ran and moved a lot faster.

Carl and Manny listened to the briefing for a few more minutes. The speaker gave an overview of the ISI's round up of suspected tangos. Then she mentioned Manny's theory that The Protectors of the Koran had planted the intelligence. Most people in the room seemed to doubt the idea. Manny and Carl just looked at each other.

Lunch was approaching, and people were getting hungry. The meeting was about to end when Manny noticed a few more orange squares appear on the big monitor behind the speaker. Carl noticed them too. They realized what was happening. ISI and U.S. Special Forces teams had been tasked with rounding up people who were tangos, based on intelligence from the computers and files found by the Marines during Manny's rescue. As Manny had suspected, many of the people were just apprehended to weaken portions of the Pakistani power structure.

As Carl had feared, the information that the intelligence was corrupted or planted didn't make it to the grunts in the field fast enough. Some of the teams were getting ambushed. Finally, some of the blue squares representing American Special Forces units started to move away from the orange squares. Then some of the light blue ones-the Pakistani commandos and ISI teams, turned gray. They had been wiped out. Manny had been transferred to the CIA in time to make sure that he was readily available for comment, but his ideas hadn't made it to the field.

The changes on big screen and on all the monitors were unmistakable. Within minutes, everyone knew that the intelligence had definitely been planted. It was one big trap. Well over 200 secret raids were being conducted all over the country. Thousands of Pakistani police, ISI, and commandos were being ambushed. The same was true for the thousands of others secretly fighting in buildings, roads, and hills all over the country. American commandos, Delta Force, SEAL's, Green Beret's, Rangers, and even Major Mosby's Marines were in the thick of it.

Back when the intelligence was being picked apart, and the operations were planned to round everyone up, there weren't enough Special Forces teams in the area, so Mosby was called on again. This time he could only bring in Echo Company, but they were holding their own. Their mission was to surround and take over a small factory outside of the coastal town of Jiwani– next to the Iranian border. The factory supposedly canned fish, but intelligence from one of the computers that Mosby recovered claimed that the canning plant was actually being used to make small anti-personnel mines for use in a terrorist attack outside a soccer game in Kabul, Afghanistan.

Everyone in the auditorium was clicking on icons and zooming in to watch the action. They had already put the word out that the intelligence was bad, but the Pakistani authorities and American Central Command had failed to act soon enough. There was nothing they could do at CIA except watch. They watched in safety and everyone had different emotions about what they saw. Some who had been in combat before felt terrible frustration to the point of nausea. Others, to whom battles were little more than life or death football games, watched with awe and amazement. Everyone, for one reason or another, felt some form of excitement.

Manny found Mosby's Marines and zoomed in on their icon. He kept zooming in until he saw the factory. Then he saw the Marines. They were all around the dull looking rectangular building. A handful of them were advancing, driving their way inside. Manny watched them throw and fire grenades. He saw the explosions from grenades and explosives being tossed by tangos. Suddenly, everyone stood up. The Marines walked toward the building, and people started walking out with their hands above their heads. Manny figured that it must have been over. The last shots might have been fired, or perhaps someone simply yelled over the deadly ground that they wished to surrender.

The people with their hands over their heads finally met with the Marines. Then Manny watched a silent black puff cover his monitor. As it faded, he saw the bodies and the black star-shaped smudge marking the explosion's epicenter. He didn't have to be on the ground to know what happened. One of the tangos didn't want to be captured so he pretended to surrender, and then he blew himself up in an attempt to take others with him.

Manny felt the bile come up in his mouth. He closed his eyes and looked away. When he looked back at his screen, he saw the Marines methodically executing all of the terrorists. They had to do it. The Marines couldn't trust anyone else to surrender without risking more lives.

Manny imagined what was happening to Mosby right then and there. Was he still alive? What must go through a commander's mind when he knowingly has to kill prisoners for the safety of his men? Manny thought about Mosby, then he realized that when it came to the lives of his men, Mosby wouldn't feel one bit of remorse. He couldn't. Manny was convinced that pity for anyone who tried to kill his men didn't exist in his being.

Lunch had come and gone, but few people had left the auditorium. Those that did simply couldn't bear to watch. Other people were coming in during their lunch break to see what was happening. Manny accepted that since it was part of their job to know what was happening in the world. One of the professional spectators pointed out that there was something happening on the east side of Pakistan that could amount to trouble.

Everyone looked. Carl spotted it and pointed at a light blue "V" with a line coming from it that was heading towards India. The room watched. A Pakistani fighter plane had either been taken or its pilot was trying to get killed. For unknown reasons, the American made F-16 Falcon flew directly over downtown Lahore, Pakistan and into India.

Instantly, the Indian Air Force reacted. A pair of yellow lines with V's at one end turned and started toward the Pakistani F-16. A few more symbols started to appear from the Yellow Indian planes. The new symbols coming from the Indian planes were missiles. The CIA's computer decided that the Indian military had become a threat, and all of the Indian symbols turned red.

Everyone in the room watched as the international incident started. A flight of four more Indian aircraft turned from their operating area off the northwest coast of India, and headed north toward the engagement. They were Russian-built Mig-29 Fulcrums. Soon the night sky over South Asia would fill with aircraft on the other side of the world.

The blue Pakistani symbol started to spin around when the first missiles came close to it. The ancient 1960's era Mig-21 Fishbeds continued to close the distance, and the renegade Pakistani F-16 fired a blue symbol-an AIM-9 Sidewinder heat-seeking air-to-air missile. The Mig-21's started to turn and spin around in an effort to elude the missile. All of a sudden, the missile symbol, and one of the plane symbols disappeared from the big screen in the auditorium. The wild Pakistani pilot had just shot down an Indian pilot, and the auditorium fell silent.

The remaining Mig-21 turned and headed toward the four closing Mig29's to its south. Clearly someone on the ground in India was directing the outclassed Indian pilot, and everyone in the room knew that if ground controllers knew about the dogfight, then it was just a matter of time before senior officers were being informed. The slower, 45-year old Mig-21 never had a chance against the more sophisticated 20-year old American-built F-16. It disappeared from the big screen as the Pakistani pilot chewed it apart with a burst of 20mm cannon fire. Another Indian pilot on border patrol was falling to Earth. Everyone in the CIA meeting room hoped that parachutes were involved, but they weren't.

The light blue F-16 turned 180 degrees and headed north. Its Pilot either had a specific target in mind, or he suspected that the Mig-21 was headed toward help.

Either way, the Mig-29's had a tough time catching up. Four more red airplane icons appeared over an Indian airbase near the Kashmir/Jammu region. They turned south and headed toward the F-16. They were more Mig-21's. Even though they were less sophisticated they did have numbers on their side, and probably assumed they had an edge in skill level. All fighter pilots make that assumption.

The Pakistani pilot started to weave back and forth over the border areas as the eight planes headed toward him from the north and south. Now more Pakistani F-16's started to turn toward the border area. No one in the room and few people in the world could have known their intentions. They might have been going to intercept the Indian planes, to shoot down their renegade pilot, or perhaps there was some other reason. No one really knew.

The wild F-16 closed with the aircraft to the north in almost no time. They fired two volleys of four missiles at him, but all went past. The Pakistani pilot never even tried to dodge them. He knew that the Indian Air Force's short range air-to-air missiles were based on old technology and that they could only find targets when fired at the rear of a plane where the engine's hot exhaust was easily detected by their missile's seekers. As long as he was within short range and coming at them head-on, the Pakistani pilot didn't need to flinch.

He knew it. The man had skills. All eight missiles streaked past him, and he let loose a missile of his own. The four Mig-21's split up and tried to evade it, but his missile was too sensitive. It detected the heat signature along the front edge of the Indian planes' wings. They were still head-to-head and closing at over 1000 miles an hour, but his missile found a target and closed on it at almost 1500 miles an hour. From 6 miles away, the Indian pilot had no time to react, and another red symbol disappeared from the screen.

The more maneuverable F-16 turned to the right and was immediately on the tail of another Indian Mig-21. Not three seconds later, the Mig-21 disappeared. It had been ripped apart by a well-placed burst of 20mm cannon fire from the F-16. Exploding fuel tanks from the falling Mig-21 illuminated the starry night sky over the Pakistan/India border.

Time was running out for the loose Pakistani pilot. The remaining Mig21's had circled around, and they were closing on his rear from both his left and right sides. The four Mig-29's had finally closed the range and joined in the spinning chase five miles above the planet's surface.

More Indian and Pakistani planes were taking off and heading for the border from both directions. Three pairs of Pakistani F-16's were headed to the dogfight from the northwest, west, and southeast respectively. Four formations were lifting off from Indian airfields. They were all Mig-21's.

When the Mig-29's joined with the two remaining Mig-21's, the Pakistani F-16 pilot decided that it was time to head back. Whatever had sparked his moment of desertion into enemy territory was long gone. It might have been that the Pakistani pilot's and ground controllers finally convinced him to return. It might have been that he had a change of heart, or had killed enough Indians to satisfy his blood thirst. For whatever reason, he headed back into Pakistan. The Indian planes followed him over the border.

That's when the real trouble started, and the crowded auditorium grumbled silently with tension. They knew that if the dogfight didn't end, it would keep growing. If it kept growing, it would eventually lead to air strikes, ground assaults, and possibly a nuclear war. The ground controllers in Pakistan and India knew this too.

The pilots on both sides knew it, but they also knew that it was their job to defend the skies of their respective nations. The Pakistani pilot had turned back, but he was still loose, and if the Indian pilots didn't get satisfaction for the planes that he shot down, then there would be no deterrent to prevent future renegade incursions. The Indian pilots had to give chase, and the Pakistani pilots had to defend their country.

When the lone Pakistani pilot met with a fresh pair of Pakistani F-16's, his two pursuing Mig-21's and the four Mig-29's with them all fired a volley of heat-seeking missiles at him from behind. He had to turn and hide the hot exhaust of his engines to evade them. The two fresh F-16's fired a pair of missiles at the Mig-21's. Both Mig-21's and both missiles disappeared. The missile that was headed for the renegade Pakistani disappeared as it passed by him and self-destructed.

The four Mig-29's fired another barrage of missiles and then closed on the F-16's. One of the F-16's disappeared from the big screen. Another Pakistani fired another missile, and a Mig-29 disappeared. The lone Pakistani turned and chased a Mig-29 that disappeared-presumably from yet another burst of 20mm cannon fire.

More planes joined the dogfight as it drifted to the south toward Lahore, Pakistan on the India/Pakistan border. Two more Pakistani F-16's and eight more Indian Mig-21's flying in two four-plane formations all seemed to collide in the night sky. They chased each other at 400-500 miles an hour. When they passed each other, the closing speeds were over 1000 miles an hour. Planes darted back and forth over the border, and missiles streamed across the black sky like twisting meteorites. Every second brought another thunderous boom as missiles found their targets or self-destructed whenever they couldn't.

Within a minute, two more F-16's and sixteen more Mig-21's were joined in the twisting furball of gray contrails drawn in the star and missile strewn sky. Anyone who lacked skill and luck were hit by missiles or cannon fire. A few ejected safely, but most fell to the ground in their twisted, torn, and burning wreckage.

Finally, ground controllers on both sides regained control of their pilots. More planes lifted off from airfields all over Pakistan and western India, but the planes patrolled well away from the border. The dogfight continued until there were only two Pakistani F-16's and three Indian Mig-21's. All total, eighteen planes were lost: six Pakistani and twelve Indian.

While the air battle flared, flamed, and fizzled, there were other deployments on the big screen. Indian naval assets in the Arabian Sea turned west and headed toward Pakistan. They remained in international waters, but Pakistani patrol boats had to hurry out of port to block the path of any incoming missile or air attacks. A few other military units became more active on both sides, mostly SAM (surface-to-air missile) units.

Carl and Manny breathed sighs of relief with everyone that had found their way to the auditorium. The Pakistan/India conflict had been avoided. The renegade pilot had been shot down in the huge dogfight. No one would know for some time why he had stolen the F-16, why he had deserted, or why he had attacked India. The answer was with his wife and two children who were falsely arrested and "visiting" the ISI interrogation halls.

Very few would even know about the dogfight. Planes had crashed all over the border area, but that was nothing new to the region. Both sides would describe the crashes as accidents, then eventually claim that the other nation had instigated a dogfight by straying into the other's airspace. Since the British left India and shattered the power structure of the region, such happenings were almost a monthly event.

Only a few ground controllers, senior military and political officials, and the pilots knew the scope of the evening's actions. In CIA headquarters, a few hundred people knew the whole story. A secret war of small-scale actions was exploding all over Pakistan. The unusually huge dogfight was only a single part of that burst in violence. By the end of their day part of that secret would be revealed to the world.

The most important events were the subtle and smaller ones. One by one, the Pakistani police and American Special Forces units operating in Pakistan started to report in. Orange boxes appeared across the country. While attention was focused on the renegade Pakistani pilot, police units were being wiped out. Even some American Special Forces units had to withdraw.

Two units ran into serious trouble. A joint Pakistani/U.S. Army Ranger task force was in a small town near the southwestern border of Pakistan and Afghanistan. Their mission was to secure the town, go into a house, and arrest someone whose name was found on one of the computers from Manny's rescue operation. It was a trap. The Pakistani police surrounded the town, and when the Rangers went into the house it blew up. The police rushed to the scene and were met with automatic weapons fire and RPG's. It was a worst case scenario-especially for the three wounded Rangers that survived the blast.

All three men were protected by their body armor. One man was badly wounded in both legs, another in both arms, and the last one was blind. The explosion deafened all three. When the Pakistani police withdrew from the village until reinforcements could arrive, the terrorist secured all three Rangers and escaped the village on horseback. They headed for the barely guarded border and the anarchy corner of southeastern Afghanistan.

Carl and Manny eventually left the auditorium for some lunch. When they returned, the situation had gotten better. Most of the units that had been sent to round up suspected tangos had finally gotten the word that their intelligence was corrupted. Missions were being re-planned, and hundreds of analysts in the building were being tasked with determining where the real bad guys were hiding.

Carl showed Manny their cubicle, and how to access the CIA's database through his network terminal. Then they went to work trying to collect and sort the clues in the database. Where were the tangos and what else were they planning besides the self-destruction of the Pakistani power structures?

Monday, February 16, 2010

Gaithersburg, Maryland

9:43pm local

After days of sad attempts to sleep followed by untold cups of coffee and caffeine pills, Manny was still unable to rest. His body cried for sleep, but his subconscious mind begged him not to close his eyes. Whenever he tried, the memories haunted him. At times, he wasn't sure if he was awake or asleep.

The real world at CIA seemed like a dream, and his dreams were too vivid to be simple memories. Manny was lost on another plane of existence, and the only thing holding him in reality was his work.

There were rules about taking work home from CIA headquarters. Manny understood them. They were common sense in the intelligence community, and they weren't too different from when he was at Homeland Security. After all, that was the Division of International Terrorist Tracking. The rules were simple: don't take anything home, don't do anything on your own, and don't discuss anything outside the building.

Reading books helped, but it wasn't current enough. After the day in the auditorium, he needed to see, hear, and read about things in almost real-time. That left surfing the internet for news, or watching more of the sensationalistic 24-hour news channels. Their sensationalism did make them more interestingeven if it was just to see how far off-base they might be on some subjects.

Coffee and caffeine were out of the question. He needed to get some sleep eventually. Manny needed a way to get so tired that he wouldn't dream. He found the answer in a bottle of bourbon. Pouring himself a full glass, he drank the entire thing in one slam as though he was back in college. When it was gone, he shook it off, flashed his eyes, and filled the glass again. Now he was ready for the TV news!

Without even noticing, Manny walked to his couch, grabbed the remote, and fell backwards to the familiar comfort zone. As soon as he hit, his cat reared up

SCOTT MALENSEK

and jumped off the couch's back. Bob was lucky. Manny had forgotten that he was there.

He surfed for a while in the hopes of finding something else to watch that would interest him, but every channel he found seemed to have a commercial. He rotated through three of the 24-hour channels and finally found the one that didn't have a commercial.

As expected, it didn't make him too happy. The past few weeks had been one shock after another, and now there was one more. There was graphic footage from a Pakistani TV station. The U.S. Rangers that had been ambushed were on their knees (even the one with the two broken legs). Masked gunmen from the village surrounded them and took turns beating the wounded American soldiers. The Rangers were still blindfolded and bloodied, but they never seemed submissive.

It was hard to watch. Already a bit on the nauseous side from the bourbon, Manny fought back the urge to vomit. Those were Americans. They were the same types of young heroes that had come all the way to Kashmir to rescue him. Here they were on international TV, being ruthlessly beaten.

The one thing that kept him from being sick was his rage. Manny knew that within hours, the whole of the world would see the pictures. Parts of the world would rejoice and vent their hatred for America, its freedoms, and its luxuries. Another part of the world would step back and hope that American retaliation would be merciful; however justified. The last part of the world would feel American vengeance.

The U.S. would demand retribution. A similar video had caused a new and weak American government to withdraw in Somalia. History showed that the entire Al Quaeda Conflict could have been avoided if troops would have stayed and made their stand then and there. In addition to the 20-20 hindsight, America still carried the anger from that conflict. Retribution would have to come swiftly and harshly to satisfy the American public.

Manny knew that it was going to be a busy morning. He finished another glass of Jack and watched to see if there was any other interesting news. The only other interesting report was that both India's and Pakistan's ambassadors were scheduled to meet with the President at the White House in the morning to discuss "…a recent rise in border disputes." Manny knew that they were going to meet to discuss the huge cross-border dogfight that he watched happen on the big screen at Langley. The rest of the news focused on the Multi-Denominational Religious Conference that would begin in the morning. All that did was emphasize how busy the next day was going to be for Manny. He finished his glass of Jack, and passed out on the couch with the TV and lights still on.

Tuesday, February 17, 2010

Gaithersburg, Maryland

11:25am local

Morning came too soon. Instead of an alarm, Bob curled up on his face to wake him up. The cat knew it was well past his feeding time. Manny had successfully slept without dreaming, but now he had to live with the hangover from hell as the price for his rest. Far worse than that, Manny was already 3 ½ hours late for his second day of work at CIA headquarters!

No one had called him. The only person whose name he really remembered was Carl. Manny hadn't even met his official supervisor yet. He rushed to get dressed, and he called Carl to apologize and tell him that he was on his way.

Thankfully, Carl understood Manny's desperate need for rest. He never even called to see if he was coming in. As far as Carl cared, Manny could get all the sleep he needed. He knew what Manny had been through, and he told Manny that he'd wait for him. They could go to lunch/breakfast together then.

Manny was thankful, but he still hurried. Bob would have to wait until he came home for food. He grabbed his overcoat, ran down the stairs, hopped in his car, and drove down the beltway like it wasn't February in Washington D.C.!

Just before noon, Manny pulled up to the security gate. He showed the guard his pass and searched for a parking space. Luck was with him for the moment, and someone who had parked close to the building's entrance had left for lunch. Manny swung in and tromped his way through the slush to the door. Inside, Carl was waiting for him again.

Both men smiled and laughed off his tardiness.

"Glad you could make it in today."

"Carl, I am so sorry. I fell asleep watching TV last night, and-"

Carl could still smell the Jack on Manny's breath. He knew that some people would have a problem with any public hint of alcohol. Others might bring him into their cliques. Either way, Manny needed to be brought up to speed before he could face anyone else. If others knew how late he was, smelled the Jack, and saw that he didn't know about the huge happenings, then both men could be in trouble with the higher-ups at CIA.

"You fell asleep in front of the TV, huh? Did you see the news, by any chance?"

"Yeah. I saw it, but that was last night. What's happened this morning?"

"Well, everybody in the country's pissed off. There're a lot of people asking a lot of questions. How were the Rangers captured? Who're the bad guys that have them? What're we doing to get them back, and what were they doing in Pakistan to begin with?"

"We've gotta keep a lid on all those round-up operations. If the press finds out about all of yesterday's screwed-up operations, they'll have a field day. What's come out so far?"

They walked back to their cubicle.

"Right now, the Pentagon's just saying that they're looking into it. Someone in the chain of command seems to think that it's actually better to look like we're confused screw-ups instead of secretly fighting little battles all over Pakistan. The President had a meeting scheduled with the ambassadors to India and Pakistan this morning. As far as I know, it's still going on. They're probably talking about the air action along the border, but they've been in there for a few hours now, and that's real unusual for this level of talks."

"Anything else?"

"Security was tight for that Multi-Denominational Religious Conference to begin with. After you came back from Kashmir, they moved it to the National Cathedral. I guess they thought that if a bomb or a plane did manage to get through, that huge stone building would be safer. Ever since that film was aired this morning, I don't think you can get within 3 blocks of the church. We expect some sort of attack anytime now, and most people are betting on the Cathedral."

"How about the ambassadors? With all that security on the north of town, how much is left at the White House?"

"I dunno right off hand, but that place is pretty secure all the time, and I don't think anyone was pulled from it. The Department of Secret Service works under the assumption that the President is going to be attacked any minute of any day at any place-especially the White House. I'm sure he's ok."

"What are we doing about the Rangers?"

"Well, as you can imagine, the folks over at the 82nd Airborne are really upset. They've got a brigade busy in the Balkans and another brigade on training rotation in Mississippi. They will throw in a battalion from the 325th Airborne Infantry Regiment out of their 18-hour ready-reserve brigade. Once we get the word on where they Rangers are, we'll drop them in. There're also the Marines down in the Arabian Sea. They got ambushed in a raid on some sort of factory yesterday, and they want some payback. I'm sure if there's any way that they can convince CINCCENTCOM (the Commander-In-Chief of Central Command), then they'll go in too. They are closer. Right now, it's up to us. We've gotta sort through the intel and decide where the tangos have our boys."

Manny fired up his terminal, logged into the network, opened up a connection to the database for Pakistan intelligence, and began searching. "Did we get any good intel from those raids yesterday?" Carl shook his head.

"I didn't see much. There was some interesting stuff, but nothing that could tell us where the tangos might've taken our Rangers. Man, those guys looked beat to hell. We've gotta find something."

Manny searched through the database. He sorted out all of the reports chronologically so that he could see the freshest reports first. Then he did a sort on the word "Ranger" and found several after-action reports for the different U.S. Army Rangers units that had participated in the raids.

He read about the team that had been decimated and captured. The Pakistanis recovered very little information when they came back in force and drove out the tangos. The only thing that seemed out of the ordinary was a digital photo of the destroyed building where the Rangers had been ambushed. In the debris-filled street, Manny noticed an American magazine. It had been blown out of the building. In the middle of the street, the magazine had fluttered open to a page with a map on it. He couldn't see the map, but he could see the name of the magazine.

Out of curiosity more than anything, Manny searched the internet and found the magazine's website. On the homepage, there was all kinds of information from the latest issue, including an article about the creative architectural layout of the Holocaust Museum in Washington D.C. Manny clicked on the article and saw the map. Then he wondered why these radical Muslim terrorists would be looking at a map of the Holocaust Museum. The predominately Jewish memorial was a good target for radical Muslim extremists. It seemed so obvious. The terrorists were going to attack the place.

He immediately called his old boss and told him the news. His boss agreed to pass the word and have a few D.C. police go and check it out. There wasn't enough information to evacuate the building, but there was enough to have the building checked out.

Manny, Carl, and countless other intelligence analysts sifted through the growing pile of data. No one could find anything that would lead to the location of the escaped terrorists and the captured Rangers. What everyone overlooked was the large amount of custom-built, armor-piercing, finstabilized, discarding sabot, .50 caliber ammunition found at the scene of the Ranger's ambush. They were ½" versions of the same types of armor-piercing rounds that main battles tanks used against each other. The Pakistanis found some small arms in the debris, but there was nothing big enough to fire the hand-sized .50 bullets. Two pieces of the puzzle didn't fit, and no one had given much thought to either one.

Not even half an hour after Manny called Homeland Security, four Washington D.C. police cars pulled up to the Holocaust Museum. Half of the officers took positions around the building, and the other half went inside. They made contact with the Museum's security officers and with the Museum's management. No one had seen anything strange all morning or in the past few days.

The museum was usually slower during the winter. It was also slowest on the weekdays. A sluggish economy and lingering thoughts of terrorist attacks at home dating back to 9/11 also impacted tourism. Three weeks of near-continuous snowfall was the final silencing factor. Attendance was far below the museum's operating minimum. Barely 200 people a day were coming through the doors. Only federal subsidies and private grants were keeping the doors open.

The four police walked through the building on normal rounds with a pair of museum guards. Outside, the other four D.C. police officers stood in the snow and watched for the unexpected. Large flakes of snow on an already white covered city made it a boring scene to survey.

Over two miles away at the White House, the President was still meeting with the ambassadors from India and Pakistan. Both men had come to inform the United States that the previous day's dogfight was an accident and that there was no need for the United States to put a hold on some foreign aid that scheduled to be sent to their countries. The talks were heated, as expected, and went on almost all morning. They took a lunch break and then toured the White House before beginning again. The last stop on the tour was the Oval Office.

Back at the Holocaust Museum, the security guards continued to make their rounds with the four Washington D.C. police officers. They had found nothing unusual in the Museum besides its macabre exhibits and the small number of visitors. The officers outside stayed at their posts as snow piled on their shoulders and hats.

Almost 2 feet of snow covered the roof of the building. Next to one of the heating units on the roof, the snow began to slowly move. Underneath the blanket of white, an exiled Serbian named Mikhail Jovanovic was stirring. He had climbed up on the roof two days earlier during the night.

Once he had found his spot next to the heating and air conditioning unit, he unrolled his thermal sleeping back and unpacked his .50 caliber sniper rifle. He slid a thermos of water into the sleeping bag with him. Then he put a few sandwiches inside, and two extra 10-round clips went in his large chest pockets. The sleeping bag's thermal lining acted as a heat barrier to prevent heat from escaping and help hide him from infrared sensors. During the silence of the snowy night, the Serbian sniper was buried and completely camouflaged.

He had spent days waiting for the moment. On the other side of Washington, every extra security guard, agent, and police officer was protecting the Multi-Denominational Religious Conference. They were almost four miles away and at least two miles beyond his target-The White House.

He watched through the scope on his rifle. It had fogged in the cold, but over time, the temperature inside equalized with that of the surrounding snow, and the condensation seemed to stay near the edges of his viewfinder. He knew from years of sniper experience outside Sarajevo that, if he could just wait long enough, the image would eventually become clear. As important as marksmanship, the best snipers possessed incredible patience as well.

Mikhail watched The White House for several days and nights. He saw the people come and go from room to room inside. He watched the two snipers on its roof change shifts. He watched as the two ambassadors entered the Oval Office at the end of their post-lunch tour.

The time had come. Mikhail and his rifle were covered in snow and surrounded by the muffled sounds of the city. The snow was his friend, but the cold was his enemy. As it was, his .50 caliber rifle had the longest range of any rifle in the world, but his shots would still be at the weapon's extreme range.

He had to plan on several different factors before each shot. The range was extreme. The cold had made his barrel tighter and his bullets larger. They would have to pass over the long, flat, and windy area known as The Mall. If the wind blew them off course by as little as 1/16 of an inch, by the time they hit their targets, the bullets would miss by a yard.

There were also the targets to consider. First he had to kill the two marksmen on the roof of The White House. He would have to fire in rapid succession to get them both, and he would have to fire fatal shots into the center mass of their bodies. There was little doubt that both were wearing some sort of body armor, so he would have to hit them in the center. A glancing hit off their sides would seriously wound anyone, but it might not prevent them from sounding the alarm.

The most important shots into the Oval Office would be even tougher. Mikhail was certain that the windows were of the heaviest bullet-resistance to protect the President. The huge half-inch-diameter bullets from his rifle could penetrate the steel armor on a main battle tank, but he wasn't sure about layers of bullet-resistant plastic. They might slow his bullets down and make them less than fatal. At the very least, the angle at which his bullets met the glass would determine which way they would be deflected, and he had to guess at that. With so many variables, Mikhail had to be more of an artist than assassin. The killer in him justified it to his soul as being the ultimate form of life and death expressionism.

The time had finally come. The two snipers on the roof walked back and forth looking through binoculars. Mikhail knew that the two men had just started their shift a few hours earlier, and that their binoculars were much more fogged than his scope. When the one on the right looked through his binoculars toward the Holocaust Museum, Mikhail took a deep breath. Halfway through his exhale, he squeezed the trigger. It was a habit that had been trained into him by the Yugoslavian Army in the late 1980's. Since then, the routine of calculating trajectories and firing at targets had become normal.

He made it look easy.

The .50 caliber rifle made a loud, booming, *crack* sound. Snow instantly fluffed off of him, and a huge yellow muzzle flash blinked over the edge of the building. His snow cover disappeared in an instant. Across the mall on top of The White House, the marksman with the binoculars exploded as the huge bullet hit him square in the heart. He never knew what hit him. Seconds later, the sound of the rifle shot reached The White House and alerted the other marksman.

In that same period of time, Mikhail aimed at the other American rooftop sniper. He took his breath, exhaled, and squeezed the trigger. Again the rifle made its familiar *crack* sound. The powdery snow blew away, and Mikhail saw two corpses on the roof of The White House. Now it was time to get what he had come all this way for. He aimed at the still unsuspecting Oval Office.

This was the moment of truth and history. He took a breath, started to exhale and-

"FREEZE!"

All eight Washington D.C. policed were on the roof. Six more armed security guards were with them.

Mikhail was not a suicidal terrorist. He was a killer, but he didn't want to die either. He let out his full breath, let go of the huge rifle, and slowly extended his cold, wet hands as the police directed. His mission was a failure.

Monday, February 16, 2010 Langley, Virginia

Central Intelligence Agency

Terrorist Tracking Center

4:31pm local

Manny and Carl were still sifting through the database. Hours had passed, and they hadn't found anything tangible. No one else in the CIA, FBI, OHS, NSA, DIA, or the various consulting firms had found anything either. The three wounded and beaten Rangers that were captured in Pakistan seemed doomed.

Breaking the monotony, Manny's phone rang for the first time since his transfer to CIA. It was his boss from Homeland Security.

"Manny, it turned out your hunch was right about that Holocaust

Museum thing."

"Really? What'd you find?"

"How about your Serbian sniper, a Barrett .50-caliber sniper rifle, and two dead marksmen on the roof of The White House?"

Carl's phone started ringing, but Manny and his old boss continued.

"You've got to be kidding me!"

"I wish I was, Manny. Someone's gotta make some sad phone calls to the families of those two dead agents. Thank God the D.C. police got to the roof just as he was about to take a shot at the President."

Manny's disbelief was incredible. A few weeks ago he was doing data entry, and now he was helping to save the President of the United States. He always knew that the data he worked with was important. It seemed like now every important piece of data was proving it's worth all at once. Now he wanted more.

"I can't believe this. Did anyone else get hit? Did they get him alive? Where is he now? Has he talked? Did we get anything from him?"

"FBI's got him over at their building. Of course, everybody wants a word with him, even the President. I was just over there, and you're one of the few people cleared to know about any of this. You and those D.C. cops really saved the day."

Manny leaned back in his chair with shock.

"As far as intel, we haven't gotten much from him. When I was there it seemed like the guy did it more for the challenge and some sort of artistic purpose. He's a psycho. There's no doubt about that, but all snipers are to a degree. We think the guy was just put up to do it by The Protectors of the Koran. There was one thing that the FBI found interesting."

Manny leaned forward and put his elbows on his desk. He was engrossed in he information.

"The Barrett .50 caliber sniper is a very rare weapon. Each one is custom made here in the U.S., but the FBI tells us that this one has an interesting past. The serial numbers were filed off, but they were able to scan it and see the impressions left in the subsurface of the metal. That gave them the serial numbers. It's a trick they use all the time in tracing unmarked weapons used in street crimes. Anyway, they did a database search on the serial numbers and found out that it was sold to the U.S. Navy. Then the Navy did a search on the serial numbers, and they said that it was left behind on a mission."

Manny was already thinking ahead.

"Do you have the numbers-'cause I can run it through here?"

Manny's old boss read off the numbers, and he typed them in as fast as he was given them. When he was done, Manny did a search on them. Instantly a report was listed on his screen. The report was classified, so Manny couldn't reveal its details, but he bent the rules a bit.

"Well, I found it. It's classified, and the Navy's right. They did leave it behind on a recon mission. I can't tell you anything more though."

"That's too bad."

"Did the Serb give out any other information about the funding for this operation? I know that's always been my focus. Let's face it, The Protectors of the Koran are a capable and well-motivated group, but without money, they're just another bunch of pissed off guys."

His old boss was still saddened by the lack of assistance from Manny, and it showed in his voice.

"No. He just said it was the ultimate challenge for someone in his 'trade.'"

"Well, remember when the Air Force sent over a UAV before the

Marines came in to get me out of Kashmir?" *"Yeah. I guess."*

"You might wanna go through those again. I'm certain that the guy that's in uniform is the guy that supplied him with at least the rifle. If I had to guess, I'd say that the guy in the suit is the one that paid for everything else. I've already run an identification match search, and nothing came up."

There was a second of silence on the other end while his old boss called up the images from the pre-raid UAV reconnaissance flight. When he saw them, he understood what Manny was trying to say. The Chinese intelligence officer in the pictures had supplied the .50 caliber rifle. It was left behind on a Navy SEAL recon mission in China.

"Thanks a million Manny. If there's anything you need from us, just let us know."

They said their good-byes, and Manny told Carl about the assassination attempt, then they both went back to work searching the constantly growing database of intelligence reports. There was still nothing. It was getting late, so Manny wrote a subroutine program that would keep searching for keywords, phrases, and combinations of both. It was an old trick he had learned at Homeland Security when he was first given the job of Database Information Input and Tracking. The program would allow him to keep searching the computer for specific pieces of information relating to those keywords and phrases even while he was at home sleeping. When it was finally written, he picked out the keywords he wanted it to keep searching for, added them to the program, and ran it to make sure it worked. It found the same pieces of information that he had been looking at all day-proving that it worked.

Manny looked at his program again. He wanted to make sure that it had all of the keywords, phrases and combinations that he could imagine. Something seemed like it was missing. He read through the long list several times and just couldn't figure it out. He looked over at Carl, and when he was finally off the phone, he asked him if he thought that he should add anything to the list.

"Hey, that's a cool idea. I hadn't thought of writing a program to do the searches. I didn't know you could run a program at the same time the database was open."

"Yeah. It's easy. All you have to do is write the program to open the database each time. If we had complete database access, then we could actually have one that just searched all day and night without opening and closing the database. It would do the same thing, but a lot faster. This way takes a bit more time, but since I usually have the searches run while I'm away, I don't need the results in real-time. I just want a report when I sit down in the morning."

Carl looked at his list of search criteria, and thought of the call he had just finished with his contact over at the FBI. He had overheard Manny's conversation with his old boss, and he knew that Manny had been appraised of the attempt on the President's life.

"Maybe you should add 'President,' 'assassinate,' and 'Holocaust

Museum?'"

"Good idea. Anything else?"

Carl thought more about his call from the FBI.

"How about cross referencing reports that include the serial number from the gun with reports that match your keywords? Can you make the search results cross-reference and search each other? That would put the reports that match the most search criteria at the top of the results list." "Yeah. I think I can do that."

Manny took a few more minutes to make the changes in his program. When it was done, he tried to run it, but there were errors. He searched through it, found them, tried to correct them, and he ran it again. Once more there were errors. Once more he found them, made corrections, and he tried to run it. This time, it worked. The top result listed an ISI intelligence report.

Manny had seen the report before, but he hadn't thought much of it. It seemed as dangerous and leading as all the others did. This time, he saw something he had missed before. Although he hadn't entered "Chinese" as a keyword for the search criteria, Manny noticed that the report mentioned the sighting of a suspicious Chinese man.

A policeman guarding a watering point in the middle of the high desert in southern Pakistan had been harassed by a large group of individuals that had stopped for refreshments. Some of them may have been on the terrorist watch list, but the policeman was termed as unreliable by the reporting ISI agent. The policemen allegedly overheard someone discussing the Jewish holocaust, a museum, and the President of the United States. The policeman thought it was odd that all three were discussed in the same conversation. When he went over the men to try and gather more information, a Chinese man wearing desert clothing rushed away. There was a lot of other information, but the most important piece was the location of the watering hole near the desert town of Panjgur.

"Hey, check this out."

Carl spun around and read the report on Manny's terminal screen. When he finished, he was robotic and serious.

"Cross reference that with other reports that mention that watering hole or Panjgur over the past 72 hours."

Manny did another search. There were three other reports from the wet spot in the middle of the high desert. There was another report that listed a large number of armed men passing through, and one of them was Chinese!

Carl asked Manny when the armed men supposedly passed through. Manny scrolled down through the report. The men had passed through town only three hours after the Rangers were captured.

Carl wasn't pleased.

"Shit. They could be anywhere within 250 miles by now. That means Pakistan, Afghanistan, or Iran."

He typed some notes into an email and sent it to his superiors to keep them informed. Then Manny had an idea.

"How about this? It says that they passed through in eight pickup trucks.

Let's try a search on 'eight trucks-Pakistan-past six hours?'"

"Might as well. Try it. See if anything comes up."

Manny typed the new keywords into his database search, and nothing came up.

"Wait a second. I wonder if anyone's made any calls from that region?"

Manny switched to a different database and searched for cell phone calls that had been detected and traced to that region. In the high desert of Siahan Mountain Range of southern Pakistan, the handful of cell phone calls stood out. The fact that any had been made from such a remote region shocked Manny. Carl was unimpressed.

Carl told Manny to do a "signal intelligence- detected transmissionsource-identification" search. There it was. One of the phone calls was from Asaad! The man on the other end was a Chinese man in Hong Kong. The location was a large hill in the middle of a 35-mile wide salt flat-about 20 miles west/southwest of Panjgur, Pakistan.

They had found it! Both Manny and Carl were convinced that the circumstantial evidence pointed to the source of Asaad's call. At the very least, Asaad needed to be rounded up ASAP. Carl typed up another email and sent it to his boss. Just as he hit "SEND," his boss walked stepped into the doorway of their cubicle.

"Carl, how's it going over here?"

Both Carl and Manny spun in their chairs. Carl introduced Manny to their superior, and there was a brief welcome to the company speech. When he was done welcoming Manny, their boss asked for an update. Carl told him about the email he had just sent and the information that they had found. Their section chief was impressed.

"Nice job, you two. I'm glad we we're able to get you in here as fast as we did. I'll run this up the flagpole and see who salutes it. That's about all we can do. You guys might as well get outta here and go home. It's getting late, and I think you've done enough for today. Good work."

Both Carl and Manny were given their 'atta-boy handshakes and then their boss left. Carl turned off his monitor and turned to Manny.

"I guess we better get outta here then-boss' orders. Why don't you let your program run, and we'll check it out in the morning? I know this great little place in Georgetown we can go to. I'll buy you one for the Prez, and you can buy me one for getting you into this gig."

Manny laughed and agreed. He setup his database search program, started it running, and turned off his monitor. Both grabbed their overcoats and headed off to Carl's favorite bar on the other side of Washington.

While they slushed through the traffic-packed streets of Georgetown, communication of their intelligence findings was "run-up the flagpole." The National Security Agency proposed the idea of another raid to get more intelligence or eliminate the terrorist organization's hierarchy. The Joint Chiefs of Staff didn't like the idea, but the NSA advisor informed the President that the person who had discovered the intelligence was the same person that had figured out the plot to assassinate him earlier in the day. Immediately, the order went out to seize or destroy everyone on that remote desert hill in Pakistan.

Tuesday, February 17, 2010

U.S.S. Peleliu

Officers Wardroom

Pre-Operation Confirmation briefing

Operation Utah

12:34pm local

For the third time in as many weeks, Major Chris Mosby received his orders to take men into harm's way. The first time he lost forty-five Marines on a rescue raid into a small terrorist-held village in the remote corner of Kashmir. His second mission was to capture a bomb factory, and he lost another twenty-three Marines. Chris wondered how many he would lose this time.

Operation Utah was a combined raid to secure a remote, 4½ mile wide, rocky, desert hill in the middle of southern Pakistan. Intelligence reports suggested that the leaders of a terrorist group called The Protectors of the Koran were hiding on, in, or near the large stone mound. The plan included a battalion of paratroops from the elite 82nd "All-American" Airborne Division, a Pakistani armored battalion, and a reinforced battalion of Marines under Mosby's command. Air support would be plentifully supplied by a squadron of F-18E Super Hornets from the U.S.S. Carl Vinson, a three-plane flight of B52 Stratofortesses from Diego Garcia, and a squadron of F-16C Falcons from the Pakistani Air Force.

The plan was complex and needed to be executed immediately. The Paratroops were already on their way, and they were being briefed in-flight. They would land at night in three drop zones around the hill-one to the north, one to the northeast, and one to the east. The Marines would be airlifted in by helicopter to a landing zone on the west side of the hill. The Pakistanis would drive in along a highway from the southeast and take up positions along the south and southeast of the hill. Everyone was scheduled to arrive in their area at midnight except the air support. They were scheduled to be on-station an hour before everyone else. With all the air transports carrying the 82nd Airborne, the helicopters carrying the Marines, and all of the air support, it was going to be a crowded night sky.

Once everyone was in place, the battle would really begin. The paratroops from the 82nd were supposed to advance up the hill from all three drop zones. It was hoped that anyone on the hill would try to escape to the other side where Mosby's Marines and the Pakistani armored force would either round up the tangos or chew them up. If anyone ran into trouble, the F18E's were available for precision close air support.

Should the operation meet with any resistance that was too fierce, then the B-52's would come in and blanket the hill with over two hundred forty-six 1000lb bombs. However, the paratroops, Marines, and Pakistanis would have to withdraw to their initial positions before the hill could be thumped by the B52's.

Mosby didn't like the plan any better than the Joint Chiefs of Staff had. He was on the opposite end of the rank spectrum, but he agreed with the people at the top. There were just too many different units from different services and nations. Each unit involved had its own chain-of-command to follow, and the senior operation commanders were too far from the action. Technically the commander of the operation was the CINCCENTCOM (Commander-In-Chief of Central Command) all the way back in Florida! The local commander of the operation was Mosby's Commanding Officer, General Shapiro-commander of his MEU (Military Expeditionary Unit), but he was going to be almost two hundred miles away back on the U.S.S. Peleliu. The senior field commander would most likely be the commander of the 82nd Airborne's battalion. He also doubted that the Pakistani commander would be eager to follow instructions from any of the American commanders. However he looked at it, command-and-control in the battle itself was going to be scattered and confused.

Mosby wasn't alone in the briefing room. Besides the different intelligence, communication, inter-service liaison officers, and his superiors, there were also the company commanders that he would command in Operation Utah. Mosby had taken Echo, Fox, and Golf companies into Kashmir. Golf was decimated, so when he took the raiding force into the factory outside of Jiwani, Pakistan, he brought Echo, Fox, and Charlie companies. On this mission, Mosby was to bring in all six infantry companies of the 1st and 2nd Battalions from the 5th Marine Regiment-Alpha, Bravo, Charlie, Echo, Fox, and Golf companies. All six company commanders were in the room with Mosby.

When the briefing was done, General Shapiro asked Mosby for his thoughts on the operation.

"Well sir, I'm sure we can do this. I am a bit concerned about the chainof-command on the field-especially with the Pakistanis being involved."

Shapiro was a professional and a matter-of-fact commander with a natural charisma that boosted his leadership abilities. He cared deeply for the people under his command, and their thoughts weren't simply valued as personal insurance for after-action inquiries.

"Chris, I completely agree with you. The fact is that we can usually say, 'We're all Americans out there,' but not in this case. The Air Force won't come in without notifying you. The Navy won't come in unless you call them. The boys in the 82nd are the best the Army's got, so I think you'll be able to work with them. I've never had anything but respect for them and from them during joint exercises. Aside from the tangos, the Pakistanis are the big unknown. The only thing I can tell you is that in a worst-case scenario, there'll be enough Americans there to get you out of harm's way."

Mosby wasn't too worried. He knew that the Marines could handle the operation. He did have some concerns about leaving the Pakistanis in place as an anvil. The Marines, paratroops, and American pilots might be the hammer, but could the tangos escape around the anvil?

"Sir, I've got great confidence in our portion of the plan, and I believe the overall strategy is good. I can't think of anything else we really need. Can you guys?"

Mosby turned to listen to the thoughts of his company commanders. Golf Company's new commander (it's former executive officer) spoke up first.

"I'll tell ya, I know our mission objective is to capture as many of these guys as possible, but if we find caves over there like we did in Kashmir, I'd like a few flame-throwers or something!"

Everyone chuckled. The Marines didn't have any flame-throwers on the ship. They were very rare in the U.S. military. Shapiro understood that they would need something special for cave fighting though.

"I'll tell you what. I think every Marine going in should bring at least one CS (tear gas) grenade. Those bastards might have masks, but it's miserable fighting with those things on, and it'll up the tension inside if you do come up on a cave."

Alpha Company's commander came next.

"How about some artillery or mortars, sir?"

Mosby nodded his head to indicate that he liked the idea of some heavy Marines ordinance on the ground. Shapiro saw the nod and agreed.

"After you're all on the ground, we'll have some batteries brought in. There's no real cover, so we'll let the battery commanders decide where they'll be most effectively located, but I'd like to see one to the west, another to the southwest, and another to the south-no...wait a sec." Shapiro stood up and looked at the map of the area.

"We don't want the Pakistanis to be any more jittery than they already are, but it would be nice to have some of our artillery available to cover themone way or the other. Let's put all three batteries over on the west side-behind our left flank and the 82nd's right flank. We'll work the details out with the artillery guys, and make them available through the forward air controller. All your support calls will go through that one frequency. Anything else?"

Everyone looked around, but no one else had any other questions or comments.

"Okay then. First wave leaves at 2200 hrs tonight. I'll have the Marine air support ready too, just in case the Navy runs out of bombs or has any problems. Good luck, and SEMPER FI!"

Everyone snapped to attention and growled out a round of, "Ooo-Yah's!" Shapiro left the room and everyone headed to their units to begin their briefings and preparations. Mosby felt much more confidant knowing that Marine support will be close by. The Navy and Air Force were the best in the world at controlling the sky, and they were very good at providing air support, but Marines always preferred Marines for support. It was a faith-incamaraderie thing.

Wednesday, February 18, 2010

27 miles WSW of Panjgur, Pakistan

11:57pm local

For yet another night, The Protectors of the Koran hid in their hill hideaway. They were on the only high piece of real estate in the area. As far as the eye could see, there was nothing but a great dry lakebed. The night's stars reflected a grayish-blue glow on its white surface. The dry lakebed was covered with a layer of white salts that had transpired up through the ground over thousands of years.

On the large rocky hill, a few pairs of sentries stood among the boulders and watched for any signs of trouble. The sky was clear and starry. The horizon was far, flat, and distant. Everything seemed fine.

The small pile of sharp rocks had a network of caves inside. Most were little more than native rocks knocked over and piled upon one another to form arches. In some place, wooden supports or sandbag walls strengthened the structure. The hill itself was too tough to allow for digging, and there were no natural tunnels. Instead, the hands of men had made the caves centuries earlier.

There were small rooms, but the tunnels were even smaller. Two men could not pass each other between the rooms, and no one could walk the entire network standing upright. Vehicles that had carried the fugitive Protectors of the Koran members remained outside under camouflage netting.

There were all kinds of rooms. Most of them were small bunkers for defending the hill. The larger rooms were dedicated to different purposes. Some stored weapons, food, and the precious water well. Others were used for sleeping, medical treatment, or prayer. Everything had to be carried or shoved through the tiny creases between the rocks that made up the cave network.

The leaders of the terrorist group were sitting in a nook at the topmost part of the hill: Ali Asmaath (a.k.a. The Ox), Munthmar Hussein (a.k.a. Papa), and Monsoor Asaad (formerly Captain Asaad of the Pakistani Secret Policethe ISI). With the three men were some of their supporters, Saudi Prince Abdualla and a nameless Chinese military intelligence officer. They all sat in front of a small laptop computer and received a secure military satellite downlink video communication.

The conversation had just begun between those at the hideaway and the remaining supporters on the other end of the satellite uplink. All at once, the men in the open-air bunker heard the helicopters to their southwest. Everyone turned and looked. Asaad had the foresight to close the laptop computer and turn it off before the satellite transmission was detected.

The flat white desert floor was bright with reflection from the pinholes of light in the night sky. No one counted, but they saw over thirty CH-47 Sea Knight dual rotor transport helicopters coming from the southwest. The exact count didn't matter. There was no mistaking it. It was the U.S. Marines.

In the desert, with few vertical surfaces to reflect sounds, there's normally a silence. Sentries all over the hill began scurrying through the caves to get ammunition to their bunkers. Everyone that passed by the three blindfolded U.S. Army Rangers gave at least one of them a solid kick or punch to vent the frustration of an impending attack.

While the tangos prepared to meet the Marine assault, the Americans and Pakistanis continued to close in. The Marines were still almost 10 minutes away, but the U.S. Air Force's Colonel Earley (commander of the 452nd Tactical Airlift Wing) was closer. Riding in the bellies of his three C141 Starlifter transport planes were four hundred-fifty paratroops from the U.S. Army's elite 82nd Airborne Division/325th Airborne Infantry Regiment/3rd Battalion/Companies Alpha, Bravo, and Charlie. One hundred fifty men and women were already standing up in the planes and ready to jump as soon as their red signal lights turned green.

The thumping of the helicopters drew closer and closer. As fast as they could crawl through the tunnels, The Protectors of the Koran armed themselves and prepared to meet the Marines in the helicopters coming from the southwest. At exactly midnight, only a few minutes after they first spotted the helicopters, the three huge 4-engine transports of the 452nd shrieked past the rocky hill from the northeast. From less than 3000 feet above the desert salt flat, hundreds of paratroopers were jumping out. They were landing on the opposite side of the hill from where the terrorists had prepared to meet the Marines.

One of the tangos near the top of the hill fired a small Russian-made SA7 shoulder-launched heat-seeking missile. It ripped through the sky toward the trailing C141 Starlifter. All three aircraft saw the launch and fired thousands of hot decoy flares to confuse the small seeker in the missile. The entire sky was lit for a hundred miles by the red flares, but it was no use. From only 3000 feet away there was little hope of confusing the missile. It slammed into the #1 outboard engine on the trailing plane's port side. The engine exploded in a white flash and a loud boom.

Thankfully, the planes of the 452nd had been modified back in the late 1990s for low-level special operations missions. Part of those modifications included self-sealing fuel tanks, explosive bolts on the engine nacelles, and a thin layer of armor. The fragments from the SA-7 shredded the left wing of the plane, damaged the #2 engine, and blew off the #1 engine. There was a small trail of fire and a huge plume of smoke coming from the plane as it continued on course at just under 200 miles an hour. By the time it was 10 miles away, the fire was already out, and the plane was heading back to Doho, Qatar. Even if everything had gone wrong, and the plane had to ditch, the 35-mile wide salt flat would have made an excellent emergency runway.

Frantically, the terrorists shuffled and crawled through the tunnels to their bunkers on the north, northeast, and east sides of the hill. As they did so, the Marines began landing. Two more SA-7's came from the hill toward the helicopters. The sky was still flickering red and white from the decoy flares dropped by the C141's. The Marines popped off a few thousand yellow decoy flares, both missiles missed. As the last CH-46 touched the ground, both SA7's exploded in the desert a few miles behind them.

The fight began. Even though there was sporadic small arms fire coming from the hill 300 meters away, the Marines assembled into their companies in short time and in good order. Alpha Company took the left flank and headed toward the paratroops to link up with them. On Alpha Company's right side was Bravo Company, Charlie Company, Echo Company with Major Mosby, Fox Company, and the decimated Golf Company on the far right side of their line.

The paratroops were all on the ground before the first Marine helicopter had fully unloaded. They also formed into companies and took their positions immediately. Alpha Company was on their left. To the right of the paratroop's Alpha Company was Bravo Company, and Charlie Company on the far right joining up with the Marines. There was far less small arms fire meeting the paratroops, and they advanced all the way up to the bottom of the rocky hill before they needed to take cover.

The Marines were running into more resistance. Small arms fire was coming from hidden places all over the southwest side of the hill. More importantly,

Mosby recognized the heavy barking sound of three .50 heavy machineguns. Those same types of guns had taken their toll on Fox Company in Kashmir, and tore Golf Company to shreds. The Marines were still in the open salt flat, so while the companies were still forming, he gave the order to advance to contact!

All 800+ Marines got up and ran toward the rocky edge of the hill. Most ran in serpentine paths to try and throw off the aim of the tangos. Some ran as fast as they could for the cover of the rocks. A few fell wounded.

Those that couldn't run fought. Caught in the open, with wounds of various degrees, every Marine that couldn't run worked their weapons to protect their buddies. The cardinal rule of the Marine Corps is to not leave anyone behind, and every single wounded Marine was carried to the rocks. Some even fired their weapons while they were being carried.

At the base of the hill, tango fire seemed to come from everywhere and nowhere. The firing ports from the stone bunkers were very small and well camouflaged. The only place that there didn't seem to be fire coming from was the southeast side of the hill.

Mosby knew why, too. The Pakistanis hadn't arrived yet, and the door was wide open for the tangos to come down the hill and try to escape or at least flank the Marines and paratroops. Golf Company on the open right flank of the Marine's line was taking fire from one of the .50 heavy machineguns, and they couldn't stretch the line to meet with the paratroops. There was no time to contact the paratroops and have them come around, so Mosby called the Navy.

While the C141's were dropping paratroops, and the Marine CH-46 Sea Knights were dropping off Marines, the U.S. Navy and U.S.A.F. had loitered high above the scene. They were five to seven miles above it, but they were there…ready and waiting. Mosby's air support coordinator got on the radio and reported the situation to the FAC (Forward Air Controller) high above them. The FAC looked through her list of available aircraft and their different weapons loads. Then she notified a trio of F-18E Super Hornets that were armed with napalm.

Down came the planes. It wasn't even a full minute since Mosby had told his coordinator to get air support to cover their right flank. He was still watching the area through his binoculars to try and find the heavy machinegun that was pinning down Golf Company. Suddenly there was a loud *FWOOSH* sound. The southeast sky exploded with twenty 1000-liter splashes of napalm. Each one hit and spread until the entire southeast side of the rocky hill was burning with jellied gasoline. The paratroops on the other side had heard the air strike being called in, but they had no idea that the flames would hit at the base of the hill and splash up to its top. Even from the other side, the paratroops watched their target silhouetted by fire. Mosby told his air support coordinator to pass on his compliments to the FAC and the pilots. No one was coming off that side of the hill.

The fireballs of napalm merged into a single orange, yellow, and black mushroom cloud. White salt flats reflected the colors of the night. Red and yellow decoy flares with crooked white plumes of smoke behind them fell on the scene.

Gray twists of exhaust from the SA-7 missiles faded into the sky toward the horizon and the stars. Yellow, white, red, and green tracer bullets streamed and bounced in all directions. Shadows of Marines and paratroops stretched out from the rocky hill into the desert. The napalm covered a full quarter of the dark rocky hill.

Then, 10 miles away to the east, the Pakistani armored column saw the light show. A sandstorm had passed over the road from Panjgur and buried it. On their way to the battle, they had become lost. Now, they were drawn to the spectacle.

With the Marines and paratroops pinned at the base of the rocky hill, Mosby surveyed the situation. The tangos had very well placed bunkers, and it had become a pure infantry battle. A few Marines had tried to get into the tunnel complex after silencing two of its occupants. The lead Marine in the small group crawled in first and was killed. They pulled his body out, and another Marine went in. He was killed. A third Marine went in with a grenade launcher. Shots were heard, then there was an explosion. He was killed, and they pulled out his body. It was apparent that getting into the hill was going to be costly.

Mosby decided to try and get his men on top of the hill to see if there might be better entrances from up there. He passed the word to Bravo and Charlie companies to take the top of the hill. As soon as each company commander gave the order, three hundred Marines began to climb as fast as they could. Alpha, Echo, and Fox companies provided covering fire into the small holes that the tangos were firing from. Occasionally a few Marines from Bravo or Charlie company found themselves either passing hidden firing ports or sticking their weapons inside and killing the occupants with point-blank bursts of automatic weapons fire.

The paratroops on the other side were having similar difficulties. Instead of trying to get to the top of the hill, they continued to try and get into the tunnel network. There were few occupied bunkers on their side of the hill, and so more paratroops were actually able to crawl inside. Wounded paratroops and dead bodies were pulled out, but there were always others waiting for their chance to get inside and help the beaten Rangers.

The fight went on for another ten minutes before the Pakistanis arrived. They brought a full company of eighteen American-built M48 Patton tanks, and twenty American-built M113 Armored personnel vehicles brought in a mounted infantry company of just under two hundred mounted infantry. They were better late than never. The Marines' battered Golf Company on their right flank was still pinned down by a well-hidden .50 caliber heavy machinegun and increasing small arms fire. The paratroops Alpha Company on their open left flank were having similar problems.

Even though he was still a mile away, the Pakistani commander could see the situation through the telescopic sights in his tank. He ordered the vehicles to spread out in a tight line abreast formation and begin firing at the gap in the American line. While Mosby was firing at a small firing port hidden high in the rocks, he saw the flashes and white-orange fireballs on the horizon. As soon as he saw them, the hill began to explode to his right. The sounds of the impacting high-explosive rounds from the tanks arrived at the same time as the sound of the tank guns firing. Each explosion seemed to happen three times. First they exploded once visually, followed by two thundering booms.

Eighteen Pakistani tanks fired at will after their first combined volley. Each tank fired between fifteen and twenty-five rounds in a four-minute period. Then the M113 APC's (Armored Personnel Carriers) pulled in front of the tanks and began advancing. While they did so, the mounted infantry inside opened the hatches on top and prepared to join the fight. Every single one of the tank commanders and APC commanders was firing their .50 M2 Browning heavy machineguns at the unoccupied quarter of the hill. Between shots, the tank gunners did the same with the turret-mounted machineguns, while loaders fed fresh ammunition into the massive 105mm cannons. All total, the onceempty part of the hill was suddenly under fire from fifty-six heavy machineguns, eighteen 105mm anti-tank cannons, and over two hundred small arms from the mounted infantry.

The impact of so much firepower reverberated through the tough stone and shook the entire hill. Mosby was knocked off the small ledge that he had been using for cover. Some of the Marines that were scaling the hill lost their balance and fell part of the way down. The paratroops on the other side scurried through the tunnels to get into the larger bunkers and rooms.

One by one and in small groups, the paratroops were infiltrating their way into the tunnels. Once inside the hill, unit cohesion fell apart. They crawled on their elbows and sometimes moved with their knees in their face, but they were advancing. Some moved toward the bunkers to help protect their fellow paratroops. Others worked their way into the hill.

Outside, as the bunkers were silenced, more and more paratroops were able to enter the tunnels, the Pakistani mounted infantry got out and connected the line between the Marines and paratroops. By then, almost two hundred paratroops had sifted their way into the pile of boulders and its maze of tunnels, bunkers, and storerooms.

The paratroops, Marines, and Pakistanis all had 1/3 of the hill to attack, but the paratroops only had 1/3 of the people that the Marines or Pakistanis had. Even with the numbers disadvantage, the men and women of the 82nd's 325th AIR were attacking hard, and they were making more progress than anyone else! Even with all of Mosby's recent actions, the paratroops were far and away the most experienced.

Major Chris Mosby tried to make sense of the scene. He wanted to go into a cave and help out, but it was his job to make sure that all went well outside. The Pakistanis were obliterating any fire coming from the entire south side of the hill. Near the top of the hill, the Marines were moving steadily. There were still countless hidden bunkers, and the Pakistanis had the situation well in-hand, so he ordered Fox and Golf companies to move up the hill.

While he was doing so, an RPG (Rocket-Propelled Grenade) exploded nearby. His forward air controller officer was severely wounded in both his left arm and left leg. A nearby machine gunner and his loader were killed, and Mosby felt a hot, sharp, searing pain in his left shoulder. He felt the warm wetness of blood in his sleeve and something heavy and chunky too. Fearing that it was some sort of bone, Mosby delicately ripped open his sleeve to examine it. Instead of bone, he found the sharp, racquetball-sized rock that had cut him.

Mosby was pissed. He could do little to influence the situation outside the rocks. His Marines were holding the base, and advancing up the hill, but there was still enemy fire coming at them hard and fast. He turned to one of his HQ staff and yelled.

"You! Get over to those Pakistanis and tell'em we need a few tanks over here. Don't come back without'em!"

Then he reached down with his right arm and grabbed the M60 machinegun from the dead machine-gunner and told one of his radio operators to load it for him. Before another belt of 7.62mm bullets was brought to him, he had the gun aimed at the spot where the RPG round had come from, and he fired repeated bursts into the small crevasse. Other Marines in the HQ and in Alpha Company joined in. Some of the Marines that were climbing up the hill tossed grenades into the hole.

Deep inside the tunnels, four paratroops entered a small 8x10x6 room. There were three other openings into the room. It was filled with boxes of food and plastic containers of water. They looked around with the aid of their night vision goggles and one of them spotted blood on the dirt floor. When he investigated, he found all three Rangers behind the boxes. They were terribly beaten and barely alive, but they had been found. Two of the paratroops started to prepare the Rangers to be moved. A third crawled back into the tunnels to get help, and the last one watched the other three openings to the room.

While waiting for help to come, the five Americans listened quietly to the rumble of the tank rounds hitting the south side of the hill and the constant ripping sound coming from all directions. It sounded like there were machineguns and automatic rifles everywhere. All around them, above them, and even in rooms and tunnels below them rifles and machineguns were working hard. Finally, three more paratroops shuffled through the tunnel to help pull out the Rangers.

As soon as the three paratroops moved the Rangers out of the room, six more paratroops came in. They divided into three pairs and went into separate tunnels to continue clearing out the honeycomb of caves. More paratroops from all three companies found the room, divided up, and followed them.

Mosby finished firing off the belt of ammunition that was in the machinegun when its gunner died. Then he made his radioman load another belt. While the radioman was remembering how to load the weapon, Mosby grabbed his M16 and fired into the nook where he suspected the RPG round had come from. Tracer rounds bounced off the surrounding rocks, but a few went inside the bunker. Suddenly there was a huge *CRACK* from a few hundred meters behind him. Dust boiled through hidden corners in the rocks and out through pores in the surrounding salt flats. The spot where Mosby had been firing exploded and huge broken pieces of rock flew everywhere. The Pakistanis had sent a tank over to help.

Mosby smiled for a moment and turned to watch the tank's commander hammer away with his .50 machinegun. It was a welcome sight. Everyone else was taken by it too.

There was at least one person on the hill that didn't like it. From yet another hidden bunker, a small burst of bullets came from a tango AK-47. Two of the three bullets skipped off the tank. The third never made it that far.

Mosby was knocked over. A 7.62mm round glanced off his body armor between his shoulder blades. If it had been an inch to the right, it never would have hit him. If it would have been an inch to the left, it would have entered under his armpit and killed him. All Mosby knew was that someone who had caught him with his back turned had knocked him down. He was pissed before, but now he was insanely furious.

Most of his Marines were almost to the top of the hill, so Mosby decided to move his HQ up there too. He told everyone to pack up and move out. Then he grabbed the M60 machinegun and started to climb without using his hands. His left arm was too painful from the rock that had cut it open, and his right arm was carrying the weight of the M60.

On his way up the hill, Mosby was shot again. He was stepping on to a small ledge on the side of a boulder when a single shot was fired from somewhere directly up the hill. The bullet hit him square in the middle of the chest. His body armor protected him, but he was knocked several feet backward off the boulder. From there he tumbled down a few more feet, bouncing off more rocks. The M60 in his right hand caught its butt on the top of a boulder and twisted around. Mosby's right arm twisted with it. He finally came to rest when his body became wedged between two boulders.

The Pakistani tank commander had seen the sniper's muzzle flash and watched Mosby fall. His response was immense. He had the gunner fire three rounds into the spot where the single shot had come from, then the tank commander fired off a full belt of .50 caliber into the dust cloud.

The HQ staff rushed to Mosby. He was alive. In addition to his left shoulder laceration, and the terrible bruise between his shoulder blades, Mosby had the wind knocked out of him, a pair of ribs were broken, his right shoulder was dislocated, and his right ankle was sprained. He was very lucky, and so were the tangos that were still alive. Mosby's rage was incessant. For a moment, his HQ staff paused to listen to him rant. It wasn't until they were ordered to remove him that they helped pull him from the boulders.

When he finally sat down with them between three boulders, he waved to the Pakistani tank commander who in turn saluted him. Mosby returned the salute, then turned around to assess the progress of his four companies of Marines. What he saw numbed him. At the top of the hill, someone had taken an empty SA-7 rocket launcher, attached a small American flag, and wedged it upright as a makeshift flagpole.

Mosby's rage melted. He shook his head, smiled, and laughed.

"Somebody's always gotta bring a flag. Okay everybody, let's get me back down this hill. Have Alpha hold position, Golf stays on top, and everyone else mops up. Better let the people in the 82nd know what we're doing, too."

The radio operator notified the 82nd's commander. Some of the paratroopers that were still outside had seen the flag. They also informed the Marines that they had recovered the Rangers and needed a helicopter to come in and get them to the ship's hospital right away. Fire from the bunkers was sporadic at best, so Mosby had the radio operator call the helicopters back to start withdrawing all of the wounded.

Paratroops were already coming out from the tunnels. As far as they could tell, they had been in all of the different rooms, and most of the bunkers. They brought out their wounded and six prisoners. Then the Marines and paratroops began hauling out the dead-paratroops, Marines, and tangos alike. Finally, not even an hour after the last shots were silenced, they began pulling out all the weapons, files, stores, and even a laptop computer.

Everything was lined up for an organized loading into the helicopters. The wounded went first: One hundred eleven paratroops, seventy-three Marines, fifteen Pakistani mounted infantryman, and three of The Protectors of the Koran. Mosby refused to leave.

Six hours later the helicopters returned and dropped off some water and MRE's (Meals-Ready-to-Eat; i.e. rations) for the Marines, paratroops, and even the Pakistanis. The dead were loaded next: fifty-six paratroops, fortythree Marines, nine Pakistani mounted infantryman, and three hundred nineteen tangos. Mixed in among them was a Chinese man wearing local desert garb and carrying a bullet in his head. Later, The Ox and Munthmar Hussein were also identified among the killed. There was no sign of Saudi Prince Abdualla or former ISI Captain Monsoor Asaad.

The Pakistanis wanted to keep the weapons and intelligence gathered, but Mosby was steadfast against it. He knew that the American intelligence services could find out much more than the corrupt ISI. Eventually, the Pakistani agreed, as long as some of his officers could accompany the spoils back to the Peleliu. He wanted to make sure that he had some sort of intelligence reports to provide the ISI. When the helicopters began returning a few hours later, they dropped off more

MRE's for dinner, and then all the captured weapons and misc. spoils were loaded. The Pakistani Executive officers from all three units went with the loot.

With the sunset, the Marines, paratroops, and Pakistanis came off the hill and moved to a featureless position nearly five miles to the south. As soon as everyone was accounted for, they called in the air strike to make certain that no one was still alive. The FAC had waited all day to call in the heavy ordinance, and so had the aircrews in the sky. During the entire engagement, only one airstrike had been called in. The Navy's planes had returned to the carrier after the battle, but the B-52's loitered for the cleanup strike. The FAC passed them the word that the time had come to make sure that no one was left alive either on or inside the rocky hill. The sun was dipping below the horizon, and the salt flats slipped into the shadows of the night. Slowly, the shadows began their slow climb up the still illuminated hill. Before it disappeared into the night, the show began.

The bombers circled down from 35,000 feet to only 15,000 feet, and then they lined up for their approach. Since there were no anti-air threats, they used the opportunity to get in some practice. Mosby and everyone else watched the first giant plane come in from their right, and drop a load of 1000 lb. bombs.

The hill exploded like a small nuclear bomb. Boulders were tossed hundreds of feet into the air and never seemed to stop climbing. Dust swelled from where the hill had been and enveloped the chunks of rising debris. Everyone cheered. Five miles away, the sound finally arrived and shook their bones. Everyone had to take a step back and brace themselves against the wind from the strong shock wave.

Another bomber approached a few seconds later. Mosby could see the bomb bay doors swing open, and another string of 1000lb bombs poured out. They disappeared into the dust cloud and more boulders climbed upward. Again dust rose and surrounded the flying boulders. This time, everyone was ready for the terrifying sound and the shock wave.

The last bomber was already approaching while everyone was still cheering. By this time, the dust cloud had risen very high, and right after the plane's bomb bay doors opened, it disappeared into the cloud. Everyone fell silent until it reappeared on the other side of where the hill had been. They cheered, and the sound of the bombs arrived as though to thank them. When it was done, every hair on everyone's body stood on end and tingled. A static charge had built-up in the air, caused by the high volume of dust thrown skyward.

There was nothing left of the hill. All three B-52's had shattered, crumbled, and dispersed the pile of stone. Dust drifted away and revealed the complete destruction. Only scattered stones, cobbles, and boulders showed that it had ever existed. There was no way for anyone still hidden in the tunnels to have survived.

Just before midnight, the last of the helicopters came to carry out the Marines and paratroops. Everyone shook hands, saluted, and even hugged the Pakistanis goodbye. As swiftly as they had arrived, the Americans left. Mosby was on the last helicopter. Almost a mile away and nearly 200 feet in the air, he looked out the window and made eye contact with the same tank commander that had probably saved his life. The commander saluted and faded away into the dust as his tank turned and headed east toward home.

Wednesday, September 1, 2010

2nd Lt. John Chamberlin-Personal Log

San Diego, California

I did it! I finally did it! All the schooling and training have finally paid off. Today I began my 1st duty assignment as a Marine officer. What exactly has it taken to get here: four years of summer jobs filling potholes for the State of Michigan, four years of studying Anthropology at Michigan State for my B.S. in Social Sciences, four years of ROTC drills and activities, two and a half months of Officer Candidate School at Quantico, Basic School (also at Quantico), 26 weeks of Infantry School at Camp Lejeune, eighteen weeks of Advanced Infantry Training, and finally-finally-FINALLY, my duty assignment! Only 28-years-old and I'm ready to lead men into harm's way. I've been assigned to the 24th Marine Expeditionary Unit-Special

Operations Capable; other wise known as the 24th MEU(SOC). Within the 24th, I've been assigned to command 2nd Platoon, Alpha Company, 1st Battalion, 9th Marine Regiment. Sounds confusing, but, it's where I get my mail, and even though it's only been a few hours, it's settling in like a new phone number.

There's a lot of new faces to learn. My boss is Captain Tom Brown. His boss is Colonel Dennis Ghetty, our battalion Commanding Officer (CO). The commander of our MEU(SOC) is General Ken Shapiro. The guy in charge of taking us where we need to go is Admiral Sam Pender.

Captain Brown is going to be tough to work with, but if there's one thing the Corps has shown, it's that the toughest challenges are the most rewarding. When I reported in, Capt. Brown was down on the well deck of the U.S.S. Wasp; our ship. He'd been arguing with one of the squids about where he could stow our gear. The sailor was new, and even though Capt. Brown only has one other MEU assignment under his belt he's an old hand. Both men lost the disagreement. It ended when a Navy Commander reamed out the sailor, then bluntly told Brown exactly where he could put our equipment-figuratively speaking. So, when my new CO walked away steaming mad, I walked up and reported for duty.

My Drill Instructors always had a way with words, especially those at Quantico, but, for some reason, Brown just didn't like me from the start. Maybe I was too clean and green. Maybe I had bad timing. Maybe it was racial. Maybe, though extremely unlikely, it was because my family wasn't destitute poor, and he obviously came from a broken home at best. How can I tell what kind of home he comes from? Four years of social sciences, his language, and his accent are pretty reliable indicators. Although I could be wrong, not many people from middle or upper class families in deep Mississippi can swear the way he does. I have to hand it to the man. He has opened up an entire new way of using nouns and verbs as adjectives and adverbs. His DI's would be enviable.

After my unsavory welcome to the unit, our Company 1st Sgt. showed me to my quarters. Here I met my counterparts, Lt. Bruce Woodward of 1st Platoon, and Lt. Hector Juarez of 3rd Platoon. Both seem professional, and, as is always the case when people meet, they were neither friendly nor hostile just attentive. I understood. The 1st Sgt. told them both about my run-in with Capt. Brown when he introduced me, but neither suggested that Brown's bite was worse than his bark. I'll assume both are lethal until further notice. The Company 1st Sgt. left without telling me where my platoon was, so I had Bruce show me. Hector was writing his wife, and seemed the more distant of my two new colleagues.

Bruce and I made our way forward and up through the ship to one of the many barracks. He stopped by his platoon's barracks. Then he introduced me to his Staff Sgt. and his squad leaders. I was too excited to listen carefully, and I don't remember their names. After a five-minute conversation that seemed to last hours, we went into the barracks next door, and I met my Staff Sgt., Bob Mamush. He'll be my right hand man from now until the end of my assignment.

Bob's a friendly, charismatic man. I can tell right away how he manages the platoon. Everyone admires and respects him. I do too. While I was at home watching the Gulf War on the TV news, Bob was eating sand in Kuwait as a Private First Class (PFC). Having just met, I didn't want to broach a possibly sore subject, but it's unusual to have a 34 year old Staff Sgt.. He should have a higher rank by now, but I'll have to find out why he doesn't later when I review my personnel files for my platoon.

My platoon, 2nd Platoon. That sound does ring.

Thursday, September 2, 2010

2nd Lt. John Chamberlin-Personal Log

San Diego, California

There was a meeting of all the battalion's officers at 0700. Colonel Ghetty introduced all the new people, like myself, then pointed out all the "old hands." I met lots of new people. I can hardly remember anyone's name, but I'm sure that they're having just as much difficulty with mine as I am with theirs. I hope. Colonel Ghetty also had some slight surprises. It seems that the 8 months of training remaining in our MEU refresher and deployment workup period has been shortened to a single combined six-month exercise. During this period, we'll be conducting some mock deployments, based out of San Diego, against various U.S. Army units on temporary assignment at Camp Pendalton.

There were smiles all around. Everyone likes to shorten his or her time away from family, and how can anyone resist the chance to show our mettle against the Army. Our game, played on our turf, well.... I almost feel sorry for those grunts.

Almost. Ghetty pointed out that this is part of a Pentagon readiness check to see how well we can perform "on the fly." We all think it's just another budget cutting effort to train two units for the price of one.

This idea was furthered when Col. Ghetty announced that "the powers that be" have decided to break up the support platoons in all MEU's. Their elements will be spread throughout the MEU on the platoon level. This kind of concerns me. It means that I'll wind up having some heavy weapons teams put into my platoon, and, even though I've been trained in how to use them, I'm not a heavy weapons specialist. It also means that I have more people to keep alive if we get into trouble.

After the meeting, I met with Capt. Brown again. He still doesn't like me. I keep thinking it's the money thing. He must think I'm some spoiled rich kid from the burbs. In truth, our family was one of the less fortunate in the neighborhood. We never went hungry or couldn't afford heat. I didn't get a car for high school graduation like everyone else, and I did have to pay for college on my own. We just had more than some, and less than others did.

Bob Mamush was with me when I met with Capt. Brown today. He doesn't like Brown either, but Bob's not the kind of guy to complain in detail about anything. He took me to Col. Ghetty's 1st Sgt. who gave me the personnel records for my platoon. He warned me though that there were going to be some changes at Capt. Brown's request over the next few days. Those changes, and the addition of as yet undetermined members of the support platoons, made my review of the personnel records almost a waste of time. At least I'd be doing something. Capt. Brown hadn't added me to the duty roster yet, so I had nothing but time on my hands. I should spend the time getting to know my platoon, but I got the feeling from the 1st Sgt. that I might as well have a new platoon any day, so why bother.

I reviewed the records anyway. As of now, Staff Sgt. Bob Mamush is my senior NCO. Sgt. Gary Anthos runs my 1st squad, Sgt. Wess Dustmann runs 2nd squad, and Sgt. Rick McGraw runs 3rd squad. Each squad has 12 people (men and women), (9) with M-16A2's and three with M249 SAW light machineguns. Each squad is divided into four fire teams, and consists of a Sgt., a Corporal, three Lance Corporals, and seven Privates.

Bob and I have put together a list of reading materials, quizzes, and drills for the platoon. Rick's squad seems to be the most professional, but Gary's squad has a significantly higher level of spirit, and I attribute this do the difference in Rick and Gary's personalities.

Staff Sgt. Bob Mamush is my right hand man. He runs the platoon. While I am in charge of it, Bob runs it. He is about 6', 170lbs., 34 years old, dark brown hair, brown eyes, and average build for a Marine. Bob, as I've mentioned before, is a veteran of the 1991 Gulf War with Iraq, and a very personable man. His rank aside, he would still be a leader through his intelligence and an amiable charisma. Everyone likes Bob. Everyone respects Bob, and when he tells someone to do something, there are no questions asked.

I have yet to see him chastise someone in front of others.

He seems to rely on the subtle humiliation factor. None the less, no one wants to be seen in front of others as the type of person who would give a good guy like Bob a hard time. I'm eager to learn from him; especially his ability to instill such influence over those he leads. I'm also eager to learn why he's only made it to Staff Sgt.

Sgt. Gary Anthos runs my 1st squad. he is 5'8", about 180 lbs., 22 years old, single, dark brown hair, dark brown eyes, and slightly out of shape. His hometown is Pittsburgh, PA, but his parents, divorced, and he really grew up all over the Midwest. Gary has a difficult, sarcastic sense of humor that is always at someone else's expense. He is has a good technical knowledge base, but is also very intelligent and well read in matters of politics and fiction (some would say they are the same). His sarcasm always offends someone in his squad, but the other's seem to appreciate his abilities and even his odd wit. Because of their appreciation, they follow him without a moment's hesitation. He is a bit on the reckless side. As much as I would like to have 1st squad follow my orders with the same zeal, I will not lead by his cavalier example.

Sgt. Wess Dustmann is the epitome of recklessness, and he leads my 2nd squad. He's 6'2", about 210 lbs., 26 years old, black hair, dark brown eyes, and even though his appearance doesn't boast it, he is very strong. Wess has Rick's technical knowledge, Gary's enthusiasm, and a wild rough and ready streak that reminds me of mountain men or the U.S.M.C. of old. He was married, but he and his wife separated shortly after. His hometown is somewhere in the southwest corner of North Carolina. Wess is a very friendly man who is extremely close to the people he calls his friends. He's very respectful to the chain of command, but his out of control R&R experiences have kept his rank at Sgt. for years. He's probably my toughest and most dangerous NCO.

Sgt. Rick McGraw runs my 3rd squad. He is 5'2", about 180 lbs., 25 years old, married, hazel eyes, and lives with early male pattern baldness. His hometown is somewhere is upper New York state. Rick is friendly towards Bob and myself, but isn't very friendly with most of his squad. In fact, I've overheard some of his members complain about his ineptitude. He hasn't shown any that I've seen yet, but I'm concerned that it will show someday. Actually, he seems very knowledgeable. His "I am the boss and that's the way it is" attitude might be why his squad has the lowest level of spirit between the three squads.

When I say men and women, I mean it. Yes, there are women in Marine Combat Units. Women have been steadily liberating themselves in the military, and this year, they finally won the right to qualify for combat assignments. It finally happened even in the rigid, traditionally chauvinistic and chivalrous U.S.M.C.. This poses a lot of challenges for a frontline leader like myself; challenges that hopefully this log will help resolve. I'm not thinking so much of the showers, latrine, uniform, or hygiene problems. I'm concerned that women might be harassed by the typical mud Marine.

Marines are known for their sense of honor, but not their sense of respect for women. Hopefully, only the most qualified women have been assigned, and they will be able to "check" improper attitudes right away. Otherwise, those women, Marines or not, are in for a tough deployment.

Of course, the other end of the extreme is also possible and equally hazardous. The male Marines could wind up trying to show off and win the attention of the female Marines. What a mess. Most of these kids are straight out of small town high school where the only degree they know of, besides a high school diploma, is an MRS. When I mentioned my concerns to Bob, he was evasive, and neither concerned nor light-hearted about the issues. I think something may already be happening in the platoon between the men and women. How could it not be?

Friday, September 3, 2010

2nd Lt. John Chamberlin-Personal Log

San Diego, California

Another day aboard ship-and what a ship it is! Capt. Brown hasn't assigned our platoon any tasks, so, while Bob has the platoon studying U.S. Army tactical manuals, I've had a lot of time to read up on the ship since I arrived.

This ship, the U.S.S. Wasp, is a direct descendant of World War II aircraft carriers and it shows. After the war, the Navy had a lot of carriers, and, since the public wanted the boys home, they didn't have a lot of sailors. There was also a worldwide movement to disarm countries. Everyone believed that any future conflicts could be settled immediately with "the bomb." During the Korean War, the Marine Corps began to experiment using the aging carriers for deploying helicopter-borne battalions. The experiments proved promising, but so did other types of large amphibious ships developed at the time.

The Wasp is a culmination and a combination of all those types. She's the size of a World War II fleet carrier. She was designed to operate helicopters, but she can operate all types of vertical take off aircraft also. I'm told we'll operate over forty large aircraft once we're at sea, but right now, there's only a few small helicopters.

We're supposed to have several different helicopters once we put to sea. We'll have some SH-60 Seahawks for detecting and attacking submarines. There will be some old CH-46 Sea Knights for delivering supplies, mail, and any other large bulky cargo. There will be a pair of CH-53 Sea Stallions for special operations. General Shapiro will have his own Huey for commanding operations from the sky. There will be a handful of AH-1 Super Cobra helicopter gunships, and maybe an old Sea King to act as an airborne radar station. We're also supposed to get some AIR-TO-GROUND Super Harriers for ground support and protection from enemy aircraft.

The most powerful aircraft will be the new V-22 Osprey tilt rotor transports. They're our ride from the ship into harm's way. The V-22's are really small transport planes that have engines at the end of their wings. Their engines rotate from a horizontal into a vertical position so that they can take off or land like a helicopter. Then they fly like a plane to our landing zone.

Unlike their World War II ancestors, the Wasp can also carry, load, and launch the smaller landing craft that carry Marines onto the beach. I've seen it done in films, but I can't wait to see the operation happen. The stern (back) of the ship has ballast tanks, like on a submarine. Those ballast tanks fill with water, and partially submerge the rear half of the ship. When the aft (rear) section is below water, a large hangar bay called the well deck partially fills. Large hovercraft (or barge style landing craft like those used in World War II) freely float out into the sea, and onto an enemy beach. If we can't fly to the enemy, we can still float to his shore.

Until we do use the aft well deck, where the hovercraft are stored, the entire level of the ship, both the aft well deck and the vehicle deck farther forward, are packed with equipment and vehicles for our MEU. Each hovercraft (called LCAC-Landing Craft Air Cushion) is big enough to carry one modified M-1A2 Abrams tank, or over a hundred Marines. We've also got artillery, trucks, missile launchers, and HUMMV's (the Corp's replacement for the jeep from World War II. It's a little bigger, and a lot tougher).

The personnel records for my platoon show that this is the first deployment on the Wasp for most. All three Corporals and three of the four Sgt.'s have been deployed on ships like the Wasp in the past. I had a few Lance Corporals who had served on ships before, but Capt. Brown pulled ten people from my platoon and the Lance Corporals with shipboard experience were among them.

The people who were pulled from my platoon were reassigned to 1st and 3rd platoon. My platoon got (10) people who I'm not sure are the cream or the crap of the crop. I definitely inherited some, if not all, of the women who seem to be causing problems in other platoons. Almost all of the minority Marines in my unit have been transferred to other units, and all of the upper and middle class Marines from 1st and 3rd platoons have been transferred into my platoon.

Now, I'm even more certain that Capt. Brown has some sort of racial or class issue with me. I hope his changes are motivated by my background in social sciences from MSU, not my family's alleged station. Either way, my challenges have increased.

I tried to spend some time with the new members of my platoon. Bob asked me to join him, and some of the people from the platoon for drinks after dinner. I didn't expect the high turnout, but almost everyone from the platoon showed up.

I think about 2/3 of the people in the platoon are under the legal drinking age. I understand the need for laws; after all, my career is devoted to defending them. I also understand a little about civil disobedience. In my mind, if you are old enough to vote, and to fight in defense of the Constitution, then you are old enough-and responsible enough-to drink. I'll never understand how an 18year-old kid is responsible enough to carry and use a machinegun but if he has a beer he's breaking the law. To that end, if the platoon wanted to have a night on the town, I was ready to help them out; as long as we kept it in the platoon.

Bob was pleased to hear my speech about frivolity, and he suggested a small bar in Tijuana, about half an hour away.

I'd been in strip joints when I was in college, but it was awkward playing the responsible Marine officer with both men and women from my platoon surrounded by half-naked dancers. Bob, along with Gary and Wess, kept the small cliques under a certain amount of control. Marines have always been a little on the rough side when it comes to bars, but some of the people, particularly those in Wess' squad (2nd squad) might actually have drinking problems. The situation wasn't helped when Wess kept challenging the entire platoon to drinking contests.

He has an incredible tolerance for alcohol. Wess bought drinks for everybody in our unit. I had to balance between being a responsible authority figure, and fitting in as "just one of the guys." After three rounds of shots in less than half an hour, I stopped. Some of the more sane people started skipping rounds. Gary tried to keep pace, but while most of his squad was able to, he passed out face down on the table. We carried him into the bathroom and left him passed out in a stall. Slowly, all of the hardcore drinkers curbed their intake-everyone outside of Wess.

Some of the women got out of control. Everyone was wound up to begin with, but the combination of age, alcohol, and sexual surroundings brought out the wildest side of our wildest women. I wasn't sure how to handle the interplatoon flirtations so I pretended not to notice. Some of the women picked up patrons that had been worked up by the dancers. Others disappeared with guys from the platoon. None of them had a problem with the dancers, and some even put on subtle lesbian acts with them! The guys from our platoon were typical, but also carried an air of big brother towards the women whenever a bar patron got too close without permission. Clearly it was not a night of Marine Corps etiquette. Actually, that would depend on your impression of Marine etiquette.

Since I still don't know everyone's name, I could only make one accurate note about the evening's romantic escapades. Bob, my platoon sergeant, must have made an impression on Portia Betts, a sweet blonde from 1st squad who definitely made an impression on him. Portia was one of the women who put on an act with the dancers. Bob apparently enjoyed this as well as she did, and the two of them slipped away in the middle of the night. I wonder how this affects Bob's ability to work with the platoon. Could he be taking the "just one of the guys" thing too far?

Saturday, September 4, 2010

2nd Lt. John Chamberlin-Personal Log

San Diego, California

I woke up this morning with the one of the worst hangovers in my entire life. We left the bar around 3am. I'd called a cab company to take the entire platoon (whoever was left at the bar) to a nearby motel. Almost half the people we started with are missing, and the combined bar tab/taxi/hotel bill has reached the limit on my credit card.

Portia spent the night in my hotel room with Bob, but nothing happened between the two of them - as far as I can tell. They got there a long time before I did. We all stayed up till sunrise talking. After the sun finally came up, I decided I had to get some sleep, but my head seemed to have barely hit the pillow when there was a knocking on the door around 9am. Privates Dave Avery, Bill Archer, and Mike Baker (all from 1st squad) were looking for their squad leader (Gary). It seems nobody remembered to take him out of the bathroom at the bar last night. I called the bar, and that they didn't know a thing about anyone passed out in the bathroom. I tried calling the police, and they, in fact, had picked him up around 11:30pm.

Bob, Portia, Dave, Bill, Mike, and I went to pick him up. He was swimming in someone's backyard pool, and, lucky for him, the neighbors called in a complaint instead of the homeowners. The police were only able to cite him for disorderly conduct and disturbing the peace-not criminal trespassing or worse. I told Gary that since I had gotten everyone else a cab, except him, I wouldn't report his little infraction. In exchange, he'd have to pay his fines. He was happy to oblige, and he volunteered to pay off the platoon's costs on my credit card bill too. I don't think Bob, Portia, Dave, Bill, or Mike will leak our little cover up-not if they ever want me to do the same for them.

By the time we got everything cleared up, it was dinnertime, and we were all hungry. We picked up some Mexican food from one of Southern California's thousands of Mexican restaurants. Afterwards, Gary, Dave, Bill, and Mike wanted to go out again, so they headed back to the motel to change and clean up. Bob, Portia and I caught a bus back to the hotel a little later so we could do some errands and make some phone calls before tonight's festivities began.

When we got back, I found most of the platoon in Wess' room. They had already started the night's drinking. Beer cans covered the floor almost ankle deep. It was impossible to walk without stepping on at least two or three. They'd already heard about Gary, and they weren't discreet about asking how I handled it. It told them, and made it very very clear that I wanted the entire issue to remain in the platoon. Capt. Brown doesn't seem to be the most flexible man in the Corps, and almost everyone in the platoon could get in trouble for yesterday night. My speech wasn't necessary. They all understood Capt. Brown better than I did, and they didn't want to listen to him anymore than I did. Bob pulled me aside for a few seconds to let me know that I had "....handled that very well." So ends the my first real test of leadership as an officer. I wonder how many other officers in the Corps' history have had similar baptisms of fire?

Sunday, September 5, 2010

2nd Lt. John Chamberlin-Personal Log

San Diego, California

I woke up in Wess' room this morning. The last thing I remember was slamming beers through a contraption of Wess' called the BBFH (Beer Bong From Hell). This beer bong was different from others. He had "found" a garden hose from someplace, and used a motor oil funnel for the intake end. He cut the end off the hose, used a hose clamp to bind the two together, then went up to Bob's room on the fourth floor. Hanging the hose out the window, we grabbed one end, held it shut with a thumb, while Wess held the funnel end. Then he filled the hose with beer (a process which took several minutes). When the hose was full, we took turns using it.

The way to use a beer bong is to put your mouth on the bottom end of the hose, and let the beer flow in as though it was a single gulp. The more beer in the hose, the faster it flows into your stomach, and the more powerful the effect. Obviously, no one was going to empty the hose (it was over 40' long and held at least a 12 pack), but we sure tried. I was one of the last to try it, and, when my stomach was full, the beer fountained out of my mouth for a few seconds before I could close the hose.

This made all those college drinking parties look tame. It's still all a foggy memory. I know I sat down on the bed after trying out the bong, and I know I woke up about 6 hours later on the same bed. This isn't history since I don't remember it, but I'm sure if it was eventful, I would have heard right away.

I was the first to wake up (except for Wess who may not have gone to sleep). He and I woke the rest of the room up, then both of us went around to all the rooms, and collected the platoon members staying in the hotel. Once everyone was together, I called the cab company again. Before the cabs arrived, each Marine was back in uniform, and ready to return to the ship.

Back on board, Bob and I took an unofficial roll call, and everyone had returned. Capt. Brown walked into the platoon's quarters just as we finished, and he gave us the pre-game speech before our exercises began. Everyone acted impressed, but I could tell they really weren't. It's comforting to see that I'm accepted by my platoon even more than their company commander is. At the same time, it also unnerves me to see any veil of discontent in the command structure.

We left port before sundown, then waited just off the coast while our aircraft were coming aboard. Most of the sailors on board were able to get some sleep, but we were up all night getting our gear ready for action.

Every unit in the MEU has specialized training so that we always have some units ready for a special situation. Some units are trained for urban combat, some are trained for rescue missions, and some are trained for demolition missions. The list goes on. Capt. Brown hasn't been able to get any specialized training for my unit, and I wonder if he may have fallen out of favor with Col. Ghetty.

Monday, September 6, 2010

2nd Lt. John Chamberlin-Personal Log

187nm WNW of San Diego, California

I went to a company briefing after morning PT (Physical Training). Col. Ghetty described our upcoming operations with the Army units up the coast at Camp Pendalton. Over the next three days, the rest of the ships in our amphibious group (called a BLT-Battalion Landing Team), are going to conduct naval exercises against an aircraft carrier battle group. This carrier group, centered on the U.S.S. John C. Stennis, is supposed to try and sink our ships before we can land and fight the Army. Our BLT is only escorted by a few Destroyers from Destroyer Squadron 6, so we shouldn't stand much of a chance. We were given a review of our General Quarters Action Stations, and then some flyers were handed out to distribute among our platoons. Everything was very smooth and professional. The only odd event centered on Capt. Brown who was late, and, despite the sarcasm from Col. Ghetty, Brown didn't seem to care.

After the meeting, Captain Brown held another meeting for all the officers in the company to discuss our roles in the upcoming exercises. It was interesting to find out that Bruce's 1st platoon, and Hector's 3rd platoon had both been trained for TRAP missions. A TRAP mission is what Marines do when an aircraft goes down for whatever reason. If the aircraft is serviceable, then a platoon accompanies a repair crew. Then, when the aircraft is capable of flying out on its own, the platoon is supposed to escort it back to friendly forces. TRAP missions are also used for rescuing downed pilots. The most famous example of this is when Marines rescued Air Force Capt. Scott O'Grady who had been shot down over Serbian territory in Bosnia. When I asked Capt. Brown what kind of training I should prepare my platoon for, he suggested it would be TRAP training, but he was evasive and unsure. I don't think he even knows what role my platoon will take on.

After our company meeting, I had Bob get the squad leaders together. He returned to my cabin a few minutes later with Gary, Wess, and Rick. I let all four of them know about my two earlier meetings, and about the TRAP training situation. Since there was no formal plan to include our platoon in the overall BLT training rotation, I asked them what they'd like to train for. Rick wanted to focus on marksmanship. Wess agreed, but also wanted to come up with some sort of physical and mental conditioning. Gary, definitely the most gung ho of the three squad leaders, wanted tactical training, even if we had to stay on the ship. We all talked over the ideas until Bob brought up a good point. Since we had become the

149

black sheep of the company, if not the entire BLT, then we could work on all of our ideas without attracting too much attention. We worked up a schedule of drills, PT, target practice, and more PT. The entire BLT is scheduled to hit the beach at Camp Pendalton within the next 100 hours. At least we'll feel more ready than we are now.

Tuesday, September 7, 2010

2nd Lt. John Chamberlin-Personal Log

340 nm NW of San Diego, California

We spent most of yesterday and last night conducting drills and PT exercises. When I finally told everyone to hit the racks, it was after 2300 (11:00pm). Of course, when I got back to my quarters, Bruce let me know that Capt. Brown had seen our training. I wasn't too surprised since we'd been rehearsing tactical drills almost everywhere except the flight deck, engineering, or the bridge. Bruce joked that I was showing off, but he knew the truth about my platoon's lack of specialization. Capt. Brown, uncharacteristically, approved of our training activities, and he asked Bruce what he was doing with his platoon. Bruce had 1st platoon cleaning weapons and arranging gear on the LCAC hovercraft (a never-ending task).

The night didn't go quietly. About 0400 (4:00am), the General Quarters alarm sounded throughout the ship. I met Bob and the rest of the platoon in the forward dining area where our action station has been assigned. The cooks had other shipboard action stations, so we hopped over the counter and kept the coffee brewing all night. We were supposed to be prepared to put out any fires in the area if there was a real attack, but since this was all a big war game, there wasn't any real fear. As soon as everyone was settled in and aquatinted with the fire fighting equipment, I left Bob in charge, and made my way to the hangar deck.

One of the elevator bay doors was open, and I could see the AV-8B Harrier fighters (from our ship) chasing, and being chased by Air Force F-15 Eagle fighter planes. I don't know if General Shapiro was informed about an Air Force presence, but I hope our Harriers made a good show of it.

Shortly after sunrise, we were allowed to stand down from General Quarters and get some more rack time. I went and met with Capt. Brown for our morning company briefing. He passed-on the news that our Harriers had been shot down in mock dogfights with the Air Force's extremely capable fighters. I asked if they had taken out any of the Air Force planes, and apparently it was an eye for an eye loss. We were happy to find out that the Air Force had taken losses from our fellow Marines in their out-classed Harriers. Capt. Brown said that the Air Force was not in the original complement of units involved with the exercise, but General Shapiro was going to make sure they were taken off the revised list immediately. The word coming down through the chain of command from Shapiro was that, once it was determined which airfield the Air Force was using, we were going to take it out. I couldn't help but smile. As I looked around, Bruce, Hector, and even Capt. Brown were smiling also.

After the briefing, I hustled back to the platoon, told everyone what was going on, and we reviewed our infantry training on airfield assaults. One of the guys in Gary's squad, Chuck Boyd, is a real computer geek, and, using his laptop (with its built in cell phone) he accessed the internet. Chuck did a subject search on "Air Force F-15 Eagle Bases" and got a list of a few thousand web sites. One of the web sites mentioned "Operation Tidal Surge 04" (the same name of our wargames with the Army!). Some Air Force computer geek had logged onto his or her squadron's website, only hours after the mock attack, and boasted about "kicking Marine butt." The website also had all kinds of information about the F-15 squadron including their current base assignment-Nellis Air Force Base, Nevada. I had Gary send somebody to get Capt. Brown. A few minutes later he walked into the platoon quarters with Col. Ghetty. We showed him the website, and they were in awe. How could someone have been so stupid! Ghetty and Brown left to pass on the word.

About 30 minutes later, Capt. Brown returned with the word from above. Our company will make the assault in the morning.

Wednesday, September 8, 2010

2nd Lt. John Chamberlin-Personal Log

90 nm W of Camp Pendalton, California

We were awake most of the night going over our plan to attack Nellis Air Force Base. Col. Ghetty designed the mission himself. It was to be more of a commando strike than an actual airfield assault. In fact, we only used three of our V-22 Osprey, escorted by four AH-1 Super Cobra helicopter gunships. As a reward for our creative intelligence gathering, our company would attack alone!

At 0500, we lined up in front of the elevator on the hangar deck. Everyone checked their gear and then each other's gear. A few minutes later, we moved onto the elevator and rode up to the flight deck. The Osprey and Cobras were already warmed up. While the sun was still a glow behind the coastal hills to the East, we boarded our aircraft.

As one, they lifted straight into the sky above the ship. We flew in a single file line a few miles behind the Cobras. While the sun broke the surface of the hills and mountains, we entered the rocky shadows. Beyond the mountains, we flew over desert, more mountains, more desert, and finally, Nellis approached.

I rode in the third Osprey. When we landed and ran out, Gary's squad was already running across the runway, and into the buildings surrounding the control tower. I followed with the other two squads while 1st platoon secured the landing zone (LZ). 3rd Platoon raced into a pair of hangars a few hundred feet away. No one had live ammo, or even blanks, but we were armed with neon orange spray paint, shaving cream, duct-tape, and even toilet paper. The offending pilots were to be captured with tape and toilet paper. Anything that would normally be

151

destroyed was spray painted. Of course, in response to the web site intelligence, 3rd platoon went around spraying "Marines kick Air Force butt!" on everything, including the F-15's. 1st platoon, having secured the LZ, then sent a squad out to find a fuel truck. We barely had enough fuel to reach Nellis, and didn't have any left over for a trip back to the ship.

I think my platoon did the biggest insult to the Air Force. While Gary's squad moved from building to building, Rick's squad cleared the buildings around the control tower. I stayed with Gary and 1st squad. Bob went with Rick and 3rd squad. Wess was on his own with 2nd squad, and, as it turned out, they had the toughest assignment of the platoon-maybe even of the entire mission.

Wess directed 2nd squad straight from the Osprey to a nearby hangar where they grabbed a truck, then drove away towards the base operations buildings. When they got there, they were approached by security guards, but Wess slung his rifle over his shoulder, pulled out a clipboard that he'd found in the truck. While he was getting directions from the guards, the rest of the squad slipped out of the truck and duct-taped the unsuspecting airmen. Wess promised to let the airmen go if they told him where to find the F-15 squadron commander. After a little realistic, though completely idle, threatening, the senior airman showed Wess how to get to the Air Combat Maneuvering and Information Center (ACMI) where the squadron was reviewing computer models of the previous evening's dogfights with our Harriers.

When they got there, Wess just marched right on in with the entire squad. It must have been an interesting scene when a dozen Marines opened the door to that dark theatre-like room filled with Captains, Majors, and even a Col., all watching movies of how they shot down Marine pilots. Later, Wess was very elusive about how he convinced these Air Force officers to follow the directions of a simple Marine Sgt. His official report says that..."although the Air Force officers were reluctant to follow my orders, I was able to change their minds in very short order. Once the authority of the situation had been redefined, there was no resistance." I wish to hell he'd tell me how he got them to follow him, but then again, maybe I don't want to know!

A few minutes later, Wess, 2nd squad, and almost twenty Air Force officers arrived at the LZ. A fuel truck arrived about 5 minutes later. I had my squad secure the prisoners in the tail end Osprey while 3rd platoon (having exhausted their supply of spraypaint, shaving cream, and toilet paper) filled the first two Marine aircraft. 1st Platoon fueled all of our aircraft, even the Cobras. When they were done, they got inside and the entire formation headed off to the west for our home at sea.

We got back at the ship about noon. Everyone's adrenaline was replaced by fatigue, and we disembarked the aircraft. While riding the huge aircraft elevator down from the flight deck, we saw General Shapiro was on the hangar deck. Col. Ghetty gave him a quick briefing, but he already had most of the details from a furious Air Force General who had called him shortly after we left Nellis. Shapiro said it was, "possibly the greatest military coup of his career."

A rapid series of debriefings followed our arrival. As meticulous as they were, the usual briefing/debriefing annoyance was overshadowed by the constant pats on the back from almost every Marine and sailor on board. The only people who didn't seem to pat us on the back were those Air Force officers. They should have been a little more-appreciative, after all, we did remove their bondage and blindfolds. As far as I know, there still 2 security guards are probably still duct-taped to the screen in the ACMI viewing room.

I think most units would have been awake all night from the excitement of the past few days, but we're exhausted. The thrill of my first command, the weekend of heavy partying, the intelligence windfall, the mission, it was all enough to put me to sleep the instant my head hit the pillow. I'm sure everyone else in the company felt the same way.

Thursday, September 9, 2010

2nd Lt. John Chamberlin-Personal Log

215 nm W of Camp Pendalton, California

I spent all morning writing and collecting reports about yesterday's action. I had all my NCO's (Non-Commissioned Officers; i.e. Sergeants) report on their actions, and the performance of each platoon member under their charge. Wess and Rick had already started theirs yesterday night, but Gary turned his in first. I suspected that his would be the least detailed, since he obviously didn't spend much time on it, but I was wrong.

He seems to have a good grasp of tactical operations. His report was extremely detailed, and supported by quotes from various U.S.M.C. handbooks. He also maneuvered his squad in textbook form. This isn't to say that he doesn't have the ability to adapt beyond "the book." During the operation, I had stayed with 1st squad, but I let Gary coordinate the actions of the squad's three fire teams.

Clearing buildings is an infantryman's most difficult task because of the ballet-like precision and timing involved. This task is made almost impossible if you have no ammunition, and your only weapon is intimidation. It was a pleasure to watch them in action. That is to say, it was a pleasure to watch from my point of view, and not from the Air Force's.

After all the reports were collected, I met with Capt. Brown. Bruce turned in all of 1st platoon's reports at the same time I was turning in ours. Hector had turned his in earlier this morning. Apparently he and his NCO's were up all night. Capt. Brown has a stack of reports to review before he submits his own, but I'm sure that, as far as documentation is concerned, he'll appreciate the effort 2nd platoon has put forward.

I haven't been continuously tracking our ship's movements. We do seem to be heading west more than any other direction. I imagine the navy wants to get out of shore-based aircraft range after yesterday. There's still no word on our "enemy" carrier's movements.

Friday, September 10, 2010

2nd Lt. John Chamberlin-Personal Log

340 nm NW of Camp Pendalton, California

Capt. Brown arranged for some live fire target practice for the entire company today. We lined up by platoon on the hangar deck, and each Marine was issued 1000 rounds. The target was a large float towed by one of the other amphibious ships in our Amphibious Ready Group (ARG). The float had a sheet, about 10' square, with a target on it held up between two poles. Each platoon was allowed to conduct several firing drills determined by the platoon commander, and supervised by the MEU's Training and Operations team, called the S-3.

The MEU has several different command sections: S-1 is general administration. S-2 is Intelligence. S-3 Training and Ops, and so on. There are so many different command teams and special units that I can never keep track. In fact, I suspect that the main purpose of the S-1 Administration section is just to keep track of all the command teams and special units at General Shapiro's disposal.

Hector's 3rd platoon was early arriving on the hangar deck, so instead of going 2nd, my platoon went last. His platoon is typical of the average marine infantry platoon. The same is true of Bruce's, except that there are no women either. It seems Capt. Brown had all the women in the company transferred to my platoon. 3rd Platoon shot off all their ammunition in about half an hour, and they seem to be pretty good shots for Marines; excellent for anyone else.

Bruce's platoon followed up and shot off all their ammunition in about 45 minutes, but the Lt. Col. from S-3 who was supervising says they scored better hits. Bruce was pleased, and privately teased Hector who took it sportingly. Then they both tried to tease me about how poorly they expected my platoon to perform. They didn't know that Bob, Gary, Wess, Rick, and I had prepared a custom series of firing exercises, or that I had the S-3 approve them while Hector and Bruce were following their textbook drills.

So while Bruce's platoon was walking off the aircraft elevator where the drill was conducted, Bob had my platoon form into three ranks, one for each squad. After 1st and 3rd platoons had formed up in the hangar, when the entire company had their attention focused on my platoon, I gave the order for the platoon to come to attention. They snapped to a rigid stance with weapons at the ready. I had them march forward, then stop at the edge of the elevator. Once there, I had the first rank (1st squad) lay down on the deck, with 2nd squad kneeling, and 3rd squad remaining at attention, I ordered them to take aim, and fire by rank. 1st squad fired a single shot volley, followed immediately by the 2nd squad's volley, then the 3rd squad's volley, then 1st squad again, and so on...

It was the same style that was used during Napoleon's time, and was even tried in Vietnam! It seems like too time consuming a procedure to use in modern combat, but the effect is still devastating. Every Marine was only firing a single carefully aimed shot, so accuracy skyrocketed. The platoon's sustained rate of fire was still intense because of the coordination between the ranks. Each Marine's magazine holds 50 rounds. We fired 150 volleys before we stopped firing. After that, we still had 950 rounds per Marine.

The other platoons were silent when 2nd platoon stood at attention and marched back into the hangar. I could see Bruce grinning with jealousy, Hector bearing both smile and sneer, and Capt. Brown alternating between watching the S-3 Lt. Col., and myself. I had the platoon about face and stand at ease, then the S-3 told us we had met our required score even though we used only (1) ammo clip per Marine! The men in the other platoons shook their heads in disbelief and frustration. We may not have pounded out the rounds, but we did the job better than they had, and they knew it.

Everyone wondered if we could handle sustained, accurate, heavy firepower in combat situations, so we had another drill planned to try something along those lines. I walked back out onto the elevator, right to the edge, then watched while I had Bob execute this maneuver. He issued as series of commands. Each squad formed into a single file line and shuffled towards onto the elevator like a police SWAT team about to raid a drug lord's house. When 1st squad was on the elevator, and the target was in their line of fire, I gave the order, "1st Squad, Commence.....Firing!" The squad stopped, and Gary, the first in line, dropped to his knee. He fired three round bursts at the target until his weapon was empty. Bill Archer, the second in line began firing immediately after Gary, and one by one rest of the squad did the same. While 1st squad was pouring out the lead, 2nd squad ran forward in single file. They arrived on the elevator next to 1st squad just as the last person in Gary's squad emptied their rifle. As soon as 2nd was in position, they opened fire. Meanwhile, Rick's 3rd squad began running forward, and acted exactly as 2nd squad had for Gary.

When 2nd squad was reloading, 1st squad was already firing again, and the three small groups continued this constant firing until Bob saw that Gary was on his last ammo clip. Then Bob shouted, "At the command, fall back by squads.....fall back!" Gary's squad stopped firing, turned, and ran back into the hangar, then he announced, "1st squad, in position!" That command stopped 2nd squad's shooting, and they did the same maneuver that Gary had. Finally 3rd squad was in position. The elevator was covered in spent brass shell casings, and the rifles were silent.

The platoon formed ranks, and I ordered them at ease. Bruce and Hector's platoons thought we were showing off, and we were, but not for so much for the applause of others as it was for ourselves. The lack of special training had made my platoon, and myself, feel sort of like forgotten outcasts. We needed to feel unique, or at least as special as any other Marine unit on the Wasp. The S-3 was talking with Capt. Brown, and, after a few minutes, I was called over to join the conversation. It seems that our drills had shown the S-3 that our entire company would be a good candidate for some urban assault training at Camp Pendalton. This would all take place after our exercise against the Army.

I went back and gave the platoon the news. They knew they'd done well, and every Marine gets a little pumped up after being able to fire their weapon in an exercise. When I walked in, they fell silent. I sat down, looked at Bob, then Gary and some of the other people in the platoon. Then I told them about how the S-3 was going to put the entire company into the training rotation for urban assault tactics after our current exercise was over. No one cheered, they all wanted to look calm cool & collected, but the quiet "nice job" and "we did it" comments between them said it all to me. I think they were actually stunned at the realization that they "showed-up" the other platoons. I also told them to check their gear and get focused because during my meeting, I was told that our company will be leading the beach assault on Camp Pendalton tomorrow morning. The silence was complete, and the focus was 100%.

This platoon is ready...now.

Saturday, September 11, 2010

2nd Lt. John Chamberlin-Personal Log

90 nm W of Camp Pendalton, California

The day began with another general quarters alarm, this time a little after 0100. We raced to the forward dining area, and went through the firefighting equipment drill, again. Then we hopped the counter and began making coffee, again.

We waited until 0430 when Capt. Brown came in. Apparently he forgot to send someone to ordered us down to the well deck and begin boarding the #17 LCAC hovercraft. In fact, we were to be ready on the LCAC by 0430! Bob shouted out a few orders, and the platoon ran back to their barracks for their gear. Capt. Brown didn't seem very honest with his sympathy. In fact, he gave me a bit of a hard time about not being prepared.

While I listened to his colorful opinion of a properly prepared Marine Lieutenant, Bob had the entire platoon geared up and moving down to the well deck where the hovercraft were all warming up. Brown and I walked back to my quarters so I could get my gear. All the while I listened to some speech about when he was a Lieutenant, and how he was always ready for the balloon to go up. It was difficult to look like I was racing to get my stuff together. I was really trying to stall the Capt. so Bob could get the entire platoon organized.

When Brown and I finally did make our way to the well deck Bob had everybody in place on the LCAC. It was loud beyond belief. All three LCAC's were warmed up, so were a dozen Screamers.

A Screamer is an amphibious armored personal carrier. It has special plates that rotate from the sides to the bottom of the vehicle to create a sleek hull. When the plates are deployed, the tracks are retracted, and a high-speed turbo diesel engine is engaged. It can race across the water, with (18) Marines, and a crew of three at about 25 knots!

Officially they're called AAAV's-Advanced Amphibious Assault Vehicle's, but we call them Screamers because of the noise they make in the water. The turbine engines on the LCAC's and the diesels of the Screamers made the well deck so deafening that Captain Brown had to finally stop giving speeches about unit readiness. I don't think he could hear himself!

Bob had already issued earplugs to the rest of the platoon while they were waiting on the LCAC-alone! 2nd platoon was the only platoon on the well deck. Each LCAC had an M-1A2 Abrams tank on it with its crew of 4, but there were no other Marines on the landing craft! Capt. Brown had pulled a fast one on us. My stunned look didn't have time to change before I noticed Hector leading his platoon into LCAC #23 in front of ours.

By 0500 the LCAC's were backing out of the well deck and onto the foggy ocean behind the U.S.S. Wasp. The Screamers came out next, and the LCAC's followed them to the beach. It took about half an hour to get to the sand, and when we did, there was already a handful of Marine units there directing us to rallying areas where Col. Ghetty would explain how and where to deploy.

V-22's from the Wasp were landing in Camp Pendalton while we were loading into the LCAC's. The Osprey dropped off a couple hundred Marines to secure a landing zone, a beachhead, and a few tactical positions like hilltops and crossroads. When I found him, Col. Ghetty was surrounded by his staff officers, his company commanders, and almost a dozen other platoon commanders like myself. He introduced us to our U.S. Army wargame referees. Every unit was assigned a referee who will be accompanying us throughout the exercise. The referees will not only moderate our actions, but also file their own reports detailing our performance. To balance any impartiality, Marine referees will follow the U.S. Army units.

Col. Ghetty told us where to pickup flyers listing the scenarios, orders, timelines, rules of engagement, and maps showing where we should deploy for each scenario. After the meeting was over, he gave us a quick pep talk, and we all went back to our units, with the Army referees tagging along. After I got back to the platoon, I passed out the flyers and repeated Col. Ghetty's briefing to my squad leaders. They in turn repeated the information to the rest of the platoon. When my poor excuse for a pep talk was done, we moved out to our first assigned position.

The beginning scenario has our company holding the line along the north side of our supply Landing Zone. Bruce's 1st platoon was ordered to hold on our left, and Hector's 3rd platoon on our right. Bruce's platoon setup on a small, peanut-shaped hill, and Hector's platoon spread out along the bottom of another hill to the right. Behind Hector's position, Capt. Brown found a quiet, shady, bamboo thicket where the Company Command Post (CP) was set up.

My platoon had to hold the road between the two hills. In front and behind us the road curved like a snake. When it reached the hills, there was a steep roadcut cliff on both sides-as high as 40' in some places. There were a few sharp 90-degree turns before the road emerged on the backside of the hills. Then, just before the LZ came into sight, the road went over a bridge that crossed a small creek. In some states I'd seen creeks this size called rivers. After the bridge, the road passed the company CP, and went then along the northern and Eastern edges of the LZ.

I feel pretty comfortable with Hector and Bruce on my flanks, so I'm setting up my platoon in a layered defense of the road and ignoring my sides. The platoon may be broken into three squads of twelve Marines, but each squad is further divided into three fire teams of four Marines each. Gary's 1st squad is spread out in the area along the base of the two hills with one fire team on each side of the road, and the last fire team further back along the road to act as either a reserve, or a covering force. Wess' 2nd squad is positioned by fire team in ambush locations along the road, and Rick's 3rd squad is positioned on the northern side of the bridge.

The intelligence report that I got from Col. Ghetty, back at the beach, says that we can expect U.S. Army units to attack with tanks, armored infantry fighting vehicles, helicopters, and infantry. Since our task force was already attacked at sea by the U.S.A.F., I wouldn't be surprised if they get involved again too. Of course, their planes may be a little easier to spot if they're still wearing our orange spray paint!

The U.S. Army has only a few types of divisions: air cavalry, mechanized infantry, armored (tank), airborne, and light infantry. It's not too hard to figure out that we're not facing an airborne or light infantry division. That wouldn't really be a challenge. The other types consist basically helicopter, tank, IFV (Infantry Fighting Vehicle) and truck units. My platoon only has a few light anti-tank weapons, and any air support or artillery support will be limited at best.

I had Bob and the squad leaders get together to go over the intelligence brief again. Gary, always up for an opportunity to fight, was eager to have his squad be the first that the Army tries to run over. Wess, equally enthusiastic, told me that he'd put all three of his fire teams in ambush positions along the tops of the cliffs that parallel the road. Rick, feigned his enthusiasm, and wants to have one fire team on top the cliff overlooking the creek, another on the opposite side of the bridge, and the last under the bridge to come up and ambush any Army units that get pinned down on the bridge from behind. Bob approved of the plan, and I really like it.

There only seem to be a few flaws with it. Since each fire team will be at least fifty meters from the next, unit cohesion is going to be difficult, and so changes to the overall plan will be equally difficult. Another problem is that if the Army chooses to hit us with a large-scale infantry assault, the platoon is too spread out to defend as a whole. We could be taken out piecemeal; four people at a time. The last major flaw, is that, if I was the U.S. Army, I would make a feint attack on the area in front of the LZ, then make an air assault into the LZ. Apart from a few CP Marines, and some logistics personnel, there's almost no defense. Our entire company would be deployed in a very bad position for counterattacking such an assault.

By sundown, the platoon was deployed, and except for a small recon patrol that Gary sent out, there was no activity the rest of the night. Even with the constant rotation of helicopters and Osprey back at the LZ, I slept well. I expect that most of the platoon did too, even if they had to take turns on watch.

Sunday, September 12, 2010

2nd Lt. John Chamberlin-Personal Log

Camp Pendalton, California

The sun rose today, and we got our first glimpse of the bad guys. A U.S. Army scout helicopter buzzed our front line. I didn't see it, but I could hear Gary and the rest of 1st squad talking on their headsets (every Marine has one built into their helmet). They were excited, and I had to tell them to cool it down. As I was cutting them off, two of our AH-1 Super Cobra helicopter gunships flew overhead and chased away the Army scout. It was tough not to get excited, and useless to tell everybody not to cheer.

I had Bob take up a good position between Gary's and Wess' squads, while I waited with Rick's squad on the cliff above the creek. Capt. Brown came over about 0800 (8:00am), and reviewed our positions. He actually likes our plan, and, before he left, he gave me a list of radio call signs that I can use to call in air and artillery support without having to go through his CP for approval. In fact, Brown told me that if he heard me call for support over the radio, he would arrange to have at least one of the M-1A1 Abrams tanks we brought ashore come forward and block the road on the south side of the creek. If all goes according to plan, no soldier will cross that creek without getting wet.

The rest of the day was spent waiting, watching helicopters fly around, waiting, digging foxholes in the sand, waiting, waiting, and more waiting. I've been on maneuvers before, but I thought for sure that an operation of this size would have been more fast and furious than those had been. It's not cheap to keep large forces like these in action, and making us sit here seems like a waste of time, effort, and taxpayer money. I had Wess and Rick send out patrols to keep an eye open, get familiar with the area, and fight the boredom.

Monday, September 13, 2010

2nd Lt. John Chamberlin-Personal Log

Camp Pendalton, California

Still no activity in the area. I went around and checked every single position in the platoon again today. I can already see the complacency setting in. The foxholes are deep, but now they're becoming comfortable! I've also noticed a good deal more flirting between the members of my platoon. This is not the type of frontline behavior that I want to reinforce, so I had everyone exchange weapons with Marines in other squads, and then I started a rumor that I was going to conduct random weapon inspection tonight. The weapon exchange took almost an hour, but afterwards, everyone was busy checking and cleaning their weapons. I had the Army referee take his MILES calibration device to each fire team and recalibrate the MILES gear.

MILES gear is what the Marine Corps and U.S. Army uses to confirm kills in wargames like this one. MILES (Multiple Integrated Laser Exchange System), is basically a two piece system: a box that fits on the end of a rifle, and a series of sensors that every Marine and Army soldier wears during an exercise. When a blank cartridge is fired through the rifle, the MILES box sends out an invisible pulsed laser beam. When that laser hits a sensor on a soldier or Marine's belt, helmet, or combat harness (suspenders), the sensor emits a loud annoying tone. That tone means that the person wearing the sensor has either been killed or wounded. It's hard to cheat since the sensor's speaker can only be turned off with a key that is possessed by an exercise referee. There are also variants of the MILES system fitted to tanks, IFVs, trucks, helicopters, and even airplanes. Those variants send and receive special coded laser signals that make sure a rifleman doesn't kill a tank, and a tank is able to kill more than one person at a time.

All the systems are integrated - as the name suggests, and all of the data is sent directly to the ECC (Exercise Command Center). At the ECC, generals can watch their units destroyed on computer screens as it happens. Data can be stored in computers, and results of the exercise can be studied for months later.

The referee is still vital to the experience. Besides being the only person able to turn off the awful shriek that the MILES sensors make, the referee can also check the calibration of MILES-equipped weapons. Calibrations ensure that the laser goes where the sights point. The referee is also equipped with the "God Gun" which can kill anything from a plane to a rifleman in order to simulate the random acts of violence that occur in frontline situations. The "God Gun" also gives the referee a little more respect from the people who are operating under his or her witness.

By the time the weapons were all re-calibrated by our referee and their new operators, it was nightfall. The platoon began their sleeping cycles according to a sleep plan that I went over with everyone while we were still on the ship. Just before midnight, when I was sure that most were asleep, I conducted my random weapon inspection and the results were embarrassing.

My suspicions about men and women in a combat platoon were not unfounded. As I made my rounds, I found several people had paired up somewhere along their service careers; specifically, Christina Boyle from 1st squad, and Chris Landas from 2nd squad. I bumped into them while I was checking the outer perimeter positions. It's an odd sight seeing a young man and woman wearing desert camouflage, and cuddling in a foxhole.

When women were first assigned to frontline combat units, I think every officer tried to prepare for this situation. Most of us expected it. After all, the average Marine in my platoon has only been out of High School for a few months. I understand being young. I still am myself, but I can't understand or permit reckless, carefree, divided attention of any kind while in a frontline situation.

To make matters worse, Christina and Chris are both the stereotypical white suburban brats that Capt. Brown envisions my entire platoon to be. She is very intelligent, extremely attractive, and overflows with confidence - not arrogance. Chris is not very bright (I wonder how he made it through High School), and, although he's a very proud person, he suffers more from arrogance than confidence. Both are from small upper middle class suburbs in the Midwest, and both have little regard for anyone else's well being. Neither strikes me as a team player. As much as I would like to bring them up on charges, I don't dare prove Capt. Brown's suspicions about my platoon to be correct. Besides, I strongly suspect that Bob Mamush, my number one man, and second in command, is at least slightly involved with Portia Betts. Bob is indispensable, and, I can't afford to put him to the coals over something that hasn't been a problem (except for this one instance). Tomorrow, I'll get together with Bob, talk the matter over with him, and then have a meeting with the squad leaders to emphasize my lack of appreciation for the foxhole love affair problem. Until then, I made sure to give both Chris and Christina a verbal lashing that neither will soon forget.

Tuesday, September 14, 2010

2nd Lt. John Chamberlin-Personal Log

Camp Pendalton, California

Well, the U.S. Army finally hit us today. About an hour after sunrise, Gary radioed that he could see dust clouds a few miles to the East. Bob put everyone on alert, and I notified Capt. Brown. About half an hour later, we heard gunfire to the south, over the hill, near Hector's platoon. The shooting rose and fell in tempo, and finally, the gaps between bursts grew longer. While our attention was split between our own frontline and Hector's ongoing action, Gary spotted an armored group coming down the road towards our position.

The Army was coming in numbers, and with great force. Three tank platoons with four M-1A2 Abrams each approached in a single file line. Behind them, another twelve Bradley IFV's followed.

Each Abrams has a crew of four, each Bradley has a crew of three+ with about six troops inside (on average). I didn't do the math at the time, but looking back, I see that roughly 150-200 soldiers hit my 38 Marines. I didn't have to do the math! When I saw twenty-four armored vehicles heading our way, I made sure to let Capt. Brown know about it. I've got a good platoon, but there's a limit to anyone's ability.

I called for air support. None was available. I tried for artillery and mortar support. They were busy too. The other units around the LZ were also under attack, and when it came to any of the help that Brown had promised, we were on a waiting list.

The Army hit 1st squad as they approached the road. Gary waited until the 1st three tanks had entered the sunken road before he gave the order to open fire. When those twelve people in 1st squad began shooting, all twentyfour armored vehicles returned fire at once. It was an awful display of firepower. Tank guns barked, the 25mm autocannons on the Bradleys began that monotonous popping of blank shells, and infantry inside the Bradleys began shooting their M-16's through their nice and secure firing ports.

1st squad lasted about fifteen minutes before everyone's MILES gear was beeping like crazy. They were completely wiped out, and only one enemy tank had been destroyed. Two of the tanks that made it past 1st squad backed out of the road cut during the fighting. The Army was going to send in the troops to clear us out.

Wess was ready and waiting with 2nd squad. The Bradley's came down the road in a herringbone formation with a good deal of spacing between each vehicle. Their autocannons and hull-mounted M-16's panned around looking for targets. Wess let the first three pass, then gave the order to open fire. Again, twelve seriously outnumbered Marines went to work.

One of the people in 2nd squad was able to shoot the driver of one of the Bradleys, and as his MILES gear began to deafen everyone in the vehicle, the column stopped. Groups of soldiers evacuated all of the vehicles and began to rush 2nd squad. Wess was hit early on, but computer records show that he took out eleven soldiers before he was "killed." The rest of the squad was wiped out in less than half an hour. The after-action reports indicate that they were able to "kill" and "incapacitate" thirty-three soldiers before their own demise around 1015hrs.

The Army rolled forward again. Bradleys led the way, with infantry walking along the sides of the road. In the rear, the M-1's crept forward. Rick and 3rd squad were their last remaining obstacles. I was on top of the roadcut cliff, in front of the bridge, with the 3rd squad's 1st fire team (Team 7). When the Bradleys came around the bend, I gave the command to open fire. The five of us forced the lead

Bradley to stop as it began to cross the bridge. The troops around it dispersed. I got two of them, and the rest of Team 7 got the remaining infantry.

The Bradleys advanced on. More infantry followed. By the time the second Bradley was in full view, we were all "dead," and Team 9, across the bridge, was firing at the fresh infantry. The tide of overwhelming numbers and force was beginning to gain momentum.

This time Tim Potter, one of the people from 3rd squad, Team 8, came up from under the bridge, ran through the surrounding infantry, into the lead Bradley, and opened fire only a few feet from the crew. Without listening for their MILES gear to respond, Tim ejected the magazine from his rifle, and loaded another. Using the partially closed rear door for cover, Tim continued firing from inside the vehicle until his own ammunition was exhausted. After he was out of his own ammunition, he began to pull from the large cache of magazines inside the Bradley.

By this time, the rest of Team 8 had come out from under the bridge and scattered among the Army troops. Another Bradley was somehow knocked out, and blocked any more reinforcements from driving to the bridge. Team 9, on the other side of the bridge, was keeping everyone's heads down.

While all this was going on, all of the "killed" soldiers and Marines stood along the top of the roadcut for the best view of the action. The referees strictly prohibited cheering or advising anyone down below, and it felt like being at a football game where you had to keep your voice down. It was frustrating enough, helplessly watching my unit decimated, but Capt. Brown climbed up and joined us. At least he had been "killed" somehow too.

Brown asked me to report on the action below, and I described our entire encounter with the Army from the beginning. He nodded slightly, and I had just started to hope that he approved when he interrupted to remind me that I was killed, and the platoon was wiped out. In fact, we had not been wiped out. I pointed to the other side of the bridge where all of 3rd squad's Team 9 was still fighting, and to the Bradley blocking the close end of the bridge with Tim still shooting from inside.

He was really the heart of the spectacle. With Team 9 backing him up from the far end of the bridge, and protected by the armor of the Bradley, the Army could only wait until he ran out of ammunition. Squads of soldiers kept trying to rush him, but Tim and Team 9 wiped out eleven attacks. They tried using several squads for covering fire, and a few rushing forward. They tried using smoke. They tried a complete charge. Everything failed, and losses were walking away quickly.

The Army could tell that Team 9 was running out of ammunition so they switched to making feint attacks, with the occasional serious push thrown in. When Team 9 was almost completely out of ammunition, they tried to get someone down to help Tim. First a lone Marine from Team 9 left cover and ran across the bridge while the others put down cover fire. He made it to the Bradley before his MILES gear went off. Then another tried made a run for it, but he never reached the bridge. The last male Marine in Team 9 wanted to be chivalrous, so he made the attempt rather than the Team's lone woman. She tried to cover him while he went to get ammunition from the Bradley, but his MILES gear began beeping before his right foot left their slit trench. Low on ammunition, the Team's lone woman switched her M-16's rate of fire to single shot, and lasted another hour before they she was eventually hit, and walked out of the trench.

When she was finally taken out, the Army troops cheered from their positions. Tim silenced them with a long burst from his M-16. By now, it would seem that the Army should be able to rush Tim and take him out by sheer numbers. They tried. Then they tried again, and they tried again! The Bradley IFV has four firing ports into which specially modified M-16's can screw into. Tim had taken all four rifles out of their ports, loaded them with magazines, and was firing long bursts from each one in a rotation to prevent the barrels from melting. After a few hours of fighting, the Army started showing signs of fatigue, but Tim was just the opposite.

The crowd on top of the roadcut was over two hundred, but he was still in the Bradley. Another Bradley still blocked the road between Tim and the rest of the Army's column. The surviving soldiers tried hiding in the poor cover that the road offered. They attacked less frequently, and with less ferocity. From on top the cliff, the rules regarding cheering were blatantly disregarded. The Army's soldiers were trying to get their surviving buddies to avenge their "deaths," and the Marines just encouraged Tim to hold out longer.

Over all the cheering, Bob pointed out to me that he thought he heard laughing. I listened, and heard it too. Every time Tim opened fire, he stopped, and laughed; sometimes uncontrollably for a few minutes. It actually seemed that he was having so much fun, that he couldn't concentrate enough to aim. The rest of the Marines noticed it too, and we all started laughing. That is, except for the army's soldiers who were slowly silenced with embarrassment. Their embarrassment mutated into rage, and they gave even more encouragement to their surviving buddies.

This time, the Army ran all the way up to the Bradley, and Tim let them approach to within about 30 meters before he came running out the back. He ran straight into the dispersed crowd of soldiers firing two M-16's from the hip. They were stunned for a split second. Then he was shot and started beeping. The dead soldiers cheered and we did, too. We cheered Tim's actions. He exemplified the Marine's *esprit de corps*.

Capt. Brown was trying not to laugh, but it was useless, and even he cheered Tim's name. For a brief moment, Brown and I were no longer adversarial. It was the first time that I felt like we were on the same team. I think he had finally come to see that despite his prejudices, we're all Marines. With the bridge free of resistance, the Army sent a tank recovery vehicle down the road to drag away the two "knocked out" Bradleys. The road was cleared in a little less than an hour, and we watched helplessly as the M-1's rolled down and across the bridge. It was a hard thing to watch; defeat.

That empty aching was pulled away about five minutes later when we heard the helicopters and watched the smoke rise from behind the hill on the other side of the bridge. The referees reported in, and accompanied us back to the LZ for extraction to the ship. On the way, we passed the smoking M-1's with flashing lights on top, disgruntled crews sitting on the road beside them. The U.S.M.C. Cobras had come in and saved the day.

Wednesday, September 15, 2010

2nd Lt. John Chamberlin-Personal Log

130 nm WSW of Camp Pendalton, California

We were airlifted back to the ship last night, and everyone was glad. It's only been a few days sleeping in the dirt, and eating from Meals Ready to Eat (MRE's), but everyone is glad to be home. It's amazing how good hot food can taste and feel. I've been on plenty of maneuvers, but it's always nice to get settled back in.

It's really got to be difficult for the Army. Not only were they defeated, but they still probably won't get a good night's sleep in a bed, or even a hot meal. The racks on the ship are VERY small, and there may only be a pad instead of a mattress, but it's like a luxury hotel for us.

Capt. Brown met with Bruce, Hector, and I before we went back to the ship last night. He said, in a matter of fact tone, that he was pleased with the results of our actions, and expected our full after-action reports before 1200hrs today. What this really meant was that he was forced to admit that we were not as inept as he expected. As a result, he was letting the entire company sleep in and take their time getting their reports together.

The platoon was put down in less than half an hour, and only a few people chose to shower before grabbing their sack time. I was in bed by 2300 hrs., and slept until almost 0800. Bob was already awake. He and the other three squad leaders were talking over coffee in the forward dining area.

It only took about an hour to get together several after-action reports. I had each squad leader document their version of the events, then note their difficulties, their needs, and their suggestions. We all went over each other's reports, then we complied a platoon report for Capt. Brown. Hector, as usual, had his reports finished, and delivered to Capt. Brown before either Bruce or I had a chance to meet with our sergeants. Brown had copies of the computer data that the ECC had gathered from the MILES gear. He gave me a copy to go over with the rest of my platoon. This was my first meeting with Capt. Brown where I didn't feel like I he expected me to apologize for something, anything. Why should I, there was only one other unit in the entire exercise that had a higher kill ratio than ours, and that was an Army helicopter troop with helicopter gunships! It was quite a feather in Brown's cap, and it earns my platoon an even share when it comes to his attitude from now on.

I went back to deliver the good news to the rest of the platoon. When I got there, as usual, my first question was where's Bob? This time no one answered. Something was obviously going on. I thought it was some sort of joke or surprise. I was wrong. Chris Landas, the young man I had caught in the foxhole with Christina Boyle a few nights ago, let me know that Bob and Portia Betts had gone someplace together. He expected them to be as busy as he was the other night. As he told me where to find them, everyone was silent, and it was obvious that his popularity in the platoon was now below zero.

I left the platoon's quarters and went to get Bob. They were on the hangar deck behind a partially dismantled Osprey. Contrary to Chris's suspicion, they were not physically involved at the time. Actually, it seems I walked in on a lover's squabble, and, I'm still not certain, but I'm pretty sure Bob was being dumped.

I've seen the look before. I've even had the look before. It couldn't be easy for him. She's one of the people he has to command, and here she is telling him that she doesn't want to see him anymore. On top of it, she's a fairly attractive blonde haired, blue eyed, well-figured young woman. At least he can take solace in the knowledge that she's as lightheaded as the Hindenburg. I don't think Bob was looking for intelligence. If he was, he must have been ignoring her obvious fault.

I tried not to interfere, but at the same time I couldn't let this situation go without being addressed. Even though I was going to have a private conversation with both individually, I wanted them both to understand that I think a break-up is a good idea. If either one has a problem with the idea, they could blame me. I don't want this to become the love platoon. They agreed, and Portia, ever the light-headed, tried to convince Bob in front of me; only to embarrass him. He got upset, and insulted her intelligence until she began crying. So there we are, in the hangar of a Marine Corps Amphibious Assault Ship, surrounded by combat aircraft, missiles, bombs, and thousands of both Sailors and Marines. There's Portia crying. I shuffled her into the Osprey, and ordered Bob back to the platoon.

Portia and I talked until she stopped crying. She had tried to break off the relationship with Bob several times before, but each time, she got back together with him because she didn't want to be alone. I told her all about the difficulties of relationships in the Corps, and about the regulations regarding men and women in the same unit. I offered to have her transferred, but she declined. I told her that if she was going to stay in the platoon, then there would be no getting back together with Bob, and she agreed.

After having dealt with the Portia problem, I went back to the platoon, found Bob, and had a talk with him in the dining area. He's in love, and he doesn't know how he's going to make it without her. I gave him the same speech about Marine Corps relationships, relationships with subordinates in the same unit, and even offered to transfer him from the platoon (a complete bluff). He didn't want to leave the platoon, so we agreed that the relationship would end this time for good. They were going to try and be friends, but that classic phrase seems doomed to fail.

With Bob and Portia all straightened out, I returned, again, to the platoon's quarters, and went over the computer data from the MILES gear. I told everyone their scores, how many kills they earned, how long they lived, and everyone had plenty of comments for each other. We all laughed with the people who never fired a shot, and lasted only a few minutes (a few members of 1st squad), and everyone cheered at 2nd squad's mysterious and lucky Bradley kill shot. The average person in my platoon, including myself, had 2 kills, lasted about an hour and a half, and fired under 200 rounds. This was completely offset by the fact that Tim had killed forty-eight soldiers (including 3 officers), survived 5 hours 3 minutes, and fired 3139 rounds from his own M-16. The other M-16's that he used were recorded as being used by the Army, and for some reason, the computer could sort out his kills, but not his rounds fired.

While we were going over the report, Col. Ghetty stopped in to see what all the noise was about. I showed him the report, and, when he asked who Tim was, I introduced him. Ghetty had Tim tell the tale about his adventure in the Bradley, and he laughed along with everyone. Then, he brought us back to reality by giving us a long speech about how the exercise is designed so that we "train the way we fight, and fight the way we train." The exercises are not supposed to be games, but rather learning experiences. He's right to assume that Tim and the rest of us didn't act the way we will if real bullets are flying. Tim was visibly sobered by the idea of how many bullets would have been thrown his way if it was a real fight. I think his pride was brought down to flat Earth, and he will probably not be as gung ho in the future.

Ghetty could feel the silence in the room more than hear it, and he was sure to take a step back and congratulate us. He said he wished that he could have seen the look on anyone of the Army's armored troops when they found that they couldn't get close to Tim's bridge. Now the incident is burned in everyone's mind as the Battle of Tim's Bridge.

When Col. Ghetty left, he motioned for me to follow him into the corridor. I followed him all the way to the Combat Information Center (CIC) while he gave me a lecture about different training programs that the platoon was perfect for. The only program that would keep us in our current company, Capt. Brown's company, is TRAP training. Training programs are a great way to get away from Capt. Brown, but I would feel like I was running from a problem, rather than dealing with a challenge. Ghetty agreed, and after our exercises with the Army were complete, we're going to be bumped to the top of the rotation. If all goes well, we'll be rescuing pilots and recovering planes by spring of 2011. Ghetty dismissed me and told me to keep the platoon ready in case a TRAP opportunity arises during the exercises. If one does come up, he would include at least part of our platoon.

I stopped by the dining area to get a cup of coffee before going back to my quarters, and bumped into 1st squad. We all had lunch together. They're in great spirits, and I feel great knowing that they've accepted me.

Thursday, September 16, 2010, 9:30am

Outside Toronto, Canada

Global Technology Incorporated-Mississauga Branch

The office's AT&T receptionist phone rang incessantly. While the receptionist was wandering the office, the Accounting Secretary manned the phones. She'd never acknowledge it, but her attitude was that she had her own job to do, and it wasn't really her responsibility. The small twenty five-person office proliferated good company morale, in general. The receptionist, fresh from a temporary employment service, still bore the less than exemplary work ethic that is notorious for such agencies. Waiting in the 20' x 25' meeting room, Walt Barber left to get a fresh cup of coffee from the kitchen. Dan Greene, his shadow, followed. On the way, as Walt glanced at the empty receptionist desk, Dan noticed his mentor's disdain and gave the accounting secretary the professional, polite, equivalent of a military dressing down. When they returned from the kitchen, on their way back to the meeting room, Walt noticed that the receptionist was gathering her things for her unplanned and permanent trip home. The attitude-laden accounting secretary was finally manning the phone. Everything had to be perfect for today's meeting.

Back in the meeting room, Walt and Dan sat down again. The branch office Operations Manager continued to come and go to take phone calls from clients. The GTI Regional Manager, the Corporate Research Manager, a Corporate Military Accounts Manager, and three Programmers remained. They made small talk about company history and events. They talked about the weather, sports, and computer programs. The only thing that they didn't talk about was current events.

Finally, after the common topical subjects wore out between the different corporate lifestyle types, there was silence. Everyone wanted to be professional in front of their superiors, and relaxed in front of their peers. Walt, truly the most senior person there, broke the silence.

"I'd like to get a few things straight before we talk about bonuses and the like. It's my understanding that the SP 2.0 software is complete and ready for beta testing. Is that correct?"

Everyone answered yes.

Walt had a question for the Regional Manager. "Okay then. How are the beta test sites to be determined?"

The Regional Manager turned and looked toward the Corporate Research Manager for an answer. The Corporate Research Manager turned and looked at the Corporate Military Accounts Manager who in turn turned and looked the programmers. One of the programmers spoke up.

"We're just now finishing up the SP 2.0 beta version. The subject of how those beta sites are chosen hasn't really come up. I don't think anyone's really championed that yet."

168

This was Dan's queue, and he followed up Walt's question.

"What kind of criteria are you looking for in a beta site?"

Each GTI employee had a comment. After the first three, Dan rose to the dry erase board and began writing them down. In essence, what they really needed was a military organization that could mimic the U.S. military in all ways. They spent the rest of the day discussing ways in which the various communication elements of the SP 2.0 software needed to be tested. Basic source to target methods of communication were listed out first, then the list of beta test operations grew until a level of extreme complexity was established. They had sandwiches brought in for lunch, pizza for dinner, and at almost 2:00am the next day, the coffee finally ran out. Minutes later the office lights were turned off, and everyone was told to go home and get some rest. Walt and Dan would have to leave in twelve hours, and no one wanted to continue the meeting through a speakerphone.

Wednesday, November 24, 2010, 1033 hours Nellis Air Force Base, Nevada

Joint Task Force Training Exercise 2010; a.k.a.-JTFTE-04

20,000 Feet above the orange and beige Nevada desert, a pair of aging F111 Aardvark tactical bombers-the last to be retired-cruised at a leisurely 450 knots in tight formation. The month-long exercise would be over after this final mission. Al Meyers and his Weapons Officer Phil Sidney had flown together thousands of times over the past two decades, but this was their last flight. When the exercise was over, the two were scheduled for individual reassignment. They were paired together back in the late 1980's, fresh out of flight school. Since then, both had received individual training periodically, but this time it would be permanent, and they knew it. Neither had much to say as each wandered through their memories of times together. They could do their jobs in their sleep, and sometimes, they had! The numbing drone of the planes engines, wind whistling into the cockpit through microscopic openings, radio garble, and instrument beeping all had become comforting and even boring.

Through the radio garble, they were instantly brought to complete attention when their wingman called.

"Wolfman, traffic at 9 o'clock, heading away. Check that. We've been made. Looks like two F-22 Raptors. They're turning right! Right turn! They're coming around hard!"

Part of Al wanted to laugh at the young 28-year-old in the other plane. The excitement was still there, but, Al kept it balanced with the reality of the situation. This was, after all, only an exercise. He answered his wingman.

"Copy, Maverick. Follow us down. It's time to get low and fast."

Both planes dove for the ground. Their wings rotated backward into a delta wing shape for easier flight at high speed. Al fired his afterburners. Their white flames stretched as far back as the length of the plane. The second F-111 followed his mentor's example. As their speed approached 1000knots, they began their pullout.

Only seconds ago they were five miles above the earth, now they were just over 500 feet. Everything was a blur except for a clear field of vision roughly 20 degrees wide, looking straight ahead. Al had gotten used to tuning his peripheral vision to watch the multifunction display (MFD) monitor that showed a radar image of the mountains ahead. A small computer plotted a suggested course and showed it on top of the images of mountains. As long as they stayed on the computer's course, all would be perfectly safe. It didn't calm Al and Phil knowing that their lives were in the hands of some programmer who might have been surfing the web years ago while writing the software, but the computer was usually right, and they were used to its idiosyncrasies. For Maverick and his Weapons Officer it was still the roller coaster ride from hell. Of course they'd brag about it later, but until then, their nomex flight gloves were filling with palm sweat.

The only consolation that Al's wingman had was that the pseudo enemy F-22 Raptors were no where to be seen. Experience had shown Al that there was no reason to take a chance by slowing down, so he kept the throttles wide open. Their target, a small white tractor trailer truck at the mouth of a valley, was almost a hundred miles away when they spotted the Raptors. As soon as the enemy planes were spotted, Phil glued his eyes on the navigation display of his MFDs. Now that they were comfortably racing over, between, and through mountains, Phil thought it was a good time to bring Al's attention closer to the target.

"Target at 15 miles bearing 005."

Al didn't even answer. He turned the plane until his heads up display showed that he was right on course. He had a choice to make. The low and fast bombing doctrine typically had the attacking plane fly over the target at a set speed to allow for the most accurate bomb drop possible without exposing the aircraft to much anti-aircraft fire from the target area. Al couldn't see the Raptors, but he knew that those pilots were of a certain performance level. They had to be good enough to qualify for a slot in a Raptor, and become a good enough Raptor pilot to qualify for this exercise. They weren't as green as... say...Maverick. With the target only 15 miles away, he rewrote his attack plan in a split second. He had to or they would fly over the target a few seconds later.

Before Maverick could react, Al chopped the power to his engine to almost nothing, popped out the airbrake, and lifted the nose of the plane 15 degrees up. The plane's wings rotated forward for better slow speed control. Their altitude climbed to 3000 feet, and their airspeed dropped to 200 knots. Maverick shot by faster than a bullet. Phil had worked with Al for so long that even though he didn't predict Al's maneuver, he wasn't surprised by it. He was ready. Phil ejected ten flares to throw off the Raptors' attempt to lock on with a heat-seeking missile emulator. Then, in the same motion, he ejected four chaff bundles to confuse any radar-guided-missile emulator. While Al leveled the plane out, Phil opened up the

Forward Looking Infrared (FLIR) targeting pod and fired an invisible tracking laser at the white truck. Al was still silent, but Phil went through the verbal checklist for their attack.

"Target acquired. Target lazed. Sweet lock. The pickle is hot."

A second later, Phil gave the immortal "Bombs away" report. The plane lightened as forty-two 750-pound bombs headed for the truck. With the plane lighter by thousands of pounds, Al had no trouble getting it back up to speed. Phil ejected more chaff and flares, and they made their way towards home.

Al was curious what happened to his wingman. Since it was their final mission, he decided to have some fun with the fighter pilots that were hopefully still chasing him.

He asked Phil for rhetorical permission. Both knew that they had to end their career together with some sort of flair besides obliterating an empty tractor-trailer in the middle of the desert. For years, Phil had known Al to be a level-headed, by the numbers pilot. Every once in a while the hot dog inside showed through, and he expected that it would on this mission.

"Since this is it for us, let's try something a little unusual."

"Works for me. What'd you have in mind?"

Al grinned through his oxygen mask. His eyes showed it, and Phil understood. He reached for something to hold onto.

"These fighter pukes have been chasing us down and making life miserable. I want to turn the tables. Let's see if we can get behind them and get some FLIR footage of their asses!"

By now the plane was over ten thousand feet in the air and flying at 850 knots. As soon as Al saw Phil reaching to hold on, he rolled the big plane on its back and looked out the top towards the ground. Right away he spotted the two F-22's chasing Maverick. The lucky kid was still alive! Al pulled back on the stick and the plane went into an inverted 45-degree dive. Halfway to the ground, he rolled it back upright and turned maneuver to get behind the overly-focused Raptors. The plane passed 1100 knots, and he eased back the throttle. It was too simple. He moved the big 30 year old plane until it was low and behind both 21st-century high-tech wonders.

Phil let go of his handholds and turned. He opened the forward bomb bay and the targeting pod rotated out again. The FLIR automatically aimed directly forward and began recording. Al and Phil remained undetected. After a few seconds, Maverick's luck ran out. He was still on the same frequency as Al and Phil, so when it happened they heard it at the same time he did. The AWAC's control plane monitoring the air battle and its invisible virtual missiles came on the radio.

"Maverick, this is Big Man. You are killed. Repeat you are killed. Climb to angles three five (35,000 feet) and RTB (return to base). Copy?"

Al and Phil openly laughed when they heard Maverick's voice answer. He wasn't a good loser.

"Copy, Big Man, climbing to angles three five and RTB. Wolfman, we are history. Thank you very much!"

Maverick's plane slowed from a ballistic speed and casually climbed while turning towards Nellis. The F-22's, on the other hand, dramatically climbed at nearly 60 degrees and began their search for more prey. All the while, Al and Phil followed behind and below both. They laughed all the way. As the F-22's began a wide lazy search circle, the F-111's FLIR pod continued filming.

Suddenly the Raptor's split up, one turning left, the other right. The one on the right dove and continued its turn while the one on the left accelerated. Al and Phil had finally been spotted, and now they were being set up. Whichever plane they chased, the other would come after them. Al rolled the big F-111 on its back again. Again, he kicked in the afterburners, swept the wings back, pulled back on the stick, and headed for the ground. Instead of rolling out of their inverted dive, Al pulled the plane all the way around in a half loop. Phil was unaccustomed to the high G forces, but low level highspeed flight was far harder on a passenger's stomach so he held his breakfast well.

The Aardvark leveled out over the open desert at 100 feet, possibly lower, but that would have required a more careful look at the altimeter. The F-22's both circled and took aim on the big plane, but it was so close to the ground that simple things like cacti were confusing their radars. They came lower in an attempt to get a clearer angle on the F-111's rear end, but Al had the plane's engines working as hard as possible. The F-22's with their hightech supercruise, fuel-efficient engines were not quite fast enough to catch up to the decades-old Aardvark and its big ugly gas-guzzlers. Al's dive had given them just enough of a lead that they couldn't get the edge. The chase lasted only a minute and a half, during which time, the F-111 made periodic jerking motions to one side or the other. The pilots of the F-22's thought the plane was dodging cacti. What they didn't realize was that Phil was giving Al minor course corrections that lead directly back to Nellis. Without asking for permission, and with complete disregard for all safety regulations at the airbase, Al flew the F-111 over the main runway at almost 1100 knots. The two F-22's, upon seeing the airbase, immediately broke off pursuit. They tried to avoid flying over it without clearance, but they had too much momentum, and both inadvertently buzzed the field. With their pursuers disengaged, Wolfman slowed down to 200 knots, requested landing clearance, and brought Al and Phil back to terra firma, Nevada Style.

By the time they were eventually detected and shot down, Al and Phil had accumulated seven minutes of film showing that the F-22's would have been killed if their F-111 had a gun or missile emulator on board. It was only their coup de tat against the Air Force's hottest plane by one of its oldest that protected them from disciplinary action for their flagrant protocol and regulation violations. The F-22 pilots did get dressed down for buzzing the field and being "shot down" by two old men in what was supposed to be an old hog of a plane. The lesser offense was drawing the more severe disciplinary action. What actions? For their safety violations and buzzing the airbase, the two pilots had to volunteer to clean up after the mess that the Marines recently left on several of the planes on the field. For their loss to Wolfman, the two hot dogs had to buy the drinks for Al, Phil, and their entire squadron, including Maverick, at the officer's club. It was a much more gentle way of losing than it would have been had the engagement been real combat.

Friday, January 21, 2011, 3:18pm

Isle of Grenada, T.K.'s Tequila Kafe'

Dan Greene watched puffy white clouds drift overhead. A gentle offshore breeze from the Gulf of Mexico side of the island moved palm leaves on trees as though they were in slow motion. The streets were empty on the sleepy island. The only man-made noise came from a bartender hanging margarita glasses and a young local washing dishes in the back of the cafe. Walt and Dan sat at their roadside table under a bright blue and red umbrella.

Both still wore the same gray suits that they wore on the morning's plane ride in. Neither had even loosened a tie.

After a few enjoyable minutes of silence, the bartender brought over a fresh round of margaritas. Walt thanked the man and asked if they could keep the tab running. They were expecting company. There was no problem.

Up the narrow and winding street leading down to the beach, a dozen oriental men came walking up. Each wore tourist shorts, souvenir T-shirts, Grenada baseball caps or sun visors, sunglasses, dark socks and custom tailormade dress shoes. Walt greeted them with a genuine smile that masked his laughter. Walt knew each man, they had all met years ago. He had spoken with each over the phone many times since then. They saw each other at state events. Some he had even met with privately.

They conversed in Chinese, the common language for all. Meanwhile, the bartender laughed at the Chinese with their dark socks. Without taking a single order, he began passing out triple margaritas of every color and flavor in the bar. Each one had the ever-important salted rim, souvenir stirrer and wooden umbrella. Walt's guests were very pleased.

Present at the apparently informal meeting were the same people Walt had gathered together fourteen years earlier. Representatives from the People's Republic of China included: General Sadis of the People's Army; General Ahr Phuhn of the People's Air Force; and General Naht Synn, Special General for

173

Internal Security and Counter Espionage. Representatives from the Democratic People's Republic of North Korea included: General Soo Pi of the Air Force and Admiral Tong Yoo of the Special Naval Operations Branch. It was the first time that they had all been together since that first meeting. Since then, all of their contact and communication had been through Dan and then Walt.

Walt and Dan removed their jackets when they were done greeting everyone. After loosening up his tie, and proposing a few toasts, Walt got down to the reason for their celebration.

"Gentlemen, it's been a long time coming."

They all happily agreed. The margaritas were taking hold.

"I am pleased to announce that the SP 2.0 software is finally complete. It is being delivered to the Pentagon for implementation next week. You will, of course, have your copies now."

Walt motioned to Dan who opened his briefcase and began passing out manila envelopes to each man. Each envelope contained six CD-ROMs with the SP 2.0 program and the battle management and data communication encryption/decipher buffer utility supplement. There was also a pamphlet on how to install the system on a PC and connect the PC to any military radio with a standard RS-232 port connection cable.

While they were passed out, Walt finished his margarita and with eye contact alone motioned each of the others to do the same. While they did so, he motioned to the bartender for another round. They were already prepped and ready. This time it wasn't because of his influence, but because the bartender knew his job well. As the tray approached, Walt put his laptop computer on the table and turned it on.

"Who will be the first to finish the transaction?"

Chinese General Naht Synn, Special General for Internal Security and Counter Espionage, handed Walt a slip of paper with a bank account routing number on it. General Soo Pi of the North Korean Air Force did the same. Walt entered the numbers into his computer. He made another toast, and waited. While they were drinking, his computer's sound card played the "chaching" sound of a cash register to indicate that the deposits were complete. Then Walt typed in a few new numbers and the money was transferred from one of his slush accounts into the accounts of several companies that he owned controlling interests in. From there, some of it was moved into his different investment portfolios. More still was sent back into some of his offshore bank accounts. Each time a transaction was completed, the computer sounded chaching, and the liquored-up crowd cheered, then drank.

It was a profitable day for Walt, but would they use the software to balance power in the world, or would they use it for their own personal profit, as he had? He felt no remorse. There was only the realization that for the first time in memory, he had done something for which there was no turning back, and there was no contingency plan possible. The closest to an emotion that he felt was the taste of irony.

A generation earlier, Walt's father had helped other men gain an overwhelming technological advantage in military intelligence. His father, Joe Barber, worked on the U.S. Army team that helped to decipher German and Japanese codes in World War II. Now Walt was handing an even bigger intelligence-gathering piece of technology over to another group of men; people who were allegedly just as concerned about world peace and the equality of man.

Of course, Walt didn't care too much about that. He was in it for the money. If there was a big deal going on anywhere in the world, Walt was always a part of it.

SECTION 2

"Warfare has always erupted at the wrong time, the wrong place, with the wrong enemy, and without enough preparation-at least from one side's perspective!"

Saturday, January 22, 2011, 5:13 PM

Andersen Air Force Base, Guam

High above the Korean peninsula, a multi-million dollar U.S. KH-11 spy satellite slipped through the vacuum of space over the DeMilitarized Zone between North and South Korea.

For years, the satellite had flown over the DMZ. Occasionally it had been sent to deviate from its course and collect intelligence data on other political hot spots. As was the case with all spy satellites, it had been delegated to monitor a certain area of the world. Its particular area of interest included Eastern Siberia, nations bordering the Sea of Japan, eastern China, Indochina, the south Indian Ocean, and central Africa.

The DMZ had been one of the Pentagon's fourteen official world trouble spots since World War II. For nearly three weeks, a low-pressure system had obscured the area. Even with its state of the art infrared sensors, radar mapping, thermal imaging, and computer enhancing electronics, the satellite was unable to see details through the thick clouds from its high orbit.

Thousands of miles away, an American intelligence officer made several calls to his commanding officers and to the commanders of various other air units. Normally, if the area had been politically quiet, American intelligence assets would wait until the weather broke, but increasing numbers of student riots in South Korea made a spy plane mission a viable concept. It was never a secret that North Korean agents usually instigated such riots. Most officers also remembered that history's greatest military upsets on the battlefield have traditionally taken place under a cloak of clouds, examples of which are everywhere: Austerlitz, D-Day, The Battle of the Bulge, and the entry of Chinese forces into the Korean War in 1952.

A few phone calls later, Al "Wolfman" Meyers and Phil "Blaster" Sidney boarded their F-117C reconnaissance plane in a closed hangar at Andersen Air Force Base. A handful of crewmen disconnected the fuel and oxygen hoses leading to the aircraft, and the blast-proof hangar doors opened revealing the setting sun behind the mountains to the west. He taxied the plane to the end of the runway, and asked the tower for clearance.

"Tower, this is Wolfman One-One requesting takeoff."

A ready response from the tower followed, "Wolfman One-One, you are clear for takeoff. Wind is 7 knots from west-southwest. Your tanker is at

angels 30 bearing 180 your current pos. Copy?"

Al answered and pushed the throttle forward with his left hand.

The engines whined. At first they seemed more like expensive noisemakers and not the plane's powerplants. The engines put through a steady and growing amount of thrust. His F-117C crept forward like an idling car. When he felt that he had achieved full thrust, Al finally let go of the landing gear's brakes. Slowly, he began to move down the runway. Eventually, when the plane reached 150 knots, he pulled back on the flightstick in his right hand. The plane's nose lifted off the ground and pointed toward the sky. Thick air rushed along the bottom of the plane and bounced it into the air. Al pressed a button and the landing gear was retracted into the fuselage. A loud thump and three green lights turned to red on one of his cockpit displays indicating that the gear was up and secure.

The landing gear was one of the few improvements that distinguished the F-117C from its predecessors, the F-117A and the F-117B. Pilots learned to fly F-117A and C versions in the two seat training version, the F-117B. Unlike the A model, which was equipped with the same relatively fragile landing gear found on F-15 fighter planes, the C model had the same heavy duty landing gear as Navy F/A-18 Hornets to enable its use from aircraft carriers. Al's plane was the longest of the three versions, having been doubled in length to accommodate more fuel, weapons, and avionics. The C version also kept the two seats from the B version, but instead of a flight instructor riding in the back, a weapons officer sat behind the pilot to operate the various armaments stored in the four weapons bays. Another enhancement of the F-117 was the replacement of the engines with the same afterburning, thrust-vectoring engines found on the F-22 Raptor. The most significant difference between C version and previous models was that the original angular design had been smoothed out to increase the absorption and deflection of radar waves, thus decreasing the potential of enemy detection.

Officially the weapons officer was designated as the WO. Most USAF personnel referred to them as "Wizzo's." Al's Wizzo, Phillip "Blaster" Sidney, quietly sat in the back seat during take-off. He passed his time by assessing the payload in each of the four weapons bays. The two bays in the back of the plane, called three and four, each held a large HARM anti-radar missile. The bay on the forward right of the plane held an M61 Vulcan autocannon capable of firing hundreds of 20mm cannon shells every second. Bay number one held the hammer with which they would try to nail down the operation, a complex photo and electronic surveillance pod.

Once he was confidant that all the systems were at the ready, Phil asked Al about the mission.

"So Wolfman, you think we'll see anything unusual this time?"

"I dunno," answered Al, "They've got a lot of SAM's (surface-to-air missiles) and triple-A (anti-aircraft artillery). If everything goes right, it should be a bore. I've been there before, you know?"

"No. I didn't."

Al and Phil had worked together for years. Their first combat assignment together had been during the Gulf War almost 15 years earlier. Since then, they had only been separated for a few weeks at a time whenever either found the opportunity for specialized training. Last year they parted company for what each thought would be their last flight. Congressional desires for a smaller, more finely tuned and cost effective military pulled them back together only a few weeks before for this mission. Phil had no idea that one of their separations included a brief tour for Al flying spy planes. He was almost in shock at the secret, but his voice remained nonchalant, as though he had known all of the time.

"What was it like?"

"They spotted us way out in the Sea of Japan from a picket line of fishing boats, then they must've scrambled fighters from every airfield in the country. There were planes all over the place. The nice thing was that since there were so many of their guys up there, they didn't fire any missiles or guns at us. I'm sure they tracked us, and were locked on a couple times, but that was only for a little while. They never had time for a good shot. That was back in an RF-4 Phantom. It was about as stealthy as a freight train. We'll see what it's like this time, though. I'm sure they've learned their lesson since then. They keep their fighters on the ground or above a set altitude to keep them distinguished from us. It's been quiet there for so long, I'm hoping for a relatively easy run. They might track us and maybe even lock on to us, but I doubt there'll be any shooting. One thing's for sure though, if it does get real tough, you can bet they're hiding something." Phil was curious.

"You think they'll ever try and invade the South again? It seems like everybody wants a peaceful reunification." Al wasn't sure.

"Who knows? You can be sure that those riots we've been hearing about didn't just start with some college kids and a few beers. They get started by the North Koreans sending down agents. It's not an uncommon thing though. Hell, I'm sure the some of our agencies do it too."

Phil remembered organized protests against the United States, and how they usually preceded some sort of U.S. intervention in the Middle-East. If they did find something over North Korea, the U.S. would be in for an interesting political and possibly a military shootout with the North Koreans. The idea was too sobering and he tried to think of other ways to pass the time. They would be in the air for a few more minutes before they met up with their mid-air refueling plane. He began to make small talk with Al on all sorts of things ranging from sports, their wives, their children, to the latest gossip at their home base in Nevada where their Colonel was arrested for drunk and disorderly during last Sunday's church services.

Still over Japanese airspace, Phil noticed on the plane's infrared monitor that there was a plane ahead. It looked like their refueler. Upon closer examination, he confirmed that it was, and let Al know. Al called the tanker on the radio and their half-empty fuel tanks were filled within a few minutes. Then Al turned the plane a few degrees to the left. They headed toward the invisible map coordinates in the sky where they would meet with the various support planes for the mission.

Although Phil and Al would be the only Americans to violate North Korean airspace, they would be backed up by nearly a hundred other U.S. Air Force personnel who would be orbiting in the Sea of Japan, over South Korea, and in the Yellow Sea. Two large E-3 Sentry AWAC (Airborne Warning And Control) planes, one in each sea, would peer through the clouds, with their immense airborne search radars looking for any threatening North Korean fighter planes or missiles. An EC-130 electronic warfare plane would fly low along the south side of the DMZ to monitor and jam enemy communications if Wolfman One-One, Al and Phil's plane, was detected. While flying in formation with the EC-130, a converted Boeing 707 JSTARS plane, capable of using special ground-scanning radar, would detect, track, record, and report any moving vehicles.

Three separate flight groups, Papa One, Two, and Three, each consisting of: three F-22 fighters, two F-117A's, and a KC-135 refueling plane, would escort the planes over both seas and the group south of the DMZ. The fighters would serve as a reserve force, directed by the two E-3 Sentrys, to intercept North Korean fighters if they threatened to shoot down Wolfman One-One. The F-117A's were loaded with a variety of air-to-ground munitions that enabled them to destroy North Korean missile sites and anti- aircraft guns if Wolfman One-One called for help. Should any or all of the planes be used, the refueling planes would ensure that no one had to worry about running out of fuel over North Korea, or running out of fuel on the way back to a friendly airbase in South Korea.

At Kunsan airbase in South Korea, two dozen American pilots waited near their planes in hangars. In the event that North Korean fighters should overwhelm the F-22's, two squadrons of F-15E Strike Eagles had been flown to South Korea. Fueled and armed, they could be scrambled and in formation with the mission support planes, in less than five minutes. Another squadron of F-15E's sat idle, but these twelve planes were loaded with both air-to-air and air-to-ground munitions. Three rescue helicopters and an AC-130H gunship also waited hoping not to be called.

In the Yellow Sea, a Marine Expeditionary Unit, code named Task Force Stingray was on routine maneuvers, and, in addition to all the assets assigned to the mission, it maintained a Combat Air Patrol (CAP) of four United States Marines Corps, AV-8B Harriers. Four more waited on the deck of the U.S.S. Wasp on alert. Below decks, thousands of Marines could be called upon for larger, more dangerous rescue operations in case everything went wrong.

"You know Al," asked Phil, "we should be meeting up with Papa One about 80 miles ahead. They'll be at about 40,000 feet--angels 40. As long as

we don't need fuel we can get down low for the rendezvous." Al differed on the side of caution.

"I think we might as well top off the tanks before we go in. Besides, it'll only take a few minutes."

There was a slight increase in gravity as Al lifted the plane up to 40,000 feet. Phil spotted the small group of planes through his infrared monitor.

"There they are. One tanker, an AWACS, two fighters, and two attack planes, Papa One. They're at angels 40 and bearing 000-as planned."

Al contacted the group and requested a refueling. He pulled their plane below and behind the tanker, a converted passenger plane. The two planes connected and Wolfman One-One topped off its fuel tanks. Moments later, when the refueling was complete, Al put the plane into a steep 60 degree dive and dropped down to 400 feet above the Sea of Japan. Once there, Wolfman One-One headed towards North Korea at 594 knots.

Phil searched the horizon with his infrared camera.

The radio called, *"Wolfman One-One, this is Papa One-One [the AWAC's plane over the Sea of Japan]. I have positive radar contact with three Mig-25's bearing 020 your position, angels 20, heading straight for you. Looks like they've spotted you somehow."*

Another call from an unfamiliar voice came over the radio.

"Wolfman One-One, this is Mama One-Two [the EC-130 communications monitoring aircraft stationed south of the DMZ]. I'm confirming your detection. Radio traffic has increased at all air bases. Repeat ALL airbases. This could be a hot one. I've scrambled the reserve force just in case."

At fifteen North Korean airfields, fighters also scrambled. A North Korean submarine in the Sea of Japan had spotted an American Stealth fighter, probably on a recon mission. Three Mig-25R's [unarmed fighter planes outfitted for electronic surveillance] flying a routine patrol had detected the emissions from the American E-3 Sentry's powerful radars to the east and west of North Korea. Several more Mig-25R recon planes were taking off and heading towards other American air groups to the east and west. They were ordered to stay within North Korean airspace.

Inside Wolfman One-One, Phil watched the map of North Korea on his monitor fill with enemy air target symbols. Part of the improved avionics in the F-117C enabled his monitor to be linked by satellite to all the information from other planes in all three support groups. Phil then plotted out new courses, avoiding North Korean air units, and passed them to Al electronically. Al turned on the plane's terrain avoidance system. A system called the GPS, Global Positioning Satellite system, sent height and location data to Wolfman One-One's flight computer. The flight computer then prevented him from flying into canyon walls, buildings, or even flat ground. If he did make a mistake, the flight computer simply wouldn't send the commands from his flight stick to the control surfaces of the plane. Previous terrain avoidance systems used a combination of radar and prewritten computer programs to detect contours of the ground and then make adjustments, but those radars acted similar to a flashlight in the darkness to anyone looking for an intruder with radar detection gear. By flying low, planes could slip in under radar beams that get blocked by rough terrain or even the curvature of the Earth.

Phil's voice came on the intercom.

"Al, we're getting scanned by a search radar at Wonson. They don't have our exact position yet, but we'd better go around to keep it that way."

"Well, we're here to get pictures of the DMZ, let's try and go south of Wonson. That'll put us between the border and the search radar."

Wolfman One-One continued to weave through valleys towards the DMZ. A loud beeping sounded in back of the cockpit.

"Al, that radar at Wonson has found us. I'm gonna jam it, but that'll only last for a while, and it'll mess up everyone's radars, including all our friendlies."

Al clicked his intercom button twice to signal his acknowledgement. He was too busy flying to make conversation. Minutes later, the radio called. *"Wolfman One-One, this is Mama One-One."* It was the JSTARS plane south of the DMZ.

"Be advised, we're seeing lots of activity on the ground around the airbases. Could be trucks with SAM's moving into position. It could also be AAA setting up."

"Al," said to Phil, "We're not going to get too many pictures from the DMZ, let's mess with these guys and head back north. They won't be expecting it, and we can get snapshots of those three bases around Pyongyang. I'm sure the electronics planes have got enough info from the beehives we've shaken up already."

Al clicked his intercom mike twice again, and Phil gave him a new course heading. While the plane banked to the right, and Phil panned the DMZ with his infrared camera--recording as much as he could.

He turned the camera forward and saw his first objective.

"Nampo airbase bearing 000 distance 55 kilometers. I'm opening bay

four."

The weapons bay on the forward left swung open and the sleek surveillance pod began recording data of all sorts. Still less than 500 feet above the ground, Al flew down one of the runways and then banked to the right turning directly north. North Korean Su-27 Flanker fighter planes were taxiing below.

Phil retracted the surveillance pod and closed the bay door. Looking forward with his infrared camera he spotted their next recon location. "Pyongyang, 15 kilometers bearing 353."

Al answered instead of clicking his mike.

"Got it. Turning to 353."

Even though it was unusual to have been so heavily detected by so many units, they were still able to fly directly over military installations and not have a single bullet fired at them. Al was feeling a little more secure with the situation.

If they wanted to shoot him down, they would have tried by now. The radio came to life with another unfamiliar voice.

"Wolfman One-One, this is Granddad One-One."

It was the AWAC's plane stationed over the east coast of North Korea.

"Be advised. There's a flight of 6, Su-27's from Nampo at your bearing 165. It looks like they're using ground radar to follow you. Mama One-Two is trying to jam their communications. Would you like the cavalry now?"

Since the North Koreans knew where he was, Al broke radio silence and answered, "Granddad One-One this is Wolfman One-One. If they get inside 20 kilometers let me know and I'll turn toward your position. They'd be lucky to get a good missile lock on us, but we can't hide from their cannons. Have the cavalry ready."

The controller on the second AWAC's called Al.

"Copy, Wolfman. Cavalry is standing by at angels 1. Your bogies are at 25 kilometers."

Al and Phil both silently hoped that the radar and infrared stealth technologies which had been built into the F-117C would prevent the North Korean planes from being able to accurately fire any heat-seeking or radarguided anti-aircraft missiles at them. If their plane's design failed, then the communist missiles would already have been approaching. They were well inside even short-range missile ranges.

Phil opened bays one, two, and three. The plane's surveillance pod, its M61 autocannon, and a HARM missile lowered into position. As they approached the North Korean capital, the pod began sucking in more data. Phil looked at his monitor and brought up a map of the current situation. All at once, five search radars, several SAM radars, and dozens of radar controlled flak guns turned on. Alarms, buzzers, and lights went off in the cockpit.

"It's a trap! They're trying to lock on to us!" cried Phil.

Al's calm voice was quick to respond.

"Okay. Sit tight. They haven't shot at us yet. They're just screwing with us. Use the HARM's and take out the first thing that does get a weapons lock on us. If they do fire, the cavalry will be here in no time."

Just then one of the flashing lights stopped flashing and stayed on. Another button on the left-hand side of both cockpits flashed red. The word LOCKON was too easy to see.

Phil selected the HARM missile he had lowered, and released it. As Al felt the plane get lighter he knew that Phil was earning his nickname, "Blaster." In a trained response to the lighter feel of the plane, Al closed his eyes so that he wouldn't be blinded by the missile's exhaust. A sensor in the nose of the HARM immediately located the position of the enemy radar which had locked-on to Wolfman One-One. The missile changed course and headed straight for the launcher's antenna. On the ground, the North Koreans turned the radar off hoping that the missile would get confused and miss its target. The emissions from the radar antenna disappeared. It was too late. The missile had saved the target's location in its microchip memory. A second later, it exploded while passing through the targeting antenna.

The Su-27's to the east, behind the American recon plane, used their infrared cameras to follow Wolfman One-One. When Phil had lowered and fired the HARM, they recorded the entire event on several different videotapes from various angles. The clarity and drama would later make any TV news network jealous.

"Wolfman One-One, this is Papa One-One. Big Brother is up and at my pos. They're ready if you need'em."

Al clicked his radio button twice as he headed towards their last objective, Sunan. Phil spotted it on his monitor, and told Al the accurate range and bearing. Only then did he have time to close the empty weapons bay and lower the remaining HARM into firing position. Several enemy weapons had locked-on them, but no one was firing. As they approached Sunan, even more radars turned on and tracked the supposedly stealth fighter.

Phil fired their last HARM at another threatening SAM radar. Seconds later it too was destroyed. A fire burned out of control where the antenna used to be. Phil closed the weapons bay.

"Wolfman One-One, this is Granddad One-One. Be advised those Su27's are at 20 kilometers. Suggest you turn to new heading 277 and meet with cavalry."

Al clicked his radio twice and turned left. The enemy fighters fired their cannons at the small black plane in the night sky. Tracer rounds raced past Wolfman One-One like six streams of shooting stars. In addition to the F-15's in the Big Brother group, two F-22 fighters that had been escorting Granddad One-One turned and headed towards the attacking enemy fighters.

By now every airbase, port, and military installation had at least four North Korean fighters orbiting above. Every search radar in North Korea was turned on and looking for the semi-elusive F-117C.

A short time later, Wolfman One-One exited North Korean air space and passed the American F-22's that were coming to its rescue. A quick dogfight erupted and Al found himself diving on an Su-27 that had followed them into South Korea. He fired a burst from his M61 cannon, and the Su-27 rolled to the right trailing smoke. Another Su-27 pulled in behind him. Al felt the plane shudder and shake as cannon shells ripped through his plane. The chase headed southeast to the safety of Kunsan airbase. Launched from almost directly ahead, an F-22's heat-seeking missile exploded a few kilometers behind their plane, damaging one of the SU-27's. All six enemy planes headed home. Only the two Su-27's had been damaged.

"Looks like we made it, Blaster. I have to admit, I was beginning to wonder. Where's the nearest friendly place to land, I think that plane on our tail might have pumped a little lead into us, and Kunsan might be too far?" There was no answer.

"Phil?"

Al turned around and saw Phil's head slumped forward over the Wizzo's control panel.

"This is Wolfman One-One declaring an emergency. Papa, give me a

vector to the nearest friendly field."

One of the AWAC's controller directed Al to a base where he landed a few minutes later. An ambulance immediately pulled Phil from the cockpit, put his limp body on a stretcher, then into the truck. It disappeared as it headed towards the base hospital. From the intakes, the exhaust port, and countless baseball sized holes in the starboard fuselage, currents of black, gray, and white smoke poured out the right side of Al's plane. Fire crews were waiting and had already begun to spray foaming fire retardant foam on the smoking plane.

Diving into the foam, a ground crewman crawled under the plane and pulled a large cartridge out of the recon pod. He ran over to a pickup truck, and dropped it in the bed. The truck raced down the runway to waiting intelligence analysts.

Al tried to be professional. He knew that in his business, people got killed. He still had questions. Why Phil? Why now? Why would Phil have to die over North Korea? Was this mission really worth it? Why did he make it back? What made him so special? Would he be so special next time? Eventually, Al's disarray turned into a need for vengeance. He wanted some payback. Whether or not his country was at war, from now on Al would be.

In a bunker on the other side of the base, a crowd of people filled a small room. The video tape and mission database was being loaded from Wolfman One-One. There was no sign of anything unusual in the North Korean positions. Critical examination of all the data would take hours, but in the end, it was a common consensus that the North Koreans were up to nothing in particular. They had, however, tried a new ambush technique on the stealth recon plane, which had almost worked. In the future they would plan better. Until then there was nothing to get really worried about.

Al rode back to the intelligence bunker once the fire was out and a security unit had taken position around the plane. When he walked inside, the bright lights temporarily blinded him. He was tired from the hours of flight and the combat. A South Korean airman handed him a desperately needed cup of coffee. There were officers from all branches of service and several countries, a handful of technicians in labcoats, and some obvious CIA types. An attractive South Korean nurse walked up from behind and told him that Phil had been stabilized. He was going to be okay! He was sure that Phil never even had a chance. He grabbed the woman by the arm, and they ran over to a truck outside. She drove him to the hospital where a few minutes later he was sitting next to his unconscious friend.

Phil had lost a lot of blood, and his face was a dull white. One of the North Korean 30mm autocannon rounds had exploded in the engine of their stealth fighter, sending a half-inch fragment through his ejection seat. Phil's lower left abdomen was punctured and scrambled inside. The fast response by South Korean aircrews and the air base's medical facility made it possible for Phil to hold onto life until his flesh could be stitched back together. He had made it. In a few months he could try and get back into flight operations.

Al's relief was immeasurable. Still, he wondered, was this worth it? He kept asking himself. There was no way to tell. His superiors would probably never tell him the results of the mission. Truly, time would tell.

Phil was not the only casualty of the night. During their escape from North Korean airspace, a pair of Jian-7's patrolling the southwest coast of North Korea, tangled with the four AV-8B Super Harriers protecting Task Force Stingray.

Saturday, January 22, 2011, 10:13 PM

Task Force Stingray, U.S.S. Wasp

Somewhere in the Yellow Sea (approximately 200nm west of DMZ)

Lying in his bed, Rear Admiral Samuel Pender tried again to sleep. While most people his age had retired years ago, he had done the impossible. He had remained in his vocation, the U.S. Navy.

His career began after graduating from the U.S. Naval academy in 1969, albeit in the lower half of his class. From the academy, he went almost directly to Vietnam to serve as the senior commander on a river monitor patrol boat. After three tours of duty in Southeast Asia, ending because of an unfortunate incident with a South Vietnamese Officer, he was sent to the North Atlantic to serve as a logistics officer on a Destroyer. Years of patrol in the North Atlantic lead to a uneventful duty assignment on a fleet oiler. By 1995, he had managed to land another uneventful duty assignment, this time on a Landing Ship Dock (LSD) as liaison officer for the Marine contingent onboard.

Around this time, other people his age began retiring, but, for the first time since his Vietnam assignment, his career began to get fulfilling again. One of Africa's innumerable revolutions had trapped hundreds of Americans in yet another African capital city while anarchy raged all around them. Marines from Sam's ship were airlifted by helicopter into the city to evacuate the American civilians. While most of the evacuation flights went flawlessly, there was trouble on one.

A storm had crept up around the small Task Force in the central Atlantic, and a flight of CH-46 helicopters was not able to land on the ships during the storm's peak. Instead, they landed outside a small factory near the coast, and waited out the storm. Rumors conveyed by the local population indicated that a rebel force was marauding in the area. The small group of civilians was only protected by the helicopter crews. Under orders from above, a small Marine amphibious group left Sam's ship to secure a beachhead around the encircled helicopter force near the coast.

A World War II-era barge type landing craft set out for the beach, but high seas began coming over the side. Soon, the landing craft was in serious trouble. While the ship's Captain searched for ideas on how to recover the troubled sailors and Marines, Sam gathered a skeleton crew for one of the rescue landing craft. He had some of the Marines on board help him rig tents over the deck, as they had done on his riverboat in Vietnam to keep VC grenades out. This time, they would keep the waves off, for a while at least. Sam sent word to the bridge that he could recover the men and landing craft. Before approval was given, Sam and his band of volunteers set off into the 20foot seas. They made it all the way to the other landing craft. Then their tent roof finally collapsed under the weight of the waves. Sam had a line passed to the other landing craft, and, using his boat to pull the flooding landing craft, the two made it to the beach.

A few hours later, the storm passed over. The flooded landing craft was pumped out and repaired. The Marines had a security perimeter established, and eventually everyone was returned to the LSD safely, except Sam who was reprimanded by the Captain for leaving with the remaining landing craft before getting authorization. The reprimand was swift and stern, and the ship's Captain promised to continue the reprimand once the civilians, Marines, and landing craft were all stowed away. It was never finished.

While the landing craft was being secured into its mooring, some sailors came running onto the deck. After a few indiscernible words, they left with the sailors that had been securing the small landing craft. One by one and in small groups, the Marines and remaining sailors left too. Everyone was in a hurry. When there were only a few sailors left with him, Sam finally overheard what the commotion was about.

While they had been risking their lives in the Atlantic, on the east coast of Africa, in Somalia, some U.S. Army Rangers had run into trouble. Sam and the remaining sailors quickly finished securing the landing craft, then they raced below deck to listen to the radio and watch the saga unfold on the ship's TV.

The Rangers were part of an allegedly international peacekeeping force that was trying to round up some Somali warlords. One of the helicopters was shot down, and the team that had gone in to round up the warlords found itself trapped in the heart of Mogadishu, a city in a state of nearly complete anarchy at the time. A rescue force was quickly put together, and soon a convoy was snaking its way through the maze of streets trying to get to the helicopter crew and the trapped Rangers. Thousands of Somalis, attracted by the shooting, came to see what was going on. Anyone that showed up with a gun in their hands was shot at by the Americans. That was all it took for the populace to rise up against the Americans. The next day, when it was all over, the U.S. Army had almost a hundred soldiers wounded or killed. No one would ever know how many Somali fell to the Special Forces guns, but reports from Red Cross and other relief agencies put the estimate between 600 and 1000 people killed, possibly three times as many wounded.

The crew watched and listened to the ordeal while it was happening. Everyone knew sailors who were off the coast of Mogadishu, or Marines that were on the ground. There was great concern, not of a battle lost, or honor faded, but of friends in harm's way. The Somali "Humanitarian Effort" was quite the media spectacle when the first Marines landed on the beach. It was even more of a spectacle when the bodies of the helicopter crew were dragged naked through the dirt streets. In the background, thousands of happy Somalis cheered and sang while dancing on the wreckage of the U.S. Army Blackhawk helicopter.

All of America was disgraced. What had been a noble effort to keep the peace was now yet another example of U.S. military overconfidence. Too few commandos had been sent. There was too little helicopter gunship support. An intelligence estimate of hostiles in the area was a complete shambles. Public relations efforts to win over the hearts and minds of the people were useless and even counter-productive. There was no real contingency plan to rescue the Rangers if they became trapped. Worst of all, requests for M-1 Abrams tanks as support units were turned down by the Secretary of Defense, not because of needs elsewhere, or even the typical budget constraints. They were turned down because the career politician decided that he could assess the situation thousands of miles away better than the career military officers could. In the end, what really mattered was the result. Hot off the heels of possibly the greatest military victory in history, Operation Desert Storm, America was already falling back into its post-war pattern of overconfidence and indolence. What pride had been gained in the desert was lost in the lawless streets of a starving nation.

Sam was sickened. He had seen the whole thing before. Vietnam was more than a historical event to him. For some reason, memories of combat tend to be not burned, but carved into one's mind. For Sam, he could still feel the hot humid air and smell the..."fertilized rice paddies." Most importantly, Sam could still see the faces of the men who died before his eyes. He didn't know all of them. Most were Marines and soldiers who were just catching a ride on his river monitor. He knew some. A few were even his friends. The dead bodies on the TV brought it all home to Sam, and, like the memories of Vietnam, another memory of war was carved in his mind, in his soul. Sam didn't consciously make a vow, but his personality was further moved toward a specific resolution. Such failures would NEVER happen under his command!

Luckily for Sam, the Captain of his LSD retired a few months later. The reprimand was never continued, and, at his next promotion review, the incident was looked upon favorably instead of otherwise. Sam made Captain, and he was assigned the same LSD on which his career almost ended. A few years of life at the helm on his first command and Sam made Admiral. Of course, it was only after everyone else his age had retired.

As part of the continuous effort to trim the U.S. military budget, carrier battle groups (CVBG's) and Marine Expeditionary Units (MEU's) were forced to share one group of escort vessels, forming one single Task Force (TF). Rear Admiral Sam Pender was assigned Task Force Stingray. Task Force Stingray was comprised of: Destroyer Squadron 6 (DESRON 6), the aircraft carrier U.S.S. Kittyhawk with its support ships (CVBG Kittyhawk), an Amphibious Readiness Group (ARG) with the 15th Marine Expeditionary Unit (15th MEU); a pair of attack submarines, a cruiser, and a long list of specialized units. Almost 25,000 men and women under his command, and not because he was the most qualified, but because he was one of the Navy's few high ranking officers with combat experience beyond flight operations in the Gulf War.

Sleep had always been a hard thing to come by for Sam. He never seemed to get enough, and yet, as tired as he might be, he could never turn off his mind long enough to relax. Tonight was no different. This time, the lack of slumber was not his fault. Admiral Pender was rousted from his attempt at resting by one of his aides. There was trouble.

Wolfman One-One, an American spy plane that was flying over North Korea, had stirred up a hornet's nest of communist fighter planes. Even though Task Force Stingray had been tasked as one of the support elements for the flight, no one in Stingray had been notified. Radar operators on an E-2 Hawkeye AWAC's plane from the U.S.S. Kittyhawk detected the rapid increase in North Korean air activity, coupled with the cryptic conversations that referred to some plane called Wolfman One-One. Only then did the National Security Agency (NSA) liaison officer on the U.S.S. Kittyhawk inform anyone in TF Stingray of the mission. While the details were being presented, and the airborne situation deteriorated over North Korea, Sam was awakened.

Wearing only a T-shirt and the tiger stripe camouflage pants he always slept in, Sam raced through the maze of corridors on the U.S.S. Kittyhawk toward the Combat Information Center (CIC). Inside, the lights were turned off, and only the blue glow from computer screens illuminated the crowded room. His aide briefed him on the situation, but it was the odd looking symbols on the screens in front of him that really told Sam all he needed to know.

North Korean fighters were patrolling close to the Combat Air Patrol (CAP) assigned to protect the U.S.S. Wasp. The Marine Harrier fighters could handle the situation. They would have to; additional fighters from the carrier Kittyhawk were busy protecting that ship, and it would take too long to launch more fighters or to get them to the evolving scene near the North Korean border. He watched and counted the seconds until the North Koreans and Marine pilots encountered each other.

Over the intercom, they listened to the different conversations. Wolfman One-One was being chased out of North Korea. It was getting shot at even after it crossed into the relative safety of South Korean airspace. Along the coast, the 30+year old Jian-7 Fishbed fighters turned and headed towards the Marine AV-8B Harrier fighters. The Marine pilots, alerted by the Navy's AWAC's plane, turned to engage. Missiles were fired, and the Migs went down in flames seconds later. The busy room fell silent. Even the voices over the different radio channels went quite, stunned. Only the fighter pilots in the Harriers spoke.

"Yankee Starbase [TF Stingray Combat Information Center], this is

Wildfire One-One (the lead Harrier pilot in the CAP flight), engaged two Jian7's. Both are down. No visible chutes. We are Bingo fuel [over 2/3's empty] and request instructions."

The crowd in the CIC smiled and professionally held back their cheers. Another flight of Harriers was ordered to take off and relieve the four Harriers on CAP.

"Wildfire Flight, this is Yankee Starbase. Relief is enroute. Return to

Starbase when they arrive. Over?"

"Copy, Yankee Starbase. We'll hang out till they show up."

Unseen by anyone in Wildfire flight, six North Korean Mig-29 fighters, flying less than 300 feet above the Yellow Sea, were moving to a position almost directly underneath the Marine pilots. Thick overcast, wandering snow squalls, rough seas, and the small size of their aircraft made them almost impossible to detect on radar. When the North Korean Migs began to their climb to intercept the Marines, the American surveillance planes finally detected them. One of the aircraft was the same E-3 Sentry AWAC's plane that had guided Wolfman One-One back to South Korean airspace.

Admiral Pender waited in the shadows of the CIC on the U.S.S. Kittyhawk. Everyone in the room knew he was there, but their attention was focused on the monitors along the wall, and the radio messages coming into the ship.

"Flash! Flash! Flash! Wildfire, this is Granddad One-One. Bandits at your pos, angels 2 and climbing."

"Wildfire One-One, acknowledged. Wildfire flight, roll right, break low, maintain visual contact, and watch for bandits. Starbase, what's the e.t.a. on our backup?"

The four Harriers rolled to the right and dropped down several thousand feet in altitude while maintaining a tight formation. A blanket of overcast clouds almost immediately enveloped them, and visibility fell to a few feet. Miles below them, the Migs, using infrared detection systems instead of detectable radar methods, locked their heat-seeking missiles onto the diving Harriers. When each of the Migs had a Harrier targeted, they fired as one. From less than five miles away six missiles raced upward at the diving Marine pilots. Halfway to their targets, the North Korean pilots fired another volley. At the last second, the missiles were detected by the AWAC's plane.

"Wildfire! Missile missile missile! They're coming up from directly below!"

The four Harriers popped out flares to confuse the heat-seeking missiles, then split into two pairs. One pair turned to the left, and continued to dive, the other turned to the right, and climbed. The first wave of missiles approached. Two passed through the pairs of Marines. The proximity fuses in two other missiles were activated and both exploded in huge black puffs with shrapnel scattering everywhere. The last two missiles tracked, followed, and detonated near Wildfire One-Three; one above, and one below. The Harrier came apart, and disappeared into the clouds.

A woman's voice sounded over the radio.

"Starbase, this is Wildfire One-Four. Three has been hit. No chute seen. We could use some help out here."

Even though her voice was calm and professional, they made everyone in the CIC tense up. Sam's eternally grouchy-looking face couldn't hide his added concern that a woman pilot was in a real combat action, alone, outnumbered, and outgunned.

Conditions couldn't be much worse for a dogfight either. The Mig-29 was one of the most maneuverable fighters in the world, capable of Mach 2+ speeds and armed with a computer-controlled, infrared sensing 30mm autocannon, over a dozen missiles; All aimed using a monocle targeting system that allowed the pilot to use all the sophisticated electronics on board without ever having to look at his controls.

The Harrier, unlike the Mig-29, was not even capable of breaking Mach1. It could carry almost as many missiles, plus two 25mm autocannons, but the electronics were nowhere as sophisticated. To fire a missile, the pilot still had to do a few keystrokes of programming beforehand. The only thing in favor of the Harrier was its maneuverability. Built around one engine with four movable exhaust nozzles, the Harrier was designed to take off and land vertically like a helicopter by aiming the exhaust nozzles straight down. Once airborne, the nozzles were slowly moved back to the horizontal for normal flight. In a dogfight, the nozzles could be turned downward, while the plane is banked on its side, to turn the plane in an extremely tight radius.

The second wave of missiles began to weave invisibly through the evening clouds; their targets visible only through infrared, mechanical eyes and digital processors. Wildfire One-One and Wildfire One-Two were both hit several times. Neither pilot survived, and the remains of their aircraft spread over miles below.

"Starbase, Wildfire One-Three. One and Two are off the air. No chutes seen. I am engaged."

"Wildfire, Starbase, help is on the way. Work towards heading 180 if possible."

"Copy Starbase. Heading towards 180."

Captain Karen Black Horse, wife of a Native American Sioux named Joseph Black Horse, was fighting for her life over the Yellow Sea. The six Migs passed her as they climbed, and she continued her dive. The closing speed between the aircraft was well over 1200 knots, and the Mig pilots were unable to see her small plane through the dark and cloudy sky. They spread out and shortly found the remaining Harrier. Four more missiles were fired. Karen cleared the base of the clouds, and saw the white streaks coming after her. She popped out more flares, and turned 90 degrees to the right. The missiles went after the flares, then impacted into the sea creating splashing columns behind her plane.

She turned 90 more degrees to the right, and, as the Migs broke through the base of the clouds, she fired one of her own heat-seeking AIM-9 Sidewinder missiles. The missile's exhaust temporarily blinded her, but unlike the Russian made heat-seeking missiles that could only track targets from behind, hers was able to lock onto the minute temperature differences along the leading edges of the Migs. Before the last Mig came out of the clouds, the missile impacted the lead Mig, and a bright flash reflected off both the clouds and the water.

Two more missiles were fired at her. Again she popped out decoy flares and turned to evade the missiles. Again they missed. The Migs kept trying to get behind her and beside her while she continued to dodge their missiles. When the dogfight closed to a shorter range, the first bursts of cannon fire from two of the trailing Migs sent tracers into the clouds above and in front of her.

Karen could see the two North Koreans bracketing her. She opened up the air brake, pulled back on the stick, and moved the throttles into the vertical position. The Harrier's nose aimed skyward, then almost stopped all forward movement while losing momentum in its climb. The Mig pilots knew they could never follow her climb without passing her so they broke off and one headed east while the other went west. Unable to head west towards the middle of the Yellow Sea, south, in spite of her pursuers, to the safety of Task Force Stingray, or east towards South Korea, Karen found herself heading straight towards North Korea.

When she saw the lights along the coast come into view, she knew that she was too close, and tried to turn back toward the south. The Migs had spread out behind her, and when she turned, tracers came at her from four different places somewhere in the distance. Slowly at first, then faster and faster as the cannon rounds came closer. Her plane was hit. She heard the pounding. The engine exploded, and the entire nose of the aircraft was thrown upward with the force of a semi-tractor trailer truck. Almost immediately, the inside of the cockpit began to come apart. As flames surrounded the canopy, she pulled the ejection handle.

A rocket motor fired under her seat. The canopy was popped off with explosive bolt, and she raced into the sky. She immediately felt the intense gforces until the seat reached it apex of flight. Then gravity disappeared. She was weightless. Finally, her parachute deployed, and she began to drift down towards the sea. Hanging in the straps, her survival gear released and dropped about 50 feet where it was held in place by a thin line. When the pack of survival gear hit the salt water, a life raft popped out and inflated. Karen hit the icy water while it was inflating. A few seconds of struggling with her parachute was followed by a swim to the raft.

Time was of the essence. Even with the survival gear, hypothermia would set in shortly. She turned on her radio beacon and hoped for a rescue. With the beacon activated, she began to relax, but it was not to be. The sounds of surf in the distance reminded her of just how close to North Korea she was. There was a quick decision to make. She could either hope that a rescue team might fly out and pick her up in the next few minutes, or she could paddle towards the beach and search for shelter. There really was no decision. She paddled.

In the CIC onboard the Kittyhawk, there was silence. The entire flight had been shot down, and only one Mig was taken in exchange. Sam had a brief message sent to the Pentagon to inform Washington of the incident. There was no sign of any survivors. After the last of Wildfire flight hit the Sea, someone eventually detected Karen's emergency radio beacon. The word was passed to the U.S.S. Wasp to prepare for a possible TRAP (Tactical Retrieval of Aircraft and Personnel mission).

Everyone in the room felt the same feelings. There was frustration, worry, embarrassment, sadness, and anger at the losses. At the same time, there was also alleviation, happiness, excitement, and anxiety over the news that a rescue might be possible. No one really knew how to feel, but everyone's heart pounded.

Saturday, January 22, 2011, 11:05 am

Channel 5 News

Just as Brook and Brett were about to kiss for the first time since Brett's wife was killed by Libyan terrorists in Chicago, and while Brook's husband was in bed with her second stepmother who was suffering from amnesia after the plane crash, their soap opera was interrupted. A network logo appeared on the TV screen with a voice announcing the reason.

"We interrupt our normally scheduled program to bring you this breaking news."

"Good afternoon, I'm Tom Wauchinski. Almost four hours ago, American and North Korean fighters tangled over the Yellow Sea. At least one North Korean plane and one American plane are reported missing. We go now live to the Pentagon where the first official briefing is about to begin."

The standard light blue curtain lined the wall behind a podium with a Department of Defense seal on its front. A turbulent crowd of reporters tangled with their equipment and debated the possible revelations of the upcoming briefing. Finally, the Secretary of Defense walked through a door on the left side of the room, and the crowd settled down while everyone tried to get back to their seats and camera positions.

"Ladies and gentlemen, if I could begin?" The reporters finally became quiet.

"Just after 10:00pm local time, about 8:00am here, six Mig-29 fighter aircraft from the Democratic People's Republic of North Korea attacked four United States Marine Corps AV-8B Harrier fighter aircraft on patrol over international waters in the Yellow Sea. One of the Mig29's was shot down, and there is no indication that the pilot has survived. All four of the Harriers were also shot down. An investigation is still underway, but I must stress that this was-without a doubt-an unprovoked attack by the Democratic People's Republic of North Korea against the United States over international waters. We will keep you informed as more information becomes available. Thank you."

The Secretary began to leave the room, but the reporters shouted a hail of questions.

One of the questions was heard above the others, and the Secretary was compelled to answer.

"What squadron were the Marines from?"

"All information regarding the Marine pilots is being withheld until the families can be notified. I'm sure you understand."

The reporter shouted back, *"So, all of the American pilots were killed? Wasn't there any sign of survivors?"*

The secretary was almost out of the room, and it was obvious that this was to be the last response to any of the questions.

"All four pilots are considered missing in action. There was a suspicious radio signal, but the weather over there is very bad right now. I'm afraid that even if someone had survived, the wintry waters off the coast of North Korea right now do not favor a successful rescue. As the water temperature diminishes, so do our hopes. There is always hope though. We have helicopters searching the area, but we're also trying our best not to provoke another attack. It is possible that she might have made it into North Korea, and that's yet another reason for restraint on our behalf. Again, I will let you now more, when we know more. Thank you."

The secretary wasn't even able to turn to face the door before the reporters, as one, shouted out the shocking question.

"Are you saying that a woman was among the Marines shot down?!"

The history of the event had so preoccupied the secretary's mind that he had made a crucial leak of delicate information to the world press, on live TV. Before he could leave the room, he had to decide how to handle his mistake.

"Ladies and gentlemen, there may be an American pilot trying to stay alive in hostile waters. If there is, does it really matter what gender they are? I don't

care. You shouldn't either. What matters here is that we conduct ourselves in a professional manner. We will do whatever we can to help any survivors of this action, and we pray. I'm afraid that prayer might be our most effective course of action right now."

The chastised reporters fell silent. The secretary's gaze peered into each one's eyes, then he quickly left the room with his staff following.

"Vulture's!" He thought to himself. "How could I have let them know about that woman getting shot down? Now it'll be like the other three men never existed. How could I have been so stupid! If we don't get her back, these guys are gonna cry havoc, and old Bill Shakespeare's 'Dogs of war' might just slip out. Oh well, it's my job to handle things like this. I guess I'll have to do it better next time."

His thoughts continued to wander and process the data from the situation. Time and again, they returned to that woman pilot, floating in the cold Yellow Sea. Still, he couldn't seem to remember her name.

Saturday, January 22, 2011 12:27 PM

Somewhere in the Yellow Sea

Karen had wrapped herself in a thin silver survival blanket to help keep in some of her body heat. The water splashing over the sides of the small raft, and the water underneath still sucked away the warmth. Her adrenaline from the dogfight was almost completely gone, and now she was suffering from extreme fatigue. Paddling wasn't getting her any closer to shore than the tide already was, and she decided to save her energy instead of paddling.

It was a good choice. The current pulled her into a small inlet near a fishing village after only a few minutes. Once she was moving up the inlet, the waves settled down, and she picked a quiet stretch of bank to quietly paddle towards. The people in the village were as still and silent as the snow falling around her.

Her flightsuit was stiff and freezing. She needed shelter. Crawling up the bank, she looked into the village to make sure no one was watching. Convinced that her presence was still unknown, she ran, with her raft dragging behind her, to the nearest building, a shed less than 50 feet away. Unsure about the usefulness of her 9mm Beretta pistol, Karen was armed only with her survival knife as she slowly opened the door. Inside the shack were all kinds of primitive farming tools that wouldn't be used until spring. Most importantly, there was a pile of straw, and several empty burlap bags that had once been used for shipping rice from the U.S. as part of its humanitarian relief effort.

She silently closed the door behind her and dove into the pile of burlap bags. Once underneath, she took off her soaked flightsuit and again wrapped up with the silver thermal blanket. Even though it was thinner than a piece of paper, it was warmer than the burlap bags, and they helped her warm up in less than an hour.

She put the flightsuit on top of a layer of burlap, then lay down on it in the hopes that she could be dressed again in a few more hours.

When the flightsuit was almost dry, and she was warmed up enough to rethink her survival plan, she tried her radio beacon again. It didn't work, so she tried her radio. The batteries were weak from the cold salt water, but she did hear some faint static, so it was working.

In a soft-spoken tone, just above a whisper, she called for help. "Starbase, Starbase, this is Wildfire One-Three, over." No response.

"Starbase, Starbase, this is Wildfire One-Three, over." Again there was no response.

"Starbase, Starbase, this is Wildfire One-Three, over." This time she thought that she heard something faint through the static. Her hope and energy brought her to a state of complete attention on her task at hand.

"Starbase, Starbase, this is Wildfire One-Three, over." No response. Her hope faded. Maybe she was just hearing things.

"Starbase, Starbase, this is Wildfire One-Three, over." A dim voice did answer her this time.

"Wildfire, this is Starbase. Copy your transmission."

"Starbase, I'm alive I'm alive."

"Copy, Wildfire. You're alive. What ship were you on when you did your first assignment?"

They were checking to see if it was really her, and she knew it. "I was on the Saipan Starbase."

A brief pause on the radio indicated that whoever found her was confirming her response. When they were sure she was who she said she was, they spoke again.

"Confirmed, you are Wildfire One-Three. Repeat. Confirmed you are Wildfire One-Three. What is your situation?"

Karen was reminded that she was in North Korea, and not some farm in a safe country.

"I'm a little cold and wet, but I'm okay. I've got a good spot to hide.

Haven't seen any hostiles. All quiet here. When can you pick me up? Over." A new voice came on the radio, one more authoritative.

"Wildfire, we're gonna figure that out. We've got your position pinpointed, and they might also if they're picking this up. We'll contact you as soon as we decide how to do it. Till then, we'll be monitoring this frequency so contact us if anything changes. We'll sign off till later. Copy?"

"Copy, Starbase. I'll call if anything changes. Wildfire out."

Help would come now. Any doubts that had taken seed since she ejected were now put away. If there was one thing she knew it was that Marines don't leave Marines behind.

Sunday, January 23, 2011, 12:43 am

Somewhere in the Yellow Sea

Admiral Sam Pender stayed in the CIC on the carrier Kittyhawk. Since their contact with Karen, the place had returned to its usual organized chaos of activity. Everyone had lots to do before any rescue attempt could be made. Intelligence on the area needed to be gathered. Options needed to be created and reviewed. Most importantly, permission needed to be given.

Sam may have been the commander of Task Force, but he couldn't send a rescue team of any sort into North Korea without authorization from higher powers. Above Sam, the Commander-In-Chief of U.S. Pacific Forces (CINCPACUSF) would have to give the order. Above him, the Chairman of the Joint Chiefs of Staff, the Secretary of the Navy, the Secretary of Defense, and a host of Washington's elite would all have to pass the word down to the Task Force. The only person who could really authorize a rescue attempt-of any kind-into hostile territory would be the President himself.

Sam turned to his right to face one of his aides.

"Wait here and get all the rescue op plans together. I want to review them all, plus their backups, in no more than two hours. Make sure everyone has their supplemental and support operation requirements ready too. I wanna keep this as simple as possible, but there is no room for failure, so whichever is the most likely to succeed is the one I will recommend. I'm going for a walk. Page me when everything's ready, and I'll meet you back here."

While he was giving the instructions to his aide, several of the officers in the CIC stopped to listen too. When he left the room, his aide didn't have to repeat a word, they had all heard, and everyone continued their planning. They knew that Sam was not one for delays. They had very little time to plan an entire operation. Scenarios had of course been developed, but not every situation, including this one, could have been planned for in advance.

Typically, the Marines would fly in, secure the area, pickup the downed aircrew, and aircraft if possible, then fly out safely (a mission called TRAP Tactical Recovery of Aircraft and Personnel). This particular situation was difficult because of the potential for more direct confrontation with North Korea, but certain aspects of it were also more favorable than a typical TRAP; specifically: the weather, the pilot's condition, and the pilot's proximity to the water. It was this last element that would be the most influential in Sam's choice as the best rescue option.

A typical TRAP mission included one or two helicopters carrying a much as a full platoon of Marines for security on the ground. In the air, a pair of helicopter gunships and numerous other aircraft provided several types of air support. A handful of variations upon this structure were submitted to Sam, along with some more unconventional choices.

One of the unconventional options was to use the SEAL team deployed with the Task Force. A fifteen-man SEAL team would board one of the two submarines in the Task Force. The sub would take them close to the shore where they would deploy while still submerged. An inflatable zodiac rubber boat would take them the rest of the way to the inlet where Karen was trapped. They would rendezvous with Karen, and take the zodiac back out to a pickup location. There, the sub would surface. The Task Force would provide air cover while Karen and the SEAL's boarded the submarine and headed for safety.

The backup plan for a SEAL-focused mission would be one of the variations on the typical TRAP mission. Three V-22 Osprey-with a full platoon of Marines on board between the three of them-would leave the U.S.S. Wasp to rescue the SEAL's and Karen if things went bad. In the event that the situation was to further deteriorate, a larger mission would be sent, this time, with nine Osprey, carrying a full company of Marines. Every air asset in the Task Force would also be ready to protect the ground units from air attack, and provide close air support.

Sam, usually wary of Special Forces operations, chose the SEAL team option. He passed his suggestion, along with some of the other options, up through the chain of command. A secure fax line, with the numbers of the fax recipients already programmed into its small memory chips, enabled the message to be sent all the way through the chain of command in one single flurry of digital faxes. Hours after the messages were sent, there were finally a reply.

TO: CINC TF STINGRAY

FROM:NATIONAL COMMAND AUTHORITY, CHAIR JCS

RE:YELLOW SEA AREA OF OPERATIONS,

TRAP MISSIONS

     EXECUTE REQUESTED TRAP MISSION AT BEST DISCRETION.

ASSETS ARE AVAILABLE AS ROUTED THROUGH CINCPACUSF VIA SECURE COMMUNICATIONS METHODS.

EXCLUDED ASSETS INCLUDE R.O.K. BASED FORCES, TF SC2, AND UNITS BASED IN JAPAN.

ALL OTHER CONVENTIONAL GLOBAL ASSESTS ARE AVAILABLE AS ROUTED THROUGH NATIONAL COMMAND AUTHORITY COMMAND CENTER-WASHINGTON D.C. U.S.A.

DETERIORATION OF POLITICAL SITUATION POSSIBLE AT ANY MOMENT.

SUGGEST MISSION CONDUCTED ASAP WITHOUT DEGRADING POSSIBLITY OF SUCCESS.

FAILURE IS NOT AN OPTION.

Sam read the transmission on the monitor as it came in. The mission was a "go". As he read through it, some of the officers in the CIC were distracted from what they were doing. They watched him for a few seconds, and he could feel their attention.

Sam turned and looked at them. The room fell silent except for the radio traffic coming in from outside sources. After a few seconds of anticipation, he simply said, "Make it happen." The room immediately turned to an increased level of activity. The unit and ship commanders were contacted and instructed to prepare for action.

An attack submarine, the U.S.S. Baton Rouge, surfaced a few miles away from the U.S.S. Wasp's port side. The SEAL team was already equipped, loaded and ready to launch. When they saw the thin silhouette of the sub break the surface, the launch left the Wasp, and took them to the Baton Rouge for their ride into enemy waters.

While the SEAL team was transferring to the Baton Rouge, aircraft were being launched. First the Combat Air Patrol (CAP) around Task Force Stingray was reinforced by four more F-14 Tomcat fighter planes, bringing the total on patrol to eight. An E-2C Hawkeye AWAC's plane was controlling and monitoring all radar, communications, and electro-sensory data around the Task Force, and it too was relieved by a fresh plane with a fresh crew. V-22 Osprey on the Wasp were fueled and loaded with Marines in case the SEAL team ran into trouble and needed help. AV-8B Harriers on the Wasp were fueled and loaded for Close Air Support (CAS) operations in case the SEAL team needed major firepower.

Aircraft with less colorful duties were also prepared. S-3 Viking aircraft were loaded with extra fuel tanks and waited to be launched as mid-air refueling planes. Anti-submarine helicopters were sent out on patrol from almost every ship in the task force in order to make sure there were no submarines watching. An EA-6 Prowler aircraft was launched to jam North Korean radar and radio capabilities.

All total, almost a hundred aircraft of various types and missions were either launched or readied to support the clandestine fifteen-man SEAL team's operation. The entire process, involving thousands of crewmen and other support personnel, was complete before the Baton Rouge was in position to deploy the SEAL's.

A little over an hour before sunrise, the Baton Rouge stopped near the mouth of the inlet where Karen hid in a shack near the shore. Two at a time, the fifteen man SEAL team left the submerged submarine through its diver's escape hatch. Once outside the submarine, they stayed underwater using rebreather apparatus that cleansed their own breath of deadly carbon dioxide so that it could be reused. Unlike SCUBA gear, the rebreathers left no telltale bubbles on the surface.

While the first members waited for the rest of the team to exit the submarine, they prepared their modified Zodiac boats for the surface. They had been attached to the sub when it was on the surface and the team was boarding from the Wasp. The first men out of the sub had to unfasten the three boats. The subsequent men had to unfold the rafts so that when they were inflated underwater, they didn't rise to the surface upside down. When the last of the team was out of the submarine, the boats were filled from compressed air cylinders. The water was less than 50 degrees, and, even though visibility was near zero, the men used no lights. Slowly they inflated the boats so that they didn't burst through the surface. All three rose together. When they pulled away from the Baton Rouge, and the men let go of their lifeline to the sub, the mission had reached a point beyond recall.

While they had been underwater, the sea felt calm and tranquil. Except for the cold, and the complete darkness, the conditions might have otherwise been considered peaceful. The SEAL's were in their element in cold dark waters, and to them, it was normal. As they rose closer to the surface, the sea conditions had gotten rougher. Once on the water, their small boats were rolled on waves almost 5' high. Everyone was lying down as the boats motored into the little cove, their eyes and weapons panning the horizon for an alert enemy.

As they entered the inlet, things changed. The sea calmed, the waves were silent. The rugged shoreline became smoother and more docile. For the alert SEAL team, the quiet was unsettling; snowflakes, waves, coastline, and no sound except for the muffled hum of the motors on the three boats. They had been trained to handle the unsettled feelings. They were professionals, and they remained focused on their mission. An American pilot was counting on them.

The shack wasn't hard to find. It was right where Karen had described it. They stopped their boats and listened for signs of North Korean movement on the shore. There was none. The shack was the closest building to the shore. The rest of the village spread out further inland, but mist shrouded its size and disposition beyond about 100 meters from the shack. Even their night vision goggles couldn't make out the entire village. Only the shack and a few other buildings could been seen. With no movement around them, they tried to contact Karen.

"Wildfire One Three, this is Shooter One One. We are in position. What is your situation? Over."

Karen heard the message, and their radio replied with a whispering voice. "Shooter, this is Wildfire. I'm still in the same location, my condition is fair. Ready when you are."

"Wildfire, this is Shooter. Skip the smoke, and use your light beacon.

We're coming in wet so direct it our way. Over."

"Copy, Shooter. Wilco immediately. Over."

From their boats, the SEAL's watched as the door to the shack opened and a bright white light began flashing from the ground outside it. The light hadn't flashed twice, and the boats began moving towards the shore, slightly faster than their entrance into the inlet, but nowhere near their top speed.

At the shore, the men slipped out and took cover along the ice-covered embankment along the water's edge. Again, they surveyed the area. A few more buildings were visible, and there was still no sign of any movement or any alert North Koreans. Satisfied that they were still undetected, the team leader made eye contact with the other members of the team. As he looked in each man's eyes, he gave a series of hand signals to issue their deployment orders. Each man nodded to show that he understood, and then the team leader began to silently communicate with the next man.

When everyone had their instructions, they watched the village for a few more seconds. There was still no movement, so the team leader issued his silent go code. With his right hand still on his M4 carbine, he lifted his left into the air and moved it forward then backward twice. Five men got up from the icy water and ran to the left of the shack. Five more men did the same, but they deployed to the shack's right. When those ten men stopped running, the last five ran directly toward it. Two of them paused outside the shack, but then cautiously continued inside.

Once in, they scanned the room and finally saw Karen hiding behind a compressor behind the door they had entered. She was wet, shaking, and aiming her 9mm Beretta pistol right at them. When she identified them, she relaxed a little by moving the pistol away from them, but she still couldn't stop shaking. The two SEAL's raised their weapons in mock surrender to show her they meant no harm. She completely lowered her pistol, and they moved to help her to her feet. One of the men pulled out a fresh set of dry clothes from a backpack. They acted nonchalant as they turned away from her while she dressed, and, after only a few moments, she was warm, dry, and ready to go.

The team leader poked his head out the door and made sure it was okay to leave. A quick look around, and an affirmative nod from one of the five men who took cover on the right side of the shack, signaled that it was as safe as could be. He motioned for Karen and the other SEAL who had entered the shack, then all three ran for the shore and their boats.

She had just started getting used to the dry clothes when she slipped and went down the embankment into the water. The team leader stopped in place, waiting to see if anyone besides his team had heard the splash. There was no sound from the village. He motioned to the rest of the team, and when he began moving towards the shore, they moved too.

At the shoreline, they all crawled into their boats one by one. As each boat filled, it left the shore at a near silent creeping speed. Even though every transmission from every team member's headset was being listened to back on the Kittyhawk, the team leader spoke directly to the CIC for the first time.

"Starbase, this is Shooter One-One. Package is on board. Cold extract in progress. Over."

He hadn't expected a response, and he certainly didn't like the one he received.

"Shooter One-One, this is Starbase. Situation has changed. We have an unidentified surface contact coming down the inlet. Probably a patrol boat. Another contact is coming down the coast towards the inlet. Advise you take dry cover until they pass."

He shook his head in disbelief. They had gotten in, picked up the pilot, but just as it seemed everything was going well, the North Koreans had two patrol boats headed their way from different directions. Since there was no way to be sure that neither one had a surface search radar, they had to assume both did. The safe thing to do was to get back to the village and hide the boats.

They turned around and headed back to the village. Back at the shoreline, the team got out of their boats and dragged them up the embankment. In the middle of the channel, through the mist, they could hear one of the patrol boats coming closer. The team members returned to the exact same positions that they had taken while covering Karen's "rescue". The five men who had originally been on the left side of Karen's shack stayed with the boats.

While they were dispersing, a searchlight came one from the unseen boat in the inlet. Another one came on from the boat farther away where the inlet and sea joined. The mist, darkness, and falling snow made it impossible for the searchlight to be of any real assistance to the patrol boats, but their mere presence suggested that they were looking for something, maybe the SEAL team.

The two boats came closer. When the second boat had entered the inlet and was less than a quarter mile to the west, it stopped. The patrol boat coming down the inlet stopped about the same distance to the East. To the North was the village, to the south, the inlet. The team was pinned between the two panning searchlights. It was obvious that they had nowhere to go.

Only the searchlights moved on the two stationary boats. Finally, the boat to the east fired a flare. Then the boat to the west did the same. The predawn morning sky flickered with white light as, somewhere in the mist, the two flares floated downward. Behind the team, beyond what they could see in the village, two more flares went up.

The team leader decided to notify the Kittyhawk of the situation.

"Shooter to Starbase. We've been made, but the situation is still cool."

Inside the CIC back at the Kittyhawk, the room fell silent. Even the intercoms and radios seemed to stop. Sam's stone face was shattered. His mouth frowned. His eyes closed, and his head slowly shook from side to side in disbelief. He knew that every plan was only as good as its contingency since no plan goes flawlessly. This was different. He could feel things getting out of control already. Even though they had taken all the precautions to ensure a stealthy mission, and to get out even if things went bad, this was different somehow. It seemed like the North Koreans knew what was going on the entire time. It felt like a trap.

Sam knew that initiative, or momentum as some would call it, was the common ingredient in all victories. Whoever had it was winning. Whoever kept it won the day. The North Koreans seemed to have taken the initiative from the SEAL's somehow, and Sam wanted it back. He got the attention of several officers in the crowded room then issued the orders that he hoped would regain their control of the mission.

"Get those Harriers in there and take out those patrol boats. Make sure the team on the ground is ready to get the hell out of there when the boats go down, and have the TRAP Marines on the Wasp get airborne just in case.

We're getting those people outta Dodge. Now expedite!"

His orders were carried out immediately. A few minutes later, the SEAL's heard a loud series of explosions where the patrol boat to the west had been. While those explosions were still echoing, another series indicated that the patrol boat to the west had been attacked by the Harriers per Sam's direct order. There was no need for a hand signal, the team ran for the boats.

When they reached them, they came under heavy automatic fire from the buildings in the village to the north, and to the northwest. They returned fire, and took cover on the embankment again.

"Starbase, this is Shooter. We're under fire. Requesting immediate hot extraction. Repeat. Requesting immediate hot extraction. Over."

In the CIC on the Kittyhawk, Sam was getting even more upset. His sadness and frustration were now manifested in pure anger, and his language reflected it.

No one needed to hear his orders. While he continued swearing, the TRAP mission Marines in their V-22's were ordered to the village.

At the village, the situation was stagnant. The SEAL's could only carry so much ammunition, and the North Koreans were throwing an almost unlimited supply at them. A quick check by the team leader confirmed his suspicions. They were running low, and he had to order his men to watch their ammunition supply.

"Shooter, this is Starbase. Shortstop is enroute to your pos. ETA eleven minutes. Over."

"Starbase this is Shooter. No good. Repeat. No good. We're running low on rounds, and there's no place to land. We need....wait one Starbase."

Now that the morning had come, the mist was clearing, and the team leader could make out a large clearing to the northeast of the village; about a mile away. If they could move along the bank, they might be able to break free of their current firefight and get to the clearing. He needed a distraction.

"Starbase, this is Shooter. There's a clearing about three clicks (kilometers) to the NE. Designate as LZ Delta. Repeat. Designate clearing, three clicks to NE as LZ Delta. Our ETA will be 30 mike (minutes). We need immediate CAS (Close Air Support) on village to N and NE. Over."

"Copy, Shooter. Designate clearing, three clicks to NE as LZ Delta. ETA will be under 30 mikes. Immediate CAS on village to N and NE. Keep your heads down. It's coming in special delivery."

The Harrier pilots had been monitoring the radio transmissions, and began dropping 500lb. bombs all over the village to the north and northeast of the SEAL team's position. When the explosions began, the North Korean firing stopped, and the SEAL's ran along the inlet's bank towards the East. About a quarter mile from their original position, they moved inland towards the clearing, LZ Delta.

The North Korean fire stopped as the bombs took their toll. An entire company, over 200 troops had opened fire on the SEAL team. That company was almost completely destroyed when the first Harrier dropped its bombs. The second Harrier killed, wounded, or buried another 23 men. The remaining 14 North Korean troops, and their lightly wounded comrades, climbed out of the rubble and the shattered buildings, only to find that the SEAL team was gone. Their officers were all dead, and all of their communications equipment had been destroyed. An alert young sergeant, 18 years old, took a flare gun from his lieutenant's dead body, and, in rapid succession, began firing handheld flares into the night sky.

Meanwhile, the first Harrier was coming back towards the village to begin a series of strafing runs. A red flare came straight at the small plane. The Marine pilot, diving at over 350 knots, mistook the flare for a Surface-To-Air Missile (SAM). Trained responses acted like instincts, and the pilot reacted by dropping decoy flares, maneuvering erratically, and breaking off the air strike. It would be foolish to wait for more SAM's, so the Marines returned to the U.S.S. Wasp.

Additional Harriers, loaded with cluster bombs and antiradiation missiles would have to come back to deal with the SAM threat. Their strafing runs would hardly be of any effectiveness anyway. The SEAL's had gotten all the help they were going to get for the next half-hour.

For thirty minutes, the situation continued to deteriorate. The SEAL's, with their rescued pilot in tow, made it to the clearing. The river was at their backs. The village was on their left of the southwest. Along the perimeter of the clearing, a leafless treeline allowed them to see when the North Koreans approached. More North Koreans did come. They came by the truckload.

A North Korean battalion commander, under the strict scrutiny of his superiors, was being driven to find the SEAL team. Hundreds of troops poured into the village and surrounding area. More patrol boats were being dispatched to the area. Battalion, Regimental, and even division level artillery assets were blanketing the clearing. The SEAL's kept their heads down and continued to notify TF Stingray of the increasing opposition. They were surrounded, outgunned, and outnumbered at least 100 to 1. No Navy SEAL had ever been captured, but this was beginning to look like there might be a first in the making.

A few minutes before the Marines arrived in their V-22 Osprey, the artillery stopped. The trucks had disappeared in the village, but the troops were surely all around. The SEAL commander tried to radio the TRAP Marines, but their signals weren't getting through. Either the radio was malfunctioning, or their signal was being jammed. For whatever reason, they could no longer send signals to the Marines. They could still receive messages, and, with the ringing of the artillery barrage still in their ears, they finally heard from them.

"Shooter, this is Shortstop. Copy?"

They tried to answer, but could not be heard.

"Shortstop, this is Shooter. Abort. Abort. Abort. We've got bad guys all over the place down here."

Their reply was not what the SEAL team leader wanted to hear.

"Shooter, this is Shortstop. Your transmission in garbled. Say again."

The SEAL team leader sent his message again, but it still didn't get through. The V-22's were in sight now, and circling the clearing.

"Shooter, this is Shortstop. We are at the LZ. Your last transmission was still garbled. Pop your smoke, and we'll pick you up."

The Marine pilots were able to figure out where the clearing was, that the SEAL's were still alive, and that their radio could receive, but not send. All they wanted to know was where to land, and how dangerous the LZ was. The SEAL's could mark the exact spot to land by using colored smoke grenades. The color would also signify how dangerous LZ Delta was. Green smoke indicated a safe landing area. Yellow smoke would indicate that the enemy might be near, and red smoke would show that they were under fire. The SEAL team commander, still huddled with the rest of his team near the bank of the inlet, threw out a smoke grenade about 75ft into the clearing. A cloud of smoke mixed with the early morning fog; red smoke.

With the red smoke spread out, all hell broke loose again. The three V-22 Osprey banked to the right and descended onto the pink cloud. As their altitude dropped, the huge rotors on the end of each V-22's wing turned upward, and they began to operate like helicopters instead of airplanes. While this was happening, the North Korean artillery began hitting the LZ again. Huge explosions made columns of black, gray, brown and white appear randomly all around. Streams of glowing yellow tracers came at the V-22's from the village and the treeline around the LZ. Door gunners in the V-22's returned fire, as did the SEAL team. Larger yellow tracer rounds came from vehicles hidden around the LZ and armed with Anti-aircraft Artillery (AAA). The AAA came in the form of the traditional flak explosions familiar from World War II, and in the form of light cannons (23mm and 30mm) firing at extremely high rates of fire; like fire hoses spurting out yellow high explosive tracer rounds.

The first Osprey landed, and Marines ran out from the back door. Normally, they would fan out behind the aircraft and return fire coming from any direction. Since the volume of fire was so high, and it was coming from all but one direction, they ran towards the inlet where the SEAL team was hiding to protect themselves, and the SEAL's. The next two Osprey made their landings, and the Marines from those two headed for the inlet's bank too. It was easy to find the SEAL team since they were the only ones in the area firing red tracers (characteristic of NATO forces).

As the Marines consolidated their position with the SEAL's, the North Koreans seemed to intensify their fire. Two more North Korean battalions were moving into the area, and two more patrol boats were coming down the inlet. Less than two minutes after landing, it was time to get back into the Osprey and leave! Squad by squad, the Marines ran towards their waiting Osprey. When the first Osprey was loaded, the Marines, SEAL's, and their rescued pilot ran for the second Osprey. The last squad of Marines, and the door gunners on the Osprey, continued to return fire from the North Koreans who were all around them.

Just before the last squad of Marines started to move towards the last Osprey, the North Korean artillery finally started hitting its mark. An Osprey exploded from a direct hit by a North Korean 120mm mortar round. The crew inside never had a chance. Additional explosions erupted around the other two Osprey, and the last squad of Marines had to run back to the cover of the inlet's bank. The first two Osprey lifted off and tried to change their positions by slowly hovering closer to the bank.

They couldn't evade the artillery either, and the SEAL's seemed to have bad luck follow them wherever they went. A North Korean 105mm artillery round exploded close to the starboard engine on the second Osprey. Shrapnel cut the fuel line, and the engine immediately caught fire. The Osprey had been designed to fly with only one engine, but it was not designed to fly overloaded with the Marines, a large SEAL team, a downed pilot, and with one engine on fire. They had to land.

Smoke poured out of the damaged Osprey, and it slowly put down along the inlet bank, in the clearing between the last squad of Marines still on the ground and the village to their left. The remaining operable Osprey also put down. Everyone evacuated both aircraft, and joined the rest of the Americans hiding along the banks of the inlet. There was no way to take everyone back to the Wasp in only one aircraft, and leaving 2/3's of a platoon was not an option for the men who could still get away. However, there was no reason for the single Osprey to wait for another lucky artillery hit, so it headed back for the ship.

Once it was clear of the inlet, a squadron of U.S. Navy F/A-18E Hornets from the U.S.S. Kittyhawk began dropping bombs around the clearing in support of the trapped Marines and SEAL's. The Hornets could remain on station for a few minutes. Then they would have to get out of North Korean airspace before they ran out of fuel, and before the North Koreans sent in more fighter interceptors. With the support of the Navy's Hornets, the extra ammunition in the downed Osprey, and the marksmanship of the Marines now on the ground, they stood a good chance of surviving until more Osprey could come in and get them out.

Back on the Kittyhawk, Admiral Pender was furious. Everything was going wrong. He knew that the North Koreans were far more formidable than Iraqis, Somali warlords, Serbians, or any other nation that the U.S. encountered in the post Cold War era. Still, he hadn't expected them to be prepared well enough to intercept one of his pilots, a SEAL team, and the rescue TRAP Marine force. One thing was extremely clear: there would be no easy way to get those people out now.

He had kept his senior commanders updated on the situation at every new development. In the Pentagon, they were able to monitor it live through the National Defense Communications Control and Intelligence (NDCCI) computer network and its new SP 2.0 system integration software. This network allowed the National Command Authority to download data from every plane, ship, and submarine anywhere in the world. They could listen in on every communication, and they could even watch video from cameras all over the globe or even from satellites above. With all this information, the President too was well informed through his aides, particularly his National Security Advisor and the Chairman of the Joint Chiefs of Staff.

General Shapiro on the Wasp had contacted Sam Pender. Ken Shapiro was a veteran of Vietnam, Grenada, Kuwait, Somalia, and a number of smaller classified actions along the way. Through it all, he remained faithful to the Corps. He had never lost the faith, but he was not known for his sugarcoating things either. That's exactly why, when it was being formed months ago, Sam had so strongly suggested that he command the Marine elements of Task Force Stingray. Shapiro had the same feeling that Sam did about getting his people out of North Korea, and he drew up plans accordingly. Those plans involved almost the entire

Marine contingent in Task Force Stingray. Sam wasn't at all surprised. In fact, just below the surface, he expected such a plan.

Admiral Sam Pender submitted the operation request to his superiors, and then he waited. They had all the information through they needed through the SP 2.0 system, and they were not at all incompetent, but it was over an hour before he got any response. When they did contact him, he was told simply that the plan was awaiting approval from the Commander-In-Chief, the President of the United States. However, Sam did not need approval to prepare to begin the operation, and that's just what he did. He had all the planes loaded with fuel and ordinance to cover the Marines, and he told General Shapiro to begin whatever preparations he needed. They wanted to be ready to begin their small invasion as soon as approval was given.

Another hour passed, and still there was no approval. Thousands of men and women in Task Force Stingray waited for the word to rescue their comrades, but no word came. At the Pentagon, the generals, admirals, and advisors also waited, but there was no word from the President. They too were getting anxious. Another hour passed, and still there was no word from the President. Finally, the President's National Security Advisor informed the Chairman of the Joint Chiefs that the President would be unavailable for the next few hours. Furthermore, the President had been informed, briefly, and they were to take whatever steps they deemed "...necessary and appropriate." It was clear that the President hadn't really been given the proper perspective on the situation, but those were his orders, and they were actually better than they could have hoped. Sam was informed, and so were several powerful members of Congress, just to make sure that it didn't seem as though the U.S. military was out of control. They were, after all, under the laws of the Constitution, ultimately answerable to the civilian leadership.

Sam passed the word, and Operation Southpaw, as it was to be called, was put in motion. Hundreds of aircraft, thousands of sailors, and thousands of Marines went into action.

January 23, 2011

British Media News Service

American Rescue Team Trapped In North Korea

There are unofficial reports coming from the Pentagon that the American Marine pilot that was shot down over North Korea has been found. This information has allegedly not been made public because of problems with a rescue team that was sent in to rescue her. Sources tell BMN that a U.S. Navy SEAL team was deployed by submarine, but that they were detected, ambushed, and are currently trapped on North Korean soil with the Marine pilot. There has been no confirmation of these reports, but Pentagon, White House, and State Department spokesmen have all refused to either confirm or deny these reports.

January 24, 2011

United Press Online News Service

~~TRAP Team Trapped~~

After the hail of questions regarding a failed rescue mission in North Korea, the Pentagon announced today that a Tactical Recovery of Aircraft and Personnel (TRAP) team was sent into the Democratic People's Republic of Korea. That team was deployed with the express mission of rescuing a Marine pilot who was shot down over international waters on the evening of the 22nd. The TRAP team was detected, and came under fire from North Korean military units in the area.

There were no further details, but unconfirmed reports from sources deep in the Pentagon insist that the TRAP team was actually sent to rescue a U.S. Navy SEAL team that had been ambushed while attempting to rescue the Marine pilot.

Monday, January 24, 2011

2nd Lt. John Chamberlin-Personal Log

Somewhere in the Yellow Sea

As thorough and efficient as my officer training has been, it did include one major contradiction. I've been taught not to get too close to the people under my command, and, at the same time, I have been told that I shouldn't segregate myself from them either. The danger of becoming too close is simple. If a situation arises where I have to give an order that requires involve one or more of their deaths, I could become reluctant and inefficient. If I don't get close to the people I command, then we may ostracize each other, and my leadership will be more likely to come into question by those I'm to lead. I've always felt that it's better to err in the former, rather than the latter. That is to say, I want to know the people I'm working with and leading. I want them to be my friends, and I'm confident that I can still order them into deadly situations efficiently as long as we all remember our jobs, and our duty as Marines.

With that thought in mind, I made sure to party with the platoon before we went to sea. Since then, I've made an effort to get close to Bob, my squad leaders, and some of the other enlisted people. I was always confident that the moment I would have to order someone to their death, I would still be able to. That day has come, and I was right.

Yesterday, some of our pilots out on Combat Air Patrol (CAP) scraped with some North Korean pilots. The North Koreans got the better of our people, and all the Marines went down. Rumor is: only one of the North Koreans got splashed. One of our pilots did manage to make it out of her plane. The cold winter waters of the Yellow Sea forced our pilot to the dry ground of North Korea. Admiral Pender sent in a SEAL team to get her out. Then the SEAL's ran into trouble. They were surrounded and pinned down by an overwhelming force of North Korean infantry and some patrol boats. Admiral Pender sent in air support to cover the SEAL's while they made their way to a clearing where they might be extracted by air. Then we were sent in to get them all out.

There was very little preparation time, but the entire company still had enough time to get assembled on the flight deck. When we got word that the SEAL's were in trouble, no one really needed to hear the orders. All three platoons ran into their Osprey, and before the last one was loaded up with Bruce's platoon, the first bird was taking off with Hector and his platoon. It was just around dawn, and the sea was as gray as it was cold. The air was thick with a combination of haze and morning fog. Visibility was poor, especially since everything was white with snow and gray from the lack of direct sunlight.

About twenty minutes into the flight, the pilots began to fly evasively, jinking the aircraft from left to right to avoid enemy fire that none of us could see. A feint thunder grew in volume as we closed in on the landing zone, LZ Delta. Our Osprey pitched its nose skyward and rotated its engines into the vertical position for a rapid landing. By now the thunderous North Korean mortar fire was overshadowing the noise of the propellers. With the nose still pointed into the vertical several degrees, the tail ramp was opened, and we prepared to exit the vehicle. There was a slight vibration, and the Osprey shuddered. I knew we had been hit, but there wasn't any real drama to the event.

The first hits that it took did startle me a bit, and I saw a similar reaction from most of the people on board. More than anything, I think we were all surprised at our lack of fear. No one got jittery, panicked, or otherwise lost control. Actually, there was a certain type of relief. Our training had been so long, and so intense, that when we finally saw the effect of tracers and shrapnel coming close, it seemed like our training was finally complete. We made it. We were combat Marines. Of course, the Osprey's hits happened in half a second, and there wasn't any time to think about their emotional effect on the platoon or myself. A look around at everyone convinced me that they were doing the same as I was. We all felt it, and later, we'd have time to figure out those feelings. Until that time, we all had to live with the odd feeling of comfort.

At just the right moment, Bob had everyone exit and form into our covering formation on the snow-covered ground around the aircraft. When I looked around, the other two aircraft had already landed, and the SEAL's were coming up from an embankment to the south about 100 to 200 meters away from me. North Korean small arms fire seemed to be coming from all around, but the company was really pouring out the lead in response.

When we were outside the Osprey, shooting at the North Koreans, I recognized that I was shooting real bullets at real people, but, like the feeling of being shot at, I don't think anyone put too much reflection into the moment. We were all concentrating on doing our jobs, and staying alive. Our rifles can only fire in three round bursts, or single shots, so every pull of the trigger meant that we were taking careful aim on a target, a person.

As for myself, I felt kind of good, maybe a little righteous, about killing people who were trying to kill me. I've never done anything worth being killed for. They probably haven't either, but I'll be damned if I'm just going to sit around and die, or offer to surrender because I don't feel like defending myself or my platoon. I've read everyone's personnel records, and some of the people in the platoon have had a few run-ins with the law, but no one, NO ONE, deserved being killed. Those weren't my exact thoughts at the moment, but looking back, that's definitely how I felt. I was NOT going to get killed, or watch someone under me get killed; not without one hell of a fight. With those feelings in our hearts, and our training in our heads, I watched North Koreans fall without an iota of remorse. Rather, I feel proud that we rose above the clamor and didn't die like sheep. Today, we were Marines.

While we consolidated, the North Koreans started to throw more mortars at us. They grew closer and closer with each shot. The SEAL's linked up with us between Hector's and my platoons. Captain Brown ordered everyone back into the aircraft, but, as the first people from third platoon reached their Osprey, it exploded from a direct artillery hit. There wasn't a lot of fire like in the movies, just a loud bang, a shock wave, dust, and a lot of debris.

I looked up and saw our Osprey waiting for us wondering if it too would get hit. If it did get hit while we were inside, that would be the end for all of us. This was the moment when I might have faltered due to friendship with the platoon, but I didn't even hesitate. I told the platoon to haul ass for the bird, and they did. Captain Brown, Hector, and the SEAL's made it into their Osprey, and, when we had gotten everyone into ours, both birds lifted off. The pilots tried to hover and maneuver their way towards the embankment where the rest of third platoon was pinned down. There wasn't enough room in the two aircraft to take third platoon back, and I wondered what was going on. It didn't matter since our bird took another hit, and we lost the starboard engine. It wasn't enough to make us crash, but we certainly wouldn't make it back to the ship. Captain Brown ordered the two birds to land again, and we all returned to our positions to fight it out. With one Osprey blown apart and another smoking and inoperable, the third headed back to the ship to bring in more help.

In the meantime, we lay on the embankment, with the half-frozen inlet at our backs, and North Koreans all around. The mission had clearly gone from bad to worse to unbelievable. No one had gotten hurt in my platoon, yet, so the issue's level of incredibility sparked me with an air of comedy for however brief a moment. It passed quickly, and, as the seriousness chilled my cold body, I had Bob do an ammunition check of everyone in the platoon.

The North Koreans are no slouches. They are well trained, well armed, and, since we're clearly in their land, they're well motivated. In a strange kind of way, I'm actually kind of thankful for all this. I wouldn't have wanted to get into combat and wondered if I was shooting the right people. Maybe it's the training. Maybe it's my character. Maybe it's the situation. In any case, I had no trouble shooting back and killing my enemy. There was no remorse. In fact, the adrenaline gave me kind of a rush, and I had to fight to maintain calm in front of the platoon.

Captain Brown crawled over to me, and, owing to the suggestion of the SEAL team leader, we decided to leave the Osprey and make our way further up the inlet. About two clicks (kilometers) farther up the inlet, there was a spot where the treeline surrounding the clearing met the embankment that we were all using for cover. That would be our first waypoint. After that, we would try to make our way to the top of a hill about three more clicks inland. Our plan was radioed back to the Task Force. The message got through, and they said that help would meet us there. They also had several planes lined up and waiting to give us air support, but they would have to wait out to sea to avoid North Korean AAA, SAM's, and enemy fighters. When we called for it, help was going to take a few minutes to fight its way in.

The company was spread out along the embankment from west to east with 1st platoon on the western left flank, and 3rd platoon to the east on the right flank. Hector and the 3rd platoon put down suppressive fire, while my platoon moved to the right of Bruce's third platoon. Then, we spread out and covered 3rd platoon while they moved through us to our right. The company continued this leapfrog maneuver until we made it to the treeline.

The North Koreans came at us in squad-sized units, about ten or more at a time. They used light infantry mortars before each attack, but since the attacks were so closely timed, there usually wasn't much from the mortars. They were well trained, and each unit operated as a team. There was little coordination between the different units, however, and their weakness was well exploited.

Every Marine in the company switched their personal helmet headset to the same channel, and, although radio discipline was difficult to maintain at times, we were able to drive them off every time. Without those headsets, we'd have been overwhelmed early on in the fight.

At the treeline, we put the platoons in single file, and advanced. My platoon took the lead. Everyone was pleased to have gotten away from the mortar fire around the Osprey, but the lack of enemy artillery meant that infantry and other ground units could hit us without the risk of fratricide. Behind my platoon, Captain Brown, the SEAL's, and the downed pilot followed. They, in turn, were followed by 3rd platoon. 1st platoon brought up the rear.

Every few minutes, a firefight would break out. When it was over, we'd walk again for a few more, then there would be another one. The cycle continued as we fought our way to the hill for hours. By the time we got there, everyone was almost out of ammo. The hill that we were working towards became less and less significant as we got closer. In the end, it was really just the highest spot on a long and wide sloping piece of ground. There was an old abandoned temple or monastery between the crest and the trees we'd advanced through. It was as good a place as any to make a stand until help could come in.

We had a lot of walking wounded, but we hadn't lost a single Marine. The North Koreans had to have hundreds of casualties. Everyone seemed to be making each shot count. Every time I looked around, a North Korean was firing, and, as soon as a Marine returned that fire, the North Korean would either go down, or disappear never to be seen again. I did see a handful of people in Hector's platoon go down, but their wounds must not have been too bad, because they always seemed to get back up. The same was true for Bruce's platoon.

I think the body armor we're all wearing is a big reason. After the U.S. folly in Somalia, when almost a hundred Army Rangers were wounded or killed, the Pentagon made lots of improvements in our bulletproof vests. Now they have more layers of kevlar, more layers of padding, and even a ceramic plate to protect our vital organs. I'm not too enthusiastic about trying out its effectiveness, but, if I have to, I'm confidant it will work.

Late in the afternoon, just before sunset, we started getting reinforcements. Like everyone else, I thought we're were finally getting picked up, but Admiral Pender has decided that we're not to be driven off. I know that everything we say is saying is being sent from our headsets back to the fleet, so there's no way he could have been misinformed about our situation. Still, it seemed like the right time to get us out, not put more people at risk. Ours in not to reason why....

Tuesday, January 25, 2011

2nd Lt. John Chamberlin-Personal Log

Somewhere in the Yellow Sea

Overnight, things have really changed. Our little TRAP mission has grown into a major operation. Yet another flight of Osprey came in to get some of us out, and the North Koreans really let them have it. Of the eight V22's that came in, six were damaged so severely that they couldn't take off again. Now there were almost two hundred Marines trapped. About an hour later, more Osprey came in-this time supported by a couple of Harriers. I couldn't see how many of either type because the North Koreans sent another infantry attack at the temple. Incredible as it seems, even with all the reinforcements, and all their casualties, they still keep attacking us. Of course, so did we. The second flight of Osprey met with AAA that was similar to the stuff that greeted the previous flight. Again, a few more couldn't take off. By now, there seemed to be smoking and burning aircraft everywhere, and the North Koreans had managed to take cover in some of them.

We had also taken our first casualties. Four Marines from one of the new platoons were killed in a rough landing. Everyone else on board was wounded to varying degrees. I watched an entire platoon leave cover and make their way, in the open, to the burning Osprey. They pulled everyone out, and made it back to cover around the temple. Some were hit from North Korean small arms during the fight.

I still had no problem shooting back at people who were shooting at me. I actually felt satisfaction and triumph when I watched them get hit and die, sometimes screaming. It was exciting. Watching Marines get hit, and some die, was quite the opposite. I hadn't looked at the North Koreans as people, just bad guys, targets, and the enemy. When the Marines were getting pasted, I felt a little shock at the reminder of my own humanity. I also felt a lot of rage. I'm sure everyone else did too, because when those people went out into the open, and we saw them buckle when they got hit, I think every Marine on the ground opened fire at the North Koreans who were shooting at them. It was the most intense fire I've seen yet.

Another wave of Osprey came in, again more went down smoking. I'm not very impressed with its ability to take damage. Of course, there's a lot to be said about the fact that only one landed hard. I guess they can take the hits, survive the hits, but expecting anything to work normally after the beating that they've been taking, well, it might be asking too much. After all, I don't know how well a tank can take damage like they've been getting.

North Korean AAA is still thick, but their infantry attacks are falling into a pattern. They only come at us about every forty-five minutes to an hour. When they do attack, and we either drive them off or wipe them out, they wait at least half an hour before they hit us again. The artillery and mortars have let up too, or so it seems. It might just be that we're already used to the explosions.

Colonel Ghetty is here now, and he's set up his CP in the temple where we've been holding out. Captain Brown has our company holding the north wall. We're lined up by platoon again with Hector on my left, and Bruce on my right. The SEAL's have joined Hector's platoon. Our downed pilot has grabbed a weapon and she joined my platoon. The rest of our reinforcements are scattered about the hill around and inside the temple.

More Osprey came in this afternoon. They met with the usual reception, and a few more litter our LZ now. One of them went down while it was making its way back to the ship. It hit with a terrible crashing sound, like a bus hitting a brick wall at a hundred miles an hour. Our platoon was the closest, so Brown gave us the word to go get the crew out. We were already moving. Bruce and Hector spread their platoons out to cover our position, and we moved forward by squads. Gary and 1st squad took the lead, Wess and Rick worked up the left and right flanks respectively. I stayed with Gary's squad.

The North Koreans saw us leave our positions, and they hit us hard. Small arms fire was followed by mortars, then a little bit of artillery. The pounding was rough. I could hear people screaming, but I didn't know whom. We crawled, returning fire along the way to the Osprey. The North Koreans knew we were heading for it, and they had it pegged with mortars. Tim Potter from 3rd squad took a direct hit, and we never saw him again. His helmet was about a hundred feet from where he had been lying. At least it was quick and clean. The Osprey crew was badly wounded, and some of the people from 3rd squad moved quickly to get them out. Once out, they dragged them back as fast as possible without standing up.

A North Korean squad found their way into a ravine about a hundred meters beyond the Osprey. When they opened up on us, 1st and 2nd platoons, back at the temple, were quick to cover us. Mike Baker from 1st squad got hit in the arm, and began to panic. The blood made the wound look very bad, but it really wasn't. Paul Gustovson from 2nd squad was closest to him, and he tried to calm him down. Gary crawled over and tried too. The three of them made a great target, and the North Koreans let them know it. Mike took another round, and his head burst open. I saw him after it had happened, but Paul and Gary were both holding him at the time, and it had to be a terrible thing for them to see up close. No one will ever forget the scene.

I can't remember how many North Koreans I've killed or wounded. I've seen blood on some, but most just fall or disappear in the smoke and mist. I've seen people die on Osprey, and I've seen some people get hit in other platoons, but this was just pitiful. It was disgusting, depressing, gross, and overall, just a bad thing to see. Getting shot at brought out a trained response, comfort. Shooting other people who were shooting at my platoon brought out a trained response; pride. Watching one of my people's heads explode took some reflection. I was stunned with curiosity and shocked by the finality of the spectacle. I was only out of touch for a second or two, but a few puffs of snow from North Korean bullets immediately brought me back to the task at hand.

Both Paul and Gary went ballistic. They went running right at the pinko murderers. There was no way that they should have survived, but they made it all the way to the North Korean position. Once there, they emptied their weapons into the enemy, then they attacked with their bare hands. A few of the North Koreans panicked and ran, but we cut them down. With their rage temporarily satisfied, Paul and Gary grabbed their weapons again, reloaded, and crawled back to the rest of the platoon. When they reached us, we all started our way back to the temple. That's when I got it.

Somehow a round skipped off the top back of my helmet, and it knocked me out cold. I'm okay now, though I don't know how. When I woke up, I was naked, in a clean bed with clean white sheets. My head feels like the bullet went in, but a nurse assured me that it didn't. I don't know where my platoon is, how I got here, or what's going on. All I know is that I'm safe, I'm alive, and I'm back on the Wasp. I also know I'm exhausted.

January 25, 2011

Central News Network Online Report

Rescue or Invasion?

Officials from the Democratic People's Republic of

Korea announced last night that an American-led invasion force had landed on the southwest corner of North Korea.

U.S. State Department officials deny the claims of invasion, but have now confirmed that elements of the 24th Marine Expeditionary Unit-Special Operations Capable (24th MEU-SOC) have been deployed in support of rescue operations in the area. There was also a repeated claim by the State Department that the American pilot who was shot down on the evening of 1/22 was operating in international waters, and that any efforts to recover the pilot were within international law.

Tensions are building on the peninsula as a second carrier task force has been sent to rendezvous with Task

Force Stingray already in the Yellow Sea. Chinese diplomats have called the movement "...a dangerous escalation of a volatile situation." However, the People's Republic of China has also begun moving additional fighter and bomber squadrons to the area, and patrols by missile patrol boats and frigates have doubled-according to sources inside the Pentagon. The Republic of Korea (ROKSouth Korea) has also put its entire military on alert fearing that the buildup of North Korean forces along the western edge of the DMZ might be a precursor to an attack on Seoul itself.

Wednesday, January 26, 2011

2nd Lt. John Chamberlin-Personal Log

Somewhere in the Yellow Sea

I'm told that the med center here on the Wasp is among the finest available on any ship, anywhere in the world. It has to be. Not many other ships perform duties that put so many of its passengers in danger. I was lucky when I got it; of course, my headache constantly tells me I wasn't just scratched either. There are at least forty other Marines in the sick bay/med center here, but there are still plenty of beds. There's also an emptiness that both relaxes and unnerves me. It's great that they're empty, but they were put here for scenarios like this one, and so they may soon be full. One of the ship's doctors gave me clearance to leave the sick bay, but not the ship. At least I could find out what was happening to my platoon, and to everyone else for that matter.

After returning to my quarters, I changed into a clean set of camouflage, and left as quickly as possible. It was eerie being one of the few Marines still on the ship. Most had departed for LZ Delta while I was passed out in sick bay. Only the support personnel, high ranking officers, and wounded are remaining on board to keep the squids company. Half of the ship may be ashore, but there is a busy pace to everyone still around. My headache makes me move slowly through the passageways, and I'm glad to be on a ship designed for fully loaded Marines to move around in. The extra wide passageways make it easy for everyone to run around me.

Most of my superior officers are ashore, so I made my way up the chain of command looking for someone to work for. I walked up to one of the conference rooms in the ship's superstructure. On my way, I caught a glance into the bridge, and the sunlight blinded me. I staggered into the conference room, and when my sight returned, I was surrounded by brass. The room was silent, and one of them sat me down at the large conference table.

Some Colonel asked who I was, and how badly I was wounded. When I had finished my impromptu after-action report, General Shapiro spoke up. I was surprised to find him sitting right next to me, and even more surprised when he said that he remembered me from the Nellis AFB assault. It was reassuring to listen to him, and my nerves relaxed. I hadn't realized how tense I had been up until that point.

General Shapiro was very interested in the North Korean heavy weapons. He asked me about the artillery, mortars, AAA, and SAM's in detail. I could tell he was frustrated by my description of how poorly the Osprey had performed. It seems that so far everything that could go wrong has. Our downed pilot's flight should never have been ambushed so successfully. Washington took way too long to authorize the rescue operation. Our SEAL team was ambushed perfectly. Our TRAP mission was itself trapped by the sudden appearance of North Korean AAA units. Then our TRAP mission's relief force was blown out of the sky by the same thing that had trapped the TRAP team. Now, Shapiro had sent in the bulk of the remaining units in the MEU, and, although they have been able to secure their own defensive position on the inlet, they too were being pinned down and trapped by heavy artillery, mortar, and AAA fire from the North Koreans.

General Shapiro asked me if I was up for duty, and I told him I was ready to go back to my platoon. Everyone looked at me as though I was trying to be a hardass, but I'm ready, willing, and able. Shapiro looked me in the eye, paused, and almost broke a smile. Then he assigned me to the conference room as one of his assistants. It's a charity position. My big contribution to the operation, from this point forward, is to keep the coffee maker operating at full speed. Everyone continued their meeting, and about half an hour later, they broke away to return to their duties.

On one of the many maps spread all over the room, I found the temple, and my company. As far as anyone in the room knows, they're still alive, and still there. I feel completely helpless and even shameful. I'm a Marine, and Marines don't leave other Marines behind. We live and die by that code of honor, yet, here I am, making coffee for the General; in the rear with the gear.

In the conference room, I'm surrounded by all kinds of staff officers, and an unusual amount of people out of uniform. There're people here from other services and government agencies beside the Navy or the Corps. I saw someone wearing an FBI jacket, another with a DEA jacket. I even overheard people talking to a woman who made no secret about being with the NSA. I'm sure the CIA is here too. Besides the maps, there's a lot of monitors and TV's hanging from the ceiling.

A few of them have live pictures from LZ Delta, or our new beachhead. The rest have news programs on CNN, MSNBC, FOX the three major networks are all running normal programs, but from time to time, they'll break in with a newsbreak. Here I am, at the very nerve center of a small war, watching it live from the battlefield, and watching yesterday's morning news back at home-also live.

No one's paying any attention to my presence anymore. I've been allowed to hear all the classified info coming into the room. The entire Korean Peninsula has gone on alert because of our little war. North Korea has called up its four-million-man 24-hour reserves, and South Korea has sent its troops to the DMZ. To keep tensions at a minimum, we're not getting any real help from units based in South Korea. In fact, we're pretty much on our own unless a full-blown war erupts. That might be our only hope since the North Koreans formally demanded our immediate cease-fire and surrender this evening. Washington has told us to obey the cease-fire as long a possible in an effort to find a diplomatic way out of this mess. The only good news is that the North Koreans have moved all of their patrol boats, particularly those armed with anti-ship missiles, back to port at best speed. No one's getting killed, at least for a little while, and, with those patrol boats backing off, the Navy's breathing a little easier.

This whole thing is so confusing, and there's so much going on, I don't really know how to feel all the time. I wish I could just get back to my platoon, and do my job. Trying to figure this out, or find some meaning in it, is really giving me a headache. Of course, I was shot in the head yesterday, and that probably has a lot to do with it!

January 27, 2011

United Press Online News Service

Marines Get In and Get Out of North Korea

U.S. government officials have been slow to provide details regarding efforts to rescue an American Marine pilot in North Korea. However, they are now being revealed. Sources in South Korea have confirmed that almost 2000 U.S. Marines are on the ground in southwest North Korea. Almost half of the Marines were sent in by air over the past few days. These same sources are indicating that the Marine force was sent into North Korea to help withdraw a U.S. Navy SEAL team that was in fact sent to recover the downed Marine pilot. North Korean military opposition is reported to be far more intense than had been anticipated.

This new information comes on the heels of a pullout of the Marine force. The evacuation of the Marines is said to have been very dangerous. Reports vary as to whether or not the Marine pilot who was shot down was ever rescued. The Navy SEAL team is rumored to have been withdrawn without loss, but the Marine force has allegedly taken serious losses-including some who may have been captured or left behind.

Pentagon officials remain reserved in their comments about the entire situation. They have confirmed that an operation was conducted to rescue the downed Marine pilot, but also cite security concerns regarding that operation as justification for their lack of clarity. It should be noted that ever since the Persian Gulf War of 1991, the Pentagon has routinely held back information regarding ongoing operations and claimed safety concerns for the troops as a reason for information blackouts. More official details can be expected as the operation comes closer to an end.

State Department officials are reinforcing the Pentagon's position about operational security. They add that not only are the lives of U.S. forces in the operation in jeopardy, but so is the entire region. Despite the rumors of an American withdrawal, North Korean military forces are still being rushed to the scene. South Korean forces have also mobilized and are building up to protect their nearby capital of Seoul. Chinese naval and air assets are being sent to the region, and even a Russian flotilla from nearby Vladivostock is putting to sea. U.S. Eighth Army units are on their highest level of alert on the eastern side of the peninsula. Even Taiwan is bracing for the possibility of a full-scale war in Korea.

Thursday, January 27, 2011

2nd Lt. John Chamberlin-Personal Log

Somewhere in the Yellow Sea

I spent most of the night watching the coffee maker work overtime, and the lines on the map go bad. Reports coming in from the Marines ashore indicate that the North Koreans have moved almost an entire corps, about 40,000 men, into the area. We've got about 2000 Marines ashore. Almost 1100 are positioned on and around three hilltops encircling a place called Beach Red Dog One. Another 800+, including my platoon, are pinned at LZ Delta.

An impromptu cease-fire fell apart after a few hours, and the lines fluctuated throughout the night. By sunrise, everything had settled down. Admiral Pender didn't want a fight to begin with, and now, all he wants, all any of us want, is to leave with honor. No one wants to run away regardless of the odds, but we can't sit and hold either. Pender knows this, probably better than anyone other than General Shapiro, and he's called for backup.

Another Battalion Landing Team (BLT) is on it way. The U.S.S. Peleliu, similar to the Wasp, is bringing them along with more escorts. They should be here in a few days. The aircraft carrier U.S.S. Kittyhawk, and her escorts, are moving north to give us better support. Her F/A-18E Hornets have already been flying CAP and close air support missions. When she gets closer, they'll be able to fly more missions and loiter over target areas longer.

She's already lost some planes, and every time we hear the radio calls, we're reminded of how we were drawn into our own little private war. So far, all of her planes have either managed to make it back to the carrier, or the pilots have been picked up by our escorts. I don't mind getting into this fight over a single pilot. The North Koreans are the ones who started it, and they escalated it by ambushing our SEAL's. The only options we had were to either reinforce our trapped unit, or surrender. Surrender wasn't going to happen, and I'm glad we hit the beach.

All I want to do now is to get back in! I'm sort of looking to avenge Tim and Mike, or get some payback for getting shot in the head. I also want to be with my platoon, to face fate with them. I don't want to hear about them getting wiped out while I sit here, in relative safety!

Friday, January 28, 2011

2nd Lt. John Chamberlin-Personal Log

Somewhere in the Yellow Sea

Another long day of watching coffee brew. I think we'll be alright as long as the coffee holds out. If it runs out, I think General Shapiro is going to go ashore himself and take things directly into his own hands. Everyone is remaining very professional, but the North Koreans are really becoming frustrating.

Back at LZ Delta, the situation sounds ugly. They keep getting hit by mortars, artillery, and now even the AAA is firing directly at them from the heights between LZ Delta and Red-Dog One. Between barrages and harassing periods of heavy weapons fire, the infantry continues to make wellcoordinated attacks. The only thing holding them back is their lack of small unit cohesion, and the absolute determination of my Marines is keeping them at arm's length.

This morning they made a strong attack between the hill on the left flank of Beach Red Dog One, and the hill in the center using light armored vehicles for close support of their infantry. When that attack finally stalled, another much stronger one came in on the right flank along the water's edge at the beach. We were able to hold them off each time, but by noon, the Marines on the hill covering the right flank of Red Dog One were stretched pretty thin. The North Koreans seemed to have had enough too, and they reduced their attacks on the beachhead to probing actions for the next few hours. At sundown, they began again. This time they hit all along the beachhead. I think fatigue, and the slippery snow and ice-covered hills are making it almost impossible for them to move with any speed or momentum.

The three hills protecting Beach Red-Dog One have been nicknamed (from south to north) Kelly, Tombstone, and Navarone. The North Koreans pushed hard on Kelly, and the Marines actually had to fall back to the base of the hill before the attack was stopped by someone who had gotten a few M1A2 Abrams tanks to the top of Tombstone. The North Koreans who were pressing down the hill towards the beachhead itself had to fall back to the top of the hill.

We tried to counterattack just before dawn. Navy F/A-18 Hornets pounded the top of the hill, and the few artillery pieces we have on the beach fired directly into the North Korean positions. We also had some more M-1's ashore. They were supposed to give cover to companies of Marines as they made a rush for the top of the hill. The attack failed.

We lost three more Hornets on the other side of the hill. Dozens of Marines have been killed, and we still don't know how many more have been wounded. The North Koreans kept only a skeleton force on the top of the hill, but they had at least thirty main battle tanks at the base of the hill on the reverse slope, out of everyone's sight. The M-1's took out every single North Korean tank as it crested the hill, but that meant that they were too busy to support our infantry companies' assault, and they got hit hard. No official order has been given to withdraw, but everyone knows it's only a matter of time now.

We've gotten word from LZ Delta that they might be able to make it through and actually help retake Kelly from the rear. If we have to withdraw from the beach, then they'll have to try, and it's a Hail Mary play at best. I hope my platoon makes it through this ok. Part of me desperately wants to get back to them. The other part of me, the rational disciplined side, realizes that I can barely stand for more than a few seconds, and I'm more likely to get someone killed than help anyone stay alive. One of the staff officers tells me that my company has taken casualties, but not as severely as most of the other frontline infantry units. I guess I can be thankful for that. I can't help wondering who in my platoon hasn't made it.

Just before dark, the official word came, we're pulling out. All heavy weapons and vehicles are to be destroyed, and the Marines will be pulled of the beach as fast as possible. They should all be off the hills by 0900, and they should all be on their way back to the ships by midnight. The Marines trapped at LZ Delta have been ordered to make for the inlet where Admiral Pender is going to try and get some destroyers and a cruiser to evac them. Very sloppy, but we wouldn't be trying something like this unless there was no other way.

January 29, 2011

Central News Network Online Report

<u>Withdrawal from LZ Delta and beaches</u>

As expected, the Pentagon announced this morning that elements of the 24th MEU had entered North Korea days earlier, but have been withdrawn. The Marines had in fact been sent in to support a Navy SEAL team that was rescuing a downed Marine pilot. The SEAL's were detected, and were met with stiff resistance. As a result, a larger Marine TRAP team was sent to support them. The TRAP team was ambushed, and a very large force of Marines was brought in by V-22 Osprey to support the mission. Another Marine force was sent to the area by LCAC hovercraft to secure a beachhead. When the beachhead was secure, the entire Marine force was evacuated. Aircraft from the carrier Kittyhawk and Abraham Lincoln supported the operation.

North Korean military units in the area rushed to the scene as soon as the SEAL team was detected. Over the past few days, while almost 2000 Marines were sent ashore, an estimated 200,000 North Korean troops closed in on the Marine perimeters. As a result of the heavy North Korean attacks, the Marines were withdrawn with some confusion, and while the Pentagon concedes that there have definitely been casualties, specific information has not yet been collected. Despite the withdrawal of the Marine rescue units, military forces all around the region continue to remain on the highest level of alert.

Saturday, January 29, 2011

2nd Lt. John Chamberlin-Personal Log

Somewhere in the Yellow Sea

Pender and Shapiro have put together an ugly plan to get our people out of LZ Delta. The U.S.S. Peleliu, a ship similar to the Wasp, is joining up with our task force. The plan is to use the Harriers and Cobra helicopter gunships to fly out to Delta, and suppress the AAA around the area. While the AAA is being dealt with, the Navy is going to fly what they're calling a modern version of an Alpha Strike. They'll have a squadron of F-14 Tomcats and a squadron of Hornets flying CAP, two squadrons of Hornets providing close air support, and two more squadrons of Hornets armed with anti-radar HARM missiles to deal with the SAM threat. The Harriers, Cobras, Tomcats, Hornets, and all the support planes will put over a hundred aircraft in the air at once. Under this huge umbrella, the Osprey from the Peleliu and the remaining Osprey from the Wasp will fly in to get our people out. They're going to use the big old CH53 Sea Stallion helicopters as transports too. The entire attack will take hours to prepare, and probably twice as long to conduct. If it works, then I'll be back with my platoon by tomorrow morning.

We've all been trained for situations like this, and I really haven't seen any outright signs of fear in anyone. Of course everyone feels it, but not as badly as one would expect. There seems to be more of a sense of anticipation, eagerness, mixed with concern for each other. It's not as though we all developed some sort of deep hatred for the North Koreans, and then decided to start attacking them. Everyone I've listened to has talked about getting ourselves out of this mess. It's not that we're afraid of the pinkos, we just don't want to see our friends get killed by them. There's not a lot of concern for one's self in anyone I've seen. Maybe too many people are looking to stick their necks out too far, for too many other people. Maybe that's how we really got into this little war.

Sunday, January 30, 2011

2nd Lt. John Chamberlin-Personal Log

Somewhere in the Yellow Sea

The battle is over, but I'm not sure anyone can tell if we won or lost. Does anyone ever? The North Koreans put up a good defense with their concentrations of AAA and SAM's. The even put up some more Su-27's and Mig-29's to mess with our air attacks. As soon as a pair of enemy fighters appeared in the area, we had to get our close air and other attack aircraft out of the area while the CAP chased off the enemy's fighters. Only when the area was clear of enemy aircraft could the Navy get back to suppressing the air defenses on the ground. With every delay, the transports had a tougher time getting into LZ Delta. The huge air umbrella existed, and it worked very well, but the leaks were substantive.

It was a costly air battle for both sides. Our pilots and radar operators estimate that almost fifty North Korean fighters, SU-27's, Mig-29's, and lots of Jian-7's, were all shot down. We lost seven more Hornets, eleven more V-22 Osprey, two SH-60 anti submarine helicopters that were out on patrol, and one of our EA6 radar jamming planes. A very expensive E-2C Hawkeye AWAC's plane had a mechanical failure while being chased, and it went down too. We also had three more shot-up Hornets crash on landing. They're write-off's. Our Harriers and helicopter gunships came out ok; a little shot up and worn out, but they're ok.

Of the 833 men and women pinned down at LZ Delta, 243 people were brought back wearing bandages (including myself), 127 bodies were recovered, and 463 are still missing. During the third evacuation flight, the North Korean infantry overran the perimeter around the temple, and the LZ's defenses collapsed. There was no way to help those on the ground except to put more Marines down, and that was completely against what Admiral Pender wanted. Even if he did, there wasn't enough time to get enough people into the LZ. Close air support wouldn't have worked either since the North Koreans were intermixed into our own people.

Marines live and die by the idea that no one gets left behind. Everyone I've seen is crushed that it went down this way. There's not a lot of talking going on.

Silence is the rule of the day. When people do speak to each other, you can feel the anger just below their skin. I'd swear you could watch the temperature on a thermometer rise when someone mentions the name of a person left behind. I almost wonder if we'll go back in again!

Around the world, the TV is showing constant news reports from government leaders. The Russians, Cubans, Iraqis, Serbs, and even the Chinese are as outraged as the North Koreans are. You'd think we'd landed on their soil. I don't think anyone will actually do anything, but the North Koreans are still mobilizing. The real fight could still be ahead, along the DMZ with South Korea.

Back in the states, the State Department has issued a travel warning to all Americans overseas, and the President is calling in our favors from friends all over the world. The British, French, Italians, Canadians, Greeks, Turks, all are sending naval units to the area in case the North Koreans come south. The list of international support goes on.

The FBI has "strongly suggested" that all Americans of direct North Korean, Chinese, Russian, and Arab descent register with their local branch offices. They say that if there are any terrorist attacks, the lists will help bring the attackers to justice. I guess they also hope that it'll make the American people feel safer too. Even the people who are supposed to register are being encouraged to do so for their own safety. Somebody seems to hope that as long as Americans can trust each other, the fear of terrorist attacks will be lessened. I hope we don't have to test their theories, but obviously someone in D.C. thinks it's about to hit the fan over here, so they're getting ready.

The fact remains that, over the past twenty years, we've made a lot of enemies around the world, and they all seem to be lining up to yell at us; as if we wanted to invade North Korea with only 2000 Marines. Do they think we wanted to get our ass kicked, or lose 700 friends for no real good reason? I wonder if political leaders actually think about what they say, or if they just take a side for future and past events based on history?

Monday, January 31, 2011

2nd Lt. John Chamberlin-Personal Log

Somewhere in the Yellow Sea

This is possibly the worst day of our lives. Admiral Pender had another SEAL team go back into North Korea last night. Since my company was evacuated to the Peleliu, I've stayed in the conference room with the General's staff. No one who was in that room last night knew about the SEAL mission, and I think it was a word of mouth chain of command that issued the orders to the team. Hardly anyone was in on it. The team's mission was to go in and locate the remnants of the LZ Delta force in the hopes that yet another rescue mission might be feasible. They found our people alright, but there is no way in hell that we're going to be able to get them out.

The SEAL's, only a five-man team, went ashore several miles from LZ Delta. On every SEAL mission, one man carries a video camera for postmission review. The tape from this mission was to be delivered to high ranking officers, the President, and a few select members of Congress. Once it was fed to Washington, it was somehow leaked to the press. The first information most people in the task force had about the SEAL mission was the on ship's TV news.

After a few hours of sneaking around, the SEAL's came upon a makeshift prison yard. There were no walls, just a wide-open clearing with thick layers of barbed wire surrounding a huddled bunch of cold American Marines. Heated vehicles with heavy machineguns and automatic cannons surrounded them. Only a few infantry guards waited outside the vehicles in the cold winter air. The SEAL's were probably two or three kilometers away, and, even though their video camera was fully zoomed in on the scene, it was hard to make out any faces. Through the snowflakes, fog, and distance, about the only thing you could make out was the fact that they were Marines, and they were not being treated well. The SEAL's filmed a North Korean officer as he pulled a Marine from the huddled mass, slapped him, beat him, and then left him motionless on the ground. The surrounding Marines rose to defend the one on the ground, but the North Korean officer waved his AKM submachinegun at the crowd, then when they had stopped, he pointed it at the Marine he had beaten, and emptied his entire clip into his body.

That was the longest one and a half-second burst I've ever seen. The network anchors all warned that the footage was graphic, but there was no real preparation for what we saw. Since I still haven't gotten a lot of information about my platoon, or even my company, for all I know that Marine might very well be someone from my unit; someone I'm responsible for. Like everyone else here, I'm stretched between complete shock, and shaped, disciplined rage. If Pender wants us to go back in, I'm going.

I'm feeling fine now after a few days rest. My head aches, but it's more of a dull pain than the throbbing pressure I felt a few days ago. I can walk fine, and, aside from a little eye/hand-coordination difficulty, I'm ready to get back on North Korean soil. Everybody feels the same way. Just let us go. We'll get it done this time, or we'll die trying. No one really had a problem with the North Koreans before. They were just people shooting at us, but beatings and cold executions, watched all over the world on the evening news, well... just let us go in. There's obviously some unfinished business to attend to a few miles to the East.

A few hours after the video went on the air, our brand new President addressed the nation. I didn't vote for the sap, but he sounded pretty good. Despite the North Korean buildup along the DMZ, and rumors of Chinese troops backing them up, the President demanded the immediate return of all U.S. captives, a demobilization of the North Korean reserves, and an immediate resumption of talks on the peaceful reunification of the two Korean countries. To back up his threat, he's sending two more carrier battle groups to the region: the 82nd Airborne Infantry Division will begin arriving tomorrow, and three more Army divisions will be sent over the next three weeks. On top of that, the aircraft carrier battlegroup Abraham Lincoln joined our task force today. We now have two aircraft carriers, two amphibious assault ships, at least four attack submarines, and about fifteen support and escort ships. Of course, there's also the remnants of our Marine Expeditionary unit, supplemented by the Pelilu's complement of Marines, about 3000-3500 Marines total.

As far as the Marines are concerned, we're still a little disassembled. The evacuation of LZ Delta and beach Red-Dog One has mixed up the Marines from the Wasp with those from the Peleliu. They're spread out between the two ships. The Navy isn't exactly organized either. It takes a day or two to add a carrier battlegroup into our formation, tie together all the digital data links, communication frequencies, and patrol schedules. They're nowhere near as messed up as we are, but they do have their work cut out for themselves.

January 31, 2011

Central News Network Online Report

<u>Battle of Yellow Sea</u>

On December 7, 1941, the United States of America was attacked by the Empire of Japan at Pearl Harbor. Almost 2000 people were killed on that day of infamy. Last night, naval and air units from the Democratic People's Republic of Korea (North Korea) and the People's Republic of China attacked the U.S. Navy's Task Force Stingray in the Yellow Sea. The communists attacked with dozens of guided-missile attack ships, hundreds of aircraft, and thousands of antiship missiles. Escort ships and combat air patrol aircraft shot down untold numbers of these missiles and aircraft, but there were simply too many. Most of the American Task Force was hit. Losses are said to be "extreme," and one

Navy official told CNNOR that, "…compared to this, Pearl

Harbor was a joke."

Among the confirmed losses are the aircraft carriers Abraham Lincoln and Kittyhawk. Each ship had over 5000 men and women onboard. Given the unusually cold winter temperatures in the area, few are expected to have survived the icy waters for more than a few minutes. However, there are reports of some survivors.

Also among the lost are most of the escorting ships that were protecting both the carriers and the two Marine amphibious assault units that were in the task force. Few of the escorts are reported to still be afloat, and the amphibious assault ships Wasp and Peleliu-each with 3000+ sailors and Marines onboard-are both said to have been heavily damaged.

When asked how such a powerful naval force could have been decimated, CNNOP defense analysts argued that any force can be overcome by sufficient numbers. The difficulty is making those numbers too high for any other nation to attempt an attack. Years of steady defense budget cuts dating back to immediately after the 1991 Persian Gulf War have taken their toll on all of the armed forces. One of the ways that the Navy has had to cope with tighter budgets has been to mothball 2/3's of its 1990 ships. This has meant that there are fewer escort ships to protect the high-value capital ships like aircraft carriers and amphibious assault craft. The number of available escort ships has dropped to the point where only a fraction of an escorting unit is tasked with protecting both a carrier battlegroup and an amphibious assault unit at the same time. There simply were not enough ships to stop all the missiles that the communists used, and now there may be as many as 15,000 sailors and Marines who paid for those same budget cuts with their lives.

January 31, 2011

United Press Online News Service North Korea Invades South Korea!

While naval and air units from North Korea and China were attacking America's Task Force Stingray in the Yellow Sea, Army units crossed the DMZ and entered into South Korea. North Korea's infamous 48-hour 5-million-man reserve force outnumbered the South Korean Army by almost 5-1. Those odds increased as another million troops poured in by rail from China. The highly-trained ROK (South Korean) Army-supported by the U.S. Army and Air Force-took a heavy toll of the advancing communists, but once more, it became a numbers game. Reinforcements are being rushed to the region from all over the globe, but in the interim, ROK and U.S. forces are withdrawing to a new line of defensive positions almost 20 miles south of the former DMZ.

January 31, 2011

The New Cleveland Press

USMC Prisoners Beaten & Film Sent To CNN

Questions of media culpability are being asked all around the United States today. Last night, a video tape was delivered to CNN and subsequently aired. The video was apparently taken by a U.S. Navy SEAL team that was sent back into North Korea to look for Marines who were still missing after yesterday's evacuation. The film shows a North Korean military unit surrounding at least 50-100 United States Marines. Despite the winter conditions, the Marines were being held in an open barbed wire area. Worst of all, the video tape graphically showed a single Marine being openly beaten and executed in clear view of everyone. The footage was showed in its complete form three times on CNN before it was finally edited. CNN representatives claim that their policy is to always review any submitted video before it is aired, but that this tape accidentally fell through the cracks. CNN is now precluding all viewing of the video with condolences to the family members of the Marines.

The video tape has also galvanized support for America's military service men and women. Civilian protests against the Marine presence in North Korea have been replaced by support rallies in small towns and at almost every college campus. It has also stiffened the resolve of Western nations to support the United States in its claims that the Marine pilot shot down was in fact attacked over international waters, and that the rescue teams sent in to recover the pilot were acting in accordance with international law.

Tuesday, February 1, 2011, 8:35 am

Berlin, Germany

Tempelhof Airport

Walt strolled through the terminal at an uncharacteristically slow pace for an airport passenger. On his way, like most stylish businessman, he spoke into a tiny, palm-sized cell phone. Unlike the other conservatively dressed passengers, he carried no carry-on luggage. His assistant, Dan Greene, followed in his shadow, bearing their only carry-on need, a custom designed laptop/secure communications computer.

While the other air travelers rushed past him to catch their flights Walt casually spoke into his cell phone connection. In perfect Korean, he questioned someone thousands of miles away; someone who clung to his every word.

"So, your units are in position, correct?"

The voice on the other end of the cell phone's connection replied.

"Oh yes. We have almost five million troops spread out in various rapid deployment centers all around the southern border. "

"How about the other units?"

"All of our aircraft are either on patrol, ready to reinforce those patrols, or on standby alert for ground attack missions. The air defense units are on full alert. The Navy has sortied almost five hundred assorted ships, patrol boats, and submarines. Except for a few token units, they're all taking up positions along our west coast, facing the American task force in the Yellow Sea."

"Good. How about our software? Is it working properly? I mean, you still have complete data on their units, correct?"

"Yes, but we can't seem to correlate all the data links into a single display. Every time we view one of their data links, the others disappear. We can look at all the unit positions, or all their patrol routes, unit compositions, status reports, even read or listen to all of their communications, but we can't do any two things at once."

"No problem, General. Are you near a terminal right now?"

"Yes."

"Fine. Are you logged into the network?"

"Yes."

"At the top of the main menu screen, you'll see the words: File, Edit, Item, Tools, Graphics, then Help. Move your cursor onto Tools. A menu should scroll down from it. On that menu, you'll see a list of all the different types of feeds that you can monitor."

"Yes, Walt. I'm familiar with this part of it, but you can only click on one data feed at once."

"True. You can set the system to overlay the different data feeds by running the setup/ configuration utility program that comes with it, or, if you only want to look at two feeds together, click on one of the options while holding down the shift key. That will save your choice in memory, and you can enter one more choice. Unfortunately, this option only works with two choices. If you want to view more data all at once, you'll have to run the setup/configuration utility on disk one of the original program, or maybe the designers left the two data feed option in your copy of the program. It was just one of their tools for designing it, and they weren't suppose to take it out. I'll also have them get you the latest upgrade version of the full package within the next twenty-four hours. I'll make sure that one of them personally installs it for you, and, if you want, my technician can also walk you through the setup/configuration program to make sure it's setup exactly as you want. I'm sorry about that, General. It should have been tailored to your needs in the first place. Will that suffice, in the meantime?"

The North Korean General immediately attempted not to embarrass the provider of his incredible software. *"No no no, Walt. Everything was fine during the install. Your people were excellent, and they set everything up exactly the way we wanted. The software is incredible. Never before have two potential foes stood eye to eye with one having so much knowledge of the other. The Americans think that their data links and communication networks give them their 'force multipliers'. They do allow them to collectively share intelligence and concentrate their firepower, but they also prove to be a most dangerous Achilles heel; thanks to your software. We very much appreciate its capabilities, and my question about multiple data monitoring was only one of fine-tuning. We can most assuredly use the system the way it was originally configured."*

"None-the-less, I'll get you an upgrade before this time tomorrow. Now, you mentioned your readiness factor. Has the son of the Great Leader made any decisions? I've tried to rally international support as best as possible, but the other world leaders would appreciate some matter of forewarning should a storm erupt on the peninsula."

"Kim Il Jong is as yet undecided on the matter. There are many of us who have already seen what is about to happen. There has been no commitment so far, but I'm almost certain that nothing will stop history's momentum. Our homeland

has been invaded. The people are outraged, and, after they have been so hungry for so long, they are seeing this as a chance at ending their suffering. The armed forces are as ready as they possibly can be. Include the international support, and the intelligence capabilities that you have provided, well, I don't know what could stop things now. I trust your confidences will be kept to maintain security and surprise?"

"Of course, comrade. Even if you were to lose the element of surprise, you would be informed as soon as orders were passed throughout the American units. On your Tools menu, you'll find an option that says DEFCON. That's the American Defense Condition level. If you click on that, you will see what level they are currently maintaining, when the last change in DEFCON level was, and what the last DEFCON level has been. If the element of surprise is lost, their DEFCON level will increase."

There was a sudden change in the General's voice. He sounded almost apologetic.

"Walt. Have you heard about our neighbors to the north?"

"Yes, General. I'm aware of their movements. In my position, I am sometimes compromised by the limitations I must impose upon myself. One of those limitations is my inability to share information that is of no benefit in sharing. That is to say that I cannot readily pass on all my news to everyone, especially if it's not done in person. I have a great deal of confidence in this secure line, but there is little substitute for a face-to-face meeting. I'll be in Hong Kong on Friday, perhaps we could arrange a meeting, maybe with a few of your northern neighbors too?"

There was a pause, longer than the satellite normally created, then the General softly spoke, almost whispered, in Walt's native English.

"We don't have that long, Walt."

Walt stopped in his tracks. The airport's bustling crowd flowed past him on both sides. He turned and looked his assistant square in the eyes, then he answered the General in a much more serious tone.

"I understand General. I will see you soon. Take care, and good luck."

Wednesday, February 2, 2011, 0218 hrs Task Force Stingray,

U.S.S. Kittyhawk

Combat Information Center (CIC)

The room was never quiet. Devoid of all light save that from the countless computer monitors and radar screens, its occupants resembled the chaos of bats in a cave. From this mission control-style room, the entire operational activities of every sailor, Marine, and otherwise family member of Task Force Stingray was observed and coordinated around the clock. Directly supervised by a Naval Commander, the U.S.S. Kittyhawk also was home for two Captains, and the Task Force commander, Admiral Sam Pender. One Captain ran the ship, another ran the ship's air wing, and Admiral Pender ran everything.

While a young Commander maintained the pace of defense operations in the Task Force, the ship's two Captains slept. Admiral Pender, ever the insomniac, nursed yet another in the endless line of coffee mugs. The Commander sat in a barbershop-style chair high and behind the numerous consoles around the room. In front of everyone, on the three walls ahead of them, three large-screen monitors showed the position of every ship in the Task Force. An outline of the surrounding coastline also revealed the location of every other ship and plane in the Yellow Sea-both friendly and not friendly. All this data, collected from sensors on escorting destroyers, frigates, and cruisers, was sent via secure data link to this room for review. Airborne patrol planes also fed their sensor data to the Kittyhawk's CIC.

The Commander watched patiently as the "not friendly" ships and planes went about their business. In the shadows behind him, Admiral Pender silently leaned on a bulkhead, also watching the big screens.

Without warning or panic, many of the "not friendly" units doubled their presence on the big screens. One of the console operators called out over his headset, and a speaker in the room echoed his voice.

"Vampire Vampire Vampire! Incoming air targets bearing zero-two three, zero-one-eight, zero-zero-four, zero-zero-one, three-five-five, three-five four, three-five-three. Suspect missile attack. Recommend General Quarters Missile."

Before the first word had finished leaving his mouth, alarms and sirens blared on every ship in the Task Force. Other console controllers began issuing orders to best defend against the attack. The Commander ordered escort ships to move into new positions and repel the missiles. Admiral Pender finished his coffee in one long healthy chug, set down the mug, and watched with arms crossed, and fingers tightly clinched. The small private war of the past few days was now crossing the line into all-out war. In a few minutes it would be over.

While the CIC's chaos was replaced with disciplined planned reaction to the sudden event, the rest of the ship's tranquility was replaced by an even higher degree of trained, disciplined, planned reaction. Men and women woke from their sleep and ran through the corridors to their battle stations. On the Flight deck, the

231

reserve Combat Air Patrol (CAP), two F-14 Super Tomcats, was launched to help intercept incoming missiles. More aircraft were brought up to the flight deck from the hanger deck on huge elevators, two at a time.

Back in the CIC, more and more targets were showing up on the three big screens. Some represented the increased CAP from the Kittyhawk, some represented Surface-to-air Missiles (SAM's) from the escort ships trying to stop the incoming missiles. More computer icons representing missiles and planes headed towards Task Force Stingray. Seconds after their detection, a handful of incoming missiles were shot down by the Task Force's existing CAP. A few more were shot down by the SAM's from escort ships. Every time one missile disappeared from the big screens, more appeared. Four more F14's were launched, then another four. A pair of F-22 Raptors on their way over the Task Force joined in the fight.

The screens were becoming confusing. Five minutes earlier, there was only the coastline, and a little over a hundred ships and planes. Now there were over three hundred missiles and planes coming at them, and almost a hundred from the ships and planes in the Task Force trying to stop the attack. The processors could handle the information, but the big screens could only show it so fast. As more and more data flowed to the screens, the image became erratic and strobed. Things seemed to jump from location to location rather than move in real-time.

Sam knew how may SAM's and anti-aircraft guns were on every ship in the task force. He added in the number of air-to-air missiles that each plane on CAP was carrying. Then he figured each ship could take one hit, so he added the number of ships to the total. It was simple in the end. There were too many missiles coming at them, and each ship was probably going to take at least 3 hits. Most were never built to take two.

It wasn't long before sheer numbers began to overwhelm the escort ships and fighters. A Knox class frigate was hit. Then the two F-22's went down. Two Spruance class destroyers were hit. The missiles were breaking through. A Ticonderoga Class cruiser began taking hit after hit. Then the Task Force's supply ships began getting hit.

Admiral Pender was knocked off his feet when the first missile hit the Kittyhawk. So was everyone else standing in the room. Even some that were sitting went to the floor. Three more hits followed in rapid succession. There were a few seconds of pause before the next series of explosions.

On the screens, there were fewer and fewer blue icons; SAM's defending the Task Force. Fewer blue ships on the screen meant that some had already gone down. Fewer blue airplanes in the air meant there was less CAP to stop the incoming missiles. The red missile icons continued appearing unabated.

Another series of explosions, fifteen total, bounced everyone around in the CIC. When they stood up, they could smell the smoke from fires around the ship, and some, Admiral Pender among them, noticed the ship's 12-degree list to port. The console operators once again rose to their feet, returned to their stations, and issued commands to unseen ships and planes in the Task Force. Only seven minutes since the first missile was detected, a calm voice sounded over the ship's public address system, the 1MC.

"This is the Captain, abandon ship, port side only. Let's be professional people, and we'll be able to come back here to collect our dues. Good luck."

As if to punctuate the Captain's voice, another explosion ripped through the ship. All around the Kittyhawk, fires and flooding were out of control. One of the fires poured down an open hatch and set off a liquid oxygen tank was used to refill the pilots' bottles for high altitude flying. That explosion cut the ship in two from the keel to the flight deck just behind the island. It was like a million cutting torches. She sank less than three minutes later.

Wednesday, February 2, 2011, 0447 hrs Task Force Stingray, U.S.S. Wasp Landing Force Operations Center (LFOC)

Everything fell quiet as a sailor entered the room usually occupied only by Marines, and reported to General Shapiro.

"Sir, Commander Blackwell has asked me to report the situation to you directly. The commies've hit the Task Force with everything they had, and we've had our asses handed to us. The North Koreans came at us with anti-ship missiles launched by patrol boats that were hidden along the coast to the east. They also used two subs to the southeast. We don't know how many bombers came at us, but those seemed to come from everywhere. To top it off, the Chinese hit us with their patrol boats and lots of their bombers too.

The General hid his shock well. Everyone suspected that their little venture into North Korea could lead to an all out war in the area, but no one ever really comprehended it.

He nodded for him to go on, and the Chief continued.

"They fired enough missiles to overwhelm our air defenses, about six to seven hundred as far as we can tell. Then it was just a matter of how much damage we could take. Our CAP made a good showing, so did our escort ships, but they were both the first to go. The pilots reported that there were so many commie planes in the air, they didn't have to worry about hitting each other."

One of the General's staff interrupted.

"How bad are the escort ships hit, or the carriers, the Peleliu-?"

The General cast a stare at his aide who understood his silent reprimand, then the Chief went on.

"The escort ships are all gone, the Kittyhawk is gone, the Abe Lincoln is burning, and the Pelilu's listing. They're trying to bring the Peleliu over towards us in case she has to be abandoned. All total, there's only four ships left on the scope, including ourselves."

"That brings up probably the most important point, doesn't it chief?" Shapiro asked. "Just what kind of shape is our boat in?

"Well sir, it's like this. We've taken six total hits, and two close calls. The Well Deck ramp is gone, but, we've been able to counterflood to keep the ship from going down ass first. Then we had a bomb or a missile skip off the deck and explode on the rebound. That trashed at least four of your Osprey and one Harrier. There's probably a lot more, but those are definitely gone. We took a hit inside the well deck too, so your LCAC's are probably messed up pretty well. They're at least trapped by the debris till we can get it cleared. We took a hit that glanced along the starboard hull, and there's a hell of a gash in our side about ten feet above the water line. Every time a wave comes along our side, the damage control guys have to hang on or be swept out. We also took two hits here in the island, as I'm sure you well know. The first hit on top and took out our aft antenna array, including the Mark 91 fire control radar. The next hit between the Sea Sparrow launcher, and the 25mm autocannon mount. The shrapnel took out most of the people on the bridge, but it's still operational.

The Captain's dead. Commander Blackwell, our XO, was down in the CIC, and he's taken command. Flooding is controllable. The fires were put out as soon as they started. We really jumped on that. All the smoke has been taken care of with our NBC equipment. Damage Control has small teams going throughout the ship doing a headcount and looking for small repairs."

The General let everything set in for a moment. Without escort ships, or carrier escort planes to protect them, the Wasp and the other remaining ships were sitting ducks. An attack of this magnitude meant that things were definitely going to go bad along the DMZ. Of course, with their direct links to the Pentagon, they higher-ups probably knew of Task Force Stingray's demise before he did. They probably watched it live and had it on tape! There was no way to make Stingray serve as an offensive unit for some time now. It was time to withdraw, and they'd need help from the Peninsula.

"Chief, that was a fine report. How's Admiral Pender taking this?"

"Admiral Pender is missing, sir. He was on the Kittyhawk when it went down. They've got plenty of survival gear on those big ships, but when the water's this cold, you only last a few minutes unless you've got a survival suit on. We've put out all the choppers we can to pick people up, and bring them here. The Kittyhawk alone had almost 5000+ men and women aboard. Add in Abe Lincoln and Peleliu with it's 3000 if they both have to be abandoned, and add in all the escorts, well... only a few hundred will make it, sir. We really got our ass kicked. Is there anything you have for us, anything you need?"

"No chief, I think it's the other way around. Let Commander Blackwell know that we're at his disposal. I might be ranking officer in this Task Force, but as far as I'm concerned, what he says goes."

Wednesday, February 2, 2011

2nd Lt. John Chamberlin-Personal Log

Somewhere in the Yellow Sea

Pearl Harbor was a joke. On December 7, 1941, the Japanese almost completely destroyed our Pacific Fleet, and almost 3000 men and women were killed and wounded. This morning, the air and sea forces from the Democratic People's Republic of North Korea, and the People's Republic of China attacked and destroyed Task Force Stingray. Words will never adequately describe what happened, they can only report the facts.

Just after 0200 (2:00am) this morning, I woke up to yet another of the never-ending general quarters alarms on the U.S.S. Wasp. Since one can never be certain of a drill's reality, I grabbed my clothes, and made my way to where my current battle station is; the forward cafeteria. Normally my platoon would be with me in case damage control was needed, but they're still on the Peleliu. There were a few stray Marines who joined me instead. We fired up a pot of coffee, and waited for the secure from general quarters command to sound from the ship's 1MC system.

The first pot of coffee was about ready when someone commented on how loud the thunder outside was. Weather at sea has always amazed me with its strength and unusual combinations. I've seen storms that make the Wasp twist like a rubber band, and I've seen storms with clear sunny skies. I've heard thunder in snowstorms back home, but never like what we were listening to. Something was wrong. I spoke with a Marine Captain who was the senior officer in the cafeteria, and he let me go out on deck to check it out.

The easiest way to the deck from the cafeteria is through the ship's central corridor. About a quarter of the way down the hall, heading aft, I went up some flights of stairs to the flight deck. The hatch was dogged, and after I got it open the wind outside flung it open. It slammed against the bulkhead from the wind. I stepped out and immediately remembered that I'd I had forgotten my coat. It was freezing cold, and the wind was running the length of the deck from fore to aft. There was a lot of blowing snow too, and I had trouble seeing the waves below, let alone any lightning. I was right. I didn't see any lightning. I saw explosions.

The first one I saw caught my attention, but I wasn't sure that it was anything other than some weird-shaped lightning until I heard the report. Then I saw four more flashes, and heard four more reports. There was no question about it. This was for real. I ran back down to the cafeteria and reported to the Captain. Then we all waited nervously for the attack to end.

About five minutes after I had reported the situation, the Wasp took its first hit. The entire ship shook like a bad car accident. The lights went out for a few seconds, and alarms screamed. When the lights came back on, we checked out the cafeteria and did a head count. Everyone was alright. The 1MC sounded and a damage control party was ordered to the well deck. Outside the cafeteria, sailors ran as fast as they could to check out the damage and begin any repairs. We stayed at our battle station and watched the action down the corridor. It was pretty exciting until the smoke rolled along the ceiling like an upside down wave through the corridor, and into the cafeteria. We all hit the deck to grab as much fresh air as possible, then we dogged the hatches to slow the smoke.

A few minutes after that, when everyone was settled down, we took two more hits in rapid succession. It started to feel like we were in a washing machine. Again the lights went out, the PA blared, and more damage control parties were allocated to sections of the ship. The smoke thickened, and it was getting tough to breathe.

We could feel the ship accelerate and turn from port to starboard, then back, again and again. They were trying to evade the missiles, or torpedoes, whatever was attacking us. I don't know how you can move a 50,000 ton ship at thirty miles an hour and dodge a torpedo moving at fifty or sixty miles an hour. What about a missile moving at 300, 400, or 500 miles an hour? I don't know if they were able to dodge anything, or not. All I know is we were hit again.

I don't remember how long it was until the next series of explosions shuttered the ship. It could have been seconds, or minutes. It seemed like hours. Four more explosions went off on the port side of the ship. The last one sounded like it hit right above us. It actually skipped off the flight deck about 100' aft of the cafeteria. This time, the lights stayed out, and the flashlights went on.

Again we did a head count, then checked the room for fires, leaks, or any other damage. Just before everyone finished up, there was knock on the hatch. In truth, somebody was banging. It was stuck. One of the stray Marines got up and opened it. The lights in the corridor were back on, and the smoke was sucked right out of the cafeteria. Outside, there were three sailors. One of them was a chief who informed us not to shut the hatch anymore. It seems the Wasp is equipped with a nuclear biological warfare air purification system. When the smoke got bad, and we closed the door to the cafeteria, someone turned on the air purification system, and vented the ship. That is, except for the cafeteria.

I wasn't needed in the cafeteria, and, since I was without even a company commander to report to, I told the Captain in the cafeteria that I was taking off. He didn't have a problem with my leaving, as long as I came back and told him what was going on. I was happy to agree, and off I went. The Chief said I should stick with him and his team, so I did.

We went throughout the deck, checking compartments for an electrical short that he and his two men were ordered to repair. We found another pair of Marines in a smoke-filled barracks, but never did find the electrical short. The chief went back to his damage control team officer to report, and I went up to the conference room where I had spent the last few days. General Shapiro and his staff were busy, but when I walked in, they all stopped and looked. They immediately got back to work when they saw that I wasn't whoever they were expected. Everyone was too busy to mind my presence, so I resumed my post at the coffee maker, again.

A Chief came in and reported the situation to General Shapiro, and it doesn't look good. Our ship, the Wasp, is pretty well beat up, and most of the Task Force is gone; simply gone. Out in the cold winter mist-covered Yellow Sea, thousands of sailors and Marines are dead or dying of hypothermia. The Chief said only three other ships survived the attack. At the time, it looked like our sister ship, the Peleliu was going down too.

Later on, I found out that the Peleliu is going to make it. She's using her LCAC's and AAAV's to search for survivors. The Abe Lincoln is still burning, but she's still making steam, so they're going to try and get her south with the rest of TF Stingray; out of North Korean and Chinese bomber and Patrol boat range. We've got a lot of air cover from the USAF in South Korea, and the R.O.K. Navy is coming over to help cover us until we're out of enemy range.

The situation on the peninsula is crazy. Terrorists or commandos are popping up all over. Everyone is on full alert, and the place is in a state of near-martial law. U.S. Troops are being rushed in, and there are reports of Chinese troops just north of the DMZ, but they might be just rumors.

The chief was right when he told Shapiro that we got our ass kicked. This is no longer a private little war between North Korea and Task Force Stingray. I imagine that a formal declaration is only hours away. You can't have 10-15,000 sailors and Marines lost at sea and not call it a war! It'll probably be on TV about the same time my company comes over from the

Peleliu tonight.

Thursday, February 3, 2011

2nd Lt. John Chamberlin-Personal Log

80nm W of the Republic of Korea (South Korea)

Well, I'm back with my platoon again. They came over on an LCAC from the Peleliu last night. Captain Brown is coming over later tonight with the rest of the company. They're pretty worn out, even more than most in the Task Force. I had Gary, Wess, and Rick meet with me once they had settled in.

Everything went as well as could be hoped for under the circumstances. They had all seen their share of action now, both at LZ Delta, and on the Peleliu. I was pretty impressed with their stamina. For the most part, they seemed tired, dirty, and not much different from when I left them at the LZ. I was a little embarrassed to have not been able to be with them, but that was erased when I realized how impressed they were that I was up an around so soon after being shot in the head! I was never hurt as badly as it looked, and everyone in the platoon obviously thought I was a goner back at the LZ. When they saw me again, they treated me like I was some sort of super-Marine.

We all got together after breakfast this morning and talked for a few hours about the past couple of days. The entire ship held a moment of silence for all the sailors and Marines lost over the past couple of days. We remembered Mike Baker and Tim Potter in particular. Mike was unusually frail and even timid for a Marine. Still, his training and his will always helped him pull through. Even when he died, he showed no cowardice. Tim Potter had been the vocal country-boy wanna-be rebel. His intelligence was never more than the GED that the Corps helped him achieve. What he lacked upstairs he made up for in heart and conviction. Tim was always looking for some sort of confrontation, and the idea of him doing anything less than measuring up to the Corp's standard is crazy. We'll miss them both. We'll also miss all the Marines and Sailors that we used to see every day: the people who served us our meals, flew us to our Landing Zones, or sailed us into harm's way. Their names may have escaped us, but the memories of our daily lives in peacetime will not.

Since Captain Brown hasn't come aboard yet, we still don't really have a direct superior officer. No one has been keeping us informed on what's happening outside our barracks. We've been able to hear all kinds of rumors. About the only thing that we know for sure is that the President has gone before Congress, and, for the first time since World War II, the United States of America is officially in a state of war. I had Chuck Boyd use his laptop computer to see if he could get any tangible news off the internet. It only took a few seconds.

Last night, while the platoon was coming over from the Peleliu, North Korean and Chinese military forces invaded South Korea. The South Koreans had detected the communist buildup, and, minutes before the attack actually began, were able to get in some very effective air and artillery strikes just as the communists were starting their attack. The communists were still able to come across the DMZ in enough force to overwhelm the South Korean and American units that were defending the border. Our units withdrew in fairly good order, from what the news reports on the internet have said, but the retreat seems without end. The North Koreans have struck with two armies along the west coast, one army attacking along the east Coast, and one army seems to have stopped in between the two prongs, probably to remain in reserve.

They also were able to infiltrate an unknown number of commandos and fifth columnists into South Korea before the attack. It's assumed that most were dropped off by mini submarines or came under the DMZ through the famous tunnels dug over the decades. Some, of course, probably took regular passenger flights from intermediate countries. They're causing a lot of trouble by attacking just about anything whether it's of military value or not.

The communist air forces have really surprised a lot people. They made an effective attack against most of the South Korean and American airfields. Our air forces are running into the same problem we did the other night. For every three communist planes they shoot down, we lose one, but they've got nearly a ten to one advantage in aircraft. At this rate, even with help flying in from the States, we'll be without air cover in just a few days.

On the lighter side of things, Wess and Rick have informed me that Christina and Chris Landas are no longer an item. Since I wasn't on the scene, I wouldn't want to call him a coward for his actions in battle, but my NCO's all tell me that Chris could always be found in the safest place. Apparently Christina took note of this, and broke off their relationship. It was not done privately, little is in a small unit, and the frustration, along with the stress of the past few days, has made Chris fairly bitter. It's for that reason that the matter was brought to my attention. I think Rick and Wess would like me to talk to him, but since Bob still has some animosity towards Portia, I can't really come down on one and not the other. I can't afford to interfere in Bob's relationship difficulties anymore than I already have. I don't think I'll intervene unless it becomes a performance problem.

Friday, February 4, 2011, 0717hrs

Kunsan Airbase, Republic of Korea

After the preemptive air and artillery strikes the previous day, the American air base at Kunsan South Korea had been the victim of three air raids: two by North Korean aircraft, and one by the Chinese. Pilots from the base had been flying around the clock intercepting those three raids and trying to stop the rain of air targets crossing the DMZ. Communist aircraft, had been unable to shut down the base's operations, but they had taken their toll. Originally, 3 Squadrons of F-16 Falcon fighter planes had were stationed at the base, 72 planes. After less than 24 hours of operation, only 52 remained operable. Almost an entire squadron had been lost between the three. The 400+ aircraft that the South Korean air force could muster did little to slow the communist planes. Their losses were similar to date. Training aircraft were now being armed, and pilots, even in shorter supply, were being pulled from all walks of life. During the night, anyone with military flying experience began to find Military Police units (MP's) knocking on their front door. Al Meyers was one of those people.

Since his plane had been heavily damaged on the spy mission a few days earlier, he had taken up a temporary residence at the base. An old friend, Colonel Madison, commanded of one of the American F-16 squadrons. When the MP's knocked on the door, Al's friend was with them. It was he, not the MP's, who suggested that Al grab the stick on one of his F-16's. Al was a bomber pilot who had always trained to fly low and fast in order to escape fighters. He'd been trained for tactical reconnaissance missions, as well as air combat maneuvering, but Al was no fighter pilot. His higher-ranking friend could have ordered him to grab the stick, but there was no need. Al was first and foremost a cool professional pilot, and he eagerly agreed.

By 0700, he was being briefed on the avionics of the F-16. Most of it didn't sink in. There wasn't enough time to train him thoroughly. Al needed to know the plane's basics; 'stick and rudder' stuff in pilot jargon. He was told how to arm the (6) sidewinder missiles, and the 20mm gatling gun. After he was told a little about the fighter's handling characteristics, they went off to the locker room to suit up.

Thanks to plenty of training and experience in other planes, Al at least knew how to use a G-suit, how to use a radio, and, most importantly of all, he knew how to use a stick and rudder.

Just before 0730, Al accompanied an airman to a hangar where his F-16 waited. It had been damaged by cannon fire from a North Korean Mig the night before, and the pilot was lucky to bring the plane home in one piece. The ground crewman worked through the night to clean and repair the plane, but until they had time to give it a complete overhaul, it was not going to be 100%. The airmen who had been up all night showed no sign of fatigue. Adrenaline was plentiful between and during the incessant air raids. They helped strap him into the fighter, and, shortly after turning on the radio, he was ordered to taxi out. A gentle slide on the throttle, and the plane was moving. Al gave the crewmen a thumbs up and a salute as a sign of thanks, but most of them were already jogging down the taxiway to another hangar with yet another damaged plane in need of their care.

His thoughts and emotions churned. What little chance would he have in a strange airplane, a strange environment, and against so many other planes? He thought about the pilot who had been wounded in the very ejection seat where he was riding. He thought about his friend who was also wounded by cannonfire a few nights earlier. He thought about the war, and whether or not it was his fault. Had he done something wrong? None of that mattered though. What mattered was that he did his job so that the men who were on the ground could stay alive; so that South Korea remained free. Idealistic thoughts, he realized, but he needed something to fight back the anxiety. That fear, wellfounded in its origins, only had to take away his concentration for a split second, and he would get killed. It was time to get to work.

At the end of the runway, his wingman, his friend, his host, would be by his side. Colonel Madison, Al's wingman, waved. They waited and listened for the control tower to clear them for takeoff.

Finally, the order came.

"Wolfman flight cleared for takeoff. Wind is from 230, traffic is inbound from 330 at angels 2. Weapons free upon visual confirmation of targets. Good hunting."

Colonel Madison knew Al's callsign, and had arranged to have their flight named Wolfman hoping that the familiar name would help Al feel more comfortable. It did. Before the tower had finished giving them their brief engagement orders, Al was already moving down the runway, and Madison had to toggle his afterburner to catch up. After all, he was supposed to be the one leading the mission.

The two planes accelerated to almost 200 miles an hour, then both slowly aimed skyward. At 2000 feet, they turned to the north-northwest. Their heads swiveled as both men searched for their inbound attackers. They crossed the DMZ in only a few minutes, and somewhere ahead, through winter's overcast sky, communist troops were attacking. In the bright sky above the cloud deck, enemy aircraft were coming directly at them. The clouds offered a small degree of safety, but only from visual detection. It didn't matter. There was so much radar energy in the air, and so much electronic countermeasure activity, that all aircraft, on both sides, found themselves relying on visual detection and communication methods.

Radio gave little more than static and garbled messages. The radar was too complex to operate anyway. They were on their own. A few tense seconds later, Madison rolled his wings to get Al's attention. When he looked over, Madison was pointing to their left. He had spotted the enemy. Al waved as a sign of understanding, and his senior wingman lead them in a slow left turn to engage their targets. Through the clouds, Al finally saw speck that might have been an enemy plane. Just as he noticed it, Madison fired off a sidewinder missile. It disappeared into the cloud cover, until there was a small glimmer then a black puff of smoke coming closer to them.

Madison gave a thumbs up to Al and then his plane exploded. There was no time to be shocked. Al rolled to the left, then pulled back hard on the stick. His F-16 Falcon turned tighter and tighter until he was completely turned around. The high G-forces left him weak, but he refused to break his concentration just because of some physical difficulty. Al knew that if he stayed in the clouds, he could be hit from any direction just like Colonel Madison had been. If he climbed out of the clouds, he would be a much better target for SAM's and fighter planes. Since his skills were in low and dirty flying, not dogfighting, he dove for the deck.

The gray clouds enveloping his plane opened up as he broke the low ceiling at about 1000 feet. Snow-covered the ground below. He knew that as long as he didn't go too far north, he would probably be able to land at a friendly airbase. A quick turn of his head to make sure no one was behind him, and Al saw his enemy. 3, Mig 21's broke through the clouds behind him. They were already spreading out to corral him.

Al moved the plane from left to right and back again repeatedly in an effort to confuse them, but it only allowed them to get closer. A missile came towards him, but it must not have locked onto his plane, because it never tracked him as he turned to avoid it. Another missile came from another one of the Migs, but the missile again failed to chase him. The last Mig was now close enough to use its guns, and tracers passed overhead. Al watched as they impacted into the snow ahead of him.

It was time for something radical. Al pulled back the throttle and extended the airbrake. He was thrown forward in the straps of his ejection seat. When the plane slowed to 350 knots, he pulled in the airbrake, moved the throttle forward until the afterburner fired. Then he began another high G-turn. The Migs were incapable of turning as tightly as the F-16, few planes were, and they went past him. He continued his turn until he made nearly a complete circle. When he heard the growling sound of a locked on sidewinder missile, he popped off one of his missiles without even questioning what it had locked on to. As he continued his turn, he lost sight of it. Through the top of his canopy, he noticed a Mig coming at him, and he fired a burst from the 20mm cannon just as it passed him from the front. Al reversed his turn, but saw no more Migs. He headed East, towards home.

A few minutes later, through the static on the radio, Al was able to make out the base's control tower.

"Wolfman flight, this is Pig Pen, over."

PigPen, the callsign for Kunsan airbase, sounded beautiful.

"Pig Pen, this is Wolfman Flight short one. No chute seen. Navigation is out. Please direct home."

"Copy, Wolfman flight. Head 053."

A small course correction, and Al was on his way back to his new home, Kunsan. Once there, he would refuel, rearm, and be given a new wingman, probably someone with less experience than him, and definitely someone with less experience than Colonel Madison. At least he wouldn't have to be the one to tell Madison's 24-year-old wife and 3-year-old little girl. No, Al would either be flying more missions, or be dead too. With only one mission in a fighter plane under his belt, and barely an hour's worth of training, Al expected his death to come soon. His professional attitude allowed him to store those thoughts until there was time. He knew that if he wasn't completely alert while flying, he would be joining Madison very very soon.

Back in the U.S.A., Congress was getting to work. Decades of partisan politics seemed to have ended overnight. The loss of Task Force Stingray at the Battle of Yellow Sea, as the press had named the event, had enraged the populace. All of a sudden, it was good politics to be a warhawk, and it was even better politics to be an effective warhawk. There were formal declarations of war against North Korea and China. Then in the same session Congress began putting together legislation to renew the draft. Over 1,000,000 men and women would be called to duty in the next few months.

Using an old law created for use in the American Civil War, Congress gave itself the authority to spend almost unlimited funds on war materials and military programs. The only provision was that the money would have to be paid after the war ended. With the fear of an even larger national debt in the near future, a movement began to cut spending on programs that were not related to the war. Social Security, Medicare, and other entitlement programs came under more than the usual amount of annual scrutiny.

A large group of senators proposed a government-brokered investment program, similar to the 401K used by many businesses. The objective of the entitlement brokerage plan was to make recipients get payment from fund managers instead of the government while, at the same time, increasing funds for the Federal Government by reducing spending and increasing funds without increasing taxes. Together with the Civil War spending law, Congress was able to flood the U.S. military with money.

American industry was about to get the biggest blanket purchase order in its history. The Pentagon needed as many planes, tanks, ships, rifles, bullets, and helmets as it could produce, and industry could now count on getting paid.

Procurement programs already underway could be stepped up overnight by adding more shifts, but, for the most part, the age old Arsenal of Democracy that had defeated imperialism in the First World War and Fascism in the Second, would again need months, even years before sufficient arms were prepared to defeat communism in the Third World War.

Around the globe, leaders chose their sides. Some would support South Korea's defense. Others supported North Korea's claims that it was defending itself. Britain, France, Spain, Turkey, Greece, and several other nations made preparations to send reinforcements to the peninsula. Russia, most of its former republics, Serbia, Iraq, and many more nations claimed to support the North Koreans. A few nations refused to comment on the war except to call it a tragedy etc.

Saturday, February 5, 2011

Washington Post

Get It Done

During the night, North Korean amphibious units landed south of Seoul, and little opposition remains in the nearby city of Suwon. Helicopter-borne troops landed along the east coast of South Korea a few miles north of Pusan. Thousands more Chinese troops left from a container ship moored at a dock in that city. Local police, Army, and reserve units managed to prevent the two Chinese groups from forming a cohesive line, but key positions in Pusan are still contested by pockets of communist troops. Radio and television stations, police stations, parts of the airport, portions of the docks, and tactical strongpoints around the city were all the Chinese could grab. Snipers are having their own private war between the skyscrapers, and only the brave walk through the streets. Only armored vehicles roam the roads, but even those prudently drive as fast as they can.

Further north, there was extremely heavy fighting along the east coast of South Korea between elements of the United States 8th Army, and the North Korean 1st and 2nd Armies. The Chinese 4th Volunteer Expeditionary Force (VEF) is supporting the North Koreans.

American forces, while initially thrown back from the DMZ, regrouped and put up a solid defense, supported by close air support from nearby USAF airfields. USAF pilots were flying a record average of 11 combat missions a day, but limited air interdiction missions by Chinese fighter bombers, and periodic fighter sweeps from communist planes are reducing the effectiveness of American sorties. Losses continued to mount among pilots and material at a rate slightly faster than the arrival of reinforcements. At the current rate of loss, the U.S. and R.O.K. air forces will be completely eliminated in less than five days. After that, only reinforcements will be able to contest communist air superiority.

Fighting along the western and central portions of the DMZ was extremely heavy, particularly north of Seoul. Communist artillery and mortar attacks occurred often, and with a fair amount of effectiveness. Frontline units are losing personnel at a rate, which would leave them below half strength within three days. As yet, infantry and mechanized ground units have attacked in force, but no larger than regimental actions have been attempted. Division and Corps-sized operations have been limited to probing actions done in force. The force of regimental sized attacks, with the full support of all the artillery, mortar, and rocket units attached to entire Corps was fearful. Overnight, the DMZ has changed in appearance. From the air it now resembles World War I Belgium, or a snow-covered moonscape.

Thousands of miles away, millions of Americans continue to cry out for revenge. Many, with the bloodless experience of Desert Storm etched in the memory, cannot understand how a defeat as terrible as the Battle of Yellow Sea could have taken place. Task Force Stingray's commander, Admiral Pender, was rescued and immediately recalled back to Washington to face a series of Congressional hearings. Those hearings were selected and put together even before his rescue two days ago. The new President, Chief of Staff, Chairman of the Joint Chiefs, and several other cabinet members have also had their actions called into question, specifically the events that lead up to the disaster at LZ Delta.

Witch-hunting has quickly proven to be a good way for warhawks to show the American people their deep involvement with the war even after it has already begun. A lack of footage from the Battle of Yellow Sea is leaving the media with an emptiness that the warhawks have only been to happy to fill with scalding commentaries, monotonous interviews, and soundbites.

The most memorable soundbite to date has been from a House Representative of Indiana. She made a comment to a network anchor during an interview about the President's indecision prior to the LZ Delta disaster. It immediately became a slogan for the people who now wait in long lines to enlist. The Congresswoman explained how the tenuous peace left on the Korean Peninsula by the Truman administration would never have occurred if President Truman hadn't relieved MacArthur. "For once," she said, "let's get in and finish the job. No more DMZ's, no more police actions, no more advisors, no more peacekeeping

forces. Let's go in and get it done."

Get it done.

Sunday, February 6, 2011

Los Angeles Tribune

The New Korean War

During the Battle of Yellow Sea, Task Force Stingray was prohibited from using one of its most valuable assets: SC-1, the Arsenal Ship. Designed as a cheap way of increasing a Task Force's firepower, the U.S. Navy built four Arsenal Ships. Each ship, loaded with 500 vertical launch tubes, can be linked to the Task Force data network. Once linked, any plane or vessel could send targeting data to the Arsenal Ship, and whatever type of ordnance was called for would be fired from the Arsenal Ship instead of from the unit calling for the ordnance delivery. This would have doubled the number of SAM's at Task Force Stingray's disposal, and defending ships could have used their own missiles to ensure the survival of their own unit. By order of the Chairman of the Joint Chiefs of Staff, SC-1, the Arsenal Ship that could have supported Task Force Stingray, was held in reserve in case a full-scale war erupted on the Peninsula.

With the destruction of Task Force Stingray, there have been only a few units capable of directing missiles from SC-1, and its entire inventory has already been used, piecemeal, and with no notable effect on the enemy's advance. By noon, the ship was on its way back to Osaka, Japan for rearming.

Back on the peninsula, North Korean forces conducted their first Corps-sized infantry attack of the war along the western side of the DMZ. After a vicious and wellcoordinated air and artillery attack, they immediately broke through South Korean units, and within hours they joined with the North Korean amphibious force that had landed south of the city of Seoul. Inside the capital, most South Korean troops have fought with iron-like tenacity, but thousands of have been routed and are fleeing to a new frontline, which is being setup south of Seoul. Fifth columnists and commandos continue to raise Cain inside South Korea. In Pusan, a few pockets of resistance have been mopped up, but the city docks and airport remain contested. Back at the DMZ, the U.S. 8th Army has been forced to fall back to a series of positions almost 20 miles to the south because of continued Regiment-sized infantry attacks. The remnants of Task Force Stingray, while trying to make their way to Osaka, came under air attack again, and the aircraft carrier Abraham Lincoln was set on fire once more.

In Los Angeles, California, a busload of Chinese businessmen were mobbed and severely beaten outside the airport. The event, caught on camera by amateur photographers, security cameras, and television news helicopters has created quite a stir around the globe. American politicians, still riding the wave of pro-war sentiment, have publicly called for restraint while encouraging the public to focus on the enemy abroad.

Overseas, dozens of nations, most having felt the broad hand of American wrath over the past few decades, have condemned the act, and used the beating as an example of a decadent society out of control. These same countries were each completely silent when it came to commenting on the Battle of Yellow Sea, or the repeated withdrawal of R.O.K., US, and other U.N. Coalition forces in South Korea.

Sunday, February 6, 2011

2nd Lt. John Chamberlin-Personal Log

Somewhere in the Yellow Sea

It doesn't look like we're supposed to make it out of this area. There's an air of frustration and borderline pessimism on the ship. Everywhere I look, familiar faces are missing. They've been replaced by tired, worn, worried, and weathered expressions. Some are survivors that were picked up over the past few days, but most, upon careful examination, are people I've known for years now. The past few days have changed everyone. I wonder if I look like a stranger too, or have I stayed the same while everyone else changed?

The low morale hasn't affected the professionalism of most people yet. Everyone seems to be doing their jobs as best they can. Nobody wants to been seen as the weak link in the chain. The only place where you can really see people who've had enough is in the med center and on the well deck. The med center is still clean and orderly, but the staff is so busy that they look like they're in a panic. If you watch for a few minutes, you'll see that they're just working at a faster speed, and the urgency of their situation makes you nervous. The well deck is where most of the survivors have been berthed. The ship was crowded to begin with, but put a few hundred extra people on board, even if you confine them to one area, and all the problems associated with close quarters creep up. There have been a few loud arguments and even some brawls, but for the most of the men and women seem stunned. There were thousands of people in Task Force Stingray a few days ago. Now, only a fraction remains alive. Most of those seem to be wondering if they'd have been better off with their friends, rather than living with survivor's guilt.

To compound the situation, we were attacked again this afternoon. At about 1530 hours Chinese bombers came in low from the East. They used the coastline to hide their approach from our radars, and we were only able to get off a few SAM's and a little AAA before they were on top of us. They went straight for our big smoldering carrier, the Abraham Lincoln.

When general quarters sounded, I met the platoon at our usual battle station, then, when everyone was settled in, and the coffee pots were working full blast, I went up to the flight deck to see what was happening. No matter where you are on a ship, when it's under attack, every plane, every missile, bomb, and bullet seems to come straight at you. Only two bombers came towards us. One buzzed us without dropping or firing anything. The other laid out a string of six iron bombs.

All missed except one that actually skipped off the flight deck. It hit near the bow, bounced, then flew over the ship's stern. The plane took a full pointblank blast from one of our old 20mm Phalanx guns, and it came right apart. The whole thing scared the hell out of me at the time, but looking back on it, it was pretty exciting. About 10 miles to the west, off our starboard bow, the carrier Abraham Lincoln took the bulk of the air attack. She was still smoking from the damage she took the other day, and when the Chinese bombers came in, I didn't see her launch a single plane. Six planes passed over her from bow to stern, just like they had done to us. Their bombs had a lot easier time hitting a big target like old Abe. I could see flashes from explosions, and smoke started billowing again. It happened pretty quickly. The entire show was over in a few minutes.

After the all clear was given, I heard that two of our remaining escorts were going to stay behind with the Abe. It doesn't sound like he's gonna make it, and I expect to see some more new faces on board soon.

Monday, February 7, 2011

Associated Press International

U.N. Forces Still Holding On

The U.S. 8th Army was able to make an orderly withdrawal to the south, but North Korean and Chinese units continue to push hard along the east coast of the Korean Peninsula. Amphibious and airborne units from Australia, New Zealand, and Britain have further reinforced the American 8th Army. Any reinforcements to the defense of South Korea are being hampered by the hundreds of Chinese still holding out in positions around the docks in Pusan harbor. All personnel and equipment have had to be flown in, and even that has been constantly interrupted by continued air strikes and fighter sweeps by communist planes.

Fighting intensified in Seoul as trapped R.O.K. army units held steadfast against an almost infinite number of communist infantry. House-to-house combat and skyscraper-skyscraper fighting has made infantry combat both a long range and a close-quarters effort. An attempt to break through the North Korean amphibious units south of the city with a South Korean armored force failed in the face of numerous anti-tank guided-missiles (ATGM's). The North Korean naval infantry came ashore with only a few vehicles, but they brought plenty of sophisticated Soviet built ATGM's and hand-held heat-seeking SAM's.

Tuesday, February 8, 2011

Washington Post

U.N. Concedes, Seoul is Lost

Small pockets of resistance are all that's left of the R.O.K. military on the north side of Seoul. North Korean units over ran Kimpo airfield, the famous sky tower, and even the Presidential Palace. Snipers and isolated pockets of separated infantry are barely slowing the communist forces. The only thing keeping them from taking the entire city are the few bridges crossing the Han River.

Bridges can only be taken by two methods. The first is for the attacker to charge across from one side, and take the enemy's side. The other method is to take both sides of the bridge at once, and then destroy the forces on it. The North Koreans may try either approach. A frontal attack across any of the bridges would be extremely costly, but with their ranks bolstered by Chinese infantry and their own reserve units, such an attack could be successful. South of the city, North Korean airmobile units could be used to make a helicopter assault on the south side of any bridge in R.O.K. hands. In either case, Seoul is surrounded, and its piecemeal capitulation is underway. How such an end comes is never of any true consequence.

Wednesday, February 9, 2011

Washington Post

<u>Japan Threatened</u>

Reinforced by units from around the world, R.O.K., U.S., and other U.N. forces in South Korea enjoyed their first day of air superiority. Communist planes attempted twelve air attacks, and tried three sizable fighter sweeps south of the frontline. All of the air attacks and sweeps were destroyed or repelled. Losses amounted to fifty-three communist fighters and bombers destroyed, another eighteen probably killed, or at least damaged. U.N. Forces lost forty-nine intercepting fighter planes, but were able to secure their airspace. Of great importance to the U.N. air defense was the use of U.S.A.F. units based in Japan. Chinese officials in New York served notice to all U.N. nations participating in the Korean conflict, that "...appropriate defensive measures would be taken at any time, without further notice, to end such operations." They went on to warn Japanese diplomats, "...such measures would be taken in the most serious manner." As a result of the diplomatic pressures from North Korea, China, Russia, Iran, Iraq, and a long list of others, the U.S. agreed not to conduct any more combat missions from bases in Japan. Support missions will continue.

Wednesday, February 9, 2011

1410hrs

Republic of Korea

Kunsan Airbase

Al Meyers was flying an average of nine missions a day in his worn-out and patched together American F-16. Finally it was getting some desperately needed maintenance, so he was on a stand down status. Still wired from days of coffee and caffeine pills, sleep was out of the question. The base commander was far too busy to trouble with his lack of official assignment problem, and, to the best of the commander's knowledge, he was all that was left of Col. Madison's squadron.

One flight with Madison, and, whether the paperwork had been pushed through or not, Al was obliged to be its sole pilot. He had been given replacement pilots with new planes from the States or around the world, but none had lasted. Since Madison's death, the squadron's pilots never numbered more than four pilots; including Al. Still in his G-suit, Al wandered the squadron facilities looking for something, anything that would tell him how to handle the situation. He couldn't just walk away. The Air Force headquarters thought it had an F-16 squadron under Madison's command, and, even with Madison gone, they still believed that the squadron existed. Al had tried looking for the group or wing commanders, but they were busier than the base commander, and couldn't be found. He'd even heard a rumor that they had been killed in a guerrilla attack.

Al wandered into the squadron offices and sat down at Madison's desk for the umpteenth time, wondering what his late friend would have wanted, what Phil would have expected, what the Air Force needed. He couldn't just dissolve the squadron. He couldn't get a new assignment, and he couldn't keep sending fresh replacements up to their deaths. Al had been thinking it over for days, and he finally came to the realization that there was only one professional thing to do. He had to find a way to rebuild the squadron in such a way as to make it a viable fighting force. Only then could he hope to give replacements a real chance. Only then could the Pentagon really assign squadron missions to more than one to four people. Only then could he responsibly search for a new assignment; one he had actually trained for!

Outside, fresh pilots and planes were arriving by the minute it seemed. Most were from other countries, but if Al was going to make his squadron work, he couldn't let that stop him. Determined, energized by the caffeine, and less than logical from exhaustion, Al went out in search of other squadrons that had been decimated. He walked down the hallway, and out to the hangar where his plane was being worked on. With every step he took, he became more and more determined to rebuild Madison's squadron by creating the 1st Composite Fighter Squadron for the Coalition, under U.S.A.F command.

In the hangar, Al spotted four men working on his F-16. He strutted straight over to the team Sergeant, a man he had worked with several times over the past few days. Still at least fifty feet away, he hollered over to him. His voice echoed amidst the thunder of artillery and the roar of jets in the background.

"Sergeant!"

The twenty-year-old technician snapped to attention out of reflex and replied just the same.

"Sir!"

"Sergeant, you've been gettin' around this base a lot lately. Haven't you?"

"Yes, sir. The war's got us haulin' ass from one corner to the other. We've even been sleeping on the wings and having our food brought over to us. I don't think any of us've seen more than three or four hours of sleep at a spell."

"Sergeant, am I the only one man squadron around here, or is that just the fluke that it seems? Don't answer! I'll tell you what I'm looking for. If there are other beat up units, I want to get them together, and form a composite unit. I'm tired of taking handfuls of fresh replacements up when we really need to be in squadron-sized flights. So, who am I going to talk to, and where are they?"

The worn out crewman had stopped what they were doing and joined into the conversation.

"Sir, you could go over to the Officer's Club and see if anyone's just hangin' out. I'm sure the bartenders will be able to point them out." Another piped up.

"Yeah, or you could go over to the tower. They'd be able to tell you who's still relatively whole or not. You might also try..."

The Sergeant knew exactly where Al needed to go, and he interrupted his teammate.

"Sir, there's a few squadrons that have pretty much had it, but you really need to talk to the R.O.K.'s. There's a liaison officer in the hangar down the way a little."

He pointed out the specific hangar to Al.

"I was just there about twenty minutes ago, but don't tell'em. I needed a few turkey feathers for your afterburner, and they've got F-16's too. I figure, we built'em, we might as well borrow some pieces from 'em too. It's not like I took 'em from a plane that was flyable. I just..."

"It's okay Sergeant. I'm glad you can handle that kind of thing. I want people that are going to be resourceful. By that I mean, you are now assigned to me. Same with your team. I'll draw up the papers tonight. Let's go find us that R.O.K. liaison, and keep the borrowing thing hush hush."

Al and his first conscript went over to the hangar and found the South Korean liaison officer. They also found three R.O.K. F-16 pilots talking to him. One was apparently upset about the lack of maintenance on his plane, but it didn't seem like the afterburners were the sole problem.

Al introduced himself, to the liaison officer, and told the South Korean his problem with the lack of command, control, or communication from his higher ranks. The South Korean pilots didn't speak any English, but they understood the scenario too well. They were having the same problem. Their squadron had been piecemeal wiped out. Their Wing Commander was missing, and now, no one was giving them orders. The only time they went up was when the base scramble alarm went off to intercept Chinese or North Korean bombers headed for the base. A few more minutes of conversation with the liaison officer, and Al had four new veteran squadron members, including three pilots with two planes.

Their next stop was the Officer's club - as the groundcrew airman had suggested. Sure enough, the place was packed with pilots who had no planes. A few days ago, the problem was too many planes, and no pilots, now it was too many pilots. In actuality, there weren't too many pilots, or too many planes. The problem was that some squadrons had aircraft come back, but pilots sent off to treat wounds. Other squadrons had pilots who brought their planes back so badly damaged that they were only good for spare parts. The pilots and planes needed to be put together. Al was more than willing to do that, and in an official manner too. He made his way through the club with his R.O.K. pilots and liaison officer. As they went from table to table, they recruited more and more pilots, 42 in all. Now, all Al had to do was find 45 planes, and create some paperwork to follow the regulations. This was not going to be some sort of pirate or mercenary unit. If he could find 45 planes, or even half that number, Al could legitimately create a composite wing! The 1st Composite Air Group.

Of course there was a lot more to do than just find pilots and planes. That was the easy part. He had to find quarters, hangars, groundcrews, missiles, fuel, and most importantly, an administration staff that could make things happen in such a way as to follow the letter of the law, but still not get strangled in regulations. His task was just beginning, but, when he took his new pilots back with him to check the status of his own F-16, Al was more than surprised. He was shocked. Word had spread around the chaotic base like a wildfire. At least a hundred, maybe two hundred people were in the hangar waiting for him to give a speech. They all wanted to be part of a whole unit, a fighting unit, a unit that actually existed, and a unit that could operate with or without command guidance.

Al saw ground crewmen, pilots, armament people, administrators, even a trio of weather specialists. They seemed to want a speech, so he climbed on top of his plane, just behind the cockpit, and gave them one.

"If I can have your attention for a moment please.... Thank you. I'm not sure what you've heard is going on here, but obviously, it's getting to be something pretty big. My name is Al Meyers, callsign Wolfman." There was some rumbling in the crowd.

"If anyone was on duty the night this whole thing started, you might have heard about my little recon flight along the border. Some would say that this whole war is my fault. It is not. This war has been going on for half a century. A lot of people seem to have been thinking that maybe it was time to finish it, and that's what we're going to do. I'm not exactly assigned to any one unit since that first night's action. My plane was pretty well trashed out, and I lost my wizzo. Col. Madison, an old friend of mine, thought I could fly, so he gave me the quick rundown on this plane. We went up, and I came back. Draw you own conclusions. Since then, I've taken up greenies right after they arrive in country. No one's lasted. If you're like me, then you realize that we don't stand much of a chance up there in small numbers. So, with that in mind, I'm trying to put together a composite unit. I'm looking for other beatup squadrons to make one, maybe two complete ones. Looking around, we might have enough people for three. I see some brass out there, I'll talk with them in a little bit. Until then, I'm not exactly sure how to organize this thing."

The crowd was silent and waited for his instructions despite what he had said. Al seemed to be the only person around with an idea, and the conviction to make it happen. Everyone else was wounded, dead, missing, or just plain worn out.

"Okay, I'll tell you what. Let's get all the pilots with planes in that corner, pilots who don't have planes on the other side of the drums in that same corner. I'll take groundcrew, armament, fuel, etc., anyone who turns a wrench to keep me the air over there by the back door. Anyone who knows how to work a phone, a computer, a map, a typewriter, or whatever, come see me right here, right now. Questions?"

There was no need to ask. As soon as someone heard their job description mentioned, they went to where Al told them. The administration people sifted through the crowd, over to Al's cockpit.

"All right then. You people are going to start getting us organized. Raise your hand when I suggest something that you think you can do extremely well. I need someone to start getting files together on these pilots. They're the one's that are going to make this unit a real thing-okay you. Now who can start cutting orders that sound like this thing was approved by the president himself?-okay you, get on it......"

The impromptu meeting went on for the rest of the day. By dawn, the 1st Composite Air Wing would be formed, with 49 planes under Al's command; 31 F-16's and 18 F-5 Freedom Fighters.

Thursday, February 10, 2011

Associated Press International

Ebb and Tide of War on the Korean Peninsula

Fighting continues in Seoul. North Korean airborne, naval infantry, and special operations units are pushing hard towards the north in an attempt to link up with their regular infantry units controlling the north side of the city. R.O.K. Units have formed solid, though well dispersed, defensive positions around the south side of the city. A weak attempt at breaking through to the besieged city was stopped by the North Korean special units to its south. American Marines, survivors of the battle of Yellow Sea, have been asked to try and support the besieged R.O.K. infantry on the southern half of the city. The R.O.K. infantry defending it have less than 9000 men, and the remnants of Task Force Stingray are estimated to be less than 3000 men and women.

On the Eastern coast of the Peninsula, communist forces attempted a helicopter assault behind U.S. infantry; specifically, elements of the 82nd airborne division. The helicopters tried to use the coastline as interference to U.N. Coalition radars, but a Canadian frigate was able to distinguish them, and twenty-nine helicopters was shot down. R.O.K. internal security units rounded up survivors within a few hours.

Communist aircraft are challenging U.N. air superiority again. U.N. aircraft were able to inflict heavy casualties on North Korean and Chinese bombers, but communist air forces are able to bring in fresh pilots and planes faster than their U.N. Coalition counterparts. Fatigue for the pilots and planes defending South Korea has begun to create nearly as many casualties as combat. The R.O.K. Air Force lost almost as many planes to accidents as it did in combat. Several U.S.A.F. Air National Guard units have been federalized for deployment to Japan as soon as possible. From Japan, they will be redeployed to South Korea before conducting combat missions against the north.

The South Korean government, reformed and operating from the city of Pusan, has sent envoys to nations around the world in an effort to form a unified coalition force. Those U.N. members currently engaged in combat on the peninsula have viewed the effort with some discontent. The proposed coalition would form a unified command under R.O.K. control, rather than that of the United States. An agreement before the conflict began made provisions for R.O.K. command of all forces in defense of South Korea, but, as units require assistance, it is the United States who is called upon again and again. Even though the United States has been stalwart in its defense, and the R.O.K. army has often been forced to retreat, a great many of South Korea's politicians and populace feel that the U.S. still has not done enough to defend their country.

This feeling is echoed in the United States by a renewed spirit of bi-partisanship in Congress; a sentiment driven by the American people's hunger for revenge. The bipartisan movement has made almost every Senator and Representative cry vengeance. Everyone wants to be seen in front of the camera, on the radio, or at least in front of their constituents as the leader of a unified American effort to get some payback. "America's sons and daughters will not die in vain" is a phrase repeated more than once every day. Budget legislation continues to pass through Capital Hill faster than it can be read in some cases. The White House and Capital Hill have promised everything and anything the Pentagon asks for. All the while, Admiral Pender sits in front of a Congressional investigation committee, seen in its entirety on all the major networks.

Thursday, February 10, 2011

2nd Lt. John Chamberlin-Personal Log

Somewhere in the Yellow Sea

Hector, Bruce, and I met with Capt. Brown today. I'm still not officially back on active duty, but we all agree I might as well be. The doctors on the ship won't have the time to certify me for a while, but as long as I feel I'm ok, Capt. Brown says he'll have me. He heard from Col. Ghetty that we're going to be sent into Seoul to backup some of the R.O.K. units holding out in the town. We're not sure of their situation, but from what we've seen on the TV news and the internet, we might as well not go. They're making the whole fight sound like it's already over. I guess that's the kind of fight Marines get sent to. Brown says that the R.O.K. guys really need our help, and that the higher-ups might just want to get some of the people off this ship before it gets attacked again. So far, only the luckiest people in the task force have wound up here. Eventually our luck may run out. I figure as soon as we're the biggest ship left on the water, we'll be the biggest target. I'd rather be on the ground than watching those bombers circling like buzzards while we've got nowhere to hide, and nothing to fight back with.

Friday, February 11, 2011 Associated Press International

TF Stingray's Marines Are Still in the Fight

American Marines, remnants from Task Force Stingray, began landing on secure rooftops in northern downtown Seoul just before dawn. Heavy small arms fire from higher rooftops around the city damaged three V-22 Osprey. Only one of the two others made it back to the ship. The last one disappeared. Marines were able to consolidate defensive positions and secure a continuous perimeter with the R.O.K. troops still holding out.

In the late afternoon, a strong attack by North Korean tanks tried to break through the perimeter in the late afternoon. It consisted of almost thirty Soviet-built Main Battle Tanks, and twice as many armored infantry-fighting vehicles supported it. Fighter-bombers from U.S. Navy aircraft that had escaped the Battle of Yellow Sea by landing in South Korea came to the aid of the Marines and the attack was decimated. Sadly, a handful of the Navy planes were lost to intense AAA and an undisclosed number of hand-held SAM's.

On the Eastern side of the peninsula, North Korean and Chinese forces attempted to airdrop supplies to their scattered troops holding position around Pusan, but without a well-defined drop zone, many of the supplies landed in U.N.-held areas. The contents of the parachutedelivered goods suggested to local U.N. commanders that ammunition is running out for their dispersed enemy. Further north along the Eastern coast, communist forces continued to push hard against American and other U.N. units. A penetration of the line occurred near where Australian and R.O.K. units met. R.O.K. internal security forces were called forward to help plug the gap. Within a few hours, the line was reformed, and the roundup of several hundred communist troops began. Large numbers of refugees moving south have also begun to hamper the resupply of frontline units.

Friday, February 10, 2011

1650 hrs.

Republic of Korea, Kunsan Airbase

Personal Log – Lt. Kim "Killer" Yoo, Republic of (South) Korea Air Force

For half a century, we've dreaded the day when the communists would restart the war. We trained and trained and trained, and it all seems useless now. In just a few weeks, my squadron has been wiped out. Our country is in shambles. Our government is on the move. The Army is trying to setup ANOTHER defensive perimeter, and even the Americans are getting pushed back. Everything we do fails. No matter what we do, or how hard we try, this nation seems doomed to communist occupation. We will never surrender, and everyone is ready to give their life to protect our country, but that simply might not be enough.

I know it's been 15 years since the United States took down Iraq in Desert Storm, but what happened? Air power was the Queen of the skies. Now, it seems useless. Our latest satellite –guided bombs and missiles aren't stopping the communists. Their old and inferior planes are almost matching us in air-to-air combat losses. The Anti-Aircraft Artillery (AAA) never even touched a stealth fighter over Baghdad, but today, every plane comes back with at least 10-20 holes. Surface-to-Air Missiles (SAM's) used to be easy enough to dodge or decoy with electronic jamming, but given the heavy losses in our aircraft, the old tricks don't seem to be working anymore.

After my last mission-the one where the rest of my squadron was finally wiped out-I was grounded by my group commander. I thought that I would at least be assigned to another squadron, but it's been three days, and no orders have come. It's a similar situation for lots of pilots. As fast as the world can send its squadrons to help us out, the units are decimated. Yesterday, one of the American pilots tried to form all of these one-man squadrons together into a composite air group. No one in the Air Force seems to remember that I exist, so I agreed to fly with them-at least until my own country calls me again.

The leader of this composite air group is an American Air Force Major by the name of Al Meyers. Rumor has it that he was flying the spyplane on the mission that essentially restarted this entire war. His callsign is "The Wolfman." It sounds like a typical, American flashy name, but in a lot of ways, Meyers isn't a typical American. He's much more reserved, professional, and has a "matter-of-fact" tone about him.

I'm not exactly sure how long this adhoc unit will stay together, but that question will not be answered by the communists. While the group's logistical staff (i.e.: thieves) are still getting their act together, Meyers organized everyone into three squadrons. Then he arranged for everyone to have at least two Sidewinders and a full load of 20mm. No one has enough fuel, but we were all bored and looking to get back into the fight, so went off on our first mission.

It was a fighter sweep over the DMZ. My English is fair, but since we have pilots from all over the world in each squadron, all of our communications are done by hand signals-a universal fighter pilot language. Meyers led the first squadron, I followed in another squadron about 20 miles behind, and roughly the same distance behind my flight was our last squadron. The American Air Force is still in charge of air traffic control over the area, but when they asked for our unit designation, Meyers just listed off all of the different squadrons that we were all from. It confused the AWAC's controller, so he just gave us the IFF code for the day, and sent us all on our firs intercepts.

Meyers took the lead squadron down from 40,000 feet to 20,000. There was a four-plane flight of Chinese H-5 bombers headed toward Seoul, so he turned the squadron to the east and pounced on them. Behind the Chinese H-5's was a full squadron of Chinese, Russia-made, Su-27 Flankers. They were trying to hide in the terrain at 3000. My squadron was still at 40,000 feet when we dove down on them. Our last squadron came in we completely overwhelmed the Chinese. We wiped them out without losing a single plane. I dropped two of the four bombers myself. It was the first total victory I've seen in this entire war, and everyone really needed it!

All of the pilots in this composite unit are not only survivors from other squadrons, they're also the best of the squadrons had to offer. The Americans say that the cream always rises to the top, but I think it's really just survival of the fittest. The communists shouldn't be too difficult for this batch of airmen. Our real problems will be dealing with all of the bureaucrats from the different nations in our unit, and just trying to understand each other.

So far, Meyers and our scroungers are doing a good job at working around all the different administrative commands. Most of the pilots seem to get along fairly well too. There are a few childish national rivalries, but I think those will work themselves out. Like the rest of the Korean pilots, I don't have time for that sort of thing. I also don't have time for the racism that some of the more arrogant pilots have brought with them. I've really only met one person who might have some real problems getting along with others, but I think he's really just upset about losing his plane to a mechanical instead of a communist. All in all, I think this unit will have some real potential if we can just get things a bit more organized-so we can fly more than just a mission or two a day!

Friday, February 10, 2011

1650 hrs.

Republic of Korea, Kunsan Airbase

Personal Log – *Lef*tenant Ian Thomas Sanders IV

It just goes to show, you can't bleeding trust technicians. "Oh, no, *Lef*tenant, sir, your plane's tip-top. Everything tight, ready for flight, off you go!" (If I ever get my hands around that rhyming yob's neck, I'll make him wish his mum was a nun.)

I suppose I'd better start with yesterday evening – it'll help explain just how neck-deep in the bog I really am. One minute I'm screaming along in full afterburner, at treetop height with the rest of my gallant RAF lads, defending this Nip-infested patch of nothing in the middle of nowhere, and the next, BOOM! Some barmy short-circuit somewhere makes a bloody cameo, and both my canopy and my WSO's ejection seat head for the wild blue. I mean, come on, we haven't even <u>seen</u> the bleeding enemy yet!

What happens next, you might ask? Well, the rest of the flight, including my now wingman-less wingman, gets sent into harm's way, and I get emergency re-routed to some godforsaken cricket pitch in the sticks called 'Coon-Song,' or some such nonsense. Twenty harrowing minutes later, I manage to pull a near-perfect landing (except for near-missing the slant-eyed wanker with the glowsticks that had the cobblers to run into my path – bloody pedestrians). No sooner do the engines spool down, though, than I'm bollocks-deep in Halon like I came down with my hair on fire.

This is no place for one of the Queen's own, let me tell you.

So, I'm being yammered at in full Sensurround by every Nip in the uncivilized world <u>and</u> his twelve brothers. I'm choking to death on Halon because, like a dunce, I removed my mask thinking there would be someone to talk to IN THE QUEEN'S ENGLISH, dammit! The berk with the glowsticks looks mad enough to try to serve me along with curried rice and boiled dog for the evening meal, and to top it all off, I've no clue of how to say "Emergency Landing – Damaged Plane" in Korean. What a marvelous afternoon.

Regardless, I'm in the middle of thumbing my holster open and deciding on firing either into the air or between someone's eyes to shut the lot of them up, when I hear someone shout something unintelligible in English. Thank God! I really didn't care that it was the Yank variety; better a colonial Philistine than these godless heathens.

I managed to extricate myself from the cockpit and was immediately led away to a debriefing by an extraordinarily young-looking USAF instrument jockey. (Some debriefing – wall-to-wall Koreans. Thankfully, one of them, presumably a liaison officer, spoke reasonably decent English.) When the usual rigmarole of how, when, and why questions had been answered, 'Airman Intrepid' led me off to a barracks for some sleep while they would try to repair my plane.

As he was walking away, I asked him what had become of my flight. He replied without turning around that they had all been killed over North Korea. Bastard! They didn't even have any news of my backseater, alive or dead. Needless to say, sleep did not come very easily at all: the spaces of wakefulness amongst the perhaps fifteen winks that I managed to grab were filled with the faces of my mates. Sorry, lads.

Today didn't start any better than yesterday had ended, either. That same mannerless wanker came and got me for morning mess, then took me to my plane. What's the matter with that, you say? Well, I'll bleeding tell you – IT WASN'T MY SODDING PLANE, that's what! Instead of my beautiful Tornado F-3, he actually had the cheek to tell me that they were going to substitute me into a frigging F-5 Freedom Fighter! Can you believe it? A plane older than I am, for crying out loud! Sure, it's maneuverable as all hell, so I'll be able to fly circles around the enemy, but I'll have to get out and bloody-well push to even get to the bastards! The gauges aren't even in English – they're in Korean! A balls-up, that's what this is – a complete farce. And when I asked just where my old plane was, he told me that it had been scrapped out for lack of parts, and the armament redistributed!

Even worse, just to add insult to injury, they wanted me to fly an F-series 2-seater with some South Korean berk as a WSO! Well, that's where I put my foot down. I asked to see the commander, and they told me that he was out flying a mission. A Yank, to boot! Even worse, he's not even a fighter pilot – he's a bomber jockey! Meyers, I believe his name is.

I made enough noise to finally get someone to answer some of my questions, but unfortunately, it was the South Korean liaison officer from last night. It seems that I've been forced to join up with some slapped-together Flying Circus of a composite unit, made up totally of pilots who were the remnants of flights like mine – either lucky or unlucky, depending on your point of view. Well, I told the Yank airman I'd fly with them if I had to, if they couldn't get me back to my unit, but I'd rather roast in hell than cart a Nip around like a bleeding rickshaw. So, he managed to find me an E-series single-seater plane, but same problem – the controls are all still in Korean.

And when I asked if he could PLEASE find me a manual or something, so I might just have a slim chance of understanding them and not get my arse blown off up in the sky somewhere, guess what? Right – it was in Korean as well. What in the hell can this sorry collection of sods possibly contribute to the war effort?

Friday, February 11, 2011

2nd Lt. John Chamberlin-Personal Log

Somewhere in the Yellow Sea

General Shapiro began sending some of our units to Seoul early this morning. Since we were the first in at LZ Delta, we're low on the list in terms of when we takeoff. For better or worse, the surviving Marines from the ships that were lost a few days ago were the first to fly out. With the addition of aircraft that were forced to escape their sinking ships, we've almost got as many Osprey as we had before Delta. The airlift carried almost 500 Marines ashore today, and the word is that they're being used as sort of a last-ditch defense in Seoul. They had a tough time landing, and had to land one bird at a time; right in the middle of downtown. They had to use a building's roof as an LZ, and not all of our Osprey made it back. Bruce heard that our guys didn't even leave the block where they landed. They were told just to hold out and defend the area as an evacuation point! I guess it doesn't matter where we get sent, it's going to be ugly.

Saturday, February 12, 2011

1444 hrs.

Republic of Korea, Taegu Highway Airstrip

Personal Log – Lt. Kim "Killer" Yoo, Republic of (South) Korea Air Force

Things continue to go poorly. The U.S. Air Force was ordered to abandon the Kunsan airbase. Engineers were hard at work blowing things up when Meyers passed the word that we were moving out. Our wing-wipers (admin and logistics people/pirates) typed up orders for us that make it look like we've been assigned to a new forward airstrip. The orders are on some U.S. Air Force stationary that the wing-wipers created, but they've also been making up some orders that assign each of us from our own Air Forces to the 1st Composite Group. My only fear is that these orders are too realistic, and I'll never be able to leave the group if I want to for some reason. The other day I was a volunteer, now I feel like a conscript.

The new airstrip is a closed portion of highway just southwest of Taegu. The area is nice and flat for a few miles around (except for the city which is right at the end of the runway to the northeast), but since we're using a road for take offs and landings, crosswinds can become tricky. The strip is being run like a racetrack. Aircraft land on the right side, taxi across the berm to the other side, then stop at any one of 22 designated areas for refueling, rearming, and debriefing. Our own wing-wipers are on their way down by truck, but the ones already here are very fast. Turnaround time is less than 30 minutes, and we don't even have time to leave our cockpits for debriefing. The intel officers just hook up a ladder and ask us about our mission while the wipers prep the planes.

It's very efficient, but I think that's because it's our Air Force that's running the show here. The Americans are great, but they're sometimes a bit slow and methodical-at least that's been my experience. Others might mistake that as laziness, but I disagree. I think that most Americans are very hard-working, but their leadership and organization can be rigid. Their leaders pale in comparison to the British who are sometimes so slow that they seem bored or even asleep. I watched and Australian bomber squadron at Kunsan for a bit, and they were pure cowboys-leaders, wipers, and pilots alike. For the most part, aggressive leadership seems lacking with almost every Air Force.

This Meyers man is a bit different, but it might be too early to tell. I've only met him, but the facts speak for themselves. While we were all sitting around waiting for orders, he actually stood up and did something. I'll give him credit for that. He's not the stereotypical American either. Most people picture them as Cowboys, Hot Dogs, or Mavericks. Meyers seems very professional. He was very concerned that (fraudulent as they might be) everyone had proper assignment orders to the group. He hasn't shown any specific commitment to doing things by the book, but I have seen him frown at some of the French pilots who were screwing around between missions. The man has a great deal of mystery about him, but he also brings a great deal of charisma to the fight. I think most people have joined the group to get back into the fight, but there are plenty of people who seem to find more interest in him than in the battle.

The communists continue to push us back all across the country. The Army's trying to set up a new defensive line near the Raktonggang River, but most of the regular units are still beat up and in a serious state of disarray. The new line will be mostly reserve divisions. While our reservists are tough as nails, everyone has quiet doubts that they won't be able to hold against the same units that shattered our regular divisions. Of course, no one talks about this openly. We all know it, but we pretend that everything is going according to plan. The entire nation is in a state of denial.

We've got an entire battalion of (South) Korean Marines protecting us, and that makes us feel a bit more comfortable. The city is being used as a rally point for an entire corps, and there are three allied divisions lined up along the river a few miles to the west (an Australian infantry division, a New Zealand infantry division, and an Indian infantry division). A squadron of U.S. Marine Harriers is parked in the field to the southeast of the runway strip, and their sole mission is to protect the road/runway. Besides the growing numbers of our group (now at 62 planes) and the Marine Harriers, there are two regular squadrons of American F-16's here. Quietly, we all wonder how much will be left of those squadrons after our wing wipers arrive and begin to borrow equipment.

Sunday, February 12, 2011

1850 hrs.

Republic of Korea, Pohang Airstrip

Personal Log – *Lef*tenant Ian Thomas "Top Hat" Sanders IV

Just when things seemed to be going our way, we've been chased out of our lovely out-of-the-way garden spot of Kunsan, and onto a mile-long strip of Hung-Lo Highway just outside of a little bit of nowhere called Taegu. What the hell? Come on, lads, even a glacier gets to move forward every now and then.

When I said going <u>our</u> way, of course, I was referring to the "First Knit-Together Air Remnant," not to the war effort in general. We seemed to be doing a bang-up job keeping hold of our little corner of Hell, but everyone's just not doing his or her part, I guess. Being booted to yet another locale is a pretty fair sign that our Allied forces are collectively getting it from the backside.

I had been busy over the last few days doing the Stick-and-Rudder-Shuffle in my ancient replacement bus, learning which of the Korean-marked gauges did what, and marking them in the Queen's English utilizing a multicolor Mini-Post-It pad and felt-tip pen (purloined from the desk of 'Airman Intrepid'). The Freedom Fighter is an interesting bird, granted, but I miss my old plane. After the move to Taegu, however, I was singled out by this Meyers at a morning briefing, and told in no uncertain terms that "playtime" was over and it was time to join one of the fighter squadrons. I attempted to protest, saying that I was simply waiting to be told where to go; he wasn't having any. The bastard turned in mid-sentence and walked away, remarking, "That's the whole idea. We don't wait on anyone's orders here. Start being proactive rather than reactive." Wanker! I'd have made a stink about it, except for the fact that he has two uncanny abilities: finding the enemy, and returning alive. He's got more luck than skill, so what's the use in arguing?

So, fifteen minutes later I was taxiing over to the staging area for the next squadron to head skyward. The only other non-rice-eater in the flight was another Yank, but nothing like our "Fearless Leader," Meyers. At least that one's relatively calm and a bit thoughtful; this tosser (who the Nips inevitably made my wingman) thinks he's in bleeding Dodge City or some such nonsense, all that "Well-them-there-Yeehah" crap to endure. What a pain in the arse!

We were sent initially to the south-southeast, simply another in a long list of fighter sweeps. We chanced upon an enemy flight of Jian-7's about thirty minutes into our sweep, though, and the fracas began in earnest. I hosed off both of my Sidewinders early in the match, missing with the first, but dead-on with the second target. I gave her as much of a goose as I could to get free of the furball and turn to re-engage, when my wingman began yammering over the radio – he was in trouble, being pursued by one of the enemy.

I began craning my neck around to try to see where he might have gone, completely disregarding my own safety, and ran afoul of the enemy myself.

261

Unnerved at first by the tracers zipping by, I threw my little F-5 into the tightest left turn I could manage without blacking out, rolling to the left and down at the same time. My tail was picked off by a missile from another member of my flight, but not before the sod got off a lucky shot, hitting me a glancing blow… you guessed it, in the canopy again, dammit! The bloody thing shattered due to the pressure difference, sucking all of my little Post-It-Notes out in a Technicolor chaff cloud and sending me limping homeward again.

Once on the ground and after being shoehorned out of the cockpit, I was surrounded by the remainders of the flight, including my Pecos Bill-esque wingman. One of the gooks was pointing at me and laughing, miming tipping a hat. When my wingman started laughing back, I asked him to translate what the Nip was saying. "He says you've got a bit of a problem keeping your lid on, buddy," the Yank snickered, clapping me on the shoulder like one of his mates. Apparently the idea got bandied around a bit, because when I returned from having a medic patch the holes that my exploding canopy had made in me, some yob had stencilled "Top Hat" on the side of my plane, along with a drawn caricature as well. I guess I'm stuck with it as a moniker, at least for the time being.

And, to top it all off, there isn't a decent cup of Darjeeling, Earl Grey, or English Breakfast to be found, neither for love nor money. I swear, if one more slant-eyed berk shoves a mug of weak, piss-colored, ginseng-laden muck under my nose and has the cheek to call it tea, I'm giving him a bleeding enema with it then and there, international relations be damned!

Sunday, February 13, 2011, 6:00 PM EST

Channel 4 News

"Good evening ladies and gentlemen, I'm Tom Waushinski, and this is Channel 4 News. The situation in the far east continues to worsen by the hour."

"U.S. and other allied forces holding the Eastern half of the frontline in South Korea were driven back to new positions south of the Raktonggang River running from central South Korea east to the coastline. On the western side of South Korea, cut-off elements of the R.O.K. army and other allied units continue to fight guerrilla actions while arranging for extraction by helicopter search and rescue teams. R.O.K Reservists began forming a new defensive line stretching from the American line along the Raktonggang River on the East, south to Tagu, then west to Kwangdshu along the west Coast."

"Sources in the Pentagon report that air combat losses to both sides have passed an unacceptable point of loss, and few air missions can be successfully flown. As fast as reinforcements from around the world arrive, fresh communist squadrons are rising to meet them. Further complicating the problem, a shortage of qualified pilots among the western nations is creating a drain on forwarddeployed air units around the world."

"During today's daily briefing at the Pentagon, officials reported that remnants of Task Force Stingrayreinforced by additional escorts from the R.O.K. and other nations around the world-is the only naval blocking force preventing communist forces from attempting another amphibious landing along the west coast of South Korea. The east coast is under the watchful eyes of a well organized group of coast watchers, and supported by aircraft operating from bases on the Peninsula and others en route from Japan."

"Closer to home, protesters outside the White House continue to grow in numbers. For decades, there have been protests for all kinds of causes. Sometimes it's seems as though it has been more of a form of entertainment than a real call for action. Since the inauguration and the Battle of the Yellow Sea, thousands have encircled the White House demanding vengeance. Down the road and up the hill, bipartisan efforts to fund the war continue in Congress...."

Sunday, February 13, 2011

Republic of South Korea

Pohang Airstrip

Al Meyers brought his rattling F-16 in for a landing. As soon as his plane's rear wheels touched down, two R.O.K. F-5 fighter planes taxied out for their takeoff. The small strip was being run like an aircraft carrier. There was no stop in flight operations, not even for a second. Ground crews rearmed, refueled, and repaired planes on the open taxiway while only the wings of the planes sheltered them from the falling snow. Drifting snow was a problem with no solution, so it was accepted and ignored. Ignored problems weren't problems.

Al had returned alone, again. He was flying an average of one mission every three hours. He had eaten three times since the invasion, and slept twice, both times for less than three hours. His plane had been damaged several times, but the holes were covered with tape, and unless something critical was broken, he had to take it back up.

The communists seemed to never run out of planes. He had shot down eleven so far. He had also lost fourteen wingmen. He didn't even have to do the simple math to figure out that things were going badly. Al could see it in the men. A few pilots had been flying from the beginning like he had, but most were new arrivals, and they still had that look of excitement. The ones who had made it through even a few missions already looked tired, even a little older. The hot shots were all gone, or new; gone meaning dead, new meaning about to die.

When Al stopped his plane at the assigned location, ten people came running at him. As soon as the ladder was up to his cockpit, the local intelligence officer got a quick rundown of his morning's work. The Intel officer, an R.O.K. air force Captain, in return told him how the war was going. The entire briefing took less than five minutes, and Al never had to leave his ejection seat. As soon as the Captain climbed down, the ground crew chief, another member of the R.O.K. air force, climbed up and started helping him out of the cockpit. Al didn't know what was going on, but he'd been through the routine countless times, so he helped with the buckles, etc., then he climbed out. The ground crew chief didn't speak English, so Al went up to the Captain for an explanation.

The Captain listened to several members of the ground crew, and then told Al why his day was over. It seemed that the rattling sound Al had been hearing was actually a 30mm cannon round that was loose in his left wing. The previous ground crew must have thought that the round went through and taped over the hole. Al's latest high-G maneuvering probably dislodged it. In any case, someone would have to open up the wing and get it out. Flying around could set it off. In a high-G turn, in a dogfight, the last thing a fighter pilot wanted was a cannon round exploding in his wing, next to his fuel tank.

It didn't even startle Al. The past few days were like one blended series of non-events. Nothing seemed real. The ordinary was gone, and a new one had replaced it. Instead of seeing familiar faces, he saw strangers. Instead of flying safe training missions, he was pushing the envelope in dogfights. He wasn't even flying the type of plane he had trained for. It was like a strange dream. After a while, the strangeness seemed real enough to accept as ordinary. The ordinary life he once knew was gone. Maybe he'd see it again someday. Until then, he had this one, and, as melancholy as he seemed, he still valued his life very highly. Others may have already accepted their impending fates, but Al had not. He hadn't fought the despair off. He just didn't feel like letting the war get to him yet. All Al cared about was a sandwich and a few hours of sleep.

Sunday, February 13, 2011

1139 hrs.

Republic of Korea, Pohang Airstrip

Personal Log – Lt. Kim "Killer" Yoo, Republic of (South) Korea Air Force

Wars are supposed to be won by the side that has the most will. After our sixth mission yesterday, we are overflowing with will. Since we came to this highway strip, Meyers has had us flying fighter sweeps over the city. There's always one squadron in the air, one on the ground (getting rest), and one on the ground in reserve. We had rotated the airborne squadrons six times, and it was our turn to wait on the ground in case the airborne squadron needed reinforcement. Meyers was in the air, and he called us up. Three squadrons of Chinese Jian 7's had come to Taegu to escort a squadron of North Korean Mig-27 bombers. It was more than Meyers squadron could handle, and the two American F-16 squadrons at our little airstrip were both up north making bombing raids of their own. Our Marine Harriers went to stop the North Korean bombers, but Meyers wanted us to come up and help out. By the time I was airborne, all of the communists were already on their way home, and I didn't even get a shot off. The turning point came when I looked down and saw a long black line of refugees heading across the river and into Taegu.

The weather has been very cold since the war started. Average temperatures are well below freezing, and that's without taking into account the wind. Occasionally it alternates between freezing rain and snow, but there's always some sort of precipitation. I had only view the weather as it related to flight conditions and combat visibility. By chance, on this mission I took the time to look down and see the effects of the war in action. Thousands of people were walking through the snow. Traffic jams and stalled cars blocked the road for as far I could see. Instead, everyone was on foot and carrying all their belongings across the bridge and into the bombed out city. I looked close enough to see that the refugees, my people, were no longer clusters of families interspersed with lost souls. The crowd seemed like one large and lost animal.

A few hours later, we were all back on the ground, and I mentioned it in conversation. Some of the other pilots immediately admitted that they had seen the refugees too. We all described what we had seen, and then there was a moment of uncomfortable silence. No one knew what to say. I didn't notice that he was listening, but from behind me, Meyers stepped forward and caught everyone's attention. We waited for him to say something-even the pilots who don't understand English. He just looked around at each of us. Then he stopped and looked right into my eyes, and simply said, "Those people are counting on us. We cannot fail them." Then he looked around again and walked away. It was an obvious statement, but one that had to be made. I'm not sure if anyone will follow this man into Hell, but he's right. We cannot fail those people. I will not, and I don't think anyone in this unit will either.

Monday February 14, 2011

Washington Post

~~Communists Advance Halted?~~

Communist forces have slowed their advances in an effort to consolidate their units and mop up the thousands of R.O.K. and other allied units operating as guerrillas behind their lines. Meanwhile, United States Marines from Task Force Stingray have been operating around the clock for several days in an effort to evacuate cut off friendly units and ferry them back to the new South Korean lines. Both sides have stopped their major operations in order to gather up U.N. stragglers.

SCOTT MALENSEK

Monday February 14, 2011

2nd Lt. John Chamberlin-Personal Log

55 nm south of the Republic of Korea

I'm back in the saddle again. Col. Ghetty officially re-attached me to my platoon today. I'd like to think that he needed my expertise in TRAP missions, but I think someone must have asked him why one of his lieutenants was just wandering around the ship. It might actually be my skills or experience that got me my job back after all. In the same speech I received from Ghetty letting me go back to the platoon, he told me to get with Capt. Brown, Hector, and Bruce to go over a mission we would go out on today. How's that for prep time?!

I jumped to and bolted for Brown's quarters. If he stayed true to his nature, the meeting would have already started without me. The corridors seem more crowded now that I have to make my way through them in a hurry. I think someone should take charge of all the survivors on the ship and get them organized. They're supposed to be in the hangar and well deck, but they seem to be drifting all over. Sure as anything, when I got down to Brown's quarters, he and Hector were already talking over the mission. At least I got there before Bruce.

Brown wasn't too upset. I only bought a disappointed glance and the sullen "It's about time" from him. Bruce showed up a few minutes later, and we went over our mission. After our unit had almost single-handedly started World War III by somehow fouling up a TRAP mission, they were going to give us another. No one really believed we screwed up, but the frustrated feeling of defeat still haunts us, and it's lingering with the smell that the survivors seem to have brought on board.

This used to be one of the cleanest ships in the Navy, but after the damage we've taken, the survivors we've picked up, the wounded we've treated, and the extra aircraft, well, every hall seems to have a unique smell. I can tell when I'm getting close to the med center by the smell of alcohol, oxygen, urine, and that coppery smell blood has. In the well deck, all you can smell is diesel, saltwater, smoke, and mildew from wet gear. On the hangar deck, I used to enjoy the smell of aviation fuel, but the place smells like its ready to explode at anytime. It's the only place I've been that doesn't smell like smoke. We've taken on as many aircraft as we possibly can, and I hear that they have to keep some in the air just to make room on the deck!

Brown says we're going to start getting rid of some of those aircraft by flying into occupied areas in South Korea, picking up surrounded good guys, and flying them to Pusan for reconstitution into front line units. The missions will be coordinated by our R.O.K. liaison team on board. They're supposed to find out where the good guys are, and direct us to them. Every Osprey on deck, and the handful of Blackhawks that have found their way here, are all going to be used. Each mission will involve company-sized teams, like ours. Every team will be brought in on at least five, maybe six Osprey, and we're told that enough extra aircraft will be sent along to pickup all the good guys, plus a spare bird-in case.

266

Brown suggests that since we're going in with so many aircraft, we might as well split out platoons up into squads so a lucky shot doesn't wipe out a third of the company.

Hector, Bruce, and I were given thirty minutes to get our platoons ready and on the hangar deck. We all left at the same time, but it took me almost an hour to get my platoon ready. Hector and Bruce beat us to the deck by almost forty-five minutes. Brown said he was getting ready to send for us, and I believe him. How they beat us, I have no idea. We were haulin' ass to get our gear together! They must have heard a rumor and been getting ready earlier. Anyway, a few minutes after we assembled in the hangar, we marched onto the elevator, and went up to the flight deck to our rides.

I rode with Gary and 1st squad, Bob rode with Wess and 2nd squad. Rick and 3rd squad rode with the company HQ staff, including Capt. Brown. Hector and Bruce had divided up too. I think Brown wanted to see if my platoon had any misgivings about my return. After everyone was on board, the birds lifted off as one and side-slipped over the port side of the ship. I was in the lead Osprey, and once we turned toward the peninsula, I couldn't see the rest of the team behind us. I did have a wonderful view of communist-held South Korea through the cockpit window.

It took us about an hour to reach the coast. Just after takeoff, we had been given a heading to fly down. As we crossed over the flat coastline below, we received our pickup coordinates, the Pyontaek Highway strip northwest of Cheongju. I didn't like the idea of being so close to a city, and I really didn't like the idea of landing smack dab on top of a military target in enemy hands. I only hoped that somebody somewhere knew what they were doing.

Low overcast, morning fog, scattered snow squalls, drifting snow, and lingering smoke from fires burning all over the country, everything masked our entry. The pilots talked it over briefly. They decided to land single file down the highway that the R.O.K. Air Force was forcing us to use as a makeshift airfield. While we descended, the pilot had the rear door opening, and I could tell the guy had seen some action. It was a simple thing that showed he was no John Wayne, but he wasn't restricted to doing things by the book either. That simple flip of a switch focused my attention on the mission, and I think it did for everyone else too. This was the same squad that had lost two friends the last time I went into combat with them, and now nothing could hold back their anticipation.

As soon as the rear wheels touched, before they could even bounce, my people were out and running for cover. Behind us, the other Osprey were doing the same thing. Where one second there was an empty highway, strewn with debris and burned out abandoned vehicles, now there were a dozen V-22 Osprey throwing snow everywhere, and well over a hundred Marines looking for targets. Another hundred-plus people came running out of hiding places everywhere. Some walked, obviously wounded. The more seriously wounded were carried on litters and on people's backs. Only a handful of armed R.O.K. Air Force security personnel covered their exits. We crammed everyone on board, then tried to fit ourselves inside, and the Osprey lifted off as one.

Above the overcast, a trio of U.S. Air Force F-16's waited with Durandural anti-runway cratering bombs. I never saw or even heard the attack, but I was told that when we left, they peered through clouds with their infrared cameras, targeted the adhoc airstrip, and turned it into rubble.

On our way to drop off the wounded, our pilot received new orders to return to the ship instead of trying to make our way through enemy territory to Pusan. The Wasp was going to get a little more crowded. In fact, it was going to get a lot more crowded.

When we landed back on the ship, we helped the wounded off the Osprey. Refueling teams were already tanking up the aircraft, and, no sooner than I assembled the platoon on the hangar deck, Capt. Brown told us to get back on board. There's a lot of trapped good guys out there, and we were going in again.

We eventually did four more missions today. They even changed out our pilots so that they could get some rest while we kept on going in. The whole day was kind of frustrating. We would get all psyched up for a hot LZ like Delta. Then, when we landed there was nothing except broken survivors. I know the North Koreans are out there, but they must be moving so fast that they don't have any time to mop up. It might be that they don't have anyone to do the job, too. Either way, tomorrow we're supposed to be doing the same thing, and I shouldn't complain. We pulled out almost 600 good guys and no one got killed. Most of them were R.O.K. units, but every flight picked up at least a few Americans. They made their gratitude very clear. TRAP missions definitely get the best smiles.

Monday, February 14, 2011

1337 hrs.

Republic of Korea, Pohang Airstrip

Personal Log – Lt. Kim "Killer" Yoo, Republic of (South) Korea Air Force

Our wing wipers finally arrived today. They had quite a tough time getting here. They say that all of the roads are clogged with abandoned vehicles, refugees, and constantly under communist air attack. Some of them are still stuck in traffic, but most of them are here, and they went right to work.

I'm not even sure if they were all off the trucks when they had struck their first deal. Apparently, the American F-16 squadrons (which have been cut in half due to combat losses already) have been ordered to fly a roaming close air support mission. The idea is that they are all supposed to load up with whatever air-to-ground ordnance and loiter over communist held territory. The communist advance has moved so fast that their advance units are separated from their support and supply columns. At the same time, we've got entire battalions cut off and fighting behind enemy lines. The Americans were supposed to be available to support our cut off units, but right now, they've got more bombs than planes. Meyers suggested that if the Americans completely load and fuel all of our aircraft, we'd help out with the close air support.

All three of our squadrons were fully fueled, and loaded with both air-to-ground and brand new air-to-air weapons. While the American wing wipers were hanging bombs under our wings, our scroungers recruited the airstrips scroungers and completely cleaned out all of the American squadrons' Sidewinder missiles (short-range, heat-seeking, air-to-air missiles used for dogfighting). It's a good thing too. Most of us were flying missions with only two missiles and our cannons. Now, we've got enough to last almost a week!

We all left the highway strip at about 0800 hours. The American F-16 squadrons lead the way to their designated positive control location-about 50 miles south of Seoul. Meyers had his squadron break up into five four-plane flights, and they spread out in a ladder formation with about 5000 foot separation between each flight. My squadron stayed together and loitered at 45,000, and our last squadron circled down in the dirt at about 1000 feet.

The Americans split up into pairs and went off to the east and west to drop their loads and help out some lost troops. None of us knew what frequency to listen to for orders (with all the radio jamming and all the different languages in our group, verbal communication would be difficult at best). So, our squadron leaders formed up on Meyers and communicated with hand gestures. He seemed to know how to contact the forward air support controllers, and soon we had our first mission.

Our squadron leaders left Meyers and headed back to join us. Everyone was told to follow Meyers toward a ground target to the east. So we kept our altitude separations and our formations, then we headed east. About 5 minutes later, Meyers rolled over and dove down. The rest of his flight followed him. Then the rest of his squadron followed him, and our squadron dove down too. The low squadron started to circle until they could see what everyone was diving toward. We were all headed down at over 500 knots, and only one man knew what the target was-Meyers.

Meyers dove down toward what looked like a normal light industrial building, and when he was down to about 2000 feet, we all saw yellow tracer rounds coming up from toward him. He toggled his release and all six of his 500-pound bombs dropped from his wings. As he pulled out of the dive and headed east (away from Seoul), the rest of his squadron started to release their bombs.

We were still starting our dive when it all started to happen, and once we were headed toward the ground, it was easy to see all the communists around the building. They were all wearing white, and some stayed to try and shoot at us, but moving targets are easier to see from a plane than a stationary one, and even those who tried to run were lined up in our heads-up displays. Meyer's squadron leveled the building and the area immediately surrounding it. Our squadron dropped 500-pounders all around where the first bombs had hit-into the fleeing communist troops. Our low squadron circled the area where we were bombing and strafed troops who tried to escape the bombs. The attack only took a minute, but when we headed back to Pohang, a full mile of snow-covered, communist-held area was black with soil, soot, and smoke.

Back at the airstrip, we found out that the American F-16 squadrons were practically wiped out. Meyers recruited their wing wipers and the four remaining pilots. Unfortunately, only three of their planes are air-worthty. The only thing that was worth salvaging from the damaged American F-16 was its cannon, and some of the engine components. Everything else was useless, or not worth the time to remove. For example, radio components are next to useless in a group that uses only hand gestures for communication, and there's so much radar jamming in the region that using a radar is almost completely futile. Some of the other Korean pilots in our group have already removed their radars in an effort to save weight and make the most of their precious fuel. We'd all like to do the same, but there just hasn't been enough time. Maybe there will be now that we've got more wing wipers.

Tuesday, February 15, 2011

2219 hrs.

Republic of Korea, Pohang Airstrip

Personal Log – Lt. Kim "Killer" Yoo, Republic of (South) Korea Air Force

Well, the Americans in our new composite group are very upset. The TV news is showing some graphic commando attacks back in America. Apparently, their FBI has arrested a lot of communist agents who have been causing all kinds of damage. Many of the American pilots are furious. Even Meyers-"The Wolfman" as they're calling him now-is showing a bit less composure than usual.

He's taken all of the American pilots and split them into a single squadron for the next few days. There will only be one American in each of the other two squadrons to act as a liason. It's obvious that he wants to lead the Americans in some sort of vengeance attack, but we dropped all of the bombs that we had yesterday.

Our normal rotation has also changed. The Americans are flying a fighter sweep along the front lines. On the ground, we still have one squadron on alert-ready to take off and reinforce those already airborne. We also have one squadron resting/sleeping/repairing. However, instead of giving each squadron a chance for rest and repair, the Americans will come back, refuel, and go back up. Only the ready-reserve squadron and the resting squadron will rotate. We think that Meyers and the rest of the Americans will try to keep this up until they run into a big enough fight to quench their blood thirst, OR until Meyers wears them out, and they're finally ready to get some rest. Either way, we're semi-grounded until they find some sort of satisfaction.

Our squadron was on the rest cycle first, and we all took the time to grab some food and get acquainted with each other. Most of the pilots in this squadron are (South) Korean, but there are pilots from several different nations. National and racial prejudices are stronger than I expected from some of the foreign pilots. None of the Korean pilots has the time for such ignorance. Our country needs us, and as much as the Americans want revenge now, we've wanted it for the past 50+ years. Some of the pilots are still professional, others-usually the ones with the most arrogance-tease each other like brothers even though they've all just met. The few who have deeper issues with nationality or race generally keep to themselves-thankfully.

Tuesday February 15, 2011, 2330 hours

Tianjian, People's Republic of China

Four miles out to sea, a British and an American submarine, modified for special operations missions, finished the slow, cold, and tedious procedure of disembarking commandos while remaining submerged in hostile waters. The maneuver had been conducted thousands of times by the American Navy SEALS, but was not quite as routine for the British Royal Marines. They had been rushed through training only a few days before. Almost a hundred feet below the surface, no light from the outside world could permeate the waters of the bay. Out of respect for the enemy's presence, no flashlights, glowsticks, or other devices were used to help the commandos find their way. Only a nylon rope between each man's fingers told him the way to his designated station. Once there, the men secured themselves to their team leaders, and their long swim inland began.

Only the team leaders knew where they were, and even that information was less than complete. Each leader used a handheld GPS (Global Positioning System) receiver to determine their exact longitude and latitude information from satellites hundreds of miles away in space. The only problem was that the depth and cold of the water interfered with the signal, and the leaders quickly found themselves relying on their instincts and best judgement. When each team felt that it had swam the right distance, they stopped and waited while their leaders would float to the surface and visually find their targets. Not a single team arrived where it was supposed to be. Seasonal and tidal forces pushed everyone miles to the south. They all knew where they were as individual teams, but their inability to communicate outside their five-man units forced each team to switch to targets of opportunity instead of their assigned objectives. Everyone hoped that someone else would find their way to the primary mission objective and take of it. No one did.

Filled with both frustration and confidence, the teams went to work on their secondary objectives, shipping targets of opportunity. Twenty-three fiveman teams began planting limpet mines on the bottom of ships around the bay and

271

moored to the docks. Everyone carried two mines, each about the size of a trash can lid and weighing nearly 50 pounds out of the water. In the water they were much lighter, but their bulk was still a substantial burden to swim with. Everyone was more than happy to clamp down their magnetic mines and attach them to bottom of the Chinese ships. Per the pre-mission briefing, every mine's timer was set for 0700 the next morning. That would give everyone plenty of time to swim back to their submarines and exit the immediate area before the communists were rudely awakened.

The mission's success would only be determined the next day when the mines exploded. For the most part, it was already a failure since the primary mission objective wasn't attacked. That objective had been the P.R.C. Naval flotillas that were recently returning for rearming after-action in the Yellow Sea against Task Force Stingray. This was supposed to be a payback for the modern day Pearl Harbor, instead, it seemed to be an example of Murphy's

first law of combat: "Whatever can go wrong, will go wrong."

Wednesday, February 16, 2011, 0649 hours

Tianjian, People's Republic of China

Walt Barber had been awake since 4:50 in the morning. The economics of the war were not as he had fully expected. He had anticipated the rapid advances of the Chinese and North Korean forces into South Korea. He had anticipated the slow response from the western nations in the defense of South Korea, and he had expected heavy damage to the American Navy in the Yellow Sea. He had, however, also expected the militant elements of the Russian military to jump at the chance for a fight with the Americans AND a paycheck. It seemed as though the Russians were asleep, along with all the other nations around the world that were traditionally rubbing the United States.

Already dressed in his businesswear, ready for the endless stream of meetings with communist party officials, diplomats, government beaurocrats, and leaders of his various business interests around the globe, Walt was ready for the day.

Without knocking, Dan Greene opened the door to the living room of his hotel suite.

"Sir, I just got a secure fax from GTI. They've detected a U.S. and a British sub right outside the port here. Apparently they're moving south, away from us, but they weren't on any of the National Command Authority position reports. They just seemed to have appeared out of nowhere. Shall I inform the group, sir?"

Walt thought for a minute. He had no military training outside of his passion for history texts. His focus had been on business, electronics, and, most importantly, accounting. All wars could historically be traced to an imbalance in economic power. Walt knew that the U.S. would never put a $500 million dollar submarine so far into enemy waters. The British could be counted on for only one thing in military history, and that was caution. What were they doing here? He looked out to the bay and pondered. When his thoughts had been gathered, he thanked Dan and pulled a palm-sized cell phone from his breast pocket. With the press of three buttons, he moved it to his ear and waited for his recipient to answer. It was early in the morning, but the voice on the end was clear, and there was a distinct bustle in the background.

"Yes."

"General Sadis? Walt Barber. I think we may have a problem."

"Mr. Barber, I hate to cut you short, but this is a very busy time for us here. Is this an immediate matter, or can we setup a meeting?"

"I'm not sure, General. I know your schedule is busier than mine, and I'm so busy I'm not sure I'll have time to relieve myself. I'll be brief. I've been getting reports that there is a buildup of less than friendly units south of the city. I'll have the specific information rushed over to you immediately, but I thought, what with your current plans being put together right now, you might want to take some extra precautions."

"Yes. Of course. I understand. This must be some sound information. I know you wouldn't have called if it was a just a rumor. I'll make some calls and see if we can't get a better picture of what's out there. Have a copy of the report sent directly to the Naval facility where our flotillas are resupplying. I'm sure they'll want to know about it. Again, I hate to be so abrupt, but I am very busy. Thank you for your help-again. Please keep me informed if you hear anything else."

"No problem General. By the way, is everything going according to your timetable, or is there something I can do to help with that venture also?"

"No. Yesterday, we were actually ahead of schedule. All of the personnel were boarded before sunset, and the heavy equipment was almost secured. This morning, one of the freighters reported a boiler problem so we've had to arrange for that ship's replacement. We've already started unloading everything to the dock, but the cranes are so busy we can only unload or load at once. We can't do both. I don't suppose you can get a 65 ton dock crane here in the next few hours?"

"What's a few hours?"

The General laughed at Walt's confidence, and then drifted back to attention as he realized that Walt was serious.

"I was joking, Walt.-"

"General, I wasn't. I could probably get a 40 ton mobile unit here within 18 to 24 hours. Can I make the arrangements?"

"No thank you, Walt." He chuckled again. *"I'm just going to have the troops do some of it the Chinese way, with muscle. I better get going. Thanks again for*

the news. I'll let you know when we're about to leave, though I'm sure you'll already know it. Won't you?!"

"Yes sir. If there is one thing I can say, I am informed."

Walt snapped the phone shut and looked out his window over the bay again. Along its closest edge, a thriving fish market continued operations in the same way it had for hundreds, possibly thousands of years. Further to the right, he could make out the pale gray shapes of over a hundred small missileequipped light frigates and patrol boats; to his left, the downtown area, and the port. Half a dozen ships were moored at piers in front of the metropolitan area, and he could count at least 23 more in the bay. One of those ships carried General Sadis, whom he had just spoken with. Most of them carried Sadis' invasion force preparing for Pusan, South Korea.

Tens of thousands of People's Liberation Army troops and hundreds of vehicles were waiting for the one ship to unload. Once its replacement was full, they would move out to the south. Two days later, they would land and end the war in one lightning strike. The thought of witnessing history in the making flushed blood through his veins. He could feel the excitement of every man waiting for his moment on those ships. Behind him, a grandfather clock, imported from England during the Opium War hundreds of years ago, began its gentle chime towards 7:00am. When it finished, Walt heard the empty silence and turned to look at it, as though it would chime one more time.

Instead, there was a fast orange reflection climbing up its gold embossed face. Walt had never noticed the clock's odd shine before. He turned around to see the entire bay fill with rising columns of flame and smoke. Almost every ship in the bay had exploded at once; at 7:00am, 0700 local. In a state of empty shock, he watched the scene unfold, and the clouds of smoke climb even higher. Finally, he heard the rumbling sound from the explosions, and the rattling of his hotel suite window from the shock wave. With water temperatures hovering in the low 40's, there was little hope of rescuing survivors. Thousands were dying. The war had just escalated to a level of loss comparable with nuclear war.

The SP 2.0 software had failed to notice the commandos, and now the thousands of Chinese were dead and dying. Walt knew that he'd have to find out why the commandos weren't noticed, and he'd have to do it soon or he might find himself on the wrong side of their General Naht Synn. A quick call to Dan, and preparations began for a flight to GTI's Canada office for some answers.

Communist response to the destruction of their invasion task force was immediate. For decades, an intricate cellular structure of intelligence-gathering teams, cross-trained in guerilla warfare, had been spread around the globe. This network was especially strong in free societies like those found in the industrialized west. Of these, the United States proved to be especially comfortable. Orders were dispatched via coded e-mail messages and web sites created halfway around the globe.

Once received, the spies put on their commando hats and went to work. What was initially thought to be normal criminal behavior quickly set a trend and law enforcement officials found themselves fighting unusually well armed, well prepared, and daring criminals. Drive-by shootings increased in ethnic urban areas, and in the thought to be havens of suburbs and small towns. Highways accidents caused by reckless drivers seemed to happen on every freeway and major intersection within hours. The most dramatic attack occurred in San Diego, California, home of the U.S. Pacific Fleet. The aircraft carrier U.S.S. Ranger burned in San Diego Bay on live TV.

That attack began at the nearby San Diego Air Sports Center. The Air Sports Center was really just a dirt patch in a small valley. Despite the fact that even small Cessna aircraft had to fly just a few feet above the contours of the valley walls, it was the hub for area skydiving. The center survived on a shoestring budget in the best of times. As a result, the runway didn't even qualify as a dirt road.

During the weekends, a dozen or so people normally took lessons and practiced their skydiving skills. However, Monday through Friday saw only a skeleton staff and a handful of visitors. Since there were no landing lights, the Center was closed from dusk to dawn returning to the area's natural desert silence.

Just before midnight, two vintage 1950 era C-119 Cargo planes approached the dirt strip. With the aid of night vision goggles for the pilots and drag parachutes to slow the planes, they swooped through the valley and onto the runway.

As they positioned themselves for take-off their engines shut down. The dust from the dirt strip reduced visibility for those without night vision goggles to only a few feet. From the small jungle separating Otay Lake and the Air Sports Center, six pairs of headlights suddenly cut shaking shafts of light through the tan fog.

Six black vans bounced their way out of the jungle to the west and up to the rear of the planes. Before they were fully stopped, a dozen men jumped out and began to unload. They pulled two long racks of rollers out from each truck and carefully slid their cargo from truck to plane.

One plane was loaded with large black cones, about seven feet high and equally round in diameter on the bottom. Nine of the large black cones were loaded into the first C-119, and, while the doors were closing, parachutes with static lines were attached.

Only one truck backed up to the second plane. Once its loading ramps were connected to the plane, all the black clad figures ran to help load the first of the four parts that made up a massive custom bomb. This truckload contained the parachute, airfoil, and fuel cells. Once loaded, another truck backed up and unloaded its part of the bomb, the outer housing. A third truck rolled up and loaded the pre-formed triangular shafts of plastic explosives into their respective places in the pie-like interior of the bomb. It was all bolted together, static lines and parachutes attached, and the cargo bay doors closed.

In the cockpits, the pilots planned their take-off while the last truck unloaded its cargo. Under both wings of both planes, a dozen ten-foot lengths of four inch wide PVC/plastic piping was secured. Each pipe had been plugged at both ends and filled with a combination of gasoline and styrofoam, which combined into a form of napalm. The truck also loaded two dufflebags into each plane.

One by one the trucks left as their cargoes were unloaded. Just before the last one was unloaded, the engines of the two planes began to come to life again. By the time the last of the four engines was roaring, the sixth truck was long gone.

One after the other, the planes raced down the dirt strip into the clear evening sky. They headed toward Lindbergh Field in Downtown San Diego. By changing altitude, heading, and IFF codes, they were able to convince anyone watching them with a radar that the planes were in trouble. Lindbergh control was the first to call and ask what the problem was.

Choi Paek, the North Korean pilot of the lead plane, responded.

"Copy tower... This is Jackson 305 and Jackson 306. We've had a small collision and we are declaring emergencies. 306's pilot is hurt and both its engines are coughing. We're gonna need a meat wagon and fire standing by."

Jackson 305 and 306 were fictitious call signs, but, by the time that was discovered, the need for the planes would be long gone.

They flew barely 200 feet above the outskirts of San Diego. As they approached the airport, the tower called to give them landing clearance and notify Jackson 305 and 306 that emergency crews were standing by.

Balboa Park passed on the starboard side, then downtown San Diego on the port side. The lights of the city, airport, and harbor lit up their target like a Christmas tree. There, ahead of Jackson 305, the copilot saw it and called out in Korean, "U.S.S. Ranger 11 o'clock low!"

Both planes opened their cargo bay doors, dropped to just over 100 feet and headed for a position just behind the Ranger. Lindbergh Field called to see what was wrong, but Choi saw no reason to answer. Besides, the planes were difficult enough to handle with both hands on the wheel, let alone while trying to turn at low altitude with an overloaded bay and one hand working the radio.

Once behind the carrier, they both banked left and lined themselves up to fly across the deck from stern to bow as though they were going to land. 306 took position behind and slightly above 305.

When the stern of the Ranger slipped under Jackson 305, Choi pulled the plane into as steep a climb as he could. The increased pitch and G forces pulled the cones out the back. PVC pipes were also released, and the loadmaster in the cargo bay pushed his dufflebag out the starboard side emergency exit.

The parachutes attached to the cones opened immediately and slowed their descent. The PVC pipes scattered like toothpicks across the deck. Upon impact their jellied gasoline contents splattered everywhere. At the same time, a parachute deployed from the dufflebag, and hundreds of small cans, the size of a cat food can, scattered along the dock next to the Ranger. Each can was in fact a homemade anti-personnel mine with a proximity fuse that was set by the can's impact with the ground.

Jackson 306 began its climb at the same time 305 did. The massive bomb that had been put together inside the plane slipped out fast, and the plane's weight difference made it a lot easier to control.

306 had let go of its PVC pipes and dufflebag at the same time Choi did in the 305 plane. The end result was that the carrier deck was covered with napalm and the dock was littered with anti personnel mines.

All of the cones set down on the deck like giant iron weights. Their parachutes automatically detached to keep them from being dragged off the 68000-ton ship. When the last one set down, there was a three-second pause. There were all the usual kinds of sounds from the city, harbor, and planes above, but to Choi and the other North Koreans, there seemed to be total silence. The quiet was misleading.

All at once, the black cone-shaped charges erupted in great columns of white flame. Each was loaded with 1500 pounds of plastic explosives. The napalm-covered deck flashed with balls of orange, red, and yellow flame.

The shape of the charges in the cones had directed the full force of the explosives downwards instead of through 360 degrees-as was the case with a normal explosion. Torch-like streams of explosive power cut through the deck of the ship. Nine holes went from deck to keel, but only one was able to actually start any flooding.

Above the burning carrier, the 20,000-pound bomb from Jackson 306 floated down via several parachutes. It swung to and fro with inertia left over from the plane. Once it was vertically facing the carrier, the parachutes detached and the rocket engine of the first component boomed to life with a bluish-white flame propelling the device downward.

It slammed into the starboard bow of the Ranger. The subsequent explosion was powerful enough to break most of the windows in the San Diego area for miles around. It also created a miniature tidal wave, which sank several small recreational craft, docked in the bay and behind Harbor Island. Lit by the fires within, a gaping hole billowing with smoke appeared on the bow as if some giant sea monster had taken a 40-foot bite from the front of the ship. Minutes later, the Ranger had a 15-degree list to starboard, and counterflooding had to begin.

Fire trucks from all over the city rushed to the dock to help put out the fires, but the anti-personnel mines took their toll. Five fire trucks were damaged and over 60 fire fighters were wounded from the innocent looking cat food cans. Two fireboats were sunk when secondary explosions capsized the little boats after they had gotten too close to the carrier.

There were only a few hundred men on the Ranger while it was docked in port, and it only a few minutes for the officer of the deck to decide to abandon the ship. By the time he gave the order, there were barely 50 men left to follow it. When they tried, they quickly found themselves unable to leave by the dock because of the mines, so they tried to jump and swim for safety. Continuing explosions from the ship deafened and killed more of them.

Jackson 305 and Jackson 306 flew back over Lindbergh Field and then above residential Point Loma. Behind them, the six men on board could already see the flames of their work reaching hundreds of feet into the air. Over Point Loma, they turned north to fly along the coast at a more comfortable 1500 feet.

Two A-7 Corsair II's, an S-3 Viking, and Two F-16 Falcons closed in on the two aging cargo planes from the south, west, and north respectively. Miramar Naval Air Station to the north was scrambling two F-14 Tomcats to join in the chase.

"Suit up! We're outta here!" Choi yelled over the roar of the engines. He turned his navigation lights off, then on again to signal to Jackson 306. In rapid succession, three men bailed out of each plane.

They opened their parachutes just over 500 feet above the houses below. Meanwhile, the planes continued north on autopilot until they were shot down not even thirty seconds later by the A-7 Corsair II's.

Each man steered his parachute to a specific address on Point Loma. At those addresses, their getaway cars were parked in the street. 4153 Mistyre Drive - Choi silently dropped into the dark streets lit only by the dim glow of a streetlight and the muffled explosions of the U.S.S. Ranger in the distance. A few feet away, he sat down in his beat-up 1981 Mazda 626, shoved the parachute into a footlocker on the back seat and drove away.

A minute or two later, he stopped at Shelter Island, less than a mile from where he had landed. From there he could watch the Ranger explode, burn, and list. He also waited for the sun to rise behind the 15,000-foot columns of black, gray, and white smoke that he had created.

While the major networks were broadcasting scenes of the burning aircraft carrier in San Diego Bay - interspersed with interviews and footage from the frontlines in Korea - another even more public act of terrorism occurred. Law enforcement units stretched to the limit dealing with the terrorist attacks were unable to provide protection from the communist fifth columnists. While there were attempts to protect likely targets such as: refineries, military posts, airports, population centers, vital industries, and transportation links, there was no way to adequately defend everything. Only eighteen security guards were protecting the Capitol Building when a rented van loaded with plastic explosives propelled itself up the steps and exploded where only a few weeks earlier the President had taken his Oath of Office. Before fire crews arrived, a camera crew was recording and selling footage to the major news organizations. The building itself wasn't destroyed, but 1/3 of the U.S. government would have to find a new place of operations for the next 8-10 months. Prudently, Congressional leaders agreed to setup shop at their Cold War emergency bunker in the hills. Within a few days, all of Congress would leave Washington D.C.

Thursday, February 17, 2011

2334 hrs.

Republic of Korea, Pohang Airstrip

Personal Log – Lt. Kim "Killer" Yoo, Republic of (South) Korea Air Force

This was a rough day for a lot of people in this makeshift group. Meyers had all of our longer range aircraft (F-16's mostly) head up north-beyond the old DMZ. Back during the 1950's, the area was called "Mig Alley" by the Americans. It's large expanse of area where communist planes-based in China-could meet allied planes based out of the old "Pusan Perimeter." Since both sides are already operating their aircraft from the same regions, Meyers thought we should go up there and see if it was a hot-spot again. It was supposed to be a simple Fighter-Sweep. Since I'm an F-16 pilot who speaks English, I was asked to go along, and I did.

We left around noon, and it took us about an hour to get up to Mig Alley. All of our aircraft were lightly loaded since we don't have any drop tanks (external fuel tanks for carrying extra fuel). In a dogfight speed kills, and speed can keep you alive. The problem is that every maneuver a pilot does creates more air resistance, and airspeed can bleed away from under a pilot's wings very quickly. The only way to regain or maintain airspeed is to increase the plane's throttle. This radically increases fuel consumption. Since we were operating at a relatively long range, and no one brought any extra fuel, Meyers had us approach at high altitude. If we ran into trouble, we could maintain a higher airspeed by diving repeatedly. It was a good plan, but like all military planning, as soon as the first shots were fired, the plan was useless.

We crossed over the DMZ, and then we spotted a dogfight that was about 20 miles to the northeast. Meyers led us toward the knot of white contrails in the sky, and everyone looked around for the communists. As we drew closer, some of the communists in the dogfight fired off missiles toward us. We all had to dodge them, and our very loose formation fell completely apart. I stayed up high, but some of the less experienced pilots dove for the valleys below in the hopes of confusing the incoming missiles.

When they dropped down low enough, we all started to get radar-guided missile warnings in our cockpits. It seemed like there were SAM batteries all around us. I saw three of the planes that went low try to dive into a valley to hide from the new warnings, and as soon as they started to snake their way between the mountains, all hell broke loose. Anti-Aircraft Artillery (AAA or triple A) started to come at them from both sides. The communist guns were designed to shoot high into the sky, but they must have been set up on their sides because they

279

were firing across the valley, and not into the sky above. Their concentrated firepower in such a small space was too much for our pilots that were caught in the valley. They were all shot down in an instant.

High above the valley, we all looked and decided that we couldn't keep diving to maintain our airspeed in a dogfight. We didn't have enough fuel to stay and fight for very long, but as long as we were close to the communist planes, the SAM batteries below couldn't get a clear shot. If we stayed, the Migs and Chinese Jians would have the advantage. If we left, we could all expect a hail of SAM's. Our only choice was to keep fighting with the communist planes and try to lead the fight south-away from the AAA flak trap in the valleys and the innumerable SAM batteries.

It only took a few seconds to figure all this out, and there really wasn't any time to make a conscious group decision. It just happened so fast, that we all knew what had to be done. We tried to drag the fight south. There was nothing else we could do.

I still don't have any idea how many planes were involved in this battle. We brought 23. When we arrived, it looked like two or three American F-15 squadrons (or their remnants) had run into an entire group of Chinese Jian 7's (copies of the Soviet-built Mig-21 Fishbed. I saw a few Mig-29 Fulcrums, but they were unusually camouflaged, and I'm not sure if they were communist or those of some of our allies. One way or the other, the sky was very busy for as far as anyone could see.

I bagged 2, Jian 7's with my wingtip Sidewinders right away. The first one took a close hit up the engine and trailed smoke until the pilot ejected. The other was hit right between the cockpit and the wings. It came apart immediately. They were relatively easy targets, and their pilots must have been novices. All they seemed to accomplish was to deprive me of precious ammunition.

With only my cannon, I tried to get close enough for another kill. It was unusually hard since everyone on both sides was busy firing missiles or dodging missiles-myself included. A few seconds later (though it seemed like hours), I was able to get on the tail of another Jian. It was headed west-toward the Yellow Sea, but every time he tired to get away, I overshot his turn and forced him back onto a westward course. This was working very well, so I stared to only overshoot his left turns. In no time, we were headed south. Back over the DMZ, I stopped playing with him, pumped a quick burst into his wing, and watched it come off immediately. There was no chute, but I was safely free of the dogfight, the flak trap in the valleys, and the multiple SAM batteries.

There were still some problems. I was still over enemy-held territory. I was alone (an easy target for communist aircraft). The rest of my group was still in the big dogfight. I was almost completely out of ammunition, and my fuel was more than halfway gone. It would have been easy and understandable to go back to Pohang, but I chose to get back into the dogfight.

It wasn't hard to find. All I had to do was climb a few thousand feet above the overcast. Off to my northeast, I could see a furball of contrails, yellow-orange glitter from explosions, and black puffs from kills. The closer I came to the big knot, the most I could discern the tiny black specs as fighters. It was hard to find a communist, so I gathered that things were going better for our side than the enemy's.

After a quick visual search, I spotted a pair of Jian 7's chasing an F-16. They were above and to my left and turning left, so I followed from below where they couldn't see me. They slowly descended, and I slowly climbed. When we were close enough, I fired a burst of 20mm at the lead plane. The burst missed, so I immediately fired another which must have hit something because the lead Jian started to trail a jet-black line of smoke. It broke off the chase, dove, and headed north right away, and while the second-and closer-plane tried to take over the chase, I pumped off a burst into it. It came apart right away.

I panned around trying to find another target. While I was searching, a few F-16's and a pair of F-15's pulled up on both sides of me and wagged their wings in thanks. I replied. Then I spotted my squadron leader's plane and pulled up next to him. He waved, and we headed back to Pohang. The planes that had pulled up next to me were from Meyer's squadron and the squadrons that were already in the dogfight when we arrived. While several of our group had been lost in this action, we actually picked up more planes and pilots than we had lost.

Back at Pohang, I had to take some down time. In the confusion of the dogfight, my plane had taken some 25mm hits in the right stabilizer and rudder. My fuel tank needed patching, and I had a hydraulic leak. The wing-wipers were amazed that I made it back, but I never really noticed any serious problems other than some sluggishness in loose turns. I thought it was just rough winter winds.

While they were taping my wings back together (they actually used duct-tape to patch several of the holes), I grabbed some food. It's a highway strip, and there are no real facilities, so we have to the same rations that the soldiers are getting: MRE's, Meals-Ready-to-Eat. They're little plastic pouches with bags and boxes of entire meals inside. Most people hate them. They seem to think that they're too dry, too stale, too bland, or too imitation. I think they're fine as long as you make sure to add enough hot sauce. Maybe it's an international taste problem, and only Koreans-like myself-truly have good taste. HA!

While I was working on my second MRE packet, Meyers came over to the 55-gallon drum that I was sitting on and sat down next to me. It turns out that he was in the F-16 that was being chased when I shot down the pair of Jian 7's. He was thankful, but not overly so. Instead, he was professional, polite, and only as friendly as a co-worker. He also told me that the intelligence officers had confirmed the plane that I shot down near the DMZ, and that I had become one of the war's first one-mission aces.

I hadn't really thought about it. The communists were not kills, scores, or even people in my mind. They were simply invading machines that brought devastation to my country and had to be stopped-at any cost. After I thought about it, I realized that I had shot down at least 30 planes already, and when I told Meyers, he said it was something to be very proud of doing. As long as the communist keep coming, it didn't matter if it was 30, 300, or 3000. I asked Meyers not to spread the word about my becoming a one-mission ace. He understood-at least I think he did. As a matter of courtesy, he shook my hand again and agreed not to tell others about my "kills" until I told him it was alright to do so. He also told me to call him Al now on.

In the past, all of my superior officers had been rigid disciplinarians. They followed the book to the letter and demanded absolute, unconditional obedience. Most were very skilled, and I respected them. I had always seen American officers as either cowboys or immovable ignorant doorstops. Meyers is different. He's charismatic in a practical and professional sense. Every day I have more and more respect for the man, and I'm glad I was able to take down those Jians.

Thursday, February 17, 2011

Republic of Korea, Pohang Airstrip

Personal Log – *Lef*tenant Ian Thomas "Top Hat" Sanders IV, R.A.F.

The long-range end of the air wing got into a real snarling clusterfuck today, but came out all right in the end. Apparently they intruded upon a dogfight that was staged directly over a SAM/AAA nest deep in the bushes – way to go, lads! (I didn't hear anyone breaking Meyers' balls about his E.S.P. being on the fritz today, though, did I? Hell no, mustn't bruise the "Magic Yank's" fragile ego.) The good news is that they came back with a few bonus planes and pilots, what was left of the squadrons who started the fight they finished up. The bad news? They're more Yanks. Bloody hell.

It looks like the Nip who gave me a hard time yesterday is getting in pretty tight with Meyers. Apparently he saved our "Fearless Leader's" bacon in the furball – I chanced across them talking over chow, swapping footnotes on the battle, sitting on the same 55-gallon drum. Blimey, they'll be holding hands next.

They share a common thread in my mind – the desire to retrieve payment from the enemy in blood for slights received personally and as a whole. Meyers and the rest of the Yanks have been busy taking back their pound of flesh for the terrorist or commando attacks back in America recently by flying "American-only" vengeance missions against the North Koreans and Chinese. Kim Yoo (that's the rice-eater's name) thinks he can kill every communist one-at-a-time, if supplied with unlimited ammunition and a steady stream of flyable planes as his get chewed to bits. His hatred is almost palpable.

I don't mean to preach here, but that's pretty damned unprofessional, if you ask me. Hate and other negative emotions have no place on a battlefield, either on the ground or in the sky. Fighter pilots especially need to keep a distance from our passions and external stimuli while in the air; we'll miss something, bungle up, and get blown out of the sky if we don't.

Now, one might ask if I could keep a professional coolness if England had been invaded: if I was fighting to keep control of my homeland. My answer? You're bloody-well right I could. It's part of what makes me the pilot I am. It's also what'll keep my arse in one piece long after those two tossers are carrion.

Friday, February 18, 2011

1051 hrs.

Republic of Korea, Pohang Airstrip

Personal Log – Lt. Kim "Killer" Yoo, Republic of (South) Korea Air Force

Our intel people have finally figured out why there are so many SAM batteries up north. It turns out that the communists do in fact have a great number of surface-to-air missiles, but nowhere near as many as we've been detecting. Instead, they've set up a network of cheap antennas all over the place. We've always thought that there was one control van and a single radar antenna for every missile battery. The communists have set up hundreds-maybe thousands-of radars all over. Only a few of them actually control the SAM's, but we can't figure out which ones. When one of their SAM batteries tries to lock onto a plane, there might be a hundred antennas that are emitting identical signals as decoys. I've seen pictures of these antennas, and each one probably doesn't cost as much as a toaster oven.

No wonder we've been losing so many planes. We can't target the SAM radars that are controlling the SAM's. They're luring us into tight valleys that are laced with AAA. The Chinese have more aircraft than the Americans and Russians combined. It's a wonder we're still here at all.

Saturday February 19, 2011

Associated Press International

Communist Tide Continues to Rise

North Korean forces penetrated the R.O.K. line of reserve units positioned north of the city of Kwangdshu on the southwestern side of the peninsula. The R.O.K. units, east of the penetration withdrew in good order, but those to the south and west were completely routed. Surviving R.O.K. units, joined by American and other United Nations forces along the east side of the Peninsula repulsed several smaller, brigade-sized attacks with heavy armored losses reported on both sides.

Sunday February 20, 2011

New York Post

Here Come the Russians

Several Russian guided-missile destroyers, frigates, and patrol craft left the port city of Vladivostock near the North Korean, Chinese, and Russian border. Officials reported the deployment as a safeguard action in order to prevent collateral damage to the city from the wandering air war and the violent ground campaign on the peninsula.

Sunday February 20, 2011

The New Cleveland Press

U.N. Forces Continue to Retreat

R.O.K. units began forming a new defensive line stretching from the Pusan perimeter, running from the south towards American and other United Nations forces holding the east/west line north of the City. There is nowhere left to run, to the east of Pusan the Sea of Japan, to the far south, the Korea Strait. Task Force Stingray's remnants and reinforcements redployed to block any amphibious attempts on the Pusan Perimeter from the Strait of Korea, and the recently-arrived aircraft carrier Carl Vinson battlegroup has moved to patrol the Sea of Japan between the Pusan Perimeter and Japan.

Sunday, February 20, 2011

1141hrs.

Republic of Korea, Pohang Airstrip

Personal Log – Lt. Kim "Killer" Yoo, Republic of (South) Korea Air Force

The communists broke through at Kwangdshu, and we're heading out. The U.S. Navy's Seabees have put together a rough strip down near Tochon. We learned of the pullout from the Marine pilots who were assigned to protect this highway strip. The Seabees made the new strip for our Marines, and we're just going to join them. It's not an easy task to carve an airstrip out of the frozen Earth. After the land is leveled, there has to be several layers of gravels and steel-mesh put down. There are also the revetments for the planes, refueling facilities, housing for pilots and wing wipers, air traffic control, security, and a long list of other requirements. It's remarkable that they were able to set up a strip for the Harriers, but now they'll have another 50-60 guests to accommodate. I can't help but wonder how we'll be welcomed?

Monday, February 21, 2011

1315hrs.

Republic of Korea, Tochon Airstrip

Personal Log – Lt. Kim "Killer" Yoo, Republic of (South) Korea Air Force

We've fallen in with thieves. These American Seabees (U.S. Navy contruction battalion sailors) are absolute thieves. Before the Marines left Pohang yesterday, Meyers had us all take off to follow them to the new airstrip. There was a flight of communist bombers that was surprised to run into almost 100 fighters. They dumped their bombs and ran away to the north before any of us could attack them. Besides that, our move went very smoothly.

We landed at Tochon and expected some verbal resistance from the Seabees, but there was none. They were just too busy getting everything put in place. We have to cram into the few tents, and some pilots have complained, but those are the same childish weaklings who don't like the MRE's. There was a shortage of parking revetments too, but we're told that problem will be solved by tomorrow night. We were pleasantly surprise to find a plentiful supply of fuel, Sidewinders, and 20mm cannon rounds. Apparently, the Seabees have a large mess hall tent devoted to gambling and other recreational activities. Whenever someone runs out of cash and needs credit extended, the Seabees have been accepting supplies in lieu of money. I suspect that at least some of the games are less than fair.

This composite group is very loose in organization. We've split up into three squadrons, and all of the planes are divided into 3-plane flights. However, the number of planes in each squadron constantly changes as pilots are lost and new people join. In an even more informal organization, all of the pilots-myself included-are really broken up into various social cliques. The younger, less experienced, more aggressive, and definitely more arrogant pilots all tend to congregate. On the other hand, older pilots and more experienced pilots tend to keep to themselves. The two wide groups are also divided my nationalities as American pilots tend to keep company with other American pilots, etc. There are also groups that cross all lines as pilots of similar aircraft share professional conversations, trade war stories, and trade performance tips with each other. How Meyers keeps it all together I don't know, but very few people have left, and our numbers are growing at least as fast as our losses.

Monday, February 21, 2011

Republic of Korea, Tochon Airstrip

Personal Log – *Lef*tenant Ian Thomas "Top Hat" Sanders IV, R.A.F.

We've been given the fucking boot again – damn these bastards! Tochon is just a dug-out flat area in the middle of even further-out nowhere than Kunsan or Pohang were. There aren't even enough "parking spaces" for all of the individual planes, or bunks for all of the men! (The "Seabees," as they're called, happen to be some of the fastest workers I've ever seen, however – they say Rome wasn't built in a day, but these sods could have constructed Milton Keynes in just under a week, <u>including</u> the concrete cows.)

Personally, I'm blasted tired of getting it up the rear again and again, chased off strips all over Korea, when all of <u>us</u> seem to be doing a bloody wonderful job of hacking and slashing our way across both the skies and the land. If they keep shoving us further and further away, into more and more isolated positions, it won't be long before they'll have driven us out of the country completely … and that would just be bad form all around.

I don't mind the close quarters that we've been squeezed into like kippers, nor the dearth of anything remotely resembling chow except for MRE's. I've been a fan of taking off into the hills on holiday since I was a wee lad; roughing it is recreation for me, not hardship. Besides, I have received a gift from God himself.

There was a group of Ghurkas here in the beginning of the construction phase, assisting with camouflage and defensive setup (who knows better how to hide than a Ghurka?). The First Composite Air Wing arrived just as they were leaving. Recognizing them for who they were, and bartering with a small diamond stud that I affected to wear in my left earlobe, I managed to secure a pound and a half of Earl Grey tea from them – part of some Sergeant-Major's "essential supplies." Ah, lovely bergamot; I never thought I would smell you again! For a moment, basking in a benevolent deity's love and the almost-holy scent wafting up from my full mug, I believed I could be no happier while still in Korea.

That is, until I found out there was gambling to be had.

I've often wondered what it is that makes American Marines of any rank, size, racial background, or job description believe (without any deep-seated basis in fact) that they are by definition better than everyone else at <u>everything</u>. Perhaps it is something that they put in the water on Paris Island, kind of like the saltpeter rumored to be in the eggs. It might have something to do with fashionably green camouflage. The government might be performing secret operations on them as they sleep. I'm not sure, but whatever the reasons behind it, this is a weakness that can be exploited.

Sit in with any group of people around a card table for long, and if one has eyes at all, one can begin to read their tells. Well, if those people are Marines, not used to hiding anything except for fear and pain, they're a bit more obvious. After two hours of playing with them and steadily winning, rather convinced I could take everything they owned, I began losing back most of it. I didn't want to build a reputation as a shark just yet, if at all; besides, I had a secondary plan. The bootnecks are rumored to be first-class scroungers and thieves, and if I can get on their good side (or win enough of their money), they might become invaluable assets if goods become scarce. We'll see. For now, simply knowing I could fleece these tossers out of their Y-fronts made me happy indeed.

Tuesday February 22, 2011

Associated Press International

Pusan Holds

Bolstered by the R.O.K. units that had withdrawn to the Pusan perimeter, the communist positions throughout the city were finally taken just before noon. Communist pressure on the perimeter continued with a series of brigade-sized probes, but there was no concentrated largescale action, and all the attacks have reportedly been repulsed.

Wednesday February 23, 2011, 6:00pm EST

Channel 4 News

"Good evening everyone, I'm Tom Waushinski, and this is Channel 4 News. I'm coming to you tonight from Pusan, South Korea where there have been dramatic events today."

"A major assault by Chinese infantry breached the Pusan Perimeter between the R.O.K. units and American forces. A mixed force from various members of the United Nations took the brunt of the attack. Thankfully, the attack was almost single-handedly destroyed by cluster munitions carried on cruise missiles and launched from an American arsenal ship that has recently returned to duty after having been rearmed in Japan. As soon as it left Japanese waters, an American naval liaison officer attached to the 2nd infantry division called in the strikes, and when the smoke cleared, there was no more resistance from the Chinese. A total of 317 Tomahawk cruise missiles were used, most armed with different types of cluster munitions that are effective against all types of targets including everything from hardened bunkers to loose infantry. The already battered 2nd infantry division then committed its remaining reserves to occupy the gap in the perimeter until the other U.N. forces were able to

rally and return to their previous positions."

"The mood around the city is that of fear and chaos. We arrived here yesterday by way of a supply flight from Seattle, via Alaska and Japan. Either because of damage or traffic, we were forced to land at a makeshift airfield that has been set up on a highway just outside the downtown area of the city. At the request of the local military authorities, we will not reveal the specific location. We hitched a ride with a small column of Australian Mechanized Infantry on their way into the city, and here we are."

"After getting into the city, we found that all of the hotels are either occupied, or have been forced empty for defensive purposes. We've taken up a new residence on top of this 30-story office building. If you look behind me, and if our cameraman can zoom in a little... over there. You'll see several wrecked vehicles on the road, and a few more near the buildings to the right and left of the others. That is all that remains of this morning's Chinese breakthrough. If we look a little closer at the city, you'll see the small roadblock that has been set up since then. We assume that other units have been moved to the area, but remain camouflaged from our view, and that of the Chinese. Directly below us, you'll also note that the streets are nearly empty except for a few roadblocks and patrolling armored vehicles. Most of the civilian population moved into bomb shelters and basements when the war started. Since then, there has been a steady stream of people from all over the country trying to get off the peninsula. Our cameras can't show it from here, but just over there, behind those two buildings, the docks are packed with civilians trying to get on a ship. Thankfully, there have been plenty of U.N. vessels bringing war materials. When those ships leave, rather than leave with their holds empty, they've been allowing hundreds, even thousands of civilians on board. I'm told that so far, every time a shipload of refugees has docked at a port,

the Koreans have been given amnesty."

"Wherever we look, the silent effects of war are present. There is some damage from artillery and air strikes and there are spots from small arms fire from the past few days of guerilla fighting on the part of the communists. What really stands out isn't the damage, but the fear. Fear has emptied this city of all its life. Now, only hidden pockets of troops remain. Nothing else speaks so strongly of fear as silence, and in Pusan today, we can hear each individual vehicle driving on the streets, even from over 30 stories up."

Thursday February 24, 2011

Washington Post

Russians Get Too Close To the Fire

North Korean and Chinese frigates in the Sea of Japan sortied out from their hiding positions along the coast and rendezvoused north of the Russian task force that had left Vladivostock a few days earlier. American electronic intelligence-gathering aircraft from the U.S.S. Carl Vinson, south of the Russian task Force, detected an increase in radio traffic between the North Korean, the Chinese, and the Russian ships. While the intelligence report was being monitored, the North Koreans and Chinese began launching cruise missiles over the Russians at the Carl Vinson battlegroup.

Those cruise missiles, like most anti-ship missiles, used internal radars to detect and attack their targets. Also like most other cruise missiles, their radars were set to turn on at a predetermined distance from their launchers. In this case, the missiles activated a few miles after they had past the Russian ships.

Over a hundred missiles were launched at the carrier and its escorts. Most of the missiles either missed their targets, were intercepted by SAM's from the escort ships, or were physically blocked by the smaller escort ships. The Carl Vinson reportedly remains unharmed. Two yet to be identified American frigates were hit, and one sank immediately. An Australian frigate was also hit and sunk, along with a Thai frigate.

In response, the American and other U.N. ships fired a volley of cruise missiles at the North Korean and Chinese frigates, and some were destroyed, but most of the American missiles turned on their homing radars before passing the Russian Task Force. Many of the missiles hit the Russian ships between the two warring flotillas. A Russian destroyer was badly damaged, two frigates were sunk, another frigate was badly damaged, and three missile patrol boats went down. Russian diplomats, along with others from nations around the globe, conveyed their fury and demanded restitution. Those nations not already joined with in the defense of South Korea are uniting in their opposition to the west.

Thursday, February 24, 2011

1359 hrs.

Republic of Korea, Tochon Airstrip

Personal Log – Lt. Kim "Killer" Yoo, Republic of (South) Korea Air Force

The communists are everywhere. We were strafed by a trio of Jian 7's about an hour after sunrise. They damaged several of the Marine Harriers, so Meyers, Al, had our squadron take off and fly high cover for the strip. We intercepted 2, 4-plane flights of Jian's heading for our strip. They tried to dump their bombs and run home, but they were all shot down. Meyers had the other two squadrons break up into 3-plane flights and take off as a sort of wandering close air support. They all headed off in different directions and searched for ground targets to attack, and there were plenty.

We're told that a new defensive line has been established along the Naktong River to a few miles to our west. I don't see it. All that I could see were communist vehicles pouring over bridges. There was a lot of tracer fire from small arms around them, but it seems like nothing is stopping them from overrunning our new home. Occasionally, I would see one of the Soviet-built tanks explode, but it seemed to happen very infrequently. I could see our new strip under mortar or artillery fire on some occasions, but it was never overrun. It seems as though the new defensive line is very static and porous.

I landed and refueled five times today. On one occasion, while I was still strapped in my cockpit and giving a briefing to the intel officer, we came under attack from some very large artillery. Everyone else ran for cover, but I was strapped to my ejection seat and couldn't get out. All I could do was sit patiently for it to end. While I waited for that one unlucky artillery round, I noticed that the Seabees never stopped working. That calmed me, and after a few more seconds, when the attack ended, I wasn't as shaken as I expected. The intel officer climbed back up the ladder and was clearly shocked that I had been so calm. If only he knew!

Friday, February 25, 2011

1413hrs.

Republic of Korea, Tochon Airstrip

Personal Log – Lt. Kim "Killer" Yoo, Republic of (South) Korea Air Force

Meyers has broken us up into six different groups now. Three of the groups will rotate as roving fighter sweep patrols. The other three will roam around doing a roving close air support. Our American Marine Harrier neighbors are being called upon to do more close air support missions than combat air patrols over the base, so we're getting some air-to-ground ordnance from them. The Seabees are "finding" other sources too. Two of the six groups (one of the fighter sweep patrol groups and one of the roving close air groups) will always be on the ground for down time. Another two will be on the ground in reserve. Whenever a flight lands, one of the reserve flights will go up to replace it. Reorganization is the norm in this group. As annoying and confusing as it can be, it shows one of the group's strengths-adaptation!

The front line is just a few miles away to the west now. We can hear the small arms and artillery fire. It's a nice switch since we can actually watch the people in our group making their close air runs on the communists. It's becoming a spectacle and even a contest to see who can make the best attack run on an enemy position. Pilots who land and have to return to attack targets a second time can definitely expect a storm of verbal taunts from others in the group. There was a time when this might have seemed sick or disgusting to compete for the most colorful ability to kill other humans, but we've all seen friends killed, and we've all seen the long lines of freezing refugees. Killing communists isn't like killing people anymore. It's like a combination of bug extermination and vigilante revenge.

I've been assigned to one of the roving fighter sweep patrols. Meyers has all of the F-5's, F-15's, and most of the F-16 clearing the skies. When he find out about the virtual cheering section that's watching the close air attacks across the river, Meyers passed the word that every aircraft is allowed to make a close air support run before landing. After he saw a few of the runs himself, he wants us all to make at least one strafing run against the communists on the west bank before we'll be given clearance to land. I wasn't sure about his demand for us all to do close air support, but after I made my first strafing run, I felt a lot better. It must have been a decent run since no one teased me about it, but it couldn't have been that great of an attack either since I wasn't complimented. Like everyone, I'll try harder next time!

Friday, February 25, 2011

Republic of Korea, Tochon Airstrip

Personal Log – *Lef*tenant Ian Thomas "Top Hat" Sanders IV, R.A.F.

Our little corner of Hell here is in danger of being completely overrun by vermin – that is, North Korean and Chinese. We're being attacked from every point on the bleeding compass, and not just by enemy fighters, either; we're being alternately shelled and probed by both foot and mechanized infantry. One thing for sure – it's never boring here in Gooksville, as one of my Yank flightmates muttered over the comm. (The tosser was immediately reprimanded by Meyers upon landing for not keeping his yap shut.)

He's one to sodding talk about keeping things in perspective – apparently he and Kim Yoo (should bloody well be "Kill Yoo," from the load of hate that bastard's lugging in and out of his bus with him) worked something out between them. He's now allowing all of our bloodthirsty "Kill-'Em-All-And-Let-God-Sort-'Em-Later" ravening maniacs a chance for retribution and/or body counts approximating those in cinema today.

The daft pillock has begun asking these berks, indeed all of us, to end flight missions with a strafing or heavier-weapons ground attack upon approach to the strip. That way, we take care of some of the Commie Nips on the west bank, and everyone gets some kind of cathartic release from it, yadda-yadda-yadda.

Needless to say, when I was advised of Meyers' latest "regulation" before my flight took off for another sweep, I told the Liaison Officer to get shanked and toggled my engines, rolling away from the little wanker while he was still on the ladder. He hit the dirt to avoid being clocked over by the wing, and I didn't think he landed well. I did manage to topple the ladder over onto him, though.

We got into a bit of a scrap to the northeast of the strip – the usual "chase and be chased" with a flight of bombers and their fighter escort. We dropped eight and chased the other four planes away, losing only two ourselves. On our return, while we were still miles away from the action below, the rest of the flight began to climb in a spiral for a speedier attack run. I separated from them and simply came in for a landing, booed heartily by the masses looking for a tastefully executed massacre.

I figured Meyers would have the bit in his teeth when I landed, and just so – he was approaching my plane as my engines were spooling down, accompanied by the aforementioned Liaison Officer, with his shoulder now in a sling. (Dislocated.) I lit a cigarette and awaited the coming storm.

He proceeded to read me the riot act, and I proceeded to tell him to go stuff himself. As far as this outlaw air force of his is concerned, I haven't seen Order One that wasn't falsified by his ever-vigilant nest of criminals that makes the Third Reich's Bureau of Propaganda and Forgery look like a nipper with a Xerox. He backed up a pace when I told him he could shove his "orders" far enough up his arse to look him in the eyes when he washed his teeth in the morning. He flushed four shades of red and tried to interrupt me when I told him

I was a pilot, not a Mongol, and if he needed to bathe in his opponents' blood at night to get to sleep, that was <u>his</u> problem to deal with. I thought he was going to raise his hand to me when I told him that if he needed someone to toss him off and kiss his arse as our fearless leader, "Kill Yoo" would be on the ground in a couple of hours, but I wasn't about to. Finally, running out of steam and words, I gave him the old "two-up" and stalked off to my tent for a nap and a cup of tea.

Bloody barbarians! I refuse to give in to the kind of amateurism and bloodlust that the rest of these wankers are basking in. The tea assisted a bit with moderating my temper, but I didn't go along with my next scheduled flight mission. Instead, I went and visited the Seabees in their gaming tent and helped myself to their loose cash. That put me in a better frame of mind, so I went back to my bunk for forty or so winks. Sod Meyers anyway.

Saturday, February 26, 2011, 6:00pm EST

Channel 4 News

"Good evening everyone, this is Channel 4 News, and I'm Tom Waushinski reporting to you from Pusan South Korea. The city of Pusan has gone from a bustling modern metropolis to a vacant city, and finally to a modern battleground. Tomorrow, or possibly within the next few

days, it will become a communist enclave."

"The Pusan perimeter was breached in a wellcoordinated attack on both the Eastern and southern coastlines. With both their right and left flanks compromised, R.O.K. and United Nations forces redeployed to the city of Pusan itself to make a last stand. Elements of Task Force Stingray recently reinforced by 3 fresh Amphibious Reaction Groups (ARG) began to withdraw United

States forces. Several naval contingents, originally sent from Europe to help defend the peninsula, also arrived in the mid-afternoon. A small vertical attack carrier battlegroup from Spain, another from Italy, and two from Britain have also begun to evacuate the city. A French carrier battlegroup based-around the carrier De Gaullejoined with the Carl Vinson battlegroup to help bolster its defense after the loss of so many escorts."

"Here in the city, we've left our 30 story perch, and made our way towards the docks where we've heard rumors of a total American pullout. At the docks, we've been asked not to show you the ships or the materials being brought in. We can tell you that supplies continue to come into the port city. The civilian population has almost completely been evacuated, and some military personnel are being moved by ship. That is all we can report. That, and the obvious fact that the entire city has become a maelstrom of explosions and bullets. There's no hiding the gunfire that you're hearing behind me, or the artillery's rumble in the distance."

"We've found ourselves this nice sandbag bunker, so we'll be staying here until the last boat leaves. Not to worry, we're told by these U.S. Army soldiers all around us, that they've spent too much time filling the sandbags, and they're not ready to leave, so we hope to report to you from here tomorrow. Until then, this is Tom Waushinski for Channel 4 News. Back to you Tawnya...."

Saturday, February 26, 2011

Republic of Korea, Tochon Airstrip

Personal Log – *Lef*tenant Ian Thomas "Top Hat" Sanders IV, R.A.F.

I was roused this morning by "Airman Intrepid," who notified me that Meyers wished to see me in the Ops Shack (Tent) as soon as possible. Still a bit miffed about what happened yesterday, I made him wait for about an hour while I changed my shorts, boiled water for shaving and for tea, stopped by the mess for a meal, and washed my teeth. When I sauntered into the Ops tent, I was informed that he had gone out on a flight mission. I told the Liaison Officer (a different one today) that I wasn't pressed, and he could bloody-well find me when he got back if he needed to talk to me.

He showed up in the gambling tent about twelve hours later, alone this time. I was in the middle of a decent winning streak with the bootnecks, but wasn't about to piss in Meyers' coffee again unless he gave me another reason to. (Mind, it wouldn't have taken much of a reason....) I jumped for first service, and gave him the lowdown on why I spoke to him in such a manner yesterday.

Air combat, in comparison to just about any other kind, is clean. All we see are airplanes exploding; not blood and brains, severed limbs, or projectile internal organs. We don't even need the pilots of the opposing aircraft to die to become an "ace;" the confirmed kills are the planes themselves. While I am indifferent to anyone else's chute except my own opening, I'm not about to take my cannon after some lucky sod who escaped a firey death, either.

I'm a fighter pilot for a reason. I don't drop bombs, else I'd be strapped into a sodding Aardvark rather than a Freedom Fighter. (Ever notice how even the name of my adopted bus seems to go a bit beyond machine-gunning foot solders?) It's just a bit contrary to my character. I've done ground attack in the past, and it didn't sit well with me. (The Queen's commission and orders from my superiors mean a little bit more to me than some Yank's preference in fighting style as well, but I kept that bit to myself.)

I as yet have no reason to hate the North Koreans and Chinese any more than any other enemy I've faced in the R.A.F. He can talk about invading a sovereign country or the mistreatment of the Navy Seals until he's blue in the face, but I'm not buying it – I've heard stories of how some American soldiers handled Vietnamese "interrogations" during that conflict. The shoe's on the other foot now, that's all that's different. I'm a professional pilot, and a servant of my government. His recent actions haven't instilled me with anything other than disgust, and I told him point-blank that I wasn't about to follow that directive, whatever the consequences.

He gave a rather lopsided half-grin, then actually apologized! Apparently, after my rather vocal display yesterday, other like-minded pilots had come up to Meyers, individually or in pairs, and had voiced their agreement with me. He said that he hadn't taken that particular point-of-view into account when he had given the order for the ground-attack passes. He also remarked that he was altering the rotation to spread those pilots who had expressed reservations with strafing ground targets evenly through the flights, and to use them as "eyes-up" while the other pilots pounded the ground. We would be the last ones to land, of course, but he would also see what he could do to offset our lighter ammo loads (not carrying any mud-moving equipment). We knocked it about for a moment longer, then I shook his hand and took my leave.

It takes a man to admit when he's wrong. I may not see eye-to-eye with Meyers, and I may not be his biggest fan, but he just may not be the complete berk I've made him out to be over the last few weeks. Time will tell.

2nd Lt. John Chamberlin-Personal Log

45 nm west/Southwest of Pusan, Republic of Korea

We've been out on 33 TRAP missions since the war officially began. Most were uneventful, and only the only action we've seen has been stray small arms fire, some light AAA and a lot of wreckage. Today, we might be going out on our last. The communists have taken over the entire peninsula, and we've been ordered to evacuate all U.S. personnel from the collapsing Pusan perimeter. We're buggin' out. Capt. Brown has called a briefing to go over our involvement with the pullout. The platoon is tired. So's the rest of the company. I'm sure the rest of our people on the ground are in even worse condition. That's what's keeping me going, and I'm sure everyone else is thinking the same thing. I overheard somebody in the cafeteria saying that we should just take our toys and go home. I say forget our stuff, let's just get our people out and get home. We were never meant to win this one anyway.

My only regret is that I was involved with starting this whole damn thing. The more I think about it, the more I wonder what I could have done differently to have kept us from being pinned down that first night; that first TRAP mission, that first blood. Thank God that I haven't come up with an answer yet. My head tells me that we did all we could to get that pilot out without anyone else's help, but my heart tells me it was my fault. Maybe that's my sense of responsibility talking. Maybe it's my training. Maybe it's simply my nature to try and find answers to impossible questions.

I know I didn't get that woman shot down. I wasn't even the guy that was originally sent in to get her out. I can't help but feel that maybe I should have been more focused, but there was nothing we could do. It was probably the same for the North Koreans too. If they had commandos in Michigan, you bet your ass we'd go after them with everything we had. That's what they did. I guess.

Sunday, February 27, 2011

0810hrs.

Republic of Korea, Tochon Airstrip

Personal Log – Lt. Kim "Killer" Yoo, Republic of (South) Korea Air Force

I'm stuck on the ground for a few days now. The communists managed to get a sniper across the river last night. He picked off a bulldozer full of Seabees before they took him out. The sniper must have reported our position, because about an hour after they dragged his body out of the bush, we started to get hammered by artillery and mortars. We lost a pair of F-16's, 3, AV-8B Harriers, and almost all of the planes on the ground were damaged. My Falcon has some holes in the nose, a flat nose wheel, and most of my avionics are damaged (worse than before).

The wing wipers are all busy repairing serious problems, so I've taken to repairing my own aircraft. Replacing the nose wheel wasn't too bad, I just used a tire jack from and American HUMMV and changed it out with a wheel from one of our destroyed F-16's. Replacing the avionics isn't an option, so I started to remove them. All of the obviously damaged ones were easy enough to remove, but now some of my gauges aren't working. I'm afraid I'm going to have to have a wing-wiper repair the damage that I might have done. At least the nose will be a lot lighter now! That means faster pitch control, faster take offs, and a bit more responsive yaw. When it's finally ready, the plane should be very interesting to fly!

Sunday, February 27, 2011

Republic of Korea, Tochon Airstrip

Personal Log – *Lef*tenant Ian Thomas "Top Hat" Sanders IV, R.A.F.

Bit of a sticky wicket last night – we got the living piss shelled out of us. There wasn't an unmarked plane on the strip, and I think we ended up losing a handful of rides. My F-5 wasn't damaged beyond the reach of duct tape, but others were, most notably "Kill Yoo's" F-16.

I don't know a lick of Korean, but the way he was yammering out loud to himself as he was jacking up the nose to replace a flat tire didn't sound like he was too happy. I couldn't resist. Before he noticed me, I turned and snuck away, returning wearing my biggest smile and loudly whistling the happiest song I could recall. I even had the cheek to wave at him as I walked by.

He stopped for a split second with an incredulous look on his face, then got even more pissed and began to give the jack everything he had, raising the volume of his bitching (I guess) ten to fifteen decibels.

I smiled for two days straight, and the pantomime description got the Marines laughing so hard I thought they were going to piss their knickers.

Monday, February 28, 2011

1346hrs.

Republic of Korea, Tochon Airstrip

Personal Log – Lt. Kim "Killer" Yoo, Republic of (South) Korea Air Force

One of the wing-wipers came over and stopped me before I completely disabled my aircraft last night. He was an American wing-wiper, and professional opinions of my mechanical abilities were clearly conveyed to me-repeatedly. The poor man worked all night repairing my attempted repairs. He also removed my radar (Apparently one of the mysterious black boxes that controls its scan rate was damaged in yesterday's artillery strike, and it was useless anyway.)

I left at dawn and joined the first fighter sweep patrol. The plane left the runway like it had an agenda! Everyone else followed. At 5000 feet, we put together a wide four-finger patrol formation and began a giant clockwise patrol around the airstrip.

One of the pilots spotted a few Jian 7's making a strafing run against the city of Taejon. We turned to the right, and headed for action. Since I knew that we were over friendly territory and there wouldn't be any anti-aircraft fire, I came in low through the city. No one expected to see and F-16 racing down the street with tall buildings on both sides, and when the rest of my flight jumped the Jians, I was in a beautiful position. Their attention was fixed above them-where the rest of my flight was attacking from. A few communist missiles headed into the sky, but I came right up from below, and found two targets for my Sidewinders. Before the first one hit, I was already able to swing my plane's nose to the left and fire my cannon at a third Jian. All three went down north of the city. The rest of my flight shot down two more. Then, we headed back to the airstrip on the southeast side of the city to rearm and refuel.

On the ground, everyone wanted to know how I was able to take so many shots at so many communists in such a short period of time. The poor wing-wiper who had been working on my plane all night was walking by, so I pointed him out to everyone and told them that I had him modifiy my F-16. It was partially the truth. His repairs had changed the weight characteristics and subsequent flight characteristics of the Falcon. Now, that poor wing-wiper is going to be busy modifying at least a dozen more aircraft. He'll be busy until the war is over!

uesday, March 1, 2011

2218 hrs.

Republic of Korea, Tochon Airstrip

Personal Log – Lt. Kim "Killer" Yoo, Republic of (South) Korea Air Force

The communists keep finding new ways to cross the river or breakthrough the northern side of our lines-the Pusan Perimeter. Meyers scrambled everyone to their aircraft before sunrise this morning because someone spotted some communist infantry about a mile from the strip. The American Marine Harriers took off vertically (a great way to waste precious fuel), and they searched the area while everyone else took their turn racing their planes down the runway and into the air to join Meyers. I was one of the last in my squadron to taxi out for take off, and while I was waiting for my turn on the runway, the all clear was sounded. I guess the Seabees went after the communists and sorted them out.

Just the same, we all hopped into the air to join Meyers. Unit cohesion fell apart for some reason, and we all roamed around the entire Pusan Perimeter-alone, in pairs, or in flights. It was probably an air traffic control officer's nightmare, but it was much worse on the few routine flights of communist bombers that were making their usual bomb runs on the port facilities in Pusan. They attacked in the same way that we were flying-along, in pairs, and in flights. It was a complete free-for-all. As soon as any communist planes were spotted, everyone jumped on them. The sky reminded me of a shark feeding frenzy.

There were so many planes in the air-our, communist, and other Coalition planes-that I never even had the chance to chase one. I saw Meyers shoot one down. There are also a few pilots who I recognized from the group that were able to down some communists. The sky was packed and very very busy.

Besides seriously overwhelming the communists today, our little composite group also took advantage of the confusion by refueling all of our aircraft. Without any sort of communication or organization, each of us in the composite group noticed the regular U.S.A.F. planes linking up with their mid-air refueling planes off the east coast of Pusan. One by one, we all lined up and took our place in line for gas. It was great. I don't think any of us have flown with full fuel tanks in days. We went through 11 refueling planes before we all started to land back at the highway strip!

Wednesday, March 2, 2011

2115 hrs.

Republic of Korea, Tochon Airstrip

Personal Log – Lt. Kim "Killer" Yoo, Republic of (South) Korea Air Force

Today was a real test for the group. The communists must have taken yesterday's losses seriously. Today, they came back and hit us with everything they had. The entire Pusan Perimeter came under multiple attacks from bombers and fighters attempting to sweep us from the sky. It was overcast, but the missile contrails and falling planes pierced the white sky and stitched it to the ground.

An entire squadron of Su-27 Flankers raided our airstrip. They dropped bombs all around the area. Our fuel and arming revetments took hits and went up right away. Some of the Seabees were lost when a communist plane strafed their tents. We all tried to get to our planes, but only a few were able to get airborne before a Peruvian Mig-29 crashed on the runway and blocked it. Our morning fighter-sweep groups were up north, and they never even knew we were under attack (radio communications have been jammed by both sides since the war started).

The American Marines were able to get up, and they made a great showing of themselves. They were able to lift off vertically. It used a lot of fuel, but at that point it didn't matter. Despite being outnumbered at least 2-1, they went up and tangled with those communists as though they had the advantage. On the ground, we were all taking cover in foxholes, slit-trenches, and in a few small trenches. I watched from a sandbag bunker near the south security gate. The communist Su-27's are much faster than the Harriers. They have better acceleration. They're more maneuverable in conventional flight, and they usually carry 5-10 times more air-to-air weapons than the Harriers. The Marines' used the moveable thrust nozzles on their Harriers to stop in mid-air, radically turn, or dive like bricks. They pulled off maneuvers that simply defied reason, and the communists must have had their heads spinning. I saw seven of the communists go down in flames, and I'm fairly certain that the Marines at least damaged all the remaining Flankers. It was something to watch.

On the ground, the (Korean) Army unit that's doing base security was busy taking potshots at the communist planes passing overhead. Pair of soldiers with an M60 machine gun was right next to me, and they were nice enough to let me have a try. I didn't hit anything, but I know from experience that when attacking a ground target, all tracer rounds look like cannon rounds. I must have at least brought a moment of hesitation to some of the Flanker pilots because I went through an entire belt of ammunition, and the tracers were all over the sky.

The Seabees spent the rest of the day clearing the runway and repairing damage. My plane took some shrapnel in the rudder, and now my rear radar detector doesn't work. Almost all of the planes took some sort of damage, and I was lucky not to have a total wreck on my hands. A 500-pounder landed in the next revetment and blew an American F-15 to pieces.

Wednesday, March 2, 2011

Republic of Korea, Tochon Airstrip

Personal Log – *Lef*tenant Ian Thomas "Top Hat" Sanders IV, R.A.F.

When I heard about Kim's escapades yesterday, I made sure that I was one of the first to approach the ground crewman who had "modified" his plane. I also made sure (thanks to a quick trip through the gaming tent) that I had enough cash on my hands to grease his palm sufficiently such that mine would be the next modified, and done posthaste. The cheeky bugger actually wrenched by lamplight for a couple hours in the dark to claim the second half of his payment quicker! (That's dedication.)

When the attack on the base took place this morning, I was already moving toward my plane with the intent of trying out her new attitude. The ready squadron had already begun taking off as I reached my plane, and I strapped in and headed for the air as quickly as possible. Unfortunately, one of the lads in a Mig-29 dicked the runway behind me, preventing any further conventional aircraft from using it. The bootnecks were able to get up utilizing their Harriers, but that was it.

We had our bloody hands full, let me tell you. What saved it for us was the Marines pulling all of their barmy acrobatics with the screwed-up thrusting capabilities of their Harriers. I've never seen anyone make those pathetically ugly planes sit up and beg like those Yanks did today. I managed to do pretty well for myself in the air, though, so I didn't feel too overshadowed by them. We ended up losing around half of the few pilots who had made it up before me, though, and three of the Harriers didn't come back either.

They had to clear the wreckage of the Mig-29 before I could land, and that got a bit dicey as well – I was pretty nearly out of fuel before I got back on the ground safely. Hell, at least my damned canopy stayed on. That's always a plus.

Thursday, March 3, 2011

1609 hrs.

Republic of Korea, Tochon Airstrip

Personal Log – Lt. Kim "Killer" Yoo, Republic of (South) Korea Air Force

We made another group effort in Mig Alley again today. This time, Meyers had his squadron spread out in a 12-mile wide, line abreast formation at 35,000 feet. My squadron followed about 20-miles behind in a loose formation that we're calling "The Gaggle" (it's really no formation at all, but we needed a term for it, and an egotistical British pilot who's not very fond of the idea accidentally coined the name.) We also had the third squadron break up into flights and come in low along the crest of the ridges-carefully staying out of the flak traps in the valleys.

The communists must have taken another beating in yesterday's raid on the perimeter. They were sending several huge groups of fresh Chinese planes to make another big bombing attack-presumably against Pusan. Their escorts didn't have any trouble spotting a few of the high squadron, and they tried to get into a fight with them. It was like a school of fish going after bait in a net. As soon as one plane was attacked, Meyers motioned for the rest of the squadron to come around and envelop the communists. While they were tangling with at least 4 or 5-to-1 odds, our squadron came in and made things a bit more even. The low squadron ran headlong into two groups (at least 8 full squadrons) of Chinese bombers that were approaching at low altitude.

The dogfight was the usual enormous, chaotic, spiraling dance of missiles, contrails, flares, invisible electronic jamming, microscopic radar-absorbing chaff, and of course explosions. I never heard a missile lock on to my plane (my rear radar detector was damaged in yesterday's air raid), and I never saw a missile come at me. I'm sure some communists took their shots, but there was so many flares, so much chaff, and so much electronic jamming that I don't think anyone's missiles locked. It was a gunfight.

Four miles below us, the low squadron ran into the Chinese bombers like two opposing shotgun blasts. It was a miracle that no one collided. I didn't have time to watch, but whenever I was chasing a target below me, I could see black smudges on the snow where planes had crashed. There wasn't a lot of time to spectate.

I was able to get good bursts into three communist Jian 7's and an Su-27. Some of my rounds might have hit an American F-16 when I overshot one of the Jians. I took some hits in the nose, but they went right through, and I never did see who shot me. There were streams of tracers going in every conceivable direction. At one point, it was like flying through a winter white out, but the flakes were 20mm, 25mm, and 30mm cannon rounds. The hardest part about seeing so many tracer rounds was the knowledge that only one in three rounds is a tracer, and that means that as many as we saw, there were three times as many rounds in the air!

When we arrived back at the strip, some American officers were there waiting. They had found out about our little composite unit, and wanted to take away some of the pilots who's squadrons were being reformed. Meyers let nine go, but the others clearly wanted to stay. The American officers were very vocal, but while they were arguing, one of the wing-wipers handed the officers copies of some authorized transfer orders for each of the pilots who wanted to stay. They were forged documents, and everyone knew it, but it was enough to buy the pilots a few more days with the group. That's all they really wanted, and the American

Officers had the ever-important documentation to show why part of their mission was not accomplished. Beaurocrats! If only we could send them to the river's edge to hold back the communists, THEN, we might be able to win this damn war!

Friday, March 4, 2011

0805 hrs.

Republic of Korea, Tochon Airstrip

Personal Log – Lt. Kim "Killer" Yoo, Republic of (South) Korea Air Force

It looks like I have to spend another day on the ground. The communists were able to get a commando team across the river somehow, and they hit us last night. We lost 21 planes to demolition charges. In exchange, the Seabees were only able to bring back a single body. Our Army security detachment seems useless, and they're off looking for the rest of the commandos right now. I'm certain that I'm not the only person who thinks we could do just as well without any security unit at this point!

My plane took some small arms fire during while the Seabees were chasing away the communists. It also has some holes in it from yesterday's gunfight in the sky. I've been told to get some rest, but the communist artillery is pounding the Army's positions along the river, and we're taking a few stray rounds here at the strip too. It's not an environment for resting. Besides, the wing-wipers just patched the holes in my plane with some duct-tape, and now it's just sitting there. No one's working on it. I could be killing communists right now, but I'm not allowed. This is insane!

Friday, March 4, 2011

Republic of Korea, Tochon Airstrip

Personal Log – Leftenant Ian Thomas "Top Hat" Sanders IV, R.A.F.

Well, I've been busy for the last few days, pulling duty after duty running fighter sweeps. Haven't had more than a 4-hour bunk in the last week, but it has yet to bother me. It tends to be one extreme or the other, as well: either excruciatingly boring, or running the full gauntlet, arse-deep in alligators the whole way.

To be truthful, I had a couple of rather harrowing near misses yesterday. Almost took a heatseeker right up the bum on the morning's first furball; as I slammed the stick as hard as I could to the bottom left and popped flares, I slid past an oncoming bogey by only a few bloody inches. He ended up taking the missile meant for me right between the eyes – poor bastard probably never knew what hit him. Then, later in the day, "Honker" got blown away less than a hundred feet from me.

About a week ago, I was griping to a Yank flightmate about how loose our formations were. I was in the middle of telling him we looked like a gaggle of dumber-than-average geese up in the air, when I felt someone tap on my shoulder. Thinking I was about to get another reaming from Meyers about being a team player, I spun around with my most evil look on my face; in fact, I was looking completely over the perpetrator.

Standing with his nose on par with my Adam's-apple was a Nip with a goofy grin on his face and a bulb-type bicycle horn in his hand. I'll be buggered if the cheeky blighter didn't honk the blasted thing in my face twice, sounding just like a sodding goose, then walked away without a backward glance! I looked at the Yank, trying to keep a straight face, and muttered, "…and that berk's been point on every formation I've flown this week." Then the two of us almost fell to the ground, laughing our fool heads off.

Funny thing was, I kept bumping shoulders with the little water buffalo-shagging git all week, not that I was trying to. Everywhere I went, I'd hear that goose-honking noise, look around, and there he'd be with the same barking-mad grin on his face, waving his arm off at me. He even popped up once in the gambling tent – I'd just told the bootnecks about "Honker" and there he was, over at the craps table, sounding his horn and waving again. The Marines about lost it there, almost falling out of their chairs laughing with me. Odd thing; no matter how often he popped up, he seemed so genuinely happy that I cracked up every time.

So yesterday afternoon, we're easing along in that blasted loose formation again, and from under me and to the left another F-5 pulls up to me tight. I look over, and it's that silly wanker again, waving and squeezing the bulb on that bloody horn like Harpo Marx. I started laughing, waved at him … and his F-5 simply evaporated. It was a goddamn miracle his debris didn't take me out as well, as close together as we were! I instinctively pulled into a diving loop, popping countermeasures as I flipped over, and there they were – nearly a whole bloody squadron of Jian-7's, firing enough missiles at us to turn the blue sky white from their contrails. The rest of the flight scattered, someone started screaming for help on the radio, and the fight was on.

Another member of the flight just happened to be beside me when I pulled out of the loop, so we paired up and flew straight into their teeth, cannon blazing. I'll admit, I was angry that "Honker" had choked on a heatseeker, but I was too busy trying to fight my way free, cover my wingman's ass, and stay alive myself to get flaming pissed at anyone in particular. Besides, how in the hell could I tell who had smoked him?

I ended up loosing all of my missiles before I could say "How about a reload?" so I ended up chasing planes around the sky with just my cannon. We shot down or drove them all away, though, with the assistance of two other flights from the active squadron that came tearing back to assist us.

After I had landed, I was walking back to my bunk when I heard a familiar yammering in front of me. I peered around a corner, and saw Kim Yoo giving one of the ground techs five kinds of bloody hell. After the berk walked away, I asked the "wing-wiper" what in the world was going on. He told me that the stupid sod was all worked up about the fact that his plane was still here on the ground, not being worked on yet due to the backlog of planes that needed fixing. I asked if it was going to be done by tomorrow morning, and he said not a chance.

Not wanting to lose such a perfect opportunity to twist Yoo's tail, I told the techie that all my F-5 needed was a little patching over the fresh holes. If he would do it tonight, and accept a bit of a monetary gift for his rapid attention to my bus, I'd greatly appreciate it. I also remarked that it would really light a fire under Yoo's arse, if the tech was so inclined. He took my cash with a grin, and I told him I'd see him tomorrow morning.

Saturday, March 5, 2011

0858 hrs.

Republic of Korea, Tochon Airstrip

Personal Log – Lt. Kim "Killer" Yoo, Republic of (South) Korea Air Force

I have to sit on the ground for at least another day! The wing wipers tell me that my engine took some shrapnel in the artillery strike the other day. They say that I lost a few turbine blades, and if any of the others have fractures, they might come loose and cause a chain reaction of broken blades in the engine. The wing-wipers showed me a handful of fragments, and I've seen engines come apart from broken blades. As soon as one breaks off, it tumbles around and breaks off two more. They do the same until there's nothing but fuel and metal coming out the back end. I do NOT understand why they can't just swap out the engine with one from some of the wrecks that we're canabilizing for parts! The hardest part is watching my plane sit in its revetment while other planes are worked on first. I can hear the communist artillery getting closer and closer ever day, and there's nothing that I can do about it except watch dust and snow collect on my aircraft while my country is being devastated. The situation is very frustrating!

Saturday, March 5, 2011

Republic of Korea, Tochon Airstrip

Personal Log – *Lef*tenant Ian Thomas "Top Hat" Sanders IV, R.A.F.

It worked! I went up to the flightline this morning first thing, before mess even, and there was good ol' Kim Yoo, barking at the same tech about his plane again. I eased up to the both of them, gave a "'Scuse me" to Genghis, and asked the wing-weenie if he had a chance to repair my plane the night before.

The tech did me proud, giving me a big smile and saying of course, right this way, yadda-yadda-yadda, and led me off toward my plane. As we walked away, I heard Yoo give a strangled scream of frustration and go stomping back toward his bunk. The minute we were out of earshot, the tech and I nearly split our bloody sides laughing. I shook hands with the lad – he said if I ever needed anything, just give him a shout. Decent sort of bloke.

Uneventful rest of the day, but I was glad for it. It was kind of relaxing being up in the quiet air, far away from the noise, shelling, and morons like Yoo on the ground below. You know what, though? I really miss Honker.

Sunday, March 6, 2011

0913 hrs.

Republic of Korea, Tochon Airstrip

Personal Log – Lt. Kim "Killer" Yoo, Republic of (South) Korea Air Force

I was finally able to get back into the air again today. After three days of being grounded for repairs, I left the runway and joined with the first fighter-sweep patrol that I could find. We circled nearby Taegu to the northeast, then we headed back over the strip to the river. We followed that all the way to the east/west frontline. The entire area looks like the Western Front in World War I-trench warfare in the middle of a moonscape of artillery and bomb craters.

Then we circled over Pusan. A few buildings were still on fire from earlier communist air raids. We also saw some buildings with gaping holes in them from large communist bombs. There are no cars in the streets-just the occasional tank, armored vehicle, or barricade. At the docks, it seemed like a million people were trying to get on a freighter that was still unloading APC's (armored personnel carriers) from some country. The weather was clear and sunny, and while it afforded a distinct view of the entire battlefield, I knew that clear weather in winter meant cold temperatures for the civilians and troops below.

We sighted a few communist aircraft, but the larger combat air patrol squadrons in the area dealt with them before we could close. I saw over 20 planes go down in flames, and almost all of them were communist. It was good to be in the air again, but the only satisfaction that I found in regards to the communists was the strafing run that I made on my final approach. I emptied my entire load of 20mm into a trenchline on the west side of the river, and it was very satisfying to see the little communist ants flee scurry for cover. Even more satisfying were all the little ant-like communists who couldn't run.

On the ground, I was hoping to refuel, rearm, and return to the sky, but the intel officer told me that I was grounded again! Meyers is having all the planes and pilots grounded so that all of our planes can be painted white. It's a very good idea since some of the planes in our group are Soviet-built. The German, Ukrainian, and Peruvian members all fly Mig-29 Fulcrums of various levels. Each one is painted differently, and in a dogfight, it's been tough not to shoot them down by accident. At the same time, in the same dogfights, the new camo will make it easier to distinguish members of our group from other Coalition squadrons.

We all seem to like the idea, but it also seems like a lot of work. Each plane has to have a tent pitched over it. Then, we have to clear the snow off the wings. The wing-wipers will do the painting, but some of the pilots-myself included-have chosen to help out so that the job can be done as fast as possible. In the cold, dry, winter air, the paint seems to dry in about 15-minutes as long as it's not put on too thick. At least it's a better cause than waiting for the wipers to get around to changing my engine!

Monday, March 7, 2011

1651 hrs.

Republic of Korea, Tochon Airstrip

Personal Log – Lt. Kim "Killer" Yoo, Republic of (South) Korea Air Force

Meyers still has the group flying in a rotation of fighter-sweep/reserve/rest. He's letting the Koreans fly as many sweeps as we want, and I was able to fly 11 fighter-sweeps today. The skies are still very crowded with both Coalition and communist aircraft, and it was difficult to chase a communist on my own. Whenever I rolled in on one, two or three Coalition planes would get between us and drop my target. I chased two or three on every flight, but I was only able to fire on four planes throughout the entire day.

It was good to get up and fight, but I don't think that airpower is making a difference right now. The communist ground offensive continues unabated. They actually managed to setup a pontoon bridge overnight, and no one knows how many communists made it across before our artillery destroyed it. I do know that our Seabees are being pulled out and sent to Pusan. The American Marine Harrier squadron (what's left of it) has been ordered to prepare an evacuation plan in case our airstrip is overrun.

After our last fighter sweep, Meyers called all the pilots together and told us that we might have to evacuate in a few days. Besides the pontoon bridge, the communists have been slipping commandos across the river in everything from hovercraft to rowboats. It seems that they want to cause as much trouble as possible to disrupt our routine operations and make it easier for a large crossing someplace.

To make matters worse, they have sent most of their armored forces to attack the American paratroops from the 82nd Airborne Division. They are holding the far eastern edge of the perimeter, but the 82nd has been in combat since the first few days of the war. They are a light division, and while they initially had some heavy equipment and anti-armor weapons, everyone expects that those precious supplies must be nearly exhausted by now. Tomorrow, Meyers wants all three squadrons to head over there and attack anything that "...smells remotely Chinese or North Korean." We don't have any more air-to-ground ordnance, but we will make good use of our cannons.

The American Marines are going to go with us too. Their squadron leader has prohibited any of his pilots from joining our composite group out of loyalty to their Marine Corps. However, they are all feeling as frustrated and angry as the rest of us. They also realize that their handful of aircraft won't be able to protect our little airstrip anymore. Besides, the American Marin pilots' exist to support the ground elements of the Marine Corps. They are very skilled in air-to-air combat, but it's not their specialty. We will welcome them tomorrow!

Monday, March 7, 2011

Republic of Korea, Tochon Airstrip

Personal Log – *Lef*tenant Ian Thomas "Top Hat" Sanders IV, R.A.F.

It was odd as hell seeing my faithful F-5 painted up snow-white. (And no, I wasn't allowed to paint the nose and landing gear bright yellow to simulate a goose's beak and legs, though the painting squads thought it was a bully idea.) I've got to hand it to those sods – they've been really putting the wood to it to get all of the planes painted. I saw some of the pilots helping as well, including the wing's favorite Mongol, Kim Yoo. Oh, yes, he was jolly-well all over helping the blokes out – and I know why. He wanted so badly to get back in the air so all of the communist bastards wouldn't be shot down before he got up there again. Wanker!

Myself, I took the downtime with a bit more grace, meaning I slept through most of it. Today, when Myers said the Koreans could take as many trips up as they wanted without waiting in rotation, I sat back and let them do it. I was only up in the air twice, while I sat in with the reserve squadron (in case I was really needed) four times.

If Meyers wants to allow the sodding Nips time to sate their passion for revenge, let them. I don't think it's all that perfect of an idea, though. What about the cumulative effects of stress and fatigue on the pilots? Doesn't that kind of balance out the need for the cathartic release of killing Commie Nips? Oh, well. I'm just bloody-well looking out for myself, and that's that.

More shite to top off the morning – it looks like our efforts here, while not quite in vain, have been ineffectual in the long run against the staggering numbers of ground forces the Commie Nips have put in the fray. As the noise from the shelling in the distance marched closer, Meyers gave us the bad news when everyone was finally on the ground – we're probably pulling out again. Damn! Tomorrow we're headed for the extreme east of the perimeter to play backup for some Yank paratroops – giving the rest of those bitten by the ground-pounding bug one last chance to hone their skills. Hmmm, with the collapse of South Korea imminent, I wonder where we'll be headed next?

Tuesday, March 8, 2011

1023 hrs.

Republic of Korea, Tochon Airstrip

Personal Log – Lt. Kim "Killer" Yoo, Republic of (South) Korea Air Force

This was a very rough morning. We all left Tochon and headed for the northeast corner of the Pusan Perimeter-on the shore of the Sea of Japan-near our old airstrip at Pohang. Meyers had us all flying in a very loose group gaggle formation that stretched from 1000 feet to 7000 feet. AAA and SAM activity was awful. We lost a few planes to both. It seemed as though the communists had moved all of their anti-aircraft weapons from the flak trap valleys in the north down to the perimeter's edge. Everyone took some sort of damage. I had a piece of shrapnel crack my canopy, and there are several new holes all over the right wing and fuselage that will need to be patched.

With so many anti-air artillery pieces, traditional artillery pieces, trucks, armored vehicles, and trenches, we had plenty of targets to strafe. The American paratroops on the ground were witnesses to a fantastic air show. Including the Marine squadron and a few other Coalition planes in the area, we attacked the communists will over 100 planes. The paratroops had some helicopters of their own join in the attack, and for almost an hour, our aircraft dove through the intense AAA and rain cannon rounds on anything that moved. I made nine strafing runs-mostly against AAA targets since they were the most satisfying. After my ninth, my Vulcan cannon was out of ammunition, but I still dove against three more AAA sites. I did it to draw fire away from those who could still attack, but I also did it to scare the communists. It was very satisfying. When I pulled out of my last dive, I could actually see the gunners running away from their quad 57mm cannon mount and trying desperately to get under cover. I laughed as I headed back to the gaggle.

While we were in the air, the communists sent almost all of their available aircraft to attack targets on the western side of the perimeter-including our new strip. Our arrival was not very enjoyable. The runway was well-cratered, and it was difficult to steer around the huge holes. Two of our planes didn't pass their ground-maneuverability test, and crashed into the craters. Both can be repaired. Our revetments were all damaged. Most of our tents are shredded now. Our ammunition bunker was destroyed, and our fuel tanks were punctured.

The Seabees are gone, so the Meyers had us all patch our own holes, repair our damage as best as possible, and refuel with as much fuel as we could before it all leaked out. Duct-tape is in short supply, and some of us will have to fly with holes in our wings. Tomorrow morning, we're all going to leave and head out to Clark Field in the Philippines. Some pilots might leave the group and join formal squadrons in their nations' Air Forces. Like the rest of the Koreans, I'm going only because there are no more airfields in the country. We've heard that the rest of the Korean Air Force is being sent to Clark Field for restructuring. I plan to report myself available for reassignment, but until I get any orders, I'll stick with Meyers and this group. We might not have stopped the communists, but I know that I was able to do more than I would have been able to do if I would have stayed at Kunsan waiting for orders.

Tuesday, March 8, 2011

Republic of Korea, Tochon Airstrip

Personal Log – *Lef*tenant Ian Thomas "Top Hat" Sanders IV, R.A.F.

Bollocks to the North Koreans! Bollocks to the Chinese as well! Everything was going beautifully regarding the attack today, as far as the worm-killers were concerned; we made our approach to the target area through the usual flak and SAM activity (a mite heavier than usual, but no great thing). The air cover for the ground-attack element, including myself, smashed our way through the twelve-or-so enemy planes flying loose cover for the Commie Nip force menacing the American 82nd. We then commenced patrolling in high corkscrews, shallow dives, and half-circle jogs – running amok in general, trying to draw some fire away from the strafing teams while keeping our eyes open for other enemy fighters.

What enemy fighters? Except for the initial slaughter, the only planes we had to defend against were one- and two-man stragglers limping back from who-knew-where. Initially, it didn't bother me too much, considering the vast amounts of flak we had to dodge, but my worries began to pile up. It started to worry Meyers a little as well, because he sent half of the almost useless covering team back full-burner to the western side of the perimeter to check on things there, but they were too late.

While we had been busy pulling the 82nd's fat out of the fire, the Commie Nips had a jolly time pulling the same ground-attack shite in the West, including our undefended base of operations. The runway resembled a moonscape, our ammo dump was now a trash dump, and the fuel tanks that <u>weren't</u> burning had leaks that the Exxon Valdez would have been proud of. Damn it all to hell! The tossers even machine-gunned the tents in the hopes that some unlucky sod would be in there still. Unfortunately, that included the tent we were using as a makeshift sick bay. Three of our lads were torn to bits as they lay helpless.

It wasn't a very nice homecoming, let me tell you. Meyers took charge, though, and got everyone busy patching themselves and their planes up, prepping to kick the dust of Korea from our feet. Our next stop will be the Philippine Islands, on some until recently deserted Yank airbase. I can only hope that it isn't going to be simply the first stop in yet another retreat across a foreign country. I'm getting a mite tired of losing the war.

Wednesday March 9, 2011

New York Post

Pusan Perimeter Perishes

Chinese forces, in bitter house-to-house urban combat, pushed their way through the outskirts of Pusan and cracked the American lines near the coast. Chinese armored forces quickly reinforced the breakthroughs, and within a few hours, the entire city was enveloped. Retreating American units, confused United Nations units, and worn out R.O.K. units were scattered, cut off, and either destroyed or captured. First estimates indicated that almost 250,000 military personnel are missing in Pusan. Almost 30,000 of them are Americans. The bulk of those killed, wounded, captured, or otherwise missing are from Republic of South Korea units.

Thousands of troops escaped the city, and the Peninsula for that matter. Twelve aircraft carriers of different sizes, types, and nationalities, escorted by almost two hundred smaller vessels, are carrying the dispersed and beaten armies, one of which now has no nation. Since members of the R.O.K. military could clearly not return home without the threat of persecution, they all qualified for amnesty in the U.S. The same was true for most of the other United Nations countries that had tried to defend South Korea. Most appealing were the offers from the Philippine government and the government of Taiwan. Both were close by, and both had ports and airfields for the naval and air units that needed new homes, and both were close to the Korean Peninsula where they hoped to return, soon.

Wednesday March 9, 2011

New York Post

Pusan Perimeter Perishes

Chinese forces, in bitter house-to-house urban combat, pushed their way through the outskirts of Pusan and cracked the American lines near the coast. Chinese armored forces quickly reinforced the breakthroughs, and within a few hours, the entire city was enveloped. Retreating American units, confused United Nations units, and worn out R.O.K. units were scattered, cut off, and either destroyed or captured. First estimates indicated that almost 250,000 military personnel are missing in Pusan. Almost 30,000 of them are Americans. The bulk of those killed, wounded, captured, or otherwise missing are from Republic of South Korea units.

Thousands of troops escaped the city, and the Peninsula for that matter. Twelve aircraft carriers of different sizes, types, and nationalities, escorted by almost two hundred smaller vessels, are carrying the dispersed and beaten armies, one of which now has no nation. Since members of the R.O.K. military could clearly not return home without the threat of persecution, they all qualified for amnesty in the U.S. The same was true for most of the other United Nations countries that had tried to defend South Korea. Most appealing were the offers from the Philippine government and the government of Taiwan. Both were close by, and both had ports and airfields for the naval and air units that needed new homes, and both were close to the Korean Peninsula where they hoped to return, soon.

Wednesday March 9, 2011

Personal Log-2nd Lt. John Chamberlin

35 nm west of Pusan, Republic of Korea

Yesterday didn't go so well. Instead of flying another TRAP, Col. Ghetty had all the infantry companies in the battalion fly in airborne reserve for the evacuating troops. Each company had to orbit the evac point for an hour and a half, then, only after another company arrived to replace them, they could head back to the ship-not for rest, but for refueling! They didn't even let us off the Osprey while they were refueled. It was a major safety hazard, but no one's been hurt-yet.

The first four companies on station didn't have any problems. They never even had to touch down. Of course, as soon as we arrived, as the company we replaced was touching down on the Wasp, we got the call. Evac point Delta One-Eight (not exactly a lucky name for our unit), was being overrun. Three Osprey had gone down, and there were Chinese armored units roaming the LZ. Why we had to go in, I don't know. None of our three platoons had any antitank weapons to speak of. Just the same, before we could really start complaining about it, the Osprey were turning their props toward the vertical, and we began our descent.

The LZ was another road, like our first evac TRAP mission's, and, like before, the Osprey landed in a single file, line astern formation. This time I wasn't in the lead bird. Capt. Brown rode that one in. I was about fourth or fifth back with 1st squad. Bob was in the Osprey behind me with 2nd squad. Third squad went in with just Rick in charge. That was fine because our command and control problems were above, not below me. Capt. Brown's Osprey took a hit from something about a hundred feet above the ground. The number 1 engine, the one on the port side, completely came off, and it cartwheeled into the ground. The rest of the company landed without incident...sort of.

As much as Brown has been a pain in my neck since I arrived in the company, I'm going to miss him. I don't know if anyone liked the man, but everyone respected him a great deal. He had a crotchety old grandfather mystique about him. I never did figure out if he had some sort of problem with me, or if he was just trying to get the best of my platoon. I never felt happy from his pride. Instead, I worked hard to spite his apparent prejudices. As much as my feelings are mixed about him, I will miss him. I think we all will.

Without our Captain on the radio to direct us, Bruce and I had our platoons deploy to the right of the road, and Hector sent his to the left. While we were deploying our units, the Osprey took off and headed out to sea to orbit in safety.

Both sides of the road had a nice treeline for cover about 50 meters from where we landed. There was a lot of small arms fire coming from Hector's side, so, Bob and I talked it over, and we moved my platoon to join up with Hector. The company radio net was out for whatever reason, so I also sent a runner, Kathy Gerrard from Wess' second squad, over to Bruce to let him know what we were doing. She never got there, and no one has seen her since. Bruce, just the same, was able to see us running across the road, and he sent a runner to us. His runner was pissed and gave me some attitude because he felt like we were deserting his platoon. Bob calmed him down a little and sent him back with instructions for Bruce to fire three green flares if he came into trouble and needed help. If he did, my platoon would come back over to that side of the road. All this was going on while tracers seemed to criss-cross the road from every direction. Up until this point, we hadn't seen any enemy armor, or any friendlies.

When we were completely linked with Hector, I went over and told him what I had done as far as Bruce was concerned. He got pissed because he had sent a runner over to Bruce telling him that all three platoons needed to be on this side of the road; poor Bruce. He decided to stay on his side of the road, and I had to put up with Hector. Meanwhile, Bob was directing Rick's third squad because Rick had been cut along the forehead by some shrapnel from a grenade. He would be ok, but I guess Chris Landas, also from 3rd squad, took the brunt of the grenade blast. We were going to have to carry out his body when we left.

I remember Chris from training back at Camp Pendalton before the war. I'll never forget that deer in the headlights look he gave me when I caught him and Christina Boyle making out in a foxhole. The two of them had been going out for about a year since then, but Christina dumped him pretty hard so she could be with Gary, my first squad sergeant. I guess after LZ Delta, she saw Chris for what he appeared to everyone. He seemed like a spoiled suburbanite who, although he had a certain trained discipline from the Corps, never seemed like the type that would volunteer for point, or take a grenade for someone. In the end, that's exactly what he did, whether he meant to or not, when he pushed Rick out of the way. I didn't see it, and I've heard different versions of the story from different people. Was he a hero, a coward, or more than likely, just somebody in the platoon doing their job, and trying to look out for a fellow Joe? I dunno.

After Bob and Rick told me the story about Chris, it was time for the Chinese to make their appearance, in style. There's no way to accurately describe the feeling an infantryman gets when he first hears a tank's diesel engine. It's not fear, or excitement. It's best described as a thought. "Oh shit." Then, when the tanks actually come into sight, that thought changes to, "You've got to be kidding me. What the hell are we supposed to do now?!"

Communications were out. We couldn't even talk between platoons let alone call for the Air Force. Those tanks were a lot bigger than the ones I'd seen in training. I remember them being the same in scale to, say, the surrounding trees and buildings, but they seemed bigger.

Hector called over to me to let me know that he was sending a runner over to get Bruce. I told him not to bother, and I fired three white flares. Bruce got the message, and immediately started coming across. The tanks came closer. There were only three of them, and they had a little infantry support, but we had the accompanying infantry well pinned down. I waited for the first cannon shot to come. It wasn't a long wait. The center tank stopped, rocked back and forth for a second or two, then fired over towards Hector's right flank. There must have been somebody over there that seemed like a good target. I'm not sure what the tank was firing at, but I do know that Hector lost some people.

Bruce's platoon linked up with Hector's where the cannon round had landed. Now our line was complete, and Hector couldn't complain. I was on his left flank. Bruce was on his right. The tanks were headed for where my platoon met with Hector's, and so Bruce had a squad try and work around to the right in an attempt to get the tanks from behind where their armor was weakest. Three more cannon rounds came towards our line, one at each platoon.

The tank commanders must have realized that if we had any real antitank weapons, we would have used them, and so they mostly used their machineguns on us. I saw a few more of Hector's people get hit, then I saw Bruce's flanking squad fire a a pair of light anti-tank rockets at the side of the tank on the right. Two of them actually bounced before they exploded. One rocket exploded on impact, and, even though the tank seemed undamaged, it stopped.

As the turret was turning to engage the squad, I saw at least twenty people jump out of nowhere and up onto the back of the tank on the left. It was R.O.K. infantry fighting with the ferocity that legend had given them. They put thermite incendiary grenades all over the back of the tank and the turret, then jumped off and headed away from all three vehicles as quickly as they had appeared. The tank's turret started turning to fire at them, then the grenades started going off. Some fell off, but some stayed and melted through the steel. The turret stopped turning. The hatches opened, and the crew was cut down by my platoon. Their bodies slumped over in the hatches. One slid off the tank like fresh fish from a bucket of ice.

The last remaining mobile tank stopped and began to back up in an attempt to get a shot at the R.O.K. troops running away. One thing that I've learned about a battlefield is that it is a very VERY noisy place. Any gun bigger than a .22 caliber, really requires hearing protection. When that protection is missing, a single round can leave your ears ringing. I fired almost 1000 rounds on this mission, and I figure most everyone else in the company did too. Throw in grenades, explosions from inside the burning tank, rockets, and the tanks engines, well…it was very noisy. In fact, it was so noisy, that I didn't hear the R.O.K. M-110 self-propelled artillery (SP) piece drive up right behind us. Louie Madill, of the quietest people in the platoon, let alone his own third squad, which was loaded with quiet people, was the first to notice the R.O.K. SP. I think he only noticed it because of the 20' barrel hanging over his head.

The M-110 self propelled artillery piece was nothing more than a howitzer sitting on top of a little M-113 APC's hull. There was no armor, or any sort of protection around the gun. It wasn't built for frontline duty. It was built to fire an 8" diameter shell at targets miles away; on the other sides of hills or cities. Here it was parked behind us, with its barrel in front of us.

Only six crewmen were around to service the weapon. The other seven must have been lost along the way. These guys were extremely cool under fire. While bullets zipped all over the place, they stood up on the vehicle, pushed in a shell, then pushed in a few bags of powder. When they were done loading the cannon, someone yelled something in Korean. They all covered up like we were going to take some serious incoming rounds.

We cleared out of the way and covered up the same way. Louie and the others around him were long gone by this time. The 8in (203mm) bore exploded. It's blast actually picked me up off the ground and threw me tumbling back at least fifteen feet. When I opened my eyes, except for the R.O.K.'s gun team, no one else was where I had last seen them.

The shell went into the ground under the tank and flipped it over like a toy. Its crew was dead from the concussion. The crew of the last tank, the immobilized one, started to turn its turret toward the SP, but about halfway there, it stopped, and they tried to bail out. They were shot with the same effect as the first tank crew. More R.O.K. infantry showed up from the woods on the other side of the road behind us. They were backed up by at least a hundred U.S. Army soldiers from the 1st Infantry Division, "The Big Red One."

Among the Army grunts was a Captain. He gave us a little bit of a situation report. From what Hector, Bruce, and I could surmise, the LZ had no clear-cut frontline anymore. The entire Pusan perimeter has been compromised, and no one, even the enemy, seems to know where the frontline is. I think the Chinese are just pouring troops into the area in the hopes that they'll be able to wipe us all out before we can consolidate another position. Since our communications were out, the army Captain called around and was finally able to get in touch with our Osprey. A few minutes later, they came in and started picking us up. Some more Osprey came in and took away the army grunts that continuously grew in number while we were waiting. We offered the R.O.K. guys a lift to back to the ship, but they refused. I guess I'd stay and defend pretty much any part of the U.S. if it was being invaded. I don't know if I'd stay after the war was lost, but I like to think I could show that kind of dedication.

SECTION 3

"History is made-for better or worse-by opportunists."

Thursday, March 10, 2011, 0805 hrs

Clark Field, Philippines

The huge American airbase bustled with chaotic activity. Hundreds of planes from around the world had escaped South Korea. Many of them made their way to the newly rented U.S.A.F. base. It had served as a Coalition rally and resupply point since the day that the lease was negotiated. No sooner had the first transports landed with groundcrew and engineers than the first damaged U.S. Navy F-18E Hornet came in for an emergency landing. The base's rebuilding and subsequent upgrade had been going on side by side with operations.

Those operations made the runways the busiest in the world. When the U.S. left after the volcanic explosion of Mt. Pinotubo years ago, the base immediately fell into disrepair. The surrounding jungle tried to rush in and reclaim itself, as did local squatters. Now it was a little piece of capitalist America again. Still, there was a lot of work to be done. The base was short of everything, except planes and pilots. Fuel, weapons, parts, groundcrew, even food, water, and shelter had to be brought in.

A normal operational structure was abandoned. Groundcrews were separated from the planes that they were assigned. Instead, the taxiway was setup to run like the pit row of a raceway. Planes from any service, any Coalition country could land and get basic service. Upon arrival, they were instructed to go down the taxiway and pick an open spot. Aircraft were lined up getting serviced. As soon as one was done refueling and rearming, it was moved away. Aircraft that were too damaged or in need of serious maintenance were moved to a second level pit row where they would be operated on by specialist, or cannibalized for parts. Ground crews were brought their meals. For weeks they were kept on an 8-hour on, 8 hour off work and rest schedule. Pilots and aircrews were debriefed at their planes by roaming intelligence officers who would run back and forth to the intelligence office for them. The same was true for mission planning officers, and even cooks. While aircrews were being debriefed, briefed, fed, or rotated with fresh crews, the planes were refueled and rearmed. Fighter planes were being turned around in half an hour. Non-U.S.-built planes were being turned around in 4550 minutes. Even the big strategic bombers from the U.S. and Britain were being turned around in less than 6 hours. Damaged planes often taxied back out to the runway with duct tape or sheet metal covering holes from AAA and SAM's.

At any given time, there was an average of 420 planes at the airbase. When Pusan airport was finally overrun, the last pilots in South Korea had to leave. Japan was denying many planes landing rights so as to ensure its neutrality in the war. Some pilots fled for Taiwan, but that was a long way away. The nearest friendly airbase was Clark Field. Aircraft seemed to land anywhere there was a runway or straight road in the Philippines, but they all aimed for Clark.

Al Meyers coaxed his South Korean F-16 through the clouds. As the plane vibrated madly from lack of maintenance, his hand seemed to juggle the flightstick. For over an hour he had watched the Philippine Sea pass below with only lone, deserted islands passing beneath him. He had only recently got into the F-16, and he was still so unfamiliar with its avionics that he couldn't figure out the new navigation system. He was still curious why the gauges and other controls on the South Korean plane were in English. Even though the plane was American-made, it would have only made sense to ship it with gauges that read in the buyer's language!

Finally, he flew over Luzon. At least now, if he ran out of fuel and had to eject, he wouldn't have to swim for it. A few minutes later, he started seeing aircraft all over the sky. Wherever he looked, he saw aircraft moving into the landing pattern. There were so many, it looked like a spinning thread of winged machines. Many moved into the pattern in formation for lack of room between landing elements. They were flying in a wide spiral shaped pattern down to Clark Field. He was never contacted by the control tower, or any other plane. With no way to request landing clearance, he had no other choice but to slip into line behind a British C-130 transport plane, and follow it down. As he got lower and lower, Al could see the planes on the field. There was complete pandemonium. Planes and personnel were everywhere. Some had crashed and rescue services were only fighting some of the fires. Others were left to burn. From 3500 feet up, Al counted fifteen columns of black smoke. There were many other crashes that simply didn't burn. Planes that were out of fuel had very little flammability.

The C-130 came in on final approach. Al was almost out of fuel. He couldn't go around and get back in line. He pulled forward, on the C-130's left side, and used hand signals to tell the British pilot that they had to land in formation. The C-130 rocked its wings, and down they went, less than fifty feet apart. The C-130 touched down first, and immediately slowed down to let Al pass by, in case he had a problem. He raced down the runway, passing the slower C-130. When the British plane stopped and was proceeding down the taxiway to pit row, Al's plane finally stopped. While he was turning to head for pit row, he noticed that an American B-52 Bomber was coming down the runway right at him. He increased his throttle to get to the taxiway, but there wasn't enough fuel. His engine shut down, and he had to coast off the runway. The big 50-year-old, eight-engine bomber turned slightly to the left, and Al was out of harm's way for the moment. While his attention was on the B-52, what looked like fifty oriental troops rushed towards his plane and began pushing it to the taxiway. They were R.O.K. groundcrewman who had made it off the Peninsula.

With smoke and jet fuel in the air, fires from burning planes in every direction, as soon as Al opened his canopy, he was mobbed by R.O.K. groundcrewman and troops-all cheering. Al was a fellow survivor, a hero, a brother in arms, and soon to be an unexpected leader.

Friday March 11, 2011

British Media News Service

<u>The Miracle of Dunkirk Revisited</u>

North Korean and Chinese naval units that had moved down the east and west sides of the Korean Peninsula the day before, made another well-coordinated cruise missile attack on United Nations naval assets in the region. While trying to evacuate the city of Pusan, most of the United Nations vessels had to move close to the coast, yet remain as far as possible from the frontline. In order to do so, they had concentrated in the Korea Strait, southeast of the city. Although diplomatic efforts hadn't fully determined where the surviving R.O.K. forces would be stationed, the certain need for the U.N. Task Forces to leave the confined area was clear. The safest way out, through the Korea Strait, constricted their defensive formations, and the ships were excellent targets of opportunity for the communists.

Using aircraft that had been redeployed to captured airfields in the former South Korea, and using the full weight of their remaining naval assets around the Peninsula, the communists saw the opportunity and hit hard. Unlike previous naval actions in the war, the communists attacked from multiple directions in a piecemeal manner. As soon as a plane could fire on a ship, it did, and as soon as a ship or submarine had the chance, it was taken. Previously, they had timed their attacks so that all their units fired their missiles at once. Piecemeal attacks were easier to defend against, but safer for the attackers. Only a handful of aircraft and small naval units were lost to the communists.

The steady attacks did encourage the U.N. forces to use more SAM's and AAA in their effort to stop the incoming missiles, and they ran out of defensive supplies faster than before. Once the SAM and AAA activity tapered off from low ammunition levels, the communists increased their attacks. The Battle of Yellow Sea, the Russian flotilla incident in the Sea of Japan, and most of the small naval and air actions throughout the campaign lasted only minutes. This wave of attacks began shortly after sunrise, and continued all day until sunset. In the end, the U.N. sank all of the larger communist vessels, destroyers, and guided-missile frigates in retaliatory strikes.

U.N. losses were stunning. The newly-arrived and overcrowded carriers from Britain, Italy, Spain, and France were all hit. Only a French and a British carrieralong with a mixed bag of escorts-still survives intact.

The U.S. Navy lost two more amphibious assault carriers, several escorts, assault ships, support vessels, and their arsenal ship. In addition, both the Carl Vinson and the recently arrived Independence will have to be withdrawn due to heavy damage.

The outrage and shock felt in the U.S. after the Battle of Yellow Sea was echoed by Europe's contributors to the battle. Even Russian diplomats, still angered by their losses in the Sea of Japan, relented in their pursuit of compensation out of shock and humility.

Friday March 11, 2011

2nd Lt. John Chamberlin-Personal Log

Somewhere in the Korea Strait

As soon as we touched down on the Wasp, Colonel Ghetty came over and told us to get some rest. He put Hector in temporary command of what's left of our company. He also told me that most of our cohesive units had been withdrawn from the Pusan area. There was still a lot of people down there, but it seems that the Task Force communications people are only receiving transmissions from small units, platoon-size or less. Even the R.O.K. guys have been shattered. Our company is to stand down while the Osprey try and go in to get anyone else out. They're under orders not to land in hot LZ's unless directly ordered by the Task Force Operations Commander. With Admiral Pender in D.C., that makes the senior officer General Shapiro. Colonel Ghetty explained that although there may be higher-ranking officers in the Task Force, we're the largest cohesive unit, and for some reason, everyone else has chosen to follow his direction instead of taking over his command.

I kind of feel good that our very own general is in charge of everything over here. He's a good guy, and we've served with him for some time now. He's also got a nice resume, and that seems to keep all the elitist officers at bay. I think what I like best about General Shapiro is the fact that he knows what to do, and when to do it. He leaves the rest to his subordinates, and rather than get in the way, he has a reputation of being helpful by being out of the loop. In OCS, we were told that one of the purposes of leaders was to get a team to provide better results than if each person were to attempt the same task individually. I think that a really efficient leader can do this, and make sure that things get done with as little micro-management as possible. Keep it simple, and keep the head honcho out of it. Generals think strategically, field officers think tactically. That's his job, and my job, in a nutshell.

After my adhoc debriefing from Colonel Ghetty, Hector, Bruce, and I put the platoons down for some rest. The ship is really really crowded, and all three of our platoons had to retake their quarters from squatting evacuees. Those squatters were relocated to the corridors with all the finesse and grace that a cold, tired, and irritable infantry platoon could gather. It wasn't a lot. Some feathers were ruffled, and I suspect that, soon, we'll be doubling up in our quarters.

Hector is steering clear of Bruce and I. I don't know if he's still upset about the poor communications at the last LZ, or if he's trying to distance himself from those he commands now. It could be that he's tired and thinking about the letters he has to write to the families of those he lost. I know I still have to write a letter to Chris and Kathy's parents. It's rubbing away at me slowly. Bruce is still Bruce, and although letter writing bothers him too, he's not letting it overwhelm him. I won't either.

I tried all night to write my letters, but they didn't come easily. In fact, they didn't come at all. Sleep did. I woke up at 0737 when the general quarters missile alarm sounded. As usual, I met up with the rest of the platoon in the cafeteria. This time, we actually had someone else start the coffee, one of the squatters. There were also people cooking right through the whole thing. Upstairs and outside, we could hear the thunder-like explosions of anti-ship missiles getting shot down by our SAM's. As the explosions came closer, we could hear the ripping sound of 20mm phalanx autocannons sending streams of depleted uranium shells at the incoming missiles. More of the targets exploded like thunder.

Meanwhile, breakfast was served. I had four scrambled eggs, six links of sausage coffee, orange juice, toast, and Bruce and I shared a fresh danish. It was good, and I was actually starting to like the idea of having extra cooks on board, then Commander Blackwell came on the ship's public address system, the 1MC.

In the past, whenever a command was given to the entire ship, it was very simple, and very formal. This time, it was different. I heard the usual speaker pop, then some feedback, and the bosun's whistle started. It was interrupted by a young man's voice. He wasn't panicked, and he wasn't calm. He was to the point. I'll never forget it.

"This is Commander Blackwell, acting Captain of the ship. If you've got a rifle, fall in on deck immediately. The missiles are on they're way, and rather than wait for it, you guys are going to put out some lead to meet'em."

The speaker popped as he turned it off. Everyone in the cafeteria looked around for another order from someone else. Bruce stood up on our breakfast tray, right on my plate, and told the entire room to get with it and get on deck! It cleared in about a five seconds flat.

On the deck of the Wasp, the Osprey were being moved around. Those that could were taking off, or fueling to leave the deck. The rest were being moved below into the hangar. There were hundreds of people on deck, maybe even a few thousand. General Shapiro was on the island, the ship's superstructure, with a microphone. He told everyone to line up along the edge of the deck and listen to some sailor that he pointed out. When everyone lined up on the deck, there were three lines of armed infantryman encircling the edge of the ship. The sailor that Shapiro had pointed out, a Lt. Junior Grade, took his place on the island, just outside the hatch to the bridge. He told us that when the missiles came close enough for the short range SAM'S and the phalanx to go into action, we might as well get off some rounds too. There was no way that a bullet from an M-16 was going to stop an anti-ship missile as big as a fighter plane, but a couple hundred might have a chance.

We were told that it was better than sitting around anyway. I looked around at the snow drifting across the wet and icy flight deck, and I remembered my nice breakfast and the warm interior of the ship. He was right though, at least we'd have something to do. I collected the rest of the platoon, and we lined up on the aft port quarter by landing spot #9. In near-complete silence, we all waited for a few long minutes. The 1MC popped, squeaked, and the Lt. JG who had been temporarily given command of several thousand infantryman announced that there was a missile coming in from the aft port quarter. It was coming right at us.

All three squads in the platoon were lined up in the same way we had done training a year before. 1st squad was laying down on the deck. 2nd squad was kneeling, and 3rd squad stood behind the other two. I told Gary, Wess, and Rick to fire by squad when the order came. As soon as I told them, with the entire platoon listening too, a flare went out from below us. Instead of going up, the flare was fired flat out to sea to show us which direction to aim. The 1MC popped, squeaked, and the order came to fire down the direction of the flare. We didn't know anything about our target's elevation, and there was no hope of visually seeing what we were shooting at through the wintry morning mist. It was going too fast anyhow.

When everyone opened fire, the sound was louder than normal, but that's like saying that the hurricane was as windy as the tornado. In other words, it wasn't a substantial difference. What was really awesome though, was the sight of tens of thousands of tracer rounds racing out from the entire length of the ship. The people lying down on the cold flight deck emptied their weapons, then when most people on the deck were reloading, my 2nd squad emptied their weapons, followed by 3rd squad, then 1st squad again. Everyone's rate of fire leveled out, and still we saw no target. From somewhere in the mist, a huge, fighter plane-sized missile appeared and silently flew overhead. Most of us watched as it seemed to move in slow motion only a few feet over the island. It disappeared in the mist on the starboard side of the ship, and I turned to look at Bob. He was already looking at me with a face that seemed to say, "Do you believe that?!" I'm sure I had the same expression on my face. After we made eye contact, we each blinked in amazement, then we finally heard the missile. Now I understand what it means when something goes faster than the speed of sound.

We all fired down the direction we were told. The people below decks were either busy bringing us ammunition or being trained in the basics of firefighting and damage control. Commander Blackwell made sure that no one was just sitting around. This was proven beyond a shadow of a doubt when Rick pointed out General Shapiro and his staff outside on the ship's island. They all had M-16's, and they were using them too. That's either a great example of a leader, or a great example of a really desperate situation; probably both.

The missile attacks went on all day, without a break. Squatters kept bringing us ammunition, food, water, coffee, and most importantly, blankets. It was a cold, windy, and exhausting affair. Some people had problems with their weapons breaking down, or barrels malfunctioning. There were some casualties too. By the end of the day, six missiles had buzzed the ship. Two fell short and drenched everyone on deck with icy cold water. Another three splashed down behind the ship. We never stopped a single one. Some of the sailors are saying that all the lead we put out may have acted like chaff to confuse the missile's guidance radars. It doesn't matter. If it did nothing else, it united every man and woman on the ship into a common objective: survival.

Saturday March 12, 2011

Air Force Weekly

Parting Shots

Anger and frustration over the Battle of Yellow Sea, the loss of the Korean Peninsula, the Battle of Korea Strait, and the countless acts of terrorism at home drove the United States to political retaliatory strike. In an attempt at saving political face in the eyes of his enemy, the President ordered a massive air attack on Korean capital of Pyongyang in the Democratic People's Republic of Korea. Thirty B-1B Lancer bombers, preceded by nine B2 Stealth bombers flew low and fast due north past the city of Pyongyang. When they reached their pre-designated positions west of the city, they all simultaneously turned and headed straight for their targets. Using precisionguided bombs, directed by GPS satellites high in the midnight sky, hundreds of tons of bombs hit the city in only a few seconds. There was little collateral damage, and no loss in American aircraft.

The city's infrastructure was gone. 2000-Pound bombs left nothing but craters and dirt in the place of their targets. Military buildings, industrial centers, transportation hubs, television and radio buildings, powerplants, and other public works were all destroyed. Even the city's statues of their Great Leader Kim Il Sung were turned into rubble. The most vengeful targets were those of the government buildings that paid the price for the terrorist attack on Capitol Hill in Washington. Diplomats and beaurocrats would need pictures to prove that their offices were once where only debris remained in the morning. Little remains of Pyongyang except residential structures. The attack serves as proof that nations no longer need nuclear weapons to destroy a city in an instant.

Sunday, March 13, 2011

Central News Network Online Report

Second Korean War: Is the War Over, is It Just Getting

While communist attacks against U.N. naval assets continue to tapered off, R.O.K. leaders in exile have decided to accept Taiwan's offer of amnesty in return for aiding in the defense of that nation. U.N. Naval commanders are moving the remnants of their task forces toward Taiwan.

The U.S.S. Wasp, sole surviving capital ship of the original Task Force Stingray, has been ordered to lead a group back to North America, by way of Europe. The Wasp's new Task Force, Task Force Phoenix, will take the heavilydamaged ships from various U.N. participants back to their home ports for repairs. Only a handful of escorts have been separated from the main U.N. Task Force on its way to Taiwan. The Wasp, and many of the other ships joining her, can still put up a good fight, but only three frigates, two destroyers, and a fresh American Aegis cruiser have been assigned to protect the wounded collage of warships.

Korean and Chinese political reaction to the amnesty agreement has been harsh. Intelligence reports hint at the possibility of a Chinese invasion of Taiwan before R.O.K. units arrive. These reports have been dismissed as rumors because of the losses in Chinese amphibious assault units to recent commando attacks around Tianjian, China. As a precautionary measure, the U.S. carrier battlegroup Nimitz has separated from the main U.N. Task Force and is racing ahead to the east China Sea.

In the United States, an enraged American public was given a new exposure to modern war. As a means of expressing its disapproval with the R.O.K. amnesty agreement, the Democratic People's Republic of Korea fired a three-stage ballistic missile called the Taepo Dong III at the continental United States. American Air Force radar operators had tracked previous versions of the missile, but their maximum range appeared to be under 6000 kilometers, barely enough to reach Hawaii. The Taepo Dong III took less than 15 minutes to reach its famous target, the city of Los Angeles, California. The warhead, containing an estimated 1500 pounds of high explosive, crashed only a few yards from the famous letters on the HOLLYWOOD sign.

The Koreans were trying to make a public statement of their capabilities. The missile's accuracy was only good enough to ensure a 50% chance of hitting the city, let alone trying to hit such a famous landmark. In fact, with the number of landmarks in the city, it would have taken a great deal more luck to miss one, than to hit one. Chance chose the HOLLYWOOD sign, but it could just as well have hit Disneyland, Capitol Records, or any number of the city's jewels.

The desired effect was reached immediately. Minutes after the missile impacted, TV news helicopters filled the air. A latent sense of fear swept the city, the state, and within half an hour, the nation. Everyone remembers where they were when John F. Kennedy was assassinated, and now they will remember how they found out about the HOLLYWOOD bombing.

Sunday, March 13, 2011

2nd Lt. John Chamberlin- Personal Log

Somewhere in the South China Sea

Chuck Boyd from 1st squad uses his laptop computer to go on the internet almost every free minute of every single day. It's annoying. Most of us feel like he's screwing around looking at anything and everything instead of working out, or doing more some of the more physical chores involved with platoon life. Every once and a while, he'll find something that's useful, but most of the time, all he has to show for it is the latest news from different world press sources, or an e-mail for someone in the platoon.

Today, he gave us the news about the missile attack on Hollywood. It really pissed us off. It just frustrated the hell out of us. The commies seriously need to get their asses kicked, and, no matter how hard we try, we can't seem to do it. What's happened to us since the Gulf War? We used to be this military machine that could win wars without hardly losing anyone. Now, we seem to get rolled over everywhere we go, and all we can do is fight to get out. This is not the U.S. military that I grew up knowing.

The word from Col. Ghetty at this morning's briefing was that our pullout from Pusan was as complete as possible. The people who were evacuated were scattered to ships throughout the Task Force, and no one is really sure of how many people were left behind. A lot of the R.O.K. military units that held out in Pusan were evacuated at the last minute, and it looks like their government in exile has made arrangements for them to be temporarily based in Taiwan. We're on our way to drop them off. The Chinese still see Taiwan as a rebel province of communist China, and their pretty pissed off at us for dropping off the R.O.K. army in exile.

Yesterday's attacks may not have touched the Wasp, but the rest of Task Force Stingray was hit hard. The new arrivals from Europe and around the globe took a lot of hits. Thousands of people went into the icy waters. We lost the last of the escort ships from the original Task Force Stingray, and now the Wasp is the sole survivor. After we drop off the R.O.K. troops that we have on board in Taiwan, we're going to lead the damaged ships out of the area, back to Europe, then to Norfolk for our own repairs. The Osprey are busy ferrying our squatters to the Japanese island of Tsushima on the south side of the Korea Strait. From there, they'll be reconstituted into units, and probably moved into regular U.S. bases around Japan and the Philippines.

There's a lot of talk about how the Chinese may try and invade Taiwan before we can drop of the R.O.K. army. Part of me, like everyone else on board, hopes they don't, so that the war will be over already. The other part of me says bring it on. I don't want vengeance for anyone I've lost so far. I don't want payback for getting hit in the head. I want to be able to stand with pride again. I don't want to be like those Vietnam vets who returned home from losing a war and got spit on at the airport. I think we all want another chance.

Tuesday, March 15, 2011, 11:00am

Outside Toronto, Canada

Global Technology Incorporated-Missisauga Branch

Dan held the front door open for Walt as they entered the office. The new secretary had obviously seen them coming and she immediately received their coats. She offered them coffee while escorting them to the branch office operation manager's office. Walt declined the coffee, and Dan copied his master's choice. They walked through the quiet 22-person office and straight into the branch manager's 10'x10' windowless room.

He was standing and waiting with an open palm. The tall casual-looking man quickly grabbed each man's hand and shook them with poorly hidden excitement.

"Mr. Barber, Mr. Greene. It's good to see you again. Can I get you some coffee?"

Dan spoke for them both with an air of disdain for the low-level underachieving manager.

"Thank you Chris, but your secretary just offered us some. Is it really that good or-" Walt interrupted his poor tone.

"Chris, I'm sorry. It's been a long flight. May we sit down for a moment?"

It was almost a rhetorical question and everyone knew it. If THE Walt Barber wanted to run naked through the office singing folk songs, he was more than allowed. He'd be cheered and applauded. All three sat down in the tiny room, and Walt tried to make small talk.

The near 40 year old manager's attention was pinned to every word, and he was taken back by the personal conversation offered him by such a powerful man. He allowed himself to feel as a peer for one brief moment. Finally, Walt brought up the war.

"Yeah, the war. My brother-in-law was killed in Seoul. Some of the people here have lost friends and family too. I used to think that we were beyond ever having another serious World War. Even if we did, I thought it would be one where someone presses a button and we'd all get nuked. This is far worse, I think. At least with the nukes, it would be quick and painless; as long as you were close enough to ground zero."

Walt didn't feel any personal guilt for selling the SP 2.0 software and enabling the communists to go to war. He sold the software for both direct and indirect profit. The direct profit from the sale was worthwhile, but Walt had hoped that by selling the software, he could offset the power that western industrialized nations were holding over those of the third world. Since the end of the Cold War they had held a monopoly of military might. There was also indirect profit to be made through his investments with the Russians and the countless companies that inevitably yield higher rewards in wartime.

Once more, Walt had led the conversation perfectly into his issue of importance.

"Yes the war. That is why we're here of course. I wanted to wait till we arrived to discuss the matter with you directly. I figure that the closer you get to the men on the line, the easier it is to find the problem with the product."

Chris, the branch manager, held his fear well, but not his embarrassment. The blood rushed from his face, and he went as pale as a ghost.

"I wasn't aware of any problems. The SP 2.0 software seems to work perfectly well. In fact, I was just on the phone with one of our buyers at the Pentagon, and she said it was working fine."

"Chris, I'll get right to it. I'm sure it's not anything catastrophic, but I took a call from one of the Korean officers who made it off the peninsula. He told me that the beta test version, the same version that's at the Pentagon, I believe, lost some of his units just before they had to evacuate Pusan. I have to say, I was fairly shocked. I thought this had been thoroughly tested before going to the beta sites?"

Chris wasn't the inept lower management underling that Dan believed him to be. His reply was detailed and immediate.

The only reason I can think that they might have lost contact with any of the R.O.K. units would be the unit strength index rating."

Walt tried to look comfortable, but his attention ruled his appearance. He was focused.

"That's the site, but what're you trying to say about a strength index?"

"The strength index is something unique to the SP 2.0 software. See, the software is really nothing more than a common computer language that translates between various machine code languages. Without any additional programming, it would simply allow one code to be viewed by another. Imagine being able to plug your TV, your car stereo, your home PC, your laptop, and your microwave oven all into a single box. Once connected to the box, you could watch TV on your laptop, or check the microwave's timer on your car stereo. All these different digital languages or codes are made interoperable by SP, but there's very little usefulness in checking the microwave timer on your car stereo. So, the 2.0 part of SP is the user interface that allows the software user to manage the information being put into that magic box. Without the interface, it would probably take thousands of people to interpret and manage all the data coming in. SP 2.0 puts all this data together in a manipulative environment. We're talking data from search radars, submarine sonars, aircraft transponders, infantry video feeds, etc."

Walt understood and knew all this, but Dan was still trying to understand the power of the software.

"Chris, is this strength index an integral part of the interface then?"

"Exactly! See, the interface is the really strong point of the software. We've created default settings that make the software weed out information that would normally be almost useless. Let's say, for instance, that an infantry division, or better yet, a fighter squadron, loses 11 out of 12 planes. That last plane hardly constitutes a viable unit, so the default interface setting takes the unit off the screen. When a squadron or any unit takes over 80% casualties, it's removed from the visible interface until it is reinforced. Those unit strengths and reinforcement levels are monitored and tracked by the accounting software, similar to the way that our office keeps track of the balance between payables and receivables. I don't need to know how each project is performing, but I do need to know which projects are active, and where our bottom line is. If I tried to keep track of every single project, I'd be swamped. I'd probably have to hire a single person to monitor each one."

Walt's attention leaked a feint smirk, and Chris knew his message had been heard. There was no problem with the software. It was the users that didn't know how to use it. What Chris couldn't possibly have known was that the users were not America's allies from Korea. Rather, they were the Chinese themselves.

The Chinese had been infiltrating over 500,000 men into Indochina over the past few weeks. On several occasions, Coalition fighter squadrons and airborne units were able to appear out of nowhere and inflict horrendous casualties on the Chinese. Tens of thousands had been killed. These numbers were further amplified by the fact that Chinese infantry officers typically chose to execute their own more immobile wounded rather than waste resources bringing them back to hospitals.

Walt hadn't told Chris or anyone at GTI that thanks to him, the Eastern Alliance was using the SP 2.0 software. They didn't need to know, and he couldn't count on anyone understanding. All they needed to do was write the software, and explain to Walt how to change the default settings. Chris showed him a few minutes later. Then Walt made some more small talk, complimented the branch office's performance in the previous quarter, and politely left to catch a plane for another meeting. Unknown to anyone at GTI, this next meeting wasn't somewhere else in Canada, or even North America for that matter. Walt and Dan were on their way to what Chris would have viewed as enemy-held territory communist-controlled Seoul, Korea.

Wednesday March 16, 2011

Washington Post

An Expensive Wake Up Call

Bipartisan activity continued in the relocated U.S. Congress, and so did new military spending. Ballistic missile defense systems, long thought to have already been in place by the American public, have finally been funded to the tune of almost $1,000,000,000,000.00. Design requests for new ships, planes, tanks, and infantry equipment have also been sent out. The LAI, League of American Industry, has been formed to organize and synthesize companies from around the country. For the remainder of the war all military purchasing is going to be funneled through the LAI. Chosen by various congressional committees, industry leaders, research teams, political representatives, and military leaders will run the organization. It will also create and maintain a series of interactive databases listing capabilities, purchases, requirements, status reports, etc.

In the meantime, while the LAI is being formed and the databases are being constructed, the U.S. Navy has been given permission to begin reactivating most of its mothball fleet. This includes six supercarriers, four World War II battleships, and almost two hundred smaller vessels ranging in size from cruisers to landing craft. The U.S. Air Force has also begun refurbishing hundreds of aircraft that had been put out to pasture in the deserts of the American southwest. The U.S. Army, having sold much of its older equipment, has begun purchasing main battle tanks from Germany in a lend-lease program similar to the one used in World War II. Small arms and other equipment production will be easier to step up, the only expected side effect being a rise in civilian gun prices.

Closer to the front, American aircraft from the U.S.S. Nimitz, supported by Royal Navy aircraft that had been deployed on the late H.M.S. Invincible, engaged Chinese airplanes over the east China Sea. Losses were similar on both sides, and the engagement only escalated once with the appearance of reinforcements from both sides. After the first and second waves of aircraft had become fully involved, both sides broke off the fight due to fuel constraints. Further air action has been limited, and little air combat action remains.

Thursday March 17, 2011

Associated Press International

~~Bad Day for Britain~~

In the East China Sea, three Chinese flotillas of light frigates and patrol craft attacked ships en route from Korea to Taiwan. Losses to both sides were minimal. The Chinese vessels were able to use the jagged coast of mainland China for protection and concealment from U.N. retaliatory strikes. Aircraft based around mainland China were also able to provide air cover from any carrier-based air assets of Task Force Stingray. Few U.N. ships were hit, but the brand new nuclear supercarrier H.M.S. Thatcher, the first of its kind from the U.K., was hit repeatedly. It's going to be withdrawn from activity duty. Her air assets have been diverted throughout the Task Force, as were the thousands of evacuees on board.

Thursday March 17, 2011

2nd Lt. John Chamberlin-Personal Log

Somewhere in the east China Sea

Call it fate. Call it Kharma. Call it God's will, or call it luck. Anyway you look at it, we are most definitely still at war. We came under another missile attack today. We didn't even have a missile come at us, but I heard the losses were bad for the Brits. They might lose their new supercarrier. All I know for sure is that I can see the smoke on the horizon when I go up on deck. I also know that there's a dozen new British Harriers on our ship.

Friday March 18, 2011

Associated Press International

Forces Regroup, But Skies Are Still Hostile

Several fresh naval groups coming into the east China Sea from Polynesia joined Task Force Stingray. Among these were two French aircraft carrier battle groups, four small South Korean task groups of destroyers and frigates, and the U.S.M.C. Expeditionary unit based upon the U.S.S. Bataan. Air skirmishes continued between mainland China, the Korean Peninsula, and Task Force Stingray. Since all sides have started running low on top of the line fighter planes, expert pilots, and sophisticated air-to-air missiles, dogfights are occurring less often. No one has been able to achieve air superiority, air supremacy, or even a tangible air presence. Civilian air traffic is avoiding the area as much as possible, and the skies appear to be disturbingly empty.

Thursday March 17, 2011

Associated Press International

Bad Day for Britain

In the east China Sea, three Chinese flotillas of light frigates and patrol craft attacked ships en route from Korea to Taiwan. Losses to both sides were minimal. The Chinese vessels were able to use the jagged coast of mainland China for protection and concealment from U.N. retaliatory strikes. Aircraft based around mainland China were also able to provide air cover from any carrier-based air assets of Task Force Stingray. Few U.N. ships were hit, but the brand new nuclear supercarrier H.M.S. Thatcher, the first of its kind from the U.K., was hit repeatedly. It's going to be withdrawn from activity duty. Her air assets have been diverted throughout the Task Force, as were the thousands of evacuees on board.

Friday March 18, 2011

Associated Press International

Forces Regroup, But Skies Are Still Hostile

Several fresh naval groups coming into the east China Sea from Polynesia joined Task Force Stingray. Among these were two French aircraft carrier battle groups, four small South Korean task groups of destroyers and frigates, and the U.S.M.C. Expeditionary unit based upon the U.S.S. Bataan. Air skirmishes continued between mainland China, the Korean Peninsula, and Task Force Stingray. Since all sides have started running low on top of the line fighter planes, expert pilots, and sophisticated air-to-air missiles, dogfights are occurring less often. No one has been able to achieve air superiority, air supremacy, or even a tangible air presence. Civilian air traffic is avoiding the area as much as possible, and the skies appear to be disturbingly empty.

Saturday March 19, 2011, 1342 hrs

Caves of the Eight Immortals

3.5 Miles NW of Changping, Taiwan

Deep inside their underground bunker, Taiwanese Air Force commanders watched seven projection screen monitors on the front wall of the theatre-like room. Between the generals and the big screens, almost a hundred junior officers answered phones, monitored radios, manned computers, and physically brought in messages from somewhere outside the cave. Despite the number of participants, the volume of data, and the pace of activity, there was little chaos. Everything was done with military order, precision, and professionalism.

Each monitor showed a separate defense zone around the island nation. Taiwanese units were represented by blue icons along with those of Coalition units. Green icons represented neutral shipping and air traffic. Unknown radar targets were yellow. After the previous day's air attack from communist China, the Taiwanese Air Force, already on full alert status, had extra air patrols over the Taiwan Strait. The entire country's defenses were on full alert. Troops were deployed along probable amphibious landing areas. TV, Radio, and government buildings were secured. Military bases doubled their perimeter patrols. Naval assets patrolled the coastline, and AAA And SAM sites were kept on the move to confuse communist mission planners.

Suddenly, the pace in the room quickened as large numbers of yellow icons started showing on the screens that representing Taiwan's western border. The generals took no chances and ordered their CAP's over the Taiwan Strait to intercept the unknown air targets. They ordered planes that had been waiting on the ground to scramble and prevent any of the unknown air targets from approaching the Taiwanese coast. Intelligence sources took crucial seconds to determine whether or not the yellow icons should be changed to red, hostile,

330

communist, Chinese icons. Still, the generals felt it safe to assume that hundreds of civilian airliners weren't suddenly flying to visit Taipei. It was doubtful too that they would fly in eleven separate squadron formations. Their instincts, their hearts, and their minds had no doubts. This was the moment that the had prepared for since they were driven from the Chinese mainland. The communists were coming.

Thick overcast from 4000 feet through to 12,000 feet made visual interception of the unknown air targets almost impossible. Without a thought, eight pairs of Taiwanese F-5 Freedom Fighters left their orbiting CAP locations, and headed towards the unknown air targets. Reserve fighters were coming from bases back on the island, but they needed time to get aloft, get to a good interception altitude, and otherwise get organized. In a type of warfare where battles can be decided in seconds, the CAP planes would have to buy precious minutes for the reserve planes. Almost at the same instant that the first pair of Taiwanese CAP planes broke through the overcast at 12,000 feet, the Air Force command center finally changed the icons from yellow to red. Long-range land-based radars had identified the incoming air targets. They were Chinese Jian-7's. No one had to tell the Taiwanese pilots, even heading straight into the Chinese formation with a closing speed over 1100 knots, they could make out the distinct silhouette of the communist planes.

The F-5's opened fire with their 20mm cannons and immediately a Jian-7 exploded. Several others broke formation, some to evade the F-5's, and others to give chase. The rest of the formations continued on course toward their targets. More F-5's broke through the clouds in pairs. Air-to-air missiles began to swirl white snake-like contrails through the air. Evasive flares and puffs of radar-decoying chaff fell from possible missile targets. Most were effective against the old Taiwanese heat-seeking missiles and the poor quality Soviet imitation heat-seeking missiles. Wherever missile met plane, a black puff appeared. Sometimes the connection was followed by orange fireballs, or black plumes of burning wreckage. Since the order was first given for the Taiwanese CAP's to intercept, eighteen seconds had passed, twelve F-5's had been lost, and thirty three communist planes were falling from the sky. From the west and the northeast, more yellow icons appeared on the Taiwanese Air Defense Command screens.

There was no more time to assemble. The reserve planes had to get into the fight. They turned and headed into the overcast, up to the clear sky above, and the furball of contrails that signified a massive dogfight. When they broke through, the first formations of communist planes passed overhead without breaking formation. A second veered slightly, but paid little attention to the fresh supply of F-5's below them. The third formation, heading directly at the F-5's, dove and fired over fifty missiles at the Taiwanese pilots. They, in turn, fired their missiles and dropped their decoys. The Migs dropped their decoys. The missiles passed each other and stitched the sky with contrails. Some missiles hit, some missed. The third formation leveled out and continued on their original course. The F-5's were falling into the Taiwanese Strait. A few more F-5's and some newer F-16's were able to get off the ground and make individual attacks, but nothing could stop the airborne armada heading for Taiwan. Only Taiwan's SAM's and some light AAA remained to be overcome.

When the communist squadrons came within sight of the island, they split up and attacked over a hundred different locations. Some targets were hit by pairs of planes with only two or four 500-pound bombs on each aircraft. Entire squadrons hit others. There was none of the precision bombing made famous in the Gulf War and latter conflicts. Cast iron bombs dropped from high altitude by fast moving planes rarely hit their target. The communists depended on statistics and quantities of bombs rather than exciting video game-style camera footage. Civilian casualties climbed into the thousands.

As the planes reformed into squadrons over the Taiwanese Strait for their trip home, the Taiwanese generals in their bunker watched the big screens with a great deal of confusion. Communist planes were falling from the sky, but there were no friendly units nearby. It looked as if some of the communist pilots were turning on their comrades and splashing plane after plane! No one in the room understood.

Less than twenty miles off the coast of mainland China, squadron after squadron of communist Chinese Jian-7's cruised comfortably towards home. The Jian-7's, notorious for their near complete lack of rear visibility, had no idea what was happening behind them. Radio silence was more than just an order for the Chinese pilots, it was a law with a fatal punishment in the People's Air Force. Anyone who disobeyed the order-the law-risked execution upon arrival. The pilots in the rear formations that were getting shot down were so well trained, that they never once called for help.

One by one, starting with the farthest back plane in the communist formation, a mixed group of Coalition planes, flying out of the Philippines, was picking off the communist pilots. Only 14 planes, a little more than a squadron, was whittling down the remnants of eleven communist squadrons. The planes were mostly Free R.O.K. F-5's and F-16's. Their pilots took their time lining up communist prey. When they were sure that a Jian-7 was going to take a full burst of cannon fire, the pilots fired their cannons, and thousands of 20mm tracer rounds out streamed-straight into the tailpipes of the enemy planes. When one plane went down, another was fired on. It took less than ten seconds to wipe out two communist squadrons, twenty-four planes. The communists continued to go down until the turkey shoot came into range of mainland China's long-range radars. Then all hell broke loose. Sophisticated communist Su-27's and Mig-29's climbed through the clouds and attacked from nowhere. Two of the Korean planes went down, then another a few seconds later. The Jian-7's, low on fuel from their day's exploits, continued back to their airfields. Another Korean plane exploded, and the mixed Coalition squadron headed for Taiwan.

The communists gave chase-firing missiles until their supply was exhausted. When they were halfway across the Taiwan Strait, the less fuelefficient Mig-29's turned and headed back to base. At about that same time, more Korean planes showed up from the north, forty planes in all. When they detected the new Korean planes, the Su-27's broke off their chase and headed back to mainland China too.

The first air battle for Taiwan was over. 108 Communist Chinese planes were confirmed destroyed. The Taiwanese Air Force lost 35, but welcomed 50 planes to the island. Among the pilots, Al "Wolfman" Meyers had brought his composite air group to a new home.

Sunday March 20, 2011

Chicago Tribune

Victory at The Battle of Taiwan Strait

Hundreds of small Chinese and North Korean patrol craft-having been rearmed at their home ports after the Battle of the Yellow Sea-made their way along the coast of mainland China. Once they were fully dispersed, they made yet another simultaneous mass attack against the U.N. ships now moving south towards Taiwan. Thousands of aircraft operated from their home bases in China to protect the fresh naval flotillas from U.N.

counterattacks. The East China Sea morning sky was crisscrossed with vapor trails from planes and missiles.

Task Force Stingray immediately detected the attack. Most of its ships were reloading their missile bays after the previous week's worth of battle. Escorts had to be rotated in the formation as SAM supplies wore out. On the monitors in every ship's CIC, it looked like wagons circling against an Indian attack in an old western. When one escort would run low on ammunition, a more well equipped one was sent to take its place. The empty escort then raced over to one of the few remaining fleet support ships for more SAM's. Cargo helicopters and Osprey had to fly low and fast to carry pallets of SAM's to ships in the meantime. The support ships were able to manhandle the weapons in record time, but only with the help of the thousands of evacuees on board.

While more and more escorts were pulled from other areas of the formation, the entire Eastern side was eventually only guarded by U.N. submarines. All air defense capable vessels were either rearming, or fighting to the west. Nothing could stop an air attack from the East. Aircraft from the remaining carriers were able to help defend against the missile attacks, but interdicting fighter and bomber raids kept most air operations busy with dogfighting.

The new communist tactic, as dangerous as it was, did allow Task Force Stingray to survive the biggest naval battle of the war to date, but not without terrible losses. One of France's new nuclear powered supercarriers was badly damaged. Hours after the missile and air attacks subsided, it and had to be abandoned and scuttled. Before it went down, French officials vowed to send their remaining supercarrier to the area as a replacement.

Despite the loss of the French carrier, the world press sees the battle as a victory because of its initially apparent hopelessness. Near the end of the attack, Taiwanese frigates and destroyers, along with exiled R.O.K. frigates and destroyers made an excellent accounting of themselves, and were credited with saving thousands of lives on U.N. capital ships.

Chinese losses were lopsided. Only a few ships were lost to U.N. counterattacks, but over 130 aircraft were confirmed shot down. Most of those planes were lost in a single hour's time. History has never before seen such a devastating loss of aircraft in such a short period of time. Even the worst day of the American bombing campaign over Germany in World War II fell far short.

3/21/11

1100 hrs.

Hualien Airfield, Taiwan

Personal Log – Lt. Kim "Killer" Yoo

The Taiwanese Air Force has been very hospitable. They arranged for us to set up our own base of operations at an airfield on the central east coast of the island, outside the city of Hualien-The City of Marble. The field is a civilian airport that was diverted solely to our use. They have sent a transport plane to the Philippines to bring out as many of our wing-wipers as possible (not many remain after Pusan's fall, and the nightlife expedites of Clark Field. In the interim period, a Taiwanese Air Force squadron of F-5 Freedom Fighters has joined us, and their support crews have been tasked with quadruple duty. I would not have expected anyone to service and supply fours squadrons in wartime with only one squadron's staff, but they are doing a marvelous job. Half of our aircraft are on the ground getting some long-awaited maintenance and/or repair. However, these Taiwanese wing-wipers are efficient enough to manage all four squadrons-at least for a while.

They are also being painted (again). This time, our group paint scheme is a pale, pastel combination of mint and pea green. While it is very unattractive, it seems to be very effective. Some pilots have argued that it hides our planes because no one wants to look at them, so the communists look away rather than vomit. They might be right, but it doesn't matter to many of us. I would personally paint my aircraft pink with large purple dots if it gave me the chance to kill just one more communist.

Meyers is shifting the group's organization once more. All three squadrons have been semi-disbanded (at least for a while). Instead, we're to fly in divisions of varying size. Each division is 3-6 planes from one nation with at least one pilot who speaks English. My division has 4, Korean F-5 Freedom Fighters, 2, Korean F-16 Falcons, and an American F-15 Eagle. There are two more Korean divisions. There would be a third, but losses in the initial engagement over the Taiwan Strait we particularly hard, and the group's Korean contingent-myself included-was heavily engaged. There are a few divisions of

Americans, a Greek division, an Italian division, and a few French divisions. We also have a specialty division of Mig-29 Fulcrums that is comprised of different nationalities.

Today, all of the pilots are getting some rest. The last days in Korea were hard. Clark Field was at least as hard-possibly harder-on many of the pilots. The flight to Taiwan was difficult, and we went straight into combat. The Taiwanese have allowed us to take over a very nice hotel nearby, and while many of us are disciplined enough to rest as much as possible whenever possible, a good number of pilots have already emptied the hotel lounge of its spirits. Taiwanese hospitality continues: as I have been writing, a civilian pickup truck has arrived and begun to unload wooden cases of whiskey. I would complain, but some of the hardest drinking pilots are our best men. Each man has his own methods, but these are not mine.

Monday March 21, 2011

Associated Press International

Biggest Dogfight Since Battle of Britain

Chinese aircraft made numerous air attacks on targets along the west coast of Taiwan. It appears to the rest of the world as though the attacks were conducted in response to the actions of Taiwanese frigates the previous day. Given that the attacks were well-coordinated, they were probably planned some time ago. Taiwanese Air Force interceptors made the raids extremely costly. For the second day in a row, over a hundred communist planes went down. This time it took the full day instead of a few hours.

Tuesday March 22, 2011

United Press Wire Service

Air War Over Taiwan Continues

There was yet another full day of airstrikes on targets along the west coast of Taiwan today. They were followed by a late afternoon/early evening helicopter assault on Sungshan Airport on the north side of the island's capital city of Taipei. Taiwanese police and infantry units inflicted moderate casualties among the Chinese airborne troops, but many survived into the night. The airport's buildings provided large structures and concealed the large numbers of troops. They also create excellent fields of fire for the communist defense.

Taiwanese interceptors attempted to prevent the air assault, but Chinese fighters were far too numerous. Losses in fighter planes are said to be unsustainable on both sides. Once more, the Chinese have lost almost a hundred planes in a single day. The Taiwanese Air Force has admittedly lost almost half of its remaining aircraft. U.N. Naval aircraft are attempting to turn the tide and intercept the assault helicopters, but a small and steady stream of missile attacks from Chinese patrol boats and light frigates is keeping them busy.

0900 hrs.

Hualien Airfield, Taiwan

Personal Log – Lt. Kim "Killer" Yoo

The Taiwanese have painted all of our aircraft, made repairs, refueled them, and even loaded them with some 500-pound bombs. Most of the planes carried 6 of the standard iron bombs. They've been having a lot of trouble with communist missile patrol boats, frigates, destroyers, and even submarines all around the island. Further to the northeast, in the northern Philippine Sea, the communists have been making air and missile attacks against all of the U.N. Coalition task forces and our Korean Navy. Meyers made a deal with the Taiwanese: if they gave us the bombs, we'd clear the area east of the island for at least one day.

Shortly after dawn, we started to launch our aircraft. My flight joined with the other Korean flights, and we headed toward the Navy task forces that were being harassed. The flight took less than half an hour, and then we ran into trouble. The communists were making a series of air attacks on the outer ring of ships that were escorting the Coalition task forces. Those same ships were busy launching missiles at the planes, firing cannons at incoming anti-ship missiles, and maneuvering to get out of the way when their ammunition ran low. American CAP aircraft were just as busy trying to intercept the communist planes. The sky was very clear. Visibility was very good. We approached in a gaggle formation at 25,000 feet, and as the 200-mile wide battle appeared over the horizon, we could see everything.

It was a confusing scene, and some of the escorts launched missiles at us by accident. It was understandable. We hadn't told anyone who we were, or what we were doing in the area. Most of our planes don't even have radios anymore. There is so much radio interference, cross-communication, jamming, deceptive transmissions, and so many different languages, that they had become useless. With the American missiles approaching at speeds in excess of mach 4, we turned to the west southwest. It was mistake.

Our new heading carried us directly into the path of the communist flotillas of missile patrol boats, frigates, and destroyers. As ships came into view, AAA tracer rounds climbed their way up toward us. Without orders, our gaggle fell apart. Everyone was hungry to kill communists, and we all chose our targets. Almost 20 Koreans dove on separate targets. Each of us waited until the last possible moment to drop our bombs. Somewhere under 1000 feet, we all let our bombs loose. Then we pulled out of our dives, and headed back for Hualien. On the way, we climbed up to our original 25,000 feet and looked for each other. Two more friends have been lost, but we all believe that we sank our targets. If it is true, then thousands of communists are also dead, and such a trade is not looked upon poorly.

Back at Hualien, the other divisions began to return. Some of the planes were badly damaged from AAA. Our Mig-29 division ran into some real trouble some American Navy F-18 Hornets that were flying CAP, so they headed home. The Hornets never came within range, but Meyers has decided to re-assign me to their division. We hope that my presence in an American-made F-16 will show friendly aircraft that the Mig-29's are not communists.

March 22, 2011

Hualien Airfield, Taiwan

Personal Log – *Lef*tenant Ian Thomas "Top Hat" Sanders IV, R.A.F.

Back in the air today, and damn! What a fight! I'm guessing the swamp-green paint scheme on our planes worked, because our losses were minimal compared to the Chinese and North Koreans'. Myself, anything <u>that</u> ugly turns my stomach, so nausea may be working for us regarding the enemy's inaccuracy.

Kim Yoo seemed in wonderful spirits again after channeling his rage through some larger than average bombs this afternoon. A few of his countrymen died, but they don't seem too sad. They rationalize that it was a bombing run, so the tradeoff of pilots to Communist sailors is hundreds, perhaps thousands to one. Myself, I shot down 2 Frogfoots (or would that be FrogFEET?) that were menacing the Coalition task force, and got a Jian-7 to boot as he was chasing one of our Mexican pilots. I waved at him as I pulled alongside after saving his bacon, and I thought he recognized me. (Can't remember the burrito-eating berk's name, though.)

All in all, I suppose you can put today down as one for the good guys. It always seems that way, though – we always seem to be kicking ass, but we get chased out of everywhere in the end.

I got a hero's welcome from the Mexicans when I entered the bar this evening; the little rat who I saved earlier even jumped up and gave me a hug … must be a Latin thing. I had a few drinks with them, but as they started to up the pace, I feigned tiredness and went to my room for a cup of tea and my rack. They were razzing me when I left and making gestures like my testicles were of smaller than adequate size, but they were all smiling, and it sounded good-natured enough. (I guess.)

Wednesday March 23, 2011

New York Post

D - Day in Taiwan!!!

Task Force Stingray answered desperate pleas from the government of Taiwan. In the largest amphibious assault since the Inchon landings during the First Korean War, tens of thousands of troops came ashore to help repel a Chinese invasion. The landings occurred north and south of

Chuangwei, a small coastal town on the northeast corner of the island nation. Marines of the 24th Marine Expeditionary Unit, the same men and women who were first into combat on the Korean Peninsula, came ashore by V-22 Osprey. More Marines and U.S. Army troops followed in helicopters and amphibious craft. Other U.N. contingents put troops on the ground too as well, including the Australians, British, Canadians, French, Greeks, Italians, Spaniards and the Turks. Free R.O.K. units came ashore too and were given a hero's welcome by locals and sightseers from all over the country.

Chinese attempts to prevent the landings were met by fierce opposition from U.N. carrier-based aircraft and escort ships. The actions on the around the "Choo" beachhead, as it is being called, did weaken the defensive support effort around Sungshan airport. Subsequently helicopter and parachute assaults were able to reinforce the Chinese infantry who had survived the night. What had once been reduced to a high concentration of communist strongpoints has rapidly become a well-organized front line. An estimated 11,000 Chinese are well entrenched and have been able to fight off all Taiwanese efforts to resecure the airport.

Wednesday March 23, 2011

1642 hours (4:42pm local)

1000 feet above the Taiwan Strait

28 miles southwest of Taipei, Taiwan

Al Meyers glanced at the multi-function display screens in the cockpit of his South Korean F-16C. When he was first put in the seat of the F-16 back in South Korea, he never learned to use the radar. There was no point with all the electronic jamming in the area at the time. The same was true of the Taiwan Strait, and he reflected on the money spent for a radar that was useless in battle. It was like air conditioning in an Alaskan car. Someday the jamming would clear up, and maybe then he'd learn to use the radar. Until then, all he saw on the screen was snow.

During the last days of fighting around Pusan, Al's U.S.A.F. F-16 finally wore out, but the South Koreans were more than willing to let him use one of theirs. In fact, no one would even think of stopping him. Besides, there were more planes than pilots, and he was much more capable of flying than the next guy. He was only able to fly two missions before his highway airfield was overrun back in South Korea. After a mid air refueling stop over the Sea of Japan, he landed at Tokyo International Airport. A few more hops from one airstrip to another, and he was on his way to the Philippines and Clark Field. Al knew that there was a lot of fighting to come, and he knew that his survival was proof of his skills. They needed him up there, and Al couldn't let them down.

Now, his new composite squadron was filled with unfamiliar faces, unfamiliar languages, and an unfamiliar home, Taiwan. They flew barrier air patrol missions over the Taiwan Strait under the direction of a Taiwanese Air Force liaison officer who was assigned to the squadron. Even though he was an outsider, Al was given the role of flight leader. Two other F-16 Falcons and an F-5 Freedom Fighter accompanied him on this mission. Thankfully, all the other pilots, having been trained by the U.S.A.F. spoke English.

Al's wandering mind was brought back to the mission by the AWAC's controller flying safely over the Philippine Sea east of Taiwan.

"Wolfman flight, you have multiple incoming traffic bearing 015 angels zero one. You are weapons free and directed to engage. Copy?"

While he answered the AWAC's controller, Al motioned to his wingmen. He directed the two F-16's to go high in an attempt at drawing any escorts away from the more important planes that would most likely stay on course. Two of his F-16's disappeared into the heavens, and, after a few minutes of waiting, he saw his targets. There were too many to count.

Helicopters, everywhere, and there were no escorts to be seen. Al locked on to one of the Chinese helicopters with one of his four AIM9 Sidewinders. When he heard the growling lock-on sound in his headset, he fired. Immediately, he locked on to another and fired. As the first missile met its target, Al and his wingman passed through the scattering Chinese formation with their 20mm cannons blazing. In one pass, the two pilots downed five helicopters with almost two hundred men on board. They circled and made two more passes using only their cannons. Eight more helicopters exploded, splashed, or both.

They turned for another attack run, but Al glanced at his wingman and saw tracer rounds exploding on his port wing. Tracer rounds also surrounded his wingman's plane. A fast glance around showed the reason. Diving down at them from roughly 5000 feet, at least twenty Su-27's were coming after them. Al and his wingman broke off their attack, kicked in the afterburners and raced for home. Al's F-16 began to outpace his wingman's F-5, and he had to slow down. The other two planes in their flight were never heard from. As the Su27's leveled out and drew closer, they all seemed to fire their 30mm cannons at Al and his wingman. Stream after stream of tracers danced around the two planes. Al watched his rearview mirrors to know which way to dodge the streams. Without warning the Chinese planes broke off their attack and turned away. Al and his wingman continued back to their airfield while unknown friendly interceptors took up the fight in their place.

Half an hour later, on the ground, Al was given a hero's welcome. He didn't know that anyone else was keeping score, but he was now the first American since World War II to make ace six times. More importantly, he and his makeshift squadrons of refugee pilots had slowed the Chinese attack long enough for Coalition forces to move into the area. Al didn't know it at the time, but his decision to go to Taiwan without orders had given the island a fighting chance.

3/23/11

1751 hrs.

Hualien Airfield, Taiwan

Personal Log – Lt. Kim "Killer" Yoo

This new division is not as motivated as the Korean divisions, but when we were finally engaged with the communists the men on my wings proved that they are very skilled. Meyers had all of the divisions in the group fly loose patrol routes around Hualien Airport. It was supposed to be a simple CAP mission where we could all become a little more familiar with the surroundings (along with radio and radar equipment, navigation is equally jammed or unreliable, so most of the time our flightpaths are based on visual recognition). When Meyers' division was around 35,000 feet, they must have spotted the air battle on the west coast of the island, because we saw them leave our CAP. We all climbed to the same altitude and saw them just as they slipped into the giant furball. As one, all of the group's divisions headed in at full military power.

I stayed with the Mig-29's per my instructions. Despite being from vastly different nations, the Mig-29 pilots coordinated their attack well (We don't really have a division or flight leader, and everything seems to be done via consensus). We came in under the crests of the mountains, but not from deep in the valleys. When we entered the furball, it was a maelstrom of contrails, missiles, tracers, Coalition planes, Taiwanese planes, Composite Group aircraft, and communist aircraft.

I tried to stay close to the Peruvian Mig-29, and before I could even look around to see what he was lining up to attack, he fired a quick burst of cannon fire and rolled left. I followed, and a Chinese Jian 7 tumbled toward me from head on trailing smoke. I tried to watch the Peruvian and see if the Jian was destroyed, but I lost sight of both. On my own, I searched for targets and found another Jian that was being chased by 2 Taiwanese F-5's off to my right. I headed over to attack the Jian from head on, and put a burst of 20mm through his plane. Both F-5's had to roll out in separate directions to avoid colliding with me as I passed between all three planes.

Then I saw 2 of our Mig-29's chasing North Korean Su-27 Flanker. They were above me and corkscrewing their way down with each of the Mig-29's taking turns chasing the Flanker. I circled around without changing my altitude and when the Flanker was about to dive past me, I was in perfect position to fire a burst into it. The plane exploded in a bright wall of yellow flame. The 2 friendly Mig-29's continued to corkscrew down toward me, and when they were pointed directly toward my aircraft, both fired. I rolled over and pulled back on the stick to dive out of their fire, and when I did, I saw the wreckage of two more Chinese Jian 7's that had been behind me falling right at me.

Once more, I rolled out of the way and looked for either communists or Mig-29's to help protect. This time, I found the Peruvian. He was chasing another Flanker, but 4 American F-16's were closing in from behind. I lit my afterburner and put my plane between the Peruvian and the Americans. When they saw that we had the same camouflage pattern, they all banked left and fired 4 streams of 20mm at the Flanker in front of the Peruvian.

The entire dogfight seemed to last forever. I fired at two more Chinese Jian 7's and damaged both. After that, I was out of ammunition, low on fuel, and hopelessly lost. The battle had worked its way from the west coast of the island into the Taiwan Strait, but Taiwan was still clear on the horizon, so I headed that way and climbed for altitude. At 35,000 feet, I found two of the Mig-29's in my division. It was the Germans. I reduced my altitude to 25,000 feet and joined them. We all waved and headed home.

Back at Hualien, we found the Peruvian Mig-29 that I had tried to protect. The airport was filling with other aircraft from our group, but while I was still trying to leave my plane, the Peruvian pilot jumped up and ran along my fuselage to help me out. He was very grateful-though highly unprofessional. That describes this entire division very well.

3/23/11

1751 hrs.

Hualien Airfield, Taiwan

Personal Log – Captain Jodl "Magicman" Steinhoff

Today was a very sad day. My friend, Dieter Von Runstadt, was shot down over Taiwan. We were flying a simple Combat Air Patrol above our new home, Hualien Airfield. Jodl spotted some Chinese planes on the other side of the island and headed toward them. I stayed with him, and the rest of our flight followed. The Chinese were flying some very old Jian 7's-copies of the Soviet Mig-21 Fishbed. Jodl dove down on them from 35,000 feet. He fired off a missile, and while I prepared to fire on my first target, the mission went to hell. At least one squadron of North Korean Su-27 Flankers ambushed us. Several attacked from our left side-south of the Chinese formation. Others attacked from a few miles behind and above. Jodl didn't even have time to pop any flares or chaff. His Mig-29 was hit by the first North Korean missile.

Our formation broke up, and the furball filled the sky. Jodl was my wingman, and since I was alone I headed for higher altitude. At 45,000 feet, I looked around and felt secure enough to begin firing into the furball. It was difficult to tell the difference between the Mig-29's in our flight and the North Korean Su-27's. Both are similarly shaped, and both have pale camouflage schemes. The Chinese Jians were much more distinctive, and I began to attack them. The furball was descending lower and lower with each plane's maneuvers, so I was very comfortable sniping Jians from 7-miles above. Sadly, it's much easier to dodge a missile attack from above instead of behind, and only one of my 6 shots found their mark. It was my last Aphid (AA-8, short-range, heat-seeking, dogfight, air-to-air missile). When it finally struck a Jian, the Chinese and North Koreans were already on their way back across the Taiwan Strait.

Dieter was a quiet and gentle man. We served together as part of the NATO, Mig-29 Fulcrum training squadron for two years. When fighting broke out on the Korean Peninsula, we were ordered to deliver our 2 planes to Kadema Air Base in Malaysia. Our planes were to serve as replacements for the ones that the Royal Malaysian Air Force (RMAF) had lost back in February. On our way, we were diverted to Kunsan, Korea because the RMAF's squadron was almost completely wiped out, and the need for replacement aircraft was urgent. When we arrived, the RMAF squadron was down to two people, and both still had their own Fulcrums. Dieter and I agreed to stay until someone could take over the planes. Our Colonel back at Laage, Germany agreed. We could not simply leave these Mig-29's for the Chinese and North Koreans to take. A few days later after sitting on the ground during several air raids on Kunsan, The Wolfman put together this composite unit. Dieter and I called Laage and asked if we could join, and we were given unofficial permission. It was hardly legal, but it was enough for Dieter and me. We were tired of being bombed, and besides, it was the right thing to do. Now, he's gone.

I will miss him. He was always smiling, and he never complained. Dieter was very professional and exact in everything he did. Tonight, I will call home and report his loss. However, if I am ordered home, I will not obey the command. This composite group is doing something positive and constructive in the face of tyranny and oppression. We let that happen once in our country, and now, I cannot sit on the ground and watch it happen elsewhere. Dieter understood that, and I don't think he was lost in vain. He knew why he was here, and he knew what he was doing.

Goodbye, Dieter. I will see you in the sky my friend!

March 23, 2011

Hualien Airfield, Taiwan

Personal Log – Miguel Cristobal Leon, Mexican Air Force

There exists a problem when a group of people fights together for the same cause – some of them will die. Most of *Los Antiguados* were together this evening in the cantina again, going over the day's events and drinking our dinner, when one of the Germans that fly the Mig-29 came in. He walked behind the bar, pulled down a random full bottle, sat down in the back and commenced drinking heavily.

I went over to him and asked what was wrong; apparently his *amigo* was killed in one of today's battles. I expressed my remorse for his loss, but it did little to assist him in his time of sadness (as few things do); he became overcome with grief and left the bar to avoid breaking down in front of everyone. I understand; it is not *macho* to show any of the negative emotions outwardly, however often and to what degree we feel them.

We drank to the memory of his *amigo*, then I looked around at my *hermanos* and told them to remember well how we were feeling at that moment. The odds of our survival through the duration of this conflict grow less and less in our favor with every passing day, as do those of every individual fighting for his or her cause. Perhaps it is fitting that we are based in the City of Marble, for as surely as the sun rises and sets, one or more of us will earn our headstones here.

I spoke with El Jefe later, and he agreed that it was something that we all had to keep in mind, but we mustn't let it distract us from our purpose. We need to be here; we have to be here. We have no other choice, because we are the only ones that can make a difference here and now. Then, he walked away toward his plane to fly yet another mission against our opponents.

I must admit, the *gringo* has a nice way with a turn of phrase.

3/24/11

0002 hrs.

Hualien Airfield, Taiwan

Personal Log – Lt. Kim "Killer" Yoo

The Taiwanese pilots are fighting like wildcats. I understand. When the safety and security of one's nation is at stake, thoughts of personal survival or rest disappear. I have spoken with some of them, and although they might speak a different language, although they look differently and have a Chinese-based culture, I have come to think of them as my brothers. Our common disdain for the communists has come to sustain and nourish us when fatigue should pin us to

343

the ground. I now know more than one Taiwanese pilot who has not slept in a week. Their faces are dark and long, but their demeanor is as candid and lucid as anyone's. Al says that if I want to fly some extra missions with the squadron that is based here at Hualien, he would understand. There are a few other Korean pilots that will join me. Al's only requirement is that I still fly with the Mig-29 division whenever they are airborne. They need a Coalition aircraft to distinguish them from the communists and keep them alive.

3/24/11

1526 hrs.

Hualien Airfield, Taiwan

Personal Log – Captain Jodl "Magicman" Steinhoff

Dieter is back! I never saw him eject, but somehow, he made it out of his plane! Dieter tells me that when his plane was hit, the missile cut it in half. The nose fell into the flames, and everything behind the intakes continued forward. He looked out and saw flames all around, so he waited before ejecting. He landed someplace called, Piekang, and a Taiwanese Army officer drove him all the way across the island to get here.

The Wolfman tells us that the Taiwanese have a captured North Korean Mig-29 for Dieter to fly. It was shot up on the first day, and the pilot decided to land at an airfield on the south end of the island near Pingtung. Our local Taiwanese interpreter/scrounger believes that he can have the plane brought up here on a flat bed truck tonight. He's already been working on it in an effort to get us some cannibalized spare parts. We'll take a look at it and see if it's repairable when it gets here.

It's very good to have Dieter back. I tried to make the call and report his loss, but I couldn't bring myself to do it last night. After we celebrate, and after we take a look at the North Korean plane, he and I can discuss whether or not to report the incident. Until then, Dieter and I are going to go out and have some beer-just to take away the soreness he has from ejecting the other day!

March 24, 2011

Hualien Airfield, Taiwan

Personal Log – Leftenant Ian Thomas "Top Hat" Sanders IV, R.A.F.

Everything's been so bloody scattered and haphazard since we left Korea. I've flown countless CAP missions and covered even more ground-ops flights than I can remember ovber the last couple of days, and the cycle shows no sign of stopping. We're busy as bees, flying all over this blasted country protecting ships, each other, and ground troops, while attacking the enemy with everything we've got.

Fatigue is starting to take its toll again, both on pilots and planes – there aren't enough hours in the day to fly as many missions as is needed, and there still aren't enough groundcrew to keep all of the planes in barely-serviceable condition. This afternoon, I had had enough for a little while, and dropped off my plane for its duct-tape regimen, refueling, and rearming. I was beat, and needed a little shuteye.

Have you ever been so tired that you can't sleep? After about three hours of dozing restlessly, I was still feeling like some yob had been beating me with a small building for a day or so. I got up and headed for the hotel pub for a couple of drinks, hoping they would knock me out so I could get some decent sack time.

There was a hell of a party going on in the pub. Most of the Mexican contingent were there, trying to hurt themselves with equal parts tequila (WHERE do they keep getting it from?) and the local hooch. The guest of honor was obviously one of the Germans, judging from the silly hat and the overabundance of alcohol he was wearing. (He looked a bit like a drowning schnauzer, but he had a silly grin on his face.) Every few minutes or so, another Mexican would dump another beer over his head and slap him on the back and shake his hand. His flightmate was sitting next to him, obviously drunk as a lord, singing something in German and waving his mug around.

I made the mistake of asking the one German, Jodl, just what the bloody hell was going on the next time he made a beeline to the bar for more booze to drink and more beer to pour on the other Kraut. He told me in broken English that the German wearing the sombrero, Deiter, had been shot down the day before, and everyone had thought he was dead. He and his buddy were having a celebratory drink or two when the Mexicans had come in.

Neither of the Germans really understood why the Mexicans had gone off the bloody hook when they had seen Deiter, but the party they had whistled up out of nowhere was quite something. (The daft bastard who I had lent a hand to a couple of days before upended a tankard of the local ale over my head as well, but it was good to see everyone having a good time, so I didn't complain too much.) Actually, their energy and emotion were extremely contagious, and soon the bar was packed to capacity with pilots and crew that had gotten drawn in. Even Meyers was there for a short while, sitting at a table by himself, occasionally chatting but mostly just sipping a drink and keeping an eye out.

I forced a couple of glasses of native whiskey down and made for the shower; sure enough, when I hit my bunk, I closed my eyes and was gone.

345

Thursday March 24, 2011

British Media News Service

Courageous Chinese Air Assault Proves Costly

Most of the U.N. troops that have come ashore spent the day re-consolidating their units which were still scattered from the Pusan evacuation. Some were able to deploy into the mountains between the "Choo" beachhead and the city of Taipei. By mid-afternoon, artillery units were in place and providing support to the Taiwanese infantry units surrounding the airport. Marine and British Harrier jump jets were tasked with close air support missions, while Task Force Stingray's escort ships fought off another day's worth of missile attacks and air strikes. Over the city, and over the Taiwan Strait, dogfights that began days ago continue to draw in U.N. carrier-based fighters and the few remaining planes in Taiwan's air force.

Two more helicopter assaults into Taipei's airport were attempted. The first suffered extreme losses when a pair of free R.O.K. F-5 fighter planes broke off from a dogfight high above and strafed the helicopters over the Strait until both pilots ran out of fuel and were forced to eject. The second helicopter assault was completely wiped out by a Taiwanese Navy frigate that had slipped between their flight path and the airport. The frigate single-handedly destroyed every last helicopter.

A Chinese parachute drop proved to be much more successful. Escorted by the remnants of three SU-27 squadrons, two squadrons of Soviet-built heavy airlift transports were able to fly low over the airport and drop out hundreds more troops. The tactic was repeated two more times between sunset and midnight. U.N. Estimates now put the Chinese strength around the airport at almost twice what it was the previous day; 21,000.

Thursday March 24, 2011

2nd Lt. John Chamberlin-Personal Log

Hill 1419, 15 miles S of Taipei, Taiwan

We're back on the ground again. An Osprey dropped us off on the beach this morning. Hector is in charge of both the company and his own platoon. The beach is a place of complete chaos. Countless units from a dozen different countries are all trying to get their act together. A lot of the units seem to be coming together for the first time since the Pusan evacuation. Too many people are trying to take charge. Officers who must have had their units shot out from under them are trying to make new ones. Hector, Bruce, and I got our orders from Col. Ghetty while we were back on the Wasp, so we walked right through the confusion.

Our initial mission was to land and move northwest on route 9 to a position near the top of hill 1419. The farther we moved inland, the more Marines we came in contact with. We walked for most of the day, then a handful of trucks that were coming down the road agreed to turn around and give us a lift up.

At the top, we're supposed to sit around a battery of Taiwanese 105mm howitzers. There was already a Taiwanese army company up here too. They had a liaison officer waiting for us, and he told us were to setup our defensive positions. They hadn't come into any contact, but they realize that this is a war, and they are definitely ready for anything. These guys are as organized as the people on the beach are confused.

There was a lot of typical overcast all day long, but in the middle of the night it seemed to disappear. We can see all the way into Taipei; about 15-20 miles to the northwest. There's a lot of fires and a lot of flashes down there. The evening was quiet except for the occasional thump and pop of explosions down below. The 105's behind us were quiet all night, but a pair of trucks showed up with some ammo so tomorrow won't be so quiet.

Friday March 25, 2011

Central News Network Online Report

Fight for Formosa:

<u>Communist Chinese Reinforce Their Airport Foothold</u>

The Chinese air campaign has shifted focus from strategic bombing of targets around Taiwan to the tactical bombing of U.N. Artillery units overlooking Taipei. Airborne troops holding the airport on the other side of the city have been hammered by the U.N. artillery since its arrival.

In the air, Carrier-based aircraft and forward deployed Harrier jump jets were able to fend off the Chinese air attacks, but were kept so busy that resupply and reinforcement parachute drops over the communist perimeter at the airport went almost unchallenged.

At the airport, a senior Chinese officer gave an interview to selected members of the world press even as artillery churned the tarmac behind him. The officer reported that the condition of the Chinese troops was excellent, and that they were waiting for the people of Taiwan to join them in their struggle against the illegal nationalist government of Taiwan. Reporters noted the apparent lackluster efforts of the Taiwanese army to dislodge the communists.

In the city of Taipei itself, Taiwanese military units focused their efforts on establishing a defensive perimeter to contain the Chinese at the airport. Barricades were put up in the streets along the northern bank of the Keelung Ho River, the airport's northern edge. South and east of the airport, in the heart of the city, buildings were evacuated and tenants replaced by wellarmed infantry. In the western half of the city, across the Tansui Ho River, mechanized units moved into concealed positions to serve as either a reserve or a rapid attack force should the opportunity arise. As U.N. forces move down from the mountains and into the city such an opportunity is expected to appear soon.

Friday March 25, 2011

2nd Lt. John Chamberlin-Personal Log

Hill 1419, 15 miles S of Taipei, Taiwan

Just after midnight, the 105mm howitzers we're guarding opened fire without warning. I was in a small slit trench with Rick and 3rd squad. When I heard those guns open up, I immediately woke up and looked down the hill toward the enemy. Turns out, that the Chinese have overrun a few of the Taiwanese Army barricades in Taipei. The 105's were called on to rain some high explosive on the reds. As I gathered my wits, I checked out my platoon's reaction. Most everyone looked like they had been awake anyway. Some did not.

I knew that having men and women in combat was going to be full of challenges, and over the past few months, I feel kind of lucky that I only had to intervene in two tristes; Chris and Christina's sexual escapade on maneuvers, and Bob and Portia's breakup. I had suspected that there were others, but 'don't ask, don't tell' equals don't matter to me. Christina and Portia seem to have only moved on to fresh meat in the platoon. When the 105's opened up, both Christina and Trent Miller, a quiet Marine from Rick's 3rd squad, could be seen buttoning up their battle dress uniforms while everyone watched down the hill for some sort of close up enemy action. Christina and Trent weren't alone. Others whose heads popped up together were Portia's and Chuck Boyd's. He's the guy in the platoon with a knack for internet intelligence gathering. Judging by the looks exchanged between Bob and Melissa Keegan (an attractive, frail Marine from Wess' squad), I would guess that Bob has decided to move on after Portia. There may be other relationships building in the platoon, and there may not be, but one thing is certain: it's going to be an interesting war.

The 105's fired through the night until about an hour before sunrise when the ammunition must have run out. More trucks arrived from the Choo beachhead around midmorning, but the guns remained silent. Overhead, the skies are clear. Normally we've been treated to the best seats in the house at the world's greatest airshow. Helicopters, Osprey, fighters, bombers, transports, SAM's and flak. Every second of the day, there's something to see. The airshow has really been distracting, but all three squad leaders have done an excellent job of keeping everyone busy with fortifying our position. There's no telling how long we'll be up here.

3/26/11

2001 hrs.

Hualien Airfield, Taiwan

Personal Log – Lt. Kim "Killer" Yoo

Due to the knife wound that I received when they were trying to get me out of my plane, I've been grounded again. A group of Taiwanese Air Force wing wipers have been busy working on my aircraft, and a new rudder has already been added. The panels that were hit by the Flankers have been removed, and the Falcon looks naked. They are very busy, and I have tried to stay out of their way.

My Mig-29 division flew a few fighter-sweep missions today. Al took my place in the division to make sure that they were not mistakenly attacked by friendly forces. They ran into the same Su-27 Flanker squadrons that we fought yesterday, and the Taiwanese F-5 squadron here at the airport has taken some serious losses. The Germans told me that they also attacked another communist helicopter assault force, and they were able to shoot all of them down. Our Peruvian pilot was less optimistic. He confirmed that the helicopters had been completely shot down, but a squadron of transport planes was able to survive and parachute-drop supplies to the communists who have taken over Taipei International Airport.

Al has arranged for me to fly a Taiwanese F-16 if the knife wound on my left side should heal before my own aircraft is repaired. I am not particularly attached to my aircraft, but I would prefer to fly it rather than someone else's. If I take someone else's plane, then there is one less pilot in the air killing communists.

Saturday March 26, 2011

United Press Online News Service

Communist Breakout Begins

After two more days of mostly successful resupply and reinforcement airdrops, Chinese troops occupying the Sungshan Airport attacked west into the city. Little progress seems to have been made. Casualties for the Chinese are estimated at nearly a thousand, while Taiwanese Army officials claim to have lost less than a hundred killed and wounded. Damage to the city is increasingly catastrophic. U.N. artillery units, supporting the Taiwanese Army's defense, are taking the brunt of the public outcry. Almost 30 city blocks have been completely leveled, and many more severely damaged from the inaccurate rounds.

Chinese and North Korean amphibious units made a daring landing on the South side of the Tanshui River northwest of Taipei. Attempts at repelling the amphibious assault by land, air, and sea all failed. Air attacks on the ships were made as they crossed the Taiwan Strait, but heavy air cover from Chinese fighters

349

based on the mainland prevented the attacks from having any real effect. U.N. ships from Task Force Stingray tried to maneuver from the east side of the island toward the Taiwan Strait, but were driven back by missiles from Chinese patrol boats and light frigates. Mechanized units from the Taiwanese Army moved from their defensive positions on the west side of the airport, but were immediately recalled by a strong infantry attack from the Chinese airborne troops who used the rubble created by the artillery as cover.

Saturday March 26, 2011

2nd Lt. John Chamberlin-Personal Log

Hill 1419, 15 miles S of Taipei, Taiwan

Another day, another air show, and more fire missions. There seems to be a lot more smoke down in Taipei today. We also saw some small artillery and air strikes conducted between Taipei and our Mountain. I'm not sure if it was our side hitting the reds, or vice versa. More troops and equipment continue to pour in from the Choo beachhead. Bob's efforts at improving the platoon's trenches and dugouts are still underway. I added to the busy schedule by setting up a routine four-person patrol every other hour. The patrols are varied so that everyone, myself included will get a chance to move down the hill and reconnoiter the terrain in front of us. I also had Gary take his squad back down to the beachhead to get some fresh water and some more MRE's (Meals Ready to Eat). We've got enough for a week, maybe two and a half if we stretch it, but water is running low.

Sunday March 27, 2011

United Press Online News Service

Marines Move to Stop the Red Rush

Chinese reinforcements poured into the beachhead northwest of Taipei. A pair of parachute supply drops to the Chinese forces at Sungshan Airport precluded another attack westward from the airport towards the beachhead. Taiwanese Army units were able to prevent the two forces from linking, but the ring around the airport had to be weakened in order to move in enough reinforcements to the shattered west side of Taipei.

U.N. units deployed into the south side of the city with United States Marines leading the way. Artillery units in the mountains have seen diminishing returns for their efforts against the advancing Chinese, most likely due to poor weather conditions and the lack of surviving forward observers in the city. In the air above the city, Chinese short range, hand-held SAM's have kept U.N. forces from conducting effective close air support operations. Communist air supply missions have been able to succeed because local SAM teams have been able to keep U.N. carrier-based aircraft from operating over the city. Pilots in Taiwan's

Air Force have been undaunted by the small heat-seeking, shoulder-launched SAM's, but losses may force their restriction from operating in the area.

Early in the morning, most of the remaining U.N. ground units were offloaded from Task Force Stingray. Repeated small-scale air and missile attacks have taken their toll on the Task Force's fuel and ammunition supplies. After recovering landing craft, the Task Force immediately moved east toward the Philippines. Ships from Task Force Stingray are expected to maintain a position east of Taiwan while rotating ships to Subic Bay for rearming and refueling.

1800 hrs.

Hualien Airfield, Taiwan

Personal Log – Lt. Kim "Killer" Yoo

It will be a few very long days before I am permitted to fly once more. Even the small wound on my left arm could be fatal in a dogfight where my body would have to pull 9, 10, 11, or more G-forces. Perhaps next week it will be closed enough to fly again. Until it becomes secure, I will have to try and find ways to occupy my time and my mind here on the ground.

I rented a car and tried to tour the area for a while as a means of keeping myself busy. This is a beautiful area. They call Hualien The City of Marble because of the large geologic deposits in the area. Streets, sidewalks, buildings, even fast food restaurants are all made from white marble. It has a mystical atmosphere. I visited the Marble Cliffs about an hour and a half north of the city-off the Island Highway out of Hisincheng. They were incredible.

I am also glad that I made the trip for tactical reasons. While I was there, I saw the Taiwanese Army setting up a SAM trap. They've put an entire battalion of air defense troops all along the cliffs. The men are armed with hand-held, American-made Stinger heat-seeking missiles. I met with the commander, and we had a very beneficial conversation. The entire nation of Taiwan is very grateful to our composite group. They are particularly grateful to all the Koreans fighting by their sides. The Taiwanese Colonel and I formulated a plan. Aircraft from the group will bait the communists and drag them into the Colonel's SAM trap. I will inform the rest of the group, and when we fly through the area, if we wag our wings, the Colonel will know that we are friendly. I will also pass the word to avoid the area unless being pursued. Hopefully, we can try out the plan very soon. The marble cliffs are beautiful, but a valley dotted by black smudges of communist aircraft wrecks would be so much more than simple visual beauty!

Sunday March 27, 2011

2nd Lt. John Chamberlin-Personal Log

Hill 1419, 15 miles S of Taipei, Taiwan

Bad day for the platoon. While Gary and 1st squad were down at the Choo beachhead scrounging for supplies, there was an air raid. Nine vintage Chinese Jian-7's came overhead in three waves. They came in at maximum speed and haphazardly dropped their bombs on the area. I'm told they didn't even come close to anything of real value, but I have to write some letters to parents who would think otherwise. The squad took cover in some overturned wooden fishing boats when the alarm sounded. After each plane had dropped its quartet of 500-pounders, everyone came out from underneath; everyone except Bill Archer, Dave Avery, and Gary. Shrapnel from a near miss went right through the boat they were hiding under, and each one took a piece. Lee Burke was hiding under another boat, and she was evacuated back to the Wasp with a piece of shrapnel in her back. There's still no word on her condition.

I put Lance Corporal Dwaine Brubaker in charge of the seven people left in 1st squad. He's a nice guy, very conservative, and lives for the image. Although he doesn't talk about it specifically, I can easily see him with a white house, picket fence, 2.5 kids, desk job, and a BMW. I wouldn't be surprised if he already has names for the kids. Still, he seems to be the most responsible one in the squad, and I'm reluctant to break up the other squads, even by one person, unless the situation demands it.

I'm not looking forward to writing Gary, Dave, and Bill's parents. I have a lot to say about each one of them, some good, some not fit for parents. I know if I dwell on them too much, I'll lose my edge and others may suffer. At least I wasn't directly responsible for their deaths. I could have sent anyone to get supplies, and the same luck that chose them also, chose the timing of the air raid and the path of the Chinese shrapnel. I saved the letters that I wrote for Mike and Tim on my palm PC. I think I might just reword those and send them as e-mail. Those letters were simple, complimentary, and overall as welldone as can be expected. If I toss in a few more nice things and maybe some personal information, I think they'll do as well as anything else. Getting a letter like that via e-mail is pretty shallow, but I think it's better to find out right away rather than after a few weeks. Maybe it'll make the chaplain's job a little easier.

Monday March 28, 2011, 6:00 PM EST

Channel 4 News

"Good evening everyone. I'm Tom Waushinski, and this is Channel 4 News. Tonight, I'm reporting to you live from the city of Taipei, Taiwan where modern mechanized warfare has taken a new and dangerous turn."

"Chinese Naval Infantry units drove through Taiwanese Army positions west of the Sunshan airport and linked up with Chinese Airborne units who had landed days before. Chinese and North Korean Special Forces units, operating in battalion strength, used hovercraft to move from the Chinese beachhead through the heart of the city. The cities' three rivers acted as empty highways to the 63 North Korean hovercraft. Taiwanese Army units and U.S. Marines were able to kill or capture most of the attackers, but an unknown number escaped into the city, and sniper activity has dramatically increased. Sizeable Taiwanese Army units still function, but a single cohesive frontline no longer exists between the four separate Taiwanese Army divisions in the city and the U.N. Forces arriving from the Southwest."

"U.N. Carrier-based aircraft, along with the remnants of the Taiwanese Air Force, have prevented the Chinese from controlling the air above the island, but the lack of numbers and low stockpiles of sophisticated air-to-air missiles is limiting their effect on Chinese air formations. Over the past few days, U.S. Marines have found themselves under air attack from enemy aircraft for the first time since World War II."

3/28/11

1521 hrs.

Hualien Airfield, Taiwan

Personal Log – Lt. Kim "Killer" Yoo

I accepted the Taiwanese offer to fly with them today. Al wasn't happy about my flying again with my arm wound still healing. He didn't say anything specific. Instead, I was given a look that comprised of concern, disappointment, frustration, and even a bit of anger. The stitches are holding fine. It has been almost a week, and I'm sure that the wound is starting to close. Normally, I would be concerned that a high-G turn would pull it open again, but I wrapped the bandages tightly with duct-tape before I put on my flight suit. That tape is strong enough to patch the holes in our aircraft, it should be strong enough to keep my stitches from tearing out.

The Taiwanese squadron was in possession of an older F-16A Falcon (They were even kind enough to put Republic of Korea markings on it for me!). The original pilot was wounded by a communist 37mm shell fragment that went through the canopy and almost severed his arm. Since then, the wing wipers replaced the canopy and cleaned the cockpit, but there is still a dark red outline of dried blood surrounded all the buttons, screens, and gauges. It is a sobering sight.

Some pilots would be disgusted by the substance. I find it to be a strong reaffirmation of my personal need to fly again. Every gauge that I glance at reminds me of the blood spilled in the name of communist aggression, and my soul is filled with both a chilling discipline and a boiling need to sweep the Earth of this red menace. They have taken my homeland. Now, they are trying to steal this island. I have even heard rumors of their infiltration into Indochina. The world is coming together to stop this cancer, and while we may be lose battles, the communist cost of victory is rising. Soon, I will make that cost too high for them.

The flight today was relatively uneventful. I flew a fighter-sweep with the Taiwanese F-5 squadron. We headed up toward Taipei. The city is covered with smoke, and a low overcast made strafing impossible. At 5000-7000 feet, the overcast cleared. I followed the F-5's on an intercept that must have been directed by ground controllers. We were still almost 50 miles away when the communist targets began exploding. Some American F-18 Hornets were in the area, and they must have shot them down from beyond the horizon.

The Americans are very good pilots, but they are prone to choose command control over personal initiative. They also prefer to make BVR (Beyond Visual Range) attacks with sophisticated, long-range missiles like the AIM-120 AMRAAM, or the U.S. Navy's older AIM-54 Phoenix. BVR attacks are very sound from an accountant's perspective, but they require lots of expensive and sophisticated equipment, sensors, missiles and aircraft. They also require highly-rained, well-coordinated, and extremely disciplined people (radar operators, communications staff, equipment maintenance personnel, weapons experts, and specially-trained pilots).

All of that effort is usually blocked by communist electronic jamming. During the fight for Korea-and during most of this action-the communists have blanketed the entire electronic spectrum with jamming. One of the Taiwanese told me that they are using the same technology that was used by radio telescopes to monitor 256 million frequencies simultaneously. The only difference is that instead of listening to all of those frequencies, the communists are broadcasting white-noise (static) on them all. The result is a complete breakdown in all communication on both sides.

Most of the time, it makes BVR attacks impossible to coordinate, and dogfighting very difficult. I believe it has also helped the communists by forcing their pilots to stay true to their specific mission orders instead of deviating to dogfight. When they do choose to fight, they usually out number our planes, and so coordination happens automatically while we are forced to fight as individuals.

Today, the Americans were able to find a way to coordinate their BVR attacks despite communist jamming. I followed the Taiwanese on two more intercepts, but on one occasion the communists turned back, and on the other occasion, the Americans must have destroyed them. I do not speak Chinese, but the Taiwanese pilots seemed unusually relieved to be back on the ground after the mission. None of us really like to fly in a sky that's being used as a long-range shooting gallery by the Americans. As the American losses have increased, they seem have become far less concerned with who they are shooting at. There is no point in complaining to them. Pilots from all over the Coalition have noticed their "Shoot first and ask questions later" attitude, and we have learned that it is best to simply to stay as far away from their ships and long-range interceptors as possible.

One of the Taiwanese pilots who speaks English told me that his squadron has already lost two aircraft to the Americans, and 4 to the communists.

There is a rumor that the Coalition naval task forces are moving away from Taiwan and into the less-constrained, northern Philippine Sea. That will make it too far to reach for most of the communist planes. It will also make it difficult for communist patrol boats to sneak up to them. Happily, it will make it difficult for the Americans to shoot any of us by accident!

After watching the Americans pick-off all of our intercepts, we finally found a flight of Chinese Jian 7's to chase down. I couldn't communicate with the Taiwanese very well, so while they tangled with a pair of the communists, I pursued the other two into the overcast. We came out right over the mountains, and all three of us had to pull up to avoid crashing into the first one. I lost one of my prey in the clouds, but I followed the other one back up through the overcast and right into the Taiwanese squadron. They had already shot down the two that they engaged, and 4 of the F-5's were in perfect position to jump on my prey. One of them fired off a snap-shot burst of 20mm cannon fire and the Jian came apart while still climbing. The flames were like a blossoming flower as they went straight up then fell over in all directions as the debris headed back into the overcast. It wasn't very satisfying for me personally, but at least there are a few less communist pilots now.

Tomorrow, I will fly the morning fighter-sweep with the Taiwanese squadron once more. We will be going back over Taipei, then west into the Taiwan Strait. I enjoy the company of these Taiwanese. They are really Chinese, but their disgust and hatred for the communists is at least as strong as my own. Some pilots paint insignias on their aircraft to illustrate the number of aircraft that they have shot down. This Taiwanese squadron has laminated photos of lost relatives on their planes where the communist kills would normally be recorded. When I asked if they recorded their kills, the English-speaking pilot showed me the stripes on the nose of his plane. Each vertical red line represented an air victory. There were 4 on his plane. The rest of the planes had at least 3 stripes, and there were two with 9!

This is one FINE group of pilots.

Monday March 28, 2011

2nd Lt. John Chamberlin-Personal Log

Hill 1419, 15 miles S of Taipei, Taiwan

Well, the airshow came home today. A pair of Chinese Jian-7's strafed our company's position today. They didn't hit anybody, and they only had time for one pass since a South Korean F-16 was right on their ass when they passed overhead. At first, when the siren went off, everyone was a little scared. I felt naked with no place to hide, and nothing to fight back with. When we saw that F-16, everyone in the company and all the guys back at the 105's jumped out of their holes cheering and waving. I doubt any of the pilots saw us, but if the Chinese did, they had to be frustrated. I think they were a little busy trying to shake the F-16.

Tuesday March 29, 2011

United Press Online News Service

Outnumbered Communist Chinese Still Threaten Entire Coalition Force

Elements of the United States Marine Corps were able to re-connect the frontline between U.N. forces, and two of the surrounded Taiwanese Army divisions. Chinese reinforcements have started to flood into the city with the arrival of three seized merchant ships bearing fresh troops from the mainland. An estimated 73,000 Chinese hold the Northwest tip of Taiwan. Nearly 2,000,000 Taiwanese troops, bolstered by almost 200,000 additional troops from around the world stand against them. Most of Taiwan's forces have been forced into defensive positions around the country. For the most part, the U.N. forces, still in disarray from their South Korean evacuation, remain at the "Choo" beachhead trying to reorganize. At no time have the Chinese faced more than 45,000 troops in, or around, the city of Taipei.

With the bulk of Task Force Stingray refueling and rearming east of Taiwan, Chinese patrol boats and light frigates made a daring raid against the lightly defended "Choo" beachhead. Casualties among U.N. troops were statistically light, but, none the less, hundreds died needlessly. All fourteen of the Chinese ships were sunk by U.S. and British Harriers based at the beachhead.

March 29, 2011

Hualien Airfield, Taiwan

Personal Log – Miguel Cristobal Leon, Mexican Air Force

Los Antiguados are all still alive despite over a week of flying nonstop missions. This evening was the first time all of us have been in the cantina at the same time since we split up into El Jefe's different groups, and we all had our stories to share, both good and bad.

One particular topic of our conversation has been the German pilots and their new pet project, the supposed repair of the North Korean Mig-29. They have been working on that thing for five days now, with hardly a moment's pause. They were both at it the morning after Deiter's party, albeit very slowly and painfully, and they break only for meals and a couple of hours sleep – both of those in shifts! When it gets too dark to turn a wrench or sort through wiring, they're banging their heads together planning out the next day's course of action.

What are they thinking - just because they can fix Volkswagens back home in the real world, it's only a few steps to aviation repairs in the war zone? What arrogance! Does every German fancy himself as an engineer? I guess so.

We started a pool, betting on how long (in weeks) it takes them to get that wreck back up in the air. Esteban feels like he was cheated – there were six slips of paper, numbered 1-5 and "Never," and he picked Week 1 out of the sombrero. That's the way it goes, but I think his dinero is gone for good. Mine's probably just as gone, though; I chose Week 2.

Tuesday March 29, 2011

2nd Lt. John Chamberlin-Personal Log

Hill 1419, 15 miles S of Taipei, Taiwan

There was one hell of an air show this afternoon. Harriers from the beachhead hit some targets on the far side of Taipei. Chuck says he saw on an internet news forum that they were Chinese freighters with supplies. All we could see was about ten or twelve black columns of smoke, a lot more fires than usual in the city, and anti-aircraft fire, both AAA and SAM's, like we've never seen before. There were a lot of dogfights to watch too. Normally, this would have been very exciting, but morale has become an issue again.

Our recent casualties have left the rest of the platoon in a heightened state of professionalism, for the most part. Some people, the ones that really show their youth, have sought refuge in each other. Christina and Portia continue to move through the platoon from man to man. They're not particularly easy women, but they act as though affairs are tallied on some sort of resume. While some men and women focus on doing their job with discipline and responsibility almost to the extreme, other men and women almost shun responsibility and show little discipline at all. I had expected the difference to be between the sexes, instead, it seems to be among the ages. Anyone under 21 tends to act as a sort of maverick partygoer; a rough and tumble Marine-male and female. People who are 21 or over seem to become more professional under stress, also regardless of sex. There are exceptions to both, but the generality can definitely be made.

Wednesday March 30, 2011

Washington Post

Restricted Submarine Warfare Begins

Chinese Naval Infantry units made amphibious landings along the coast, specifically: Fangyuan (west-central Taiwan), Yungan (southwest Taiwan), Fanglio (southsouthwest tip of Taiwan), and Taimali (a small picturesque fishing village at the south-southeast tip of Taiwan). Taiwanese Army Units began deploying to the areas immediately. Reservists who were initially charged with defending against invasion were overrun faster than expected, and intelligence data for counterattacking elements of the Taiwanese is sketchy at best.

Since the capitulation of South Korea, U.N. diplomatic efforts aimed at ending the war have stalemated. Chinese and North Korean officials, backed by a collage of U.N. countries hostile to the United States, see no reason to cease their operations. While the fighting intensified on and around Taiwan, U.S. and British submarines were called into the area from around the world. As they appeared on the scene, they immediately started searching and tracking their prey. At 0000 hours, 3/30/05, they received their orders to attack Chinese and North Korean vessels at will. Between Vladivostock and Vietnam, the submarines went to work. Within a matter of hours, 43 communist ships, ranging in size from light frigate to a Soviet-built heavy missile cruiser, all went to the bottom.

At the same time that the submarines were let loose upon the seas, orders went out from the Pentagon to shuffle the dwindling number of carrier battlegroups around the globe. Almost two thirds of the carriers that the U.S. started the war with are either badly damaged or have been sent to the bottom. While the damaged capital ships are being repaired, while old ships were refitted, and while new ships were being designed, the remaining carriers will have to cover more territory. Some areas will be left completely without the benefit of a carrier's rapid response capability.

The same shifting of forces is happening within other services, and other nations. Fighter squadrons around the world are being moved to the east. Ground forces, the most difficult to move of any military unit, are being left on their own until air and naval reinforcements can be brought to protect them. Dwindled down by decades of budget cuts, weakened by lowering recruitment quotas, and slashed by immense losses in battle, once mighty militaries are finding themselves stretched thin and playing on a less than level field.

3/30/11

1606 hrs.

Hualien Airfield, Taiwan

Personal Log – Captain Jodl "Magicman" Steinhoff

After several days of tinkering, Dieter and I have finally brought his new aircraft back to life. The North Korean pilot who was flying it must have been as dumb as a cow. The avionics are shot to hell from direct hits in both the forward and aft avionics bays. We spent most of our time trying to figure out the Russian wiring scheme, and although we still don't have everything working perfectly, the engines do fire, and the controls do respond-at least on the ground. The North Korean pilot probably panicked when the right engine shut down. The hinge on the aft inlet ramp to the starboard engine was hit by shrapnel. It came loose and closed off the intake. After having 20mm hits immediately in front of the cockpit and immediately behind his seat, the pilot was certainly shaken, then when the inlet ramp fell into the intake and shut off the engine, he must have feared for his life. None of us have figured out why he didn't eject. Maybe he was afraid of ejecting and preferred to bring it in. None of that matters now. Tonight, Dieter and I are going to take his new plane in the air and try to figure out the last of the wiring mysteries.

March 30, 2011

Hualien Airfield, Taiwan

Personal Log – Miguel Cristobal Leon, Mexican Air Force

I suppose I owe our German compatriots an apology – tonight the Korean Mig-29 took off from the runway. I would never have believed it, but Pepe came running into the cantina shouting "They did it! They did it! It flies!" Esteban choked on the beer he was drinking, then got up and did a little victory dance before demanding his money from all concerned. (There were a few grumbles, but everyone paid.)

When Jodl and Deiter entered the cantina a few hours later, they entered to a round of applause from everyone there. When they found out about the pool we had going, though, Jodl was a bit miffed that "Never" was one of the options offered. All in all, they took it pretty well. Perhaps they should start working on the problem of why the Coalition forces keep getting beaten so badly around the world!

3/31/11

1200 hrs.

Hualien Airfield, Taiwan

Personal Log – Captain Jodl "Magicman" Steinhoff

Dieter's new Mig-29 is working very well. Our friends on the neighboring Taiwanese ground crew staff have been helping us put the plane back in order. They've also re-painted it in the same colors that the rest of our group are using-a pale green. Dieter says that the only real problem is an uneven power shift between the two engines as soon as the landing gear touches down. I think we'll have to disable the intake ramp activation mechanism in the rear wheels. His port engine probably strains since it's not getting as much oxygen as the starboard side that doesn't have an intake ramp. The ramps are designed to protect the engines from sucking in debris on unimproved runways. Here in Taiwan, the field maintenance people are extremely effective, and we really don't need to use the ramps. A simple wire-cutting operation should take of the matter as long as we don't accidentally clip the brakes by accident.

The other day, the Chinese began making amphibious landings all around Taiwan. The landed at Fangyuan in west/central Taiwan, Yungan in the southwest, Fanglio near the south-southwest tip, and Taimali near the south-southeast tip. The Taiwanese F-5 squadron is going to make a few raids on the closest Chinese beachhead near Taimali. The Wolfman offered our services as escorts, and so tomorrow, the entire group is going up. Dieter is looking forward to getting back into action, and so am I. This time, I will try to watch out for him even harder, and I'm sure that he'll think twice before diving on a target that looks too simple.

We'll be flying in our All-Fulcrum division, but we will be accompanied by a South Korean pilot and his F-16 again. The Korean's name is Kim Yoo-callsign Killer. He crashed his Falcon the other day, but the Taiwanese Air Force has loaned him one of their older A-models. It's a lot slower, and a lot heavier, but we've flown with him in the past, and he is very skilled.

Yoo is a quiet man. He speaks English (it seems to be the most common second-language for us all), but he rarely converses with anyone. He is very clearly bitter about the fall of South Korea to the North Koreans and Chinese ("communists" as he calls them). Yoo's had his callsign before the war, but it seems a very astute choice for today. He's not particularly tall, short, slender, large. In a crowd of Asian pilots, he melts away in front of my eyes. The only real distinguishing characteristic is his cold demeanor. Nothing seems to give him any joy or contentment except discussing the details of a communist pilot's death. He's a unique man, but he has also proven to be an excellent pilot. The Wolfman assigned him to our division so that other Coalition pilots don't mistake us for Chinese or North Korean. It's a simple enough role, but he stays with us (most of the time), and on more than one occasion, he's watched all of our six o'clocks! Yoo is so eager to fly that he flies with our division, AND he flies with as many Taiwanese fighter-sweeps as he possibly can. He must sleep as often as he laughs!

Wednesday March 30, 2011

2nd Lt. John Chamberlin-Personal Log

Hill 1419, 15 miles S of Taipei, Taiwan

With all that's going on around us, we are bored. Overcast skies today blocked out the airshow. The 105's were packed up and moved out before sunrise. Even the sound of other artillery has faded. Occasionally we hear a plane fly low overhead, or some booms and pops of gunfire in Taipei. Maybe we've grown used to the war already. Maybe there really is a lull. Whatever it is, we're getting bored. The flow of troops and equipment from the Choo beachhead has trickled, and I had Rick send part of 3rd squad down to check it out. Could it be that we're as committed as possible? I would think that there should be some more reinforcements coming. Maybe they're holding the bulk of our units back someplace, like the Philippines, for an invasion of Korea.

Thursday, March 31, 2011

2111hrs local

U.S.S. George Washington CVN

Eastern Atlantic Ocean

The Moon looks bigger at sea. It calms the massive ship's roll, and it takes the edges off the waves. White foamy crests remain, but their splashes ripple silently toward the horizon. Under the giant, white orb, their swells fade into a stillness of sight, a silence of sound, and a numb feeling from the ship's nuclear reactors.

Lieutenant Chuck "Spider" Snyder stood on a catwalk along the starboard side of the flight deck. From 80 feet above the Atlantic, he looked down and peered into the depths. Was there an ancient Viking longboat below him? Maybe there was a sunken U-boat, or more likely a U-boat victim. Leaning on the largest warship that had ever been built, he couldn't help but wonder about the wreck-strewn bottom of the Atlantic Ocean. It was night, and he could still see deep into the water, but the mysterious, invisible fathoms below were conjuring images in his mind.

The unique smell of an aircraft carrier on the high seas filled the wind and pounded his face with warm spring air from nearby Africa. At any given moment his mind was forced to choose between smelling jet fuel, oil, smoke, burnt tires, hydraulic fluid, solvents, Atlantic salt water, or Saharan dust. The concentrations were so strong that he started to get a gasoline & ocean taste in his mouth.

The U.S.S. George Washington-a Nimitz class aircraft carrier-was on its way to the Mediterranean Sea. There was trouble in the Arabian Sea, the Philippine Sea, the Yellow Sea, the Sea of Japan, the Taiwan Strait, and the Korean Strait; instead of going to America's latest war zone, however, his ship was being sent to the tight confines of the Mediterranean Sea. Since the days of the Barbary Pirates in the late 1700's, the United States had maintained a naval presence in the crowded warm waters of 'The Med.'

Decades of military budget cuts had trimmed the number of American ships-especially aircraft carriers. Normally, five of the ten vessels were back in their homeports for refit, rest, or repair. One was always used for training, and the remaining four were tasked with protecting America's national interests all over the globe.

While the military equipment and programs were sacrificed in favor of domestic entitlement programs, the soldiers, pilots, Marines, and especially the sailors had found that their international assignments were increasing in size and number. Conflict seemed everywhere. The Pentagon classed these deployments as "Conflicts other that War," but they were often fought as major battles.

More than 5000 men and women were on board the George Washingtonthe "GW." They were well-trained for war, but Chuck Snyder and his fellow shipmates weren't going to the front lines. In times of crisis, the first question an American President always asked was "Where's the nearest carrier?" In this case, it wasn't the George Washington.

Chuck's ship was in Norfolk, Virginia when the President asked his fateful query. They had completed their last 6-month Atlantic patrol back in January. Soon after their arrival, trouble had sparked in the Pacific, and all of the available aircraft carriers were sent to the region. Normally, Chuck and his shipmates would have enjoyed a few months of duty based in Norfolk, but in a time of war (even an undeclared war), the ship was sent back out.

Their mission was to replace the 6th Fleet's U.S.S. Theodore Roosevelt. The "TR" had been ordered through the Suez Canal, the Red Sea, and into the Arabian Sea to join with other carrier battlegroups that were taking part in another one of the Pentagon's, "Conflicts other than War." Most of the press called it 'America's Latest War.' What it amounted to was a Mediterranean without an American carrier battlegroup. Given the unstable nature of the countries along the inland sea, the U.S.S. George Washington was given its sailing orders post haste.

None of that was too important to Chuck. His thoughts were of the moment. The past was too difficult, and the future was too complex. At the moment he was 80 feet above the Atlantic Ocean, bathing in moonlight, floating in the warm equatorial moonlight, and getting hypnotized by the glimmering glow of the waves. Everyone joined the navy for different reasons; Chuck was there for the experiences, the moments.

He felt alone for while and at peace. Protecting him from miles awaybeyond the horizon in all directions-were hidden escort ships. There were 5000+ people on his ship, but almost 10,000 people were protecting him. They were on the two cruisers, three destroyers, two frigates, and two submarines of the U.S.S. George Washington Carrier Battle Group (GW-CVBG).

Even on the moonlit deck of the carrier, he wasn't alone. There were a few deckhands on the flightdeck: Red Shirts, Yellow Shirts, Purple Shirts, Green Shirts, White Shirts, and Silver Suits. Each color designated their tasks. Red Shirts dealt with explosives, ordnance, and crashes. Yellow Shirts moved the planes around like taxi drivers. Purple Shirts handled fuel. Green Shirts controlled the launching catapults and arresting wires for landing. White Shirts took care of safety, medical, and oxygen needs for the pilots and crewman. Lastly, the silver suits put out fires. Some of the planes were parked in readyto-launch positions, but there were none scheduled to launch, and nothing was expected to land for another hour and half. It was a rare moment of peace on the flight deck.

That peace was broken by the addition of a few new arrivals. Three pilots from the ship's U.S. Marine Corps squadron were making their way up through a nearby hatch. Their boisterous frivolity broke the peace. Chuck thought that their long flight to the ship would have sapped some of their energy, but it hadn't. They were Marines, and there were untold reserves of strength burned into each one at Parris Island.

One of the three, the shorter Marine that was clearly the center of attention, stopped in his tracks when he saw Chuck. He turned toward the other Marines, put both hands on his hips, and started shaking his head.

"Aw, shit! Another squid. Dammit, guys. You'd think we could find someplace on this tub where we wouldn't bump into one of these waterjockeys."

Chuck looked around. He wanted to seem surprised at the young Marine's arrogance. Chuck knew he was alone on the catwalk, but he also wanted to make sure. If the Marines wanted to go beyond insults, then the odds were three-to-one against him. Chuck confirmed that he was alone, so he simply made eye contact with the loud Marine and slowly turned his back toward him. The young pilot was not amused.

"What's this?! Are we disturbing you...squid?"

Chuck looked back over the side for one last look at the moon-just to burn in the memory of the previous moment. Then he turned and walked over to the Marines. His hand was stretched out in greeting, but his face was blank and professional.

"Welcome aboard, gentlemen. I'm Chuck Snyder. I don't believe we've been introduced."

The other two Marines smiled professionally and shook hands with Chuck as they introduced themselves. The one pilot that had been so arrogant moments earlier changed his tone. Now, he tried to be both as professional as Chuck, and as cordial as his two companions.

"Hi-ya, Chuck! I'm Lieutenant Codie Custer. Callsign's 'Cowboy.'

Chuck's eyebrows flared in curiosity as he shook the pilot's hand and replied, "Wow. Custer. That's quite a name. Are you related to the general?"

The smile on Codie's face said it all. Chuck had erased the Marine's arrogance and offered to turn it into personal pride for the man. The change should have been natural, but Codie wasn't the sharpest knife in the drawer, and veiled by his grin, it confused him a little.

"Yes I am. Let me tell you, there's nothing but pure-American hero in these veins. I was born at The Little Big Horn, and he fought at the Alamo."

The other two Marines laughed, and Chuck pretended not to notice the fumbled boast. Codie never noticed his mistake, and his friends' laughing confused him. He laughed too, but he had no idea why.

Chuck kept his cool face as long as he could, but seeing the other Marines laughing in his peripheral vision was breaking him down. He was going to start laughing soon, so he tried continuing the conversation to hold off the breakdown.

"Well, Codie. I'm sure I speak for all of the squids on this ship when I tell you that we're thrilled to have you guys on board."

He continued with a sad attempt at mimicking Codie's Texas drawl. "You know, we've got women-folk in this here Navy, and I think each and every one of them's gonna sleep a tide bit better knowing that the United States Marine Corps' very own Codie 'Alamo' Custer is on board to keep them safe." He finished with a smile and a deep look into Codie's eyes.

The other two Marines completely broke down in laughter at Codie.

"It's 'Cowboy,' squid! Not 'Alamo!' I was born at the Alamo, but it's not my callsign!"

The deafening silence of the sea and the laughter of his friends trimmed Codie's whine. When Chuck eventually broke down too, Codie figured out that all three men were definitely laughing at him. He was pissed, and it showed, but there was nothing he could do. Chuck saw it, and decided that it was time to leave. He nudged past Codie and his two friends, then through the hatch and back toward his quarters deep in the ship. The laughing continued in the distance until the dull hum of white noise from the ship's nuclear reactors faded the sound of the Marines.

Chuck made his way through the crowded corridors of the ship. He snaked his way through a group of crewmen that was doing some painting. Then he went down a few more stairwells and down another corridor to his quarters. Inside, his roommate was busy playing video games; flight simulators, of course.

Lieutenant Rick "Ironman" Meyers was glued to his laptop computer. He had been playing the same mission since he arrived, and he still hadn't completed it. Every time he came close to succeeding, he'd try some sort of stunt and get killed. Chuck wondered what kind of a pilot he would prove to be in the real world instead of the virtual world.

Rick Meyers was from the same Marine squadron as Codie Custer. He had come aboard in advance of the squadron to arrange for their berthing and orientation to the ship. When Rick landed, he met with the airwing commander who was very busy and told him to "find someplace to stay and have your squadron commander meet me as soon as he comes aboard."

The GW's airwing commander was a short man in stature and in temperament, but he was large on authority. If he wanted Rick to get lost, then Rick decided to get lost. He introduced himself to the first naval aviator that he met, Chuck, and petitioned him for quarters. Chuck's roommate was in one of the squadrons that were missing, so the two became roommates.

The GW's rush to sea meant that two of the squadrons in its airwing were unable to join the ship. They had been sent to bolster another carrier battlegroup in the Philippine Sea. Both squadrons were of 30+year old F-14 Tomcats.

The Tomcats were supposed to be the GW's airborne umbrella against potential air attacks. The airwing commander shifted one of his F/A-18C attack squadrons to the airborne umbrella role. The other F/A-18C Hornet squadron would have to do double duty in both attack role, and Combat Air Patrol (CAP) i.e.: airborne umbrella roles. A few hasty and curt phone calls to the Pentagon had arranged for the Marine squadron's deployment to the GW as a replacement for the Hornet squadron that had been shifted primarily to CAP. He hoped that another squadron might come aboard later, but forces were being stretched to the limit in America's newest war.

After settling in as Chuck's roommate, Rick Meyers was given a tour of the GW. The inside of the U.S.S. George Washington was typical of PostWorld War II supercarriers. Fully loaded for a combat deployment, the ship displaced almost 100,000 tons of water. If the ship could stand on its stern, it would be taller than the Empire State Building. Most portside visitors didn't even recognize it as a ship from the docks; it was too big to comprehend.

Supercarriers had always been compared to floating cities. Nearly 5000 men and women lived, worked, and fought from a single ship. Each carrier contained several cafeterias, community bathrooms, barber shops, dentist offices, libraries, stores, and even a closed circuit TV station. The big joke among the hundreds of thousands that sailed on them over their half-century of existence was always that the ships had bowling alleys. Some did, but they were always makeshift and never part of their design.

The most important aspect of the supercarriers could be found in their four cavernous hangars and on the 4-acre flightdeck. Though the ships often varied in size by as much as 80 feet and several hundred tons, each one carried roughly 100 aircraft of all shapes and sizes. Half of the ship's 5000-strong crew were actually pilots and aircrew for its airwing contingent. In addition to providing the international muscle that the supercarriers so often flexed, the airwing was also responsible for protecting the GW and all of its escort ships.

Chuck had been on the GW's previous deployment. That 6-month sailing had been extended by 3 ½ months so that the GW could replace the U.S.S. Abraham Lincoln in the North Atlantic. It was his first deployment, and now he was showing someone else around on their first time at sea.

During his first time out, Chuck had come to know the inside of the ship intimately. The notorious winter storms of the North Atlantic had seen to that. Swells as high as 40' tossed icy waves over the ship's bow at times. There was a four-week period where not a single aircraft could launch or land because of either heavy snow, ice on the deck, or poor visibility. When he was able to fly, it was rare that he could see the front of the ship while landing on the back end! A rush trip to the Mediterranean so shortly after coming home from an extended deployment might have upset many crews, but Chuck was looking forward to the warm weather and comparatively placid waters.

Like any new arrival to a carrier, Rick was lost in no time. He learned that there were two corridors running the length of the ship. If he made his way to either corridor, and up to either the hangar deck or the flight deck, then he could find his way back to their room, the "Dirty-Shirt Café" (a cafeteria for naval aviators and pilots only), or the row of squadron ready rooms. With that key knowledge, he could learn about the rest of the ship from wandering. It wasn't that Chuck was a poor tour guide, but Rick wanted to get back to his games.

He was obsessed with computer flight simulators. Rick grew up playing with his family's PC. First he learned how to draw pictures with a mouse, then he learned to play pre-school games. By the time he could read, he had already discovered the joy of the flight simulator, and was mastering games as fast as his parents could buy them. Reading had opened a new world to him, and before he was in Junior High School, he was already learning his third programming language. In High School, he wrote his own flight simulator. College computer classes were a joke to him, and he made a fair amount of spare cash by doing homework for the slower members of his classes.

His career options seemed awesome. He was granted scholarships to almost 20 different universities-including the Air Force Academy and the Naval Academy. With his love of flight simulators, he seemed destined to become a fighter pilot, but the regimentation of a military life wasn't appealing to Rick. Instead, the stereotypical high-life of a skilled computer geek in the 21st Century-especially the financial opportunities-had transfixed him on the almighty dollar. He chose a little-known college in Northern California that specialized in teaching the skills needed to write video games. Then there was the code incident. Two years into his collegiate career, Rick was arrested for almost 2500 counts of software piracy. That incident, and in particular the presiding Federal Judge, changed his life's path.

Back in High School, he had written a video game program that saved information from the last time someone played it. When the same player would try and play against the computer again, the computer would read the information saved from the last time it played that specific opponent. It would then analyze the saved information and choose tactics that were different from those that the player had faced the previous time. It was a primitive form of artificial intelligence, but the more a player tried to go against the computer, the better the computer would play. This also meant that the player had to play better each time.

Rick was too young to know the intricacies of copyright or patent laws. He never even tried to maintain the rights to the software. Instead, he passed it around to anyone who he thought wanted a more challenging flight simulation experience. Through the exponential capacity of the internet, his program spread wildly. Thousands of his fellow computer geeks and geek wannabes soon installed it in their computers.

The trouble stemmed from right in his home. His mother's second husband was living with them at the time. He and Rick got along famously since he worked at a software company. When Rick's mother divorced his stepfather, it was a bitter scene. The stepfather did a complete reversal toward Rick and filed the piracy charges against him as a method of hurting his former wife.

The Federal judge didn't understand the case. He also didn't comprehend the intricacies of what made computer software work. The last nail in his coffin was that the judge couldn't accept that Rick was capable of writing the small program back in High School. Rick was given a choice: military service or 14 years in a Federal jail.

The Navy and Air Force had heard about the court case and were waiting in the courtroom with open arms-too open for Rick. He was still independent enough to resist the drooling dogs of modern war. Instead, he called the local Marine ROTC commander and asked for his help. The Marines were definitely the most stringent of the three services offering flight opportunities, but their professionalism appealed to him in the face of the overeager Navy and Air Force recruiters. Besides, his uncle was a pilot in the Air Force, and he was as stale as a loaf of month-old bread. That was not the lifestyle for Rick!

As soon as he signed, he was given the oath of allegiance. Three days later he was sent to basic training-the same basic training that every Marine (pilot or infantryman) had to go through. Two years after the court case, Rick graduated from his computer college and headed off to flight training. Another year and he found himself landing a brand new F/A-18F Super Hornet on the deck of an aircraft carrier headed for the Mediterranean Sea!

When Chuck came back into their quarters from his peaceful interlude on the catwalk, Rick was in his natural position-at his computer. Most pilots and naval aviators felt more comfortable in their cockpits. Rick wasn't particularly uncomfortable there, but playing flight sims and editing their code was as calming to him as needlepoint, painting, or writing was to others.

Chuck was about the same age as Rick, but the video game fascination still stunned him. He really wondered about Rick's state of mind because every time that he saw him playing games, Rick seemed to be losing. Chuck couldn't understand how someone could be so obsessed with a game that he always seemed to lose.

"Rick, now don't take this the wrong way, but… do you ever win at these games? You always see to be getting killed. In fact, I haven't even seen you survive more than a few minutes. If you fly anything like you do in those games, I gotta tell ya man, I am NOT going up there with you!"

Rick laughed and tapped the 'P' key on his keyboard to pause his current game.

"You're missing the point. Sure I play for fun, but I use these games to try out different tactics. I re-wrote the AI (Artificial Intelligence) for the enemy pilots. The computer saves data from the last time that I played, and then interprets it and re-directs the AI so that it gets tougher and tougher each time. Let's say that I break right six out of ten times after I close with an enemy fighter. The next time

I play the game, the computer's AI pilot knows that I have a 60% chance of breaking right, and it acts accordingly. Now, take that to the 'N'th degree. If the computer detects that I'm flying an F-16 Falcon against an Su-34 with a tighter turn radius at low speeds, and I've already broken to the right, then it'll also determine the best attack maneuver for the AI pilot to counter me and my simulated F-16. It's like I'm fighting the ultimate omniscient bad guy."

Chuck understood. No modern fighter pilot hadn't grown up playing flight sims. While he was impressed that Rick knew how to re-write the game's code so that it became more challenging the more that he played, this degree of difficulty was simple sadistic. It was nuts. If you played often enough-like Rick-then the computer would be impossible to beat. It would just have so much information that it would actually know what you were going to do before you did!

"So, once more, why are you playing this again and again. You're just making it closer to the impossible. You'll never win." Rick's face became unusually serious.

"Well, I'm constantly having to fly and fight harder and harder each time, but that means that I'm learning to fight better and better each time too. I may get killed, but it's better to in the virtual world than in the real world. Right? Besides I've got another idea that I've been working toward. Check this out…"

Rick quit the current game that he was playing and went back to the sim's main menu screen. There, he logged out as the pilot, and typed in "AI5" as the new pilot's name. Then he started another scenario. As soon as he clicked on the "Start Mission" button, the screen went black. Both men sat and looked at the motionless monitor, but only Chuck was confused.

"What happened?"

"The computer's fighting itself."

They exchanged stares, and then Rick explained.

"See, I flew a few missions as a pilot named 'AI5.' Then, once the game's AI program had gotten use to my flying technique, I went back to the main menu and re-classed the 'AI5' pilot as a wingman. That means that the computer had to create a new database for the AI5 pilot and run a new baseline AI program based on the data that I had created under the name AI5. In other words, there's a second artificial intelligence program running-and learning-in the game. If there's no real person in the scenario, then the two artificial intelligence programs fly against each other. They learn from each other's habits, and eventually create an unbeatable bad guy." Chuck shook his head in disbelief again.

"Once more, why are you doing this? I mean, you know right off the bat that you're not gonna win."

"Yeah, but I'll learn, and so will the computer. So, if anyone else ever plays against me online, I'll have already played against the toughest possible opponent. I have to admit, I have been known to login to an internet game or two and make other people fly against my little friend AI5. No one survives, and all it does is piss other people off, but I love it. It's like teaching your kid to be the best football player in the world, and then letting him loose on a field of little leaguers. It's hilarious."

Now Chuck was beginning to understand. Rick's artificial intelligence program was his goal. That's why he played. Most people played sims for the fun, or the experience, but Rick played to perfect his program. He wanted to make it as ornate and intelligent as a masterpiece painting. But Rick was a new kind of painter; someone who's art was bits and bytes, zeroes and ones, on and off switches, micro-circuits and lines of code. His masterpiece was only tangible in cyberspace. There was more.

"AI5's not just fun to let loose on the internet, but there are some real possibilities too. Imagine an intelligent missile or drone that anticipated every move its target might make milliseconds before it happened. What would happen if you had a missile on your tail that knew when you were most likely to drop chaff or flares, and it switched its guidance systems just before you made your defensive move? I'll tell you exactly what would happen, you'd get your ass kicked! That's really what you see happening every time you see me getting killed in these games. Maybe someday, someone will want my AI5 for their missile guidance system? Who knows? By then, the program will REALLY be good!"

The screen came back to life with an after-mission report. AI5 had beaten the computer's AI. In fact, AI5 had been assigned to fly a 1950's era American F-86 Sabre against three state-of-the-art F-22 Raptors. The Sabre was detected early-on, but was able to dodge all of the missiles fired at it. One by one, it closed with the Raptors and picked them off with its miniscule armament of six .50 caliber machineguns. The Raptors hadn't even been able to get a single bullet hole in the Sabre.

Rick read the report to Chuck and explained what had happened. Then he pulled up another screen that showed a 3D trail of how the dogfight had happened. It was a twisting squiggle of lines that only fighter pilots could understand as organized aircraft maneuvers. Chuck was impressed, but the futility of fighting an all-knowing foe still masked any sort of entertainment or satisfaction from him.

Next, Rick opened up a window that showed the game's flight history for the pilot that it knew as AI5. The AI5 had accumulated over 2000 hours of simulated combat-flying time. The average mission lasted only 3 and a half minutes. It had been shot down 318 times, but had amassed a total of 9102 aircraft in almost 700 missions. Those numbers wouldn't seem so impressive, except for that fact that the simulated opponents were all other AI programs that no human being alive stood a fair chance in fighting. Rick smiled as Chuck read the screen over his shoulder.

"I've been letting it play at night while I sleep, and sometimes while I'm in briefings or when I'm flying the real deal. Little AI5's been getting awfully smart. Right now, the database file is almost 100 meg, and it's getting a little sloppy. I need to get a faster memory chip so that it can access the database and make decisions faster. Sometimes it's been getting shot down because it can't get enough data to make a really informed decision."

Chuck could see that Rick's hobby meant a lot to him. Flight sims were more than video games to the average 21st Century fighter pilot, and Chuck wasn't as impressed by them as most, but he could see that Rick was really into his extracurricular activity. Besides, he could understand the reasoning that if the sims were hard enough they could teach him better airborne tactics. That made him wonder what Rick had actually learned from his virtual ultraace program.

"Are there any tactics that you've found help out against your program that might be useful in the real world?"

Rick jumped at the chance to further justify his programming habit.

"Oh yeah! Check this out…" He clicked on a few on-screen menu buttons and opened up a tactics page for his fictitious squadron of virtual AI pilots-including AI5. "This screen shows all my AI pilots that I've had working together. I wanted to see what kind of small unit tactics worked best against the toughest AI programs. I haven't really put it through the paces as much as the individual AI's, but sometime I will. Ultimately, I'd like to take the best AI's and form them into a squadron artificial intelligence program. That program would allow all the sensor data and database intel from all the different AI programs to work together. The programs would act like a pack of wolves. It's almost impossible to fight and win against a single AI, but the chances do get better if I bring along a few wingmen. Now, in a small unit vs. small unit sim, using individual AI programs for the enemy pilots, if I place the individual AI programs and the master squadron AI program for the players, the players will always win."

He called up another screen that showed a few grids and a 3D box on the right hand side of the screen.

"Check this out! I based the squadron AI program on two things: multiple data input sources, and multiple output choices. Lots of data comes in from the individual AI programs in the virtual squadron. This data includes all the sensor data from each AI's plane-just like our planes can download radar data from the carrier or other planes, and then see them on our own monitors even while our radars are off. The individual AI programs also send in their situational assessments. If an AI pilot determines that an opposing helicopter is most likely going to turn to the right based on previous experience stored in its database, then it sends that data to the squadron AI program. The squadron AI program collects all this data, assesses it, and creates what I call a play."

Chuck was really seeing the potential now. The computer could actually create new small unit tactics, try them out, and then flight leaders could find out what tactics are most likely to work in different situations.

Rick continued. "See, first it creates a new file and numbers it. Then the squadron AI program chooses a formation. It numbers the aircraft in the formation. Then, it assigns flightpaths to the different aircraft in the formation based on its assessment of all the data from the individual AI programs and their simulated sensors. Finally, it issues special commands to the different aircraft. This is also based on the assessed data from the individual AI programs."

Chuck was starting to get confused. "Why not just use the same commands that we use in the air? Otherwise, to use the tactics that the program comes up with, we'd have to learn a whole new way of doing things."

"When we're in the air, and I'm the flight leader. If we spot a pair of bandits (hostile aircraft), then I have to radio the rest of the flight like this...Bandits bearing two-nine-zero, angels 10. One has lead. Two cover. Three and Four go to 2 o'clock, angels 25. That means that in our four-plane formation, I'm gonna take the lead bad guy. My wingman will cover me, and our other two planes in the formation are going to go straight ahead, a little to the right, and up to 25,000 feet. Right? Okay. Well, the computer's squadron AI program would just send a single command to the other individual AI programs. That command might be something like '030256.' Instantly, all the other planes would understand what was happening and what they were supposed to do."

Chuck thought the idea of using pre-planned plays like a football team was nice, but not very appropriate to a field of battle as fluid as a modern dogfight. When closing speeds between aircraft and missiles normally break

2500 miles an hour, there was no time to think 'what was play #030256?'

"I dunno, Rick. That's a lot to try and remember in a shootout."

"I admit, there's way too much for a person to remember. Some of the squadron AI's plays are super complex. They involve one plane popping chaff, another dropping flares, a few planes doing both in different combinations, and as many as 18 planes getting orders all from a single number like that 030whatever. Hopefully, a few really good and common plays will stand out, and we can just learn to remember those. After all, how many plays does the average high school football player have to remember and instantly understand? There's probably at least 100. If I can find 100 plays that are common, simple, and proven effective by the squadron AI program, then I've just given us a little bit of an edge in the real world."

Chuck knew he was right, but that was a lot of work to do for a little bit of an edge.

"How long have you been working on this, Rick?"

Rick shook his head and smiled. "I've been playing with flight sim AI's since I first learned to program back in junior high. This particular AI program's only a few months old, and the squadron AI is something that I've dabbled with for years, but only really gotten into over the past few months."

Chuck was impressed. Rick's fascination with flight sims went far beyond hobby. It was enough to be a profession! "Did you ever think of selling any of this? I would have thought that the Pentagon, a defense supplier, or-at the very least-the gaming industry might be interested!"

Rick's smile faded, and he turned back to the monitor. Then he paused and thought about the question for another moment. "Hphh! That's how I wound up in the Corps."

Friday, April 1, 2011

1314hrs local

U.S.S. George Washington CVN

Eastern Atlantic Ocean

Under the hot equatorial midday sun, the GW conducted the last series of training flight operations before passing Gibraltar and entering the Mediterranean Sea. The newly arrived U.S. Marine Squadron, VMFA-251 "Thunderbolts," had been ordered by the ship's airwing commander to conduct a round of navigation exercises along with a few simulated airborne threat intercepts.

The Marines were launched in pairs and sent to different locations around the ship. Each pair was assigned an escort ship to protect. Each of the escort ships was at a different point on the compass and anywhere from 15 nm (nautical miles) to 40 nm from the GW. What they didn't know was that Chuck's entire squadron, VFA-136 "The Knighthawks," had launched two hours earlier.

Besides having the home-field advantage of operating at sea, Chuck and the Knighthawks had the element of surprise on their side. None of the Marines knew what was out there. All they knew was that there were supposed to fly to some invisible coordinates and then take orders from either their assigned escort ship, or the head honcho of all air-defense flight operations, the AAWO (Air-Attack Warfare Officer) back on the U.S.S. George Washington-callsign "Big Man."

Over 200 nm to the east of the GW's battlegroup, The Knighthawks landed at the British airbase behind the Rock of Gibraltar. There, they refueled their F-18C Hornets and waited. The planes weren't equipped with any extra fuel tanks so that they would be as aerodynamically clean and ready to dogfight as possible. Even with the element of surprise, they needed every possible advantage to beat the Marines.

The Marines were flying the F-18F Super Hornet. Almost identical in designation, name, appearance, and mission, the two planes were greatly dissimilar in capability. The Super Hornet looked almost identical to the older Hornets that The Knighthawks were flying, but it was completely different under the skin. Its wings were larger and carried more fuel. The engines were newer, more fuel efficient, had better acceleration, and a much higher top speed. There were far better avionics for detecting other aircraft or surface targets. There was a second seat for a Weapons Officer (WO or "Wizzo") who could focus on identifying targets and guiding weapons, leaving the pilot to focus on flying and dogfighting. Most importantly, the plane had a layer of radar-absorbent stealth coating all over its skin, and the few exterior enhancements had been made in conjunction with the latest in stealthy design. Those minor angular changes reduced the Super Hornet's appearance on radar screens by almost 100%!

The Knighthawks were radioed by the AAWO back at the George Washington. The Big Man told them all that the last of the Marine Thunderbolts had launched, and that The Knighthawks could begin their attack at their convenience. The planes that had been fueled taxied to the runway and began taking off. Once airborne, they circled and joined into a formation. Another twenty minutes passed before the last of the planes was fueled and left the ground to join the formation.

Almost 300 miles away and 30,000 feet in the sky, a U.S. Navy E-2C Hawkeye AWACs (Airborne Warning And Control aircraft) continued to loiter in a 10-mile-wide circular pattern. Mounted on the top of the twinengine turboprop aircraft was a 24'-wide spinning disc. The plane's huge radar and electronic sensory antennae were concealed in the disc. Those antennae and three crewmen in the plane's fuselage comprised one of the U.S. military's most important weapon systems.

They could detect and track hundreds of airborne and surface targets through the plane's complex electronics suite. They could also electronically transmit that data to any American aircraft, ship, submarine, or headquarters on the planet. In return, the plane could also receive the same types of radar and electronic sensory data from other fighter planes, ships, submarines, and even satellites! With the combined sensory knowledge of the entire U.S. military in their three consoles, the plane's crewmen were tasked with directing all airborne intercepts of surface and airborne threats to the George Washington Battle Group. The only people with higher tactical authority were the AAWO, the Battlegroup commander, or the admirals back at the Pentagon. Whenever someone sailed or flew into the airspace monitored by the Hawkeye, they were at the mercy of those three controllers sitting in front of the three consoles.

Protecting the Hawkeye was a pair of the U.S.M.C. F-18F Super Hornets from the Thunderbolts. Two more pairs cruised in formation with the three planes. Those four planes would act as a mobile reserve force for the Hawkeye to send as reinforcements toward any of the patrolling pairs of Marine Super Hornets or to help stave off a large attack on either the Hawkeye or the GW.

When The Knighthawks were all airborne and organized, they were still over Gibraltar. The E-2C Hawkeye was hundreds of miles away and almost six miles in the sky, but it watched their every move. All four of the reserve F18F Super Hornets flying were sent to intercept the Navy's F-18C Hornets. The Knighthawks had conducted endless exercises and actual operations in conjunction with the Hawkeye, and they knew that they had been detected. This was going to be a stand-up fight.

The only people who would be even slightly surprised were the Marines. They still thought that they were alone in the sky with only the Hawkeye to keep them company. For all they knew, they were just going to be ordered to intercept fictitious enemy aircraft.

The Navy's Knighthawks circled and spiraled their way high up into the sky. In a dogfight, speed kills. Whoever has the most speed and the tightest turning radius tended to be the victor. Despite the presence of two General Electric F404-GE-400 turbofan engines with a combined thrust of over 32,000 pounds, the best way to get speed was to fall-or dive. That meant that to get into an optimal fighting position, the Knighthawks would need every foot of altitude that they could scrounge.

The twelve planes leveled off at 45,000 feet. Then, they broke into three four-plane formations and headed straight for the George Washington. The lead formation kicked in their afterburners and began burning fuel at 12 times the normal rate. The second flight accelerated to full power, but didn't kick in their afterburners. The last flight would come in at a leisurely 500 knots.

Coming straight for the Navy's planes, the four Marine Super Hornets wanted to intercept their quarry and show how fast they could move from one patrol location to another. They lit their afterburners and raced for the intercept point that Hawkeye had directed them towards. The closing speed between the four Marines and the four naval aviators in the lead formation from The Knighthawks was almost 2500 miles an hour! (high-powered rifle bullets travel at about 1500 miles an hour.)

A little more than 10 minutes after leaving Gibraltar, the Hornets and Super Hornets spotted each other. All the planes had turned off their radars. The Marines didn't think there were real adversaries in front of them to detect, and The Navy's Knighthawks didn't want the Marines to detect them anymore than the Hawkeye already had. Not even two seconds after spotting each other, the Hornets and Super Hornets passed each other with less than two miles apart.

The Navy's Hornets never flinched and just blew right by the Marines. The Marines broke into two pairs. One pair banked and turned to the left, and the other to the right. When they were almost in position to give chase to the Navy's Hornets, the second flight spotted the Marines and lined up to shoot them down. Once the naval aviators were locked on to the Marines, they popped flares to signify air-to-air missile launches and declared that the Marines were dead. In reality they were very very unhappy, but they were alive. Had it been a real engagement, the Marines would have been blasted from the sky.

From the E-2 Hawkeye came an unintentionally sensual voice from one of the controllers. *"Thunderbolts one, two, three, and four, you have been killed. RTB (return to base i.e. the U.S.S. George Washington). Thunderbolt one-seven and on-eight, you have traffic inbound at angels four five. Engage at will."*

Chuck laughed in his oxygen mask. He was flying in the last formation of F-18C Hornets from The Knighthawks. He knew that the E-2 had watched them from the moment of their take-off, but he'd never heard the new controller's voice. Her phone sex tone had to be incredibly insulting to the Marine pilots that had been shot down by the second formation from his squadron. He glanced to his left and saw his squadron commander, Commander Don "Dutch" Shultz. It didn't look like he was laughing.

Dutch had a reputation as a cold leader. He was extremely detailoriented, had an aptitude for monitoring several different situations at once, and seemed a bit on the snobby side to some. He was a tall and slender man of poor posture and average looks. He spoke with a somewhat delicate or even feminine tone. Some people mistook his sexuality and misjudged him greatly. While he did prefer to read magazines about interior design and woodworking, Chuck knew him to be a gentle and purely cold man. His wife, a well-known New York ballerina, must have known him in a different light.

Chuck's glance toward Dutch must have sent psychic signals to his commander. Dutch looked back and shook his head at him. The message had no words, but it was clear. Chuck's attention had drifted, and Dutch wanted him to return his focus to the mission at hand-specifically the two new Marine threats that were coming toward them at 2500 miles an hour! His focus returned.

The lead flight of Hornets that had blasted past the first flight of Marines by using their afterburners decided that they needed more speed. Their mission was to simulate an attack on the GW. Normally, the escort ships patrolling the area miles around the carrier would have been a major problem for any incoming aircraft or missiles, but they had been ordered to stay out of the exercise so that the Marines could get some practice. That meant that the only real threat to the Navy's Hornets was the Marines' Super Hornets. To continue with their air combat edge, the lead flight began to increase their speed beyond simply using afterburners. From 45,000 feet, they began a long and shallow dive. Their speed grew rapidly to over 1200 knots.

They kept their radars off to make detection a little more difficult for the Marines. Instead, they homed in on the GW using its Instrument Landing System (ILS). The same electronic emissions that would guide them in toward a successful landing on their ship would guide them in to target the ship. Best of all, there would be zero margin of error. The ILS was designed to bring planes on to the ship in zero-visibility within inches of the optimum landing spot (the #3 landing wire).

The same voice that had announced the simulated deaths of the first four Marine pilots-and their backseat Wizzos-now ordered the rest of the Marine squadron to leave their patrol stations and attempt to intercept the diving Navy F-18C Hornets. There was no way it was going to happen. For the Hornets to successfully complete their mission, they only had to get within 20 miles of the carrier. At 20 miles, a real attack aircraft from an unfriendly nation would be able to launch its anti-ship missiles and head for home. Once those missiles were launched, there was no reason to attack the airplanes, and the Marines would have lost their little wargame. Given the rate of their increasing speed, the distance between the first flight and the nearest Marine Super Hornet, and the closing rate of all the planes, there was no way to stop the Navy's first four planes.

It didn't take long for the controller on the E-2 to figure this out. She changed her orders and told all but the nearest two Marine fighters to change course again. Now, they were to head for the second and third flights of Navy Hornets. The Navy was going to get their first four planes to the 20-mile mark, but if the Marines followed her orders and performed as expected, then there would be no more successes for the Navy's Knighthawks. From five different points on the compass, pairs of Marine Super Hornets turned and accelerated toward the second flight of Navy Hornets.

The second flight of Navy Hornets lit their afterburners and began to dive toward the GW. They were going to make a run for the 20-mile mark too. In a few seconds their speed passed 1000 knots (almost 1200 mph). The closest pair of Marine Super Hornets turned on their radars and immediately locked-on to the Navy planes. A series of flares were fired from the Marine planes, and the Hawkeye controller came on the Navy radio frequency.

"Knighthawk five, six, seven, and eight, you have been killed. Repeat, you have been killed. Descend to Angels 3 and RTB."

Dutch knew that if the planes had been destroyed and there had been no radio conversation to indicate a dogfight, then they must have been lost to a simulated air-to-air missile attack. He motioned to Chuck and the other two remaining Knighthawks. They were to turn on their radar jamming devices, drop to 1000 feet (the lowest allowable altitude for the exercise), and accelerate to the 20-mile mark. All three planes wagged their wings in acknowledgement and then dove steeply form 40,000 feet to 1,000 feet. On their way, they lit their afterburners and in an instant passed through the sound barrier. A few seconds later, and they passed mach 2.

As each of the four planes began the monstrous effort to pull out of the dive, their radar warning indicators began to light up in their cockpits. The Marines were searching for them, and the sophisticated radar detectors told them that the nearest Marine planes were coming from their forward right-to the northwest. The radar warning systems also told them that the Marines were almost within missile range. For good measure, Dutch popped out a pair of chaff bundles.

The little canisters of chaff would show up on the Marines' radar screens, but they would look like small fuzzy clouds instead of definitive targets. That's almost what they literally were. Each bundle was shot from the back a Hornet by a small explosive about the size of a shotgun shell. When the bundle was blown off the airplane, it exploded behind it. The bundles were just long thin fibers of plastic with a radar reflective coating. When scattered into the wind, the Hornet's wake, and by the bundle's explosion, the fibers were too small to be seen by the naked eye, but to a radar, they looked like either millions of small targets or one huge target. Either way, those four chaff bundles confused the Marines' radars for just another second.

In that time, Dutch, Chuck, and their two wingman were able to travel another mile. As they drew closer to the invisible 20-mile mark, the miles were important, but the time was even more so. The mathematics of the situation were that in less than 60 seconds, the Navy could win or lose the exercise. The same was true for the Marines.

"Knighthawk one-one, you have broken the exercise hard deck. You are destroyed. Climb to angels 3 and RTB. Repeat, Knighthawk one-one, you have broken the exercise hard deck. You are destroyed. Climb to angels 3 and RTB."

The radio message from the E-2 controller was ominous. Chuck's wingman had slipped under 1000 feet, and he was now considered dead. Dutch wasn't pleased. Even with the oxygen mask on, all three of his pilots knew it. It might have been a knot of turbulence in the air. It might have been pilot error. It might have been an error on the part of the E-2 controller. None of that mattered.

What did matter was that only three planes remained in their squadron. According to the rules of the exercise, at least half of the squadron had to make it to the 20-mile mark to declare victory. The first four planes had made it, so the Navy wasn't going to lose, but if at least two more planes didn't make it to the invisible line in the sky, then the exercise would clearly be a draw.

The white plasma plume from their afterburners stretched as far behind as their Hornets as the planes did in front of the flames. Their engines were pouring out fuel as fast as it could be done, and their turbines were spinning at their redlines. Every iota of speed and power was being used.

Dutch could sense that the closest Marines were probably about to lock on to their planes. He was about to pop out another bundle of chaff when his radar warning system began to alarm again. More Marine Super Hornets were closing in from straight ahead to the west, and another pair from his forward left to the southwest.

Dutch popped out two more chaff bundles. Chuck and his wingman saw the bundles ejected through their peripheral vision, and they did the same. As they had done before, the small clouds confused the Marines' radar, but this time, it had less of an effect. It also allowed the Marines to determine where they were. By looking at their radars and imagining a course from the first set of chaff clouds to the new ones, the Marines could guess with great accuracy where the Navy pilots were. There was far less confusion, and the Marines continued to bore in on the Navy Hornets to block them from reaching the 20mile mark.

Rick "Ironman" Meyers was leading the flight of Marine Super Hornets that was headed straight for the Navy Hornets. By looking at his radar screen and determining where the Navy planes were, he figured out when they would begin to merge or pass each other. He and his wingman were both flying with their afterburners on, but they were still in level flight at just over 5,000 feet. When Rick decided that they were about to come within visual range of the Navy planes, Rick and his wingman began a pre-planned maneuver. Both planes began to roll as fast as they could. Rick spun his wings clockwise, and his wingman went counter-clockwise. They also began to pop out thirty decoy flares from each plane.

When they were in visual range of each other, Dutch, Chuck, and his wingman saw what looked like two sparking egg-beaters coming right for them. Flares were going in every direction. Dutch knew that the flares couldn't be simulated missile launches, but Chuck's wingman recognized them as just that. He broke formation, popped out a pair of chaff bundles, and climbed frantically. He climbed straight toward the two Marine Super Hornets that were closing from the northwest. Not even a second later, there was a pair of flares high in the distance, and the sexy sounding E-2 controller announced the Navy pilot's demise.

Dutch and Chuck knew that the two Marine Super Hornets ahead of them were just trying to confuse them, but the flares and the violent roll rate of the planes forced them to climb and avoid a possible collision. Their speed brought them a mile into the sky in almost no time. That altitude also drained a few hundred knots of airspeed.

Rick and his wingman passed underneath with lightning bolt speed. A few miles behind the Navy Hornets, they stopped rolling and popping flares. Both pilots slowed down, turned to their left, and headed for Dutch and Chuck. The pilots who had just shot down Chuck's wingman dove and headed for them too.

Dutch finally picked up the GW on his radar. He locked on and popped out a flare in a simulated attack. Chuck did the same, then he listened while Dutch reported it to the E2 controller. She waited a long second, then replied. *"Knighthawk flight, confirmed kill on target. Knock it off and RTB. Congratualtions Navy!"*

The Marines had lost, but it was very close. Dutch and Chuck throttled back and slowed down to land on the GW. They were the last from their squadron to land, and everyone was waiting. They thought that it was a victory, but Dutch wasn't very happy. They had come too close to losing, and he let everyone-even Chuck-know that he expected better from them. He was actually disappointed.

After a short frowning speech, everyone headed for the squadron readyroom for the official debriefing. They were all covered with sweat and flight gear, so some stopped by their quarters to change. Normally, they kept all their flight gear in their locker room, but The Knighthawk's prep room-or locker room-was being re-modeled.

The airwing commander waited patiently and silently in The Knighthawks' ready-room. He sat in an ornate wooden rocking chair that Dutch had made. There were similar chairs in Dutch's stateroom and in the ship's Captain's lounge. Each was a further attempt by Dutch at improving his hobby's skills, and the best chair was in his stateroom. They were known throughout the ship as symbols of high stature, and the airwing commander seemed to be rocking in his throne.

Dutch was one of the first people to get into the ready-room. He didn't bother to change to clean up. When he arrived, he and the airwing commander spoke privately in a corner for a few moments. Then the airwing commander sat back down and returned to rocking while Dutch stood at his side. They both waited for the rest of The Knighthawks and for the Marine squadron, The Thunderbolts.

After almost twenty minutes, the Thunderbolt squadron commander came in. Just like Dutch, he privately met with the ship's airwing commander, and then they both spoke with Dutch. During their subtle meeting at the front of the room, pilots, wizzos, and naval aviators filled the large cushy chairs and lined the rear of the room.

The Marine squadron commander motioned for the last Marine into the room to close the hatch, and then the low-level mutter of conversation faded into total silence. The airwing commander sat rocking until the only sound in the room was the creaking of the chair. Then he stood up and began a chastising tirade that none of the naval aviators had seen before. The Marines hadn't heard the likes of it since their initial boot camp training, and no one understood all of the swear words that were used. The airwing commander actually managed to curse in no less than five different languages!

It continued for over an hour. He reviewed each and every airplane's activities. There was something nice to be said for each person. Then he was able to complain about at least three different things for every pilot, aviator, and wizzo too. It was a dressing-down that covered everything from a person's appearance in the briefing to the order in which they worked the different modes on their individual radars. When the airwing commander finally finished, there was nothing left for either squadron commander to add. Both simply frowned.

Technically, The Knighthawks won the exercise, but the Thunderbolts were able to them. None of that mattered to the airwing commander. In his mind, the U.S.S. George Washington was steaming into Harm's Way, and the people under his command were unable to protect it. He worried about what they might face on their trip into the 'Med.

There was trouble all over the world, and the 'Med seemed like one of the few quiet spots. Still, he worried. France, Spain, and Italy were secure and stable countries, but that's where the safety ended in the great inland sea. Greece and Turkey were both NATO members, but the history of conflict between the two could be traced back across thousands of years. North Africa and the Middle East had always been volatile, but since the Iranian Islamic Revolution of 1979, The 1991 Gulf War, and the 2001 War on Terror, they were a powder keg waiting to explode. No two countries combined were a real threat to the GW, but if all the Islamic countries ever did join together in the much-vaunted and long-awaited Jihad, then the GW would be a big fish in a small pond. He worried a great deal.

The ship was heading into the tight confines of a body of water where the escort ships would have to operate closer to the carrier. In the great expanse of the Pacific or Atlantic Oceans, they might be 30 nm from the carrier, but in the Med,' they would have to be 10-20 nm away. That meant any incoming air threats would have less distance to cross between the escorts and the carrier. Less distance meant that there was less time and fewer opportunities to intercept any threats. Worst of all, the airwing commander knew that he didn't have enough planes to do the job, and the Marines that he brought on board to try and do the job simply weren't ready for the challenge.

While he rocked in his chair, the silent and crowded room contemplated the same thoughts. Everyone knew the situation. Everyone knew the dangers. While everyone also hoped to magically acquire the skills needed to protect their ship, some of the escorts were already passing Homer's 'Pillar's of

Hercules'-Gibraltar-and entering the Med.'

Friday, April 1, 2011

The New Cleveland Press

Macedonian Liberation Front Seizes Town in Yugoslavia

While conflict seems to be raging all over the world, the flash-point region of the late 20th Century is starting to smolder once more. The last remaining republic of the former Yugoslavia, Macedonia, has become the object of a massive terrorist attack. Reports from European peacekeepers indicate that the border village of Lasnjovica was entirely taken over by as many as 100 terrorists from the Macedonia Liberation Front based in Northern Greece. Details remain loose and without verification. The only certainties are that the Macedonia peacekeeping force is mobilizing for an attack, and local authorities have denied access to the village.

While Lasnjovica has little strategic importance, this attack is seen as a confirmation that the Macedonia issue that flared near the end of the Kosovo Conflict of 1998 has not faded away completely. Macedonia's large Greek population, its Orthodox Christian heritage, and its historical ties to Greece have left many to believe that the two countries should be re-united. Serbian leaders are strongly against losing their last fellow republic in the former Yugoslavia.

Leaders in Kosovo, Bosnia-Herzegovina, Croatia, and even Albania also prefer that Macedonia remains part of what was once Yugoslavia. Most feel that if it were allowed to stand alone as a separate nation, it would be too weak and a regional power vacuum might invite larger nations into the area. Greece is one of the nations that people are specifically worried about.

Added to the danger that the area poses is the Turkish possibility. Turkey and Greece are both still at odds over the 30-year question of sovereignty over the island of Cyprus. While many Greeks would like to re-unite with their pre-World War II province of Macedonia, Islamic Turkey opposes the potential spread of Orthodox Christianity into the long-oppressed, smaller, Islamic countries like

Kosovo and Bosnia. A direct move by Greecewhile unwarranted and unlikely-would trigger a yet another popular religious war between the two nations.

World War I began with the assassination of Archduke Ferdinand in the Balkans. World War II saw the shattering of nations and the resulting rise of nationalism in the Balkans. The Cold War and the Soviet occupation saw those same nations pinned in poverty and infected with yet another idealism. After the fall of the Soviet Union, and the subsequent disintegration of its former nation statesincluding Yugoslavia-all forms of civilization fell apart in the Balkans. Once more death camps, rape camps, mass executions, and ethnic cleansing ran rampant. Now, with the seizure of a single village, the world is being reminded exactly how tenuous the situation is in this small and critically important province.

Saturday, April 2, 2011

0513hrs local

U.S.S. George Washington CVN

Western Mediterranean Sea

Chuck had managed to drag Rick away from his computer briefly for some dinner. They both had CAP duty in the morning, but sitting in the small stateroom was monotonous after only a few minutes. Rick didn't have too much of a problem with it, but as was the case on his last cruise, Chuck found himself either in one of the hangar bays or on the side of the flightdeck. He just couldn't stand being cooped-up.

Since the airwing was shorthanded, everyone was ordered to at least stay in their flightsuits even when they were off duty. If any air threats were detected, the pilots and naval aviators would be able to save precious minutes by spending less time prepping for flight. Anti-ship missiles and aircraft normally flew at 500-1500 knots, and would most likely be detected within 100 nm. That meant that if threats did come at the GW or its escorts, there would be only a few minutes to try and shoot them down. Clearly trouble was expected.

Rick continued-as always-to babble about his flight sim experiences and his most recent programs for them. His quest for the ultimate computer AI player seemed like it would never end. Rick had also re-named his favorite AI program. Now, instead of calling it AI5, Rick was referring to the program as "Bob."

Chuck tried to pay attention, but they had only been together for a few days, and he was already getting tired of the flight sim topic. It was interesting, but Chuck's glass was about full of computer conversation. He was almost to the point of talking about politics, religion, or anything that would ignite a two-way conversation.

That's when Codie "Cowboy" Custer walked in. He was full of piss and vinegar. There was a line of people waiting to order their food, but Codie managed to drive most away with his obnoxious statements about the food's quality. He had the audacity to claim that Marine MRE's (Meals-Ready-toEat, i.e. field rations for troops) looked and smelled better than the private Navy cafeteria burgers. The sailors and chiefs behind the counter weren't impressed, but a few of the Marine pilots and wizzo's laughed heartily. He drew plenty of attention, and spotted Rick and Chuck sitting together.

He feigned discreetness and called to his friends over his shoulder. "Check it out, you guys. It looked like that squid we met on deck the other night's trying to learn our tactics from Ironman." Then he turned, reached over the counter to help himself to a plate of burgers and fries, and walked over to join them.

"Hey Ironman, what's going on? 'You telling these squids how to fly?" He turned to Chuck and continued. "What's your name again? Fly, no-Spider right? Look Spiderman, if you wanna know how you guys could've done better yesterday, then you're gonna have to ask the best. That's me. Ironman's good, but he's a little computer-happy at times. I think he spends too much effort watching his screens and not enough time with the stick. Don't get me wrong, he's good, but I'm the best. How can I help you?"

Rick was embarrassed by his fellow Marine's behavior. He wasn't typical of the squadron or the Corps, and he was giving both a bad name. It wouldn't have been as bad if there wasn't a taste of truth to the cafeteria banter. Rick believed that he did spend too much time working his plane's avionics, and as far as Codie's flight skills were concerned, he was at least one of the best.

Chuck looked across the table. They had everyone's attention in the small cafeteria. Rick was red-faced and quiet. Codie was spread over a chair that he had turned backwards and was sprawled all over. He glanced into the eyes of both Marines for a long second, then he replied.

"Look Alamo, uh…I mean Cowboy, we don't really call each other by our call-signs too much outside of the cockpit. I can't speak for Rick here, but you can call me Chuck. As far as the flight lessons are concerned, I don't think anyone in the entire Navy is going to be asking some Top Gun wannabe like you for help. Besides, we may not have kicked your ass the other day, but at least our squadron was able to achieve most of its objectives. I know I made mine. Did you?"

Codie was infuriated at the public humiliation. As soon as Chuck had called him Alamo, he could hear others telling the story of how he had verbally jumbled his words the other night. That further burned into his ego.

"Well, Chuck, I thought we got most of you guys. I know I bagged two of you myself. Were you one of the first four that came in balls-out with nothing but speed and prayers to keep you from getting picked up and pasted, or were you the one with that fag squadron commander of yours watching over you on your way in?"

Codie's homophobia was apparent. Dutch was a gentle and sometimes delicate-sounding officer. He was definitely not a wild womanizer, but everyone who knew him knew that he was not homosexual. Even if he were, anyone who ever had to take orders from him or was dressed down by him knew that it would have been irrelevant. Dutch was possibly the coldest and hardest man Chuck had ever met.

"Codie, I won't ask, and I won't tell about you or your preferences, okay. Yeah, I was the one that came in with Dutch, our squadron leader. Where were you? The only Marines we saw were Rick and his wingman. They were the ones that almost blew it for us. Wait a minute…you must be Rick's wingman, right? That makes sense, since he clearly has a better grasp of tactics."

Codie was pissed and leaned forward to get closer to Chuck's face. It was a pale attempt at intimidation.

"Look squid, I was the guy that took out half of your second flight. Ironman, Rick for you, did a nice job of scaring you guys on your way to the 20-mile mark, but I'm the guy that was comin' in on your six o'clock. Another second or two, and I would've popped a flare right up your ass!"

He took a primitive chomp out of one of his hamburgers and spit it out. Then he turned over his shoulder and aimed a shot right at the enlisted cook that was flipping burgers behind the counter. "This tastes like crap! I see you flippin' these things, but are you letting them cook at all?!!"

The officer in charge of the cafeteria came out to see what the commotion was about. He was a Lt. Commander and out-ranked Codie, but that wasn't going to deter him from his tirade. Chuck had gotten him so mad that nothing was going to stand in his way-almost.

"Is there a problem here, Lieutenant?" asked the Lt. Commander. "Because if there is, you can take it up with me. In fact, lemme get you another burger." He reached down to the tray and picked one up. As he handed it over the counter, he pulled back and examined it more closely. Codie thought for a second that the Lt. Commander was about to see how raw the meat was and join him in his rant about the food quality.

"I'm sorry Lieutenant. You're right. The cook hasn't been serving you fully prepared burgers."

He un-wrapped the fast-food style package and lifted the bun from the burger. With a loud snort, whisp, and spit, he pewced on the paddy. Then, he wrapped it back up, handed it to Codie, and cleared his throat. Everyone was silent.

"Here you go Marine. I'm sorry about that. Try this one."

Codie's face went ashen. His eyebrows lowered in disdain for the Navy as a whole. The Lt. Commander stared blankly into his eyes. There was a feeling of a bar-room brawl about to begin.

The Lieutenant Commander added to the fire's fuel. He seriously wanted to see Codie eat the contaminated burger, and he was standing on his rank to see that it happened.

"C'mon Marine. I insist. This burger'll be a lot more satisfying than the one you spat out. It's much more worthy of a Marine Lieutenant like yourself."

Codie held the defamed burger in his hand. He examined it, then stared back at the Lieutenant Commander. Finally, he dangled it between his index finger and thumb, and he dropped it on the floor. With a final glare at Chuck and Rick, he walked out of the Dirty-Shirt Galley.

The room's low-level drum of conversation returned instantly. The Lieutenant Commander patted each of his staff on their backs, and walked back into the kitchen smiling. Rick, like the rest of the Marines, shook his head in denial. Chuck stared at the hatch that Codie had left through, then he gently closed his eyes and put his head down.

He knew that Codie was more than just a jerk. He was one of eighteen Marine pilots that would protect the GW, its escorts, and even Chuck's life if they ran into trouble. The thought of his life in Codie's hands was too scary to deal with, and he was trying to ignore it.

Sitting in the back corner of the room, with his back to everyone, Dutch had remained quiet and uninvolved. There was no ill-feeling about Codie's homophobic and personal remarks, but his thoughts about safety were the same as Chuck's.

Saturday, April 2, 2011

United Media Network

Macedonia About To Fold

European peacekeeping forces entered the town of Lasnjovica during the night, and their findings have the entire region in an uproar. Lasnjovica had been allegedly taken over by the Greek-backed Macedonian Liberation Front terrorist organization. Local authorities spent hours preparing for a military assault on the town, but when they finally arrived there, it was empty-save for at least 800 bodies. The terrorists reportedly left a video of their execution and a message for the Macedonian authorities, but it has not been released to the press.

Reaction around the globe has been swift, but predictable. While the executions were condemned across the political board, a few countries did express an understanding that may even veil their support. Greek officials in particular have speculated that the town may have been harboring Muslim extremists bent on conducting terrorist attacks in Athens.

The idea of terrorists fighting terrorists has international appeal, but has also created a split in the global condemnation-particularly from predominately

Muslim nations in the area. Kosovo, Bosnia, Croatia, Albania, Turkey, Syria, and most of the Middle Eastern nations have expressed frustration at the continued aggression against Muslims in the Balkans.

Popular sentiment echoes the official responses from those same governments. Crowds immediately took to the streets in Tripoli, Cairo, Damascus, The Gaza Strip, and Istanbul. In Kosovo, British peacekeepers had to be called in to break up a crowd that was threatening the Greek consulate. Already, some Mullahs are calling for a Jihad against Orthodox Christians.

Saturday, April 2, 2011

British Media News Service

War Fever Spreads to Europe

One of the gaps created by the shift in American carrier battlegroups has left the entire Mediterranean Sea without significant U.S. naval presence for several days. Marauding forces are emerging from several places in the former Yugoslavia, most likely Serbian backed, and they are allegedly attacking U.N. peacekeepers in Bosnia and Kosovo. Regular Serbian army units are moving into Macedonia under the pretense of putting down a possible insurrection. NATO and U.N. military response to the multitude of actions in the region has been completely defensive and poorly coordinated between nations.

At sea, most of the U.S. and British submarines operating between Vladivostock and Vietnam continue to maintain radio silence. Chinese and North Korean media reports claim to have sunk 18 submarines since the unrestricted submarine warfare erupted in the region. Intelligence reports indicate that as many as 37 more communist ships have been sunk since the first days actions. In addition to the large number of ships lost at sea, a lone British submarine was able to penetrate Hong Kong harbor and sink an additional 14 ships. The confined space and the sinking ships made it impossible for the submarine to escape. While trying to get out it was damaged, forced to surface and surrender. The crew and officers are being paraded on TV for all the world to witness.

Saturday, April 2, 2011

2nd Lt. John Chamberlin-Personal Log

Hill 1419, 15 miles S of Taipei, Taiwan

Chuck downloaded some news footage of captured British sailors on display in Hong Kong. They looked pretty beat up. I'm reminded of the people we left behind in Korea, and the wasted TRAP mission that got us into this war. I know it wasn't the mission, LZ Delta, or any one thing that rolled us into war. If I have to blame anything or anyone, it's fate. Only fate, or God could have orchestrated such a twisted chain of events.

Hector got word from Col. Ghetty that a heavy weapons platoon will be joining up with us over the next few days. We should be getting some mortars, heavy machineguns, maybe even some anti-tank and anti-aircraft teams. We sure need it. I'm a little concerned about getting strafed again.

Sunday, April 3, 2011

2nd Lt. John Chamberlin-Personal Log

Hill 1419, 15 miles S of Taipei, Taiwan

This morning, Colonel Ghetty came up and gave the entire company a big briefing. It appears that there's trouble brewing all around the globe. The U.S. military is being stretched to its limits, and the rest of the world knows it. There's no way to hide it. In the Middle-East, Iraqi forces have stepped up their probes of the U.N. No-Fly zones. Iranian military units fired on a French freighter passing through the Straits of Hormuz; the entrance to the Persian Gulf. There have been some air skirmishes in over the Mediterranean between U.S., British, French, Greek and Italian aircraft on one side, and Libyan, Syrian, Turkish and Algerian planes on the other. It wasn't a single big dogfight, but it sounds like everyone's patrol planes have widened their rules of engagement. On the Balkan Peninsula, terrorist attacks and sniper activity are increasing everywhere. No one really seems to know what's going on inside Russia or any of its republics. There's so much activity, and so little coordination, that Colonel Ghetty says it's like large-scale anarchy.

We were all starting to think about how lucky we are to be on this mountain, and I think that's exactly what he wanted. After about an hour's worth of doomsaying, Ghetty paused and began his briefing about our situation. The fleet has been forced back into the Philippine Sea, behind Taiwan. The only air cover we can count on is from a Marine air group of Harriers back at the Choo Beachhead. Artillery support is being ordered only with approval from division HQ. That means we might as well forget it. By the time anyone at division HQ gets the word from us, the fight will be long over. We were originally supposed to be guarding a Taiwanese artillery unit that left a few days ago. Now, we're supposed to stay here and keep the road on our right flank from falling into Chinese hands. We haven't seen any unusual activity close by, other than the slowing of reinforcements moving from the beachhead to toward Taipei. Colonel Ghetty says that we might soon.

Tuesday, April 3, 2011

American Internet Media Services

British Peacekeepers wounded in Cyprus Mortar Attack

United Nations peacekeeping forces have been on the island of Cyprus since the mid 1970's, and their mere presence has been deterrent enough for the island's extremist citizens of both Turkish and Greek descent. There have been instances of tension and even violence during the U.N. presence, but the cease-fire line patrols went largely un-scathed until yesterday.

While conducting a routine patrol between Greek and Turkish neighborhoods, a British Army unit came under well-directed mortar fire from the Greek neighborhood. Three U.N. peacekeepers were killed, and nine more were wounded-two seriously. A reserve force raced to the scene and was met by automatic weapons fire from the Turkish neighborhood.

When the British were finally able to extract their wounded and recover their dead, a larger force-almost the entire contingent-went looking for the perpetrators. All that was found were some empty bullet casings on the Turkish side of the cease-fire line.

British officials report that the lightly wounded were immediately stabilized. The more seriously wounded were then flown back to England for further medical treatment. Both are reported in serious condition.

In response to the attack, Britain has sent a small task force to the island. Code-named Task Force 5, the flotilla consists of a destroyer, two frigates, and three troop ships with over 1100 Royal Marines on board. They are expected to arrive off the coast of Cyprus within the next three days.

Turkish and Greek officials had identical responses to the attack. Both sides denounced them, but both privately expressed an understanding of the escalation of violence. There is a growing undercurrent of discontent with the U.N. presence on Cyprus. The U.N. would prefer not to be there, while Greece and Turkey seem to be mutually in favor of settling the island's affiliation once and for all. This situation has become further exacerbated by the tensions on the Balkan Peninsula over the Greek-sponsored terrorist attacks in Macedonia.

There is very little talk of peace, and even less genuine belief that diplomatic talks could settle the issue. Even in a time when the entire world seems to be in conflict, it appears as though both sides are walking toward war with very little reluctance.

Tuesday, April 3, 2011

American Internet News Service Not in MY Shopping Mall!

Israeli government officials report that the Bat Rabin Shopping mall was attacked just after 11:00am local time in Jerusalem. At least eight Palestinian gunmen jumped from two stolen cars and ran into the central food court. Israeli soldiers on security duty stopped two, but were then killed themselves. Immediately, the gunmen closed off the exits to the mall and began executing civilians. At least 43 people were killed and over two hundred were wounded. Among the dead were six Americans; two from New York City, and four from Detroit, Michigan. The State Department has not issued names yet.

In retaliation for the attack, Israeli warplanes attacked suspected Palestinian terrorist targets on the West Bank. Several F-16 fighter planes and Apache gunships were used in the attack. There is no official word on Palestinian casualties. Israeli officials claim that the strikes were surgical and only a few known terrorists would be among the dead, but reports from Palestinian sources say that hundreds of civilians in nearby homes have been killed or injured.

Tuesday, April 4, 2011

American Internet News Network NATO Strikes the Balkans Again!

In only its second major use of NATO military force, American, British, French, German, and Italian aircraft are once more bombing in the Balkans. The first major use of NATO military force was during the Kosovo Conflict of 1999, and this time the focus of NATO airpower is on Kosovo's neighbor, Macedonia.

Years of unrest and terrorist attacks were building to a crescendo when the NATO hammer was swung. Unlike most previous military assaults, there was no warning or political brinkmanship preceding today's strikes. Military officials claim that the element of surprise was the key to their surgical missions against Greek-backed terrorists in the small country.

There is no word on casualties. However, reports indicate that the bulk of the airstrikes employed Global Positioning System-guided munitions against urban targets. Typically, urban targets have the highest collateral damage, and precision-guided munitions-however accuratehave never been able to avoid civilian deaths.

As expected, Greek diplomats had not been informed of the attacks despite their NATO membership. Their responses illustrate a sense of shock and deep-seated anger within the Hellenic government. Turkish government officials applauded the action and said that "…terrorism knows no religious boundaries, and it will be fought regardless of location, affiliation, or religion."

Monday, April 4, 2011

2nd Lt. John Chamberlin-Personal Log

Hill 1419, 15 miles S of Taipei, Taiwan

Since we've been up here, Hector has been holding morning briefings for Bruce, myself, and the platoon sergeants of all three platoons. Today he let us know that Colonel Ghetty wasn't just pumping our morale yesterday. Hector pulled out a small palmtop PC that someone had brought up from the beachhead. The palmtop was connected to the U.S. Military Command, Control, Communications and Intelligence Digital Network (U.S.M.C.C3IDNalso called the MilNet).

On his palmtop, he showed us how he can access all kinds of information, including enemy locations, and various friendly unit locations depending on access level. Of course, Hector's clearance isn't that high, so most of the units he showed us only showed us that a unit was in a location, but didn't describe the unit. For example, we could easily see the frontline by looking at the positions of friendly ground units, but we couldn't see the name or composition of any unit except for our three platoons and Colonel Ghetty's command post. It was unnerving to know that anyone with access to the MilNet could see all this information, but it was also comforting to know that, as long as Hector kept updating his palmtop, the higher-ups would know what kind of situation we were in. Maybe artillery support from division HQ won't be as hard to get as we thought?

Hector's little digital wonder doesn't just show us where the friendly units are either. He showed us where Chinese units were pushing on the Taiwanese ground units all around the island. It was the first time I had really seen an overall view of the island's fight. There are Chinese everywhere. I counted seven different beachheads, and two areas under control of airborne units. I could also see where there were Chinese ships east of the island, between our fleet and us!

The little 3 x 5 liquid crystal display also showed air activity all over the place. In fact, there was so much air activity that it was hard for Hector to explain what was going on, so he disabled that control, and millions of blips seemed to disappear from the screen. After he turned off the radar display, I think everyone in Hector's bunker looked over their shoulder, out the passageway, and at the sky above. None of us saw a thing.

The point of Hector's little presentation was to illustrate, in no uncertain terms, that the bad guys were coming. We may not see them with our eyes, but someone tied into the MilNet has, and they sent the info for anyone with clearance to see.

Thursday, April 5, 2011

7:11am local

Global Technologies Inc.-Medical Research Division

Dearborn, Michigan

Dr. Joshua Silverstein walked through the heavy glass door and into the lobby as usual. He passed the familiar security guard who recognized him and then watched in awe as he passed. Everyone in the building-and most people in the company-knew what had happened to Dr. Silverstein the day before. No one had expected him to come in after his grievous loss. He continued to his cubicle. Down the hall, past the administrative assistants, past the accounting office, and through the maze of cubicles, he walked. Every step passed another shocked co-worker, but no one said a thing.

He meandered his way back to his familiar cubicle. There, he found his lab coat, and while his computer was booting, he donned it as usual. On the outside, Dr. Silverstein seemed saddened by his recent loss but no more than anyone else would have been. No one could see that the heart of this normally jovial and warm man had been burned away with the rest of his family.

His wife and three daughters had been visiting some old college friends of his wife's in Israel. They had been there for three weeks and would have returned in two more days. Before they left, Mrs. Silverstein and the girls wanted to buy the good doctor some souvenirs at the local mall.

Security was unusually tight for the Americans, but it was normal to the Israelis. Everyone had to be searched. Every bag had to be searched. Every Arab needed to show identification. Cement barricades-disguised as flower planters-blocked the open-air mall from any potential car-bomb attacks. Armed soldiers with assault rifles stood at the ready to oppose any terrorist attacks.

They were not enough. In the middle of the afternoon, while Dr. Silverstein was asleep at home in Michigan, yet another nightmare began in the Holy Land. Later described as Palestinian extremists, a group of eight men jumped from two cars near the mall's entrance. Two were killed immediately by the Israeli soldiers standing guard, but the rest fought their way into the mall. The guards were all killed or wounded within a few seconds.

Then the killing began. Systematically, the gunmen took positions near the exits to the mall. As shoppers frantically tried to escape, they were murdered by automatic weapons fire. The entire scene was recorded on bank machine cameras, a victim's video camera, and the security cameras at the entrances and exits to the mall.

Dr. Silverstein received the call in the middle of the night. It was a little past 4:00 am. Like anyone else, he worried. Occasionally he broke down and wept-profusely. Every fifteen minutes he was called and given more information by the State Department.

While he waited, the news hit the 24-hour cable-news stations. They showed all the different camera angles. They showed the most graphic portions of the footage. They showed the bodies scattered about the food court. They showed two little girls-his girls-crying next to the body of their mother-his wife. They showed a gunman casually walk over and execute each of the girls. He didn't need to wait for any more calls from the State Department. After seeing the footage, it was he who gave the State Department official the news.

Dr. Silverstein's loss was total. He didn't know the fate of his third daughter. He didn't know if his wife was wounded or dead. There was no doubt as to the fate of his two oldest girls. Before long, his crushing sadness melted in frustration, then rage. The more he suffered, the more his soul evaporated.

With absolute hatred in his blood, and unencumbered by his soul, Dr. Silverstein became determined to find revenge. The news reported that Israeli police had killed all of the gunmen, but he was undeterred. Inside his head, under the veil of a migraine headache, he decided that others would pay. He wanted vengeance brought upon those who had ordered the attack, those that had trained the gunmen, those that had armed them, those that had housed and fed them. He wanted the people that had motivated the gunmen to pay. He wanted the violence to end once and for all.

The more he thought about it, the more he realized the reality of the situation. Those eight gunmen would have been recruited, trained, supplied, and overall aided by hundreds, maybe thousands. He knew that suicide attackers didn't commit themselves after a few days of rhetoric. He saw the deaths of his children, and probably his wife too, as just another example of how Palestinian people were dedicated to the destruction of the Jewish race. As was the case in NAZI Germany, some-possibly even many-of the Palestinians didn't fit the profile, but Dr. Silverstein had been convinced that most Palestinians-most Arabs-wanted to kill every Jew.

His family's deaths were proof to him. Cyclical violence in the Middle East wasn't based on religious rights. He believed that it had become a matter of racial prejudice. Until one side or the other could be beaten down like the Germans had been in World War II, there would be no safe place for a Jewespecially in Israel.

Dr. Silverstein's vengeful rationalization continued unabated. He made arrangements for a flight to Israel through the State Department official that kept calling. Then he prepared to go to work. There were some things to be done there before he left. Vengeance would be his. It would be swift, silent, and deadly, but he needed to go to work first. Only there could he put in motion his plans for the silent genocide of all the Arab people.

"Josh?"

Dr. Silverstein's department Operations Manager poked his head over the cubicle wall. His face was alternating between pale green and white. It was obvious he knew, and that he didn't think Dr. Silverstein had heard about his family.

"I already know, Doug. Thanks. Listen, I hope you don't mind, but I've gotta go over there and…take care of things. The State Department's got me on a flight this afternoon."

Relieved but still saddened, Doug Kowalski's shoulders slumped and his head cocked to the left.

"I'm real sorry, Josh. We all are. If you wanna come into my office we can talk about it." Dr. Silverstein shook his head. "Alright then. If there's anything any of us can do, just let us know, OK? I think we'll just hold off on the PAR (Production Authorization Report) until you get back."

Dr. Silverstein's computer finally finished booting and logging into the local network. He needed to get to work right away if his plan was even going to get off the ground. He stood and patted his boss on the right shoulder.

"Thanks, Chief, but I can get it set up for you guys before I go. That's why I came in. Besides, I can't stay at home. C'mon, what am I going to dowatch TV? It'll be a while before I do that again. I need to keep my mind busy for a while."

His boss nodded in approval.

"Okay, but don't overdo it. No pressure. Just get done what you can, and we'll hold off until you get back if we need to. We'll all give you some space, but if you wanna talk, don't hesitate, just come on in."

Dr. Silverstein nodded, thanked him, and sat back down to work secretly on the computer-as fast as he could. On the other side of the cubicle walls, his boss was discreetly motioning to each and every person. He wanted to hold an impromptu hallway meeting to let everyone know that Dr. Silverstein needed some time and space to deal with his tragedy.

All the while, Dr. Silverstein's computer screen flashed through pages and pages of data. It was the blueprint for his company's new pharmaceutical for the treatment of Cholera, called ZP9. Instead of attacking the disease directly, ZP9 was designed to genetically change people's bodies so that they could reject the trace chemicals that Cholera bonded with to infect them. As long as people's immune systems could be genetically manipulated by ZP9, then Cholera would never be able to take hold again. It was the great hope for many in the Third World, and after ZP9, a long line of new gene-altering drugs would follow.

Dr. Silverstein continued to search the database. He was looking through tests done on human subjects-particularly those of Arab descent. Each time he find one, he marked it and continued his search. After half an hour, he had assembled a few hundred. Then, he did a database search on the common genes among them. He cross-referenced the list of gene variables, and eventually found 163 bars-DNA markers-that identified the genetic differences between people who had Arab blood in their veins and people who did not.

His Doctorate was in Genetic Research and not medicine. There was no Hippocratic Oath to make him think twice. His soul was gone anyway. All that remained were happy photos, and the image of his 3 and 4-year old daughters, screaming and dying on their mother's body, under the food court table, on the other side of the planet. He had watched it on TV, but the sound struck at his core. Hormones, instincts, and the absence of love drove the once-loving parent and healer toward his evil deed.

Dr. Silverstein typed a few strokes into his computer and found the Production Authorization Report for ZP9. It was the final blueprint for his company's labs to begin making the drug. The drug, ZP9, actually contained nine separate gene-enhancing drugs to make three changes in people's genes. Dr. Silverstein called up the genetic formula for the ninth drug and made a few changes. Instead of targeting all human immune systems for genetic manipulation, he changed its formula to effect only people whose DNA contained the 162 bars of DNA code that were specific to people of Arab descent. Then, instead of having the ninth drug block Cholera-bonding chemicals, he deleted a few other bars of

DNA that hadn't had their uses determined. All that was known was that bars in that region of the human genome controlled people's natural life-expectancy. A few more keystrokes, and he was done.

Dr. Silverstein met with his Project Manager and gave him the go-ahead for production. He accepted more condolences from his office and lab coworkers. Then, he logged out of his computer, turned it off, removed his lab coat, and headed for the airport.

He knew that over the next few weeks, as thousands of people were inoculated with his new drug, a large portion would die. He also knew that genetic mapping was too small to trace. Accountability was irrelevant to the man without a family, and the thousands-possibly millions-of people who were going to die were all going to a greater end. They would apparently die of old age or natural causes, but he figured that in a year or two, there would be no more Middle Eastern problems. For the man whose conscience had been murdered with the rest of his family, it was genocide justified.

4/5/11

1042 hrs.

Hualien Airfield, Taiwan

Personal Log – Lt. Kim "Killer" Yoo

I flew with the Taiwanese F-5 squadron again. The weather is still overcast from 1000-5000 feet. Above 5000 feet, the sky was bright, clear, and sunny. As we flew over Taipei, the overcast was muddied with a dark grey/brown hue from all of the smoke below. I watched a television news broadcast that showed fighting in the city, and it didn't look bad enough to color the clouds with smoke. However, that was last night, and the tide of battle has been very fluid. Conditions might have changed since the broadcast.

Air activity was unusually quiet. The F-16 that I am flying is very old, and while its radar is out-dated by at least 15 years, it does work. There were too many targets to track at once, and the computer kept erasing contacts to report new ones. Almost all of these radar contacts were Coalition, so it didn't matter too much. Since the skies were almost vacant of communist aircraft, and both the radar and radios were working, one could easily hope that the communists have suspended their air operations. This is good news to most people, but many of us find it to be both disconcerting, and unsatisfactory. We are pilots, and we are tasked with killing communists. As long as the overcast remains, we cannot attack their ground forces effectively, and as long as they keep their planes on the ground, we cannot fight them.

After we landed, the Taiwanese pilots put in a request for permission to attack the Chinese airfields in mainland China. It is an incredibly bold and brave notion, but I fear that the more conservative high-level commanders will not support it. Even if they did, the diplomats would probably prohibit an attack on mainland China since it would inevitably prove as a propaganda bonus to the communists-particularly if one of us were captured on the mission!

Tuesday, April 5, 2011

2nd Lt. John Chamberlin-Personal Log

Hill 1419, 15 miles S of Taipei, Taiwan

There's still no unusual activity around here. The once heavily-traveled road only sees an occasional passer by now. There's about as much traffic coming back to the beachhead as there is going out from it. I think that means that we're holding our own. Bob and I have the platoon camouflaging their positions and identifying range reference points for better accuracy if we get rushed.

At Hector's meeting today, he showed us a new trick he learned on his palmtop. If he's looking at the unit display map, with the airborne unit display off as usual, and he switches to the report unit data menu, he can change the area displayed from Taiwan to anywhere in the globe. We were able to watch what was going on in the Mediterranean with almost no trouble at all! It was actually pretty boring until Bruce suggested turning on the air unit display, and then we could clearly make out planes chasing each other around. As hightech as the scene was, it reminded me of the days when TV was new and you'd see pictures of entire families watching a small black and white screen as though it were a window to another world. In Hector's bunker, the six of us watched with the same intensity and awe, but this time, it was real life, and it was another world, another battle.

Friday, April 6, 2011

0850hrs local

Eastern Mediterranean Sea

35,000 Feet above the royal blue Mediterranean waters, a grey, twinengine turboprop E-2C Hawkeye AWACs plane loitered at 180 knots. Inside the tiny fuselage the three radar and electronic surveillance operators were glued to their monitors and control panels. Behind and below them-40nm away-the U.S.S. George Washington was cruising along at 19 knots. In a 30 nm arc, the escort ships raced along and then slowed down to listen for submarines. Recon helicopters did the same as they hurried from one end of the battlegroup to the other to identify fishing boats, cargo ships, and pleasure craft. Commercial airliners criss-crossed around the circle of ships and helicopters. Foreign air forces patrolled the skies over a dozen nations-some with pairs of planes and others with entire squadrons. The radar screens watched everything with a God-like view, and the screens were crowded.

Far away from the Hawkeye-code named Big Eye Five-Five-a pair of U.S. Marine Corps F/A-18 F Super Hornets was flying east just south of Cyprus. Codie "Cowboy" Custer was the flight leader for the two aircraft. Their mission, Combat Air Patrol, was under the direct supervision of the controllers on Big Eye Five-Five.

In the warm and crystal-clear spring air, the four crewmen in the two Super Hornets panned the skies with their eyes looking for trouble. They'd only been in the Mediterranean for a few days, but already they had flown eight CAP missions. This one had been completely uneventful for the first two hours. Finally, a sensual, phone-sex voice broke the radio boredom.

"Alpha-Whiskey Lead, this is Big Eye Five-Five. Climb to angels onefive and identify air contacts designated tracks Bravo-Lima-five-two-five, Bravo-Lima-five-two-six, and Bravo-Lima-five-two-seven; bearing three-zeroseven at angels ten. Copy?"

Codie answered back with a succinct "Copy, Big Eye Five-Five." Then, he glanced to his right and saw that his wingman was nodding. He had heard the radio call also. That was all that was needed. The two Super Hornets turned left and climbed two miles higher into the pale blue.

On the other side, of Mount Olympus over the waters separating Turkey and Cyprus, three Turkish F-16 Falcon fighter planes were on patrol. The sleek grey shapes slipped through the high altitude air with almost no air resistance. The spaces where bombs, missiles, and fuel tanks would hang under their wings were almost vacant. Their only armament was the three M61 20mm Vulcan gatling cannons mounted on the left side of each plane and a pair of small missiles at their wingtips.

Codie turned on his ultra high-tech radar and immediately detected all three Turkish F-16's. The AN/APG-73 detected the number of spinning turbine blades deep in the intake of each plane. Then it searched through its database and determined that the planes were American-built F-16C Fighting Falcons using Pratt & Whitney F100-PW-220 engines; one in each plane. The information on Codie's center cockpit computer monitor (MFD-MultiFunction Display) labeled the previously unidentified symbols. The radar also interpreted its microwave returns and determined that the three planes were only armed with 2, AIM-9 Sidewinder, heat-seeking, air-to-air missiles on each plane. As soon as he read it, he punched a few keystrokes and sent the information to both his wingman and back to Big Eye Five-Five. Then, he continued on course to get a visual identification.

When he was less than 20 miles away, Codie finally spotted the Turks. The sleek, grey planes looked like transparent specs at first, but experience had shown him that they were more than just smudges on his helmet visor. He recognized them. They were coming in from his right side and it looked like they were going to pass in front of him, so he led his wingman into a lazy left turn toward an invisible intercept point in 3D.

Soon they were clearly visible. The left side F-16 wagged its wings in a salute of friendliness. Codie and his wingman did the same, then all five planes joined into a formation. Both Codie and his Turkish counterpart radioed back to their controllers that they had identified the mysterious military aircraft near their assigned patrol locations. Then, the Turkish pilot held up a hastily written sign. It was his email address! A round of polite and nostalgic picture-taking followed, but soon Codie and his wingman were called back to work.

Nikki-Codie's controller back on Big Eye Five-Five called him.

"Alpha-Whiskey Lead, this is Big Eye Five-Five. If you boys are done playing, there's more work to do. Climb to angels three-three zero and identify air contact; designated track Charlie-Bravo-three-three-five; bearing twozero-seven at angels three-three-zero. Copy?"

"Oooo, baby! I love it when you talk so strongly. We're on our way. Just gotta say goodbye to these guys from Turkey. I'll tell'em you said hi."

Codie's less than politically correct or professionally permissible reply didn't go unnoticed. Nikki was a beautiful woman, but in the post-Tailhook era of the U.S. Navy, she had been surprisingly well-treated. Such blatant chauvinistic behavior was unusual…but not unheard of to her. With an emphasis on the letter "P" she changed from an inviting voice to a threatening tone.

"Alpha Whiskey Lead, what is your name Pilot?"

The U.S. Navy doesn't have pilots. The men and women who fly their aircraft separate themselves from regular airmen and women because they have to have special skills and abilities to take off and land on moving ships in high seas. It is almost offensive to refer to them as pilots. Instead, they are called naval aviators. The Air Force, the Marine Corps, and the Army have pilots.

Codie recognized the change in tone. He knew how to fly a plane and how to work the most complex avionics. He knew how to get along with younger pilots who were still young, dumb, and full of-bravado, but he didn't have a clue about how to speak with a woman; let alone a woman in an authority position.

"The name's Cowboy, ma'am. Now, I bet you've got the nicest tail feathers in the Navy, so don't go getting 'em all ruffled. We're on our way to that new contact, and I didn't mean any disrespect by calling a sexy-sounding lady like yourself a baby. We okay now?"

Almost 200 miles to the west of Codie, in the cramped confines of the E2's fuselage, Nikki shook her head in awe. The other controllers stopped what they were doing. They glanced at her, shook their heads in disbelief too. Then they went back to work guiding other patrol fighters to identify the seemingly endless stream of unidentified aircraft and ships around the battlegroup.

"Get to work, Alpha Whiskey Lead. Big Eye Five-Five-out."

Codie laughed in his oxygen mask. He thought the whole snippet of conversation was cute. Somewhere inside, he even danced with the idea that Nikki was interested in him-beyond professionally. Social graces were his least capable skill. He went back to work; flying from one side of his assigned patrol sector to the other. There were a lot of ships and planes to be identified.

With Codie and his wingman heading away, the three Turkish F-16's began a slow turn to the left. They headed south, and Nikki was content to watch them even as they headed straight toward-then directly over-the divided island of Cyprus. No one knew it, but they were on an electronic intelligencegathering mission. Their orders were simple: fly over the island and see if anyone tries to shoot at them. Codie had failed to notice the conformal electronic emissions monitoring pod mounted under the fuselage of one of the Turkish F-16's. Even though he hadn't, Nikki had tracked the planes, and a recon mission was the only good explanation of their technically illegal flight path.

Wednesday, April 6, 2011

2nd Lt. John Chamberlin-Personal Log

Hill 1419, 15 miles S of Taipei, Taiwan

The platoon is becoming a little bored with digging, camouflaging, and the other routines of daily life on the mountain. I had Bob send some of them down to the beachhead to scrounge for fresh food. He sent four people from Rick's third squad, but they came back after about three hours. They had been turned back by a roadblock on the road down. The entire area's security has been tightened. A Taiwanese military police squad came by looking for a sniper that we never heard fire, but I think they were just checking to make sure that the people Bob sent down to the beachhead made it back and that the road was still covered.

Hector says that the situation in the Mediterranean is getting really bad. U.S., British, French, German, and Italian airstrikes in Macedonia have reopened centuries old wounds between Greece and Turkey. It's making everyone a little nervous since there are both Greek and Turkish forces helping us fight off the Chinese. So far it's been trouble-free over here, but in the Med, there's already been a lot of rioting and even a small skirmish between patrol boats in the Aegean Sea. Sometimes it seems like the entire world is coming apart.

Saturday, April 7, 2011

1610hrs local

U.S.S. George Washington CVN

Eastern Mediterranean Sea

While alarms began sounding, the ship's 1MC speaker system popped, whistled, and squealed to life. Before a word was spoken, every human being on the ship dropped whatever they were doing and headed to their battle station.

"General Quarters. General Quarters. All hands, man your battle stations. Traffic routes: forward and up to starboard, down and aft to port. Repeat. General Quarter. General Quarters. All hands, man your battle stations. Traffic routes: forward and up to starboard, down and aft to port. Missile attack portside, aft. Missile attack portside, aft. Move it people!"

It repeated twice, and finally, as each of the ship's critical operating departments was secured for battle and reports were sent to the bridge, the alarms ended. Pilots and naval aviators scurried into their flight gear and then to their ready rooms. Everyone was ready for a missile attack.

As fast as they could be brought on deck, F/A-18's-both Hornets and Super Hornets-were launched. Controllers on the patrolling E-2C Hawkeye radioed orders to the pilots and naval aviators. Pairs, trios, and quartets of planes were sent to patrol areas at maximum speed. Once there, they searched and waited.

It was just a drill, but it could have been very real. When everyone was allowed to stand-down and return to their normal duties, the reality sank in. Drills were commonplace, but with tensions around the globe at their highest since World War II, and within the tight confines of the Mediterranean Sea, this drill carried significantly more importance than most.

Saturday, April 7, 2011

The New Cleveland Press

Cyprus Protests Turn Violent

While the rest of the world is struggling to end the conflagrations in East Asia, the Eastern Mediterranean seems to be trying just as hard to erupt into conflict. Greece and Turkey continue to politically fire across each other's bows with torrents of legal positioning, argumentative rhetoric, and sometimes even blatant propaganda.

Greece has ordered its embassy and all consulates in Turkey closed. All of their staff are being ordered back to Greece. Turkey has followed suit by cutting off its direct diplomatic ties in Greece.

Both countries are arguing in the world press about the sovereignty of Macedonia and Cyprus. Greece seems determined to end the stream of Islamic terrorist activity in neighboring Macedonia. At the same time, there is undoubted support for the Greek minority on the divided island of Cyprus. Turkey is at the very least understanding of the Muslims in Macedonia, and rumors persist of direct support for the Islamic terrorists that have been active in that country. According to British Intelligence, there is definitely a direct connection between the Turkish Cypriots that attacked British peacekeepers and the Turkish Intelligence Agency.

Saturday, April 7, 2011

The Akron Times

Charity Survives

In the midst of all the chaos in the world, one American company is still trying to fight disease instead of soldiers. Detroit-based World Pharmaceuticals Incorporated (WPI) has been sending a brand-new drug for the treatment of cholera to overseas relief agencies at no charge. The new drug, called ZP9 by its manufacturer, has been engineered to genetically build up a resistance to the common pestilence of the Third World. It is the first inoculation to use genetic therapy as a preventive treatment, and it may even lead to the elimination of the common disease. The drug has been sent to medical relief agencies all over Africa, South and Central America, the Middle East, and Central Asia. Military activity has hindered shipments to Asia. World Pharmaceuticals Inc. is a division of Global Technologies, Inc.

4/7/11

1547 hrs.

Hualien Airfield, Taiwan

Personal Log – Captain Jodl "Magicman" Steinhoff

Our Taiwanese scrounger has really come through for Dieter. He "found" an old American 20mm external gun pod. An American company had loaned it to the Taiwanese Air Force for trials on their Cobra helicopters. Now-just as Dieter's 30mm cannon was beginning to jam on a regular basis-he has some heavy firepower. The 20mm rounds aren't as large, explosive, or high-velocity as our 30mm rounds, and the external pod can't be linked to either the infra-red/laser fire control system or the helmet mounted targeting system. However, it does have a variable rate of fire, and it uses the same type of cannon round as the rest of the NATO aircraft in the group. Everyone in our All-Fulcrum division has always enjoyed the helmet-mounted targeting system, and the infra-red/laser fire control system, but in this war the only thing more difficult to find that spare parts or fuel, is 30mm Soviet-made ammunition. We're all very jealous, and we've all taken to harassing the Taiwanese scrounger for external cannon pods of our own.

There is a serious ammunition shortage right now. The Taiwanese claim that they can't find any more AMRAAM's. Even the older AIM-7 Sparrows are in short supply. Almost all the late model, short-range AIM-9 Sidewinders are gone, and the group is down to using "B" models that date back to the American experience in Vietnam! Our division has a few more AA-8 Aphids left, but I don't think there are any other Soviet-made missiles on the island right now. The Wolfman says that there are some dual AIM-9 Sidewinder wing racks in the group, and when we completely empty our supply of Aphids, he will have the Taiwanese weld them to our racks. I'm not sure how they'll be wired to our aircraft since even Dieter and I have yet to figure out the Russian wiring, but The Wolfman is confident that we can overcome this supply problem.

Whenever someone sounds doubtful to him, he's quick to remind all of us that if we're having supply problems, so are the Chinese and the North Koreans. It's a good point, and I think everyone can agree with him. After all, we are all officers, and we should all understand the military thinking. The Wolfman is very astute to the strategic perspective of this war-an amazing thing given our near complete lack of intelligence data.

Thursday, April 7, 2011

2nd Lt. John Chamberlin-Personal Log

Hill 1419, 15 miles S of Taipei, Taiwan

It looks like the war is coming back to us. There was an early morning artillery attack that hit the company pretty well. Bruce's squad took some heavy casualties: four KIA, nine wounded, six of whom had to be evacuated. I told him and Bruce about my e-mail form letter for the KIA's families and they both gave me a hard time. They said it was too cold, but after I showed it to them, and asked them to improve on it, they both reluctantly made copies of it.

My platoon came out of the barrage okay. Bob said some shrapnel hit one of the people in Wess' second squad, but I checked, and she was only scratched. There wasn't even any blood to speak of. Nobody really cared too much since the woman that was hit is probably the least liked person in the platoon. Her name is Maria Jansen.

Most women in the military have always had one of a few stereotypes attached. They tend to be classified as lesbians, sluts, bitches, babes, or one of the guys. I guess women will only really be integrated into the military when society in general fully integrates them. Anyway, Maria fits four out five of the categories. Her promiscuity in and out of the platoon has lead me to believe that she may be a lesbian, probably is a slut, most definitely is a bitch. Because of all this, she is usually regarded by the most chauvinistic of the platoon as one of the guys-at least until she verbally abuses them enough. Her most obvious quality is that she is a major wimp. I have no idea how she got to become a Marine. The shrapnel that hit her cut her left wrist with such little force that it was actually stopped in her BDU sleeve!

Friday, April 8, 2011

2nd Lt. John Chamberlin-Personal Log

Hill 1419, 15 miles S of Taipei, Taiwan

More artillery hit us again today. No one took any hits-not even a scratch this time. If anyone felt any pain, it was because the entire company is digging their trenches and bunkers a little deeper now. The blisters are adding up.

Bruce thanked me for the e-mail KIA form letter today. I don't think that Hector understands its value yet. A few more letters, and he will.

There was more bad news from around the globe today. During Hector's morning briefing, we watched Turkish jets flying combat air patrols over the Greek section of Cyprus. Greek naval units are in the area, and we all wonder how long it will be until the two NATO members go at it. Bruce pointed out that, off the coast of Taiwan, a pair of Greek frigates have been moved to a position northeast of Taipei, while a trio of Turkish frigates, along with a destroyer, have moved to the southwest corner of the island. On top of that, the Chinese seem to be avoiding both forces.

Tuesday, April 8, 2011

0335hrs local

U.S.S. George Washington CVN

Eastern Mediterranean Sea

Codie had been on Combat Air Patrol (CAP) duty for almost 8 hours. Before he left the GW, he'd spent another hour and a half prepping for the flight, checking his aircraft, and waiting in line to launch. He had another hour of CAP duty, then he could return to the ship, debrief, change, and get a few hours rest. The ship was short-handed, and now he and the other Marine aircrews in his squadron were feeling the fatigue. Worst of all, they had just started the operational phase of their deployment, and Codie knew that he would have to keep up his schedule for at least another 6 full months!

Starlight reflected off the surface of the warm Mediterranean Sea almost 7 miles below him. In his cockpit, a dim green glow from both his heads-up display and from one of his Multi-Function Displays (MFD) turned the night into day. He could peer through the tiny electronic screens, and the low-level light from his surroundings was electronically amplified. The only drawback was the monotone green. After a few hours in the cockpit, it became dull and boring. He had been sent to identify the occasional fishing boat or yacht, but it was a quiet night. At least one of the controllers on the AWACs plane was Nikki.

Codie was supposed to lead his flight where ever she ordered. Her job was to detect potential threats, then send the Marines to check them out. Somehow, in Codie's mind, she was giving him a quiet night on purpose, and he saw it as a matter of her affection. Of course, there was no reality in his thoughts. It had in fact been a slow night, and Nikki was just doing her job. Besides, like most women, Nikki found Codie obnoxious, distasteful, and allaround unattractive. The only thing that he had going for him was that he was pilot, but back on the GW a woman as attractive as Nikki could have her pick of Naval Aviators. Why would she want to get involved with a Marine pilot-let alone one like Codie? Such thoughts passed the time for the Marine pilot, but Nikki was so unimpressed by Codie that they never even entered her mind.

She truly had zero interest in him. After all, there was work to be done.

Codie's radio sounded.

"Alpha Whiskey Lead, this is Big Eye Five-Five. Descend to angels onefive and identify air contacts designated tracks Lima-Lima-two-two-one and Lima-Lima-two-two-two, bearing one-eight-zero. Copy?" Codie snapped his mind back to the real world.

"Copy, Big Eye. We're on our way."

He rolled his plane to the right and gently dove. As he corkscrewed downward from 7 miles in the sky to 3 miles, the green glow of the MFD in the center of his cockpit slowly showed 5 tiny specs. He closed in, and the specs grew into aircraft. Codie recognized them as Turkish F-16's.

"Big Eye from Alpha Whiskey Lead, looks like we've got five Turkish Falcons over here. I'm on their 6 o'clock, and I don't wanna spook 'em by accident. I'll uplink my video.....now."

With a few keystrokes, Codie digitally connected the video from his lowlight level MFD through the airwaves to Nikki-150 miles away and 4 miles above him. She relayed the images back to the GW where they were recorded.

"Alpha Whiskey Lead, this is Big Eye Five-Five. Uplink established. Maintain your distance and continue monitoring. Await further instructions." Codie and his wingman were almost 20 miles behind the Turkish pilots. The passive stealth design of their Super Hornets, with their thin layer of radar-absorbent material, made it very unlikely that they would be detected. Visual detection from such a long range would be just as unlikely during the day, and even less at night. They kept their guard up, but they also felt very safe.

Then it happened. Before Nikki could spot it and call them, Codie and his wingman watched the Turks begin firing missiles. Bright white exhaust from rocket motors raced from under the wings of the Turkish F-16's and toward some unseen location far away on the horizon. The missiles were heading away from

Codie and his wingman, but they knew that with their payload spent, the Turks would soon turn around and spot the Marines.

Codie and his wingman were close enough to use hand signals. They made a plan and initiated it. Immediately both planes turned around and began to dive to pick up speed.

"Vampire, Vampire, Vampire. Alpha Whiskey Lead from Big Eye FiveFive. Missile launch. Climb to Angles three five and come right to zero zero zero."

Nikki had finally detected the Turkish missile launch, and determined a course of action. Codie and his wingman were already diving to sneak away from the scene as fast as possible.

"Big Eye, this is Cowboy. We're already leveling off at angels one. Did ya'll see that video?"

Nikki watched her radar screen as the Turkish planes turned to their right and started following Codie and his wingman.

"Alpha Whiskey Lead from Big Eye Five-Five. Confirmed. Video received. Bogeys inbound at your six o'clock. Repeat. Climb to angels three five immediately."

Codie smiled in his oxygen mask. She was making a mistake. He knew it. His weapons officer in the Super Hornet's backseat knew it. So did his wingman and his weapons officer.

"C'mon, Nikki. We appreciate your worryin' about us, but if we climb, we'll either lose airspeed and they'll spot us, or we'll burn too much fuel and you'll have to send a tanker out here. Either way, those boys behind us are gonna know they've been seen. How 'bout we stay low till we get back on our station?"

"Alpha Whiskey Lead...," Nikki replied, *"remain on course and at your angels. Bogeys are working toward your northeast. Good call, Cowboy."*

Codie was in seventh heaven, and everyone knew it. Mediterranean waves glimmered with starlight as his mind took the compliment and amplified it a thousand times. Nikki had added fuel to the tiny romantic flame that he carried. Still, Codie knew that he wasn't conversationally comfortable with women, so he retreated into his professionalism so that he might hide his excitement.

"I got ya, Big Eye. Thanks for the update. What's the word on those vampires (missiles)?"

The excitement of the moment wasn't her first as an intercept controller, but her near-mistake was unusual, and it left her disarmed. Her by-the-book professionalism evaporated.

"It looks like the Bogeys made a strike on a ship that's been sitting off the northwest coast of Crete. We've been watching it for a while, and you can make your own assumptions, Cowboy."

"(She called me by my name!)" He thought.

"I understand, Big Eye. Where do you want us next?"

"Cowboy, it looks like the Bogeys are far enough to your east now. Take your flight up to angels three-five, and continue on your current heading. I'm sending you some waypoint coordinates through the uplink. When you get them, head for it. That'll be your new positive control point. Once there, just hang out, keep your eyes open, and await further instructions. Big Eye Five-

Five out."

Codie watched his MFD. The new coordinates were received, and he headed for his new patrol location. Nikki looked at her radar screen and watched a dozen missiles from the Turkish F-16 airstrike impact in succession. Their target: a less than subtle Greek fishing boat covered with radar and radio antennas; a spy ship. A few hours later, Nikki would watch as Greek helicopters circled the sunken ship's location. None of them would stop to try and pick up survivors, and the U.S. Navy would deduce that there were none.

Saturday April 9, 2011

The New Cleveland Press

Gene Therapy Goes Horribly Awry!

An American miracle drug, ZP9, could be the source of a strange series of deaths in the Middle East. The new genetic therapy drug is manufactured by an American-based company, World Pharmaceuticals Incorporated (WPI), which has provided the drug to several different children's relief organizations. On the surface, the effort seems to be a public relations-oriented program. Some skeptics of genetic therapy-based drugs hint that WPI's charity program is simply an attempt by the company to build a case history for the drug and speed its approval by the FDA.

Following a series of unusual deaths, authorities in several Middle Eastern nations have halted the use of ZP9. Only a few days after some children were innoculated with WPI's latest creation, many subjects began to exhibit a series of symptoms more commonly found among the aging. Some have suddenly developed arthritis, heart disease, and other geriatric disablties. A few children have already died as a result of their apparent side effects. Amplifying the situation is the fact that only children of traditionally Arabic descent-particularly Palestinian-seem to react to ZP9 inoculations.

Official response from WPI is that they are looking into the matter, but they stand by the drug's usefulness. Officials also claim that it was thoroughly tested, and that the idea of a drug affecting only a particular race is "as ignorant as it is offensive."

Authorities from the Centers for Disease Control (CDC) have offered to assist and coordinate investigations between the various nations effected. To date, almost 100,000 children have been innoculated with ZP9. Estimates suggest that as many as 200 might have already died as a result of the alleged side-effects, and as many as 5000 children may already be showing symptoms.

Saturday, April 9, 2011

2nd Lt. John Chamberlin-Personal Log

Hill 1419, 15 miles S of Taipei, Taiwan

Hector's palmtop is really nice, but it doesn't show any small units. While we've been watching the position of companies, battalions, brigades, divisions, corps, and armies, small squads and platoons have been moving up towards us. Right after this morning's briefing, we came under sniper fire from below. It's a brave, determined or deliberate man that opens fire on an entire company from below. We rained down so much lead that I'm sure we got him. Then again, the patrol Bruce sent out to find the body came back up with nothing; not even a blood trail.

Sunday, April 10, 2011

San Diego Times

Allies Heading Home

The remnants of several United Nations naval task forces in the Philippine Sea, the South China Sea, and around Taiwan will all find themselves with fewer escort ships today. Greece and Turkey, at the urging of United Nations officials, have sent their naval elements home from the Asia Theatre. The United Nations officials declined official comment, but with the cameras off, there was great speculation that the two nations might be going to war. No one in the United Nations wanted to see the chaos that might result of both countries had vessels in the same U.N. task force.

Officials from Greece and Turkey are applauding the U.N. insistence. Both countries seem to want a fight, and there is speculation that both countries were about to order their ships home even if the U.N. had not asked them to leave the Asiatic flotillas. Turkey is withdrawing three American-built, Perry class, guided-missile frigates (re-named Gaziantep class), and four American-built, Knox class, guided-missile frigates (re-named Muavenet class). Greece is withdrawing three former American Knox class guided-missile frigates, and two former American Adams class guided-missile destroyers. The loss of the Greek and Turkish ships will seriously impede the U.N. Asiatic task forces' defensive abilities and lessen their deterrent value against further regional aggression from the People's Republic of China.

Sunday, April 10, 2011

2nd Lt. John Chamberlin-Personal Log

Hill 1419, 15 miles S of Taipei, Taiwan

Colonel Ghetty came up and gave a briefing to each company this morning. The Greece/Turkey thing is getting out of hand in the Mediterranean. All Greek and Turkish units have been "asked" to leave the area. The fear around the globe is that if, although it seems like when, they should start shooting at each other, one side or the other may decide to accept aid from either the communists, or any of the Arab nations that have been getting out of line.

The Coalition that we've been working with since the Korean thing exploded has been unexpectedly strong. I think a lot of that can be attributed to the heavy losses early on the Battle of the Yellow Sea, on the Peninsula, and the communist terrorist activities around the globe. It's weird that the worse we seem to get beaten, the stronger our resolve, either as a Coalition, a country, a platoon.

Monday, April 11, 2011

The New Denver Times

Allied Aircraft Tangle With Turks

British and Italian fighter-bombers were intercepted and attacked today by aircraft from their former ally, Turkey. After days of precision airstrikes against suspected terrorist bastions in Macedonia, Turkish aircraft suddenly attacked a strike package of British Tornado bombers that were being escorted by Italian fighters. One Turkish F-16 fighter was shot down, and two Italian Tornados fighters were lost.

NATO officials have not yet publicly spoken about the incident. However, sources inside NATO command headquarters in Belgium have told the American Internet News Network that there is a great deal of speculation as to why the Turkish planes were even in the area. Some believe that they may have been there to prevent Greek interdiction in the airstrike campaign. Turkish and Greek officials remain tight-lipped about the incident.

4/10/11

0850 hrs.

Hualien Airfield, Taiwan

Personal Log – Captain Jodl "Magicman" Steinhoff

We accompanied the Taiwanese F-5 squadron on two more raids of the Chinese beachhead near Taimali. They lost another pair of planes to AAA. We lost two, American F-16's. Kim Yoo took a great deal of AAA and small arms fire, but he managed to bring back the F-16A that he borrowed from the Taiwanese Air Force. The Wolfman's plane took a great deal of small arms fire, but his reputation grew even greater.

Our division was flying low cover over the Pacific coast. The Taiwanese F-5's were strafing a truck column heading north on Highway 54-toward Hualien and our airfield. The AAA was intense, and the sky was filled with tracers. We all knew that most of the tracers were from small arms fire (not very effective against our aircraft), but there was no way of telling which glowing, yellow streaks were from a rifle, and which were from a cannon. Two of the Taiwanese F-5's went down in a row, and they broke off their attack to consider a different approach.

Kim left our formation and headed toward the coastal highway at very low altitude. He was so low that a plume of water vapor was pulled up from the waves and followed him all the way to the coast. The Chinese were quick to spot him, and splashes of water burst all around him while tracers came at him from straight ahead, and from 45-degrees to his left and right. We climbed to get away from all the AAA that was missing him and coming toward us! About 3-miles away, Kim began to fire bursts of 20mm into the center of the truck column. He didn't pull up until he was almost over the coast. When he did, we could see the sparks of small arms rounds hitting his plane from almost 10-miles away. Everyone thought he was a dead-man. He lit his afterburner and climbed into the overcast and safety.

While we were busy watching the show and searching for Chinese or North Korean planes, The Wolfman dove out of the clouds and flew down the length of the column from south to north. His plane trailed sparks from small arms ricochets even worse than Kim's had. He was firing bursts of 20mm and continued his dive until he was under the telephone and power lines along the road. Then there was a large orange flash from the tail of his plane, and his engine began to streak thick black smoke.

Everyone in the group and the entire Taiwanese F-5 squadron followed him back up to Hualien. It was a long and tense 10-minute flight before the airfield came into view. The Wolfman turned toward the Pacific, then headed back toward Hualien and lined up for his approach. When he dropped his landing gear the smoke suddenly stopped, and his plane lost airspeed. It started to stall just as the his rear wheels touched down at the end of the runway. Emergency

407

vehicles met him on the taxiway, and we circled until the runway was clear. After two passes, his canopy was open, and he was heading back to his sandbag revetment like nothing had happened.

Kim landed next. The Taiwanese F-5's followed, then the rest of our group in division order. The All-Fulcrum division had the easiest task of the day, so we had burned the least fuel, so we passed up our place in the pattern and orbited as a CAP until everyone else was down. On the ground, Dieter and I went down the taxiway that The Wolfman had used so we could stop and see if he was okay. He was outside and showing the Taiwanese groundcrew what had happened with the international language of hand gestures.

The underside of his plane lined with 1'-2' long black and grey streaks. They had been caused by all of the Chinese small arms hits, and it gave the F-16 a strange, hairy appearance. Thousands of bullets had missed his cockpit by feet-even inches. He was bothered at all. As we taxied past his revetment, Deter and I both looked at each other in disbelief. The Wolfman was smiling and laughing like he had just won the World Cup. He's clearly lucky, but I'm not sure what else drives him. Whatever it is, the man is overflowing with destiny. It seems like nothing can touch him, and I think he knows it.

Monday, April 11, 2011

2nd Lt. John Chamberlin-Personal Log

Hill 1419, 15 miles S of Taipei, Taiwan

Well, the war took another turn for the worse last night. Hector sent a runner over to get Bob and I just before sunrise. We met up with Bruce and his platoon sergeant in Hector's bunker as soon as we could get our gear. I thought that maybe Hector found out that the Chinese were coming up the hill. His news wasn't anywhere near as local, but maybe just as bad. He was watching his palmtop, and showed us how an American Navy unit in the Aegean Sea had launched aircraft into Macedonia, a former part of Yugoslavia like Croatia, Slovenia, Bosnia-Herzegovina, or Kosovo. The American aircraft had been met by Turkish aircraft, and both sides lost units. We couldn't tell if the aircraft were helicopters or planes, and we couldn't tell what type any of the ships involved were. We were able to guess that the U.S. Navy units on the palmtop were probably a carrier battle group since the American aircraft seemed to originate from them, and there were far too many planes for them to have come from a ship like the Wasp. All of us wondered how this was going to change the war.

Sunday, April 12, 2011 1414hrs local

Eastern Mediterranean Sea

The Turkish reclamation of Cyprus was hours behind schedule. It seemed impossible to anyone involved that the Greeks were still unaware of the invasion task force's presence, and everyone waited for an attack. There were eight large LST-style amphibious assault ships in the formation, and the 2500 Turkish soldiers were all on deck with their eyes wide open in search of threats. Surrounding the small task force were a dozen guided-missile frigates (FFG's) and another dozen guided-missile patrol craft (PCG's). Below the surface, the entire Turkish submarine force of 15 vessels kept a wide perimeter free of any opposition ships.

The plan was for the first wave to land at dawn. The landing craft would then begin making round trips to a nearby Turkish naval base on the other side of the Gulf of Antalaya. At the base, two more commando brigades and six infantry brigades were waiting to for their ride to the island. If everything went according to plan, there would be over 30,000 crack Turkish troops on Cyprus within 48 hours. Of course, not everything was going according to the plan. In fact, the operation had slipped almost 9 hours behind schedule, and the first wave wasn't even in sight of the island yet!

Commander Don "Dutch" Schultz was accompanying a four-plane flight of Marine Super Hornets on the afternoon CAP-eastern sector Four Charlie. He wasn't in command, but he did outrank everyone else in the flight. He also had more experience flying CAP missions than the entire flight combined.

They had been airborne for over two hours. As usual, it was an endless cycle of flying from one location to another and identifying objects that the fleet's E-2C Hawkeye had detected. For the Marines, it was as monotonous and repetitive as a game of musical chairs at a stadium. They never seemed to run out of contacts that needed investigating, and it seemed like they all had to checked-out right away.

In the Hawkeye, 150+miles to the west, Nikki was on duty again. She had the same monotonous feeling that the Marines had, but she also had to deal with the tight, dark confines of the Hawkeye. In addition, she had the burden of watching the entire Eastern Mediterranean Sea on a single monitor. At any given time, there were thousands of freighters, airliners, fishing boats, patrol boats, pleasure craft, and cruise ships. The computer was nice to have since it sorted everything out and saved the identification data on each radartracked target, but sometimes her screen just looked like a big bunch of static.

She was getting tired of being on duty all the time. Most of the GW's crew worked 12-18 hour days. Since the Med' was such a confined space, the GW was keeping two of the ship's four E-2's airborne at any given time. Another one was always on standby, and that left only one crew with any downtime at any given moment. Nikki and her crew were working 6-hour days, but they had been sent up three times in each of the past four calendar days. She was averaging only two hours of sleep, and the plane's twin turboprop engines were droning her into a stupor.

Nikki rubbed her eyes and leaned back from the screen to stretch her back. It only took her a moment to regain her focus. When she looked back at the screen, she saw the new radar contacts. Over fifty new surface contacts were all bunched together and headed for Cyprus. She didn't need a lot of experience to know that it was probably an invasion force.

"Bravo Sierra Lead, this is Big Eye Five-Five. Remain at angels three five and come to heading zero seven three. Identify multiple surface contacts. Proceed with caution. Probability is high that contacts are on a high alert. Repeat: proceed with caution. Copy?"

Back in his F/A-18 Hornet, Dutch listened to his flight leader acknowledge the orders and direct the flight accordingly. They made a gentle turn to the right, and headed for the contacts' location. A few minutes later, Dutch heard the Marines talking amongst each other. They were excited and eager as they reported back to Nikki in the AWACs. The contacts were indeed an invasion task force, and there seemed to be ships everywhere. Dutch couldn't help but chuckle when he heard the Marine flight leader give a detailed report about the class and capabilities of the American-built, Turkish amphibious assault ships. Marines definitely knew about amphibious assault ships!

Nikki answered back while she uplinked the information back to the GW's Combat Information Center.

"Copy, Bravo Sierra Lead. Contacts identified. Home Plate (the U.S.S. George Washington) has been updated. Return to station immediately. Bogeys inbound to your pos. Probability is high that contacts are on a high alert. Advise that you expedite back to your station. Copy?"

The Turks had obviously spotted the U.S. Navy patrol and sent up their own aircraft to prevent any American intervention from going un-checked. The flight made their slow, lazy, and peaceful turn toward their original holding location, and as a squadron of Turkish F-16's started to close in from only twenty miles away, the Marines-and Dutch-accelerated away from the scene. The Turks would leave them alone, but from that point forward, Nikki and the E-2 Hawkeye crews would keep a close eye on the Turkish invasion force.

Back at the GW, the word was sent back to Washington about the Turkish invasion. From there, NATO coordinating officers from every member country were notified-including the Turkish representative. There were some choice words expressed between the Greek and Turkish representatives, but after a few minutes, they raced off to notify their respective countries about the invasion's detection.

As the Turkish landing craft anchored and loaded troops into smaller, shore-assault craft, the Greek response closed. There wasn't enough time to move their guided-missile frigates and patrol craft into position, so the Greek Air Force was called upon to do the job. They would do so with a maximum effort.

The Greeks loaded up and launched most of their air force. A strike package of 126 planes was sent. It consisted of 2 squadrons of American-built A-7H Corsair II attack planes, 2 squadrons of American-built A-7E Corsair II attack planes, 4 squadrons of American-built F-16CG Falcon fighters, and a squadron of American-built F-4E Phantom II fighter-bombers. Escorting the strike package were another 92 fighters: 3 squadrons of American-built F-5 Freedom Fighters, 2 squadrons of French-built Mirage F-1CG's, and 2 squadrons of French-built Mirage 2000's. It was a powerful attack force, but there was a long way to fly before they could attack the invasion task force, and the entire trip kept the attack force close to Turkish soil-and Turkish airbases.

Nikki's shift was going to end in an hour, but when she heard the controller sitting next to her report the detection of the Greek attack force, she suddenly wanted to stay on station. Being able to watch the battle on the radar screen was a rare and important opportunity. She passed her thoughts on to the aircraft commander. A few radio requests later, and they were given permission to stay on station. A mid-air refueling plane would be sent to keep them aloft, and the regularly scheduled E-2C Hawkeye would still keep to the schedule. Apparently, there were others who wanted to take advantage of the opportunity too.

It wasn't long until the Greeks started to run into trouble. The two leading squadrons of Greek F-16's were detected by the Turks, and a Turkish squadron of F-16's was sent to intercept them. Nikki instructed Dutch and the Marine flight that he was with to move within 100 miles of the battle to observe. Then she and the rest of the people on the pair of E-2 Hawkeyes watched the show.

The first few squadrons of Greek F-16's snaked through the islands in the Aegean Sea at 10,000 feet and a casual 450 knots. The Turkish F-16's left their airspace and headed toward the Greeks to intercept them. Just before they came within visual range of each other-with about 40 miles between them, the Turks launched a barrage of American-made AIM-120 AMRAAM air-to-air missiles. The Greeks detected the missiles immediately and began defensive maneuvers. As they did, the Turks closed.

A 600-mph dogfight followed. It lasted for almost a minute and a half. The aircraft from both nations were identical in shape, manufacture, and even paint scheme. It was impossible to tell friend from foe, and losses were heavily. Both nations lost over half of their planes involved.

None of the Americans could believe what they were watching. Nikki and her crew watched the battle on their radar screens from almost 200 miles away. So did the other AWACs crew in the regularly-scheduled E-2C patrol that was also airborne. Back on the GW, the CIC was crowded with spectators. The same was true in the Pentagon's National Command Authority Center and in the White House Situation Room.

Dutch and his flight of Marines had a front row seat. They could actually see the huge 25-mile wide dogfight. They watched the aircraft, missiles, contrails, decoy flares, explosions, tracer fire, and explosions. The planes looked like metal specks. The missile and aircraft contrails seemed like kite string. Explosions, flares, and cannon tracer fire reminded them of Fourth of July fireworks during the day.

With both sides bloodied, the remnants of the Greek and Turkish squadrons broke off their engagement and headed for home. Their planes moved as fast as they could burn fuel, but the battle for Cyprus had just begun. More squadrons were on the way. This had just been a clash of first-wave escorts and available interceptors.

The main attack was coming. Three squadrons had been mauled: two Greek and one Turkish. Now, three more squadrons of Greek F-5 Freedom Fighters were zipping across the surface of the Aegean. As they passed Crete, the Turks sent up more interceptors. This time, they sent another squadron of F-16 Falcons, and two squadrons of F-5 Freedom Fighters.

The Greeks flew at an extremely low level. Two planes disappeared from the American radar screens as they crashed into waves. The low level meant that the precision guidance radars on the Turkish planes took longer to lock-on to their targets. There was no long-range missile barrage this time.

The squadrons met over the open sea, and the planes began chasing each other in another dogfight. The strenuous maneuvering at extremely low level brought down four more planes in crashes. There were also some collisions as the agile fighters twisted and turned at high speed. As had happened in the earlier combat, both sides were flying identical American-made aircraft. There were no Greek F-16's in their formation, so the Turkish F-16's were the easiest to identify and target by all the Greek planes. That Turkish squadron was almost wiped out. As its numbers fell, and the identical F-5 Freedom Fighters from both countries began to target each other, fratricide became the big danger, and losses mounted.

After almost two full minutes, the Greeks headed for home. They were low on fuel, and they had taken a beating. Besides, their mission was complete. They had drawn the Turks away from the coast and out into the middle of the Eastern Mediterranean.

More planes were on the way. The main strike force of Greek aircraft was now airborne. There were 126 fighters armed with Harpoon anti-ship missiles, AGM-88 HARM anti-radar missiles, and AGM-65 Maverick guided missiles. The strike force was escorted by 7 squadrons of fighters; 92 more planes.

Back on the GW, the combined radar, sonar, and misc. intelligence data showed something disturbing. The Turkish aircraft that had been flying CAP over the invasion task force were now moving out toward the open sea. They were also moving closer to some of the GW's escort ships: the guided missile destroyer U.S.S. John Paul Jones, and the cruiser U.S.S. Antietam. Both ships were ordered away from the scene and closer to the GW. Instead of having a ring of protective escorts 50 miles away, the GW was now pulling in some of those escorts. Less distance meant that there would be less reaction time to any threats. It was a risk, but safer than leaving them out in the open sea with hundreds of fighters, bombers, and missiles getting tossed around.

There was something else that bothered the people back in Washington. It wasn't just the U.S. Navy, the Greek Air Force, and The Turkish Air Force in the air. Syria, Libya, and Egypt were putting entire squadrons into the sky. At the moment, they were still over their own respective sovereign nations, but they were clearly forming up for something. The word was sent back to the GW to be on the alert and take any appropriate preventative measures.

As soon as the GW got the message, every available aircraft was launched. Helicopters lifted off from the GW and from all of the escort ships to search for any possible submarine threats. Refueling aircraft went up next in case there was some sort of emergency. All of the remaining Marine Super Hornets and the Navy Hornets were launched immediately following. Finally, the S-7 Viking planes were sent out to patrol against any threatening surface or subsurface contacts. If anyone tried to attack the GW, they'd get a hot welcome.

It took almost 45 minutes to get all of the aircraft airborne and into patrols. During that time, the Greek strike force passed low over the island of

Crete. When they had crossed the southern beaches, the airborne armada turned left and headed for Cyprus. They remained at low altitude, but they also accelerated to 650 knots.

The Turks had anticipated the attack, and their airborne CAP was already in position. Outnumbered and four miles above their targets, the Turks had to dive fiercely to get into good combat position. As they did, the Greek spotted them. Most of the escorting squadrons slowed and climbed to engage the diving Turks.

In seconds, missiles were drawing white lines across the sky. When the gray specks were close enough, tracers arced down or raced upwards. There were tiny black puffs at the end of some of the white missile contrails, and a few scattered about the rain of glowing yellow and white tracers. Decoy flares glittered in the distant horizon, and another dogfight ensued. Many of the Turkish pilots kept their eyes on the prize and dove right through the escorting Greeks who had come up to block them. Those who continued their dive immediately found themselves being chased by those same Greeks who had climbed past them.

When interceptors and CAP aircraft were pulled back down to wave top height, the strike force was in danger. A fast, running, and twisting chase began. It took the Greek strike force another minute to get within range of their Harpoon missiles. Finally, some of the planes in the strike force launched the heavy missiles and joined the escorts in repelling the pursuing Turkish CAP. A few more seconds, and the Greeks who were carrying the AGM-88 HARM anti-radar missiles were within range. They fired the radar-homing rockets and also joined the escorts.

From that point on, the Turkish CAP was defeated. They had been patrolling all day, and they didn't have enough fuel left to dogfight. The Greek escorts had eroded their original numbers for almost 5 full minutes, and now their opposition was growing in size. Worse than that, for the Turks to leave the engagement area, they had to fly through the Greek strike force to get back to their airbases.

The last group of Greek attack planes closed on the Turkish invasion task force. Four squadrons of F-16's, armed with precise and devastating AGM-65 Maverick missiles were all closing on the Turkish ships. In front of the attacking Greeks were a wave of powerful Harpoon anti-missiles, and a barrage of HARM anti-radar missiles.

The Harpoons cruised at wave-top height. When their computers detected that they had passed a pre-programmed point of potential target contact, the missiles' radars automatically came on. Each missile searched 45 degrees to the left, then 45 degrees to the right, looking for targets. Whenever a ship was detected, a Harpoon's built-in computer searched its database for a similar radar detection file. It then decided if the detected ship was a fishing boat, a rock, an American aircraft carrier, or a Turkish ship. In seconds, all of the Harpoons had found their targets and turned toward them. As each one closed to within a few miles, they climbed straight into the sky and dove down on top of the targeted ships. Surface-to-Air (SAM) missiles from the escorting Turkish frigates and patrol craft stopped a few. Cannon and machine gun fire stopped more. Most, however, hit their targets.

While explosions were splitting the small ships into pieces, and fires from unspent rocket fuel leftover in the Harpoons raged out of control, another wave of missiles closed. The AGM-88 HARM missiles detected the radar antennas on the Turkish ships: search radars, navigation radars, and especially defensive weapons targeting radars. Even if a radar was turned off, each HARM could store the location of the antenna and attack it. Their goal was to leave the Turkish ships without the ability to target the F-16's that were following.

None of the HARM missiles were stopped. The invasion task force had been too disabled by the Harpoons. The HARM's slammed into radar antennas on every ship, and left the remaining vessels virtually blind and defenseless. The only defensive weapons left were the small arms and shoulder-launched heat-seeking missiles that the invasion troops carried. They were immediately brought to the deck of the landing craft.

Only a few hundred made it topside before the Greek F-16's and their Maverick missiles reached their launch point. The sleek gray planes were still at wave-top height when they closed on the burning perimeter of Turkish patrol craft. Just as they passed through the columns of smoke, the Greek pilots pulled back on their sticks and aimed into the sky.

It only took a second to climb up to 5000 feet. They leveled off there, and began to pick their targets. The weather was clear and sunny, and the only thing obscuring targets was the thin veil of smoke from a few badly burning ships that had taken hits from the Harpoons and HARM's. The Greeks could see everything. They aimed their missiles through the MFDs in their cockpits, and as each one was locked onto a target, it was fired. A few small tracer rounds zipped back from the transports, but there was no time for the Turkish troops to set up an effective and organized defense. They were in chaos, and as the Greek F-16 pilots passed overhead at 700mph, they could see it. They also saw their Mavericks impact into the American-built Turkish landing craft. Hundreds died. Thousands were wounded, and-as quickly as the attack happened-it was over. Only the flickering flames, the choking black smoke, and the cries of the wounded remained. Even the warning sirens and horns on the ships fell silent.

The Greeks banked and turned to their right to leave the area. Their exit route took them over Cyprus, and as they passed the Turkish landing beach, each of them saw defeat. Below them, most of the Turkish commandos had been able to land. Some more small arms fire and a pair of shoulder-launched heat-seeking missiles chased the Greeks as they headed back out to sea to begin their roundabout course home. They had come and dealt a heavy blow to the Turkish navy, but the Turkish army was already ashore.

On the southwest side of the island, the Greeks joined into squadron formations at low altitude. They were to low on fuel to even strafe the Turks on the beach, and as they tallied their losses morale sank. Over half of the planes sent into the attack were missing. One quarter of the Greek Air Force had been lost in just under two hours. During that same period of time, they had expended almost every precision-guided munition in their inventory. The cost in fuel, weapons, and aircraft was over a $1 billion (U.S.). The cost in experienced pilots was far greater; 103 Greeks had either been killed, were missing, or had been captured.

Near the eastern edge of Crete, the Greek formations split up and headed back to their individual airfields. Their Army and Navy would have to take up the fight now. It would take days-even weeks-to mobilize the Army, but the Navy was already on its way.

Greek guided-missile patrol boats were racing toward the remnants of the Turkish task force at over 30 knots. Frigates and destroyers were on their way from all over the Aegean Sea. They had planned on meeting off the northeast coast of Crete, but it was not going to happen.

Word of the Greek strike against the Turkish invasion task force must have spread rapidly throughout the Turkish command structure. Back on the GW, reports started coming in from the GW's escort ships and patrol aircraft that unknown submarines were suddenly becoming active all over the Aegean and Eastern Mediterranean. The monitors in the GW's CIC and back in Washington all showed new submarine icons.

A few minutes later, a helicopter from the U.S.S. Antietam identified a torpedo's distinctive sound. The torpedo was headed for a Greek frigate that was only ten miles away. Efforts were made to notify the Greeks through their coordination officer back in Washington, but it was hopeless. Less than ten minutes after the torpedo was detected, it exploded underneath the small Greek ship. The explosion lifted the ship out of the water. It snapped in the middle as the weight at the bow and stern of the ship exerted more force than it could structurally handle. The aft end of the frigate capsized immediately and slipped ingloriously beneath the waves. The remains of the forward section began to drift as they burned.

More submarine and torpedo contacts were reported back to the GW. They were scattered throughout the area. The rough undersea terrain and the collage of islands made them difficult to detect, for all but the most expensive and sophisticated search equipment (i.e., U.S. Navy sensors). In fifteen minutes, several more Greek ships were sunk. Then, the situation became much worse.

Nikki was the first to spot it on her radar screen. The Egyptian aircraft that were loitering off the coast of Alexandria started to head north-toward the GW and the battle area. Then the Syrian aircraft started coming in from the east. Finally, a squadron of Libyan aircraft began making the trek from the southwest. The GW's battlegroup was about to be in the center of a combat zone involving five other countries, and no one really knew who was a combatant or who wasn't. It wasn't even clear if there would be only two sides to the battle, or if there were going to be six!

Nikki ordered the Marine squadron to form up on the northeast edge of the carrier battlegroup's perimeter. Then she had Dutch return to his squadron and moved them into a position near the southern edge of the battlegroup.

Another Navy squadron of Hornets was directed to the southeast side of the battlegroup, and the last squadron of Hornets was sent to cover the north and northwest sides. Senior commanders back on the GW ordered the carrier's escorts to move closer to the GW. Then, every radar and sonar in the group was turned on and readied for action.

The waiting game began. American ships and aircraft moved into position over the next ten minutes. All the while, the Libyans, Egyptians, and Syrians came closer, and the Turks patrolled at the northeast edge of the battlegroup's perimeter. There were also small skirmishes and dogfights going on between scattered pairs of fighters from Greece and Turkey.

Finally, the American squadrons were in place. The Libyans were the first to be seen. It was a single squadron of ten Mig-25R Foxbats; unarmed, Soviet-built recon planes. When they saw an entire American squadron of F/A-18 Hornets in front of them, they harmlessly and nonchalantly turned back for their home airfields.

Then the Syrians spotted the Marine squadron near the battle area. The Syrian squadrons-two groups of fourteen Mig-23 Floggers- turned back too, but not before a pair of them dove down to the surface to visually inspect the damage to the Turkish invasion task force.

The Egyptians turned back without ever coming into visual range of the Americans. They had been given their own E-2 Hawkeyes as part of President Jimmy Carter's Peace Initiative settlement between Egypt and Israel. The AWACs planes must have detected Dutch's squadron and notified the 36 American-built F-4 Phantom II fighter-bombers. No one ever even saw them in the crystal clear sky.

Every few minutes, another submarine contact or torpedo contact was reported back to the GW. The Turkish submarine group that had been positioned around their invasion task force was hard at work. The eight submarines of the Greek Navy were working hard too. There seemed to be contacts all over the area. For the two American Los Angeles class attack submarines that were escorting the GW, it must have really been tense.

Unknown to almost everyone involved, the two of Syria's three submarines had taken the opportunity to take out some "targets of opportunity." A Syrian submarine Captain sank an Israeli submarine. He later claimed, "Its threat potential in such an environment was determined to be greater than the threat of political incident."

The Israelis wouldn't discover their loss for some time, but one of their corvette commander had the same thoughts about a freighter that was heading for Beruit, Lebanon. It was known to be carrying small arms, and it was served a Harpoon IIc anti-ship missile courtesy of the Israeli Navy. The ship sank in Syrian waters less than an hour after their submarine was lost in the same area. Slowly, the military forces all seemed to suspend their operations. Libya, Egypt, Israel, and Syria all pulled their ships, submarines, and aircraft back into their territorial areas. The Greeks had taken serious losses in the air, on the sea, and under the waves. The Turks continued to ferry troops to Cyprus with the few ships that remained seaworthy. By midnight, they were mostly in control of the island. Only pockets of resistance remained in the mountains.

Tuesday, April 12, 2011

2nd Lt. John Chamberlin-Personal Log

Hill 1419, 15 miles S of Taipei, Taiwan

Up until now, the standing order of the day has included a complete shutdown of all digital communications below the platoon leader level. No one ever gave us the order directly, and even the meticulous Hector never mentioned it. It just seems like common sense since any transmission would give away our position to the Chinese.

Every Marine's helmet has a built in radio to communicate on a platoon radio net. This allows the squad leaders and myself to understand what exactly our people are doing. Typically, each Marine is tied into their squad's net, and the Squad leader is tied into the platoon net. All of these transmissions are voice activated, and very feint, thus making them difficult to monitor, track, or pinpoint.

417

On top of the headset comm units, most of the people in the platoon have palmtop PC's for sending e-mail back home, etc. As long as we're in a combat zone, no one logs onto the internet, or turns on their helmet mounted comm unit as a safety precaution. The Chinese may have a difficult time pinpointing the transmissions, but it could be done. We don't want to take that chance.

Today, we couldn't take it anymore. During Hector's morning briefing, I suggested sending some people to the top of the hill and letting them all log on for three minutes. We needed news, and we needed to check our e-mails. Hector thought it was a good idea, and I had Bob pick some people out. He sent them up as soon as they were ready. He sent almost all of our platoon nerds/geeks/slackers to the top after they had collected everyone's outgoing emails, including mine. When they were ready, they walked away, up into the clouds.

Each person had an even share of e-mails to send and receive. After they finished, each one was assigned a few different web addresses to check out. This way, we could take care of all the e-mail, and get lots of news from different sources, all in a few minutes time. It worked. They entire company welcomed them back like they were heroes. They were. They had the mail!

They also had the news. Turkish troops had landed on Cyprus and quickly overran the U.N. Peacekeepers that have been keeping the Greeks and Turks apart since the mid 1970's. The U.S. Navy group we watched in Hector's bunker was a carrier battlegroup centered on the U.S.S. George Washington. All the Turkish planes that intercepted them over Macedonia were lost, and the Carrier suffered no losses. The Turkish, Chinese, and Communist Korean governments have are calling themselves the Eastern Alliance, and they've sent out invitations around the world for others to join. Yesterday's World War predictions were correct.

At least we all got mail from our families, friends, etc. I actually got a letter from Mike Baker's dad thanking him for the nice e-mail I sent him after Mike was killed at LZ Delta. I felt kind of guilty for not having known Mike better, but I also felt lucky to have not know him better. It's strange how many emotions someone can feel at once about the same thing.

Wednesday, April 13, 2011

2nd Lt. John Chamberlin-Personal Log

Hill 1419, 15 miles S of Taipei, Taiwan

Today was absolutely uneventful. It had all the typical routines, headaches, and characteristics that everyday seems to have, but completely lacked anything unique. Everyone dug their homes deeper, and tried to make them as comfortable as possible; a task made much easier by the warming spring temperatures. Everyone ate their MRE's. I went to my morning briefing at Hector's again. The situation all over the globe still make the U.S. look like a small band of firefighters trying to put out forest fires thousands of miles apart. Everywhere you look, you hear about some new dogfight, naval skirmish, or terrorist attack. Each day, we call up more reserves, and come up with some new way to supply the war with materials. The routine daily patrols went out, but found nothing and came back, tired, frustrated, and bored. On the one hand, it's nice not to have to struggle to survive everyday, but on the other hand, waiting for the fight to come to us creates an unusual combination of extreme boredom and tense anticipation.

Friday, April 13, 2011

8:07am local

Global Technologies Inc.-Medical Research Division

Dearborn, Michigan

Dr. Josh Silverstein was sitting in front of his computer with his eyes transfixed upon the data. He was reading the latest field data that had been emailed from some of his company's representatives in the Middle East. It was obvious that his personal attempt at genocide was working far more effectively than the media knew. Over a million children had already been given ZP9 inoculations. According to the computer, 349,765 had already died of unusual side-effects. An almost equal number were also having unusual reactions.

Josh felt no remorse. In his mind, he was removing a scourge that had plagued humanity for millennia. He also felt no joy or even satisfaction. With the loss of his wife and children, Josh was devoid of sentiment. Even revenge was absent. Instead, he saw his mass murder as a task or chore that needed to be done.

He had seen the morning news, and had already heard the public reports about ZP9. Josh figured that he'd be arrested soon, and his modification to the ZP9 code would be discovered. Nothing would stop his arrest from coming, and he really didn't care about it. He did want to finish his "work," though. So, while he waited for someone from his company, his parent company, or some government agency, Dr. Josh Silverstein began modifying all of the genetic therapy drugs in his research database. It was an immense task that one man alone couldn't complete in month, but it was his best hope of continuing his perceived planetary purge. It was also a convenient way of masking the onceunique changes that he had made to ZP9.

Sleep had been hard to find since the gory murder of his family. Every time that he closed his eyes, he saw the bullets and blood from the TV news report. Coffee was his new best friend. Despite his immense amount of "work," Josh needed some more caffeine. He turned off his monitor and headed for the office kitchen.

Josh walked down the corridor of cubicles, rounded the corner to the left, and stepped into the kitchen. The receptionist was walking out, and he noticed that the coffee maker was full with a fresh pot. He knew that the receptionist had probably made the fresh pot, and that she was notorious for making what was called "mega-strong" coffee. Normally it would have deterred him from filling a cup, but Josh needed the java, so he grabbed a styrofoam cup and topped it off. Without even waiting for it to cool, he slammed a healthy swig.

With his cup still raised, Josh's boss, Doug Kowalski, leaned into the room from the hallway.

"Josh? Ya got a minute?"

Dr. Silverstein finished his swallow of mega-strong and nodded.

"Let's go in my office."

The two men walked a few feet down the hall to Doug's office. It was a tiny, stark, windowless, 10'x10' room. There wasn't even enough room for Josh to sit down while Doug closed the door. Both men pulled in their bellies while Doug squeezed his way behind the desk. It was a familiar dance to both men.

"So, how're you doin', Josh? Is there anything we can do?"

"I'm fine, thanks. It was…tough to go over there and see them like that, but it'll all be over soon. My sister-in-law is making the memorial arrangements today. I'll let you know when it is, and maybe you could pass the word to everybody around here for me. Okay?"

Doug's eyes squinted in serious sincerity, and his head bobbled in gentle approval.

"Sure thing, Josh. I'll take care of it. Listen, I don't know if you've had the time to watch the news, but there's quite a fuss being made about some alleged symptoms that seem to be tied to ZP9. Some of the relief agencies are refusing to use it. They want concrete proof that the drug's not causing the symptoms. Have you heard about any of this?"

Dr. Silverstein knew he was a terrible liar, and so he tried to tell as much of the truth as possible.

"Yeah, Doug. I saw something about it on TV last night. I went over that design for months. I even reviewed it before I left. I don't think there's anything to it. Don't get me wrong, it's a terrible thing if it is true, but I was just over there in that region the other day, and I have to tell you, everything is a conspiracy to those people. It doesn't matter if you're a Jew or and Arab. If there's something even remotely wrong, then it's a conspiracy. In fact, I'm going through all of our stuff right now. I want to triple and quadruple-check all of the differences in our designs. If there is a problem, I'll find it in the next few days."

Josh's boss leaned forward and put his elbows on his desk. Then he folded his hands and bowed his head. Their eye contact and friendly atmosphere faded for a moment. There was a pause while he thought of what to say.

"Josh, the CDC's been looking it over, and they have some questions. They're…uh, a bit too technical for me, so I'd like you to handle it. Okay?"

Dr. Silverstein felt his inevitable meeting with the authorities coming closer and closer, and he hoped that he would have time to finish his chosen task.

"Sure, Doug. Not a problem."

Josh's boss leaned back in his chair, nodded in satisfaction, and then stood up. Josh followed suit as his boss moved to squeeze past him and open the door.

"Okay, Josh. I'll give those guys a call and ask them to stop by so that you can work things out with them. I imagine they'll come right over, so you might want to finish up whatever you're immediately working on, and get ready for them. Thanks again."

The two men shook hands in the hallway next to the kitchen. It had been an unusually brief conversation, but Dr. Silverstein understood very well what was happening. He knew that he'd probably be arrested in a few hourspossibly even minutes. With that in mind, he fought the urge to run back to his corner cubicle, and instead managed to move at a casual working pace.

Once there, he knew that he wouldn't have enough time to continue modifying all of the company's genetic therapy drug design database files. It would have taken forever anyhow. He needed something faster. Sitting at his desk, he felt the minutes-even the seconds-ticking away. Maybe he could automate the search and modification. Josh identified the string that he had modified in ZP9 to make it artificially age people of Arab descent. Then he created a new database search program that would systematically access all of the drug design database files in the company's file server.

If a drug design database file met certain criteria that Josh wrote into his search program, then that same search program would modify the drug design database file to include the changes. Those changes would be random, but similar to the ones that he had written into the various drugs that consisted of ZP9. The search program would have to run for a few days to search the company's entire file server, but at least Josh knew that his work would be complete whether he attended to it directly or indirectly.

Dr. Josh Silverstein finished writing his search program moments before the receptionist paged him. He didn't have time to test it, but-like most scientists of his generation-he was very computer savvy, and he had confidence in his abilities. After being paged, Josh saved the program to the company's file server and then started running it. He logged his terminal off from the server and walked to the lobby to meet with the investigating authorities.

Friday, April 13, 2011

The New Cleveland Press

Greece Concedes Cyprus to Turkey

Yesterday, the largest naval engagement since the Falkland Islands War took place in the Eastern Mediterranean Sea. A naval task force from Turkey was sent to land troops on the island of Cyprus. The invasion force was in direct violation of the 1974 United Nationsbrokered peace agreement between Greece and Turkey.

Aircraft from Greece rushed to stop the invasion fleet. Reports indicate that losses were heavy to extreme on both sides. In response to the Greek air attacks, Turkish submarines began a coordinated campaign of unrestricted submarine warfare in the Aegean and Eastern Mediterranean Seas.

By midnight both the Greek and Turkish navies had suffered grievous losses. However, the Internet News Service reports that Turkish commandos are already in control of the populated areas of the island. Defense Department sources confirm that sporadic air and naval attacks continue in the area, and the American military forces in the region could be called upon to intervene. State Department officials announced that "…although no one has made any declarations, it is very clear that a state of war now exists between Turkey and Greece."

Friday, April 13, 2011

Dearborn Daily News

Genocide Cannot Be Justified

The FBI today arrested local genetic research scientist Dr. Josh Silverstein. Charges are still being determined, but it is believed that Dr. Silverstein is connected to the international outcry of tampering claims made by various Middle Eastern health officials and government agencies. According to Justice Department sources, Dr. Silverstein may have altered the design of a genetic therapy drug that was being distributed by overseas medical charity organizations.

The alterations in the some of the chemicals in the drug are causing the drug to not only create a level of immunity toward the symptoms of cholera, but also to weaken the immune system in people of Arab descent. The drug in question is called ZP9, but the CDC is examining all genetic chemical formulas that Dr. Silverstein has been working on over the past few years. Hundreds of thousands of Middle Eastern children have either died or are in serious condition as a result of the free inoculations provided by the drug's manufacturer. Company officials at first claimed that there was no connection between the apparent side effects and deaths of so many Middle Eastern children, but in the wake of Dr.

Silverstein's arrest, there has been no further denial or any other official comment from the company.

The suspect drug, ZP9, was designed by Global Medical Technologies here in Dearborn, Michigan. It is a subsidiary of World Pharmaceuticals Inc., which is primarily owned by Global Technologies, Inc. All of the companies involved are major campaign contributors to Senators and House Representatives from both parties, and while calls are already being made for Congressional investigations, the question of bias is expected to loom large in the appointment of committee members.

Dr. Silverstein recently returned from Israel. The U.S. State Department funded his trip so that he could positively identify the remains of his wife and two daughters. Palestinian gunmen executed all three during a terrorist attack at an Israeli shopping mall a few weeks ago. Dr. Silverstein had seen them executed on the now familiar security camera videotape of the attack. That tape has been aired repeatedly on television stations around the world, and it is assumed to be a source of motive for his alleged crimes.

Local reaction has been a mixture of shock, surprise, outrage, and on occasion, sympathy! It is no secret that the largest concentration of Arabic people outside of Cairo, Egypt is here in the Detroit area. As expected, the thought of one man attempting single-handedly wipe out the Arabic culture has drawn a combination of shock and anger in all communities, but among the local Arab community in particular. Also in a state of shock are local Jewish leaders and groups, but there is an undercurrent of understanding and compassion for the Jewish-American scientist who allegedly acted in revenge for his family's killing. Such odd sentiment remains private, but the undercurrent is very apparent in many conversations. So far, police have managed to keep protesters away from the investigation, but as more facts are revealed and the community waits for justice, law enforcement officials concede that violence is expected.

Friday, April 13, 2011

The New London Daily Chronicle

Mid-East Rages Over Genocide Allegations

American Justice Department leaks about the ZP9 scandal spread throughout the Islamic world yesterday. What apparently began as a genuine report that the FBI was investigating possible criminal intent in the ZP9 modification and subsequent deaths, soon evolved into an alleged report of a CIA attempt at genocide aimed at ArabMuslims.

Violent protests seemed to spring up within hours. In Cairo, Egypt, the American embassy was surrounded and fire-bombed before Egyptian police could disperse the crowd. In Beruit, Lebanon, an initially peaceful protest turned into a riot at the American University. In Gaza, an American TV news crew was taken hostage, blindfolded, and paraded through the streets for almost an hour before Israeli soldiers drove away the crowd and rescued them. Similar protests seemed to occur in almost every Muslim nation-particularly those in the Middle East.

State Department officials have issued an alert to all Americans abroad. They are to being told to be as cautious as possible, change their routines, stay in large groups of other Americans and westerners, be aware of the local authorities, and to always keep their passports on their person. Americans in Middle Eastern nations, especially predominately Islamic countries, are being advised to leave at once.

Thursday, April 14, 2011

The New London Daily Chronicle

IRAQ JOINS EASTERN DEFENSE

After years of anti-Western banter, the government of Iraq announced yesterday that it plans to join the defense alliance organization that was created by The People's Republic Of China and The Democratic People's Republic of Korea (formerly North Korea) a few weeks ago. British and American diplomats have expressed a great deal of frustration at the move. Some fear that, in the wake of the recent public outcry about the American drug ZP9, Iraq's move might encourage other nations to join the socalled Eastern Alliance and form a more unified front against Western nations. With tensions between Greece and Turkey peaking over the Macedonia question and still simmering from the 1974 Cyprus issue, Turkey is expected to begin making efforts to join the Eastern Alliance within the next few days. Such a move would also involve Turkey's withdrawl from NATO, removing the only Muslim nation from that stalwart defense organization.

4/14/11

2310 hrs.

Hualien Airfield, Taiwan

Personal Log – Lt. Kim "Killer" Yoo

After almost a week of repairs, I was finally able to return to the air and fight in my own aircraft today. The Taiwanese are not very pleased with me. I believe that they partially blame me for the heavy damage to the Falcon that they loaned to me. Some of their wing-wipers were not very helpful in repairing my aircraft, and they left the work before finishing. I do not understand why, but it does not matter.

Their morning fighter-sweep patrol left before I could join them even after I told them to notify me when the flight was ready to leave so that I could have joined them. I was forced to wait for the Mig-29 division to wake up, recover from the last night's drinking, and finally leave for the mid-morning fighter-sweep patrol. We headed south to harass any communist air support that might be protecting the communist advance north from the Chinese beachhead near Taimali.

There was a squadron of Mig-29's flying 25,000 feet over the Pacific. We approached through the mountains at low altitude. The communists did not fire on us, and I assume that they did not recognize us as Coalition aircraft despite the presence of my F-16. I fired the first shots with my wingtip Sidewinders-the only missiles I could get mounted by the Taiwanese. Both struck separate Mig-29's, but they both flew on. The Mig-29's in my division climbed behind me, and the communists dove. Meyers has suggested that the communists must be having a missile shortage similar to our own, and I believe he is correct since none of the communists fired a single missile at any of us. I was able to fire a burst of 20mm into one of the Mig-29's that I had hit with a Sidewinder earlier. It came apart immediately, and I watched as large pieces of it passed by my plane on the left side, the right side, below me, and above me. The other Mig that I had hit with a Sidewinder took a few 30mm rounds from one of the Germans. As the communists passed me, I pulled over into a loop and dove back down after them. Our Peruvian Fulcrum pilot had remained at low altitude, and I watched him shoot down 2 more Mig-29's with a few quick bursts of cannon fire.

Almost all of our cannon kills were done with head-on attacks at high-speed. This division is not the most professional, but they are very skilled aerial marksmen. I am privileged to fly with them. Together, we are able to kill many many communists.

Thursday, April 14, 2011

2nd Lt. John Chamberlin-Personal Log

Hill 1419, 15 miles S of Taipei, Taiwan

Colonel Ghetty came up and visited the company again today. He says that Iraq has joined Turkey, China, and Korea in the Eastern Alliance. Israel is getting nervous that all the Arab nations will join up too. That would leave them naked in the desert without sunscreen. It also makes everyone wonder how many Islamic republics of the former Soviet Union will join up? Ghetty doesn't seem overly preoccupied with the global situation, though. The longer I know him, the more I respect his ability to view things. He does a nice job of not looking at things too tactically, or too strategically, all the while watching everything.

Closer to home, the Chinese are continuing to slug their way through the mountains of Taiwan. Our fleet is still too weak to come in closer and give us some real protection. They're still out to the east in the Philippine Sea. Reinforcements from the States haven't shown up, but we are getting new types of MRE's, and ammo is still plentiful. Ghetty says that the Taiwanese Army, backed by the forces that we were able to pull out of Korea, has been able to slow the Chinese advance considerably, but they've been unable to stop it completely. He compared it to a time when he was helping hold back the Mississippi during the 100-year flood in the early 1990's. "...as soon as one leak gets plugged, another dike starts to overflow three or four miles away." The map he showed us was evenly split into four areas: friendly, enemy, unknown, and untraversable. We're not at a stalemate by any stretch of the imagination, but we have definitely reached the high water mark for this island.

Friday, April 15, 2011

2nd Lt. John Chamberlin-Personal Log

Hill 1419, 15 miles S of Taipei, Taiwan

Most of the time that we've been on this hill, except for a few days when we first arrived, it's been cloudy. In the first few days, we were able to watch dogfights, missiles, flak, and all kinds of air activity as far away as Taipei. It seems to alternate between fog, drizzle, light rain, overcast, then the cycle repeats again; everyday. Today, the sky was crystal clear, and, from just after sunrise to just before sundown, we didn't see a single plane, missile, or tracer round in the sky. Most of us think that everyone's running low on high-tech weapons like air-to-air and surface-to-air missiles again. Everyone's probably running low on planes and pilots too.

There was a lot of smoke and fire from areas south of Taipei, and a few in the city itself. The city has been left in ruins. It reminds me of what Seoul, Pusan, Grozny, Sarajevo, and Beruit looked like after war had passed through their streets; modern buildings without windows and pillars of rubble hanging over the cluttered streets. Pock marks from all sorts of caliber on everything.

Hector suggested we have someone from each squad looking out for aircraft at all times. Bob and I agreed and made it happen. Bruce said he would too, but we only saw one person on watch in his platoon at any given time. The only time there might have been two people in Bruce's platoon looking out would be during shift change. I think he's starting to feel the effects of the boredom.

Saturday, April 16, 2011, 4:73 PM

35000 feet above the Sea of Japan

GTI Flight 1030

A blanket of overcast lay below the small corporate jet. The Earth was out of sight and out of mind - at least for little while. Dan Greene was busy mixing drinks from a mini bar hidden in one of the plane's small cupboards. General Naht Synn sat in one of the large and luxurious reclining chairs surrounding a conference table. At the head of the table immediately to his left, Walt Barber was busy pouring him a cup of tea.

The whisping sound of the planes three jet engines was well muffled by the thick insulation of the plane, but in its wake there was silence. Most of the time the silence and the big comfortable chairs made for a very relaxing flight. Sometimes, in moments of conversation where much was at stake, Walt had learned that the silence could be used as an intimidating weapon. As the plane's defacto owner, he could feel at peace with the silence, but it had been known to make guests on the plane uneasy in the past. Such nervousness sometimes played well on in-flight financial negotiations.

General Naht Synn was not the kind of man to lose confidence in any situation. He knew why he had come on the plane, and he was completely focused. While Walt quietly prepared his tea, and Dan readied Walt's bourbon, Naht Synn was intent. If anything, his fixed gaze at Walt had inadvertently conveyed his lack of mental peripheral vision.

The plane had taken off from Pyongyang, Korea with little more than the usual pleasantries. There was talk of the war, the weather, world politics, and a tinge of gossip about common associates. When Dan left the room, and Walt began his little prolonged silence trick, the Chinese general knew that the moment of purpose was at hand.

Dan finally returned with the bourbon, passed it to Walt, and took his seat. He sat at the conference table, but as far from the two as possible. His ego kept him next to Walt, but his head still knew that he was not much more than a glorified secretary.

The general opened the conversation.

"Walt, on behalf of the Chinese people, and socialists everywhere, I want add my thanks to you-our unsung hero who has made all of this possible." He rose his cup of tea in toast and continued.

"Without the investment that you and your associates have made in our nation, without the communications software that you provided, and without your tireless work, the people of the entire world may have never been able to free themselves of the capitalist noose that the west has held for so long. Thank you."

Walt and Dan raised their bourbons high and sipped in sync with the general and his tea. Then Walt politely thanked him for the thanks. It seemed to bounce back and forth, again and again. Finally, after a few minutes, it ended, and Dan rose to get some snacks. Naht Synn wanted to cut through the fat in the conversation. Walt opened the door for him.

"You make my actions sound so noble. You should know, I spent months going over my decision about selling the software. In fact, I'd always thought of myself as a good American. When I found out about the software contract, and how it would destabilize the world's military balance, I realized that the United States was no longer really a government by the people, for the people, and all that. I don't think I was really convinced until I met with Kim Il Sung and Deng

Xioping just before the Gulf War. I had been fortunate enough to counsel both on financial matters before that day. They were good men, with good ideals."

Walt paused for a few seconds, but not too long. Then he continued.

"I miss them both. I think that's the day you and I met, wasn't it, general?"

He seemed genuinely saddened at the memory of the two late socialists, but the politically savvy Chinese general had spent too many years in the counter-intelligence business. He could smell illusion in the same manner that a sailor could smell the sea from miles inland. He could imitate it too, but preferred the Chinese military style of a powerful frontal attack.

Naht Synn delicately sipped his tea, returned the cup to its saucer, and ever so slowly leaned forward. The plane's engines were so quiet that even Dan, now sitting at the far end of the conference table, could hear Naht Synn's wool dress uniform rub as he leaned forward.

"Mr. Barber...Walt, you and I seem to have so much in common. We both come from successful, moral families. We both studied at American universities, we both have found our calling as silent heroes for the people."

He sat silently for a few seconds and let his thoughts sink into Walt. Naht Synn knew that Walt was waiting for him to finish his speech. Both men knew that this speech would be the heart of the general's request for the meeting.

The plane began a slow wide turn to the west while it began it's descent into Taiwan International Airport. Beams of sunlight crawled their way across the cabin from front to back. Naht Synn, on the right hand side of the plane still had his eyes locked on Walt. Dan moved closer to the two men. He just had to get into the conversation even if it was on the peripheral; even if the sun was directly in his eyes. The general continued.

"Walt, sometimes I have to wonder? Why do I find myself interrogating downed pilots in muddy bunkers during artillery barrages? Why is it that I'm in a uniform, carrying a gun? Why do I have to wear gloves to keep blood from getting under my nails while you probably get a regular manicure?"

The general's tone was getting more methodical, more rehearsed. Dan unconsciously clinched his fists to hide his recent nail job, but he also made a mental note to cancel Walt's for the following day. His mentor was focusing on every single aspect of the general's speech for some clue of what was coming, and how to appropriately respond. In his mind, he rewound and played back every word, every syllable.

"Walt..."

The general paused and leaned back slowly with his uniform making even more noises this time.

"Walt, I know. I know why you were so willing to sell your software. I know your secret. I know that even though you and I are alike in so many ways, I know that we are not some sort of patriots-not both of us."

It was time for verbal poker. Walt was normally a winner, but this time, he was up against a professional. If ever there was someone who might have been able to claim that they had figured him out, Walt knew it would very well be a professional. Someone who's career had been forged from unveiling secrets; exposing lies. Of course, that same course of schooling would have also made someone like Naht Synn a very good liar too. Was he telling the truth? Did he really know something, or was he trying to bluff? Walt knew only one thing, if he tried to answer any allegations too quickly, it would give the impression of some sort of guilt. Naht Synn was a dangerous and powerful man. Walt had always known that, so he let the general continue while he put on his best full house face.

"I know about your investments, Walt. When I was first asked to attend that meeting back in January of 1991, I pulled your file. It was clean; too clean. I had never known Kim Il Sung to have met with someone who hadn't needed to cross the line at some time or another. Secrets always reveal themselves – eventually - and our records almost always at least have a clue to a person's 'lack of innocence'. I'd seen children with far more footprints in their path. When I started asking around about your relationship with Kim Il Sung and Deng Xiaoping, I found out that you were the one who had facilitated the massive grain shipments during our unfortunate harvests the year before. China could have survived. We may have lost a million or two people, but that has often been the Chinese way. Korea would have fallen to the capitalists for certain. One would think that this would have made you a patriot, but I'm a suspicious man, Walt. I could taste your mettle just by reading your file."

Walt's poker face was perfect. Dan's was close, but it hinted at his hidden uneasiness. Even though Naht Synn was staring into the blackness of Walt's pupils, into his soul, the general could still feel Dan's walls cracking.

"I asked myself why a quiet, successful, western businessman would discreetly arrange for those massive shipments of grain. Boatload after boatload came to the docks. Hundreds of thousands of people would live because of your generosity. Why? There was no previous example of philanthropy, and there hasn't been one since. Why?"

Naht Synn was getting close. Walt knew it. He kept his poker face, but Naht Synn was beginning to see his soul.

"It was all about motivation. Surely you're a capable man. No one else in the entire world was able to get those ships through the Taiwan Strait."

The general leaned back the rest of the way, rested his arms on the sides of his chair and laughed.

"I remember how you scared the Chinese Nationalists and they sent out their ships to block the grain. I remember how the Americans even moved in one of their carrier battle groups to stop them. That's when I noticed that there was more to the story than just grain. The Americans would never let themselves be seen as the open advocates of a starvation campaign, even against communists. So why were you, an obvious and avid capitalist, trying to feed the poor communists against the will of your own people, and why were your own people trying to stop you?"

There was silence. The plane's descent grew greater.

"What could you gain? That's when I began to have your finances examined. I quietly had a team put together just on the chance that you weren't really a socialist saint."

Air speed reduced.

"It took a few months, but we finally got onto the trail when one of our people noticed that a company, of which you owned a controlling share, owned a controlling share in a construction company. The construction company had recently been awarded an operations license to build a new factory outside of Beijing. We looked into the company and found that it had been getting its materials and labor for free through one of the People's Liberation Army's engineering battalions."

Flaps were deployed, landing gear came out. Naht Synn knew he was hitting home and peeling away layers of the onion that was Walt Barber. Still, the flight had been too swift, and he wanted to finish just as they landed.

"Eventually Walt, we found that your web of holdings was allowing you to trade food for business permits in my country, in Korea, and all over the globe. You've been setting up and running capitalist operations in communist countries. You've been exploiting the low labor rates of our people to bleed hard currency out of the west. You've been profiteering from tensions between east and west. Not only did you profit extremely well from the SP 2.0 software, but I think you baited us with it just to make the tensions heat up and make your investments more valuable!"

After a brief silence, the plane's wheels barked on touchdown with the runway.

"You and I both know people who have disappeared for far less than what you are doing. However, I have to admit, without you, none of this would have been possible. The Chinese Nationalists would still be an issue. Korea would be divided. The west would still hold three-fourths of the world by their short hairs."

Dan had sat and listened carefully as their ad hoc plan for international war profiteering was being unveiled. Typical to his demeanor, Dan interrupted in an attempt to defend his superior; his idol, his mentor.

"General, I'm sorry you've put so much effort into pursuing some sort of wrongdoing, but I can assure you that no one has-"

In one fluid motion, the general reached into his right front pocket, pulled out a palm-sized snub-nose .38 caliber revolver, put it to Dan's head and jerked the trigger. As fast as the weapon had been drawn, it was put back in his uniform while Dan's lifeless body slumped back into its huge chair gushing blood from a gaping hole in the back. Walt didn't even flinch. He almost expected it the instant that Dan opened his mouth. General Naht Synn had never given even the slightest impression of being anything less than completely ruthless.

"Walt, I'm sorry, but I just couldn't take anymore of that ass-kissing 'yes man.' Besides, this helps to prove my point. The software that we've purchased has a flaw that I don't think can be remedied."

Now Walt's expression finally turned to concern. He forced himself to ignore Dan's death, the flowing sound of blood from the chair to the floor of the cabin, and the smell of gunpowder in the plane.

"General, I'm not sure I understand the flaw. Could you elaborate so that I can have it looked into?"

In the plane's cockpit, the pilot and copilot had glanced back to see what had happened. Since then, they kept their conversations with the control tower very quick and quiet as the plane was directed towards one of the battle damaged airport gates.

"Look Walt, the software allows us to know exactly where every man and woman in every active Coalition force is as long as the unit that they're assigned to is connected to the U.S. MilNet. That part works fine. The problem is that units that are taken from the MilNet due to losses of over 70% no longer appear. The same is true for most special operations teams; especially ones with less than 25 or 30 people involved."

Walt understood. In fact he knew about the problem already. He also knew that some entire countries, like Vietnam, didn't have any of their forces tied into the MilNet. Still, in order to keep a bargaining edge with the Chinese, he was going to let them stumble into Vietnam. Tens of thousands of Chinese would be killed, wounded or captured all because he needed to be able to bargain with the likes of Naht Synn. He wondered if Naht Synn knew about Vietnam's lack of integration into the MilNet?

"General, I believe the software is working as advertised, but there will always be the unexpected in a war. There might be some sort of way to convince the Coalition that they should attach all their units to the MilNet, but it might take some time to come up with a good answer. Do your people have any ideas?"

"Actually, Walt, they do. It's been suggested that since the software is working, but its usefulness diminishes with every battle we win; every unit we decimate below 30% strength. It's been suggested that your usefulness has come to an end. In fact, some have said that you might be a security risk. At the very least, there are some questioning why we should allow you and your capitalist associates to rake in profits derived from either the People's Republic of China, or the Democratic People's Republic of Korea?"

If the general had come to kill Walt, he'd have done it by now, and both men knew it. Naht Synn wanted something from Walt. They both waited. Naht Synn waited for Walt to beg for forgiveness. Walt waited for Naht Synn to stop playing around and make his request, whatever it might be. They waited until the plane finally arrived at the gate. Then, as even the pilot and copilot remained seated and silent, Naht Synn spoke.

"Walt, we're going into Indochina. You probably already know that. Make no mistake about it, you are being allowed to live, but your life is no longer footloose and fancy-free. As we continue to reduce Coalition units below 30%, your software will be less and less useful. You're going to need to do something, anything, extraordinary to justify your existence. Good day, Walt."

The general rose and headed for the door. Walt followed. The copilot raced to open the door and drop the stepladder. No one on the plane wanted to be the next Dan Greene. General Naht Synn left the plane without even shaking hands or nodding goodbye. It was clear to Walt that even though he was secretly one of the world's first trillionaires, his life had no value. In fact, he was on borrowed time. When the Chinese moved into Vietnam, they'd find out just how valuable he and the SP 2.0 software really could be. He'd see to that. Still, he needed to reinforce his value, his survivability with the communists.

Sunday, April 17, 2011

2nd Lt. John Chamberlin-Personal Log

Hill 1419, 15 miles S of Taipei, Taiwan

We're still on the hill, and the world is falling apart. Colonel Ghetty came up again and joined us for Hector's morning briefing. He tried to explain what's happening in the Mediterranean, but it was too confusing, especially the Balkans. Greece and Turkey are officially at war. Greece has invaded Macedonia. Turkey has taken Cyprus. Syria, various groups in Lebanon, and Libya have all recently attacked shipping in support of Turkey. Greece has threatened to close off the Black Sea if Turkish ground forces try to cross over the Bosporus Strait or the Dardenelles. Israeli military forces on are the highest alert, which is even making the Egyptians nervous. The only carrier group in the area is the George Washington. So far, working closely with NATO air forces, they've been able to scare off any would be attackers-so far. Here in Taiwan, the word is that reinforcements have joined up with Task Force Stingray. Everyone's hoping they'll move back into the Taiwan Strait and block Chinese reinforcements. There was even a tinge of hope in Ghetty's voice when he told us about the scuttlebutt.

It wasn't until he told us about the possible reinforcements that I realized just how low morale has become. Everyone seems fine, but I think we're all just putting on aires or trying very hard not to let the stress of the situation get to us. I know I still feel a sense of guilt about that first TRAP mission, LZ Delta, Mike, Tim, Kathy, Gary, Bill, Dave, Lee. I can deal with each one by thinking about them rationally and logically. Combined, I guess they have built up some stress. Bruce must be have had the same thought I did at the same time because he immediately turned to me and reminded Colonel Ghetty that we haven't had any days off this year. He wasn't whining, but it was an important reminder to make after 4 1/2 months. I'm sure the grunts are feeling the wear and tear too, but they tend to get closer and make their own R&R whenever they can.

Sunday April 17, 2011

Washington Weekly Magazine

Balkan Fuse Is Lit Once Again

The fall of Cyprus to Turkish commandos, and the recent naval engagement in the Eastern Mediterranean Sea proved to be 'just cause' enough for Greek decision makers. With two huge political and military defeats in as many days, Greece ordered its troops into the troubled province of Macedonia. The move was an expected response to the recent terrorist attacks in that neighboring nation, and the probability of the Greek invasion was most likely the cause of Turkey's move into Cyprus.

Macedonia was recently a province of the former Yugoslavia. Following the disintegration of that nation, Macedonia found itself in turmoil similar to that of the other former Yugoslav republics. Serbian-backed insurgents, Islamic extremists, nationalist revolutionaries, socialist militants, bandits, gangs, and local militia groups all threatened to move the former province into the same chaos as Lebanon, Afghanistan, Bosnia, Somalia, Rwanda, or Kosovo.

After the failure of international intervention in all of those national self-destructions, NATO learned its lesson and moved peacekeeping forces into Macedonia. Their presence had been based on the condition that they were to be removed in stages over a period of years. However, as the unit-strength of the NATO forces diminished, insurgency began to rise. Regardless of the situation, NATO forces planned on continuing their timetable for withdrawal, and terrorist attacks grew in number and intensity as a result. As long as NATO was planning to leave the beleaguered country, and as long as anarchy continued to rise, the intervention of Greek forces was not only expected, but is also considered necessary by many in the world community.

Monday April 18, 2011

1112hrs local

U.S.S. George Washington CVN

Eastern Mediterranean Sea

On an aircraft carrier, night and day lose some of their meaning. That's particularly true on a carrier that is short-handed. Such instances leave the naval aviators and pilots on board doing even more than their usual share of duties. Flight time increases. So does the experience of the pilots, as well as their fatigue levels.

Chuck had just come off of a 14-hour patrol. The Mediterranean sun made his cockpit unbearably hot. As usual, his ass felt as though he had been sitting naked on a 2x4 over a bucket for half of a day. After arriving on the ship, there was another hour of debriefing, aircraft maintenance review meetings, and the like. Finally, he skipped eating and hit the sack. It was the middle of the day for most of the crew, but for him, it was just as noisy and lit as any night.

The smells of an aircraft carrier were amazing. Everything had the scent of jet fuel, hydraulic fluid, burning tires, and hard-working men and women. Chuck was deep asleep when he awoke to an unusually pungent odor. He opened his eyes and saw that his darkened stateroom was filled with smoke.

Only his new roommate's desk lamp told the story.

Rick was soldering. Under the dim glow of the lamp in the windowless room, he had covered the room's smoke alarm, put a wet towel at the bottom of the hatch, and was hard at work doing something with a soldering iron. The smell pinched in the back of Chuck's throat. He snorted and tried to clear it, and an apologetic Rick discovered that he had awoken his host.

"Oh man, I'm sorry. I didn't mean to wake you up."

Chuck rubbed his dead-tired eyes and rolled over to get a better look at what Rick was doing. With a grumbling mumble, he asked him.

"What the hell are you doing? It's almost midnight."

"Rick chuckled. "No it's not. It's almost noon. You just got in. I tried to be quiet, man. Guess the solder woke you. I didn't think it was that bad, but when you've got your face in it, well…you get used to it."

Chuck's eyes regained their focus, and he saw what Rick was working on. On the floor next to the small desk where Rick was soldering was a 1' diameter, 2' high nosecone to a HARM missile.

"Uh, Rick…what're you doing with that warhead?"

"Oh no. It's not a warhead. This is just the guidance system for a HARM."

Chuck huffed and tried to get a better answer.

"That's not what I mean, Rick. What…are…you…doing?" "Well, have you ever met a Petty Officer named Richardson?" Chuck shook his head.

"It doesn't matter. He's a red-shirt. We were talking back in Hangar 2 by my aircraft last night. He's into sims too, and when I told him about Bob-my AI5 program-he thought that it was a monster idea."

Chuck's eyes searched around looking for an explanation about Rick's

'monster idea.'

"You okay, Chuck? I'm real sorry about the smell."

"Nah, I'm fine. What's your 'monster idea' again? I forgot."

"I mentioned to him that someday I hoped Bob-AI5 that is-might be used in a missile-guidance program. He thought it was a great idea. So, we talked some more, and he gave me this spare guidance system to try it out with."

Chuck paused while he tried to come to grips with the reality in front of his eyes.

"Now, Rick, you mean some Petty Officer just gave you this multimillion-dollar missile-guidance system cause he thought your flight sim program was cool?"

"Nah. These HARM's are all pretty old. This one goes back to just after Vietnam. They were only a few hundred-grand when the Navy first bought'em, and this guidance system isn't even worth that. What happened was that they had a HARM with a faulty rocket motor. They had taken the guidance system off when they were searching for the problem, and when they found it in the rocket motor, the manufacturer wanted it back. The Navy was just too lazy to return the guidance system, and the manufacturer was too lazy to make a stink. So, this thing's been down in the hold taking up space for years."

"How do you plan on having it track something? I thought HARM's just had passive antennas for locking on to radars."

"Well, this one USED to have one of those. I took it out and put in an old laptop that I had. It's just got an old 56k modem/cell phone, but that'll do the trick. The funny thing is I have to call the missile to establish an uplink between the old laptop and the computer in my plane. The big question is, 'Do I have enough minutes left in my cell phone account?!'"

Chuck flopped backwards into his rack and closed his eyes. It was so screwed-up, twisted, and typical that the story was believable. The hardest part to comprehend was that no one had tossed it overboard!

Rick was on a roll, and even though it was obvious that Chuck wanted to get back to sleep, he continued.

"Check this out. I took a 23-pin data ribbon from a guy I met outside of CIC. I connected that to this converter bar, then I spliced these wires to fit into an old

RS-232 plug. That plug fits right in here. Now all I've gotta do is remember how this old DOS system works, and I'll know how to input my program."

"What're you gonna do with it if you're able to upload your program? D'you really think that it'll work? I thought the AI5 program was based around flying an aircraft. How're you gonna come up with the flight characteristic stats for a HARM missile?"

Chuck's eyes were closed, and silence followed his possibly projectending questions. He opened them to see what Rick thought. It was just in time to see him flipping through a three-ring binder; the maintenance and technical specification manual for an AGM-88 HARM missile.

"Here it is Chuck…the acceleration rate, max-weight, top speed, and…yep, turn radius. I can input that into my laptop. I'll just create a fictitious aircraft and give it those specs. Then I'll make Bob fly it for a few days until he gets the hang of it."

Chuck huffed and flopped backwards. In less than a minute, he was sound asleep in his disbelief. Rick was truly reaching deep into the bucket of super-geekdom. In the small 12'x12' stateroom, one pilot was crashed out while the other burned away with his soldering iron.

Tuesday, April 19, 2011

9:51am local

Federal Building

Detroit, Michigan

It may have been April, but the weather outside was reminiscent of February. An Alberta Clipper cold front had brought a late snowstorm down from Canada. The temperature had fallen from 61F to 10F overnight. The sunrise was blocked by the typical low-level overcast sky. A heavy, lake-effect snow was being carried over the state from Lake Michigan to the west. In Detroit, it was accelerated by 23mph winds, and the crowd outside the Federal building felt like they were being sandblasted. The storm wasn't a common occurrence, but it was far from unheard of also.

Inside the building, a small army of security officials protected one man. U.S. Marshals, State Police, Detroit Police, FBI agents, ATF agents, and more than a few less-public agencies guarded Dr. Joshua Silverstein. He had committed the worst crime in American history, and attempted to commit one of the worst atrocities in world history. There was no specific American law against genocide, so instead he was being charged with "an undisclosed number" of homicide charges (security officials didn't want the growing number of deaths to become public.). Dr. Silverstein had tried to wipe out an entire people. The effects of his mass-poisoning plan were still being felt. People were dying in the Middle East, and they would continue to die until someone discovered the extent of Silverstein's tinkering with different genetic drug formulas. Even while he was in court, his still undetected database program continued to identify new drugs and alter them.

Outside the Federal Building, covered in snow and frostbitten by wind chill, hundreds of reporters and news crews waited for the American mass murderer to come out. They all needed their five seconds of film, and everyone prayed for a soundbite. Silverstein's acts were as despicable and disgusting as possible, but for many of the media members, that had to take a back seat to getting some sort of an edge to their story.

Around the reporters and news crews were thousands of protesters. Detroit's Arab and Muslim populations were the largest in America, and if Dr. Silverstein's plan worked-which it was-their lives were in jeopardy. The only reason for their deaths seemed like their race or religion. Instead, Silverstein had condemned all people of Arabic descent to death because of what he perceived as their permissiveness. There was no doubt that he was motivated by revenge for the brutal deaths of his family, but his rationalization was that violence in the name of Islam, Zionism, or simple cyclical violence all could have ended if people weren't so permissive.

The entire world was guilty of that crime, but Silverstein had confessed to the FBI his belief that only Arabs could have convinced other Arabs to break the circle of violence. Argumentatively, he had pointed out that every other group of people had tried in one fashion or another, and each was dismissed as either being Zionist-backed or Zionist-serving. There was even a moment of dark irony during his FBI interrogation when he described his "work" as a "Final Solution" to the Middle Eastern problem. Adolf Hitler had used the exact same words when he ordered the death and destruction of Dr. Silverstein's people.

Josh was going to die. He knew that. There could be no other result to his arrest. The best that he could hope for was consecutive life sentences, but living was empty for him without his family. They were everything to him. Whenever guilt or remorse had creaked open a doorway to his soul, he remembered the special moments with his wife-before and after they had been married. He also remembered the butterfly kisses from his little girls, their tiny hugs, and the sugar-sweet sound of their synchronous chorus, "We love you, Daddy." Those images, feelings, and harmonies had been torn apart by the repetitive footage on his TV. Instead, there were new sounds of their terrified shrieks, new images of their bodies going limp when they were executed, and there was a bitter bile taste in his throat from watching it happen. Any remorse or guilt vanished in the face of such fuel for vengeance.

His court appearance began at 6:00am. It took hours to read the charges to him and explain the evidence to the judge. It was just the first hearing in a long line of attempts at justice, but it finally ended. As the parade of uniformed and disguised security left the court, Josh was stopped so that a bomb squad heavy bulletproof vest could be put on him. While a few people muscled it around him, the senior FBI agent (the same man who had interrogated him earlier) stared him in the eyes. In a calm and professional voice that seemed to assume cooperation, the FBI agent asked for Dr. Silverstein's help.

"Josh, we're pretty sure we've found all of the different formulas that you were working on, but I'd like to show you a list, and have you pick out any that we may have missed. Would that be okay?"

Silverstein's face was blank and cold. Whenever he verbally heard details of his "work," he blocked off possible incursions of remorse or guilt by remembering his family.

"I'm sorry. I can't help you," he replied.

The courtroom had cleared behind them except for a few groups of police at the rear doors to the room and the agents who were about to finish putting on Josh's vest. His FBI agent turned and asked the other agents, "Could you gentlemen give me a minute alone here?" They stepped back and watched the perimeter of the hall and the other groups of police on guard duty.

When their attention was fully and purposely focused elsewhere, the lead agent continued his impromptu discussion with Josh. He tried rationalization.

"Look, Josh, we all understand about your family, but what you've done is as wrong as it possibly can be. Right now, you're the one doing the killing. We just want it to end. You've made your point. The entire world sees that, but now it's time to help us out and make it all end. It's up to you."

Josh's mind had traveled back to a toy store that they had all visited months earlier. There was no response at all. He was blank, and now the FBI agent tried a new tack.

"Okay, Josh. Let me put it to you this way. What you've done is put the entire nation at risk. All kinds of countries around the world are supposed to be helping us out right now, but instead, even some of our long-time friends are calling for justice. Everyone has some sort of thing that they'd like us to do to you, but very few people actually know that we need your help. There have been some suggestions from Washington concerning your fate. These court appearances are going to be just that-appearances. So, I've been unofficially ordered to gain your cooperation at any price. If that means that something happens to you, well…that's just the way it'll have to be. Every minute that you delay, more people will die, and international hatred for our country will grow. So, here it is: either you look at this list and tell us if we've missed any of your 'work,' or I'm gonna have the guys take off this vest. Then, you and I are going to walk out to that prison bus alone."

There was silence. Josh's efforts to block out any regret had been emotionally draining. He was becoming more and more depressed. Consecutive life sentences would be the worst of his fates, and he selfishly prayed for an end to his own personal grief. He didn't even try to answer, and the FBI agent wasn't bluffing.

There were no orders to kill Josh, but it had been suggested that if he didn't cooperate, the government's responsibility to protect him would become limited at best. The lead agent had the vest removed, and they headed out of the Federal Building. The sound of the crowd outside drew closer, and everyone looked into Josh's eyes to see if he had changed his mind. He didn't.

At the outside doors, the lead agent stopped and gazed into Dr. Silverstein's soul. His eyes were as blank and empty as his heart. Through the doors and down the stairs, the Detroit Metro Police had set up a police line to keep the media back. Behind the hundreds of pretty people with microphones and/or cameras, thousands of protesters chanted for Josh's death, "Nazi-Jew! Nazi-Jew!"

The agent looked at his subordinates and the representatives from the other agencies and addressed them.

"Okay, you guys, it looks like Metro has a good perimeter prepped. You two go ahead and hold open the door. You, go ahead and get my door on the left passenger side. The rest of you help out with the perimeter."

There were no objections from anyone-including the U.S. Marshals who were really in charge of Josh's security operation. Everyone headed out to their tasks. When they were in position, Josh and the lead FBI agent walked out.

When the doors opened, the crowd went nuts. Their chant faded into a roar of slurs. Members of the media frenzy began shouting questions from 50' away along the sidewalk and from around the waiting prison bus/van. Josh took his first steps forward, and debris of all kinds began to fly at him from the crowd.

He walked down the twelve steps to the sidewalk and was deafened by the media/crowd. There were questions of all kinds. "Did you do it?" "How could you do such a thing?" "Are the drugs still dangerous?" "What do you think will happen to you?" "What did you hope to accomplish?" Finally, a single reporter who did her homework asked a question that Josh couldn't resist. "Did you alter the drugs because of what happened to your family?"

Josh snapped. He stopped, turned to face the reporter, and began waving his cuffed wrists at the camera.

"You're damned right I did it because of them! I watched my wife and little girls executed on your station! I watched it over and over again, and no one did anything. Not a damn thing! The next day, someone else lost their family in the same tit-for-tat attacks. It doesn't matter if you're an Arab, a Jew, an American, or whatever! It doesn't seem to matter who you are, the Arabs will never let the Jews live in peace, and Jewish revenge will never stop. Well, I found a way! I'm putting an end to it. The Arabs had their chance to accept

Israel. Now they have to pay the price for their decisions!"

The reporter's jaw dropped, and she fell silent. So did most of the media members in the immediate area, but the crowd went crazy. He had tried to blame his gross sin of vengeance on an entire race-the entire crowd. Now they wanted blood.

A reporter at the top of the steps drew Josh's attention.

"Dr. Silverstein! Do you really think that you can wipe out an entire race by yourself? What good can possibly come from the death of millions?" Again he waved his cuffed hands in front of the cameras.

"Peace will finally come to that region, and YES, I can do it myself! In fact, I'm doing it right now!"

He waved at the Federal Building doors and ranted onward.

"Just inside those doors, the FBI tried to get me to help them identify all my 'work.' I can't! What I've let loose is changing medicine! All over the world, it's changing the water, the air, and even the food! One person can do it, and the world had better wake up to the fa-"

A shot rang out from somewhere in the crowd. The security detail that had stood by idly while Dr. Silverstein verbally drew fire was now piled on to his body. It was a vain attempt at finally appearing professional. Other police ran into the panicked crowd. Thousands of reporters, TV crews, and protesters ran in every direction.

Then more single shots rang out. They were followed by the steady booming sound of police shotguns, and the small burping sound of submachine guns from other security members. Heavier automatic weapons fire started to bark from all directions. In the chaos, people were tripping, falling, and being shot.

A small band of white supremacists had melted into the crowd and opened fire in an attempt to seize Dr. Silverstein. When they opened fire, some of the people in the crowd who were carrying concealed weapons (pistols mostly) also opened fire. The police thought that the crowd was attacking them, and the crowd thought the opposite. In the middle were the six white supremacists that had spread out into the crowd before the shooting.

One of the supremacists, wearing blue jeans and a black vinyl coat, knelt down next to the prison van and began picking off police officers who were firing into the crowd. He shot nine before someone spotted him and killed him. Another white supremacist dove through the pile of panicked media members around Josh and tried to pull him from underneath the FBI agents and U.S. Marshals. He was pinned down, subdued, and arrested.

Three of the other four supremacists were shot and killed by U.S. Marshals and Detroit Metro police. One was killed at point-blank range by a shotgun blast to the head. All the while, a TV camera that had fallen to the ground transmitted the scene.

The last of the supremacists escaped into the panicked crowd. She was arrested an hour later by Toledo, Ohio police at a checkpoint as she headed south on Interstate 75. When asked why they had started shooting, she told investigators, "…even though he was a Jew, Silverstein held the key to a racially pure world."

At the end of the day, the scene outside of the Detroit Federal Building was that of a war zone. An army of ambulances was scattered in the surrounding streets. Wounded protesters and members of the mass media littered the street as though on a battlefield. Debris of all sorts was everywhere. So were drops of blood and spent brass from the storm of people who had been shooting. Over a hundred people were killed. Four times as many were wounded. Among the dead and lifeless was the limp body of Dr. Josh Silverstein. His orange prison jumpsuit was spattered red with blood, and he was finally on his way to true judgement.

The FBI had assumed that Dr. Silverstein knew exactly what he had done, but he did not. All that Josh knew was that he had written a program to automatically search for chemical formulas, determine if they targeted people of Arab descent, and then modify those formulas. Josh had no idea if the program was even working.

It was. The program was searching, identifying, and modifying so many formulas that the FBI couldn't determine which questionable formulas had been tampered with. It was like trying to find a single leaf on a tree. Only luck could intervene.

A few hours after the Detroit Massacre, their luck changed. The FBI was able to determine that many of the formulas had been changed in the same manner. Then they discovered that almost all of the questionable formulas were changed after Dr. Silverstein's arrest. That led them to determine that a computer program was running and changing the formulas. No other human was capable of was searching, identifying, and modifying so many formulas so fast. They wrote an adhoc trace program, isolated Josh's, and stopped it from running.

The artificial genetic aging disease that Dr. Silverstein had imbedded into the drug formulas was easy enough to identify. It's spread was stopped by halting the distribution of the drugs. However, the effects were impossible to correct. It would be weeks before a retro-gene therapy drug could be developed, and people would be understandably reluctant to use it. In the meantime, hundreds of thousands were dying in the Middle East.

Those who were not dying, protested. Dr. Silverstein's last words had been caught on tape by a hundred TV cameras. They were broadcast around the world. Anti-American protests and riots of all sizes erupted all over the world, including inside the United States. Ironically, the safest place to be suddenly became Israel. It seemed that there was finally a greater threat to Arabs than Israel; The United States of America.

Wednesday, April 20, 2011

1533hrs local

U.S.S. George Washington CVN

Eastern Mediterranean Sea

America's Latest War (the same "Conflict Other Than War" that had drawn all the available aircraft carriers away from the Mediterranean) was now overshadowed by a rising death toll among civilian Arabs, claims of American genocide, and graphic footage of the Detroit Massacre. Even the war between Greece and Turkey went unreported by most major global media outlets. Everything was publicly put on the shelf while the 'ZP9 Crisis' evolved.

In reality, America's Latest War in the East continued unabated. The war between Greece and Turkey was still being fought all over the Eastern Mediterranean Sea, and the U.S.S. George Washington was firmly in 'Harm's Way.' World politics as seen by even the most well informed news junkie didn't hold a candle to what was really going on. At any given moment, the GW was tracking a missile, plane, or torpedo from either Greece or Turkey. A full-fledged shooting war was going on in the Aegean and Mediterranean Seas, and few knew.

The GW had been on and off of a General Quarters readiness level for the past few days. It was wearing on both the aircrews and the ship's crew. Everyone was pulling extra duty shifts. That meant more aircraft flight time and less time for routine maintenance. Less maintenance meant more aircraft breakdowns, and soon, Chuck's squadron of aging F/A-18C Hornets was down to only two available planes. All of the others were finally grounded for maintenance.

So, Chuck picked up some chicken sandwiches from the Dirty Shirt Café and brought them back to his stateroom. Rick and one of the other Marine pilots had linked their laptop computers together and were playing yet another flight sim. Chuck passed out the sandwiches, sat on his rack, and watched as a crowd of people passing by the stateroom started to gather in the hatchway to watch.

In the simulation, Rick was flying a Vietnam War Era Phantom F-4C. His opponent was flying a Super Hornet identical even in call sign to his regular "real-life" ride. To make it more interesting, Rick had re-written the program's code so that his Phantom used the same flight characteristics as the HARM missile that he was modifying (almost every pilot and naval aviator on the GW knew about the project by this point). While Rick had the advantage in speed and turning, his acceleration rate was zero, his visibility was very poor, and he had no weapons. The only way that Rick could win was to ram his opponent; his squadronmate.

The crowd of onlookers grew until the small room was packed. Chuck could no longer see the small monitors just a few feet away. He had to rely on the "ooo's" and "ahh's" of the crowd to know what was going on. Every few minutes their cheers would reach a crescendo as Rick would win a round. In the drone of the crowd, the muffled activity of the flightdeck above, and the hum of the ship's

THE X-MAS WAR

reactors, Chuck heard Codie Custer's annoying voice in the stateroom. It seemed as though every Marine pilot on the ship was there.

Finally, the Marine squadron commander came by and broke up the crowd. The ship was on alert, and it was definitely not the best time to be having any sort of spectator sport on the ship. Chuck was pleased. It meant that he would be able to get some seriously needed sleep-or so he thought. Codie wanted to stay and babble. He even tried to make conversation with Chuck as though he, Rick, and Chuck were best friends. It was quite the arrogance role reversal.

Before the crowd was broken up, the Marine squadron commanders had watched a few minutes of the simulated aerial ballet, and it had intrigued him. He and Rick gabbed about it for an hour. Asking Rick any flight sim question was a guaranteed way to waste at least 30 minutes. An hour and a half later, they finally decided to go down and continue their conversation with the people down in the protected ordnance area-where there were more HARM missiles.

Rick took his laptop, and Chuck was just about able to get some sleep. It felt like he had just closed his eyes when the General Quarters Missile alarm went off. He sprang out of his rack, threw on a flight suit and ran to the ready room, but there were no planes ready for him to fly.

It was different for Codie and Rick. Their newer aircraft were less worn and needed less maintenance. Both young men prepped in their ready room/locker room and were briefed as quickly as possible by the ship's Combat Information Control CAP Coordination officer. There were no maps to look at, or stats to review. They were just told the situation over one of the ship's MCs (speaker-phone), and then they all ran to their aircraft. The GW's escort ships had been tracking missile launches from both the Greeks and the Turks for days. This time, the some of the missiles were coming dangerously close to the GW.

In just a few minutes, their Super Hornets were lifted to the flight deck on giant elevators. They taxied out to the four catapults, and a dozen more fighters were thrown into the air. Codie enabled his aircraft's computer data uplink and was immediately connected to every radar and sonar in the small fleet. He saw the threatening missiles right away and-since he was a flight leader-he contacted his AWACs controller.

"Big Eye Five-Five from Echo Foxtrot Lead, I have positive uplink established, and I'm showing multiple vampires bearing 045, heading....looks like 210. That's a bit close to Home Plate. Advise?"

Back in the E-2, the three controllers were busy. They had to track all of the Greek and Turkish missiles, and they had to direct all of the newly launched U.S. Navy fighters. Those three crewman would direct the defense of the GW and its protective task force. One of them was Nikki, and she was too busy to stand on a lot of protocol.

"Cowboy, this is Big Eye. Come right to heading 045. I'm sending you coordinates for your positive control point. When you get there, you are free to

engage any air targets within 15 miles. Repeat: you are free to engage any air targets within 15 miles. Do you copy?"

Cowboy was well on his way already. He was a Marine, but he knew that if the Navy was putting everything it could into the air, then the threat was real. When Nikki told him that he was authorized to be shooting things down, his hunch was confirmed. It also flushed his blood with electricity. He was finally going to get to use some of his extensive training!

A glance at his radar screen told him that there were already at least a dozen missiles in the air. Most of them were going to come close to the U.S.S. Antietam, an AEGIS guided-missile cruiser. Codie wasn't familiar with the ship's specific capabilities, but he did know that the AEGIS missile defense system for which the class of ships was named had a reputation for extreme effectiveness. In truth, the Antietam carried almost 100, long-range surface-toair missiles. It could fire them all in just a few minutes, and send each one after a separate target. There were also chaff rockets, close-in defense cannons, short-range surface-to-air missiles, multi-role artillery, electronic jamming capabilities, and a long list of other defenses. It was the most formidable vessel in the battlegroup besides the GW itself.

Codie headed for his positive control point. It was almost 5 miles behind the Antietam. If any missiles did escape the cruiser's long reach, Codie would have to attack over it to assist. The sky was lightly overcast and featureless. He watched his radar screen instead of peering out the canopy of his Super Hornet.

On the small MFD, he saw Turkish missiles heading from a flotilla of frigates toward a Greek squadron of patrol boats. The Greeks were responding with a stream of anti-ship missiles toward the Turks. At the same time, both navies were launching surface-to-air-missiles (SAMs) in an attempt at preventing the opposing anti-ship missiles from striking home. It was a line of radar contacts coming and going in both directions. Whenever a missile was hit by a SAM, it fragmented and the number of radar contacts grew exponentially. Back in the American E-2C Hawkeye, Nikki was getting confused. Codie was beyond that.

In the Combat Information Center (CIC) on the GW, the radar screens were packed with lines and icons. It was almost impossible to metally keep track of all the threats. The only thing that was clear was that the missiles from both sides were traveling at hundreds-even thousands-of miles per hour. The Antietam was too close to the action, and it could only escape at 35-40 mph.

The Greeks and Turks knew this, and the ships on both sides tried to maneuver toward the Antietam. They hoped that the presence of an American ship would deter the other side for just a few moments. There were also veiled hopes of letting the American ship stop the streaming, long-range, anti-ship missile crossfire. It was not a concept that thrilled the Captain of the U.S.S. Antietam.

Captain Cora McRaven commanded the U.S.S. Antietam. She had come to the ship less than two months earlier, and in some people's eyes she had yet to prove herself. A cruiser command in the U.S. Navy held a great deal of prestige for most people, and many of her peers were more than a bit jealous. Cora wanted to show everyone that she was the right person for the job, and now she would get her chance.

The ship had been cruising at a comfortable 15 knots before the shooting started. When it did, she anticipated the missile crossfire, put the ship on a General Quarters Missile alert, and turned the ship around. When she was headed away from the crossfire, she had the Antietam speed up to 33 knots.

The winds and currents were against them, however, and they were only making 25.

The smaller Greek patrol boats and Turkish frigates were closer to islands. The small rocks that jutted from the sea also blocked the wind and current. They were making almost 40 knots in the same direction. Only four minutes into the Greek/Turk missile exchange, the crossfire began to overtake the Antietam. Over the ship's intercom, Cora stood on the bridge and heard the call that she had been waiting for with both dread and eager anticipation.

"Vampire! Vampire! Vampire! Bridge from CIC, we've got multiple inbound air contacts coming in from the southeast. It looks like those Turkish Harpoons are locking on to us by mistake!"

Everyone on the bridge tensed. Everyone except Cora was wearing their U.S. Army surplus steel helmets (painted Navy gray, of course). Cora kept her ship's baseball cap on and tugged at its brim in frustration. Then she grabbed the ship's intercom system and went to work.

"CIC, this is the Captain. Go to Weapons Free status. Take down anything and everything that even remotely looks like it's coming our way. Then pass the word back to Home Plate (the U.S.S. George Washington). Let 'em know that we're under fire. I want both helos up and away from us too. I want blip enhancers running on both, but task one with ASW (submarine hunting). We don't wanna get surprised. They could be driving the fox to the hounds."

She hung up the telephone for the intercom and worked the panel to switch it over to the Engineering section.

"Engineering, this is the Captain."

"Captain, this is Chief Petty Officer Dillard-Engineering Div 1."

"Chief, we've got inbound air contacts and lots of 'em. Get everything you can out of those turbines. The wind and waves are in front of us, and the vampires are behind."

"Captain, I've got-"

She hung up the intercom and yelled at the nearest sailor.

"Helm, come to new heading 250!"

The ship started to heave to the side as it began its tight turn to the left. With her mental battle clock running, Captain McRaven knew exactly what was going to happen and when.

"Okay everybody, close your eyes, and either get down or look away."

Just then, the forward vertical missile launch (VLS) tubes began to burst with flame. One after another, SM2ER rockets were being thrown into the sky from the VLS tubes immediately in front of the bridge and at the far aft end of the ship. The sound was deafening, and no one even tried to talk or listen except between the launches. It was like being 50' away from a 1/10 scale Space Shuttle launch.

With the Antietam now turned and headed toward the GW, Codie was less than 5 miles in front of the ship when it started launching its air defense missiles. He instinctively rolled, turned away, popped out decoy chaff and flares, then dove for the waves. A few feet above the water, he leveled out and looked in his rearview mirror. The Antietam was enveloped in smoke and bristling with spike-like, white contrails from 11 missile launches-and counting. As he watched the dots of orange flame from missile exhaust climb then dive toward the far horizon, Nikki called him on the radio.

"Cowboy this is Big Eye Five-Five. Help's on the way, but you've gotta cover those guys behind you. You are weapons free. Engage all hostile and unknown air contacts coming in your direction."

Codie was thrilled. In a second, he readied his weapons, climbed, and turned back toward the ship. Hs radar screen was filled with air contacts. There were so many dots that it started to look like snow or static. There was no time to reply and confirm his orders from Nikki. He had to get busy.

When he was almost turned around, Codie locked on to a Turkish Harpoon anti-ship missile that was headed for the Antietam. As fast as he had locked on to it, he fired one of his AMRAAM medium-range air-to-air missiles. Again, he locked on to a Turkish Harpoon and fired off another AMRAAM. The Antietam passed below and to his left, so he did a tight turn to the right to avoid being accidentally shot down. When he came around again, he individually attacked two more Turkish Harpoons. A glance at his radar screen showed that many of the Antietam's SAM's were hitting their targets. The influx of Turkish missiles dwindled, but it did not stop. He turned right into another circle. As soon as he could, he attacked two more Turkish missiles with his last air-to-air missiles.

His Super Hornet was now unarmed except for its 20mm cannon. For him to use that, he had to fly into the now three-way crossfire, turn around, get behind a missile and chase it down like a target drone. That would also put him heading straight into the teeth of the U.S.S. Antietam. Instead, he circled around the cruiser and headed back to his positive control point. Timing was everything, and as the Antietam had steamed away from the crossfire, she had drifted directly under Codie's coordinates. He circled a few miles to the north instead.

"Big Eye Five-Five, this is Cowboy. I'm down to my 20 mil. Better get some help out here like it was yesterday. There's a world of hurt coming our way."

Just like the other two controllers in the E-2, Nikki was being pulled in every direction.

"It's coming, Cowboy, but we've got trouble all over the place. Advise you descend to Angels three. I've got a pair comin' in from your west at Angels five, and they are weapons hot. Contact U.S. Navy Eagle Flight 221 and 222."

Codie looked up from his radar display just in time to watch four AMRAAMs fly 1000 feet overhead at mach 4. They came out of the overcast horizon to his west, passed above and between himself and the Antietam, then disappeared toward the horizon to the east. He descended to 3000 feet before they were gone.

"Eagle 221 from Echo Foxtrot Lead, check your angels. I am at angels three above the cruiser."

"Copy, Echo Foxtrot Lead. This is Eagle Flight. We're coming over right now. Keep your head down."

Codie looked above and ahead of him, toward the horizon where the AMRAAMs had come from. A second after he heard their radio reply, he spotted the pair of F/A-18C's. They were at full afterburner, and white-hot plumes of burning jet fuel plasma stretched out behind each plane. In a steady succession, they took turns launching missiles at the unseen Turkish Harpoons at the far horizon. Below and to his left, the Antietam was still firing off its 100-missile stockpile. As he watched the radar uplink on his cockpit tactical display screen, Codie wondered if that would be enough. He didn't have to wait long for his answer. Aircraft were moving at 1000 knots. Missiles were passing each other in five different directions at over 3000 knots. Threats were resolved as Turkish Harpoons were either shot down or missed. When they missed, they continued on toward their accidental target-the U.S.S. Antietam. Those that evaded destruction closed the distance at almost a mile a second, and most were less than 80 miles away when they were launched!

Back on the Antietam, things were about to be taken to the next level.

The bridge intercom buzzed, and Captain McRaven picked up the phone.

"Bridge from CIC…"

"Go ahead, CIC. This is the Captain."

"Cap'n, we're down to 20 birds left. The guns are ready, but we've got to turn to bring them to bear."

"Understood." Cora slammed the phone back down and began connecting to work the panel so she could connect with the Engineering division. Another SM2ER missile launched from the forward missile bays, and the bridge was again illuminated in a flickering yellow light. Captain McRaven turned and shouted into an 18-year old sailor's helmet.

"Helm! Left full rudder! Come to new course, 180!"

Cora put the phone back up to her head and called Engineering.

"This is Chief Petty Officer Dillard-Engineering Div1."

"Chief, this is the Captain. Report."

"Cap'n, we're already running at the redline on all of the turbines. They're designed to cut off at a certain point so that they don't sieze up. We could find ourselves paddling."

"Chief, if you don't find a way to get some speed, we WILL be paddling. Just make it happen."

Cora slammed the receiver down and tugged at the brim of her cap again. The officers down in the Combat Information Center (CIC) were good. The ship was phenomenal. Everything was working exactly as planned. All she could do was maneuver the ship to bring the point defense cannons to bear. Now, it was up to the people guiding the missiles and working the keyboards down in CIC. She had already given the commands to maneuver the ship, so the only thing she could really do was wait.

Seconds seemed to take hours. Then, everything happened in rapid succession. First she could hear the chaff rockets firing to decoy the Turkish Harpoons, but she knew that they were American-made, and they wouldn't be so easily decoyed. Then she heard the aft and forward 5inch cannon turrets begin their vain effort to stop a missile. Another SM2ER lifted off from the forward VLS and blocked out all the rest of the sounds. When it faded, she heard the ripping or tearing shrill of the two 20mm Phalanx cannon turrets. They were the last ditch defense. Now, there was nothing between the Antietam and an incoming missile except luck and lead from the cannons.

The ship's 1MC intercom blasted throughout the cruiser. "BRACE! BRACE! BRACE!" Everyone tried to prepare themselves for an impact. There was just enough time to make sure that both feet and both hands were firmly planted, and then the first missile hit.

The Turkish Harpoon had approached at wave-top height. The Antietam had fired a pair of SM2ER missiles at it, but both had overshot. A follow-up pair was launched, but the Harpoon initiated its terminal popup maneuver and began climbing just as they would have shot it down. Near the apex of the climb, one of the AIM-9 Sidewinders from the Eagle Flight missed it. From over 5000 feet up, it dove straight down on to the ship. The 5-inch cannons tried to shoot it down, but the angle was too steep. The Phalanx turrets streamed armor-piercing, depleted uranium sabot rounds at it in shrinking cone-shaped patterns, but all they managed to do was knock off a small fin. Already diving, there was nothing that could stop the missile. Even if it had been hit, its debris would have still rained down on the American cruiser.

The missile hit between the hangar bay and the aft vertical missile launchers. There were still several Tomahawk cruise missiles and ASROC anti-submarine missiles in the aft VLS bays, and they went off with the Harpoon's warhead. Unspent fuel from the Harpoon and the missiles that were in the VLS bays instantly turned the aft portion of the ship into a giant blowtorch. No less than 1/3 of the crew was killed instantly, and almost everyone else-including Captain McRaven-had been wounded.

The U.S.S. Antietam was mortally wounded, and more missiles were coming. Codie was watching his radar screen when the missile hit, but the resulting flash, fire, and ink-black column of smoke caught his attention in an instant. He knew about the other missiles, and so did Eagle Flight. They were all out of air-to-air missiles, and there was almost nothing that they could have done, but that didn't settle their stomachs one bit. Each of them was filled with anger, frustration, and the resulting nausea of seeing so many dying so horribly.

"Echo Foxtrot Lead from Eagle Flight Lead, I have tactical command here. Get back to Home Plate, re-arm, and get back out here ASAP. Copy?"

Codie was being sent back to the GW, but he was too filled with emotion to comply.

"That's a negative, Eagle Flight Lead. I'm staying to cover those guys down below. I will not leave them out here unprotected."

"Very well Echo Foxtrot. Then you are instructed to take out the next vampire that comes this way."

"That's another negative, Eagle Lead. I'm down to the Vulcan, but maybe I can scare any bad guys from trying to round up our crewmen. Why don't ya'll go back and load up. I'll wait."

Codie's emotion was blocking his sense of logic, and even his trained responses to such a situation. Something else his playing with his thought process. The voice from Eagle Flight sounded familiar. He couldn't place the name or the face, but while he thought about it, the mystery was solved.

"Echo Foxtrot, uh...if you're down to just your Vulcan, why don't you go forward and try to take down some of those targets with it? We'll follow you."

"Big Eye Five-Five, this is Echo Foxtrot. I know you're pretty busy right now, but, uh, could you explain to Eagle Lead that I'm the senior tactical commander here?"

Nikki was busy enough before the Harpoon hit the Antietam. Now, she had to direct rescue helicopters from all over the battlegroup, and she had to deal with Codie.

"Cowboy, you are not the senior TACCom (Tactical Air Commander on the scene). Eagle Flight Lead is also the VFA-136 Knighthawks squadron leader. You need to RTB immediately. Do you copy?"

Codie turned his Super Hornet and drifted closer to the pair of USN Hornets that were circling the smoke from the Antietam. When he was close enough, he saw the writing on the side of the lead aircraft's cockpit: Don "Dutch" Schultz- the very same man that he had insulted in the Dirty Shirt Café a few days earlier. Dutch looked over at Codie, pointed at him, and then pointed down to indicate. It was a clear sign that he wanted Codie to land back on the GW and rearm. Codie just shook his head negatively.

"Copy, Big Eye, but I can't leave these people out here. Eagle Flight Lead, if you really wanna try and take out some more of those 'poons (Harpoon missiles), then I'm right behind you."

Dutch turned and headed out toward the Greek/ Turkish missile crossfire to the east. His wingman followed, and so did Codie. All three aircraft were virtually unarmed. In front of them-unseen beyond the overcast horizon-their radar screens showed well over 300 missiles in the air. There was no telling which ones were anti-ship missiles, which ones were SAMs, and which ones were air-to-air missiles from the GW's other planes.

They raced away from the burning U.S.S. Antietam at over 700 miles an hour. It was still in view when the first missiles began to zip across the sky. Codie saw what looked like a white telephone pole approach from his lower right, climb over him, and dive over his left shoulder. Dutch and his wingman immediately rolled left and dove behind the missile. It was close enough to clearly identify it as a Harpoon. The missile continued on its course, and it never turned to attack the Antietam. While they chased it, Codie spotted another missile and chased it. When he was close enough, he was able to identify it as some sort of SAM. More missiles were coming and going from every direction, and the three pilots split up to chase them all. Finally, a Harpoon passed Dutch's plane from almost head-on. There wasn't enough time to turn and chase it, so he called Codie.

"Echo Foxtrot, we've just had a Harpoon blow past us. Can you take it?"

Codie looked around. He saw Dutch, and then he saw the Harpoon. It was headed for the Antietam. Immediately, he rolled left, turned, and dove for the white rocket. The Harpoon fell to wave top height as made its attack run. Codie was still a few miles behind it, but he fired a burst of 20mm cannon fire anyway. The shells splashed ahead and to the left of the missile. He corrected and tried again. He was closing on the slower missile, and this time the range was much closer. His cannon shells splashed all around the Harpoon, but nothing happened. He leveled out and closed in behind it even more. Meanwhile, the Antietam drew closer, and it opened fire on the missile with the forward 5 inch multi-role cannon and the 20mm Phalanx autocannon. Yellow tracers zipped past him and huge splashes from the 5 inch cannon seemed to lift the waves right out of the sea. Still the Harpoon closed. Codie fired another burst at it, and finally hit it. Some of his cannon shells knocked off a few of the control fins. When it tried to initiate a terminal pop up maneuver by climbing for altitude, the missile lost control, corkscrewed into the air, and splashed 6 miles off the starboard beam of the Antietam. Codie climbed a little and did a victory roll as he passed through the smoke rising from the American cruiser.

" *"Echo Foxtrot from Eagle Flight Lead, well done. When you're done playing around, perhaps you wouldn't mind coming back over here and giving us a hand? There's another pair coming in from 145 (southeast), angels 3."*

Codie turned and headed back toward the missile crossfire. The overcast was burning off, and the sky seemed laced with twisting white and gray contrails from all of the anti-ship missiles, SAMs, and air-to-air missiles. Large black smudges dotted throughout the braid of white lines showed where some of the anti-ship missiles had been stopped.

In less than two seconds, Codie spotted Dutch and his wingman chasing down another pair of Turkish Harpoons headed for the Antietam. With the ship still behind him, it seemed as it they were coming right for him. There was no time for any of the pilots to warn the other. Dutch and Codie both fired at the missile from opposite directions. Large white splashes snaked around the small target. Together, they fired almost 600 20mm cannon rounds at the Harpoon, and they all missed. Thankfully, they missed each other too.

The Harpoon passed under Codie's right wing, and Dutch flew over it still chasing and still firing. His wingman broke off the chase when they came too close to the Antietam, and it began to fire its defensive cannons at the missile. Dutch stayed with it. White-hot tracer rounds of armor-piercing depleted uranium sabot rounds spun a spiral around the Harpoon. Then, the computer controller on the Antietam's Phalanx gun mount closed the diameter of the spiral and tried to narrow its fire at the missile. Dutch fired off another burst from his 20mm cannon and nailed the Harpoon. It was still at wave-top height when it exploded, and its debris flew forward like a shotgun blast. Most of it hit the Antietam, but at least the warhead and rocket fuel-the most dangerous parts of a Harpoon-had been detonated before hitting the ship. Dutch pulled up and turned to the right to rejoin the intercept.

Codie and Dutch's wingman had already given chase to a pair of Harpoons and both escaped them. Thankfully, the missiles never locked on to the Antietam, and they continued to head toward the Greek flotilla. Dutch joined them in formation, and they all looked at their radar screens to determine which missiles to try and attack next. While they were doing so, a pair of U.S. Air Force F-16's from Italy arrived over the Antietam. They were fully armed, fully fueled, and eager to defend the ship. After briefing them on the situation and letting them know that some missiles had been shot down, Dutch led his wingman and Codie back to the GW for refueling, re-arming, and a formal de-briefing. They hadn't prevented the Antietam from getting hit, but they did keep it from being sunk.

Back on the burning cruiser, Captain Cora McRaven wasn't so certain about her ship's fate. The bridge was filled with a choking black smoke, and the fire from the missile's impact was still burning deep into her ship. There were reports of stress fractures, and subsequently innumerable small leaks seemed to appear all over the ship. The Antietam was almost completely unarmed, and there was still very little word about the status of the ship's turbine engines. It was just too soon to tell if they were going to make it.

Out of the 350+ officers and crew that had set sail with the ship, over 1/3 were still unaccounted for, and almost everyone had at least some serious bruises and scrapes from the impacts. The Harpoon that Dutch had shot down had even sent a piece of shrapnel across Captain McRaven's left cheek. It was nothing that a dozen or so stitches couldn't mend. She would live, but it crushed her heart to know that so many of her crew-people who trusted her with their lives-would not. Survivor's guilt struck hard. She loved her ship, and the crew was a second family to her. She knew each of them, and she knew each of their families. The guilt only subsided when she occupied her thoughts with ways to get the ship back into a fighting capacity.

Thursday, April 21, 2011

2nd Lt. John Chamberlin-Personal Log

Hill 1419, 15 miles S of Taipei, Taiwan

It looks like we'll be getting some action soon. Hector let Bruce and I know that the Taiwanese Army has fallen back out of the city of Tainan on the opposite side of the island. That means that five of the Chinese beachheads have linked, and now they can concentrate on driving north towards us. Hector has set up a patrol schedule from his platoon to make sure that there is always a fire team observing from the high point on the mountain behind us. I think he's got a feeling that the Chinese that are driving north along the east coast might come up behind us. It would be pretty tough if tens of thousands of Chinese suddenly appeared back there, and no one wants to be a prisoner. We're not even sure that they keep prisoners after they've been paraded.

Monday, April 25, 2011

2nd Lt. John Chamberlin-Personal Log

Hill 1419, 15 miles S of Taipei, Taiwan

Hector let us send another communications patrol to get e-mail and check the news. The news was still filled with confusing and sometimes contradictory reports of what's happening in the Med, or here in Taiwan. The only significant news was that the OPEC members that are also members of the Arab League want to double the price of oil. Only Mexico, Venezuela, and a handful of others want to keep prices where there at. Even Saudi Arabia wants to raise the price! None of this mattered to the platoon of course. All they cared about was their mail. Everyone seems pretty happy for the most part. The only mail that we're getting now has been the occasional e-mail/news run. Regular mail that goes through the Post Office takes weeks, and it hasn't at all come this month! We have been getting Global Parcel Service packages though. One of the times that Chuck Boyd (our nerd/geek/slacker from 1st squad) went up and did the e-mail/news run, he ordered a new pair of boots from an online surplus store. They came in today, he did have to pay special shipping charges to get it to a combat zone, but he was extremely happy. Now everyone is making their purchasing requests. I wonder who will come first, the Global Parcel Service delivery truck, or the Chinese?

Wednesday, April 27, 2011

2nd Lt. John Chamberlin-Personal Log

Hill 1419, 15 miles S of Taipei, Taiwan

Having hardly fired a shot the entire time that we've been up here, the word from above has come down. We're pulling out of Taiwan! Task Force Stingray and all the air force units nearby have been unable to gain significant control over the island, or so we're told. As a result, all Coalition forces will be withdrawn from the Choo Beachhead ASAP. We're to remain here and cover the troops as they fall back.

To help us out, the small reinforcements that we were promised weeks ago, have finally arrived. Each platoon got a .50 caliber heavy machinegun team, a TOW anti-tank missile team, two 60mm mortar teams, and a Stinger III team. Now I have to think seriously about ammunition, its supply, and all of the different types. We've also been given a bunch of M249 SAW light machineguns. In fact, I've got enough to give one to every fourth person in the platoon! The reinforcements help offset the bad news that we're pulling out, but what really raised Bruce and my spirits was that Colonel Ghetty gave us each our own MilNet palmtop PCs-just like Hector's. It's as much a status thing as well as a tactical force multiplier. Okay, the truth is, it's just cool to have.

Thursday, April 28, 2011

2nd Lt. John Chamberlin-Personal Log

Hill 1419, 15 miles S of Taipei, Taiwan

What a fireworks show there was last night! Task Force Stingray started pulling people out from the Choo Beachhead, and not long after, the Chinese hit hard. I went up to the top of the mountain and watched it for a while. I could hear the Chinese artillery fire from the west, go over our heads, and land all over the beachhead down below to the East. I watched the Harriers takeoff and intercept

Chinese fighters, bombers, and attack helicopters. I watched Chinese anti-ship missiles skim above the waves and meet SAM's from Stingray stop some of them. I saw the ones that weren't stopped by SAM's find their targets on the beyond the horizon. Through all this, there was AAA everywhere. There was so much AAA that I went back to the platoon and made sure everyone was under cover to protect them from the rounds as they came back down. The show lasted all night. When morning came, there was too much smoke to see very much, but I could hear it. It never let up. What should have been a minute or two long engagement has gone on for hours at full intensity.

When I was back with the platoon, I saw friendly troops pulling back down the road to the beachhead. The MP's at the roadblock weren't slowing them down at all. Most of the people I saw were U.S. Army and British Army. There were some uniforms I didn't recognize, but most were familiar. A lot of equipment was moving out too. In addition to the orderly withdrawal or displacement of infantry, there's a lot of tanks, artillery, and trucks leaving too. Unlike the Korea evac, we're taking our toys home this time!

Friday, April 29, 2011

2nd Lt. John Chamberlin-Personal Log

Hill 1419, 15 miles S of Taipei, Taiwan

The retreat continued today. It was still very orderly, but the pace was noticeably faster. There was a lot of overcast today, so we didn't really see any air activity. We did hear the jets and helicopters all around us, though. A platoon of Taiwanese Army troops came off the road and said that they were supposed to stay up here and cover our retreat. Hector hasn't gotten any orders for us to pull out yet, and Colonel Ghetty hasn't stopped by in a while. Until then, we're not leaving.

Hector, Bruce, and I agreed to put all the 60mm mortars in the pits that the Taiwanese artillery used when they were up here. I had the M249 SAW's spread out, along with all the ammo. I gave my TOW anti-tank team to Hector to help cover the road up the mountain. Bruce kept his to cover our left flank. I put my .50 caliber heavy machinegun team near the crest of the mountain. From there, they'll have a 360-degree field of fire. They can also serve as lookouts, and they'll be almost impossible to take out without some sort of heavy support. The Taiwanese infantry have spread out around the company in three, four, and five-man groups. They seem like real hardasses.

Saturday, April 30, 2011

2nd Lt. John Chamberlin-Personal Log

Hill 1419, 15 miles S of Taipei, Taiwan

Another bad day for the good guys. The procession of units retreating down the road has fallen to an intermittent trickle. About 1530 (3:30pm local) this afternoon, there was a quick artillery barrage against the Taiwanese MP roadblock. At the end of the barrage, the Chinese fired lot of smoke shells. We couldn't be sure if it was gas or smoke, so everyone rushed into their CBW (Chemical Biological Warfare) suits. No sooner had we covered up than the .50 caliber team on the crest opened fire. The Chinese were coming straight at us.

The mountain's face is extremely steep. The 60mm mortars had trouble hitting their targets, but the debris they loosened amplified their effectiveness. Boulders rolled down into the reds, thinning their ranks and taking away their cover. The SAW's that I dispersed out really rained lead down on the enemy troops that had survived the .50 cal and the mortars. The Taiwanese troops in our line proved to be very deliberate, accurate, and effective with their fire. They probably learned fire control discipline the hard way weeks ago.

The attack on my platoon continued, off and on, for about an hour and a half. It was a little longer for Hector's platoon, and a little less for Bruce's. After sunset, Colonel Ghetty came up and gave us the word to hit the road. The remnants of that Taiwanese Army platoon will take over our entire company's position and cover us. Hector, Bruce, and I agreed to let them have the .50 calibers and the mortars to help out. They were happy to take them and gave our mortar and .50 cal crews their small arms so they wouldn't be defenseless on the beach.

On our way down, we stopped occasionally for rest. It's surprisingly difficult going down the mountain after sitting on top for so long. During one of these rest stops, I let Chuck use his palmtop PC to get on the internet and check the news. It wasn't good.

In the Med, the George Washington carrier battlegroup got caught in a crossfire of anti-ship missiles between a Greek flotilla, and some Turkish vessels. The Greek missiles missed, but the Turkish missiles hit a few of our escort ships. None were lost, but the rules of engagement demanded that the Turks be fired upon as an immediate response. Most of those were sunk by the much more accurate American Harpoon III missiles. In response, terrorist attacks in the U.S. seem to have increased, and both Syria and Libya have begun making small probing air attacks on the George Washington group. The small probing attacks have little chance of succeeding, but they do wear down the men, the ships, and most importantly, the ammunition supply.

Saturday, April 30, 2011

1035hrs local

U.S.S. George Washington CVN

Eastern Mediterranean Sea

Chuck's Hornet had finally been repaired. He prepped for a routine patrol, and headed out to the hangar deck to pre-flight the aircraft. While he was walking around, looking in the landing gear wells, peering into the engine intakes, and tugging on the AIM-9 Sidewinders that were mounted on each wingtip, he noticed something out of place.

Next to his Hornet was a Marine Super Hornet with four HARM missiles mounted under its wings. Rick, Codie, their squadron commander, and Don were all kneeling next to one of the missiles and talking. It was an odd pairing of personalities, and the aircraft's payload was unusual given the previous day's hostilities. There was no reason to have anti-radar missiles mounted on any of the aircraft, and there was certainly no reason to mount four of them on a plane that was assigned to defensive CAP duty.

Chuck walked over without drawing any attention, so he asked them what they were doing. Codie-always the blowhard in any group-was the first to answer.

"Hey there, Spider. Look's like your roomy here's gone Super-Geek on us. He and some of the red shirts came up with the idea of turning these unassigned HARM's into some sort of super-long-range missiles."

Chuck had considered the possibility that the HARM loadout might be Rick's AI5 program experiment, and he wanted to see if it would work too.

"Hey Don, I'm next up for CAP duty. Would it be alright if I tagged along and made sure Rick here was safe? My wingman's aircraft could still use some more downtime on the starboard engine, and I'd rather have Rick up there with a fully prepped bird than one that might give out at any time." Don thought about it for a second, then agreed.

"Okay, I'll clear it with the Air Boss and the Air Group Commander. Just remember, this area's a combat zone now. Make sure you get a downlink from the E-2 before you do any live-fire testing. Follow the rules of engagement, and don't launch any of these. You're only authorized to flight test their payload stability. They're so old that no one ever tried to hang four of 'em at once on this new bird. Take it easy. Got it?"

They nodded and followed their aircraft to the elevator as they were taxied for lifting to the flight deck. A minute later, they climbed into their planes, taxied to their respective catapults for launching, and soon, they were in the air.

Chuck called the E-2 Controller to let them know that he and Rick were available for CAP duty. The controller told him that they could help bolster the CAP over the stranded U.S.S. Antietam. He also informed Chuck that a trio of tugboats from Italy were on their way, and that the U.S.S. Stark was near the Antietam to provide assistance, fire support, and an evacuation point if necessary.

While he heard the reports, Chuck led Rick toward the Antietam. It wasn't hard to find. A dense black line of smoke rose from the surface. It was still on fire. A few minutes after the controller had finished giving order to Chuck, they passed over the burning ship. Between the forward bridge structure and the aft gun turret, it looked as though a giant monster had taken a huge bite out of the ship. The Harpoon had set off several attack missiles still in the aft VLS bays, and their combined explosion was twenty times as damaging as the Harpoon. It was a miracle that the ship was still as intact as it was.

Through the smoke, they looked down inside the ship. Small fires glowed and flickered off the water that had filled the gap of twisted and torn wreckage. On the port side of the ship, the damage reached the waterline, and the ship seemed like two pieces joined together underwater instead of a single vessel.

Right next to the Antietam was the smaller American frigate, the U.S.S. Stark. It was pouring water into the gaping hole from fire hoses. Along with the fire hoses that were working feverishly inside the Antietam, the smoke was slowly turning gray. By the end of the CAP duty, the fires would be out completely.

They circled above the two ships at 10,000 feet. The E-2 controller sent them an uplink, and they were able to see that the war between Greece and Turkey was still very hot. Via the uplink, their radar coverage extended from Italy to Israel and from Athens to Cairo. Hundreds of aircraft and ships were transiting, patrolling, or attacking. A few pairs of Turkish F-16 Falcons were raiding a Greek island in the Aegean Sea. An equal number of Greek F-16's were trying to stop them. Small dogfights seemed to be the rule of the Aegean, and through the middle of it, civilian aircraft and jetliners from a dozen nations made the routine transit unopposed-right through the battle zone.

While smoke still billowed from the U.S.S. Antietam, the gravity of the situation was further brought home when the E-2 Controller called Chuck.

"Charlie Delta One One (Chuck) from Big Eye Three One (the E-2 Controller), you have traffic (possible targets) bearing 110 at angels 2. Identify probable hostile target tracks. Proceed with caution immediately. Copy?"

Chuck glanced at Rick off to his right. He nodded to let him know that he had also heard the message. Then Chuck called the E-2 controller while he and Rick began their right turn toward the potentially hostile aircraft. "Copy, Big Eye. We're on our way."

The two planes headed to the southeast, and the stunning scene of the Antietam was left behind them, but only physically. Chuck and Rick were cool professionals. Neither had the stereotypical balls-out, hair-on-fire, hot shit attitude that fighter pilots were notorious for having. They were a new breed; calm, cool, calculating, professional, and extremely intelligent. Both young men knew that the fate of hundreds of sailors on the Antietam was in their hands. The ship was out of sight, but not out of their minds.

Rick's Super Hornet had better avionics than Chuck' vintage F/A-18C. He also had the four modified HARM missiles that he was itching to try out. Rick had renamed them BOB missiles after the pet name that he had given to the artificial intelligence program that he installed in each one. The flight characteristics were fine with four of the huge missiles hung under his wings, but now curiosity was taking over.

"Spider, I think I should hang back a bit with these beasts under my wings. If we get into a scrap, I'll either be worthless, or I'll have to drop 'em into the sea."

"Good call, Ironman. I'll take the lead and come in low. You go high. Besides, you'll be in a better position to help if it does turn into a scrap."

The two planes separated. Rick climbed to 22,000 feet, and Chuck descended to 2,000. Higher altitude also meant that Rick's superior sensors would have a longer detection range, since they wouldn't be blocked by the curvature of the Earth. He would also have the advantage in a possible dogfight because he could pick up speed while diving in from above. In air-toair combat speed, is directly correlated to one's survival.

As he passed 15,000 feet, Rick's radar started to pick up targets. First, he spotted two trios of F-16's coming in from the south-the same ones that he and Chuck had been ordered to identify. Then, he noticed two pairs of F-16's that the E-2 had not detected approaching in from the northeast. It would take a few more precious seconds to determine the nationality and/or the weapons being carried by any of the radar contacts.

"Spider from Ironman, I'm picking up the bogeys to the south, but there are six of them, not two or three. There are also four more coming in from the northeast. You better come back up here and join on me. Copy?"

Chuck answered with a simple "Understood. On my way." Then he turned around sharply and climbed to join on Rick's wing. While he did so, Rick's radar figured out that the F-16's to the south were Turkish, and the ones to the northeast were Greek. All of them seemed to be carrying large missiles under their wings. It was too soon to determine what kind. He sent his report to the E-2 and uplinked his radar data so that it appeared on the E-2's screen as well as the GW's CIC monitors and even those back in Washington D.C.

While he was pressing buttons to send data, the Greek and Turkish F-16's merged and began shooting missiles at each other. The ten planes began spinning a circular web of contrails to the east of Chuck and Rick. Occasionally, a missile left the furball in the sky and headed either north or south. Rick's radar determined those stray missiles as Harpoon anti-ship missiles and HARM anti-radiation missiles used to take out tracking and targeting radar antennas. None of them seemed directly targeted against the Antietam or the Stark, but while Chuck and

Rick watched the scene from 20 miles away, some missiles began to leave the dogfight and head for the American ships.

No orders were needed. All total there were eight missiles diving for their wave-top approach to the Antietam or the Stark. Chuck locked on to one after the other and began firing his air-to-air missiles. One by one, he let loose four medium range AMRAAM missiles. Then, as the anti-ship missiles came within 10 miles, he fired off his pair of AIM-9 heat-seeking Sidewinder missiles, and he dove down to try and shoot down another with his cannon.

While Chuck raced from 15,000 feet to 500 feet, Rick had a choice: he could either jettison his four modified HARM missiles/his BOB missiles, or he could try them out. It wasn't really a choice for Rick. He was eager to give them a try, and he was just as confident that they would work perfectly. Flying straight and level for another moment, Rick maneuvered a cursor on his radar screen. Then he selected all of the anti-ship missiles and sent that data to his missiles. Finally, he connected the adhoc data transfer uplink in each missile to his Super Hornet's computer. Everything was ready, so he turned toward the approaching anti-ship missiles and fired all four BOBs.

The Super Hornet was pointing down toward the Harpoons as they skimmed across the surface of the sea. When the four BOBs fired, they launched all at once, and Rick's plane felt like it went from 700 miles an hour to reverse! Almost a quarter mile in front of him, as his Super Hornet started picking up speed again, Rick watched as all four missiles corkscrewed away in different directions. One even nose-dived and went underneath him.

Chuck's air-to-air missiles had taken out three of the eight Harpoons. He was closing on the lead missile, so he fired a burst of 20mm cannon fire. Unlike Codie and Dutch's first attempts the previous day, Chuck's burst struck home and shredded the white missile. It blew up into a million pieces, but it was still far enough from the Stark and the Antietam for all the debris to splash down harmlessly. With the explosion temporarily blinding and shocking him, Chuck pulled up and headed vertically into the sky. Flight instinct had taken over to prevent him from flying into the cloud of dangerous shrapnel.

Flying straight into the sky, he looked over his shoulder. From there, he could see the dogfight still going on behind him. He also saw even more missiles than he had originally. While disappointment started to settle, however, some of the missiles began to collide. All four of Rick's BOB missiles had found their targets. Only one more approached the Stark and Antietam. Thankfully, it malfunctioned, was damaged, or simply met with a chance wave. In any event, it crashed into the sea miles away from either American ship.

Rick was ecstatic. Chuck was just as happy. The E-2 controller had directed some of the other CAP pilots from Rick's squadron to the area, but they didn't arrive until after the last BOB had struck home. They all waved their wings as they passed him and headed to Rick and Chuck's patrol station.

Then Chuck and Rick were ordered back to the GW. They were out of missiles. Their fuel was lower than expected, and a larger CAP unit had arrived to replace them. It was time to go. On their way back, Rick dove down to 1000 feet and passed over the Antietam and the Stark. There was no victory roll, but it was clearly understood.

A few minutes later, both Chuck and Rick were back on the deck of the GW. There were no congratulatory crowds, or cheers from the deck hands. The only congratulations they received was from a small crowd of red shirts in the hangar deck. They were the ones who had supplied Rick with his guidance systems and eventually the entire rockets for his BOB missiles. As Rick and Chuck rode down the elevator with their planes, Rick was busy shaking hands with all the red shirts.

Chuck started to head toward the mission de-briefing room when his squadron leader, Don, appeared on the hangar deck. Don walked with a purpose, and when he found the crowd of red shirted enlisted sailors, he burst through them like tidal wave. Face to face with Rick, Don began a verbal lashing that scattered the red shirts and then anyone within hearing distance. He was furious that Rick had fired the untested experimental missiles. The only saving grace was that they had actually worked. Don was adamant about the fact that if they hadn't, the Stark and the Antietam might very well have been hit and even sunk by friendly missiles that had been made by some guest Marine in his spare time. The dressing down was fast-paced, loud, furious, and completely one-sided.

Everyone cleared away from the scene. Rick was handling it better than most people would have. The only two people who remained in the near vicinity were Chuck and Codie. Chuck waited for Rick because he knew that his squadron commander would be looking for him next, and if he had to look for him, Don would only get even more irate.

Codie watched in awe. He had the utmost disrespect for Don, but that was changing. Up until that point, Codie respected Don's rank, and his flying ability, but not the man-the leader. He had assumed that Don's normal, delicate, formal, and soft-spoken demeanor were indications of weakness. At first he had thought that Don was even a homosexual, but those thoughts were quietly dismissed as he was told about Don's famous and glamorous wife. Now, as he watched Don yell in the same manner as a Marine Drill Instructor, Codie began to finally respect the man. It was becoming clear that Don was no wimp.

Sunday, May 1, 2011

2nd Lt. John Chamberlin-Personal Log

25 nm E of Chuangwei, Taiwan

We made it down to the beachhead before sunrise. The place is just miles of debris. There's hardly anyone or anything left intact. There's a lot of stragglers who look lost. There's also a lot of Taiwanese Army people dug in and looking for a good fight. When we got within sight of the water, an LCAC came roaring out from behind the remnants of what appeared to be a fishing village. All three platoons got on board, along with some stragglers, and two British trucks loaded with supplies. By the time the sun was fully up, we were on our way back home to the Wasp. Colonel Ghetty says there's only a few more Coalition units left to pull out, then the Taiwanese will get picked up. They wanted to be the last to leave as badly as everyone else wanted them to be. They really are hardasses.

I think everyone had a little bit of a rough time getting their sea legs back, but its nice to be home. We kicked some U.S. Army squatters out of our old quarters, and dumped off our gear. No one bothered to stop for a shower; we all went straight for the cafeteria for some hot chow. The cooks didn't recognize us, but after a while they remembered who we were, and they microwaved some pizzas for us. We got to jump to the front of the line. There were a lot of upset people behind us, but my people were very, very happy.

Tuesday, May 3, 2011

2nd Lt. John Chamberlin-Personal Log

45 nm E of Chuangwei, Taiwan

We spent most of the day at General Quarters thanks to repeated missile attacks from Chinese light frigates and patrol boats in the area. I never saw any of the missiles. I think our air defenses are getting better at dealing with the air attacks, but the sailors are telling me that the missile attacks haven't been as big as they used to be. It seems the Chinese have given up their massive onslaught attacks in favor of continuous small attacks. Either way, we get to hang out in the cafeteria, which is a hell of a lot more accommodating than that hill we were on a few days ago!

Colonel Ghetty came down and hung out with us for a while. I think he came mainly for the coffee, but Bob and I were able to get the latest news from him. The Arab League is standing firm behind Turkey in the Med. The George Washington Carrier Battlegroup over there has now come under direct attack from the Turks, but so far, it's been nothing serious. Back home, the price of oil has skyrocketed, and it's really hindering our industrial capacity to reinforce and resupply us. Closer to us, the Japanese have been directly threatened by the Chinese and Communist Koreans, so we're no longer allowed to use their bases. That's really going to set back our operations. It should also lead to a faster restructuring of our Philippine bases, Clark Field and Subic Bay. There's a lot of rumors about Chinese military movements in Indochina, but that's been going on for decades, maybe centuries. Bob says that one of the people in Wess' second squad heard from a British sailor on board that we're going to Vietnam! Colonel Ghetty said that the rumors about Chinese incursions into Vietnam were disconcerting, and we should try not to feed the rumor mill. Of course, he didn't sound to doubtful about the rumor, so maybe there's actually some truth to it? Ghetty didn't have any specifics about reinforcements, resupply, or our next deployment, but he really didn't need to give us any.

Wednesday, May 4, 2011

2nd Lt. John Chamberlin-Personal Log

15 nm E of Chuangwei, Taiwan

Bad news today, about 1000 hrs (10:00am local), there was a big commotion to get all our aircraft in the air. The Choo beachhead was collapsing. Chinese infantry made it to the top of the mountains overlooking the beach. As soon as they were able to get some mortars up there, they sent in the infantry. It must have really been clear that the fight was ending because the Taiwanese Army actually called for the evacuation. When I left them, they gave me the impression that they were ready to die in their positions. For them to have asked for a pickup, it must have been pretty bad.

The Wasp is incredibly overloaded right now, so none of them were brought over here. I did see an Osprey go over to a tiny Australian Perry class frigate and unload at least fifty people. It was like watching a clown car at a circus. They just kept coming out and coming out. I have no idea how they all fit inside, or where they went in that little frigate, but they did it. I imagine that's what the entire evacuation has been like. Get the people off the beach and to the Task Force. Then, get a ride from the Task Force to the Philippines. At least we'll be leaving this area soon. I wonder where we'll go next; maybe to the Med' to back up the George Washington carrier battlegroup. I hear that another carrier battlegroup, based around the U.S.S. America - fresh out of mothballs - is going to the there already.

Wednesday, May 4, 2011 Central News Network Online Report

Fight for Formosa:

Coalition Driven Into The Sea!

Once more the combined military force of an international Coalition have failed to halt to advance of aggressive communist forces. After several days of fighting, the Taiwanese Army has been overrun. The last units to maintain a cohesive defensive perimeter were serving as a rear-guard for the evacuating Coalition forces near the "Choo" Beachhead. During the night, communist forces broke through Taiwanese Army lines and compromised the Choo Beachhead. The struggle for TaiwanNationalist China-has ended in debacle.

SECTION 4

"True heroes aren't made. They're cornered!"

Wednesday, May 4, 2011

United Press Wire Service

Generation Failure

By UPWS Columnist Richard Hedd

American history is filled with battles won and lost. While most of America's struggles have begun with defeats, somehow, all of the previous guardians of democracy have managed to pull together and rise to the occasion. From the Boston Massacre to Yorktown; from Fort Sumter to Appomattox; from the Lusitania to the Argonne; from Pearl Harbor to Berlin; from the Tonkin Gulf to Kent State; from Kuwait City to the Highway of Death; and from the World Trade Center to Kabul, the American spirit has always emerged from smoke-filled battlefields of defeat to final victory. A special standard of social strength in the face of inequality, evil, oppression, and terror has always filled our flag with a strong wind and displayed those stars and bars for the world to see.

Today, a new generation has been handed the colors. Like the Minutemen of the revolution, they are being asked to face an enemy of far greater numbers. Like the valiant army of Abolitionists in the Civil War, the children of the Baby Boomers are called upon to free a people living under the absolute tyranny of a few. Like the liberators of Europe, they are being asked to protect America from yet another "ism." America's youth, the so-called Generation X, has been thrown into a quagmire of conflict, but the struggle has the justifications of all the previous wars in the past combined.

Given such a history, and given a cause that surpasses all of those in the past, why do today's flag-bearers fail so utterly? They have lost every battle on the land, in the air, and on the sea. They have now permitted two nations-two close allies- to fall under the yoke of dictatorial communist regimes. Where are our heroes? Where are the men and women who have always risen to the occasion? America wonders if Generation X is truly a slacker generation, and if so, why?

The parents and grandparents of these apparent slackers have done everything to make this country stronger, and now, when our children are tasked with toeing the line, they have clearly come up short. In the 1960's we protested, fought, and died to end an unjust war so that they wouldn't have to fight again. In the 1970's we introduced the microchip and advanced every form of technology exponentially. In the 1980's every parent worked-sometimes two or three jobs-so that our children could be economically secure. In the 1990's we worked even harder for their educations. In this century, as the keys to the seats of power have been handed over to a new age of caretakers, this group has cast aside all established social, artistic, political, and particularly family values. Now, when faced with international turmoil, we should not be surprised that today's youth are at the very least lackadaisical in their convictions.

However, this is not the failure of a single generation. We Baby-Boomers must recognize that during those developmental years, we may have failed en masse. In general, this new generation has been given all the tools needed to advance as a social group. Perhaps, that is our utmost failure. It might very well be that America's new guardians have been too well-provided for in the past.

Earlier generations have been challenged by History, but this generation has not. There has been no monarchy to rebel against. The chains of slavery were broken well over a century ago. Europe has been liberated-twice! The fight to end the Vietnam War and bring down the militaryindustrial complex was won a long time ago. This war is the first historical challenge that these kids have seen. Instead of asking why they keep losing, should we be asking, When will they start winning?

Richard Hedd's opinions are his own. They do not necessarily represent those of The Scott-Smith-Stevenson Publishing Co., Fleet Inc., BRFC LLC, UPWS, its contributors, or its advertisers. Comments and rebuttals to Richard Hedd's column should be addressed to...

Thursday, May 5, 2011, 11:17am

U.S.S. Arizona Memorial

Pearl Harbor, Hawaii

The tropical Hawaiian Islands had always been synonymous with gentle ocean breezes, warm sunshine, and seas so blue that they melted into the sky. Anyone who ever thought of the paradise on Earth, conveniently forgot about the errant storms that could whip across the chain of rocks with little regard for their presence at all. One such storm approached. Ships came racing into the small harbor to escape the monstrous waves that the Pacific was mustering. The storm was still over two hundred miles away, but there was no longer any hint of blue in the skies. Instead, the dawn's overcast was turning gray. Shadows of more ominous black clouds were rolling from one horizon to the other; and still... the seas matched the heavens.

Walt Barber stood on the stark white memorial looking at the names of the sailors who were entombed below him over half a century earlier. The Japanese had attacked without warning, or so the victors had written. He was intelligent enough to understand that the previous World War, like its predecessor, and like this one, was all about economics. They were not about national security, not the interests of any nation, but rather the financial interests of hidden elite powerbrokers, aristocrats, power mongers, kings, kingmakers, and bankers - his class. In truth, they were his ancestors. Even with 1000+ names staring down on him, the thought pacified any shred of guilt for his deeds.

Walt turned and watched the ships come in from the storm. The memorial was empty of tourists and visitors on such a poorly promising day. He stood alone as it began to drizzle. He listened to the sounds of the harbor, and finally Walt heard the footsteps for which he had waited. They seemed to take forever, but patience was one of his best virtues. Exactly as the footsteps were within the perfect distance, Walt turned and embraced the man he had been waiting for.

"Walt, it's good to see you. Is everything alright? This place seems very inappropriate."

Their embrace ended, and the two men shook hands firmly. Generals needed the firmest handshakes and the warmest greetings of anyone Walt worked with. General Alexov Kryuchov was out of uniform, but his age, experience, wisdom, and rock-like conviction showed his rank none-the-less. Walt respected the man, and he needed him. Everything Walt had done for the past twenty years could be traced to Alexov. Secretly, both men felt as though they were using the other. At the same time, each wondered if he had been duped into a dangerously Machiavellian game.

Alexov looked around for Dan. "Where's your shadow?"

Walt sighed. He smiled and looked Alexov square in the eye. They killed him, Alexov; right in front of me. I knew he was a kiss ass, but that's no reason to blow the man's brains all over my plane."

Alexov understood. The Chinese and North Koreans that Walt had been dealing with were not exactly an understanding crowd. In fact, one of the few things that they had in common with each other, aside from the obvious ideological views, was a pure ruthlessness. Walt had done something to upset someone, and he wasn't dealing with forgiving people. There was a good chance that he was in serious trouble.

"Alex, I always wonder if this is the right thing to do. Few things console me when I think of the tragedies all over the world that I've enabled to occur. I take heart in the knowledge that fate dictates such things. I try to always understand and prepare for all contingencies, in all situations, but sometimes I feel as though I'm only fate's pawn. Still, there is nothing really wrong. I can deal with these re-evaluations. I wanted to meet with you to talk about an idea that one of my people submitted."

The former Soviet and Russian general was curious and concerned. Walt had always seemed like a man who had no cares, no concerns, and no remorse. There had been thousands of lives ruined by his international war profiteering, but Walt apparently never really grasped the lethality of the Chinese until Dan's death. Some had considered Walt to be too cavalier in his attitude, but this was obviously not a man without his doubts. Alexov asked him to elaborate, and he casually continued with his normal tone slowly coming back.

"Alex, the Chinese are responding to everything that's happened so far like children in a candy store."

Walt could tell that the general, though fluent in English and well versed in American culture, didn't understand his analogy.

"All they needed was a little taste of victory, a taste of war without fear of the U.S. military. Now they can't get enough of it. If wars and battles are really decided by initiative and momentum, the Chinese might soon get out of control. Someone else will need to stop them, someone who would never be tied into the American/Coalition MilNet, someone who could not be detected by the SP 2.0 software."

Alexov thought for a few moments. Walt was very correct in his concern towards the Chinese. Once a military war machine gets fueled and starts moving, if it isn't stopped, the war could be over far too quickly; too quickly for them to make their profit margin.

"Walt, I agree. Something will eventually have to be done. Maybe now is the time. I can only think of a few nations that would never agree to being tied into the MilNet. It's a very lucrative force multiplier. There is only one that seems to fit all the necessary criteria."

Walt tried not to appear thoughtless, but there was no way around it - Who?

"We can't encourage any sort of action that would enable the Coalition to get a mainland foothold against the Chinese, but the two sides must be kept in conflict so for us to make anything out of this deal. There is only one country, other than Mother Russia of course, on the Chinese border, that could stand up to the Chinese. One country."

Alexov smiled and grinned, but Walt was confused and concerned. His strength - unyielding confidence - was passing. They seemed to pass second by second with the wind from the oncoming Pacific storm.

Friday, May 6, 2011

2nd Lt. John Chamberlin-Personal Log

85 nm E of Chuangwei, Taiwan

We're going home. Part of me is overjoyed that the war is about to end for us, at least for a while. Most of me, and I think its the same for everyone in the platoon, wants to fight on. We're not hungry for blood, or vengeance, but no one, no Marine, wants to leave a job incomplete. It goes against our grain.

Captain Blackwell (formerly Commander Blackwell of the Battle of Yellow Sea) announced over the 1MC intercom that we're to lead a splinter task force of damaged ships home. We're going to go through the Indian Ocean, the Red Sea, the Suez Canal (if possible), the Med, and finally across the pond, the Atlantic Ocean to home. It would be a lot faster and easier to just head for Pearl Harbor, San Francisco, San Diego, or the Panama Canal, but there's some unusual difficulties. We have to provide cover for all the damaged ships in Task Force

Stingray. Most of them are foreign: Italian, Spanish, British, French, Australian, New Zealand, and the list goes on. Some of them will get repaired at Subic Bay courtesy of the U.S. Navy, but the South Korean and Taiwanese ships will get priority after the larger U.S. vessels.

The grand tour will also serve as a show of force to the Arab League. Our little task force may be small and damaged, but it could still cut them off by sea in no time. Another problem is that lots of ships have been disappearing in the South Atlantic. Submarines are suspected. A carrier battlegroup based around the U.S.S. Forrestal (another ship fresh out of mothballs) has been sent to find the cause. I'm sure there's a few of the Navy's hunter/killer attack submarines down there too! After a few more supply and redeployment flights from the Philippines pass through the Wasp, we'll get the Task Force together and head west. They're talking about Sunday the 8th.

Monday, May 9, 2011

2nd Lt. John Chamberlin-Personal Log

155 nm E of Chuangwei, Taiwan

We haven't quite left for home yet. Chinese air attacks have slowed the pace of flights from the Philippines. We haven't been able to load up supplies for the trip, and we still have a lot of evacuated troops on board. I'm starting to wonder what the Pentagon's big plan for this war is. We have thousands of veteran troops on board. There are tens of thousands, maybe hundreds of thousands back in the Philippines. The ship's closed circuit TV showed a news report stating that over 1,000,000 troops are being deployed to active Army divisions around the globe. What are we going to do with all these people?

Monday, May 16, 2011

1352hrs local

U.S.S. George Washington CVN

Eastern Mediterranean Sea

Don had arranged for Rick and Chuck to be grounded indefinitely. Everyone who even heard about the punishment knew that it would only last until the battlegroup needed pilots again. When that happened, no one would care about the BOB missile incident.

Hostilities between Greece and Turkey began to subside. The inventory of expensive, American-made, precision-guided munitions was running low. The same was true for the valuable fighter-planes. In the wake of the "accidental" attacks on the U.S.S. Antietam, the United States was not expected to replenish either side's stockpile of modern weapons or planes. Instead, a cat-and-mouse game took over in the Aegean and Eastern Mediterranean Seas.

Instead of flying routine CAP operations, Chuck and Rick were condemned to playing flight sims in their stateroom, supervising squadron maintenance readiness, and watching TV. The flight sims kept Rick occupied, and a few additional beaurocratic tasks did the same for Chuck. When they watched TV, the punishment really hit. Every minute of every day, there were all kinds of news reports about the American scientist who had killed thousands. The footage of his assassination was played again and again. The same seemed true about the resulting Arab riots all around the Mediterranean. The rest of the ship was busy preparing for any eventuality that might result from the latest breakdown in East/West relations. Everyone had something valuable to do-except Chuck and Rick.

Finally, the moment that they had trained for arrived. Out of the puffy white clouds along the coast of Turkey to the east, a direct attack on the U.S.S. George Washington began. The Turks had been badly mauled by the Greeks, but they still were able to secure Cyprus. When the Greek attacks began to seriously impede Turkish resources, they had tried to get the United States to block some of the incoming missiles. That resulted in the near sinking of the U.S.S. Antietam. Now, with international support for America waning in the wake of Dr. Silverstein's genocidal revenge plot, the Turks saw an opportunity to gather Arab allies in their war. The Americans threw their political hat in the ring with the Greeks. Given thousands of years of animosity, it was guaranteed to draw an attack from the Turks. The moment was ripe for Jihad.

At 0600 hours, the Turkish Air Force commander passed the word to his subordinates. They were to launch a maximum effort against the U.S.S. George Washington. A victory against the Americans was sure to bring other Arab nations to their aid against the Western-backed Greeks.

The attack was delayed until the afternoon. A Turkish fishing boat-operated by an intelligence-gathering branch of their Navy-was having a difficult time locating the American battlegroup. It wasn't until almost 11:30 am that the carrier was spotted and its location reported back to Istanbul. Then, events happened rapidly. Aircraft were loaded with almost all of the remaining precision-guided munitions, and then they taxied out to await their orders. The airfields furthest to the north and east put their planes up first. As they joined into low-level formations and headed for the GW, the airfields closest to the Americans began to put planes into the sky. They all joined into a line-abreast formation that stretched for over ten miles from one end to the other. Then they descended as low as they could in the hopes that the nearby waves would confuse the American radars.

None of it worked. The American-intelligence collecting agencies were all too heavily tasked. None of them warned the GW of the attack. Instead, it was Nikki, the E-2 controller who was the object of Codie's affection. She was on a routine patrol and noticed the Turkish squadrons gathering in the north and far east. All of them headed for the airfields closest to the stricken Antietam. Nikki directed a pair of Marine Super Hornets to reinforce the four Navy Hornets that were protecting the Antietam, the Stark, the rescue helicopters, and the four tugboats that were preparing to bring the Antietam back to Italy for repairs. The Marines arrived over the scene and extended the battlegroup's radar coverage. When they sent Nikki their uplinked data, she immediately noticed the Turkish line of attack planes coming. Nikki saw the six-mile-wide formation, notified the GW, and then brought in reinforcements from all over the Mediterranean.

Back on the GW, a hundred different alarms began to sound. Chuck and Rick ran to their ready rooms to find out what was happening. Not even five minutes later, they were in their planes and getting thrown off the GW's catapults as fast as possible.

The GW's flight operations staff was busy. They were scrambling every fighter into the air as fast as they could throw them down the deck. First the fighters were launched. Using all four catapults, 36 Hornets and Super Hornets were launched in less than 10 minutes. Then another E-2 was sent up to widen the radar coverage and act as a reserve in case something should happen Nikki's plane. Finally, anti-submarine S-3 Vikings and rescue helicopters were launched.

Elsewhere in the battlegroup, each individual ship prepared for battle. Almost every escort ship had at least one-usually two-helicopters for antisubmarine and rescue patrols. Eleven more helicopters went into the air.

There was a lot of command and control activity. The GW and the E-2 controllers were busy directing air traffic. In the CIC onboard the GW, the escort ships were tasked with defending patrol sectors that circled the carrier like rings on an onion. Some of CAP Hornets and Super Hornets were sent to assist the escort ships. Others were sent forward to intercept the Turks. An entire squadron was held back to defend the GW.

As the air umbrella went up over the GW battlegroup, the Turks closed in at just under 700 miles an hour. Sophisticated, interlinked, American radars had no trouble discerning the F-16's from the waves. Some of the radars had been designed to detect things as small as periscopes in heavy typhoon-like seas. During this attack, the waves were only 2-4 feet high. The F-16's were small and stealthy, but Nikki and the rest of the combat controllers of the GW battlegroup saw them coming, and they were ready.

It was clear that the Turks either intended to attack, or were playing an incredibly dangerous game with the Americans. There was some indecision as to how the Turkish threat could be dealt with, but there was no time to argue. They were closing at over a mile a second. Since it was easier to apologize than to ask for permission, the battlegroup commander decided to attack them as they came into range.

The American frigate U.S.S. Stark was still guarding the heavily damaged U.S.S. Antietam, but when the Turkish line of F-16's came within 35 miles, the Stark opened fire. The forward Mark 13 missile launcher carried up an SM2 missile from the storage bay below deck. Then it spun in the direction of the Turks, and fired. Both the Stark and the Antietam were smothered in white smoke from the missile's exhaust. When it faded, one could see that the Mark 13 had already reloaded and was aiming to fire once more. The launcher fired another 33 missiles at the incoming Turks, and spikes of smoke contrails spread like evil fingers from the ship's bow toward different points on the eastern horizon-beyond which, the Turks continued to approach.

In the Turkish formation, havoc was about to hit hard. One by one, the planes closest to the Stark were exploding as the SM2 surface-to-air missiles from the Stark found their targets. The F-16's were bursting into orange and red flames then disappearing in black puffs loaded with expensive debris. At the same time, 4, three-plane formations of Marine Super Hornets were firing their 48 AMRAAM air-to-air missiles into the formation. Then Rick and Codie's squadron attacked at over 1000 miles an hour from only 1000 feet above the water. Their missiles closed on the Turks at almost 3000 miles an hour. When the range between the Turks and the Marines was down to 10 miles, the Marines fired off 48, AIM-9 Sidewinder heat-seeking missiles. It was all happening too fast for the Turkish pilots' eyes to see the missiles and send the impulses to their brains. All that they could do was to continue their attack.

The Stark and the squadron of Marine pilots had blown a gaping hole in the middle of the Turkish formation. Half of the Turks passed north of the Stark and the Antietam. The other half passed south. While they were doing so, Don and Chuck's squadron of U.S. Navy Hornets opened fire on the northern group, and another Navy squadron worked on the Turks to the south. Each squadron carried another 48 AMRAAM missiles and 48 Sidewinders. The devastation was almost complete.

Less than 20 Turkish F-16's survived the attacks from the Stark, the Marines, and the Navy Hornet squadrons. Those that remained were scattered and confused. They were also inside the range of the remaining escort ships. Of the GW's escorts, the Stark was the weakest in terms of air defense. The most powerful had been the Antietam. Now, a pair of Arleigh Burke class guided-missile destroyers (the U.S.S. John Paul Jones, and the U.S.S John McCain) both opened fire. Each was equipped with the same VLS missile system found on the Antietam, and fountains of SM2ER missiles leapt from their forward and aft missile bays. A few hundred feet above each ship, the missiles arced over and headed for the Turks.

Less than 10 minutes after the decision was made to engage the Turks, the attack was over. The Turks never came close enough to use their anti-ship missiles, and the U.S. Navy had destroyed every single F-16. Not a single Turkish pilot survived. A few had managed to eject, but their approach was too low, and there wasn't enough time for their parachutes to open. It was completely one-sided, and-for a few hours-it was a complete secret to the rest of the world. Only government officials in the United States and Turkey knew that the attack had happened.

Tuesday, May 17, 2011

1230hrs local

U.S.S. George Washington CVN

Eastern Mediterranean Sea

The secret battle of the Aegean Sea did not remain an enigma forever. In fact, it was born into public knowledge hours after it occurred. One of the Turkish pilots had a brother-in-law who was a reporter with deep government connections. When the world was finally informed, the one-sided battle had a slant from the losing side. Most people were told that the Turks were ambushed. American Defense Department officials provided video and radar data, but it was dismissed as Hollywood propaganda. The Americans had totally won the 15-minute battle, but they utterly lost it in the public's eye. Even some Americans had their doubts (there's <u>always</u> some).

If the public hadn't found out about the battle, there would have been a great deal of support for holding back on any retaliatory action. However, when the cat was let out of the bag, it was decided that the United States had to strike back. The American media alleged that the public demanded it.

Don gathered his naval aviators in their ready room and briefed them on the day's operation. They were to provide escort to a squadron of U.S. Air Force F-16's that were already leaving their base in Italy and were headed for the Turkish coast. The Air Force had been tasked with attacking the four primary airfields that the Turkish pilots had attacked from. Off the coast of Turkey, they planned on rendezvousing with Don's squadron. Then the F-16's would split up and drop Durandural rocket-assisted, anti-runway bombs. Each plane carried two of the huge bombs, and it was expected that the resulting craters would prevent any further Turkish air strikes. The Navy's escort planes just had to loiter off the coast and assist as needed. No one in the room failed to see the futility of the air strikes. After the previous day's battle, there was no more viable Turkish Air Force left. Still, Washington wanted to attack, and they would oblige.

When the details of the briefing were finished, the naval aviators finished prepping for their flights. One by one they filtered into the hangar deck and inspected their aircraft. The maintenance crews had worked through the night, and every Hornet was ready. Just after noon, they were taxied to the ship's four catapults and shot off the deck by giant steam pistons.

After his recent performance both in training exercises and in combat, Don reassigned Chuck to be his wingman. It was an honor, but Chuck thought that Don might have made the move to keep a better eye on him after the incident with Rick and his make-shift missiles. He saw it as a continuing punishment.

The Hornets formed into three, loose, four-plane formations. While they formed up, they circled and climbed to 30,000 feet. Once everyone was in position, and they were all at the right altitude, they headed for their rendezvous point. It was over 200 miles to the northeast, but the flight only took them a little more than 15 minutes.

From the west, the Air Force F-16's faded into sight. They were right on time. Don slowed his squadron down, and wagged his wings to the Air Force pilots. Their squadron leader wagged back, then they split up and headed toward their individual targets. As the F-16's slipped out of sight as smoothly as they had appeared, Don's radio called.

"Delta Lead from Skymaster one-on- zero."

It was a controller from an Air Force E-3 Sentry AWACs plane-a larger version of the Navy's E-2 Hawkeye.

"Copy, Skymaster. Go ahead."

"Delta Lead, identify multiple contacts bearing zero-one-three, at angels three-five. Copy?"

"Copy, Skymaster. Wilco."

Don switched to the squadron radio frequency and ordered one of the three flights to investigate and identify the radar contacts. When he switched back to the inter-unit frequency, the radio called for him again.

"Delta Lead from Skymaster one-one-zero."

"Copy, Skymaster."

"Delta Lead, identify multiple contacts bearing one-three-four, at angels three. Copy?"

"Copy."

Once more, Don switched frequencies and sent off another 1/3 of his squadron to investigate radar contacts that the Air Force couldn't discern with just radar. Despite decades of ultra-high-tech research and electronic evolution, the best way to determine friend from foe was still a good pair of human eyes.

When Don switched back to his inter-unit frequency, he heard the Air Force controller ordering elements of other escorting squadrons to check out radar contacts. He knew that there were lots of squadrons involved in the air strikes. Some Air Force units had been tasked with attacking Turkish airfields. Others were doing escort or search-and-rescue patrols. There were even planes from other countries involved, like Britain, Italy, and Germany. When America wanted to exercise its military might in Europe, it called on its NATO allies, and that meant that hundreds of planes were probably involved.

He also heard reports coming in from the F-16 pilots as they made their attack runs. There was a great deal of AAA (anti-aircraft artillery; i.e. "flak"), and there were reports of SAMs. Cursed with the burden of experience, the F16 pilots already knew how to dodge, decoy, or evade SAMs. There had been hard-learned lessons in countless earlier missions through the years-at the expense of many pilots. Now, those lessons were saving lives. Everyone survived. As individuals and in small groups, all of the F-16's that Don had been ordered to escort appeared and joined in formation with him.

The two flights of Hornets also returned. The first flight had identified a Syrian patrol flight of six Mig-23 Foxbats (Soviet-built fighter-bombers). The second flight of Hornets flew a hundred miles to the southeast and identified an Egyptian patrol of F-16 Falcons, then they headed back to rejoin their squadron. Everything had gone according to plan-for both sides.

Wednesday, May 18, 2011 0455hrs local

Eastern Mediterranean Sea

75nm Southeast of Crete

American and NATO air strikes continued against targets in Macedonia and Turkey. On the evening of Saturday 14 May, 2011, an intense cold front passed over Germany, Italy, and the Balkan Peninsula. It then collided with the warm and wet air over the Eastern Mediterranean Sea. Only the sudden brutality of the weather halted the airborne attacks.

Back on the GW, flight operations had to be temporarily cut back. Wind gusts over 50 knots and waves averaging 20 feet made it even more dangerous than usual to launch or recover aircraft. The CAP that was airborne was kept aloft when the U.S. Air Force sent a KC-10 mid-air refueling plane from Germany, but any reinforcements or replacements were told to wait until the storm subsided.

The conditions were too rough for most attacks anyway. Standard iron bombs would drift off target. Aircraft wouldn't be able to attack at low level because of the wind and waves. Precision-guided munitions would have far too difficult a time distinguishing waves from ships, and laser-guided munitions were useless in such dense cloud and rain cover. It would be a useless and suicidal operation for any nation to attempt an attack against the U.S.S. George Washington and its supporting battlegroup. That's what the conventional military wisdom suggested.

Growing anti-American sentiment within the nearby Arab nations dictated that any attack against the genocidal Americans-any Jihad-would be worth the attempt. In difficult flight conditions, and in near-total radar blindness, hundreds of fighters and fighter-bombers lifted off from airfields around the Mediterranean. Libya, Egypt, Syria, and a handful of scrapedtogether planes from Turkey all headed into the sky. Just before dawn, they headed toward the American battlegroup.

Fishing boats were used to track the American ships. An aging Egyptian E-2 Hawkeye confirmed their locations just before the worst of the storm enveloped them. The final straw came from a North Korean intelligence operative who had mysteriously and conveniently provided accurate longitude and latitude positions

474

for each American ship in the battlegroup-even the fleet supply ships. It was a windfall report for the coordinated Arab attack forces.

Libya had a long-lasting grudge against the United States, and despite the weather, it sent most of its Air Force against the GW battlegroup. Over 130 Mig-23 Floggers of various types headed out to sea. They were accompanied by 45 French-built Mirage 5 fighter-bombers. Serving as escorts were 53 aging Mig-25 Foxbats-mach 2.8 capable, Soviet-built fighters. There were another 36 ancient Mig-21 Fishbed fighter planes as well. All of the aircraft were old. Their pilots were ill-trained, and their weapons were of questionable reliability. Still, each man was eager to strike a blow in the name of Islam and in defense of Arabs everywhere.

Egypt had been testing the reaction time of the GW's CAP for weeks. Each time that its reconnaissance planes were met by American Hornets and Super Hornets, an Egyptian E-2 Hawkeye monitored and recorded the activity. Now, in the pre-dawn hours of a once-normal Wednesday morning, the Egyptians joined the fight against the Americans. They sent 137 Americanbuilt F-16's into the attack. They were joined by 48 Mirage D's, 72 Mig-21 Fishbed fighters, 24 American-built F-4E Phantom fighter-bombers, and 18 state-of-the-art, French-built Mirage 2000 fighter-bombers. It was a formidable force of foreign planes, but the Egyptian pilots had come to know them as their own, and it was in their own aircraft that they went to defend Egypt and Arab people everywhere.

Syria had the most battle-beaten air force of all the attackers. For decades, the legendary Israeli Air Force had beaten them senselessly. Each pilot was looking forward to an opportunity to reclaim its pride. Given that their target was a symbol of American invincibility and due to the increased anti-American sentiment in the post-ZP9 days, the Syrian pilots looked forward to this attack even more than if it had been against their old foe, the Israelis. Joining the attack from Syria were 170, Mig-21 Fishbeds, 134, Mi-23 Floggers, and 36 cutting-edge Mig-29 Fulcrums. They had thrown everything they had into the fight. Though they were lacking in skill, and no one's military technology was equal to the Americans, the morale of the Syrian pilots was atmospheric, focused, and aggressive.

Weeks of sporadic attacks, meager defensive actions, NATO bombing, and a few large-scale battles with Greece had left the Turkish Air Force in ruins. Only 24, F-16's were patched up and thrown into the fight. Still, they had a trick up their sleeve that would amplify their effect in the threedimensional battlefield that was the Eastern Mediterranean Sea.

Allah blessed his holy warriors with a break in the weather. Not long after the airborne armadas were aloft, the storm began to subside. First the coastal areas cleared, then, as the swarms of Arab aircraft made their way toward the GW, the clouds, winds, and waves subsided along their flight paths. With the fading storm, the Americans finally detected the encircling aircraft, but there wasn't enough time for the GW to launch all of its interceptors. That morning, Allah favored the Holy Warriors of Islam.

475

The Libyans had the farthest to fly. They left a few minutes before the agreed-upon launch time, and so they approached the GW Battlegroup first. At the extreme south-southwest of the formation, a pair of Marine Super Hornets was flying CAP over an American destroyer, U.S.S. Ross. They were 35 miles away from the GW, and when the Libyans were finally detected, they had already made it within range of the Ross' SM2ERII surface-to-air missiles.

It was only a few seconds after the formation was detected that the decision to fire was made back on the GW. Radar contacts were increasing in number, and it was clear that a major attack was beginning. During the next two minutes, the entire fleet went into action. Navy Hornets were launched to bolster the 12-plane CAP that was spread out around the GW Battlegroup. Every ship in the formation went to General Quarters readiness. The Ross opened fire, and the two Marine planes that were overhead went to work.

The leading edge of the Libyan assault was the 53, Mig-25 Foxbats. Normally, the planes were limited to a reconnaissance role. They had been designed as high-speed, high-altitude interceptors, but carrying missiles under their wings slowed their performance. The Libyan Foxbats closed at over mach 2. It was as fast as they could fly since they had been specially fitted to carry 4, AS-11 Kilter anti-radiation missiles. While the Ross fired off a few SAMs at the Foxbats, the Foxbats launched a wave of 212 missiles in return.

As the Libyan launches were detected, the Ross switched to targeting the AS11's instead of the rapidly disarming Foxbats.

Still, the Foxbats closed. As long as they approached the Ross, they could detract SAM launches at their missiles by using themselves as bait for the Ross. The Marines took advantage of the situation. Each plane shot down 5 Foxbats, but then they had to head back to the GW to rearm. The Marines had fired all their missiles and emptied their 20mm cannons.

Flying at 2008 knots-roughly 2100 miles an hour or half a mile every second-the 212, AS-11 Kilter missiles closed in on the Ross rapidly. The Ross was loaded with 70, SM2ERII surface-to-air missiles. A few of the missiles smashed into 11 more of the Libyan Foxbats, but the remainder were aimed at the AS-11's. Of those, a high percentage found their targets, and 48 of the 212 Kilters were shot down. The Ross was also armed with a pair of 20mm Phalanx autocannons, and a 127mm multi-purpose cannon. When the rain of AS-11 missiles was within 4nm, the guns opened fire, and 7 more missiles were stopped. Small rockets loaded with radar deflecting chaff were launched to confuse the missiles, and 4 more missed the Ross.

Finally, the ship was out of ammunition and out of luck. Out of the 212 missiles, 146 remained. They began to impact. The first missiles hit the different radar and communications antennas. When those were destroyed, and their emissions ended, the remaining missiles lost their guidance and impacted haphazardly. Many went past the ship; over, under, in front, or behind it. Statistics were with the Libyans. Over 98 of the missiles hit the ship. Each had a relatively

476

small warhead, but the effect of so many missiles hitting so quickly was devastating. Each one hit with enough velocity to pierce the hull and superstructure before exploding inside. Over 200 of the 325 crewmen and women were killed. Everyone was wounded. After 20 seconds of impacts and near-misses, the ship was dead in the water with a pale gray smoke venting through the countless holes. The ship was shredded and damaged beyond any hope of repair.

The Mig-25's circled clockwise around the ship once, and then they headed back to their bases. Their job was done. Now, it was time for the second wave of Libyan aircraft. Old, slow, and very low-tech, 36, Libyan Mig21 Fishbed fighter planes approached the mauled U.S.S. Ross and headed for the U.S.S. George Washington.

Between the Mig-21's and the GW, an American frigate awaited. The U.S.S. Elrod was 12 miles southeast of the GW, and the Libyans were approaching at low altitude from south-southwest. Their flight path didn't bring the Libyans over the Elrod, but they were close enough for the frigate to fire its complement of 40, SM1MRII surface-to-air missiles. Unlike more modern ships, the Elrod was a product of 1970's era design, and its launcher was a mechanical beast instead of the rapid-firing VLS systems on newer destroyers and cruisers.

The Elrod had enough time to launch 11 missiles before the Libyans spotted the white spikes that were contrails from the Elrod's approaching SAM's. They looked like thin white fingers reaching up from a single point on the cloud-dotted horizon. When the Libyan Mig-21 pilots spotted them, they turned slightly and climbed to attack the Elrod.

At 2000 feet, with their noses aimed high, the Libyans opened fire. Every plane carried 4, UV-16-57 rocket pods. Each of the pods were loaded with 16, 57mm unguided rockets. All total, the Libyan Mig-21 pilots fired 2,304 rockets high into the air, then they turned and headed back to their airfields.

The rockets were small. They had been designed for infantry support and destroying small targets like trucks and tanks. None of them could do much damage to even the smallest of ships-like the Elrod. They didn't even have the range to reach the American frigate. Coupled with their lack of guidance systems, the rockets had the effect of shooting a shotgun at a building from 20 miles away. It was a harmless attack, but an expensive one for the Americans. There was no way to determine that the rockets were short-range and unguided. As a matter of precaution, the Elrod had to fire the remainder of its multi-million dollar SAMs at the $100+ rockets.

It was a futile effort. The Elrod only had 29, SM1MRII surface-to-air missiles to shoot down 2304 rockets. The rockets were far too small to accurately target, and most of the Elrod's SAMs missed. The few that were able to find their targets did so just as the Libyan rockets were running out of fuel and falling into the sea. Now, the Libyan ruse was over, and the Elrod was almost completely out of ammunition. It would take day-maybe weeks-before replacement SAMs could be loaded. In the meantime, the Libyan Mig-21's were headed for home, and the GW's E-2 Hawkeye reported that another wave of attack planes was coming.

This wave of Libyans was almost twice as big as the two previous attacks combined! Soviet-built Mig-23 Floggers-130 of them-were racing over the Mediterranean's white-capped waves at 500+ miles an hour. Every plane carried 2, AS-7 Kerry missiles. One squadron of Mig-23's turned and closed on the Elrod. From 5 miles away, the Elrod's 76mm turret and its 20mm Phalanx autocannon both opened fire. At the same instant the muzzle flashes were first seen coming from the Elrod, a dozen Libyan Mig-23's fired 24, AS7 Kerry missiles. The rockets ignited and they streaked toward the tiny ship. Steered by tiny joysticks on the Libyan pilots' flight controls all of the missiles found their marks.

The Elrod boiled with explosions. After the first few, the rest were mere flashes of orange and white under a blanket of black smoke, gray steam, and white seawater. The Floggers passed overhead a few seconds later. As they turned to return to their airfield, the squadron leader reported that the ship had broken apart and sank. There was barely any debris, and no survivors would ever be recovered.

The other 116 Libyan pilots continued toward the U.S.S George Washington. Since the attack had started, only a few minutes had passed. For some it was an eternity. For others, it was just enough time to scramble to their aircraft and launch from the GW. The Marine CAP aircraft that were scattered in 200 miles around the battlegroup were now arriving-just in time to meet the Libyan air armada.

Codie was among them. He had been patrolling over the destroyer U.S.S. John Paul Jones. It was over 100 miles north-northwest of the GW. When the word about the attack spread throughout the battlegroup, Codie and his wingman both turned and headed south before they received any orders to do so. They were already over the GW, when Nikki called and told them specifically where to go.

"Big Eye Five-Five to Cowboy Flight, bandits, bandits, bandits, bearing two-zero-five at angels two. Expedite. There are no friendlies in the area. You are weapons free. Engage targets at will. Recommend you expedite, Cowboy. They're in some real trouble down there."

"Copy that, Big Eye. Yeeeeeeeeee-haaaaaaaaa! We're already on our way, darlin.' Now you're gonna see what Marines are made outta! Cowboy flight, engaging."

"Cowboy from Big Eye Five-Five, this is serious like a heart attack. Get it together and get it done fast. We're sending everyone in, so once your wings are empty, get outta there. Copy?"

"Ahhh, don't you worry your tail feathers, girl. We've got this. Cowboy flight out."

As Codie and his wingman passed over the busy flight deck of the GW, their radars locked on to the Libyan Mig-23's. Both planes immediately started picking targets and opening fire. From less than 20 miles away, both Super Hornets launched 4, AMRAAM medium-range air-to-air missiles. The selfguiding

AMRAAMs were equipped with their own built-in homing radars, and they pursued their designated Mig-23 targets at over mach 4. When the range closed to 15 miles, Codie and his wingman each fired 2, AIM-9 Sidewinder short-range, heat-seeking, air-to-air missiles. Then they aimed their planes toward the center of the Libyan formation for an attack with their 20mm cannons.

The 116 Libyans all popped out bundles of radar-confusing chaff in an effort to confuse the AMRAAM missiles, but it was useless. The Americanmade missiles were too sophisticated, and 8 Libyans found themselves on the deadly end of finger-like missile contrails. Those eight orange flashes and black puffs of smoke were followed by an impressive fireworks display as the remaining Flogger pilots popped out decoy flares in the hopes of evading the Sidewinder missiles. It was the same result as with the AMRAAMs, and 4 more Libyans were shot down. Codie and his wingman were aces now, but neither cared one iota. They were boring in on the Floggers at full-afterburner speed.

The Libyan formation looked like a line of gnats when Codie and his wingman first saw them. Both Marines were flying head-on and too fast to accurately hit any of the Mig-23's. They passed through the formation, pulled up, and at the top of a loop they rolled over and dove down behind the Floggers. The Libyans continued to press their attack. Their cockpit design prevented them from easily seeing Codie or his wingman, and even though they knew that the Americans were behind them, they did nothing to evade.

Codie lined up behind one of the Migs and fired. A few hundred 20mm rounds ripped through the thin skin of a Mig-23. It trailed vapor, then burst into flame and spun into the sea. It splashed as he was lining up another. His wingman shot down one, then Codie followed suit. It was a turkey shoot, but there were too many turkeys. After he and his wingman had shot down 12 Libyans with missiles, and 6 more with their cannons, the Marines had to head back to the GW. They were out of ammunition. The problem was that the remaining Libyans were headed for the GW too!

The large formation drew closer to the American carrier. Behind the GW, almost 5-miles north-northwest, the Antietam's sister ship, the U.S.S. Anzio, opened fire. It could have earlier, but to prevent the accidental downing of the Marine fighters as they made their attacks, the Anzio held back its wrath. Nikki directed Codie and his wingman to climb from 1000 feet to 30,000 feet to distinguish them from the Libyans. When they passed 5000 feet, the Anzio's fire control computer assessed that it was clear enough to begin launching missiles. From both the forward and aft VLS bays, streams of SAMs arced into the sky, bent down toward the horizon, rocketed over the GW, and reached out to the oncoming Libyan formation. They launched in pairs, and by the time the first American SAMs found their targets, 34 missiles had been launched. They continued until all 88 had been fired.

The Libyans were losing planes left and right-literally. Their formation was being whittled down to stragglers, but the Arab pilots-filled with racial and religious anger toward the Americans-flew headlong into their fate. It was the Light Brigade's charge in the 21st century.

When the first Libyan Mig-23 reached the 10-mile mark, the GW itself opened fire. Both starboard Mark 91 missile launchers lashed out with all 8 modified AMRAAMs that were in each one. The flight deck was as busy as a NASCAR pit row. Right next to the parked and taxiing aircraft, rockets were firing off in defense of the ship. Another 16 Libyans died before they reached the attack range of their own missiles.

Eventually, some of the Libyans came within 7-miles of the GW, and they were able to launch their own missiles. Short-range, point-defense missiles and Phalanx cannons on the GW tried to stop the 96, Libyan AS-7 Kerry missiles. The Phalanx' were able to detonate 4, and the short-range missiles stopped another 32, but the Libyans were able to steer 60 into the carrier.

Explosions boiled all over the ship. Most of the crewmen and women on deck were killed by the explosions, shrapnel, or secondary explosions from aircraft that had been hit.

It was hard to miss a target as big as an American Supercarrier. Still, the Libyans acted on the side of caution and aimed for the center of the ship. Fore and Aft, there were only a few explosions, but the center of the flight deck and the ship's island were turned into punctured and twisted, burning wreckage.

The Libyan's passed over and around the GW at low altitude. When they did so, the port-side missile launchers and Phalanx mounts were brought to bear with great effectiveness. Of the remaining 48, Mig-23's, 36 were brought down. The Anzio continued to launch its missiles throughout the attack, and it single-handedly shot down the remaining Libyan squadron.

Codie and his wingman watched the action from 6 miles above. Both were speechless. Their ship was burning, they were out of ammunition, and despite their best efforts, the under-rated Libyan Air Force had beaten them. With the thrill of battle replaced by the throat-choking grip of defeat, he called Nikki for instructions.

"Big Eye Five-Five from Cowboy Flight, request re-direct to nearest landing opportunity for refuel and rearm."

Nikki was getting similar calls from all the Marine pilots and all of the Naval aviators that managed to get off the GW before it was hit. No one knew what to do or where to go. Thankfully, only the Marines seemed to immediately need a place to land. She directed them all to the U.S. Air Force base in Aviano, Italy. Then she called the base and requested that mid-air refueling planes be sent to meet the Marines. It was a long trip. She also requested additional aircraft for CAP, and a mid-air refueler to come out and tank-up her plane.

Worst of all, Nikki knew the truth. The radar on the GW was knocked out along with all its communications equipment, but with the weather clearing, she could see the new storm on the horizon. Another 45 Libyan aircraft-Mirage 5's-were approaching the battlegroup from the southsouthwest. To the southeast, hundreds of Egyptian aircraft were approaching. The situation was going from awful to inconceivable.

The Libyan Mirages approached at 3000 feet, and they passed over the smoldering wreckage of the U.S.S. Ross at 450 miles an hour. Libyan morale was high before the attack. After the Ross and Elrod were destroyed, the pilots were euphoric. That feeling increased with the reports of hits on the GW, but when the Mig-23's were all shot down, the Mirage pilots lost hope. The venting U.S.S. Ross restored their morale. Quietly and privately, each man rejoiced in Allah's good will.

In an instant, their glee was shattered. Chuck and his squadron had made it off the GW just before the AS-7 Kerry missiles began impacting. Don was the last one off the deck. When his landing gear thumped as it left the deck, he had looked back and seen the first missiles hit. The memory was burned into his psyche, and he remembered it well as he gave his squadron the order to attack. All at once, every plane fired off its 4, AMRAAM and 2 Sidewinder missiles. The Libyan Mirages (with 2, 1000 lb bombs hanging under every wing) were shot down during a 3-second ripple of explosions. None of them survived.

Around the fleet, and back at home in Washington D.C., commanders watched their monitors and saw the naval aviators defend their ship. There was the satisfaction of revenge, but as Don led his squadron in a circle high above the GW, the Egyptians finally commenced their attack from the southeast.

There was little hope of stopping them. Don's squadron had nothing left to fight with except its cannons. The U.S.A.F. reinforcements were 15 minutes away, and there were only two more Hornet squadrons in the air to defend the GW. Not a single escort ship would be able to assist in its defense. The Anzio was almost completely out of defensive ammunition, and the GW's missile launchers would take hours to reload. The Libyans had effectively punched a hole in the GW's protective ring of frigates, destroyers, and cruisers.

Nikki directed the remaining Hornets toward the Egyptians. She had Don take his squadron up to 10,000 feet and ordered them to loiter over the GW. They only had their cannons, but they were better than nothing. Besides, as a result of the attack on the Antietam, they had already shot down missiles with their cannons a few times. She hoped that they might be able to do it again.

The Egyptians came in force, and they came all at once. Almost all of the planes were flying low and relatively slow. They cruised toward the GW at 400 miles an hour and under 2000 feet. Leading the attack were 18 Mirage 2000's. They carried 2 huge Exocet anti-ship missiles each, and when they were 30 miles away from the GW, each of the Mirages launched their missiles and turned for home. Immediately behind the Mirage 2000's were 72, Mig-21 Fishbeds (carrying 2, 500lb. bombs each). A few miles behind the Mig-21's were 137, American-built F-16 Falcons-each carrying 8, 500 lb. bombs. Farther behind were the Egyptian heavy-hitters: 24, F-4 Phantoms carrying 240, 1000 lb. bombs, and 48, Mirage

D's carrying another 192, 1000 lb. bombs. The math was simple. There were just too many planes for the Navy's Hornets to handle.

The two squadrons of Hornets that launched off of the GW before Don and Chuck's squadron went to work. They used 96 AMRAAM missiles to shoot down all of the Exocet missiles and 60 of the Egyptian F-16 Falcons. The combined speed of the Hornets and the Egyptian aircraft was over 2000 miles an hour. Soon after their AMRAAM missiles were gone, the Hornet pilots started firing their shorter-range Sidewinder missiles. Given the high closing speed, the forward angle of attack, and the intrinsically stealthy shape of the F-16's, many of the Sidewinders were unable to find their targets. Still, another 18 Egyptian pilots were brought down. Armed with only their 20mm cannons, the Hornet pilots split up into pairs and began dogfighting with the Egyptians. The planes were technologically similar, but the American naval aviators were far-better trained, and their experience showed. Only 9 Egyptian pilots managed to drop their bombs on the GW, and they were all shot down on their way back to Egypt.

For almost a full minute, the 500lb. bombs rained down on the U.S.S. GW. Concealed by dense black smoke, fires and explosions rocked the ship as 11 of the bombs found and hit the GW. The remaining 61 bombs missed and threw pillars of seawater 200' into the air all around the huge ship.

A full ½-mile of the Mediterranean Sea was blocked out by highexplosive splashes, smoke, flame, steam, and thunderous explosions. Somewhere inside the confusion, an entire American supercarrier was obscured from all view. Onboard the GW, thousands of sailors were fighting for their lives. Whenever a bomb hit or missed, the concussion knocked everyone on the ship off their feet or into a bulkhead.

Only a handful of Hornets had any ammunition left to deal with the remaining Egyptian planes, and no one had enough fuel to make it to Aviano. Each naval aviator decided to stay and try to deter the remaining Arab attackers. As the two squadrons of F-4 Phantoms closed in on the GW, the Hornets pounced on them from 5000 feet above. Most of the planes were unarmed, but the Phantom pilots had no idea, and they had to evade. A few of them were shot down by the last remaining rounds of American 20mm shot from the Hornets. The rest dropped their bombs aimlessly around the GW. None of them hit, but the effect of 240, 1000lb. bombs impacting in the Sea around the ship was huge. To those onboard, it seemed as though they were in the middle of a nuclear explosion. So much water was thrown into the air that the fires on the flightdeck were temporarily extinguished!

The last of the Egyptian attackers were close enough to the Phantoms to see the ensuing dogfight. Some of the Hornet pilots (still unarmed) turned and headed straight at the 48, Mirage D's. Too close to break off their attack, and with still enough time to evade the Hornets, the Mirage pilots climbed and tossed their bombs high into the air toward the GW. Only 4 of the 192, 1000lb. bombs hit the ship. Once more, the crew was subjected to explosions equal in size to a small nuclear weapon. This time, the fires were almost completely extinguished. In fact, the supercarrier took on so much water that it began to list-10 degrees to port.

On their way back to Egypt, the Hornet pilots repeatedly dove and chased after the Phantoms and Mirages. The Egyptian pilots thought that their skills were keeping them alive. They never suspected that the Americans would try to defend their ship even while unarmed.

In Nikki's E-2 Hawkeye, there was no time to reflect on the bravery of her fellow naval aviators. The battle took place less than 20 miles away and 6 miles below. She was sitting high above the action, and now she had to watch it unfold yet again. The Syrians and Turks were coming. Nikki tried to convince the Hornet pilots to head for Aviano, but it was useless, they wouldn't leave. Some of the pilots-the ones lowest on fuel-started to say their goodbyes before ejecting. By the grace of God and the professionalism of the U.S.A.F., help arrived.

All the way from Aviano, a squadron of F-16 Falcons had raced to the GW's aid. At the far edge of Nikki's radar screen, she could see 4, KC-10 midair refuelers headed to the combat area. The naval aviators flew toward them to refuel, and the F-16's headed for the Syrians and the Turks. More U.S.A.F. fighters were on their way.

The Syrians and Turks met north of Cyprus, and they joined into a combined formation. First to attack the Americans were the Turks with 24, F16's. They closed in on the stranded U.S.S. Antietam and its protector, the U.S.S. Stark. Armed with standard, 500lb., iron bombs, the Turks came in low and fast. Despite being outnumbered two-to-one, U.S. Air Force F-16's dove down and scattered the Turkish planes. A dogfight followed, and since both sides were flying nearly identical aircraft, no one could tell friend from foe. Having experienced that kind of engagement earlier with the Greeks, the Turks jettisoned their bombs and headed for home. Neither side inflicted any losses on the other, but some of the Americans' precious air-to-air missiles were fired in vain. That meant fewer shots at the Syrians.

As the Syrians left the relative safety of the Cypriot Strait, more U.S.A.F fighters were closing in on the U.S.S. George Washington. This time it was a squadron of F-22 Raptors that had been on patrol over Bosnia. They had passed a squadron of F-15 Eagles that was leaving Aviano. Help was definitely coming.

Nikki directed the first squadron of American F-16's toward the 170, Syrian Mig-21's. The U.S.A.F. was outnumbered more twelve-to-one, but they attacked with ferocity. Their volleys of AMRAAM missiles brought down 41 of the Syrian planes. The range closed in seconds, and a barrage of short-range Sidewinders found another 22 Syrian victims. Unlike the Libyan and Egyptian pilots, the Syrians knew when they had been out-classed. The remaining Mig21's dumped their bombs and tried to head for home. Unfortunately for them, the U.S.A.F. Falcon drivers had their blood up, and they shot down 18 more with just their cannons.

While the Mig-21 pilots scurried for home, they passed through an undaunted formation of 134 Syrian Mig-23 Floggers. In hot pursuit, the American F-16's that were chasing the Mig-21's also came upon the Floggers, and a dogfight immediately began. The F-22 Raptors raced into the scene from high altitude and began picking off the Floggers that continued toward the GW. A handful managed to slip out of the dogfight and fire off a few AS-7 Kerry missiles at the Stark and Antietam. Before their missiles hit the

American ships, the pilots were killed by the American Raptors. That left the AS-7 missiles without any guidance, and they went wildly in every directionexcept into the ships.

Fatigue was taking its toll, and the U.S.A.F. F-16 pilots were getting worn down. They had been dogfighting for nearly ten full minutes when the F15 squadron from Aviano arrived. The F-15's joined the fight and brought down most of the remaining Syrians. A few tried to press their attacks, but they were taken down by the last remaining rounds from the Phalanx turrets on the Stark and the Antietam. Only 19 Floggers escaped back to Syria.

Nikki was still directing aircraft when the pilot of her aircraft ordered the radar turned off so that they could refuel from an Air Force KC-10. Even with the radar off, she still received data via digital uplinks with all the American aircraft in the area. Three more squadrons of F-15s and F-16's from Aviano arrived in the next 15-minutes. More were on their way from Germany and England. The Navy and Marine Hornets and Super Hornets were already well on their way to Italy. There, they would debrief, refuel, rearm, and head back out to the GW.

Smoke, steam, and pockets of flame still flickered across the sea around the U.S.S. George Washington. The ship had taken 11, 500lb. bomb hits. They had hit just as the ship was trying desperately to launch all of its aircraft, and the flightdeck was loaded with fueled planes.

Every single remaining airplane and helicopter was damaged beyond repair. Many took direct hits by the Egyptian bombs, spilled their fuel all over the deck, and added to the destruction. Only the Hornets and Super Hornets were safe, but none of them could land back on the GW. There was no steam available to the arresting gear systems, and so safe landings were out of the question. All four of its catapults were also out of commission. The flight deck around the island and the center of the ship was warped by heat. The Landing Signal Officer (LSO) station was destroyed.

Every sailor on the ship had some sort of injury. Most were limited to bad bruises, concussions, or blunt force injuries. There were 813 burn and shrapnel victims. Another 423 people were missing and presumed dead. They had either been direct victims of the 11, 500lb. bombs, or they were blown overboard by the shock waves from the roughly 400 near misses!

Those same explosions also saved the ship. When the first bombs hit the GW, the flight deck was crowded with fueled and armed aircraft waiting to launch. By the time the last of the 500-pounders had landed, every plane was on fire and most were already exploding. Nearly all of the deck crew were killed. Despite the heroic efforts of firefighters, the blaze was completely out of control-until the near misses. Concussion blew air away from the flames, and the towering fountains of seawater that surrounded the ship nearly wiped the deck clean. Many of the planes, most of the smaller debris, and a large number of sailors were washed overboard. The fires were almost extinguished by the first series of near-misses. The second series left only minor blazes on the deck. In the hangar and below decks, every sailor knew their job. They had all been trained as firefighters, and while many perished as a result of the smoke and flames, thousands lived.

The huge GW had taken the best that the Arab countries in the Eastern Mediterranean Sea could throw at it, and it had survived. Many factors had contributed. The skill and heroism of the Marine, Navy, and U.S. Air Force pilots brought down most of the attackers, and drove away an equal number of them. Technology, training, and commitment on the part of the escort ships resulted in the loss of the Elrod and Ross, but all the escorts had done their jobs against far superior numbers. Blind courage and trained responses on the part of the GW's crew kept the ship afloat even while tens of thousands of pounds worth of high explosives rained down on and around them. Most importantly, luck had scattered the 400+ bombs around the ship instead of into it. The American supercarrier-sometimes called a "Bomb Magnet"-had taken the brunt of four nations all at once, and it was still on the water.

Repairs were already underway. All of the near misses had sprung countless leaks in the hull. Every crewman was busy either fighting the fires, plugging leaks, clearing debris, or rebuilding damaged areas. By nightfall, the ship would be watertight once more. The fires would be out. The flight deck would be cleared. All of the arresting gear would be repaired, and two of the four catapults would be functional once more. The engines never stopped running. The island had taken many hits, but other than losing most of its external sensors (radar, etc.) and most of its communications, the ship's command and control staff were no more wounded than everyone else on the ship. Not long after the attack, basic radio communication was restored, and by the next morning, it was expected to have all communication systems in working order.

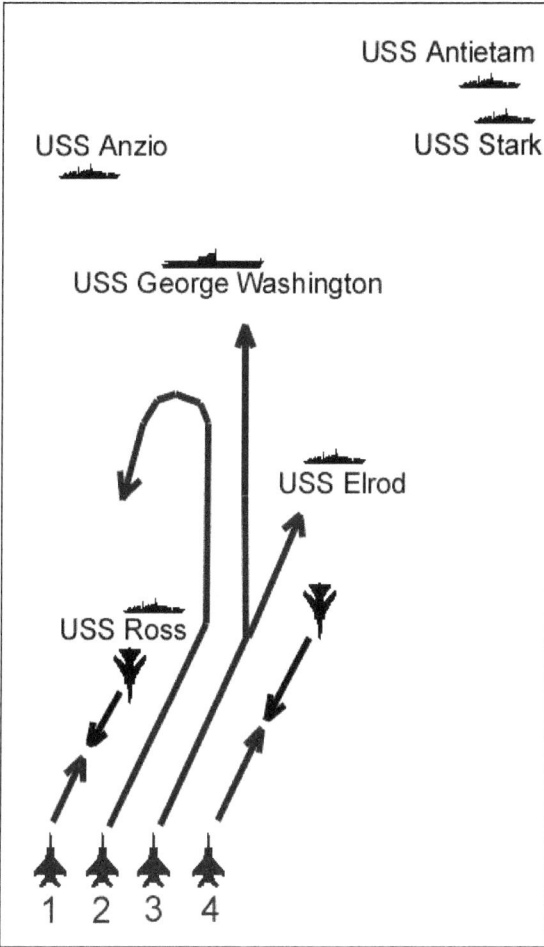

LIBYAN ATTACK ON THE USS GEORGE WASHINGTON CARRIER BATTLERGROUP

1.) 53 , Mig-25 Foxbats attack the USS Ross with AS-11 Kilter anti-radar missiles.

2.) 36 , Mig-21 Fishbeds - armed with 57mm ground-support rocket pods as decoys, attacks the USS Elrod.

3.) 130 , Mig-23 Floggers attack the USS Elrod and the USS George Washington with AS-7 missiles.

4.) 45 , Mirage 5's attack the USS George Washington and are all shot down by F/A-18 Hornets from the American carrier.

EGYPTIAN ATTACK ON THE USS GEORGE WASHINGTON

CARRIER BATTLEGROUP

1.)   18, French-built Mirage 2000's attack the USS George Washington with Exocet missiles.

2.)   72, Mig-21 Fishbeds attack the USS George Washington with 500lb.

bombs.

3.)   137, American-built F-16 Falcons attack the USS George Washington with 500lb. bombs.

4.)   24, American-built F-4 Phantoms and 48, French-built Mirage D's attack the USS George Washington with 1000lb. bombs.

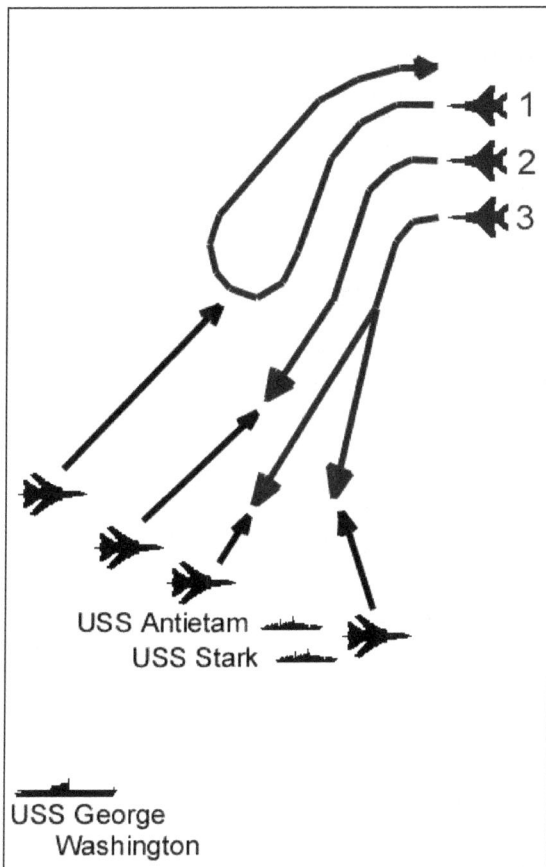

TURKISH AND SYRIAN ATTACK ON THE USS GEORGE WASHINGTON CARRIER BATTLEGROUP
1.) 170, Mig-21 Fishbeds attack the USS George Washington with 500lb. bombs, but they are turned away by a single USAF squadron of F-16 Falcons.
2.) 170, Mig-23 Floggers attack the USS George Washington with AS-7 Kerry missiles.

3.) 24, Turkish F-16 Falcons attack the frigate USS Stark that is protecting the damaged cruiser USS Antietam.

Wednesday, May 18, 2011

2nd Lt. John Chamberlin-Personal Log

Somewhere in the Philippine Sea, E of Taiwan

At dawn this morning, we finally left the Philippine Sea for home. We've got all our evacuees off the ships. We're all loaded up with fuel, food, and ammunition. All of the Marine units on the various ships have been reconstituted and joined back together, and the ships have been repaired to the point where they're considered by most to be seaworthy.

Everyone's glad that we're not going through the South Atlantic. The ship's TV news reported that the U.S.S. Forrestal took three torpedoes and a missile while looking for hostile subs down there. It's still afloat, but the picture on TV showed it listing at least 10 to 20 degrees to starboard. There's no word on enemy sub losses. Under the water, the war is silent and invisible. The only losses and deployment that are ever reported occur on the surface.

In the Med, the George Washington carrier battlegroup came under a well-coordinated air attack from Turkey, Syria, Libya, Algeria, and even Egypt. Most of her escorts are gone now, and the ship's headed for Italy all aflame. A few days ago, the Washington tipped off the Greeks that a Turkish amphibious group was headed towards mainland Greece. The Turks were ambushed and almost completely wiped out. Right after that happened, some of the members of the Arab League joined the Eastern Alliance with China and communist Korea. I'm sure the Washington was ready for the fight, but numbers are numbers. If the enemy fires enough missiles at a carrier battlegroup, and there aren't enough missiles or planes to shoot them down, then a certain percentage of the anti-ship missiles will connect with their targets. What the navy needs is an unlimited supply of SAM's to stop enemy aircraft and anti-ship missiles. I'm sure that's what they had in mind with those arsenal ships, but even those get emptied too fast.

I'm not sure how we're going to get through the Suez Canal now that Egypt has turned hostile. There's enough aircraft, missiles and Marines in this Task Force to take it by force if we need to, but that could get kind of rough. I'm sure someone higher up has a plan.

Thursday, May 19, 2011 0944hrs local

Aviano Air Base, Italy

Only a few of the Hornet and Super Hornets from the GW had been lost in the morning's combat. The rest made their way back to the American air base in Italy. There, the pilots and naval aviators waited anxiously while their aircraft were refueled and rearmed. Debriefings were quick and conducted by Air Force intelligence officers while the aircrews were still in their cockpits shutting down their planes. A few questions were asked. Then the U.S.A.F. officer would run to the next plane and repeat his gentle interrogation. All that the Marine and Navy officers had to do was to summarize their own activities during the morning and add any input that they thought was pertinent.

Of the 36 Super Hornets and Hornets that had left the GW (12 Marine and 24 Navy), 29 landed in Aviano. The rest were either lost in combat or missing and presumed lost-probably as a result of either damage or lack of fuel. Don, Chuck, Rick, and Codie survived the battle. Like everyone else from the GW, they were lined up and waiting on the flight line for the order to return to the combat area.

While Air Force groundcrews worked feverishly to prepare and repair the planes, a small crowd had gathered. Marine pilots and naval aviators were bonded in blood. Everyone had killed at least one Arab. Some, like Codie, had killed several. In just a few hours, many had gone from greenhorn to ace.

The flightline was a beehive of activity. Air Force vehicles raced back and forth bringing fuel, weapons, ammunition, tools, loading equipment, and the ever-important duct-tape. There wasn't enough time to properly repair most of the planes. After all, reinforcements were arriving from all over Europe on their way to do CAP duty over the GW. High-pitched jet engines drowned out all but the noisiest of dropped wrenches and loud Air Force Master Sergeants.

The crowd of people waiting to return to the GW was buzzing with conversation. Everyone had seen the escorts lost, the GW's hits, the nearmisses, and the incomprehensible number of aircraft in the sky. Only one person had ever trained for such a situation, and he did it on his own time.

In one of the infinite number of scenarios that he had created for his flight sims, Rick had tried to see what would happen if a large number of planes attacked the GW. Specifically, he wanted to see how the ship could be defended if an enemy attacked with more planes and missiles than there were surface-to-air missiles in the battlegroup. He designed the scenario after he was told that his squadron was being assigned to do CAP duty for the GW. It was a worst-case scenario, but trying it to survive and succeed against the impossible was Rick's style.

His results were far different from what had happened. In Rick's scenario, all of the escorts were picked off and the GW was sunk. The only real difference was that he had combined the airborne attack with simultaneous attacks by several submarines and patrol boats armed with antiship missiles. While everyone else was sharing their personal combat experiences, Rick, his squadron commander, and Don were all trying to figure out what would happen next. During their discussion, the rest of the crowd slowly started to eavesdrop, and the reality of war set in with each of them. Soon, Rick had everyone's attention.

One of the Marine pilots who had been bragging about shooting down six Libyan heard Rick's discussion and asked him about his theories.

"So, you're saying that this was NOT the big attack? Even though the Air Force is putting over a hundred planes over the GW, you still think there's more to come."

"Oh, yeah. C'mon. Let's face it. They wouldn't have gone head-to-head with an entire carrier task force unless they were ready to use everything they had."

Another Marine piped up from the crowd and interjected.

"What else could they have left?! My Wizzo and I dropped four of those bastards. Everybody tagged at least one or two. Hell, there must've been thousands of them up there. Now tell me, what else could they have left, Rick?"

"How many of you saw anti-ship missiles? Did anyone hear of any subs being detected? Think about it. Each of those countries knows that by directly attacking an American aircraft carrier, they're basically declaring war. So sure, they're gonna throw everything they've got into the game, but does anyone really think that they'd go in without a backup plan?"

Rick stunned everyone into silence. Then Codie spoke up.

"Okay. So what's their next move?"

"I have a tough time believing that the generals and admirals in all those Arab countries were just going to sit back and let their air forces do all the work. I also have a tough time believing that their political leaders would try to take on the U.S. with only one branch of their armed services. Given that, I think they'll hit us with every sub, ship, and remaining aircraft that they've got. Who knows what their armies will do, but I wouldn't want to be an Israeli right about now."

"Hey! Where were the Israelis when they were taking off?" asked a voice in the crowd.

Don had the answer.

"They probably watched 'em take off, but decided not to help us so that they wouldn't provoke those anxious armies that Rick's talking about."

Everyone thought that over for a few moments. They excitement of the earlier dogfights was now completely replaced by the awe of the latest war. Rick's speculation was logical, and it made each pilot and naval aviator even more anxious to get back to the GW. They figured that the U.S. Air Force would need all the help it could get.

As usual, it was Codie who broke the silence. He left the crowd and walked back to his nearby Super Hornet. Four Air Force mechanics were busy working on it. One was taping over a hole in the wing. Another was fueling the plane, and the last two were trying to load fresh 20mm cannon rounds. Pointing rudely at the overworked men who were loading the cannon rounds, Codie tried to exude authority, and he did so loud enough to embarrass his crowd of peers 100' behind him.

"You two! What's taking so long?! I gotta get back to my ship, so how 'bout y'all pick up the pace?"

One of the airmen, a Tech Sergeant, stopped cranking on the ammunition feed mechanism, picked up a ¾" wrench, and threw it toward Codie.

"Hey, asshole! How 'bout you shut the hell up and let us work? That is, unless you wanna pick up that wrench and give a hand!"

It was gross insubordination, but the sergeant was tired. He and his crew were given a month's worth of tasks and then told to complete them before sundown. The automatic feed on the 20mm re-supply loader was broken, and they had to crank the long belt of heavy cannon rounds into the plane manually. It was extremely hard labor. They were physically drained, and Codie's plane was only the fourth of almost forty planes that needed the ammunition. They had work to do, and Codie was just getting in the way.

In an unusual incident of common sense, Codie identified the real situation and did the right thing. He picked up the wrench and called to the crowd of loitering pilots.

"Hey, you guys! Are you gonna stand around and shoot the shit all day?! Let's get to work! These guys need our help too!"

With that, he walked over to the ammunition loading crank, pushed aside the Tech Sergeant, and began cranking in his own rounds as fast as possible. The rest of the pilots scattered as they ran to help out the other Air Force ground crews in any way that they could.

Two hours later, all three squadrons were given clearance to take off and head back out to the GW to bolster the U.S.A.F. CAP that was already there. Several Air Force mid-air refuelers were circling above the GW. There was also a huge E-3 Sentry AWACs plane that had replaced Nikki and her smaller E-2 Hawkeye. The two Navy and one Marine squadron patrolled uneventfully for the rest of the day. There were a few recon flights from the different Arab countries, but they were driven off without firing a shot. Everyone refueled twice, and then-just after sunset-they were allowed to land back on the GW.

During the night, the Air Force maintained and rotated its CAP over the GW. Helicopters picked up survivors from the Ross and searched for any from the Elrod. Tugs continued to move the Antietam closer to the GW for protection. The Destroyer John Paul Jones moved in from the north too. Throughout the night, U.S. Navy four-engine, turboprop, P-3C Orion subhunters also scoured the area in search of any submarines.

Most importantly, the Air Force brought in a squadron of B-2 Stealth Bombers. Each one headed for a different Arab airfield. They were loaded with satellite-guided Durandural Mk3, rocket-assisted, cratering, anti-runway bombs. Without warning, and with precise timing, 18 airfields were hit; from Libya to Iraq, and from Bulgaria to the Sudan. In the morning, repair crews would be busy all over the Arab world. Their pilots would have to sit and wait to try another attack on the GW.

There were other raids too. The United States Air Force was running low on precision-guided munitions. There never seemed to be enough. Instead, standard iron bombs were used to make a series of strategic deterrence raids. Since the accuracy of the weapons was limited, planners opted to strike larger targets. B-52 Stratofortresses and B-1B Lancer Bombers flew to the region from bases in the United States. Just before sunrise, every Arab port and naval base around the Eastern Mediterranean Sea was hit. Several large ships were sunk, and countless small patrol craft were lost. The bombers leveled loading and unloading facilities, warehouses, docks, air defense centers, forward command and control centers, and dry dock facilities. In one night, the U.S. Air Force set the Arab world's maritime capacity back decades.

Underneath the waves, the two American Los Angeles Class attack submarines that were assigned to escort the GW went to work. When they received news about the attack on their battlegroup, both submarine captains were also given authority to attack any participating Arab warship or cargo ship carrying war materials. One of the submarines had been tracking three Egyptian submarines, and it immediately attacked and sank all three. The other American sub had tracked a Libyan submarine. It was hunted and sunk less than an hour after the American captain received his war orders. Now, both were searching frantically for enemy submarines in the area. Besides Rick, there were a great number of people along the chain of command who believed that the Arab submarine threat was very real.

Rick's thoughts about Israel were correct too. Fearing an all-out ground attack by its Arab neighbors, Israeli leaders decided not to inform the Americans about the air attack on the GW. The next day, Arab armies waited for orders to attempt another invasion of Israel, but no order came. Each Arab nation's leaders decided not to press an attack against Israel while in direct conflict with the Americans. After all, it was the Americans who had attempted to wipe out the Arabs, and there was a strange sense of brotherhood with with the Jews who had been victims of genocide themselves. Both sides still held animosity toward each other, but there was a new respect also.

Friday, May 20, 2011

1350hrs local

U.S.S. George Washington CVN

Eastern Mediterranean Sea

The GW had lost most of its aircraft. Its crew was almost completely exhausted. Everyone had some sort of wound, and nearly half of the ship's complement was dead, dying, badly wounded, or missing. The ship had taken on so much water from near misses and fighting fires that it was listing 11 degrees to starboard. None of the radars were operable. Its defenses were either damaged, inoperative, or out of ammunition. Landing operations were manual and very hazardous. Everything had a burnt metal stench to it, and there was a general feeling of defeat on board. Things were so bad that the ship's food was starting to seem like the highlight of the day.

Rick and Chuck were lucky enough to still have their stateroom in good shape. The smell of smoke and fire-retardant foam was still heavy in the air, and everything seemed to be covered with a thin layer of soot. After they landed, both men made their way to the quarters and passed out regardless of the distractions. They were drained.

It seemed as though they could sleep forever. Each planned on staying unconscious until ordered to do otherwise. Five hours after he hit the rack, Rick was shaken awake. The lights were out in the stateroom, but the door was wide open, and a beam of light from the corridor reflected off the thin veil of remaining smoke in the room.

"Meyers. Are you Rick Meyers?"

"Yeah," replied Rick as he coughed and tried to open his eyes. "Who the hell are you?"

"Rick, my name's Manuel Luka. Uh, you're supposed to come with me." "What're you talking about? Who's Manuel Luka? What's going on?" "You can call me Manny. Everybody else does. I'm with Central

Intelligence. Let's go outside and talk."

"You're joking. Who are you? Who put you up to this?"

"I'm serious, Lieutenant. Let's go. Don't make me tell you again."

Rick sat up and looked at the man in civilian clothes. Since the ship was in such grave danger, he knew that there was no way a civilian was still on board. With that, he decided that the man might actually be telling the truth.

He climbed out of his rack and followed him into the corridor.

"What's all this about Manny?"

"Rick, you're coming back to Seattle with me."

"Excuse me?! I'm in the middle of a war out here, ya know? We need every stick jockey we can get. In case you haven't noticed, this ship took a beating, and most of our escorts are gone. I'm one of the few left to guard this tub, and you wanna take me outta the fight. Sorry, Manny. That's just not gonna happen!"

The civilian looked around to see if anyone was listening. Sailors were busy all around, but everyone was either too busy working or too dazed to care about their conversation.

"Of course I know about the war, dammit! I know more than you, and let me tell you it's a hell of a lot worse than you might imagine."

Manny looked around again. No one was nearby, and no one seemed to have heard him slip about the GW's fate.

"Look, Rick, we know about those HARM's that you modified the other day. We need to know exactly what you did and how they worked. We know it had something to do with an artificial intelligence program that you wrote."

Manny pulled out a piece of paper from his inside pocket, handed it to Rick, and continued.

"Read for yourself. Your orders are to come with me immediately. Bring your gear. You're not coming back here."

"This sucks, Manny."

"I know it does, Rick, but it's for the best. Besides, I'm probably saving your life. You're gonna make it off this boat alive and intact."

This time it was Rick who looked around. He found an empty stateroom and dragged Manny inside. Then he closed the door, put his face right up to the civilian's, and asked him to clarify his last remark.

"Rick, you and everybody onboard knows that we lost the Elrod yesterday. The Ross is useless, and the Antietam is lucky to be afloat. There's no hiding that fact that we lost some escorts. What only a few people know is that this entire battlegroup is almost completely out of ammunition. There's hardly any SAMs left on any of the escorts or on this ship. There's also an acute shortage of air-to-air missiles in the area. Just a few carts is all that's left. Didn't anyone think it was odd that you guys aren't taking full loads of missiles up when you take off?"

Rick thought about it. The escorts and the GW probably were at least low on ammunition. There were certainly plenty of targets in the air a few days earlier, and none of them could be ignored. Every incoming missile and plane had to be targeted. It made sense, but he couldn't figure out the air-to-air shortage on the GW.

"What the hell happened? I thought we had a huge supply on board?"

"I can't give you the details, but just understand that budget cuts limited the number that the Navy could buy over the past few years, so in times of crisis, the admirals have been borrowing from the Air Force. It's sort of a cashand-carry kinda thing. From what I understand, the Air Force has been using up their supply over the past few years too. Hey, it's really just a matter of being pulled in every direction and only equipped for a few. I'm sorry. Like I said, be glad you're getting outta here in one piece."

"What about my aircraft? These guys're gonna need every bird they can have!"

"I met with your squadron leader earlier. He says that he's got a plane that needs an engine replaced, so he's gonna assign that pilot to your aircraft. Everything's all set. Grab your gear, and we are outta here. The Air Boss has orders to launch us before anyone else, and I've got a Greyhound waiting on deck."

With a final "This sucks," Rick went back into Chuck's stateroom and packed his gear. Chuck woke up and asked what was happening, but Manny was outside the hatch. Rick knew it, and he knew that he probably wasn't allowed to talk about his new assignment. It was understood that anything involving the CIA, NSA, or any other spy-type agency just couldn't be talked about. Instead, he just told Chuck that he couldn't talk about what he was doing, or where he was going, and Chuck understood-at least a little.

His only reply was, "Okay...well, good luck to you. I guess."

Rick knew the truth about the situation, and he hid it well, but not completely. The shook hands, and said goodbye.

"No, Chuck. It's me who needs to wish you luck. Fly fast, turn tight, fire first, live longest, and remember: stick and rudder. Okay?"

"Okay, Rick. Take care of yourself too."

Five minutes later, Rick was in the air and headed for the United States. Right after the transport plane carrying him left, the ship was called to General Quarters readiness. The Arabs were coming again.

Once again, Chuck ran through crowded corridors and up sailor-packed ladders. He rushed into the squadron briefing room as other naval aviators from The Knighthawks were arriving. Don was already there and speaking into one of the ship's communication phones. Unlike drills or earlier alerts, this time, he slammed the phone. Then he told everyone to get prepped, get to their aircraft, and get launched as soon as possible. Unit cohesion would take have to take a backseat. The GW's air wing commander wanted every jet jockey in the air immediately.

All that Don told his squadron was that the Arabs managed to move a few missile patrol boats along the coast of Turkey. Others had moved out to sea from Libya and Egypt. The U.S.S. Anzio had launched a few Harpoons and Tomahawks at the Libyans, and the U.S.S. John Paul Jones (10 miles north-northwest of the GW) had already launched against the Egyptians. The Stark and the Antietam had fired against the Syrian and Turkish ships to the east. Unfortunately, the Arab ships all carried either Harpoons or longer-range Soviet-made anti-ship missiles. No Arab missiles had been detected, but it was obvious to the leaders of the GW battlegroup that they were probably already in the air.

By the luck of the draw, Chuck's F/A-18C Hornet was on the deck. He quickly zipped on his g-suit, grabbed his survival vest, and ran to the stairs that led to the flight deck with his helmet in hand. Over the past few months, Chuck had come to know his way around the ship fairly well, and he climbed out of the stairwell just a few feet from his plane. He put on his helmet and survival vest. Then, a yellow-shirted crewman helped him into the cockpit, and he began powering up the plane.

When his engines were ready, he taxied to the number 4 catapult. Some of the flightdeck crew connected his plane to the catapult's shuttle while other set the valves and dials for the proper weight of his plane. If they gave it too much steam, the catapult would rip the nose gear out from under Chuck. Too little steam meant that he would roll off the front of the aircraft carrier, drop into the sea, and be run over by the huge ship. The seconds ticked by as the still unseen Arab anti-ship missiles closed. Finally, with a thumbs-up and a salute, his Hornet was thrown off the deck of the GW and into the air.

"This is Spider Five-One-Three. I'm feet wet and ready for tasking." *"Spider Five-One-Three from Big Eye Five-Five, come to heading 090. Expect Vampire traffic after passing Sector Nine. Weapons free status as soon as you pass over friendly ships. Search and destroy all vampires. Copy?"*

"Copy, Big Eye. I'm on my way."

Chuck rolled his plane to the left until it stood 90-degrees on its wing. Then he yanked back on his flightstick and pulled the plane into a tight turn until it was heading toward the suspected Syrian and Turkish missiles. Once pointed in the right direction, Chuck leveled out, accelerated, and climbed to 7500 feet. At that altitude, his radar would be able to detect any missiles from 40 miles away, and he would still be low enough to dive on any (most antiship missiles attack at very low altitude).

On the way to his patrol area, Chuck listened to his radio and heard another pilot as he was directed to follow him.

"Big Eye Five-Five, this is Cowboy Six-One-Seven. I'm feet wet and ready for tasking."

"Cowboy Six-One-Seven from Big Eye Five-Five, come to heading 090. Expect Vampire traffic after passing Sector Nine. Weapons free status as soon as you pass over the Antietam and Stark. Search and destroy all vampires. This is all about you Cowboy. Make it happen. Do you copy?"

Chuck shook his head in anticipation. Cowboy was coming to join him. That meant things weren't going to go by the book, but he knew that at least he'd have a skilled pilot. No one really enjoyed Cowboy's company. He was socially challenged by his own lack of people skills, but no one could argue with his naturally aggressive aerial combat capabilities.

Visibility was excellent, and a few minutes after he left the deck of the GW, Chuck spotted the U.S.S. Antietam and its escort, the U.S.S. Stark. A pair of tugboats led the mini-procession toward the relative safety of the American aircraft carrier. As they passed below his left wing, Chuck's search for incoming missiles increased in intensity. Whatever his radar detected, he had already been ordered to shoot down.

While his head spun on a swivel, and his eyes looked high and low, Codie "Cowboy" Custer pulled into position about 100 yards off his right wing.

"Hey there! Is that you, Spider? It's me, Cowboy! You ready to take down summa those scaaary Vampires?"

"Copy, Cowboy. This is Spider Five-One-Three on your wing. Listen, your radar's got better range and look down/shoot down capability. Why don't you go high and see if you can spot anything?"

"Roger, that Spider. I'm up, up, and away!"

Codie climbed quickly-all the way up from 7500 feet to 25,000 feet. Chuck found a bit of comfort in knowing that-even though they were still well within radio range-at least he didn't have to look at Codie for a few minutes.

A few seconds after Codie climbed, his sophisticated radar looked over the curvature of the Earth and began to detect incoming missiles. The Syrians and Turks had indeed launched an attack. Codie pressed a few buttons to digitally send the data back to Nikki in the E-2 Hawkeye. From there, the data was shared with every plane and ship in the battlegroup via the digital datalink network.

"Spider, this is Cowboy in the sky. I've got Vampires closing in on our position. I'll stay up here and keep an eye on 'em while you move out and take 'em down. How's that sound?"

"That sounds good, Cowboy. I'm on my way. Spider Five-One-Three is engaging."

Chuck gently climbed to 10,000 feet. Then he used the data from Codie's plane to lock on to the closest missile. As soon as his AMRAAM missile acknowledged that it knew where the incoming missile was located, Chuck fired it and began the procedure again. He had launched from the GW with only 2 AMRAAM's, 2 Sidewinders, and a full load of 20mm cannon rounds. It was only a few seconds before his missiles were gone.

"Cowboy Six-One-Seven from Spider Five-One-Three, I'm outta

missiles. Take your shots, and let's get to work."

While Chuck had been locking on to the closest missiles and firing, Codie was miles above him locking his missiles on to the incoming vampires.

As soon as Chuck announced that his missiles were gone, Codie let his fly too.

"They're on their way, Chuck. Let's go get 'em!"

As Codie started to dive toward the incoming missiles, Nikki called him.

"Cowboy from Big Eye Five-Five, remain at your current angels. Do not engage vampires. Repeat, do not engage vampires. We need your radar high in the sky to spot them all Cowboy. I've got two more aircraft coming to help you guys. Stay high!"

"Awww, c'mon baby. I wanna go play too. Send one of those fresh planes up high to spy."

Nikki was furious. She was far too busy to deal with Codie's hot rod antics. The Syrians and Turks had fired 80 missiles toward the GW from the east. The Egyptians had thrown another 32 missiles from the southeast, and the Libyans launched 24 from the southwest. Nikki knew that the GW's escorts were almost out of defensive missiles. Every ship in the battlegroup would have to depend on Nikki's direction of the intercept CAP aircraft-like Codie and Chuck. She didn't hold back in her response to Cowboy Six-One-Seven.

"Dammit, Codie! This is a major attack coming in from three sides. You will listen to me and do exactly as I order, or I will have you brought down. Do I make myself clear?!"

Meanwhile, Chuck's missiles connected with 4 of the 80 incoming targets from the east. Codie's AMRAAM's and Sidewinders found their targets a split-second later, and 8 more anti-ship missiles exploded below. Codie's radar picked up the remaining eastern missile attack, and soon everyone's radar screens showed the same target-rich clutter as his.

They were only 15 miles from the Stark and Antietam. Still, not a single surface-to-air missile was fired. There were none left. Both ships had managed to reload their Phalanx 20mm cannon turrets, and their multi-purpose cannon turrets were ready (though relatively useless against anti-ship missiles).

Chuck dove down, turned back toward the Stark and Antietam, and then he lined up on the lead surviving missile. Two quick bursts from his 20mm cannon blew the missile to pieces. He climbed, turned, and dove down behind another one. Chuck fired another burst. This time, he damaged the anti-ship missile. It crashed into the sea and exploded into a huge column of water.

Codie looked at his radar and watched as two more U.S. Navy Hornets approached the area. He knew that they'd arrive shortly, but not in time to help stop the next few incoming missiles. Despite his orders, Codie rolled his plane over, pulled back on the flightstick, and dove toward one of the missiles. Immediately, Nikki noticed his altitude changing, and she called him.

"Cowboy from Big Eye Five-Five, return to your station, Codie!"

"Sorry, Big Eye. There's no time to talk about it. Spider's doin' what he can, but our backup's not gonna make it here in time to stop them vampires. I'll go back up as soon as that help does get here. Is that a deal?"

"Cowboy! You get your ass back up there right now. This is not a negotiation!"

While Nikki pleaded with him to remain at altitude so that he could still relay distant radar data to her, Codie was already aiming at a tiny Turkish Harpoon missile. A quick burst of 20mm cannon fire, and it exploded. While remaining at 1000 feet, Codie turned around. As he completed a complete circle, he spotted another missile, lined up behind it, and fired. It exploded into an orange ball of flame, then a black puff of smoke, and finally a million shards of debris and splashes.

Chuck was too busy trying to shoot down another incoming missile to notice Codie's activity. After some effort, he was able to get into position and fire his last remaining rounds at the vampire. It veered sharply to the right and disintegrated from the violent force of the maneuver. He had damaged it, but no missile was built to take hits. It was only then that he realized Codie's presence.

Since he was completely out of ammunition, and Codie seemed to be doing a nice job, Chuck fired his afterburners and climbed to 30,000 feet. There, he connected his radar to the digital data uplink network and sent information back to Nikki. Other than the 49 remaining missiles, there were no more vampires approaching.

"This is Spider Five-One-Three. Big Eye Five-Five, I'm outta ammo, so I'll take high cover now. Cowboy Six-One-Seven is busy engaging. What's the sitrep on that backup?"

"Copy, that Spider. Thanks for going high. Backup's only a few miles west of the Antietam right now. They'll be in range any second."

While Nikki was talking to Chuck, Don "Dutch" Schultz and two more Hornets from Chuck's squadron came within missile range.

"Cowboy Six-One-Seven, this is Dutch. You're blocking our shots. Go high and get outta the way. We're firing….now."

At that instant, 18, AMRAAM and Sidewinder missiles started zipping off the wings of Don and the two other members of The Knighthawks. At speeds over mach 3, the missiles passed the Antietam and Stark in less than a second. Each one headed for its individual anti-ship missile target.

Codie was closing in on another missile. A faint blur of white high above the horizon told him that Don wasn't exaggerating. The white blur was a blend of contrails from all the air-to-air missiles that had just been fired. Codie didn't know whether or not the missile he was chasing had been targeted, so he immediately pulled back on the stick, lit his afterburners, and climbed as fast as he could.

Two of the Sidewinder heat-seeking, air-to-air missiles that had been locked on to anti-ship missiles detected the white-hot plasma plumes of Codie's afterburners. The missiles changed their targeting priorities to Codie's plane. Both turned and climbed toward him from slightly below a head-on aspect.

Codie spotted one of the missiles just before it hit his plane. He popped out decoy flares, and the missile missed him. The other missile approached from beyond his field of view. Its proximity fuse detonated a few yards below and behind his left wing. Shrapnel cut off his left rear stabilizer, punctured his fuel tanks, and heavily damaged his left engine. Immediately his aircraft trailed smoke and flame.

Chuck was still above him and saw the incident. He tried in vain to warn Codie, but his pleas fell on deaf ears.

"Cowboy! Eject! Eject! Eject! You're on fire! Get outta there!"

Instead of seeing Codie's canopy depart from the plane and his ejection seat race into the air, he heard Codie's voice on the radio.

"That's a negative there, Chuck. Everythin's okay here. I've got it under control."

His voice was calm, cool, and unusually professional. While he was talking, Codie switched fuel tanks, shut-down the damaged engine, and began to regain control of his crippled aircraft.

"Big Eye Five-Five from Cowboy Six-One-Seven, declaring and emergency. I'm requesting re-direct back to the GW and immediate landing position in the pattern. Over."

There was no response from Nikki. While she had been directing the battlegroup's defense against missile attacks, the controller next to her had failed to notice a pair of Syrian Mig-29 Fulcrum fighter planes. At extremely low altitude, they had flown out of western Turkey, through the Aegean archipelago, and even between the GW and the U.S.S. Anzio! When they were within 15 miles of Big Eye Five-Five, the two planes climbed up-showing themselves much more clearly on radar-and attacked. Each Syrian plane carried 4, AA-11 and 4, AA-12 air-to-air missiles. In a matter of seconds Nikki went from conductor of a modern naval battle's orchestra to victim. The E-2 Hawkeye's pilot never saw the missiles that struck his valuable command and control aircraft, and no one would ever find more than a trace of its existence.

It happened so fast that no one knew what had happened. All of a sudden 75% of the radar data that was being transmitted to all the ships in planes in the GW battlegroup simply disappeared. Codie tried several more times, butlike everyone else-he never heard from Nikki again.

Instead, the air traffic control officer-the Air Boss-answered Codie.

"Spider Six-One-Seven, you are cleared to land. Wind is 8 from zerofour-three. Speed is 33. Heading three-three-three. Approach at will and contact on final. Copy?"

Codie was busy trying to keep his plane from exploding, crashing, or going out of control. All that he could muster was a simple, "Copy" as a reply.

Don and the two other members of The Knighthawks split up and engaged the remaining anti-ship missiles. They were coming in as a stream from due east. Each naval aviator busy chasing down and strafing the small aerial targets, but there were too many. They shot down 9 more, but their ammunition was exhausted, and they had to climb to get out of the way before the cannons on the Stark and Antietam opened fire.

Don was the last to climb. He was pursuing a missile that was headed for the Stark. There was just enough time to fire off the rest of his ammunition. When it was gone, the missile was still flying, and he had to climb to get out of the way too.

He took a few seconds too long. The Phalanx turret on the Stark opened fire just before he pulled back on his flightstick. The comparatively tiny white turret ripped out hundreds of 20mm depleted uranium sabot rounds and tracer rounds, and all of them came right at Don. A few cut through his Hornet as he tried to climb out of the way.

Don's engines both slowed down as sabot rounds and broken turbine blades went wild inside. He couldn't shut down both of them, so he throttled them back until there was just enough speed to stay airborne. Inside his cockpit, lights and alarms seemed to be the only things still in working order.

Every possible warning light was on, and he couldn't discern the different alarms by sound because there were simply too many. His engines both spun down and virtually shut off. His communications were gone. The digital instruments and controls in the cockpit weren't responding. All he could do was steer. For a limited time, it was stick and rudder time.

Don's plane was climbing over the Stark and toward the Antietam. Among the cacophony of alarms, he never heard the stall warning, but he could feel it. He dropped the plane's nose and spiraled around to his right to regain airspeed and stay aloft.

As he did so, the missile that he had been chasing impacted the American frigate. Don's plane was riddled with even more fragments-larger ones. A large piece of aluminum cut right through his fuselage, into the rails on the back of his ejection seat, and jammed it. Don felt the thuds as pieces cut into and through his plane, but there was no time to assess the specific damage.

As smoke filled the cockpit and altitude continued to decrease, he tried to eject. His canopy blew off as soon as he pulled the black and yellow-striped handle between his legs. Solid rockets under his seat fired, but the debris kept his seat in the falling plane. Plasma from the rockets cut like blow-torches, and in less than a second, the cockpit was turned into a molten mass around him.

Eventually, the rockets loosened the jammed piece of debris, and the ejection seat left the plane. Don was tossed into the air, and the seat separated from him as the parachute opened. In a fluid motion, he swung one way as the parachute inflated. Just as it had a full canopy of air, he hit the water. Neon green dye from his survival vest trailed across the tourquoise blue waters. Don was still conscious enough to remove his parachute harness as it pulled him beneath the waves. Activated by the saltwater, a small life vest around his neck inflated and brought him back to the surface.

He watched his plane fall flat into the sea less than a mile away. Then he looked around and saw the Stark a few hundred yards away. It was burning, but at least it was nearby. He hoped that it would survive the attack so that someone could come and rescue him soon. His legs and hands were badly burned from the ejection seat's rockets, but-for the time being at least-the pain was lost to adrenalin.

Don also had a front row seat for the next scene in the modern naval battle. The two Hornets who were with him had left to chase down the Mig29's that shot down the Hawkeye. Now, the Stark and the Antietam were undefended. Both were almost out of ammunition. The Antietam was still in poor shape from the missile hit it took days earlier, and the Stark was smoking from a 30'-wide hole behind and below its bridge.

The Syrian and Turkish missiles continued toward their targets. Don bounced helplessly on the waves and watched them fly past. One slammed into the Stark amidships, then another. Another one missed the Stark and hit the Antietam aft on the port side. More were coming.

A large, Soviet-made missile (about the size of a small airplane) plowed deep into the center of the Stark and split the ship into two pieces. The forward section slipped to starboard, lay down on its side, and slowly sank. The aft section lifted the propellers high into the sky. It bobbed and swayed from side to side until a pair of smaller Harpoon missiles hit it. It disintegrated in the blasts and disappeared.

The damaged and virtually unarmed Antietam was all that stood between the GW and 23 more Turkish and Syrian anti-ship missiles. Don watched the Phalanx and multi-purpose turrets open fire. He heard the tearing sound and the booming of the guns, watched their tracer shells fly overhead, and winced at the whistling sound of travel. All around the horizon, as aircraft, air-to-air missiles, surface-to-air missiles, and anti-ship missiles exploded, Don heard thunderclaps.

Some of the rounds from the Antietam hit an anti-ship missile close enough for Don to feel the concussion. Shrapnel and debris splashed all around him. He felt the heat and smelled the flame. Somehow, he survived without further injury.

The Antietam kept up its rate of fire, but there were too many missiles. Two passed by the Antietam. A few seconds later, Don heard them and watched as two more hit the port side of the Antietam. It fell over on its side and slipped slowly beneath the waves.

Don wasn't alone in the water. Almost 500 people had been on the Antietam and the Stark. Half of them went down with the ships. The other half were swimming.

The tugs that had been towing the Antietam cut their lines as the ship went under. They were in the wrong place at the wrong time. The Turkish Harpoon missiles were equipped with computers that could determine whether a ship was a warship or a civililan ship (specifically a frigate, a cruiser, a landing ship, a carrier, etc.). The Soviet-made Syrian missiles were not as choosy. Whatever their radars detected, the missiles attacked. All of the tugs were blown to bits by the huge Syrian missiles and their thousands of pounds of high-explosives. None of their crewmen survived.

Chuck missed the action over the Stark and Antietam. When Codie was given his landing clearance, he followed him to see if there was anything he could do to help. There wasn't, and Codie would never admit it, but he felt a lot less lonely with Chuck on his left wing.

Codie lined up on the carrier announced that he intended to land, and made his approach. With one engine destroyed, the other damaged and overworked, part of his tail missing, and leaking fuel rapidly, Codie put his mauled Super Hornet down on the deck as though nothing was wrong. His tailhook caught the number 3 arresting wire, and he went from 200+mph to 0 in just a hundred feet.

Flight deck fire crews surrounded his plane and when he looked up, everything was already white with fire-retardant foam. Someone pulled the exterior emergency canopy release lever, and his canopy opened. Foam rushed inside, and Codie felt someone undoing his ejection seat harness. Before he knew it, he was out of the plane and being carried through the slippery foam that blanketed it. It was no surprise to Codie, but everyone else seemed stunned that he had actually made it.

Chuck circled above the GW. The battle was raging as far away as 100 miles in every direction. To the southwest and southeast, most of the anti-ship missiles had been stopped. A few miles to the north, the Mig-29's that had shot down Nikki and the E-2 Hawkeye had themselves been brought down. To the east, the Antietam and Stark were gone, but the larger Soviet-made Syrian anti-ship missiles were still coming.

A few miles off the coast of Turkey, 52 miles away from the U.S.S. George Washington, a four-ship flotilla of Syrian Nanuchka I&II corvettes watched their huge SS-N-9 Siren missiles close on their target. Almost nothing remained to stop them. The missiles silently slipped past the remains of the Antietam and Stark, and the sounds of their jet engines followed a few seconds behind.

The largest remaining radar in the GW battlegroup was actually on the George Washington itself. Deep inside the ship's Combat Information Center (CIC), commanders were watching the Syrian missiles close on their ship. Hornets and Super Hornets that had been sent to deal with Libyan and Egyptian anti-ship missile attacks were ordered back to the GW with the utmost speed, but it was too late. The first of the Syrian Sirens was already 5nm off the starboard aft quarter. Only one of the ship's 4 Phalanx turrets could be brought to bear on the missiles, and as luck would have it, that turret was the only one that had been damaged in the earlier air raids on the ship.

The first SS-N-9 Siren plowed into the GW on the aft side of the ship's island. The explosion sheered the emergency break-free bolts that held the island in place. They had been designed to snap and release the island if the ship listed more than 15 degrees. Instead, the entire mass of the island was moved forward almost an entire foot!

The next missile hit the #1 arresting wire's starboard side mount. The wire came loose and whipped wildly across the deck like a steel sawblade. Crewmen tried to get out of the way, but the deck was slippery from all the foam that was used to put out the fire on Codie's Super Hornet. Several were cut in half. Others lost limbs. All the cuts were as clean as if they'd been done with lasers.

Two more Sirens came in and hit the aft end of the ship. One completely blew apart the starboard Phalanx and SAM turrets. The other hit in the same spot, but just above the waterline. Cracks from the explosion went well under the ship and it began to take on water.

Another missile came in and hit the #3 elevator while it was in the down position. The lift cables were sheered by debris, while more shrapnel went directly into the hangar. Half of the elevator jammed in place while the other half dropped into the sea and dangled. Inside the hangar deck, fireproof doors started to shut and split the deck into four separate hangars. In less than 30seconds, they were sealed tight and the compartments were filled with fireretardant foam.

The dangling elevator pulled the ship to starboard, and the turn was further tightened as the starboard propeller shaft had to be slowed. It had been damaged by the hit above the waterline, and it was threatening to rattle itself loose from the ship. The Captain tried to keep the ship going forward so that it would present a smaller radar signature for the incoming missiles, but there was too much drag to starboard.

As the GW began to drift to one side, the incoming missiles saw the ship's 2-dimensional view grow as its profile slowly came into view. Now they had a much larger target area. Another missile hit just above the waterline near the middle of the starboard side. Instead of puncturing the ship and exploding, it hit the side, exploded, and continued forward leaving a large 15'x75' gash in the side. Two more missiles hit the same area, went into the gash, and exploded deep inside the ship.

Finally, the last missile hit the ship. It went into the aft hangar through the opening where the #3 elevator had been. Inside the ship, it hit the fireproof doors, blew through, and exploded in the #3 hangar. Once again, the aircraft inside caught fire and burned.

The U.S.S. George Washington was in serious trouble. It already had an 11-degree list to starboard. Now it had a deep gash that was almost 100' long and went right to the center of the ship. The reactor rooms were compromised, and the ship's engines had to be taken offline. Half of the ship's interior was in flames, and the fires in the #3 and #4 hangars were spreading. There were spotty radar and communication abilities since the earlier attacks. With the engines offline, there wasn't enough steam to run the remaining arresting gear or the catapults. On top of it all, the battlegroup's admiral was missing and presumed dead, the ship's captain was badly wounded when the island was knocked off its mounts, and the air wing commander was still in the air.

The U.S.S. Anzio pulled along the starboard side of the GW and began to help fight the fires with its hoses. It also launched its two helicopters to try and pickup survivors from the Antietam and Stark. The cruiser's captain also radioed the air wing commander to head for land. It was clear that the GW was out of the fight. The destroyer U.S.S. John Paul Jones pulled along the port side and did the same.

Above the ship, the air wing commander could see the damage. He collected the aircraft that had made it off the ship and led them to back to Aviano. Chuck went with them. The U.S. Air Force had an E-3 Sentry moving into the area along with another wave of reinforcement/replacement CAP aircraft: 3 squadrons of F-16's and a squadron of F-22 Raptors.

While the GW was relatively dead in the water, the Italian Navy went into action. They sent 3 frigates, a destroyer, 11 tugboats, and a squadron of Tornado F-3 fighter planes. The Greeks arrived with one of their last frigates, and even the French sent a squadron of Rafale fighters for protection. With the renewed presence of the U.S. Air Force, and the added reinforcement of international protection, the ship was at least temporarily safe from further Arab attack. The day's naval battle was over, but the fight to stay afloat was just beginning.

Codie had been rushed to sick bay after his emergency landing. When the missiles began to hit, he checked himself out and went to the flightdeck to see if there was anything that he could do to help out. He climbed up the portside aft ladder and arrived just in time to see the arresting wire cutting into everyone and everything as it was knocked loose by one of the Syrian SS-N-9 Siren missiles.

That same missile had impacted and exploded not too far away from where he had come out of the ship, and pieces of the flight deck went into his left shoulder, arm, and torso. Bleeding from cuts, abrasions, and lacerations all over, his flightsuit turned red, but Codie still had the strength to pull wounded crewmen off the flightdeck and back down to sickbay. He was on the deck for each impact, and on several occasions he led groups of sailors through the flames, foam, smoke, and debris in the #4 hangar to get to sick bay.

The inferno in the ship's hangar deck took the fire out of Codie. There was no time to fantasize about Nikki. There was no point in being a hardass in front of sailors who were fighting the fires as though it were hand-to-hand combat. He would never be the same Cowboy again.

Friday, May 20, 2011

2nd Lt. John Chamberlin-Personal Log

Somewhere in the South China Sea

There were a lot of disappointed people today. The Task Force is continuing eastward, but we're not going home anymore. All around the globe, things are going bad for the Navy.

Another big air attack hit the George Washington today. The TV report says that the ship is dead in the water with an 11-degree list to starboard, and the fires are still not under control. NATO aircraft from all over Europe have been dispatched to fly cover until French, Spanish, and Italian naval escorts can get to the area and coordinate their defenses.

In the South Atlantic, the U.S.S. Forrestal took an unknown number of torpedo hits and rolled over with almost 5000 people on board. The escorts are trying to pickup survivors, but even if they picked up everyone, there's not enough space on board to put all those people. Everyone's hoping that there are enough friendly subs in the area to pickup people and get them to safety.

Sunday, May 22, 2011

The New Cleveland Press

Peace Conference Held In Rome

The U.S. Secretary of State arrived in Rome today to take part in the Italian-sponsored Peace Conference. While most NATO countries and non-NATO European nations have also sent representatives, the big surprise has been the presence of delegates from Egypt, Libya, Syria, and Turkey. Those four nations have been in a technical state of war with the United States or the past few days. Given the ferocity of attacks on U.S. Navy assets in the region, it was widely assumed that formal declarations of war were immediately forthcoming within the next few days.

Pentagon officials have been adamant that despite losses to the U.S.S. George Washington carrier battlegroup, the American Navy has seriously eroded the military capacity of all four nations. Claims of destroyed Arab aircraft losses have been in the hundreds. It has also been claimed that most of the small craft in the Arab navies have either been sunk or seriously damaged. U.S. Air Force officials have played the traditional edited bomb damage video and photos at press conferences.

However, military officials from all four Arab nations have completely disclaimed all losses. Given the presence of Arab delegates at the Peace Conference, however, it might be presumed that at the very least there have been substantial Arab military losses.

Tuesday, May 24, 2011

2nd Lt. John Chamberlin-Personal Log

Somewhere in the South China Sea

Another day at sea, and no one seems to know where we're going. We continue to head East, and the speculation is all over the place. Sometimes I wonder if there might be a rumor out there that we're going to Mars! Colonel Ghetty is as elusive as ever, but I don't think he even knows. Hell, Captain Blackwell may not even know. The only thing that I'm certain of is that we've been told to turn in our winter gear and exchange it for summer. This has only fueled the rumor mill.

Even more fuel was added by the latest TV news broadcast. The George Washington has come under major air attack several more times since it was initially set ablaze, but NATO land-based aircraft have done a nice job protecting the boat. A bunch of Italian tugs are trying to get it to port, and the TV footage is spectacular. On top of that, the Egyptians have officially closed the Suez Canal to all use by ships bearing the flag of any Coalition nation. The U.S. is at the top of that list. We can't go around Africa because the submarine threat in the South Atlantic is still very real. The latest rescue total from the Forrestal sinking is just over 3500. I wonder where they put all those survivors.

Tuesday, May 24, 2011

1640hrs local

U.S.S. George Washington CVN

Central Mediterranean Sea

Four days after the devastating anti-ship missile attack, the fire in the #4 hangar was still burning. Most of the communication systems had been replaced. The GW's engines were still out of commission, but the Italian tugs were moving the ship toward Italy at a remarkable 7 knots. Of the ship's 5000+ crewmen and women, everyone was wounded in some fashion. Seriously wounded, killed, and missing totaled over 3200 people.

Many of the survivors from the Stark and Antietam were recovered by helicopters from the John Paul Jones and the Anzio. They were brought back to the GW for further evacuation, but the seriously wounded were being taken back to Italy and Germany via a helicopter ferry to Athens.

One of the first people brought back to the GW was Don. Because of the burns to his feet and legs, he was immediately taken to Athens by helo, then Aviano, and finally Ramstein, Germany by U.S. Air Force planes. Don would spend months in surgery, recovery, and rehabilitation, but eventually he would be able to walk and fly again.

While in Italy, he was visited by Chuck and the remnants of his squadron in the hospital. As was the case with most Americans in the region, Chuck felt defeated. They had taken an entire carrier battlegroup into the Mediterranean Sea, and in a few days that symbol of American power projection was reduced to a handful of ships. They had lost thousands of sailors, Marines, and even some U.S. Air Force personnel.

Don was in an incredible amount of pain, but he still had the strength of character to tell his naval aviators that their contributions to the battle were critical. He tried to remind them that they had been ambushed by four countries, and they had survived. More than that, they had practically destroyed the air forces and navies of all four attacking nations!

Some of the naval aviators understood his logic. The Arab navies had been completely destroyed by American Harpoon missiles and the GW's two escort attack submarines. Their Air Forces were nearly wiped out by SAMs from the

GW's escort ships and by U.S. Navy and Marine CAP aircraft. Punitive air strikes by American B-52's, B-1's, and B-2's had severely damaged or destroyed all of their airfields, naval facilities, many of their refineries, and a sizable portion of their commercial port facilities. While the GW might have been seriously damaged, and all of their escorts were either sunk or out of ammunition, the battlegroup had almost single-handedly fought a war with four countries, and it had won.

Chuck was one of those who understood Don's logic, but he felt something else. Everyone seemed to have lost friends in the attacks, and there was a tinge of vengeance in the air. Chuck had lost familiar faces too, but he had never been particularly close with anyone on the ship-except Rick, who had escaped the last and most devastating raid. Chuck had always been slow to befriend others, and his craving for revenge was not really about lost friends. It was about the possible friends who had been taken from him. The latent anger, the strange paradox of victory and defeat, survivor's guilt, and the ego-fueling pride of his aerial victories all coupled together and changed Chuck.

When Don was finally taken away for his flight to Germany, the old Chuck left too. Instead of being a simple, reserved, and professional naval aviator, his persona changed as though by a light switch. He overflowed with confidence that bordered on arrogance. He clearly had less respect for those who had never been under fire before. There was an underlying sense of commitment that could only come from vengeance or a deep prejudice. Where once there had been a man who kept to himself, Chuck now tried to bond with his fellow naval aviators by joining in their conversations instead of just listening. The combat had clearly changed him.

Friday, June 3, 2011

The New London Daily Chronicle Armistice Signed In Rome!

After only a few days of meetings, the Italian Peace Conference in Rome has already reached a few monumental agreements. There will be peace in the Mediterranean, and a formal NATO-based coalition military force will maintain that security. The Rome Accord of 2011 is scheduled to be signed at a special ceremony tomorrow evening.

The Rome Accord will bring all NATO and United Nations forces around the world together under a single banner. It creates a joint-international logistics, command, and control structure for maintaining peace in the Mediterranean. It also provides funding and logistical pledges toward U.N.-based military operations around the globe-including those taking part in America's New War. Most importantly, the Rome Accord will be signed by all four Arab delegations-effectively ending their undeclared war with the United States. In return, the American Secretary of State has committed $200 billion in reparations to the families of those who were killed as a result of the ZP9 massacre (roughly $20,000 per family).

One of the shortest, most vicious and expensive engagements in human history has ended.

Saturday, June 4, 2011

2nd Lt. John Chamberlin-Personal Log

Somewhere in the South China Sea

The war is officially the Third WORLD War now. Diplomats from all over have been gathered in Rome for the past few days to discuss the spreading conflict and the growth of the Eastern Alliance. They decided that a formal alliance of nations would be needed to contest and deter the Eastern Alliance. The Italian government, still very upset about its naval defeat in the Korea Strait months ago, led the effort and drafted the Rome Accord of 2011. In essence, the Accord reinforced the legal footing to the Coalition forces that have been fighting since the invasion of Korea.

Defending Korea was clearly authorized by United Nations mandates going back to the 1950's. The defense of Taiwan had less legal basis since the Chinese always considered the island as part of China. In reality it was, but the same argument for its existence could be used to disavow the legality of the Communist Chinese government. Taiwan was part of China, but it was part of the original democratic China. Communist China, created by revolution, was less of a legal China than Taiwan. None of this matters now that the island has fallen, but the legal right for Coalition forces to intervene did force the need for the Rome Accord. Now we are part of an official and legal Coalition of forces. I don't think it will have any bearing on stopping a Chinese missile attack, but it might deter more countries from joining the Eastern Alliance.

It took several days for our Coalition allies to decide whether or not to sign the Accord. Even our signatures aren't binding yet. It still has to be ratified by Congress back in the states. There are a lot of officially neutral countries around the world, but they all lean towards either the Eastern Alliance or our Coalition now. The consensus is that most will make their orientation official in the next few weeks.

Once more the ship's TV reports were impressive I'm not sure whether it was a driving force, or a deterrent, but, while diplomats debated the terms of the Accord, they could see a distant column of smoke from the George Washington. At night the horizon glowed from its flames.

How long can a ship burn?

SECTION 5

"My enemy's enemy is my friend."

Monday, June 6, 2011

2nd Lt. John Chamberlin-Personal Log

Somewhere in the South China Sea

Geneeral Shapiro held a briefing of all the Marine officers on the ship in the hangar just after lunch. The rumors about Vietnam were correct.

The CIA has been watching troop movements in the area for the past few days, and now they've seen Chinese and Vietnamese aircraft dogfighting. The Vietnamese air force, although well-trained in their own style of airborne guerrilla warfare, probably won't last too long. Human assets, meaning spies, report that the Chinese have made movements into Indochina all along their border from Myanmar to Vietnam. Some reports indicate that only small units have infiltrated south, but other reports claim that entire corps is on the move. Our little ragtag task force has been ordered into the Gulf of Tonkin to help give the Chinese something to think about besides invading all of Indochina. The rest of Task Force Stingray will be a few days behind us. Until they catch up, the Wasp is going to be the biggest ship in the area. Everyone's a little nervous. So many carriers have gone down lately that we have to wonder just what chance we're going to have? It's not that we're that badly damaged. It's that we're so badly out numbered and outgunned.

Thursday, June 16, 2011

2nd Lt. John Chamberlin-Personal Log

Somewhere in the South China Sea

General Shapiro and Captain Blackwell held an unprecedented meeting of all the officers on the ship. We all met on the flight deck about midafternoon. Intelligence on the situation in Vietnam has been sketchy and rumor-oriented during the entire affair. It must be pretty bad. The Vietnamese government has petitioned for membership to our Coalition. If the other Coalition countries accept them, and it is expected that they will, then the U.S. Marines will be going into Vietnam-as allies! Shapiro and Blackwell called the meeting to make sure everyone was in the right frame of mind should those orders come. We are to remain professional and not to treat the Vietnamese as either old enemies, or superior victors. It will be up to the first Americans in Vietnam to set the tone of our new relationship, and it is supposed to be brotherly. I guess it makes sense, and it's probably workable. It reminds me of the former Union and Confederate soldiers who united to fight the Spanish in the Spanish American War. The U.S. Marines and the Vietnamese Army have fought the same battles, and we should be able to respect each other's capabilities better than anyone else on the face of the globe. If we go in, I think we can create that brotherhood.

Wednesday, June 22, 2011

2nd Lt. John Chamberlin-Personal Log

Somewhere in the South China Sea

General Shapiro held another shipboard officer's meeting this morning. We're definitely going into Vietnam. A contingent of British Royal Marines supposedly flew in to Haiphong while he was briefing us. They're going to reconnoiter the dock area, and prepare the way for our ships. Most of the Coalition troops will simply be offloaded from their ships at the docks. In traditional form, we're going to hit the beach. Someone higher up thought it would be safer and it gives a stronger impression of our force.

The Vietnamese Army is estimated at nearly 3 million strong, and they're asking for help. All the Coalition can send for now is this beat-up, tattered and torn, semi-worn-out task force. The British are putting in 1500 Royal Marines and roughly 2000 Ghurka infantryman. We're throwing in almost 6500 Marines, the remnants of the 82nd airborne division (about 1800 strong), and, we're told, about twenty various Special Forces teams. There are also another 2500 R.O.K. troops, and 1200 Taiwanese Army refugees. Not even 20,000 troops to help out 3 million? Later on, the U.S. Army is going to fly in an entire army of new divisions fresh out of Advanced Infantry Training (AIT). The Brits and some of the other Coalition members have made similar pledges, but the unspoken feeling around the ship is that we'll probably get run out of Vietnam before they get here.

Thursday, June 23, 2011

2nd Lt. John Chamberlin-Personal Log

Somewhere in the South China Sea

The closed circuit TV's on the Wasp today really drew a crowd. There was a TV news broadcast of U.S. Marines from our Task Force walking down the streets of Haiphong. They were flanked on both sides by columns of Vietnamese troops, but all around there were crowds of cheering people. It was unreal. Two countries that had hurt each other so badly resulting in decades of tension, were marching side by side into harm's way. We were all a mixture of emotions. At any given time everyone felt the years of tension, the relief of finding a strong new ally, and that strange sense of brotherhood. The one emotion that coming to the surface for me was anticipation. Maybe now, with veteran forces from South Korea, Taiwan, and Vietnam all at our side, maybe now we could finally make a successful stand.

Friday, June 24, 2011

2nd Lt. John Chamberlin-Personal Log

Black River Valley near the city of Hoa Binh

(about 50 miles west of Hanoi, Vietnam)

I can't believe we're here. We flew from the Wasp straight to the combat area just before first light. The Vietnamese sent an English-speaking liaison officer to every Coalition battalion that came ashore. Colonel Ghetty passed information from the liaison down to all of us before we left the ship.

The Chinese have been shelling the entire northeastern border of Vietnam for over a week now. Meanwhile, Vietnamese Army outposts reported thousands of Chinese coming through the jungle-covered mountains along the northwestern border. The Vietnamese, arguably the finest jungle fighters that the world has ever known, slowed the Chinese advance, but with only a few thousand scattered troops on hand, they were unable to stop it. As soon as the shelling along the northeastern border ended, Chinese mechanized units drove into Vietnam. The Vietnamese Army was never structured for anything other than jungle warfare, and they took heavy losses. That's when they called on us. All of the Coalition heavy weapons units are being deployed outside Haiphong to defend against the Chinese mechanized units that are approaching from the northeastern border. The lighter units, like ours, are being deployed here, on the western border, in the Black River Valley.

On this side of Vietnam, the Chinese are using the river as a highway through the jungle, and the Vietnamese have coordinated their attacks to make it seem as though that's the way to go. Vietnamese reinforcements are supposedly moving into the hills on along the sides of the valley, while we move to create a roadblock along the Chinese water highway. If all goes according to plan, in three days the Chinese will run into us. We'll force them to try and flank us by going over the hills, and the Vietnamese will mow them down. That's the plan.

Our company is deployed along the treeline of the jungle behind a clearing with a now abandoned village in it. We've got two more companies of U.S. Marines on our right flank, and another one on the left flank between the Black River and us. On the other side of the river, the British have deployed about half of their 2000 strong Ghurka contingent. We don't know where the other Ghurkas have been deployed.

Heavy fire support? There's a few light mortars that have been collected by the different Marine companies on this side of the river, but they won't be too much help, and they're under Colonel Ghetty's direct command. Air support would be a surprise since Task Force Stingray only has a few squadrons of Harriers, and those will probably be needed to protect the ships and the Osprey as they move troops and Marines around. As far as Hector, Bruce, and I can estimate, we'll be on our own.

The heat here is unbearable. The jungle's shade only provides enough cover for the mosquitoes to feel comfortable. It seems like only yesterday we were in the snow of South Korea, now we're in 100-degree triple canopy jungle with poisonous snakes, strange animals, a billion types of insects, and tigers (if you believe the rumors). At least the ground is soft and easy to dig.

Bob got right to work on getting bunkers and trenches dug. I'm not sure the rest of the platoon liked getting a workout so immediately after leaving the air-conditioned Wasp, but they're Marines. They'll adapt. A Vietnamese liaison officer came around with Colonel Ghetty and had lots of praise for our fortifying ability. He was a little cocky when it came to how well we were camouflaged in comparison to the invisible Vietnamese troops at the tops of the hills. Other than that though, he seemed ok. In fact, apart from those we flew over on the way here, he's the only Vietnamese we've seen all day.

Saturday, June 25, 2011

2nd Lt. John Chamberlin-Personal Log

Black River Valley near the city of Hoa Binh

(about 50 miles west of Hanoi, Vietnam)

Three days nothing! About 1500 hrs, a Chinese recon patrol came down the river and was ambushed by one of our Special Forces teams operating in the area. The entire valley heard the firing for almost 20 minutes. Then there was that eerie after-action silence. The Green Beret team brought back the bodies so that they might be able to use the same ambush location again. We're all sure the Chinese will come looking for them soon.

Sunday, June 26, 2011

2nd Lt. John Chamberlin-Personal Log

Black River Valley near the city of Hoa Binh

(about 50 miles west of Hanoi, Vietnam)

More e-mails to families at home tomorrow. Peter Goodman and Terry Meade were both killed when the Chinese came down the river. The platoon on our left flank spotted a small barge with a platoon-sized force on board. When they came too close, the Marines opened fire. Everyone on the barge was killed, and the boat, freshly vented with 5.56 NATO bullet holes, took on water as it floated past the Marines. No sooner had it passed than everyone along our line started to take small arms fire from the treeline on the other side of the vacant village.

Hector had our company hold fire for a while, and the companies to our right followed suit. The Chinese must have thought that they could flank the Marines who had sank their barge. They came out of the jungle right at us. Hector passed the word that we weren't to open fire until they spotted us, so we waited. I'm sure it was only a few minutes, but it seemed like hours.

I don't know if they spotted us, if someone literally jumped the gun, or what happened, but Bruce's platoon, on our right flank, opened fire first. The Chinese who were caught in the open were killed immediately. No sooner did they go down than more came out of the jungle beyond the village. It seemed like for every Chinese that was hit two or three more came out of the jungle! After a few minutes the entire field between our line and the far treeline was littered with bodies. What had started out as a platoon ambush grew into a battalion ambush. It was like a shooting gallery. They couldn't get close enough to put down accurate fire on our positions, and they had no place to hide. While they kept coming, I began to think about ammunition.

Even if every Marine hit a Chinese with each three round burst fired from their weapon, we'd be out of ammunition in less than two hours. I could tell Bob was thinking the same thing since he started telling for everyone to watch their ammo. I heard other Sergeants echo the order up and down the line.

My estimate turned out to be close to the mark. After about an hour and forty-five minutes, I heard someone in the platoon scream "I'm out! Throw me a magazine!" It wasn't what I wanted to hear. The Chinese dead had become so thick that their reinforcements were piling bodies two and three high for coverright out in the open - body barricades! Roughly half an hour later the Chinese got to about 50 meters and the grenades started flying faster. It reminded me of children throwing rocks at each other, but these rocks exploded. Pete and Terry were sharing a bunker and a grenade rolled in. There was no time for them to get out or even look at each other. More grenades came at us, and more Chinese came out of the jungle. The grenade duel didn't last even five minutes.

It ended when Bruce came running over with a few people from his platoon. They'd been overrun and the Chinese were all around. I yelled over to Bob, but he already knew and was firing at some of them who were running past our right flank. Hector had his platoon sergeant come over with a squad to help out. Bruce was able to get some people together and, with the help of Hector's squad, they were able to retake their positions after a few worrisome minutes. The Chinese that were driven out of Bruce's positions went running out of our treeline for the alleged safety of theirs on the other side of the village. In a domino style effect, other groups of Chinese, caught in the open, joined the ones fleeing as they were passed. Their attack collapsed. We had to let them go to conserve our ammo.

They didn't hit us again the rest of the day, but we're pretty sure they're not going to leave without trying their hand again. Thank God Colonel Ghetty was able to get us some more ammunition before nightfall. Otherwise, it was going to be a long restless night.

Pete had been a constantly happy guy who always did his job professionally with pride. He rarely complained, and never without very good reason. Terry was the big strong man of the platoon. He gave the appearance of being a few cans short of a six-pack, but he was one of the platoon's best weapons technicians. If you had a problem with your weapon, any weapon from knife to machinegun, Terry could fix it. Both were married and will be missed.

Not that any one person's life has more value than another's. In addition to Pete and Terry, we did take two other casualties, wounded, but possibly even more severe. Harry, our platoon point man and general lucky charm was wounded in the face by some grenade shrapnel during the "rock/grenade fight", and Bob, my right hand man, took a bullet in the right buttocks. I think he was as pissed off as he was embarrassed. He'll return to duty in a few weeks, but he'll never be the same indestructible guy. He'll also never live it down. He's always telling people to "Bunker down, or you'll get that pretty ass of yours shot off." He's never going to hear the end of the ribbing!

Monday, June 27, 2011

2nd Lt. John Chamberlin-Personal Log

Black River Valley near the city of Hoa Binh

(about 50 miles west of Hanoi, Vietnam)

It turned out to be a long and restless night anyway. The jungle gets louder at night than it is hot during the day. After sunset it's nice and cool. Every living creature wants to get out and look for food or fun. Overhead, the Chinese kept a constant vigil of parachute flares in the sky. We watched through our night vision goggles as their medics crept into the field and pulled out the wounded. They've got some night vision too, but we've got one for every Marine, which makes a night assault on their part virtual suicide. In addition to the noise, there was the occasional sniper fire from both sides, and an ambush here and there. All night long, there couldn't have been fifteen full minutes of peace. Thank God for our med packs and the miracle of caffeine pills.

As soon as the sun came back up, and the river's mist had lifted, we heard the bugles again. The Chinese still use them to give orders to their frontline troops. I'm not sure how effective a communication method it is, but it's always put me on edge. Not long after they sounded, the Chinese came across the field again. This time, a few mortars supported them, but the effort was lackluster at best. Actually, it was kind of a relief to know that they're as low on heavy firepower as we are.

The Chinese came on slow, walking at first, then a slow run after they passed the village. It looked like two companies coming straight for Bruce's platoon and my platoon. We let them get a little closer than we did yesterday, when we finally opened fire, the Chinese hit the dirt and returned it as best they could. The element of surprise wasn't as effective as it had been before.

While we pinned them down and slowly picked them off, the Chinese sent a battalion sized force against the Ghurkas on the other side of the river; just under 1000 troops. It was a bad choice for them.

The Ghurkas are volunteer Nepalese tribesman. For decades they've fought with great distinction in the British Army, and they are some of the finest troops in the world. Most countries even consider them a Special Forces unit. They are highly trained, highly motivated, but most importantly they bring with them qualities that only a third world background can: unlimited endurance, and a stomach for what most westerners would call savagery.

The Ghurkas let the Chinese come to less than 50 meters of their position before they opened fire. When I heard them open up, I looked down the line and saw what was going on. The Chinese were cut down like wheat. Those that remained fired into the jungle with panicked aim. More Chinese came from the far treeline further up the river. After the majority of the first wave of Chinese had been hit, the Ghurkas ceased fire and waited for the next wave to get just as close to their position as the first wave had. Their camouflage was extremely good, and the Chinese must have thought that their foe had retreated. They ran up to take the Ghurka position. Again the Ghurkas cut them down and picked them off. By this time, I think a rational commander would have tried a new tactic against the Ghurkas. The Chinese tried the same wave style attack three more times before the Ghurkas fell silent for the day.

On our side of the river, the Chinese had hit our line fairly evenly. Each platoon had to fight off about 100-150 Chinese. Thankfully, unlike the Chinese commander facing the Ghurkas, my opposite stopped his attack after his first wave was mauled. He's a smart man. After all, we were a lot more conservative with our ammunition today. Everyone wanted to make sure that they weren't left empty-handed if the Chinese made it to the jungle again.

The clearing in front of us is starting to become quite a scene. Overhead the sky is pure blue. The jungle is deep green with plants of every shape and size imaginable. The village still looks the same as when we arrived. It might even be the same as it's been for a thousand years. The knee-high grass flows in the light wind like waves on a lake, and with each breeze it seems to reveal more hidden Chinese bodies. I think if this continues for a few more days, the wind won't have to blow to show the corpses. There will be enough to cover the grass entirely.

I took another casualty today. Melissa Keegan was hit in the left armpit by a Chinese bullet. It hit right where her bulletproof vest couldn't interfere. The bullet went right on through and came out her other armpit. She was a nice person, slim, very attractive and intelligent. She wasn't known for complaining, and she did her job well. Melissa had been hanging out with Bob for the past few weeks, and I think the two of them had consummated their relationship, though neither ever toyed with the other around the rest of the platoon. Everyone's pretty sure they had something going on, but no one seems to be able to confirm it. I do know for sure that Bob is going to be hurt when he finds out, especially since he couldn't be here when she died. I was never to close to Melissa, but I never had a problem with her. I respected her as a person and as a Marine above other men, even other Marines. We will all remember her.

Tuesday, June 28, 2011

2nd Lt. John Chamberlin-Personal Log

Black River Valley near the city of Hoa Binh

(about 50 miles west of Hanoi, Vietnam)

The Chinese came at us again today, just before sunrise. This time, they floated down the river on a bunch of barges lined with sandbags. My platoon, and the rest of the platoons to our right, didn't really have a shot at any of them as they came down. Hector had his people hold their fire until the Ghurkas opened up. They actually waited until the first barge was about to pass through our line, then the Ghurkas tossed grenade after grenade over the sandbags. Colonel Ghetty had the mortars join in and add to the Chinese confusion.

The first barge passed right on down the river without a living soul left on board. The second and third barges let loose a torrent of small arms fire at Hector and the Ghurkas. It was impressive and daunting, but again handfuls of grenades over the top took it out. Three more barges came down the river. One ran aground in front of Hector, the second one in front of the Ghurkas, and the third tried to pass on down the river. After three and a half hours of sporadic fighting, the remaining Chinese made it back to their own treeline. Four more barges came down, but they all ran aground by their own positions in the jungle.

Our front was quiet except for sniper fire and a few stragglers that wandered past our guns. I don't know how many men the Chinese have lost so far, but it's at least in the hundreds, maybe even thousands. Everywhere you look, between our positions and the Chinese treeline on the other side of the village, there are Chinese bodies, and they are REALLY starting to smell! I'd send out a burial detail, but I absolutely refuse to put one of my people in a sniper's sights to bury enemy dead. Of course, if the smell gets much worse there may be volunteers-if it can get any worse!

Wednesday, June 29, 2011

2nd Lt. John Chamberlin-Personal Log

Black River Valley near the city of Hoa Binh

(about 50 miles west of Hanoi, Vietnam)

The Vietnamese plan seems to have worked. The Chinese hit us with another attack around lunchtime. They left the Ghurkas alone this time, and they hit us with about as many men as they did yesterday. Today, when we opened fire, they immediately hit the ground and began making their way back to their treeline.

Something seemed strange. Hector and Bruce came over so that we might try and guess at what was happening. About five minutes into our conversation, all hell broke loose up and down the valley. All along the tops of the ridges, we could see smoke coming through the top of the jungle canopy. The only place that seemed quiet was the no-man's land in front of our positions.

We think the main Chinese force finally made it down to our position late yesterday. When they heard about the difficulty getting past our line, we figure they decided to try and go over the ridges past our flank, just like the Vietnamese had hoped. Bad idea. Thousands of well-trained Vietnamese Army troops hold that high ground, and they are not ever going to leave. They'll have to be killed off, and none of us think that's going to be too easy. It wasn't easy in the '60's and 70's for us, and it won't be for the Chinese in this new millennium.

The fighting hasn't let up in volume since it started six-and-a-half hours ago. In fact, for the first time since Korea, we're seeing friendly air support in action. On average, every five minutes, a Marine or British Harrier comes through the valley and pounds a hill. When I say pounds, I mean completely destroys. These planes are flying overhead with cluster bombs, napalm, 2.75" rocket pods, 25mm cannons, and 500-pound bombs.

I even saw one drop a string of four 1000-pounders. When just one of those goes off, its shock wave compresses the humid Vietnamese air into a bubble of compressed steam that flashes up through the trees. As soon as the steam goes away, trees fly up through the new clearing. It's like it happens in slow motion.

I don't know if the Chinese will continue their flanking effort, retreat back up the valley, or if they'll hit us harder tomorrow. I do know that even though we hardly fired a single round, this was the most explosive day I've seen in months!

Thursday, June 30, 2011

2nd Lt. John Chamberlin-Personal Log

Black River Valley near the city of Hoa Binh

(about 50 miles west of Hanoi, Vietnam)

All night long the Chinese never let up on their attempt to go over the ridges. It's still going on right now. This morning, they tried to hit us again. This time, they came out of the treeline so close together that it seemed like they were shoulder to shoulder. There were hundreds, maybe thousands of them. They walked-weapons at the ready. It was a bit daunting so we had absolutely no intention of firing until they made it to at least 100 meters in front of us. They never made it that close.

When they made it to the village, all hell broke loose. Explosions blew huge gaps in their ranks, and small arms fire, coming from the village, starting picking off the remaining Chinese who were still searching for targets. Somewhere in the grass and in the village, a friendly unit had crept in and taken up position to intercept the Chinese. Hector came running over to make sure Bruce and I were still where we were supposed to be, then he made sure the rest of the companies to our right were still there. They were.

I watched through my binoculars until I finally saw one of our mysterious friends,. It was some of the missing Ghurkas that had been wandering through the Chinese jungle looking for a fight. There couldn't have been more than a hundred of them, yet the Chinese were falling in groups of ten and twenty. During the night, the Ghurkas had managed to sneak close to the Chinese treeline, plant claymore mines all over the place, then take up position in the village. Not a single Chinese or Marine noticed; even though we all have night vision goggles!

No one needed to do it, but the companies to our right opened fire and supported the Ghurkas as best as possible. We did the same. The Chinese attack on our line lasted about half an hour before the bugles sounded their retreat. I'm not sure if the Ghurkas stayed or left the village, but we haven't seen or heard them since the firefight. I'm glad they're on our side.

The mountain and hill fighting continues unabated, but now there are some new players in the game. I saw a few flights of U.S. Navy F-18E Hornets make air strikes on the hills. That means we've got a carrier somewhere off the coast now. The F-18's can carry about twice as much ordinance as the Harriers, and the Chinese have to be feeling it. As the carrier gets closer, we'll all be seeing more appearances from the F-18's. Soon the Chinese will have to either retreat, or hit my platoon with everything they've got. After today's incident with the Ghurkas, I think they might try the retreat option.

Friday, July 1, 2011

2nd Lt. John Chamberlin-Personal Log

Black River Valley near the city of Hoa Binh

(about 50 miles west of Hanoi, Vietnam)

Now the U.S. Army has arrived. While our little frontline was waiting for another early morning attack, we heard engines behind us on the river. Hector sent a patrol to see what it was. The U.S. Army, carried like Marines in landing craft and sampans, came up the river and dropped off an entire battalion. About 1000 troops are supposed to take over our position. We held the line for days, and they come in to take the victory in what will probably be the most important part of the battle; the final Chinese attempt to breakthrough before they have to retreat. We're to take up new positions along the bank of the river in case the Chinese try to get some more barges down past our line. None of the Marines are happy about it.

These soldiers coming off the landing craft are as green as green can be. They've got brand new jungle-pattern BDU's, and I don't think I saw a dirty knee on a single one of them. Their weapons are still factory black, and they still walk in step with each other. I was embarrassed when the Ghurkas and the Vietnamese Liaison officers looked at them.

When we found our new positions, Hector, Bruce, and I had a meeting to discuss our situation. All three of us agreed that the Army units might not be able to hold off the Chinese. Hector had his platoon sergeant hang out as a liaison officer with one of the Army's company commanders who had taken our positions. If the Chinese look like they're going to break through, we'll find out right away. On Hector's word, we'll run back to our old positions and back up the green grunts.

The fight for the valley's high ground seems to never let up. Air support is starting to clear holes in the jungle, and we can even see some of the fighting now. This also means an increase in enemy sniper fire, so we try not to move around too much. It's been a busy day, and everyone is starting to show some fatigue from the past week. While my people are tired, wearing torn uniforms, and eating MRE's from 1999, the Army is right now cooking their meals on their portable solar-powered stoves.

Saturday, July 2, 2011

2nd Lt. John Chamberlin-Personal Log

Black River Valley near the city of Hoa Binh

(about 50 miles west of Hanoi, Vietnam)

When people use the phrase overkill, I now have an image to personify it. The Army got their teeth cut in a typical morning Chinese attack. As usual, the Chinese bugles sounded, and they came across the clearing in two waves. They're a little more cautious now, after the Ghurkas ambushed them the other day. Army soldiers are only trained to be marksmen out to 100-yards, while Marines are trained to hit targets at 500-yards. As a result, the Army waited until the Chinese were right on the treeline before they opened fire. The Chinese hit the dirt, and took cover behind the smelly bodies of their comrades left over from days before. Then they returned fire as best they could.

The Army troops held their ground, but when the Chinese finally fell back for their positions, the Army called in a trio of Apache helicopter gunships to prevent the Chinese from settling back in. One of the gunships took a hit from a hand-held heat-seeking SAM and went down in the Chineseheld jungle. The other two backed off and unloaded every weapon they had. 128, 2.75" Rockets each went into the jungle loaded with thousands of nailsized flechettes. Hundreds of 30mm rounds from the cannons mounted under the helicopters' noses topped off their delivery. Then an entire company of Army soldiers moved out of our treeline and ran towards the Chinese. The Apaches apparently had little effect on the well-entrenched Chinese, and the Army's company was met by very heavy small arms fire. They had to crawl back to our treeline for the rest of the day. Incredibly, no one was badly hurt.

Around mid-afternoon, while the Army's dogfaces were still licking their wounds, Colonel Ghetty came over and told us that there was to be a ceasefire, effective immediately, until at least sundown tomorrow. The Vietnamese had positioned two reinforced corps along the tops of the river's valley (5 infantry divisions on each side). Last night, Royal Navy Marines were airlifted to a position farther up north in the valley that linked the two Vietnamese corps.

The Chinese woke up to find themselves surrounded. Coalition forces poured in firepower their left, their right, in front of them, behind them and from above. After a while they sent over a negotiating team to arrange the cease-fire and discuss their possible surrender.

When the platoon heard this, they were evenly split between stunned silence and jumping with joy. Myself, I'm one of those stunned with silence and awe. We've had our hats handed to us at every battle so far in this war, now, after only a few day's fighting, we might be part of the war's first victory - and a decisive one at that!

Sunday, July 3, 2011, 1410 hrs

Phuc Yen Airfield NW of Hanoi, Vietnam

Surrounded by dense jungle, Phuc Yen airfield was as peaceful and tranquil as its setting. A cease-fire with the Chinese was holding, so the aging Soviet and Chinese-built Vietnamese aircraft were finally getting the maintenance attention that they craved. Only the occasional falling wrench or short periods of air compressor usage interrupted a symphony of the afternoon's jungle noise.

Working on his forty-year-old Jian-7, a groundcrewmen was the first to notice the mixed formation of fighters overhead. They flew in a tight "V" formation and circled the field silently at less than 2000 feet. As they passed, the sound of their engines followed at least a mile behind. Other groundcrew and staff were alerted, but the formation circled peacefully so, with the cease fire still apparently still in place, no one felt the need to run for cover or even scramble an interceptor. After two full circles of the field, the planes peeled off from their formation and began landing without clearance or warning.

The first plane, an American built F-16 with South Korean markings, touched down and rolled to a stop in front of the small field's control tower. Immediately Vietnamese troops surrounded the plane and took aim on its pilot. Three more South Korean F-16's followed, then four South Korean F-5 Freedom Fighters and four South Korean F-4 Phantoms. After the last plane in the squadron had pulled to a stop, another squadron, twelve Taiwanese F-5 Freedom Fighters, began circling in "V" formation.

By this time, Al Meyers, the pilot of the first South Korean F-16 that landed, had made his way out of the plane. He attempted to introduce himself to the senior officer in charge of the troops. He was calm, and even joyful while he ignored the nearly fifty AK-47 assault rifles pointing at him. No one understood his English, and he had never even heard Vietnamese spoken before. More troops arrived in trucks from unseen locations in the jungle. One by one the other pilots came over and stood with Al. One of the Taiwanese pilots tried to communicate with the Vietnamese officers that had gathered. When one of them responded in Chinese, their dialogue finally began.

Al, through his Taiwanese interpreter, explained that he was in charge of what had become the Coalition's 1st Composite Interceptor Group. He had hooked up with these pilots after they had been forced out of their homelands, and now they were looking for a new base of operations.

The Vietnamese officers only knew of a strict chain of command within military organizations, and they looked at the new pilots as though they were deserters, mercenaries, or pirates. Only when a Vietnamese pilot came through the crowd and joined the conversation did things get better. The Vietnamese pilot, accustomed to flying outdated and worn-out fighters left over from the Vietnam War of Independence, immediately started examining Al's high-tech, 1980's era F-16. The first thing that the pilot noticed was the kills painted on the side; all forty-eight of them! Minutes later, after a few words from the Vietnamese pilot to the rest of the crowd, Al and his ad hoc unit were welcomed as heroes.

They had been driven out of South Korea in March. Al and his R.O.K. pilots barely made it to the Philippines. Then they fought their way into Taiwan only to be driven back to the Philippines again a month later. In the "PI" Al's notoriety spread among pilots who had planes but only occupied homelands to fight for. His piecemeal unit became an evening news side story, but there was nothing more than routine patrols out of Clark Field. When they found out that there was another fight brewing in Vietnam, Al and his pilots voted to join it. Their unit wasn't directly under anyone's command, so they created some fictitious requisition papers, secured weapons, fuel, and flight plans, then they simply left. A few hours later, they found a new home in the middle of the Vietnamese jungle.

Monday, July 4, 2011

2nd Lt. John Chamberlin-Personal Log

Black River Valley near the city of Hoa Binh

(about 50 miles west of Hanoi, Vietnam)

Sometime last night, the cease-fire ended. I woke up this morning, and while making my rounds, several of the platoon's members told me that they heard firing up the river. These reports were backed up by a close call with a sniper just before noon.

I was walking over to check on third squad, and, as I passed through a bamboo thicket, I noticed some of the thin trees coming apart. I hit the ground and rolled back the way I had come before I heard the shot. It took so long, I think I can assume that it was from a good distance. That was the whole event. I tend to walk a little faster and a little more crouched over now.

The cease-fire's cessation was also reaffirmed when the Chinese made yet another morning attack. This time, there was no preemptive artillery or even a mortar barrage. I was still making my rounds, about to go back to my dugout for the morning NCO meeting, and I heard someone up at the Army's frontline shout out. A few more shouts, and the small arms fire began. Then came the bugles. Usually, there are only four or five from the other side of the clearing. Today, there were so many, it sounded like the four or five were echoing in a steel drum. They were everywhere.

I looked over towards Hector's platoon, and he was looking right at me. Neither of us had to say a word. We both nodded at each other, and sent our platoons back to their original positions. The Army was going to need our help. My platoon was up and running as one group, with me in the middle. We made an excellent target, but only stray rounds came through the jungle. I got up to my old command dugout and surprised an Army 2nd Lt. whose platoon had relieved me. I thought he was going to shoot me when I jumped in, but he never had the chance to turn his gun around. His look went from stunned surprise to complete relief. Even though the Army was bloodied the other day, we're still the toughest vets around. Actually, I don't feel as tough as I did when I first took command of the platoon, but, as a unit, I'm sure we look the part. Now we could act it too.

The Army's soldiers were firing away at the oncoming Chinese with serious resolve. Rightfully so, since none of us had ever seen that many troops in such a concentrated area. There were so many of them that they came on like it was Waterloo, Yorktown, or Gettysburg. Walking in rows, almost shoulder to shoulder, thousands moved forward. They even came forward at a slow walk. Each one stopped occasionally and fired a full clip from his or her assault rifle then they advanced while reloading. I think they must have known their chances of survival were slim at best, but every commie must have wanted to make damn sure that they went down firing. They put out a lot of lead. Brass shell casings were coming out so fast that there was an aura-like gold glimmer around them. We didn't wait for them to close the distance, and the Chinese quickly learned what the fourth of July means to Americans; fireworks!

Bunched up as they were, even the Army couldn't miss. They fell like dominoes. Every three-round burst from every M-16 brought down at least a pair. As long as you aimed in the general direction you couldn't miss. How someone could order so many people to sacrifice themselves is beyond my understanding. Thousands were falling, and still they came forward, row after row after row. Each one advanced through the field of knee high grass. The field was wet with blood and littered ankle-deep with week-old corpses covered with empty brass bullet casings. Still they came. It was unsettling.

I had never seen anyone panic until today. As Marines, we're trained to the point, some would say, of insanity. We have had it burned into our minds that we can kill anyone with anything, at anytime. We fight for our country, the corps, etc., but more than anything, we fight for each other. With that thought foremost in my mind, I have never even thought of running away and leaving my platoon to die. The thought actually puts a bitter taste in my mouth just trying to envision such a situation. I know that no one in my platoon would desert for the simple reason that we care too much for each other. Up until now, I could never picture someone running out on their friends like that. Today, I saw it happen.

Bullets were raining through the jungle like a horizontal lead hail. They were ripping through thick leaves, snapping twigs and branches, splintering trunks, and even dropping some saplings. When the Chinese got to within about 30 meters, I saw a soldier on my right side run past me, heading for the rear like there was some sort of safe spot back there. A few seconds later, I saw two more from my right side. I pointed them out to the young Lt. in my bunker. He glanced to the right, then went right back to shooting at the Chinese. I could tell he was a little nervous, and I tried to calm him down. It didn't do any good, he yelled at me and told me to shut up. It was really no big deal to me if he wanted to get uptight, but we needed everyone at his or her post if we were going to make it through.

I told him that if any more of his men ran, my people would probably shoot them just to make a point. This really pissed off the greenhorn, and he tried to threaten me. While all this is going on, I was just lying there, shooting commies and talking no louder than I needed to in order to be overheard through all the firing. When he threatened me, I stopped firing and asked him if my platoon should pull back to join up with the rest of his gutless wonders. He was really pissed off now, and he was about to swing at me when another soldier from the right side of the bunker came running by. We both looked at the panicked young lady, then the Lt. looked at me to check and see if I was going to shoot her. I wasn't, but Wess came running up from behind, dragging one of the soldiers that had run away, and pointed a .357 magnum at the 18year-old Army girl. She went back to her foxhole just as fast as she had left. The young man Wess was dragging ran with her. When I turned to my left to see what he thought, the Army Lt. was busy shooting at Chinese again.

Wess came into the bunker. He apologized for leaving his squad, but said that if he didn't stop the soldiers from running, the rest of his squad was ready to "…take 'em out, use their ammo, and heat our MRE's with their --- solar stoves." The Lt. kept firing, and the mood between both platoons, though much more tense, was also more focused on the Chinese attack; an attack which was turning into a grenade fight.

When the Chinese were close enough, our people started throwing grenades at about the same time they did. It was like children on a playground with rocks. Each grenade threw more and more dirt, dust, debris, and smoke into the air. It was only a few seconds before no one could see a thing. If we stayed in our bunkers, we were perfect targets for the grenades. If we retreated, the only chance we had at finally winning a battle in this war would be lost. I told everyone to get out of his or her bunkers and fight hand to hand. We did, and the Army, at least the platoon that was with mine, went with us. Bruce's platoon, reduced to a little more than squad size now, waded into the sea of Chinese too. Hector's platoon waited to pour out a little more lead, then they left their bunkers and our entire frontline became a lightning-quick bar fight with machineguns and grenades tossed in for effect.

When my M-16's magazine was empty, rather than take time to reload, I swung it at the first Chinese I saw. It hit him square in the face, and he hit the ground back first. I'm not sure if I killed him or knocked him out. I picked up his old Type 56 assault rifle, and another Chinese came at me with his bayonet. The Type 56 was set to fire full auto, and I emptied the magazine into the oncoming commie before I had the time to release the trigger. There was no time to reload, so I grabbed a fresh rifle from the Chinese I had just killed. This time I didn't even have time to stand up before I saw more Chinese coming at me, two of them. I turned and used the rifle to flip the ends of theirs upward and over their heads as they ran past. When they turned around, someone else shot them both. More Chinese came through the smoke. I followed the same pattern of shooting someone, taking their rifle, and using it to shoot another. I don't know how many I killed, but there was no sense of remorse or sadness, only an enraged desperation to survive. Despite how young they seemed, we weren't out here butchering school children. These people were trained and motivated to kill us-nothing less, and nothing more.

After what felt like an eternity, they finally stopped coming. An individual here and there still appeared every second or two, but they were immediately dispatched. Some got off the ground and ran at us. Most just ran or crawled away as fast as they possibly could. My BDU's were shredded from bayonets, bullets, close calls with grenades, and Chinese hands. Every part of my body bled from countless small cuts and abrasions, but there's nothing serious. Through it all, I never lost my cool, and as soon as I was sure that the Chinese were pulling back, I shouted out for everyone to gather weapons and ammo then get back to their positions. I told one person to take a roll call and find out what our status as a platoon was. I had three others check on the other platoons and the Army's status. The Army's medics were already treating the wounded; Army, Marine, and Chinese. I knew that as long as the Chinese didn't hit us again in the following 1/2 hour, we would be fine. They didn't, but I think we could have handled them if they did. While I was taking care of the cleanup, the Navy got into the action; yes, the Navy.

None of us heard them coming. We had to actually see them drive by before any of us even knew that they were there. It was like watching a parade on the river. Thirty-six river patrol boats (PBR's-Patrol Boat River) went by with about 20 meters between them. Each boat has only a ten-man crew, but they're armed with four twin .50 caliber machinegun mounts, a dual Mk19 full auto grenade launcher, and two 5.56 gatling guns. I think two or three of these boats could have stopped this attack, let alone what thirty-six could do to a weaker follow-up attack!

The rest of the day was fairly quiet. There was a lot of artillery and close air support action up-river. We watched flight after flight of fighters and bombers fly up into the valley. The Chinese must have really gotten a beating somewhere. For the first time ever, I saw B-1 Lancer strategic bombers used as close air support.

They came through individually, but talk about power! They're silent as they approach, but when they pass, the sound makes your heart rumble when they fly over. I only saw one actually drop its load, 84, 500-pound fragmentation bombs: "Snakeyes." That particular strike was about four miles away, and I'd swear we could feel a breeze from the shock waves. Nothing bigger than a cockroach could have survived that. They say you only have to get close with nuclear weapons and hand grenades. Add strategic bombers to that list.

Monday, July 4, 2011

1410 hrs

Phuc Yen Airfield NW of Hanoi, Vietnam

A worn out South Korean F-16 circled above the small airfield and its surrounding jungle. Out the right side of his canopy, Al Meyers glanced at the ancient silver Vietnamese Mig-21. The pilot's father had flown the same plane against American pilots during the Vietnam War of Independence in the 1960's. With all that Al had seen in recent months, this scene still seemed the most unusual and difficult to grasp. His radio brought him back into reality.

"Wolfman Flight, this is Tower Red 33. Copy?"

"Copy, 33. Go ahead."

"Wolfman Flight, this is Tower Red 33. Radio intercept indicates possible traffic (enemy aircraft) 75 kilometers bearing 263, altitude unknown. Number and type is also unknown. You are ordered to proceed and confirm contact. Reinforcements are preparing to leave now. Copy, Wolfman flight?"

"Copy, 33. We're on our way. Have our reinforcements hold over your position until requested. We don't want to swat a fly with a hammer. How do you read 33?"

"Wolfman Flight, this is Tower Red 33. We read you load and clear.

Reinforcements will hold over your position until requested. If needed, call for Hammer Flight and say number of reinforcements requested. Good luck. Tower Red 33 out."

Al turned his F-16 and headed west toward the unidentified aircraft. His

Vietnamese wingman banked and turned with him. Inside that silver plane's cockpit was a smiling twenty-two-year-old pilot, filled with the same confidence and cockiness that fighter pilots of old were famous for. Al's moment of disbelief was no where near as powerful as his wingman's, though for different reasons. He had listened to his father's tales of glory and excitement during the war against the Americans. He had learned to respect the American pilot's skills, their courage, and their technology. When, at the insistence of the airfield commander, he was ordered to fly as the American's wingman, it was like nothing he had ever imagined. At one time the Americans were mortal enemies of the Vietnamese people, but now they had tried so hard to be friends. Most Vietnamese believed that if it wasn't for the capitalists running the government in the U.S., America could have been Vietnam's greatest ally long ago. Today, in the face of a powerful Chinese invasion, they were.

In keeping with Al's bomber pilot past, and his new experiences as a fighter pilot, they remained under 1000 feet. Both were confident that they wouldn't hit any power lines or skyscrapers in the third world country's rural backlands. Al never took the time to learn how to operate the avionics in his fighter plane, and the Vietnamese pilot was strictly schooled in not using his Mig 21's radar. Instead, he had been trained to rely on the guidance of groundbased radar and it's presumably more intelligent commander. In this war communications between ground controllers and pilots had been spotty, so Al and his wingman would rely more on visual identification than anything else. Communication was through hand signals since neither pilot spoke the other's language. It wasn't the best way to conduct an intercept, but Al was a bomber pilot, running a multinational fight unit with little experience, and only a few minutes worth of training. With all that going against him, the current mission would have been a sad reason to start complaining. After all, it didn't matter how it got done, but a job had to be done.

Only two minutes after they left the airfield, Al spotted a pair of Russianmade Mig-27 ground attack bombers. Another look around, and Al saw three more pairs in front of the first one. His Vietnamese wingman spotted another pair on the right side of their small flight. All the Russian-built Chinese planes were flying from the south to the north, towards the Black River Valley where Al had heard that an entire Chinese Corp had been surrounded. Somehow, the Chinese managed to get their bombers into Vietnam and circle to the south side of the valley. That way, after they made their strikes, they could race straight back into China where enemy fighters were probably waiting to intercept any Coalition planes chasing after the bombers.

There wasn't enough time to get reinforcements into the fight, but here was their chance to stop the bombers before they got to the valley. If the Chinese pilots followed typical bomber tactics, they'd aimlessly drop their bombs as soon as they were attacked to allow them to get away. Then they'd kick in their afterburners and run for cover under a line of Chinese fighters at the border.

Al motioned to his wingman, and they moved into position behind the trailing pair of enemy bombers. His wingman stayed with him and waited for Al's first shot. Soon his headset growled signifying that his Sidewinder missile was locked on, and Al let it fly. A split second later, his wingman fired two more heat-seeking missiles at the other plane. All three of his rockets spiraled to their targets, and two of them hit. Both Chinese planes exploded, one immediately broke apart, the other corkscrewed into the jungle. The third missile went past both planes, locked onto one of the bombers in the next pair forward, and tracked on it. The missile's proximity fuse detonated its warhead just behind the unsuspecting plane. Shrapnel ripped through its engine and control surfaces. The bomber trailed smoke. Its pilot ejected while the plane drifted to the left and collided with its wingman.

After three shots and four kills, the American and Vietnamese pilot moved up towards the remaining Chinese bombers, none of which had been alerted to their presence. Al listened for his missile's familiar growl and fired again when he heard it. His Vietnamese wingman was out of missiles, so he fired his 23mm cannon instead. While Al's missile spun towards his target, tracer rounds from his wingman zipped between the two planes and alerted their pilots. Both bombers dumped their bombs. Al watched the tracers miss and immediately knew the bomber's next moves. He fired his afterburner while they were still dumping their bomb loads. At least those bombs wouldn't hit their targets. Technically, the mission was already a success, but Al thought he could get a little more.

He gained on the accelerating bombers and started aiming his 20mm gatling gun. His wingman was slower to react and his older plane was even slower to accelerate. Still, Al had never learned the fighter pilot rule of sticking with your wingman, so he went on. The Vietnamese pilot would catch up as soon as someone tried to turn. When he was within four kilometers, Al began firing bursts from his 20mm gun. The first four bursts missed, but made the bombers start jinking from side to side as a matter of evasion. This cost them airspeed and let Al's wingman catch up. He joined in the cannon fire, and soon the bombers were caught in a crossfire of American 20mm tracer rounds, and Vietnamese 23mm rounds. Both were hit at the same time by rounds from both planes. Al and his wingman could share the credits. They searched for any remaining bombers, but they were long gone. Al motioned to his wingman: nice job, let's go home.

At the airfield, they landed without warning. Al didn't want to wait for the tower to give him instructions, he was getting tired, and such protocols from a backwater airfield seemed pointless. He landed and noticed that every plane at the airfield was waiting to takeoff. While he taxied to the tower, his wingman circled and eventually landed. They both pulled up while groundcrewman and ran over to assist them. Along with them, an obviously disgruntled Vietnamese officer, some of his staff, and some of Al's administration people.

I perfect, accent-free English, Al was about to get reprimanded.

"Major, I do not know how you are accustomed to doing things, but in this country, in this air force, and most importantly, in this airspace, pilots will follow the orders given them by their ground controllers without question or deviation. Is that clear, or do we need to re-discuss your presence at this facility?"

Al's slow to build fatigue was finally reaching the point of carelessness, though he didn't realize it. His reply bore the attitude of someone who knew how valuable they were to the war, someone who was now an ace ten times over, someone who just didn't feel like getting reprimanded in public.

"Look, I don't know who you are, or what you think your doing here, but I'm making a difference. I'm not sitting around in some safe underground bunker watching over people who are doing nothing but watching radar screens that have been left blank by Chinese ECM. Instead, I've been up there dropping bad guys from the skies of your country. I'm ten thousand miles from anything resembling home, and I'm kicking ass for you and yours. So are all of the other people that I've brought here. If you've got a problem with me or any of my people, let me know, and we can find someone else's country to defend. Its seems like there's plenty of fighting going on around the globe right now. I hear the Mediterranean is very busy and the people there might be a little more appreciative. Now, would you like us to pack, or should we stay and drop some more planes?"

The Vietnamese officer, red in the face glared into Al's eyes. He had never in his military career been dressed down-not even for a uniform infraction. Now, in front of his men, and the men of the man whom he had tried to chastise, the Vietnamese officer was being put in a box verbally. He was stunned silent. As Al was helped out of his cockpit, he started up again.

"Look, I want to stay here and continue making a difference. I'm sorry if you don't understand our ways, but there was no time to get any reinforcements to the scene in time, and, quite honestly, I'm awfully tired lately. I should have called for landing instructions, but I was lazy. After this greeting, I can assure you, I won't make that mistake again. There's lots of work to be done here, besides, I hate Italian food, and the only things that the French can do is drink wine, make love, and lose wars. Your pilot and I just dropped six Russian-made Chinese bombers. We did it in one pass with only four missiles. What-da-ya say we get some coffee and try to figure out whether or not it was ever possible to get the other two?"

Al was offering the Vietnamese officer a way of saving face. Would he take it? He had to. The officer nodded his head and reached out to shake hands with his new friend.

"Major Meyers, you're absolutely right. We are both tired. If you land appropriately in the future, I will not take issue again. As far as your intercept... how about I get you a cot. You can catch some sleep in the radar control room. It's quite cool down there, and, as you said, it's relatively quiet due to all the electronic jamming."

Everyone was smiling. No one liked a tense situation, and Al was able to defuse it well. The Vietnamese officer went on as though nothing had happened. In a way, he felt proud to know a man that would stand up to him in public and still not take advantage of his error in judgement.

"So, you say you got six planes today? That makes your total... what? Fifty-four?"

"No, my wingman, your pilot got four of them. I'm an even fifty now. In fact, it was quite a thing. Your man here got one with a missile, then he dropped two with another missile, and he finished up by bagging another with his gun. Not something you see everyday. The kid's got some skills."

531

Al's wingman didn't understand a word, so the Vietnamese officer translated. When he was done, the young pilot hugged Al and began jibbering in Vietnamese. All of his comrades joined in, and Al was able to drift away for some sleep. In the background, another patrol flight was taking off, and all of the planes that had been waiting to act as reinforcements were being powered down. Their pilots ran over to Al's wingman to hear the story. The kid had all of the attention of a pro athlete, and Al's report was the reason. Now everyone wanted to fly with Al. All he wanted was sleep.

Tuesday, July 5, 2011

2nd Lt. John Chamberlin-Personal Log

Black River Valley near the city of Hoa Binh

(about 50 miles west of Hanoi, Vietnam)

The Army took some serious losses yesterday, so, Hector, Bruce, and I decided to stay here at the treeline in our old positions. As long as the Navy has those PBR's going up and down the river, nothing's going to happen along the bank. Besides that, our company's pretty beat up, and we probably couldn't stop any real push against us even with the Army's "help." Hector's platoon is down to about half strength. Bruce's is down to little more than a single squad. Somehow, my platoon came out of yesterday's fight without heavy losses. I think everyone has some sort of cut, scrape, or bruise, but only two people have to go home in bags.

Rudy Johnson and John Kowalski were in the same bunker. A grenade must have rolled in and taken them out. There were some Army guys in there too, but we're not sure how many. It was a real mess. We had the Army collect the remains, then I had the rest of second squad fill in the bunker. No one wanted to use it again anyhow.

My canned e-mail letter home just doesn't seem to work for these two guys. Almost everyone else that I've lost was single, but Rudy and John were both married. Rudy was only 21, but he had three little girls at home. John had gotten married about the same time that I came to the platoon last fall. I wrote each wife a new e-mail. I also sent some digital pictures I'd taken of the platoon when we were in Taiwan and some from here in Vietnam. The e-mails were tough to write. I didn't spend too much time with Rudy or John, no less than anyone else, but not as much as I have with my NCO's or some of the people who have become my friends. Still, I had to spend a few hours writing, and each e-mail turned out to be very lengthy. I wish I would have spent more time with them. They were nice guys, very dedicated, very smart, and all around good eggs. It shouldn't have happened to them, but at least it was quick. The condition of their remains left that much without a doubt.

When this morning's mist lifted from the field, I couldn't believe the sight. The Chinese were clearly the real losers yesterday. Bodies were everywhere. I tried counting, but it was disgusting. I lost count sometime after two hundred. There have to be thousands out there by now. There are spots where they're piled high enough to kneel behind. All night, we heard people moaning, crying, and calling out in Chinese to their comrades. Thank God the artillery and air strikes upriver increased when the sun went down. What really unnerves me is that I was able to sleep so soundly with so many people in so much pain so close by.

I made my rounds this morning, sat and had coffee with several different groups of Marines and soldiers, yet no one else seems to have any problem with the cries of the dying in the field. There's still a lot of hostile feeling towards them since they were after all trying to kill us yesterday. That said, a part of me still wishes that someone felt some sort of remorse. Our only reaction to their cries is frustration, and occasionally, a soldier will yell out to them "Die already!", "Sucks to be you!", or countless other niceties. I'll take the verbal taunts and terms of endearment over ammo wasted on wounded enemies anytime.

As far as remorse for myself, I just keep in mind that one of the people out in that field could very well be the same person that took out Pete, Terry, Rudy or John. Just the same, I probably have more remorse than Rudy's daughters do. How would they want me to feel? What would they want me to do? I think they want exactly what I'm doing-nothing.

Tuesday, July 5, 2011

0547 hrs

Phuc Yen Airfield NW of Hanoi, Vietnam

The Vietnamese squadron took off before Al's composite group. They circled the field over the remote jungle in a tight echelon formation. When the last Mig-21 was airborne, Al taxied to the runway, turned, and made his way down, faster and faster, until his wheels finally came off the ground. He climbed only to about a 1000 feet, then circled the field behind the Vietnamese formation. Three dozen more fighters left the ground after the Wolfman. They joined on his wing and formed into 4-plane echelon formations until the last formation was just ahead of the Vietnamese, and the 4-miles-wide circle was complete. Then, Al dropped his flight down to 500 feet, slowly turned to the right, and headed north.

"Tower Red 33, this is Wolfman Flight. Heading 345 at angels 05. Operation is a go. Monitor this frequency as best as possible. See you in an hour or two. Wolfman out."

The other formations peeled off from the circle and followed Al to the north through the prehistoric looking valleys, over timeless villages, and reflected in innumerable subsistence rice paddies. They were on a modern hunting expedition, a fighter sweep. Some of Al's veteran South Korean and Taiwanese pilots had come up with a plan to fly north into China, and taunt the Chinese Air Force into a huge dogfight. Hopefully, they would take out some of the Chinese fighters, but the big goal was to keep them from being able to cover any of their bomber attacks. If the Chinese fighters were busy dogfighting, or refueling on the ground after a dogfight, then the bombers would be have no safe cover to flee to. Hopefully somebody could get in and take them out. If not, their bombers would at least have to fight their way through Al's group to get back into China.

The train of mixed fighters headed north for almost twenty minutes. Without looking for a map, everyone knew exactly where the Black River Valley was from the smoke, AAA, and SAM contrails. Al wondered why anyone was even trying to use SAM's anymore. Most of the Chinese and Vietnamese SAM's were so old that modern fighter planes only had to carry an ECM jammer to throw them off. The heat-seeking ones were fairly easy to deter with flares. Newer, more effective SAM's were still out there, but they had become so expensive and so scarce that they were usually held back to protect major strategic targets like ports and military industrial areas. Of course, even the most useless missile still had its effect. No one would ever take the chance that a missile coming after them was a waste, so every one had to be treated like it was the most high-tech and efficient thing ever built.

Al was still thinking about the tactical usefulness of SAM's when he noticed a flight of Vietnamese Mig-21's on his left starting to climb. His entire group had been flying extremely low to avoid detection. Al knew that if a fighter pilot felt he had been detected, then he would race for altitude. In a dogfight, altitude can be turned into speed, and speed is what keeps pilots alive. One of the Vietnamese must have felt that the fight was about to begin.

Al motioned for everyone else to climb high. He lit his afterburner, put the beat up F-16 into a nearly 60 degree climb, and visually searched for enemy aircraft. As he passed through 5000 feet, Al spotted a line of fighters circling around the base of a mountain off to his lower left. They were flying in groups of three at about 1000 feet. It was like they were taking off, or at least waiting for something. Maybe there was an airfield on the other side of the mountain? Al looked carefully at the opposite side of the jungle-covered rock as he passed 10,000 feet.

Sure enough, there it was, and it was huge. A full-scale airbase had been carved out of the jungle. The red earth was still wet from a small cloud of evening rain that had passed by in the night. It reflected like a mirror. On the runways were hundreds of Chinese-built Mig-21 copies (called Jian-7 by the Chinese) a few Su-27 copies and some other misc. aircraft including a huge Russian Condor transport plane, one of the largest planes ever built. It had target written all over, but Al wasn't falling for the bait.

One of his Vietnamese flights did. All four planes in one of the formations on Al's left rolled over, turned below him, and dove for the airbase. A second later, the carousel of Chinese fighters circling the base all started to climb in separate directions. On the ground, Al could see people, vehicles, and planes starting to move in every direction. The element of surprise was lost.

Al leveled out his climb, and the rest of his group did the same. From almost 20000 feet, they all started to line up on their targets. Al figured that most of the Vietnamese rookies would go for the closer Chinese fighters, so he took aim on a Su-27 that was heading away from the far side of the mountain. When his missile growled with the sound of a lock on, Al let it go. The others in the group took this as the international command for open fire, and forty more white contrails zipped toward their targets. A trio of SAM's came up from the base, but while nearly a hundred missiles and planes were in the air, there was no hope for a lock on. It was too confusing for the radar seekers on the SAM's. Some AAA came up from five different locations, but these Chinese hadn't taken the time to work out a crossfire plan so they just tried the spray and pray method. It was useless except for maintaining Chinese morale.

Al had guessed right about the Vietnamese pilots, and their missiles found targets first. Five enemy fighters popped into black puffs of smoke around orange-white fireballs. Two more followed, then seven more. As the Coalition flights passed over the mountain and turned for another pass, there were parachutes and airplane debris raining into the jungle.

Now the Chinese responded. A pair of Vietnamese planes went down from missiles. Another exploded between two Taiwanese F-5's. Both trailed smoke and tried to head south. There was no time to see if the pilots were alright. All of the Chinese, Vietnamese, South Korean, and Taiwanese planes were now meshed in a twisted death chase. Their exhaust, missile contrails, and tracer rounds spun together to leave the classic furball design painted in the sky. Each plane was chasing one while another moved in on its tail. Now speed would keep a pilot alive, but maneuverability would get other pilots killed. Whoever could turn the tightest would get the most kills. After only a few seconds, most of the less maneuverable Chinese and Vietnamese planes were going down. Black lines fell from the furball of contrails in the sky to the jungle below.

Still more fighters, better fighters, came up from the airbase. The Chinese had most of their best planes, the short-range Mig-29 and long-range Su-27 copies, still on the ground when the Coalition group attacked. They tried to takeoff as many as four at a time, but the remainder of the Vietnamese group that had dove on the airbase was circling at low altitude, and they picked them off before they could get up to either combat speed or altitude. It was like shooting fish in a barrel. Everyone was close enough to their airfields, so fuel wasn't a real concern. Al's only real worry was ammunition. He had already fired off his four missiles, and his cannon was down to only a few hundred rounds. Of course, he had also put five more Chinese into the dirt. Everyone else had to be running as low as he was, so he thought it was time to break off the attack.

The Vietnamese were doing a good job of keeping the Chinese on the ground, but they had yet to close the runways. It was too bad no one brought any bombs, Al thought to himself. There was always tomorrow, but that wasn't good enough for the Wolfman.

Al left the furball, and the three planes in his flight followed. Two more flights saw him leave and figured that he was up to something good, so they left too. The remaining Coalition planes had the Chinese who remained in the furball very busy. Al lined up on a row of Chinese planes still parked on the ground, and fired two bursts from his 20mm cannon. Two more planes burst into flames. The other Coalition planes saw his tactic, and began strafing the entire airbase. Al never saw it, but a South Korean F-5 behind him took a lucky hit from some of the random AAA and exploded. It hit the ground at a shallow angle and slid down the runway into the huge Condor. The Condor, capable of lifting over 400 troops or 150 metric tons of cargo, was loaded with something extremely explosive, and it went off like an atomic bomb. Its shock wave compressed the moisture in the air and a white misty bubble flashed past his canopy from behind.

At the same instant, not even 1000 feet off the ground, the combustion in Al's engine was blown out. Then he lost power. There was a strict procedure on how to restart the engine in midair, but Al never learned it. He'd have to try it as though it was on the ground, but the plane was still moving. More air than usual would be rushing in, and it might blow out the engine again as fast as it got started. Al pulled up the nose until it was almost vertical. Then he rolled the plane over on its back. As the plane lost airspeed, he began trying to start the engine the same way he would before a takeoff. After some frantic switch flipping and button pushing, the engine whined to life, but he was about to stall the plane for lack of airspeed. If it stalled, he would drop like a rock from barely 2500 feet. Directly above an enemy airfield was a bad place to eject, so Al went for the desperate measures.

He fired his afterburner, pulled back on the stick, and snap-flipped the plane just before it completely lost airspeed. He was still falling, tail first, but at least he had some forward movement to provide lift under his wings, and his speed was increasing. It was a race to see if his speed would increase faster than his altitude decreased. As the ground came up, with his nose almost 45 degrees high, Al dropped the landing gear to cushion his impact... an impact that never came. As soon as the afterburner's flame touched the ground, it pushed off the Chinese runway, and the plane into went into the air. When his airspeed finally passed 150 knots, Al lifted the gear and climbed back into the sky. Most would call it an impossible move, but it was also something that would have gotten even the best fighter pilot court-martialed for trying to practice in peacetime. During a war, over an enemy airbase, about to fall straight down, any pilot might have tried it.

Al joined his formation circling the mountain. The Chinese had been swept from the air, and the base was left in ruins. He had lost 6 planes from his composite group: 3 Taiwanese, and 3 South Korean. Their Vietnamese escort had lost another 6 planes. With their losses totaled, they headed back to Phuc Yen Airfield.

The planes landed in order of damage and fuel status. The least flyable went in first despite the danger of wrecking and blocking their own runway. Al landed second to last, followed only by the Vietnamese squadron leader. When he climbed out of his cockpit, Vietnamese troops, groundcrewmen, and other pilots ran up to his plane. Everyone was ecstatic. They attached the usual ladder while he shut off the plane's systems and began to unstrap himself. This time, no less than four people climbed up to help him out. Once on the ground, a firetruck that had pulled up began spraying water over the ground. The Vietnamese officer that had tried to chew him out the other day cleared a path at the bottom of the ladder, and the crowd's volume fell as he began to speak, on everyone's behalf in a language that Al could understand.

"Major! We watched the whole thing! That was incredible! Do you know, that today, because of your efforts, the Vietnamese Air Force shot down more planes in this battle alone, than it has in any other engagement in history?! We are in your debt."

Al was being shaken and patted to the point of confusion.

"What about our losses? Did those two planes make it back?" He asked.

The officer pointed to a hole in the treeline at the far end of the runway. "They're over there. Both pilots survived. They're in the infirmary now. Their planes are lost though, I'm afraid. No one else came back. Were those the two you're talking about?"

"Yes. I want to get over there right away. Just to check on them, ok?"

"Sure, sure, sure. How about you, we saw you disappear off the radar, then appear out of nowhere? What happened?"

Al told the officer the story while he translated it for the crowd. Everyone listened seriously, and shook their heads in disbelief when he was done. There was silence. Having told the tale, Al finally realized how unbelievable the maneuver was, and he was proud. The officer looked him dead in the eye searching for a hint of fighter pilot exaggeration, but there was none. He glared in stunned disbelief until a Vietnamese speaking Taiwanese pilot who had seen the event made his way through the crowd and confirmed it. Two of the Vietnamese pilots did the same.

The officer, convinced at last, saluted and rose to attention. "You are truly a hero of the People's Air Force. It is an honor to serve with you."

Al felt it was time to clear up any misunderstandings about their relationship. "Whoa. I'm here fighting for Uncle Sam. The rest of these men are fighting for their own reasons too. Unless we start getting paychecks from your government, let's just consider us allies, not new recruits for the people of Vietnam. Ok? As far as the hero thing goes... I was just trying to stay alive. They say heroes aren't made, they're cornered. Well, I was cornered in a bad way! What do you all say we move this little party over to that hangar where our admin people can debrief us and get all the facts together. Maybe some of you will get the gun camera footage and data tapes together so we can watch some movies tonight?"

Everyone smiled as the Vietnamese officer translated and issued orders. Within a minute, the crowd was dispersed to their tasks, and the pilots made their way to the debriefing hangar. Hours later, the day's tally of losses was Coalition 12, Chinese 122 in the air, and an additional 74 on the ground. Every pilot who went on the mission was now an ace; some, like Al, were several times over. The day's effect was so enormous that rumors spread throughout the country in hours. The next morning, the press started arriving, along with lots of brass from Vietnam, the Taiwanese Air Force in Exile, the South Korean Air Force in Exile, and the U.S.A.F. Everyone wanted to see what this ragtag composite group was made of, where it came from, how it was surviving, and what was making it so successful.

Wednesday, July 6, 2011

2nd Lt. John Chamberlin-Personal Log

Black River Valley near the city of Hoa Binh

(about 50 miles west of Hanoi, Vietnam)

Colonel Ghetty came by and checked on us this morning. He's starting to show a little wear and tear. Like most of us, he hasn't shaved in a few days, and I couldn't help but notice that he could probably use a haircut too. He didn't wear much in the way of gear either, just a belt with a 9mm Beretta pistol, a map case, and a bandoleer of magazines for his M-16.

I'm sure we all look a little battered. I know my BDU's have small holes, bloodstains, mud, dust, and who knows what all over them. Still, I've made it a point to try and look my best; my most orderly. I've tried to pass that on to the rest of the platoon too. Everyone still has on their full web gear, their packs are in their bunkers and dugouts, and their BDU's are in as good of shape as possible.

Hector and Bruce's platoons look a lot like Ghetty does. Maybe I've been too meticulous about our appearance, but I think it gives the platoon something to take a little pride in. We don't look anywhere as professional as the Army's fresh soldiers do, but we still look like pros. Ghetty seems a lot more informal with us at our meetings too. I guess his other companies are as shot up as we are. We seem to be in the best shape of them all. I think he might try and make Hector a Captain to formalize his command. He really kissed his ass about the past few days' action.

I'm not jealous. Hector deserves it. I do wonder if Captain Brown would have been wounded instead of killed, would he have recommended Hector as his replacement? There's no telling with Brown. I think maybe Hector will get his promotion, we'll combine his platoon with Bruce's, and maybe, according to Colonel Ghetty, we'll get a heavy weapons squad with some .50 caliber, grenade launchers, anti-tank guided-missiles, a hand-held SAM team, and some light mortars. It would sure be nice. We could have used them yesterday.

I was surprised that he wasn't at all upset about our change of position. It was only a few hundred meters anyway. In all truth, I think he was pleased. It's pretty obvious that the Army was never going to hold without us. They seem to be more reliant on support units than we are. Our infantry training had the same essentials, but I think our company has come to realize that artillery, mortars, air support, gunships, naval support, whatever; none of it can be counted on. The only thing that we can count on is each other. They'll catch on.

Ghetty says that there's another cease-fire in effect. That might explain the lack of sniper fire or distant rumbling from up river. He also says that the very fact that the Chinese Corp commander in the valley is willing to discuss surrender shows that it's only a matter of time. The war around the world seems to be tuning down a bit too. There's terrorist activity everywhere, along with dogfights and naval skirmishes, but there are so many that even the press has stopped trying to keep track. They list them out, give body counts, and then go into more in-depth stories about the bigger battles or most important fights. Yesterday was big enough to count as a battle, we're told, so maybe we'll be in the news.

In case the cease-fire doesn't hold, Ghetty had a few people bring over some infrared beacons. They look like generic 3-pound coffee cans, but they'll project an invisible beam of infrared light into the sky whenever they receive a coded signal from a friendly attack unit; including friendly planes, tanks, ships, or even other ground units. We were told how to use them, and that we would get our own special radios to toggle other units' beacons in a few days if these worked. It's a double-edged sword. The beacons will keep us from getting any friendly fire as long as other friendly units use the special coded toggle radios. The down side is that if the Chinese figure out the code, they can toggle our beacon at anytime, and it'll be like standing up in the open saying, "Here we are! Come bomb us!" Hector, Bruce, and I all had our beacons put in the field in front of us. At least if there's a problem, we won't take too much damage. I hope.

Wednesday, July 6, 2011

2205 hrs

Phuc Yen Airfield NW of Hanoi, Vietnam

Al woke up in the radar operations center after having slept for almost eight full hours. The radar room was so far underground that the cool earth on the other side of the cement walls kept the room a comfortable 65 degrees even with all the heat generated by the various radios and other electronic equipment. There was quite a commotion going on in the usually library-like bunker. Chinese radar jamming had inexplicably ended, and the screens were covered with Coalition aircraft. A few small dogfights along the northern border with China were sideshows to the continuous flood of Coalition bombers moving into the Black River Valley. They were lined up from the Eastern edge of the radar screens to the Valley, and back again. There were hundreds. Task Force Stingray had moved off the coast, and fighter-bombers from four carrier groups were pounding the Chinese into submission. A ceasefire had been called during the night, but Al found out from the senior duty officer in the room that the Chinese had tried to get a flight of transports to the valley for an airdrop. Some Vietnamese fighters from an airfield closer to the border intercepted them. Despite losing ten of twelve planes, the Vietnamese were able to shoot down twenty transports.

When the flight was first detected, the word must have gone out through the Coalition command that the cease-fire was over. For the past hour and a half, hundreds of planes crossed the coastline and began pounding the Chinese. There were some specific targets that took repeated attacks, but it seemed as though most of the planes were either picking their own targets or just dumping bombs in non-friendly areas of the valley. Any civilians had either left or been killed long ago. As long as pilots knew where the friendlies were, they could dump their bombs in between the good guys.

If the Chinese thought that an airdrop would get their people out of the trap, they were wrong. Now, they'd be bombed like nothing anyone had seen in almost 40 years. Every Vietnamese in the room understood what was happening to the Chinese. The last time a ground unit had been attacked so intensely was 1968, when they had tried to overrun the U.S. Marine base at Khe Sahn. History seemed to love irony, particularly in Southeast Asia.

Thursday, July 7, 2011

2nd Lt. John Chamberlin-Personal Log

Black River Valley near the city of Hoa Binh

(about 50 miles west of Hanoi, Vietnam)

It looks like our beacons worked. Last night there were more artillery and air strikes than I've ever seen or heard before. When the sun rose, the entire valley was white with the usual morning mist, this time laden with smoke and cordite from explosions up the river. The attacks ended between 0600 and 0630.

When the haze finally thinned to the point where we could see the other side of the clearing, the scene was unreal. It was like we were transported to somewhere else on the planet, or in the solar system. The field has been completely churned over from explosions. It looks like a large construction site where only the earthmovers have had a chance to do anything. The village is totally destroyed. Erased might be a better description. When the artillery and air strikes ended this morning, there wasn't a sound from up river-not a sniper, not a bird, not even a noisy jungle beetle. Everything is dead. If I had to paint a portrait of what I imagine hell to look like, this is it.

About 1200, I watched through my binoculars as a group of three Chinese climbed out of the tangled and ripped-up jungle. No one shot at them. They had a white sheet tied to a small broken sapling. They moved towards a spot in our line close to Colonel Ghetty's HQ; hopefully by accident. A few Marines went out and talked to them for almost fifteen minutes, then everyone went back to their lines. Shortly after that, a runner came over from Colonel Ghetty and told Hector to pass the word that it was over. The Chinese had been ordered to surrender unconditionally to the nearest unit. The ones in front of us were going to come out as soon as possible. The jungle was pretty torn up, and they were having a tough time of it. Word passed quickly along the line. Meanwhile, Chinese troops came out of what was once a jungle in small groups. They came over individually, and in groups of as many as twenty. We searched them for weapons or anything useful, then had them line up along the river for pickup.

The Navy sent some PBR's and barges up to take them somewhere away from the line. I assume that somewhere, there's a bunch of engineers building a holding area as fast as they can. I'm told that in the Gulf war, fifteen years ago, they just used pits with barbed wire. Here, any hole dug deeper than three or four feet seems to fill with water, so there better be a lot of barbed wire.

Friday, July 8, 2011

2nd Lt. John Chamberlin-Personal Log

Black River Valley near the city of Hoa Binh

(about 50 miles west of Hanoi, Vietnam)

It looks like this fight is over. Since yesterday, over 3500 Chinese surrendered to our Company. Actually, they surrendered to the Army company that's in our old positions. We've been moved back to the riverbank to guard prisoners. The Navy's been running up and down the river since shortly after the surrender. We didn't have any barbed wire until one of the Navy barges brought up an entire pallet's worth. Hector had the prisoners clear a holding area and set up the wire with their bare hands. It seems cruel now, but at the time, it just seemed efficient, like it was the right thing to do. The best part of the whole thing is that Hector had a few burial details put together, and the bodies in the field are finally being taken care of. Most are being put in craters from the other night's bombing. It doesn't matter where they go, just so long as they get buried and we don't have to breathe any more of that smell!

The prisoners are really beat-up. Some of them just sit there and babble to no one. All of them have some sort of wound or another. The most seriously wounded get loaded onto the barges first every time. The first few loads were almost all stretchers and stretcher-bearers. Since then, they've been evenly split between seriously wounded, lightly wounded, and walking wounded. I still don't know where they're all going to wind up going, and I suspect that the Vietnamese won't be the kindest of hosts.

Colonel Ghetty says we're going to stay here for a few more days. The Army's been flying in thousands of troops since this Vietnam thing started. As soon as they can, they're supposed to send over some replacements for us. After that, no one knows where we're going. We're all a little afraid to mutter it, but maybe we'll go home, or at least get some leave. It's been seven months. There's no way to hide the fact that the entire company needs some rest, recovery, and reconstitution. As a company, we're down to about half strength. Our equipment has been through a frigid North Korean winter, a damp and wintry station in the mountains of Taiwan, and now the jungles of Vietnam, the latter being the harshest. We'll see.

Sunday, July 10, 2011

2nd Lt. John Chamberlin-Personal Log

Black River Valley near the city of Hoa Binh

(about 50 miles west of Hanoi, Vietnam)

We are out of here! The Army sent another green company fresh from the States to replace us. We have to stay through the night, but after that, we are outta here! Everyone's really been excited, especially after Wess brought up something I hadn't thought about in a while. We haven't had a place to spend any of our pay in over half a year! The Corps uses electronic direct deposit for all Marines while they're at sea or in the field. We've been both. I know I've got some money to spend! God help us all when this platoon finds a bank.

Sunday, July 10, 2011

7:21 am

United Confederation of Russian Republics, Moscow

In the same unobtrusive building along the Moscow River where he had met with Defense Minister Dimtri Yazov before the Soviet Union's collapse, Walt Barber waited for his Russian confidant, Alexov Kryuchov. When he finally entered the office, Alex looked around in puzzlement. There was no gorgeous receptionist, no smooth talking Dan Greene, and no lights. Only the dim glows of a computer monitor in an office further down the hall. He investigated and found Walt who was hanging up his phone and looked unusually distraught.

"Walt? Is everything alright? Where is everyone?"

Walt sat down at his desk and watched the monitor. The only other light came from the reflection of the building's outside lights on the blowing snow outside the office window. Walt turned around, saw that it was Alexov, and rose to embrace him. In a glance, his hanging face changed along with his obvious mood. The businessman's stoic face turned to a rare and genuine smile.

"Alexov! How good it is to see you. I'm sorry. I apologize. I completely lost track of the time. Is it really time for our meeting already?"

Alexov had never seen Walt shaken up. He thought it might be a trick of some sort, so he smiled and played along. Little did he know, the ruse was real.

"Actually, Walt, I'm a little early. Sorry if I disturbed you. What's going on? Where is everyone?"

"The receptionist doesn't come in until 9:00 or later some days. With a body like hers, I don't question it. I can't afford not to have her here. It's not like she really does anything of any value. It's the other perks that I don't want to lose, if you know what I'm trying to say."

Both men smiled. She was, after all, more of a call-girl than a receptionist. She was also the right woman for the right job.

"So what's going on then, Walt? Have they figured out your little scheme?"

Walt sobered. His little war profiteering scheme, his plan to finance or own controlling shares of military industries on both sides was hardly a little scheme. Alexov knew that trillions were at stake.

"Naht Synn, the Chinese Special General in charge of-"

"I know who he is, Walt. We went to school together in North Carolina back in the 80's. So he figured it out. He was bound to. If anyone was going to do it, he's the one. How much does he know?"

"He's building a pretty thorough diagram of my holdings, and he knows about the 30% strength setting in the SP 2.0 software, but any more than that...I just can't be sure."

Alexov thought about the situation. If Naht Synn knew that certain Coalition units not tied into the MilNet wouldn't show up on their pirated version of the software, and he knew Walt wasn't some sort of patriot, then why was he allowing Walt to live? More to the point, Walt knew it too, and his deep well of confidence was drying up fast.

"What's he want with you, Walt? You wouldn't still be around if he didn't need you for something. He must have just been trying to make a point, but why?"

It was the source of Walt's nervousness. What did Naht Synn want from him.

"I don't know, Alexov. He said I could live as long as I proved myself useful, but I'm not sure I know what he wants specifically. I'm not sure he even knows. He may just be thinking that I could prove to be a good wildcard in the future. In any case, I'm safer here in the snow than back east in a jungle or something."

Alexov walked over to Walt's liquor cabinet. He filled a glass full with vodka, and another with some Kentucky bourbon.

"Well, the cold's not too bad this year. There have been far worse. Besides, I'm a Russian, and we don't exactly handle 120 degrees of beating sun and steaming jungle too favorably."

Both men drank regardless of the hour. They looked out over the river at the white Parliament building almost lost in the dark and cloudy morning. Each considered the situation silently. Both took another series of hard swallows from their drinks, then they sat down at the two guest chairs in front of Walt's desk. What to do? Alexov was particularly curious.

543

"Walt, do you have any idea where they might move next?"

After thinking for a moment, Walt took another heavy drag on his beverage and answered.

"I was just looking at the latest strategic view on my copy of the software."

He pointed to the 26" flatscreen monitor on the side of his desk. It showed the location of all the Coalition forces in the world in blue, and their combat strength as an orange icon at the top left of each. He sat back down at his desk and used his mouse to highlight the area between India and Alaska.

"It looks like a lot of the smaller Coalition garrison units in northern Indochina are reporting skirmishes. Some have even had their combat ratings reduced. If I had to guess, I'd say they're going to go around Vietnam and hit it from the south. I seriously doubt that they want to have another incident along the Black River."

Alexov drew closer to the screen, and asked Walt to zoom in on the area around the Chinese/Myanmar/Laotian border. Then he pointed to an area where there were no Coalition units at all. Walt sat and listened to Alexov's professional military opinion.

"Look here. The Coalition doesn't have any large or healthy units in this area, or in these two. If I was a Chinese decision maker, I think I'd try and attack down these undefended river valleys, use them as supply routes, then turn back to the East. They could cut off the Vietnamese in the north, hold the mountainous areas and let the Coalition come at them through the Cambodian jungles. If they were really ambitious, they could drive west at the same time and link up with Bangladesh, threatening India."

Walt could see the potential for a huge offensive, but Indochina offered the worst terrain in the world for fighting. As soon as that thought crossed his mind, he realized its dichotomy. It might be the worst terrain in the world, but it would be far worse for a mechanized, industrialized, high-tech army. This would be an infantryman's fight. More than that, it would be a knife fight! That's where the Chinese would have a clear advantage. It had to be where they would strike next. Now how could he use that information?

Alexov wondered why they should even try?

"Walt, I know I'm much more financially secure since you and I began investing in these 'specialized' markets 15 years ago. I think I can safely assume that you've done even far better than I have. We could just sell out; take the money and run. I'm sure that we could buy ourselves some nice out of the way hideouts, hire lots of 'receptionists', and simply disappear. We could buy ourselves some nice little islands, but… just not too warm for me, huh?!"

Walt normally wouldn't think of selling out. Selling his assets went against his grain, but things were changing. Maybe it was time. After all, he could probably buy an entire country; nothing too big, but most of Central America and

Africa could seriously be bought up piecemeal. The thought was fun, but then he realized that spending hundreds of billions of dollars would inevitably draw attention. He couldn't hide. He couldn't buy his security. He'd have to make it, and Walt would do almost anything to anyone to keep himself alive. Walt was many things, but never benevolent at his own expense.

"No, Alexov. The kind of money you, I, and several others have accrued just doesn't hide very well. It could be a lot of fun, but somehow, I think we're going to have to either make ourselves indispensable, or completely forgettable. It's a dangerous path either way."

"Well, permit me to make a suggestion?"

Alexov thought about what Walt had said, but now it was time to brainstorm. Walt's attention began to light up like the morning sun. There had to be a way out of this.

"What if we could become indispensable, but still be far from either the Chinese' or the American's first concern? More to the point, how far are you willing to go?"

Walt didn't even hesitate. He was still infected with a greedy compulsion, a responsibility to make money, but he also had to survive. He had to continue-at any cost. While he pondered, Alexov finished his drink, poured another, and asked the big question.

"What if the war escalated? Both sides are fully committed. Both have already endured the worst casualties in generations, it's getting as bad as a..." Walt interrupted him.

"I'm willing to do just about anything, but I don't see how either side can get more involved. I mean, if they do much more damage to each other, they might as well pull the nuclear trigger! Nobody wins then. There's nothing left. You just give the order, and somebody blows up the world. You know the drill, Alexov. What are you getting at?"

"Look, Walt, nobody knows about the nuclear trigger better than myself. Remember, I was the one with his hands on the Dead Hand launch system when the Soviet Union fell! I know!"

Alexov felt himself getting excited. He stopped, caught his breath, smiled and lowered his tone.

"Walt, I'm not saying somebody should put 40 or 50 thousand warheads in the sky to save our asses. I am saying that China doesn't have that many. If they did, and they used them, they could receive retaliatory strikes from the Americans in a hundred-to-one ratio. Still, there might be another way. Let's say that the Chinese had more warheads and delivery vehicles than they do right now. With your funding, and my connections, we can get them as much as they want, really."

Confused, saddened, sickened, and hopeful, Walt poured some more drinks while Alexov continued.

"There's not a lot of control on the hardware in Russia. Let's face it, Kazakstan, Uzbekistan, Tajikistan, and several more of the new republics have a tough time controlling their population, let alone their borders and the old high-tech weapons sitting in storage depots."

Walt's attention was galvanized. Alexov could see it, so he let the brainstorm rain.

"Let's say that they pull the nuclear trigger. Once both sides do enough damage, they'd get stuck, bogged down. No one would be able to mount a major operation without becoming a target. Meanwhile, I keep getting you the weapons, you keep selling them, the Chinese keep buying, and the both sides keep your factories humming. The best part of the deal is that nukes and nuclear defenses would be worth their weight a million times over in gold to the Chinese. All we have to do is get someone else to sell them. While nukes are being tossed around, we all move down on their arrest lists; indispensable and invaluable."

Walt was shocked, but he hid it well. The idea was disgusting. Either side would kill tens of millions of people, maybe even trigger a complete nuclear exchange; all for the chance at saving a few unknown billionaires and trillionaires. Of course, Walt, Alexov, and any of the other nouveau riche weren't the type of people who would really care about anyone else's life. If their lives were at risk, why shouldn't everyone else's be too? Walt sat for a few moments and considered the idea further.

Since the day he first met with Deng Xioping and Kim Il Sung during the Gulf War, Walt had played with the idea that he might be letting loose a dragon on the world if he sold the SP 2.0 software. There had even been moments when he thought that perhaps he might have done the wrong thing. They were fleeting moments. His conviction, his plan, his greed had never failed to keep him on course.

"Alexov, if we we're to encourage such a situation between the Eastern Alliance and the Coalition, how would we go about it? I mean, let's face it, the Chinese know as well as everyone else does that the United States has enough warheads to wipe them off the map. Then there's the whole problem of Chinese missiles going over Russia and the North Pole just to get to the U.S. I'd have to convince them that they could get away with it. That might just be a little too complex."

Alexov had anticipated Walt's question. Only someone who'd had their finger on the button, nuclear codes in their hands, and a national crisis in the background could be prepared to answer it; someone like Alexov.

"Look, I can give a heads-up warning to certain Russian generals so that all the Chinese have to do is give a phone call's warning, and Russia won't get involved. There may be a lot of people north of China that get seriously nervous as nukes are tossed over their heads, but that'll be inevitable. As far as an edge over the U.S., all the Chinese have to do is tickle one of the American DSP satellites, and they can cut the available American response time in half, maybe more."

Walt was more comfortable with the conversation the more that they talked it over.

"What do you mean 'tickle'? You're starting to sound like one of my software computer nerds."

Alexov laughed at the similarity. He wasn't a particularly husky man, but as a Russian soldier, he was a far cry from a stereotypical weakling nerd.

"It's kind of like this. The Americans have four Defense Support Program, DSP, satellites in a geosynchronous orbit along the equator. They all use infrared sensors to watch for missile launches of any kind. The four satellites can only detect them while a missile's boosters are throwing the warhead delivery vehicles into space. There's so many things moving around on and just above the planet that trying to detect a missile launch with radar is like looking for a clear piece of sand on a beach. That's why they use the infrared sensors. Once tracked into space, there are dozens of satellites that can follow the objects."

Walt's attention was back in its usual flinch-free poker face, and Alexov understood that he could follow his conversation. Alexov continued.

"A nice little remnant of our Soviet Union is some laser research technology that's just sitting around looking for sponsorship. Some of the last things that we were using were these ground-based sites for what we called 'tickling'. We would track an American satellite and fire a laser at it from the ground. The lasers were always too weak to do any harm, but we could then monitor their communications and determine just how important the satellite was to their defenses. If there was no reaction, we could safely assume that it was just a regular commercial communications satellite. Sometimes we'd take MTV off the air, and sometimes we'd stir up a hornet's nest over at their NORAD. The best part about it was all the laser really did was point a very very bright spotlight at something very small and very optically sensitive; something like an infrared detection satellite."

Finally, Walt understood. He could finance the Russian laser research program himself, albeit discreetly as usual. Alexov could arrange for the Russian's to look the other way as nuclear missiles flew over their country. The Chinese, if pushed, could find the possibility of a limited nuclear engagement all too enticing. Once in the fight, Walt, Alexov, and all the silent investors that had made the war possible could fade away while the Chinese were preoccupied with making sure the U.S. didn't wipe them off the map.

The only thing he needed was time. Luck wasn't even considered. Walt always believed that a man-made his own luck.

Tuesday, July 12, 2011

2nd Lt. John Chamberlin-Personal Log

Cam Rahn Bay, Vietnam

A few of the Army's UH-60 Blackhawk helicopters took us to a Vietnamese Naval Base in Cam Rahn Bay around mid-morning. The base was actually built by the U.S. almost forty years ago. Since then, it's been used by mostly Russian ships, some Vietnamese patrol boats, and even a few visits from the Chinese when times were more peaceful. Now, it looks like the Americans are back in force. All of the Marines that came ashore weeks ago, including the British Royal Marines, have been sent here for R & R.

The local populace remembers the days of old, and a shantytown has sprung up along both sides of the base-along the beaches. There are black marketeers, car dealerships, drug peddlers, electronics stores, food markets, computer stores, tailors, bars, brothels, casinos, and banks. Small portable Automatic Teller Machines (ATM's) have been set up at the gates of the base. There's at least a hundred near the south entrance alone. We can actually pick which bank's ATM we want to use. Most people just pick the one with the smallest line, and there are thousands of Marines and troops in line at any given time.

The Vietnamese Army sent some troops in to try and keep order, but everyone thinks that the locals must have paid them off. No one cares what goes on outside the base as long as no one gets hurt. There's a lot of money flowing outside those gates.

After we were assigned our barracks, an officer from the base came in and passed out requisition cards. We are allowed one full set of gear for each requisition card. Everyone is going to get the latest in equipment. All that we have to do is make sure that our old gear gets either turned in or destroyed so as to keep it from getting sold outside the base. After everyone gets their new gear, we'll all get to use the showers.

The company's down to half strength, so Hector let me put all the young women from my platoon in his barracks. His people are going to stay with Bruce's platoon. I suspect that, instead of being welcomed for privacy, the women will be using their quarters as a sort of company party center. Everyone's just in that kind of a mood, and we won't be allowed off the base until Tuesday thanks to some sort of beaurocratic snafu. It turns out we weren't expected here until then, but the Army relieved us a little early. Don't get me wrong. No one is complaining!

Wednesday, July 13, 2011

2nd Lt. John Chamberlin-Personal Log

Cam Rahn Bay, Vietnam

This place is a mad house. I feel like a farm boy in New York City for the first time. The base is packed with thousands of Army, Navy, Air Force, and Marine personnel. There are small groups of people who seem to be wandering aimlessly. Thousands of others seem to be walking fast or even running to get wherever it is they think they need to be. With so many people in such a hurry, I wonder if our platoon is supposed to be somewhere? Some units are marching or jogging in formation, and others, like mine, are just sort of hanging out watching it all happen.

I would have thought that last night, our first night out of the bush, would have been cause for a real party, but I checked the beds just before midnight, and everyone was sacked out. There were a few couples that didn't even try to hide their affection, but with all that we've been through, I'll let it slide until it becomes a problem. After all, these people, these kids, have been playing the high school-like dating game since I first arrived. Trying to stop it has been like trying to pull magnets from steel. It just isn't going to happen. It hasn't created much more than a few unique situations in the field. So far, no one's been hurt in combat because of a dangerous liaison. If anything, it's helped keep morale....up.

This morning, when reville sounded at 0530, everyone woke up, got dressed, and headed for the mess hall. I imagined that with thousands of people here, the food would be mass-produced, and lack quality. Inside, there's no line. We just seated ourselves, and someone actually came over to take our orders! The menu, yes menu, has food for all three meals, so if we felt like bacon and eggs for dinner, or pizza for breakfast, it was there for the taking. Price? Free. When life in the Corps is rough, it's really rough, but when it's good, it's great.

I ate with Hector and Bruce. They had breakfast, but I had a 16oz. steak dinner with real peppers and onions, baked potatoes, salad, soup, rice, a beer. Everyone over 21 is allowed one beer per meal as long as the mess commander doesn't have any trouble.

When I got back to the barracks, the platoon was checking out their new gear and debating what cherished pieces of the old they wanted to keep. Boots were the first to go, then chemical gear and BDU's. Some accepted the new rifles, but about half still want to keep their M-16A3's. It seemed like no one wanted the new pistols, but for everyone one that didn't, there was someone who wanted to carry two; old-west fashion. There were only a few of those left over. Ponchos were turned in too, but only because all of ours have been ripped, not because of some new improvement in the replacement ones.

Hot new items on the equipment list: new weapons, new helmets, new chemical suits, new flak jackets, new socks, and new boots! The new boots are all leather like the old ones. These go higher up the leg to help keep them on in deep mud, or on long marches. They're comfortable, like good running shoes. Our old boots had two metal plates built into the sole to prevent punji sticks from going through, but these new ones use kevlar and a plastic/rubber mesh for a sole. It looks like they took fibers, knitted them together in the shape of a boot print, then heated them up and stamped them into a solid.

Our new socks are a lot lighter than the ones we've been wearing. Socks are perishable supplies in the jungle environment. I think there might have only been one or two people in the platoon who hadn't worn out their jungle pairs of socks and had to switch to their winter ones left over from Korea. Heavy winter socks, in thick leather boots, in a jungle...it's a miracle we didn't all come down with trench foot. Next to marksmanship, footcare is the most important thing for all infantrymen.

Our new flak jackets are incredible. The old ones weighed almost 25 pounds, but these aren't even ten! The old ones were kevlar with padded composite ceramic plate inserts that could stop any bullet up to 9mm. They could also severely limit the amount of damage from even something as powerful as a rifle bullet. The new ones are also made of a kevlar based material, but the manual that came with them says that the padding is built-in and the plate inserts are now some sort of aluminum/boron/ceramic compound. The book says that these new vests, or jackets as some would call them, can stop all small caliber bullets including most rifle rounds. I hope we never have to find out, but this war's got a long way to go, and I'm sure someone is going to wind up field testing their new body armor.

If we run into a chemical or biological attack, we're a little better equipped than we were before. The old chemical suits went on over our BDU's, and they were hot as hell to wear for more than a few minutes. The new ones go on like long underwear, and they're supposed to be worn all the time. When they're on, they look like a black spandex jumpsuit. I thought they would be hot as hell too, but they have a built in pattern of small tubes, not even 1/16 of an inch thick. These tubes collect condensation from the temperature difference on both sides of the suit. They use a capillary action to pull the cooler moisture from the bottom of the suit to regulate body temp. It was the most incredible thing. It's like wearing an air conditioner. There is one major flaw in the design. After they've been on for more than a few minutes, say half an hour or more, they are extremely difficult to take off. This is only a real issue when it comes to routine bathing, or more importantly, the call of nature. The trade-off will only be measured in the field, so I haven't pitched my old suit yet. I'm going to see if I can stash it from the base supply officer who's re-equipping the platoon.

Our new helmets are something else too. They look like something from a science fiction movie. Like the flak jackets, they're made out of new plastic based materials, so they're about half the weight of the old covers (helmets). They old ones had a clip on the front to mount night vision goggles, and clips on the side to mount our platoon radio communication headsets. The new covers have both units built in.

There's also a monocle built into the helmet. It flips out from the area that covers the right ear. As soon as the monocle reaches a certain point, a small pinhole projector shines onto the little eyepiece. The result is a one-eye night vision piece. The light that comes into the helmet is converted into a digital projection on the monocle. The built-in comm system has no microphone hanging down in front of the wearer's face like the old one. Instead, it has four tunable microphones, each about the diameter and thickness of a dime. They're actually better at picking up my voice than the old ones were. I think it's because they're on the inside of the helmet. Outside noise has a more difficult time reaching them, and they're able to filter out ambient sounds much, much better. At the rear base of the helmet, there's a computer cord connection port. It allows us to connect our palm PC's to the helmet. When they're connected, all images that come in through the monocle, and all sounds that are transmitted through the radio, can be sent to the palm PC. From the PC, data and images can be sent to anyone on the globe with access to the network. I can also use the monocle as a small monitor for the palm PC. The monocle isn't even 2 inches square, but when I hooked it up to my palm PC it was like looking at a drive-in movie screen.

Speaking of palm PC's, everyone in the U.S. military is being given their own. I'm told that there's going to be a class here at the base for everyone that hasn't learned how to use one yet. Of course we'll all have our own level of clearance, but that will automatically be setup by the main computers that tie all the palm PC's, radars, etc. together. It's a dangerous proposition to allow such easy access to all our intelligence, but it's also an incredible force multiplier. With everyone tied together, they're saying that a 10-man force will be able to call in the firepower of 10,000. We'll see.

Our little Christmas in July lasted most of the day. It was interspersed with trips to the mess hall for more steak, burgers, pizza, and BEER. As night fell, I began to wonder if tonight would be the big 'cut loose, let it all hang out' night for a party, or would it be tomorrow when we were allowed ATM access to several months worth of pay?

It was tonight.

Thursday, July 14, 2011

2nd Lt. John Chamberlin-Personal Log

Cam Rahn Bay, Vietnam

Yesterday's party began just before dinner. Everyone was finally settling down with his or her new gear, and the crowd was slipping away. When I came back from dinner, no one was left in the men's barracks. My suspicions about the women's barracks proved correct. It was party central.

Wess and some of the guys in second squad managed to "find" a few cases of beer behind the mess hall. A good portion of Hector and Bruce's platoons were informed of the gathering, and the barracks was packed. It was shoulder to shoulder with a beer in everyone's hand when I walked in. Instead a fearful silence, I was greeted by cheers and raised beer cans. It's difficult to complain about that kind of appreciation. To keep that good relationship with the platoon, I smiled, waved, and opened one of the beers that had been passed to me. I was only going to have a few sips when I noticed that I had finished it already. Still thirsty, I had another.

The half a dozen women in my platoon were acting it up in varying degrees. Portia and Christina were both with their companions of the week. The previous night, I had seen Steve Burns from first squad in bed with Portia, and Louie Madill from third squad was in bed with Christina. They actually make nice couples, but given Portia and Christina's track records with relationships, there's not a lot of hope for Steve or Louie.

Most of the guys in the platoon seriously dislike Maria, so she was trying, to no avail, to pickup someone from either Hector or Bruce's platoon. Late in the evening, she was able to pickup one of Bruce's guys, but only after he was completely beered up. Maria's relationships always seem purely sexual. She's like a macho guy with a Casanova complex. The only problem is that she's also a real bitch. I've had to talk her down a few times when she was grossly insubordinate. I think she's a victim of her attitude and a slave to her libido.

The other night, I also saw Melissa Warner and Dwaine Brubaker sleeping together, but I let that slide too. They're kind of a cute couple, and neither one is really the promiscuous type. I think they're relationship is a little more serious than the ones that Portia, Christina, and Maria are known for.

Lori Lawrence, normally a very conservative, straight-arrow-type Marine, really comes out of her shell with a few beers. I only saw her have a few drinks, but she was leading a sing along, sort of karaoke-like, but without the music. Late in the night, someone saved us all by bringing a radio to the gala. Lori and Karl are very close, like Dwaine and Melissa are. I think they're good examples of healthy relationships in contrast to those of the other women, girls, in the platoon.

Amy Broady, like Maria, is slave to her sex drive, but as rough as Maria's attitude is, Amy's is amiable. She's an incredibly nice and friendly young woman who is among the most intelligent people in the platoon. She's also fairly attractive, though a little short and stocky for a runway model. It makes no sense to me why she is as provocative as she is when she has so many other things going for her. During the day, when we were figuring out how to use the new equipment, it was Amy who figured out how to connect the palm PC's to the helmet. Then at the party, she went nuts with the beer, and she was all over an entire squad of men-and women! Through the night, I saw her making out with Marine after Marine, some of whom were women. Amy's not an overt lesbian, but she's obviously flexible and active in her orientation.

The guys in the platoon, and those from the other platoons that had joined the party, didn't act any better than the women; after all, it takes two to tango (except in Amy or Maria's case). If I had to guess, I'd say that one out of three guys at least had a chance with the few of the women on hand. Those that didn't went wild on beer. Wess had only "found" a few cases of beer, but more and more made its way to the party. Everyone was fully loaded up in no time at all.

The drinking contests slowly evolved into one, lead by Wess. He had everyone bidding on who could do the most pushups. The highest bidder was put to the test. They always failed, and had to drink the number of beers bid as a punishment. No one ever made it to the number of pushups or the number of beers bid. When the victim passed out, the bidding would start again, and another victim went to the floor.

Sometime in the middle of the night, I went back to the men's barracks and passed out. Before I left, Bruce and Hector had joined the party sometime in the night. Hector disappeared, but Bruce agreed to remain and make sure no one got hurt. I woke up this morning feeling like I'd been run over by a tank.

It's amazing how sunny Vietnam can be in the summer. Thank God for our Marine-issue eye protection. We're supposed to wear these special sunglasses to protect our eyes from dirt, debris, or non-lethal lasers. I've never met anyone who was hurt by a laser, but that doesn't mean they're not out there. It just means that no one I know has been hurt by one. In the past, these special sunglasses have kept dirt and debris out of my eyes, but it's the morning summer sun in Vietnam that's the real problem.

I went to breakfast with Bruce. He's in much worse shape than I am. That brought my morale back up to where it needed to be. My appetite wasn't really there, but a steak dinner platter for breakfast and another beer brought me back to operable condition.

After breakfast, the entire company went to a pre-leave briefing. We learned all about the local culture, sights to see, and how we're supposed to behave. It lasted for about an hour and a half, then we went back to the base outfitter. Everyone was issued a brand new set of Marine dress blues. The platoon really looks good.

Every single person knows that they've fought hard and fought well. When a Marine makes it through basic, they know they're something special. The moment we all heard those first rounds hit that Osprey back at LZ Delta, we all knew we were proving our mettle. After Korea, Taiwan, and the Black River Valley, there's not one bit of doubt. We're Marines; heart and soul.

We checked out of the base around 1000 hrs, and there were already long lines at all the ATM machines. I had to wait in line till almost 1130 until I finally got my chance. The machines were set to only allow a withdrawal of no more than $1000/day, U.S. My balance is over $35,000! When I left San Diego, I barely left $1000 in the savings. I put all the rest of my money in my online investment account. I hadn't really checked to see how those investments were doing, but sometime I'm going to have to transfer my savings into my online account. I guess all that overseas and combat pay really added up.

I went with second squad down the main drag of the shantytown. One by one, or in pairs, the entire platoon split off and disappeared into the communist country's new center of cancerous capitalism.

I wandered around with Barry Gallagher. He's a very nice guy from Southern California; the stereotypical muscular, blonde hair/blue-eyed, surfertype who says "dude" a lot. Most of the people in the platoon are nice to him, and he's nice to everyone without exception. His only flaw is that he is a serious born-again Christian. There's nothing wrong with being a born-again, or having any type of religious interest. Like anything else though, it needs to be taken, or more importantly shared; within reason. I find too much of almost anything to be a turn off.

I like Barry. While everyone else was enjoying a wild and spend-thrifty time, I spent the day in theological discussion with Barry. My own religious background is also Christian, but I never really had it forced on me, and I never really picked it up. I've certainly prayed plenty in my life, but I don't know many people who are as devoted churchgoers as Barry. I've heard it said that there are no atheists in foxholes, and it's true. Everyone says a quick little prayer here and there when they're being shot at. Still, Barry is the only one in the platoon that prays out loud. Most everyone else is like I am. We all believe in God or some sort of formal religion. We all practice it when convenient. None of us, except maybe Maria and one or two others, would dare call themselves atheists, but religion just isn't in the forefront of most of our minds. Survival seems number one, then it seems to be a toss-up between beer and sex for most people. As awkward as Barry may be, he's an intelligent and gifted conversationalist. At least it was an opportunity to dwell on thoughts other than trying to keep everyone alive. That was refreshing.

I went all day with $1000 in my pocket, and the only thing I wound up buying was a paperback book, and a necklace for Barry. The necklace is handcarved wood of different types with a small cross every twelve beads. Barry doesn't have his own bank account. All of his pay goes directly into his sister's. I'm not sure why, but I'm sure there's a good reason. When I saw the necklace it seemed like the right thing to do, and it was a good way of changing the conversation from one type of bible discussion to some other aspect of his religious lifestyle.

We got back to the base before dinner, and I met up with Colonel Ghetty outside the mess. He had just finished eating, apparently alone, and he asked if I'd like some company. I didn't realize he felt so close to any of the officers in the battalion, but into the mess hall we went. I ordered a pizza, and he even found room for a slice. Ghetty got us a few extra beers (rank has its privileges), and we hung out for a few hours. We talked about how nice the base is and how rough it was in the jungle. We talked about the view we had in the mountains overlooking Taipei, Taiwan, and how cold it was in Korea. We also talked about Captain Brown.

Ghetty says that Brown actually liked me. Before I arrived, he and Brown reviewed my records, and Colonel Ghetty made the remark that it looked like they were getting some spoiled white kid. Brown took that to mean that Ghetty wanted me taken down a notch, and so he did. Thankfully, it didn't take to long for them to recognize that my personnel file only showed where I was from and what little I had done in the past. When they saw what I could do with a platoon, they knew they had something, or so I was told. Anyway, it was good talking with the

Colonel. He doesn't seem so big and powerful anymore. Now he seems more like someone who's chewed the same dirt I have.

After about an hour or so, he had to leave to meet some old Air Force friend of his in the Officers Club, so we both went our separate ways. As we were walking out, I made it to the other side of the street when he yelled over to me. I thought maybe I'd forgotten something in the mess, or maybe I'd walked away with his hat instead of mine (happens to me more often than one would think). No, that wasn't it at all.

Colonel Ghetty asked me what time tonight's party was supposed to start! He said he had a lot of fun at last night's. I don't remember seeing him there at all, but I guess it does explain why the party didn't get broken up. Anyway, I yelled back to him that I didn't know when it was going to start, but I expect it to be well under way by ten or eleven. We both laughed. He said he might stop by, and then we went our separate ways.

I got back to the barracks and changed into my new BDU's. Having worn BDU's almost exclusively for the past year, I feel pretty uncomfortable in a dress uniform now. I think most people these days feel the same. Everyone who's staying on base for any real period of time is wearing the BDU's of their service. Even the sailors here walk around in them. Who knows? Maybe it's turned into a fashion thing?

After I changed, I hung out in the barracks reading my new book. It got boring real quick. The stories I used to find exciting now seem slow, artificial, and staged. After all, they are. I think I'll have to find a new genre. Barry was hanging around the barracks too, and he suggested his bible. Some other time, maybe. I thanked him, told him I couldn't possibly take his, and, as I got up to go check if the party had started, I asked him if he'd like to join us. He didn't think it was a good idea, but he thanked me anyway. Barry's a genuinely good guy.

I went across the street to the women's barracks (a.k.a. party central). Karl and Lori decided to double date with Dwaine and Melissa at the base theatre. The other women were already beered up. Portia and Steve were all over each other on Portia's bed. The same was true with Christina and Louie. It was almost like they were having a contest to see which pair could be the biggest exhibition. I walked over to interrupt them, half for fun and half to make sure things didn't get too out of hand. I made it about three paces then I noticed the smell.

I have no problem with marijuana in the civilian world. It may be a controlled substance, but, to me, it's kind of like the policy we have with homosexuals; don't ask, don't tell. After all, unlike other drugs (cocaine, alcohol, LSD, or PCP for example), you never hear of someone getting violent because they were smoking pot! None-the-less, in a combat unit, there's absolutely no place for it. Anything that takes away your edge or slows you down will definitely get you killed.

On my way to go kick somebody's ass, I had to stop for a minute and remember that no one in the platoon was old enough to be drinking. Yet I was allowing that to flow freely. The Corps takes a harder stance on controlled substances like marijuana, so I had to also. I followed the smell.

The barracks is broken up into ten enclaves with eight bunks or racks in each, just like back on the Wasp. I walked by two empty enclaves and couldn't believe my eyes when I came upon the third one on my right side. Amy Broady, Maria Jansen, Tom Franks, Peter Giovanni, Kurt MacDougall, Dan Farnum, and Wess Dustmann were all crammed in the small area. Maria, Tom, and Kurt were taking turns on a plastic water bong that had to be three feet high. Dan and Wess were sharing a bottle of Whiskey. Amy and Peter were doing their best strip tease acts for the little crowd.

No one even noticed me. I had to stop for a second and try to gather my senses. There have been some out-of-hand parties with this platoon, but this definitely crossed the line.

I was at a definite loss. Where to begin, who to yell at first, whether to yell at all or to try and be the nice Lieutenant and get each one of them alone. I was stumped; not to mention how would Colonel Ghetty deal with the issue. He could arrive at any minute, and there was no way to hide that smell! I thought for a few seconds, then I turned to action. When in doubt, blast it out. I left the party and went back across the street to the men's barracks. In the stack of crates with our new gear, I pulled out three tear gas grenades and went back to the party. Barry saw what I was up to and followed me with three more grenades. We both looked at each other and smiled like the pranksters we were. I had Barry to wait by the front door until I threw my grenades in from the back. Not even a minute later, I came by the back door and, one by one, threw in all three of my grenades. Barry had the front door opened just enough to see when they went in, and his followed before the last of mine went off. One grenade would have been enough to make it near impossible to stay in the barracks. Six was overkill in the worst way.

Portia and Steve came running out first. Both looked terrified. Christina and Louie came out right behind them. All four were embarrassed as they tried to get their BDU's buttoned up. Then came the real fun part. A small crowd of people that had been walking down the street began to gather around the smoking tin building. Since the tear gas first started coming out, almost a hundred had gathered.

Wess, my wild and wooly squad leader came out next, bottle in one hand, pistol at the ready in the other. He was so drunk that he tripped on a stone outside the door, hit the ground, fired his weapon in the air, and passed out-all in one spinning motion.

Tom and Kurt came out next. They tripped over Wess. Kurt fell face down next to him, and Tom landed between the two. Both remained semi conscious and rolled around moaning in an effort to get back up. Gravity was not cooperating. Tom must have felt the world spin one too many times, and he vomited all over the back of Kurt's head. A nice pile was forming. A few more seconds passed with no one leaving, then Dan came out, a little disoriented, but marching, some would say stomping, over the others into the crowd. He tried not to step on anyone, but he was lucky enough to be walking. There was no way he could manage to avoid planting one boot squarely between Tom's legs and another on Kurt's vomit-covered ear. This lead to a near loss of footing on the rough, slippery, moving surface, but Dan recovered and moved into the crowd.

Maria came out after him whining and complaining with every step. She was furious and demanded to know who the culprits were. She shut up the instant she saw me dangling grenade pins and staring at her. We were all laughing out of control, until she came out filled with emotional fire, then we calmed down, knowing how direct and personal her temper could be. When she finally shut up, the laughter quickly grew back to its buckling roar.

The only people that remained were my two amateur strippers. The crowd grew as we waited and waited. It seemed like it took forever, but then we finally heard the familiar-choking coughs coming toward the door. With the street filled with the growing crowd, Peter (completely naked except for his boots) came out the door with an equally bare Amy in his arms. He had her bent over his left shoulder in a fireman's carry position. Her bare butt aimed straight at the crowd who cheered as though he had rescued her. Amy would have reacted, but she passed out just before they cleared the doorway. Peter looked around at the crowd, smiled, walked directly over to me and saluted. "Sir," he said with pride, "I think we all made it out ok, but Private Broady here seems to be wounded. If it's ok with you I'd like to take her back to the barracks and try to revive her." I leaned over and told him quietly in his ear to get her in our barracks, get her dressed, and, yes, get her revived. I knew the by "reviving" her, he really wanted to get a little more friendly with her. It was fine by me as long as they were a little more subtle.

Colonel Ghetty came out of the crowd as Peter, still naked as the day he was born, was walking into our barracks. I thought for sure he was going to read me the Riot Act, but instead, he took the grenade pins out of my right hand and shook it firmly. He said he did the same thing once, but he had only used a single purple smoke grenade. It may have been a better idea. Some of the old purple smoke grenades left a nice residue, and everyone must have come out looking purple. Ghetty said he smoked his platoon because they had overslept passed morning PT, and he wanted to know why I did it tonight. I was as vague as possible, but couldn't lie; not only because of his position or our time together today, but I simply didn't want to. I explained the unique situations that men and women together in combat units can create. I told him that I had found such situations to actually be beneficial to morale rather than detrimental, but sometimes the platoon needed to be drawn back a bit. There are still lines of conduct that can't be crossed. Thankfully, he didn't ask what line had been crossed, but Peter and Amy's appearance might have lead him in one direction.

There was no mention of the Marijuana. One by one, throughout the night, I made sure to explain my lack of tolerance on the topic to each of the people that I saw in Amy's enclave. Most wouldn't remember, so I pulled them from the off base permission list at the gate, and tomorrow, after they try to get off base, I'm sure they'll come to me for an explanation. The tear gas will not be the extent of my punishment either. I will not become a toothless tiger for the platoon to play with. As much as I want them to continue liking me, I have to make sure that they stay sharp, disciplined, and obedient.

The rest of the night, there were small parties at different barracks, warehouses, hangars, and halls around the base, but our company area was quiet.

Friday, July 15, 2011

2nd Lt. John Chamberlin-Personal Log Cam Rahn Bay, Vietnam

I woke up this morning in the usual condition; on the edge of reality. Having gone for so many months without a single drink, my tolerance seems gone. A solid steak breakfast and another morning beer got things going today.

Most of the platoon, and the rest of the company for that matter, has started feeling the effects of several days worth of hard drinking. The few conservative types (the more responsible people in the platoon) have been able to practice tolerance in their drinking, and they've been faring much better than the rest of us. There are two exceptions to the hard drinking rule: Wess Dustmann and Dwaine Brubaker, my second and third squad leaders. They wake up at reville, and before I can get my second boot on, they're out leading people in morning PT. Thank god somebody's out there instead of me.

Bruce, Hector, and I met with Colonel Ghetty today. He tells us that our parties have to officially end since someone above him made mention of last night's tear gas fun. I asked if this was a permanent decision, or if we could slowly pick them back up after a few days or weeks. He looked around to make sure no one was listening, then leaned over and told the three of us that we wouldn't be here much longer. Colonel Ghetty met with General Shapiro yesterday after he and I had dinner. We're going to ship out at the end of the week.

Ghetty took us outside for some privacy before he gave us the details. It seems that we stopped the Chinese cold in the Black River Valley and captured a reinforced corp? There's almost 100,000 Chinese in Vietnamese prisons. Two other Corps tried to break through to them from the north, but the Vietnamese slowed them down, the Brits stopped them, and everybody got a piece of them before they went back into the mountains on their own soil. Given our recent history of only one victory, no Coalition force is looking to chase the Chinese onto their own turf.

When the Chinese realized that they couldn't break through, they tried to draw pressure off of their trapped guys. They sent several divisions South into Laos, Thailand, and Myanmar. Now, those countries need our help. The deal is, they'll supply us with food, but we have to pay for all the rest of our supplies.

It's not that these countries are ungrateful. They simply can't afford to keep an expensive modern military war machine like ours in business. The

U.S. Army sent its replacements directly into Laos, and that's why it took so long for us to get relieved in the Black River Valley. We also made a bunch of small airstrips and firebases all around the border.

In Thailand, the Australians and New Zealanders backed up the Royal Thai forces. They hit hard and fast, but they just didn't have enough people to stop the Chinese flood, and they were overrun. After a few days, they were able to reform and occupy good defensive positions in the mountains. That made it too costly for the Chinese to send in reinforcements.

In Myanmar, the Chinese ran into a well-organized national defense force. It bent like a reed in the wind, then snapped back with pent up force. The long coastal country started out with an estimated 350,000 men to stop the flood, but most of those have always been tied up fighting insurgents. When the Chinese came across the border, they found a network of garrisons and mobile units that caused serious damage. It seems that the Chinese have pulled back into the mountains along the northwest border connecting Myanmar, China, and India together.

Smaller, more elite units remain scattered all over Indochina. For better or for worse, most of the Southeast Asian nations have their defenses structured around fighting guerrilla wars. With that in mind, Laos had a militia self defense force of over 100,000 men. Thailand set up its Army to be about 150,000 strong, but it has several police forces numbering almost 100,000 men. Add to that a special corps of about 20,000 Thahan Phran; huntersoldiers who are trained to hunt down insurgents. They're the largest organized sniper unit in the world, and they're good too. Myanmar built its Army, Navy, and police into units that can counter any of the 43 different insurgent groups anytime, anyplace, and in any combination. Now that we're backed by frontline Coalition forces from around the globe, and almost 100,000 U.S. Army troops in each country, the Chinese are taking some losses.

Where does that leave my platoon and I? Colonel Ghetty says that we're going to go back to the ship, move to a position off the coast of Myanmar, and wait to act as a Ready Reserve Rapid Reaction Force. The way he described it, it sounds like it'll be like a quiet shore leave on the ship. He doesn't want us to spread the word yet, so we won't. I will, however, start getting my own things in order since even Colonel Ghetty doesn't really know when we'll bug out to the Wasp.

After our meeting, I met with all of last night's rabble-rousers and told them that they had blown the entire party scene because of their behavior. I also gave them all a speech on the Marine Corp's policy regarding marijuana and other controlled substances. Maria gave me a hard time about not letting them get high, but letting everyone get drunk instead. I told her that it was all a bad idea, and that I'm willing to look the other way when it comes to a lot of things, but right now the Corps is tougher about drugs than it is about drinking, so I have to follow suit. I also reminded them all that if I really wanted to get tough about the issue, I could have had them all arrested. They might not have been sent home, but they sure could have been reassigned to some unit still in the jungle. She shut up, but I can tell she's as irritable about the whole thing as she ever was.

When I was done with all of them, I had a separate talk with my squad leaders. I wanted to reaffirm to each of them that they are as responsible for the conduct of their squads here at the base as they are for their conduct in the field. I told them that I had a meeting with Colonel Ghetty about last night's antics, and he was displeased. I also mentioned how Ghetty was spoken to because of our platoon's actions. It all rolls downhill, and I made sure that the lowest level commanders felt the momentum.

As we were walking out of the barracks, we got a welcome surprise. Bob's back. My right hand man, who'd been shot in the ass almost three weeks ago, is back. He got the mandatory razzing about his wound, and we all had a lot of fun about it; including Bob who I thought would be tired of such comments by now. He does walk a little slower than the rest of us, and now he sort of has a John Wayne-style strut. Everyone is very glad to have him back.

I told Bob about how the rest of the fight went in the valley, and about the happenings here on the base. He had seen pictures on the news from the valley, but he said that none of them illustrated it quite as well as my description. When I told him about the tear gas fun last night, he almost hit the floor laughing. There was a good deal of disbelief until Chuck Boyd, the platoon computer nerd, showed him a picture that he took with his digital camera. The picture was a little dim, but its composition was perfect. Chuck took it at the precise moment Peter came out with Amy over his shoulder, with a pile of drunken and stoned bodies beneath him. We all had great ideas of what to do with the picture, but Chuck has already put it up on a web site he made for the platoon. Now anyone in the world, including our Chinese counterparts, can view the picture and find out all about our little unit's mischief.

It's good to have Bob back.

Saturday, July 16, 2011

2nd Lt. John Chamberlin-Personal Log

Cam Rahn Bay, Vietnam

With Bob back in the platoon, it was almost inevitable that there would be another party. I was over in the women's quarters with Bob and my squad leaders. Amy gave us all beers that were leftover from the previous nights. A few more people from the platoon came over and also grabbed some of the left over beer. We sat around talking about the Valley and the parties here on the base. I didn't want it to happen, but at about the same time that I felt that the little gathering was turning into a party, Colonel Ghetty showed up with two other Colonels and a couple of flats of beer. Rank has its privileges. Hector, Bruce, and the remnants of their platoons showed up a few minutes later. More and more people I had never seen before started to show up. In no time, the barracks was shoulder-to-shoulder. Again, someone brought a radio, and the music changed the atmosphere into a party. Ghetty had thrown caution to the wind and created a last hurrah.

It went from a small group of people in the platoon sitting around talking over a few beers to a monster party. The number grew and grew. The people that were coming from around the base soon couldn't get in, and they just started hanging out in the street drinking beer and joining similar gatherings at nearby barracks. Everybody on the base seemed to have heard that there was a big party going on, and they all wanted to be part of it. It was immense. With all the different uniforms from all the different services from a dozen countries, it was like a military Mardi Gras.

People drifted off late in the night, but the party didn't end until the mess hall opened. Then the crowd moved over there. Everyone feasted and got fresh beer, then it was over. When I left the mess, it was still crowded. It was shoulder-to-shoulder in some places, but on the other side of the door, out in the street, the base was quiet. It was business as usual. I went back to our barracks and passed out in my rack. I was too tired to do a bed check, and there was no way I expected Hector and Bruce to hold a normal morning briefing.

When I woke up, it was almost 1300 hrs. The entire platoon, though in a complete daze and operating in slow motion, was gathering their gear. They were nice enough to let me sleep as long as possible. Colonel Ghetty himself had come by before noon and gave us our marching orders. We had to be down by the docks, ready to catch a ride back to the Wasp by 1530 hrs. It took some work, but I got my gear together in time. Bob did a roll call, and we were waiting by the dock at 1515. Hector and Bruce brought their platoons by a few minutes later. At least everyone had the same hungover look about them; everyone except Colonel Ghetty, of course.

In the Black River Valley, I noticed that the months of battle had taken their toll on his appearance. Today, after several nights of heavy celebrating, Colonel Ghetty was wide awake and in spit-shine condition. He looked like new! I felt even more sick than I already was.

SECTION 6

"Warfare's gone digital, but mud is still mud to all us grunts."

Sunday, July 17, 2011

2nd Lt. John Chamberlin-Personal Log

10nm off the coast of Cam Rahn Bay, Vietnam

Well, we're back on the boat again. The Wasp has been through quite a few small air and gunboat attacks since we left. They've moved a few hundred welders, electricians, and other repair teams on board. Our quarters have been taken over by some electricians from New Orleans, Louisiana.

No one told us where to bunk, so we're stuck living on the rear well deck. The entire company is packed in here. We get all the noise of being on the flight deck, but it's all compacted into half the space and kept inside the ship instead of being let out into the open sea air. This kind of thing would never be done in peacetime, but the ship's loaded with over twice the normal capacity of people and barely half of the vehicles we'd like to have. Everywhere you walk, you have to step over, under, or squeeze past somebody else. They've set up MRE distribution tables outside of the mess hall and head so that if we can't get into the mess, we at least get to eat something. I hope we're not here too long.

Monday, July 18, 2011

Correspondence

From Corporal Matthew McKenna,

U.S. Army, 1ˢᵗ Infantry Division, 12ᵗʰ CHEM To Jonathon McKenna (*Grandfather*) Yo Gramps!

I called mom yesterday, and we talked for almost 15-minutes. I'm sure she was surprised to hear from me, but there are telephones all around me, so it wasn't that big a deal. It sounds like she's doing okay. Sorry to hear about your heart. Mom says they put in those stints all the time now. She told me not to worry too much. I won't. You're a hardass, and we all know it, so I'm sure you'll be fine. Good luck with it anyway.

I know mom's worried about me over here, but it's not at all like you see in the movies or on the news. After all that work going through basic training and AIT (Advanced Infantry Training), they assigned me to this CHEM unit. I signed up to fight, but instead, I'm learning to take water, soil, and air samples, install wells, and place electronic sensors. I tried to convince mom that it's about as safe as it can be, but I don't think she believed me.

The main job of my unit is to detect chemical attacks, but we also handle all the environmental tasks. That means that we put in wells and record the rise and fall of the water table to detect enemy tunnels being dug under the lines.

We also place and monitor all the electronic sensors in our division's area. It's a big job, and it's incredibly boring, but at least I'm not stuck pushing paperwork across some desk. So far, all of the fieldwork that we do is still in the rear. There's just no reason to go forward when the sensors are already monitoring what the Chinese are doing behind their own lines. Most people just see us as just another group of REMF's (Rear-Echelon Mother Fuckers).

I'm sure you remember those types from your Vietnam days.

I better go. There's lots of work to do, and all of it's boring. Hope your stint goes well. Say hi to everyone, and try to make sure mom doesn't worry too much.

Take Care,

Matt

Tuesday, July 19, 2011

2nd Lt. John Chamberlin-Personal Log

Somewhere of the east coast of Vietnam

The ship began moving out last night. People are still coming by helicopter. I was a little surprised to see so many U.S. Army infantry coming on board. The R.O.K. expatriates have a lot of Marines in their cadre. I wonder why they weren't brought on board instead.

I would have liked to have them. Those guys are tough as nails. They're the types to chew glass and spit sand. I don't have such nice things to say about our own U.S. Army, though. Back on the Black River, they didn't seem to really have the stomach for a fight, maybe due to shortcomings in our wartime training back home.

I can't compare it to the wartime training that the Marines are getting since I haven't gotten a single replacement yet! Someone in the chain of command is not getting the news about our company. We're barely at 40% strength. Of course, if I get replacements that are as poorly-trained as the Army's troops were at the Black River, I don't want them!

Wednesday, July 20, 2011

2nd Lt. John Chamberlin-Personal Log

Somewhere of the east coast of Vietnam

There was a general quarters alarm this morning. The scuttlebutt says that the Chinese have been trying to get a squadron or two through the Task Force's defenses for the last couple of days now. It should keep us on our toes, but when the alarm sounded, the only people that really moved to their action stations were the sailors and some of the old hand Marines. Our entire platoon hustled to get through the crowd and make their way to the mess only to find that it's someone else's action station now.

When we were hustling, the rest of the Marines and soldiers around the ship seemed to just hang out. There were a lot of nervous and confused faces, but no one had been given any instruction on how to handle the alarms. The new guys don't seem to have any fire or damage control training. This ship seriously lacks discipline.

When the alarm ended, Bruce, Hector, and I found Colonel Ghetty. We talked about the problems we saw during the drill, and he was as surprised as we were. He's going to make mention of it at tonight's commander meeting, if they still have those. Ghetty's the kind of guy that if nothing gets done about the lack of alarm preparation; he's going to take matters into his own hands. It'll be awkward for the Navy to accept the fact that a Marine Colonel will be telling people what to do during an attack, but if they've got a problem with him, they'll have to sort it out at the Colonel level or higher. That's where decisions get made, and things happen.

Thursday, July 21, 2011

Correspondence

From Corporal Matthew McKenna,

U.S. Army, 1st Infantry Division, 12th CHEM To Jonathon McKenna (*Grandfather*) Yo Gramps!

I got a letter from mom saying that you already had your heart stint put in! Sorry-I didn't know. I just got your letter. It was postmarked after the letter I sent you, so everything must be going okay. I'm glad. You gotta take care of yourself. Okay?

I didn't know you were in the Big Red One back in Vietnam. What a coincidence! I was talking with unit's top dog. His name is Colonel Dick Saunders (no one dares to joke about his name, though). Colonel Saunders says he thought my name sounded familiar. He was with the Big Red One back in Vietnam-just like you. I doubt if you two ever met or remembered each other. Do you know him?

Colonel Saunders is the head of the environmental surveillance portion of the 12th CHEM. He's a nice guy-for an officer. Sometimes we go out to lunch together. I can't tell you where our headquarters is, but we're outside a pretty big city, and so we can go into town for lunch almost every day. Colonel Saunders sort of keeps the unit rolling. We have a bunch of Captains who administrate the different types of work we do. The Lieutenants run things in the field, and we grunts get to do all the work. One day, I'll be putting in a well with one Lieutenant, and the next, I'm placing field microphones for another. It gets kind of confusing, but Colonel Saunders seems to always know who's where and what they're doing. That's more than I can say for some of these ROTC/George Patton wannabes.

I'm not close enough to see the front yet, but I hear that we'll be doing some well installations closer to our lines in the next few weeks. Right now, I can see the smoke coming from the valleys to the north, and I can hear the artillery, bombs, or mortars. Sometimes, we even hear a faint echo of gunfire, but it's unusual. I'll be careful when I do go closer to the lines.

Don't bother to tell mom about our trip toward the front, she'll just worry.

Say hi to everyone, and take care of your ticker!

Take Care,

Matt

Thursday, July 22, 2011

Correspondence

From Corporal Matthew McKenna,

U.S. Army, 1ˢᵗ Infantry Division, 12ᵗʰ CHEM To Jonathon McKenna (*Grandfather*) Yo Gramps!

I'm okay, but I thought I'd let you know that we had a little accident today. One of our Lieutenants was supervising a drilling operation.

See, before you can put in a groundwater monitoring well, you have to drill a hole in the ground. It's usually about 12"-24" wide and about 50'-100' feet deep. Then we put in some special PVC pipe called "screen" with a plug glued on the bottom end. The PVC has thousands of slits cut in it to let the groundwater flow into it. Most of the time, we have to screw a bunch of screen pipes together since they come in 10' lengths, and the holes are usually much deeper. After the pipe is in place, we dump in small gravel to fill the space between the pipe and the walls of the hole.

A few hours later, groundwater has seeped into the pipe, and we can take a measurement. We try to check the groundwater levels in all the monitoring wells every week. That gives us enough data for a good computer model. Then we watch the levels, and when the computer notices that some of the wells in the area are lower than they should be, we know that the Chinese might be digging a tunnel in the area closest to those monitoring wells.

Well, this Lieutenant was not very sharp. Most of them have a lot to learn when they come straight from ROTC or even West Point, but they usually seem to come around and learn how things are done. We have one that's a real moron, and two that are on power-trips. Out of fifteen Lieutenants and four Captains, that's not too bad-especially with the crunch that the war's put on training them back home. So this one LT has his face right next to the ground and right next to the drilling auger while it's cutting into the mud. It hit a small rock-they call them cobbles, and the auger started to jam. For some stupid reason, this LT tried kicking the auger free while it was still under tension. Somehow, it actually freed itself from the rock. It probably cut through it. Well, when the resistance from the rock was gone, the auger took all its tension and jerked back into digging deeper. The LT's foot was right there, and it came right off.

I saw the whole thing happen. Everyone tried to tell him to stand back, but he was too bossy to listen. Now he's going home without his right foot. It's stupidity at its worst.

It was also kind of sick. When we pulled the auger back up, his foot had been ground up like hamburger in a boot. Blood from his leg was still everywhere, and we were all throwing up for at least an hour. It's a sad thing when someone gets hurt, but now I think about something else whenever I hear about people getting wounded. I think about the guys who have to clean it up.

I guess you don't have to be in the trenches to have injuries. Definitely do NOT mention this to mom. I only mention it to you because I know you've been there. We've always tried not to talk about it before, but now I know why. Unless you've actually seen it, you can't understand it, and you can't stop thinking about it.

Take Care,

Matt

Sunday, July 24, 2011

2nd Lt. John Chamberlin-Personal Log

Somewhere in the South China Sea

I was right about Colonel Ghetty and the alarm training. They are still having commander meetings twice a day, and he brought up the lack of preparation last night. From what I gather, he was told that if he thought there was a need, then he should be the one to address it. With Colonel Ghetty, that's not being passed the buck, that's being given a green light.

The well deck is always the noisiest place on the ship, but this morning, Ghetty stood up on an Army M-1A3 Abrams tank with a megaphone, and he began his campaign of emergency organization. All officers, including myself, were called up individually to report their unit's condition, their place in the chain of command, and to report on what their unit is supposed to do during an alarm. That took most of the morning. Then, he put together a makeshift emergency staff from the different services and began drawing up a training schedule. By lunchtime, Colonel Ghetty had volunteered a seemingly lost U.S. Navy Lt. Commander to conduct fire training for the hundreds, maybe thousands of new Marines and soldiers that had arrived on the ship. The Lt. Commander protested, but it was no use, Ghetty had been given the green light from Captain Blackwell at last night's meeting. Any trouble, and the green Lt. Commander was given a good description of how fast he could see himself on an Osprey headed for a new assignment in the jungle. Colonel Ghetty can really make things happen when he wants to.

Monday, July 25, 2011

Correspondence

From Corporal Matthew McKenna,

U.S. Army, 1st Infantry Division, 12th CHEM To Jonathon McKenna (*Grandfather*) Yo Gramps!

Thanks for the letter. I knew Vietnam was ugly, and it sounds like you saw a lot more than a simple injury. I'm not having too much trouble sleeping right now. Maybe it hasn't set in yet. Maybe it isn't bad enough to stick with me as a war memory. It's probably that I'm just so tired! From the time I get up at 5, go through Colonel Saunders' PT routine, get some chow, get to a monitoring site, do my rounds, go to three other sites, and then get back to the Operations Center, I've already been at work at least 14 hours. Then I have data to submit, gear to clean, and only then do I get to grab some more food.

We're eating MRE's at the sites to save time. They're not too bad. Some people hate 'em, but everything I've had so far has been okay-as long as you use the hot sauce in the packages! They do get to be kind of dry and filling, but we burn calories like there's no tomorrow. It's hard work. It's 100% humidity because it rains at least part of every day, and a day in the 90's is like a cold spell.

I guess you can see by the postmarks on my letters that I'm in Myanmar. You might know it by its old name: Burma. They changed the name back in the 80's, I think. New governments like new names.

They tell me that it's a lot like Vietnam. Most of the country is heavy jungle, mountains, flood plains, rivers, and rice paddies. Except for the war, the place is a paradise. There are so many different plants and animals that it sometimes seems like another planet on a sci-fi movie.

The people are incredibly diverse too. There are real Indian tribes in the high hills. In the valleys and along the terraced steppes, there are dirt-poor villages. Closer to the cities there are suburbs just like back home. The cities are HUGE! I thought they'd be a joke, but there are tall glass-lined skyscrapers, highways, airports, even fast-food restaurants (thank God!). I wish the war was someplace else. It would make this place perfect.

Even the rain seems to leave around noon. Okay, it does rain everyday, and that does wear on those of us who have to work in the mud, but it only stays long enough to water the plants, and they're incredible. I actually saw a purple and red jungle the other day. There were no green plants, just trees, vines, and bushes that were purple and red. There are flower farms that do nothing but grow stock for florists, and it makes the air smell sweet for miles around. It's worth the rain to see those kinds of things.

Poking out of the thick, triple-canopy jungle and the mirror-like steppes of rice paddies, everywhere you look, you can see the gold-covered spires of temples. They're thousands of years old, covered with real gold, and they never seem to tarnish. At sunrise and sunset, they reflect the sun while it's changing colors, and it's like something from a fairy tale. It's also weird seeing these ancient temples and brand new glass skyscrapers sitting side by side and reflecting a red-orange sunset!

Maybe the reason that the LT's injury isn't bothering me too much is because I'm surrounded by this incredible scenery? Well, I'm gonna get some rack time while I can. Take care of yourself, and say hi to everyone.

Thanks,

Matt

Tuesday, July 26, 2011

2nd Lt. John Chamberlin-Personal Log Somewhere in the South China Sea

We came under another Chinese air attack this morning. They always seem to hit us. When the General Quarters alarm went off, things were different. Our Lt. Commander has only had two days of training for all of us here in the well deck, but things have definitely changed. Instead of no one knowing where to go, or what to do, thousands of people, Marines, soldiers, sailors, and civilian workers dropped whatever they were doing and ran full speed to their action stations. It looked like complete chaos, like a riot without fighting, or like a free-for-all track race with no lines and no direction. Still, it only lasted a few seconds. In under a minute, everyone was where he or she was supposed to be.

Our new action station is on the catwalk at the edge of the flight deck by landing spot #9 on the aft port corner of the ship. It's a long way to go from where we are on the well deck, but at least we're outside, and the weather is balmy. If we were off the coast of Alaska, I might feel differently. The view is beautiful too, and we really get to watch a lot of action since we can look all the way up the length of the ship. As locations go, I think this is the best. Lots of people, myself included, come up here just to look out at the scenery for something to do. I think Colonel Ghetty knew that and maybe that's why he gave the nicest location to one of his own commanders. I'm only surprised that Bruce or Hector didn't get it.

Bruce is farther forward by landing spot #7, and Hector is just below him at the aft port aircraft elevator.

Once again, the Chinese turned back as soon as they let go of their antiship missiles. As usual, there weren't enough missiles launched to get through the wall of SAM's put up by our Task Force. There's still a lot of work to be done as far as training. I don't think any of the new people will stay calm for a long when confronted by the black smoke that a shipboard fire generates. We'll need to train more realistically or at least a lot more often before they react under such pressure. These small-scale attacks are bearable, but the tension, the speed, and the lethality of those major engagements are a lot to handle.

Thursday, July 28, 2011

2nd Lt. John Chamberlin-Personal Log

Somewhere in the South China Sea

We had a general quarters submarine alarm this afternoon. There was some confusion in the well deck when the alarm sounded. Lots of people didn't know if there was a different procedure to follow. We just went back up to the catwalk, and watched the scene. Helicopters from all over just seemed to appear out of nowhere and started circling off to the north. Over the next few hours, their search pattern narrowed and came closer. We watched them drop sonobouys all over the place.

The buoys dropped into the water, then sank to a designated depth. A float connected by wire to the buoy can transmit audio signals from the water's depth to the helicopters above. Joined together by digital network, the helicopters can receive data and sounds from any buoy dropped by any chopper. After enough time and enough sonobouys, I can't see how a sub could survive.

Of course, add to that the silent unseen war under the water as the enemy sub also had to evade any friendly subs accompanying our Task Force. I'm sure there's more than a few. I heard that every carrier battlegroup always has at least two Los Angeles class attack subs escorting it. Our Task Force currently has five carrier battlegroups in it, plus the remnants of three others. We've also got a few smaller Task Forces of surface ships from South Korea and Taiwan escorting our main Task Force. No one will tell us exactly how many ships are out here, but the rumor is that there's well over a hundred.

Monday, August 1, 2011

Correspondence

From Corporal Matthew McKenna,

U.S. Army, 1ˢᵗ Infantry Division, 12ᵗʰ CHEM To Jonathon McKenna (*Grandfather*) Yo Gramps!

Thanks for your last letter. It sounds like you enjoyed Vietnam's scenery as much as I do this place. There's a rumor that we're sending troops in there too. I don't think they'll redeploy our division though, and we're so dug in along the frontline that I don't think they'll transfer our unit to another one.

Right now, the Big Red One needs us, so they'll probably keep us.

There isn't too much new over here. I still haven't even seen the frontline. The only way I know how the fighting is going is by the number of Medevac flights going back and forth over the hills to the trenches on the other side. We have had a few air raids, but when the sirens go off, we take cover, and then there doesn't seem to be any damage when we come back out. I'm doing a lot of well monitoring. We think we might have found our first tunnel. Tunneling under trenches and fortifications goes back before castles. They think it's one of the ways that the North Koreans were able to come through the DMZ so fast. After all, the South Koreans have been finding North Korean tunnels since the end of the first Korean War-some big enough to drive two tanks through side-by-side! Well, now we think we found one, so we've set up six more monitoring well sites with ten wells at each site. It'll take a few days to determine the normal groundwater level in all the new wells, but when we've got it figured out, we'll narrow down the location. Then, we get to call in the Air Force and its deep-penetration, bunker-busting bombs! That should be a real show!

Tell everyone that I'm doing well and that I'm safe. Take care of yourself too. How's that stint doing? Do you feel like you've got a lot more energy now?

Take Care,

Matt

Tuesday, August 2, 2011

1st Lt. John Chamberlin -Personal Log

Somewhere in the South China Sea

Colonel Ghetty had some promotions and awards to pass out today. Hector, as expected, was promoted to Captain and officially given command of the company. Bruce's platoon will absorb Hector's, bringing that platoon back up to full strength.

I was made a 1st Lieutenant and given a few commendations for valor. There were a few promotions among the enlisted personnel as well. Everybody wound up with something. For example, we all received commendations for our actions on the Wasp at the Battle of Korea Strait, and most of us were given commendations for our actions in the Black River Valley. I don't think we earned the Korea Strait commendations just for shooting our guns in the air, and the Black River Valley didn't make many heroes; just survivors. Some deserve much, much, more, and others, well...somewhere in the chain of command my after-action reports might have been misinterpreted or even ignored. All total, my platoon received: 6 promotions (including mine), 115 Battle ribbons, 69 Operation ribbons, 71 commendations for valor, 8 Bronze Stars, and 15 purple hearts; 12 posthumously.

It was a good day for everyone. Even those who are no longer with the platoon were alive and well in our memories.

Wednesday, August 3, 2011

Correspondence

From Corporal Matthew McKenna,

U.S. Army, 1ˢᵗ Infantry Division, 12ᵗʰ CHEM To Jonathon McKenna (*Grandfather*) Yo Gramps!

I'm glad you're feeling better with the stint. Should you really be back to work so soon? It seems like most people your age are golfing-not working! Just make sure to take it easy, Okay?

We're close to finding that tunnel now. The computer has it narrowed down to a mile wide area, but it needs more data. So, we sank a few more wells, and in a few more days, we should have it triangulated. We're also putting some U.S. Navy sonobouys into a few wells. They're supposed to be really sensitive, but no one knows how well they'll do underground. In the ocean, their technical specs say that they can pickup metal-on-metal sounds at up to 15-miles. We've got them within 2-3 miles of the area where we think the tunnel is, so we might be able to pickup something.

It's nice working with new technology and different services. When I was first assigned to this unit, I was disappointed that I wasn't going to fight. Now, I'm glad I'm here. I like what I'm doing, and I'm one of the best we've got-much better than most of the officers. Colonel Saunders says that whoever finds the tunnel will get a meritorious service medal and a week in Bangkok. I wanna be that guy!

I've also started to get a feel for the frontlines, and I'm not as eager to go up there now. There are a lot of medevac flights going over those hills lately, and no one tells good stories about the trenches. I'm still anxious to see them, but I have a new respect for my REMF duties. I'm lucky to have them, and I know it.

Don't push yourself too hard. Say hi to everyone. Tell mom I'm okay. Try to stay away from all those commando attacks we hear about happening back home, okay?

-Matt

Thursday, August 4, 2011

Correspondence

From Corporal Matthew McKenna,

U.S. Army, 1ˢᵗ Infantry Division, 12ᵗʰ CHEM To Jonathon McKenna (*Grandfather*) Yo Gramps!

We did it! We found the Chinese tunnel that we've been looking for! Colonel Saunders-the guy that says he remembers you from Vietnamcompiled the computer data from all our groundwater monitoring wells, and he narrowed down an area where the tunnel might be. The U.S. Navy Lt. Commander who brought the sonobouys did the same thing with their recordings. They went over some geologic maps and determined where the best place to dig a tunnel was, and then they called in an airstrike.

It was great! We all sat around our HUMMV's and drill trucks and waited for the Air Force to come in. About an hour after Colonel Saunders called in the strike, 2, F-15E Strike Eagles called him back on the radio. They were too high to see, but he gave them the longitude and latitude coordinates, they typed the data into their bombs' computers, and then they dropped 'em. There were 4, 5000 pound bombs, and when they hit, it seemed to lift the entire hill! The dust and debris from the explosions covered the hill in a few seconds, and as the dust started to blow away, the hill seemed gone. Finally, we could clearly see all four craters. Each was about 100' across, and maybe 50' deep. Where the bottom two craters connected, there was a third, smaller crater that Colonel Saunders says was the caved-in tunnel. I was worried that some of the Chinese might come pouring out, but he says that until the tunnels are complete, the only people inside were probably engineers, and none of us have any doubt that they had to have been killed in the airstrike. Later today, a company from the 9ᵗʰ Engineers is going to go over the craters and search for the tunnel. I could see it clearly through the binoculars, so I don't think there will be any doubt about finding it.

What about the week's R&R in Bangkok? It's typical. The information that finally helped Colonel Saunders triangulate the tunnel came from the Navy Lt. Commander and his sonobouys, so he gets to go. Saunders could tell that we're all unhappy about his choice, so he's recommending us all for meritorious service ribbons. It's not much. It's not cash, and it's not a week in Bangkok for sure, but at least we get something out of the deal. Even without the medals, everyone is thrilled that all the drilling and monitoring finally paid off. A lot of people in the division are looking at us a little differently now. We're not the same REMF's that the accountants and logistics thieves are anymore. It's nice having the respect of our peers.

How's everyone back home? We heard that the FBI made a big break in the terrorist/commando thing. Were all those attacks really caused by terrorists, or was it Chinese commandos like everyone's always thought? Say hi to everyone for me!

Thanks,

Matt

Thursday, August 4, 2011

1st Lt. John Chamberlin -Personal Log

Somewhere in the South China Sea

After yesterday's awards, my platoon and Bruce's platoons walked around the ship like superheroes. There were plenty of people who had more awards and much more impressive war records. However, none had to wait as long as we did. Since we didn't have any dress uniforms, we all had to wear our new rank and medals on our BDU's. We looked extremely impressive. So much so, that the green Army soldiers on board barely even teased us about the look. Normally, we tease them, and they tease us about anything and everything. It's an inter-service rivalry that helps everyone try to be their best. There was no disputing it, we were royalty yesterday.

With that in mind, we had to pay the piper today. General Shapiro had taken a little heat from Captain Blackwell about the way we looked. Blackwell always seemed to be a level-headed officer, but I think we might have made his crew look less than appreciated, and they do deserve all the appreciation they can be given. This ship has been through hell, and its one of the few ships to have gone all the way from the Battle of Yellow Sea to this point in the war. She's taken her licks, and the crew's seen her through. They say it rolls downhill, and since we're at the bottom, our two platoons were ordered from the top to swab the flight deck. There were very few flight ops today, so we had at it.

Usually, a few hundred sailors get the pleasure of making sure that the flightdeck is clear of loose debris. Today it fell on a little more than fifty Marines. We started at the bow after morning PT, and by lunch, we hadn't even reached the island amidships. The sailors were loving it until Blackwell saw what had become of his previous day's comments. Maybe he felt sorry for us, or maybe he just didn't like the way the sailors were grinning at the effort we were being put to. Maybe he just wanted to make sure it got done. In any case, after lunch, Blackwell himself came down from the island, grabbed a trashcan, and joined us. A few other sailors followed his act, and soon there were hundreds on deck.

It wasn't just a menial task. If so much as a loose bolt gets sucked into a Harrier's engine, the fast moving intake blades can chip and shatter. If one blade breaks, its pieces can break into two more, and so on until it comes apart or explodes or both. The deck was clean and clear in less than an hour after Captain Blackwell came down to help. As much competition and teasing that goes on between sailors, Marines, and soldiers, there's a strange sibling-like bond that joins us all. Whenever we see one in trouble, the others do what they can to come to their aid. I think we all just like being part of a team that can take share pride when things are getting done. It works on the ship, and it works in combat.

Friday, August 5, 2011, 8:33pm

Isle of Grenada

T.K.'s Tequila Kafe'

Alexov waited at a small curbside table with three empty chairs. Around him the small bar and grill was only half-full. Of the people who were dining and drinking, at least half looked like locals. The late summer heat tended to be the slowest time of the year for tourists. There were the occasional cruise ships, but they always came and went in 48 hours or less.

In front of him, Alexov was on his third bottle of vodka. His plane had landed ahead of schedule, and waiting in the Caribbean sun had left him thirsty. Only his tolerance kept him completely vertical.

Out of the back of the café, his waiter approached with a tray of appetizers and margaritas for another table. In his right hand was another bottle of vodka for the Russian. Alexov smiled and wondered which he would see first, the bottom of the fresh bottle, or Walt who had called the night's meeting? While focusing on the bottle, Walt stepped out from behind the waiter.

He reached around the waiter's head and stole one of the margaritas. Raising the glass in toast, he welcomed his co-conspirator.

"Alexov! It's good to see you. Here's to your health!"

"Walt, I was beginning to wonder if something had happened to you."

"I'm sorry, my plane was late, and it looks like yours must have been early. Look at all of these dead soldiers! What time did you get in?"

"We landed just after 1 o'clock, but the time change and the jet lag are taking their toll. Last week I had a meeting in Halifax, then another in Vladivostock two days later. I can't tell if I'm coming or going right now. How about you?"

"We landed about half an hour ago, but there was some confusion at the airport about our flight plan. It was really just one of those unexpected cash tolls that you expect from a small airport like this one. We had a little rough weather over San Salvador too. I apologize for not sending word, but it looks like you've made yourself comfortable. Anyway, Naht Synn isn't due in until 10:00. I would expect him to be late given the difficulty in getting here, but he's a very resourceful man, so… who knows. Maybe we'll see him a little early?"

They both tapped glasses and drank. Alexov startled Walt with his lucidity, but it was a welcome surprise. They'd both need all their wits about them as soon as the Chinese general arrived.

"Walt, I think we should emphasize our plan as a sort of last-ditch option to our friend. I thought about it, and I've even asked around a little. No one seems to know what their end game is going to be. Ever since Korea, our friends have been running with no end in sight, and no idea of how to find one. It's like once they were let out of the gate, there was no stopping them. In fact, there really wasn't one until Vietnam."

"You're absolutely right, Alexov. It's not as though they lost any truly valuable assets in that valley. Let's face it, they've got hundreds of millions of troops, and losing a few hundred thousand is like a drop in the bucket. If they had an end game, it was probably a series of tentative objectives like Korea, Formosa, and Indochina for its food and resources potential. Now that the Coalition is making a real fight of it, they'll have to start thinking of exit strategies other than total victory with conventional arms. In fact, I think the only way that they'll be able to achieve a total victory is through unconventional means. Your idea's probably the last choice they'll have, but it is a decent one."

Down the sloping street Walt spotted someone coming up the sidewalk. It was a single figure, an Asian dressed in blue jeans and an untucked plaid button down shirt. As the man drew closer, they noticed that it was General Naht Synn.

In perfect English, with all the exuberance of a long lost friend, he called out.

"Alexov! You old sunofabitch! How the hell are you?! I haven't seen you since school!"

The two men shook hands and embraced while Walt stood and motioned for the bartender to bring over more drinks.

"Walt, I didn't know you and Alexov knew each other. You know he and I went to North Carolina State together? Sometimes it seems like yesterday. Other days it feels like another lifetime! Where have the years gone?"

The small reunion was starting to draw attention. All three men were out of their normal day-to-day uniforms and dressed in casual Caribbean attire, but their volume was unusual.

"Ming, it's good to see you. It's been far too long. Drink with me and we'll toast-"

Naht Synn's attitude had been the complete opposite of the last meeting with Walt. He was jovial and gregarious, but now he started to sound stern again.

"Alex, there's no way I'm going to drink with you. I don't care how many bottles you've already put down, I know how you can be. I'll tell you what. I'll have one to toast with, and that's it. Okay?"

Perfectly timed, the waiter returned and passed out another round of drinks. Alexov was falling behind. He had yet to finish the bottle that he was winding up when Walt arrived. Now there was a fresh bottle and a tall blue margarita in a genuine, brushed aluminum, "TK's Tequila Café" stein. Naht Synn turned and shook hands with Walt.

"Walt, it's good to see you again. I hope there are no hard feelings from our last meeting. I simply couldn't take Dan anymore. Every time he opened his mouth, I felt like that parasite was sucking your blood and trying to get mine too. I hope you're not too upset?"

Naht Synn actually apologized to Walt. It was an incredibly weak apology by anyone's standards. It almost sounded like a neighbor was apologizing for breaking a borrowed lawn mower. Still, it was an effort from a man who never needed to ask for forgiveness. This was unproven ground to Naht Synn. Walt feigned remorse and forgiveness.

"Let's just forget the whole thing and have a good time, shall we?"

The general replied with his impression of someone who was sincere in their remorse, then all three sat down and relaxed.

Still thinking about how he had fallen behind in his drinking, Alexov ignored the glasses, and finished off his first bottle while tipping it back and holding on to its neck with only two fingers. Walt and the general watched in slight disbelief. It was a sight to see. When the bottle was dry, Alexov stood up, threw it out into the street and raised his blue margarita high.

"A toast! To long lost-"

Walt pulled him back into his seat and told him to shut up. Naht Synn's early arrival and his own late arrival had changed the scenario that Walt wanted to create for Naht Synn.

They sat down and tried to look a little more discreet. Tropical tourist islands like Grenada were probably the last place in the world that the CIA might be looking for any of them, but there was a good chance that there would at least be one informant on the island. They all looked around cautiously. The rest of the patrons had gone back to their beverages.

"I'm sorry, you two," Alexov pulled back their attention. "I told you before. Russians don't like this kind of heat. I should slow down a bit. Really though, I'm fine. In fact, I'm better than fine. I'm fantastic."

He smiled, and when Naht Synn grinned back, he knew it was safe to continue.

"Ming, I'm sure you realize that I'm not here by accident."

Naht Synn was serious again. There were no accidental meetings, no accidents whatsoever when Walt was involved. What could Walt gain by reuniting him with Alexov? The answer was on its way. Alexov was coming around quickly, and unless someone had seen the volume of his consumption, they could never guess it now.

"Walt and I met back in '91 when the Soviet Union collapsed. We've kept in touch from now and then ever since. He never wants to lose one of his contacts, I guess. You know how he is."

They all laughed, even Walt who was already finishing his red margarita. His laughter was more because of Alexov's use of Naht Synn's first name, Ming. Walt had never heard it used before, and to have seen it spoken without a gun being drawn was a confirmation of their meeting's good tone.

"I know, Alexov. Walt and I met around the same time. Ever since, I get these calls about new investment ideas. The tips are good, but where I work, investing is sometimes not considered…. politically correct."

Again the meeting broke into light laughter. Naht Synn, who hadn't planned on drinking, relaxed a bit and began working his drink in unison with Walt. Alexov had ceased his vodka and margarita operations, but now it was Walt who would carry the conversation.

"Gentlemen, please. My investment ideas are almost always sound, and I hope you don't feel too pestered. After all, that is why you're both here tonight."

Walt looked around and casually leaned forward a bit to speak just a bit more privately and not draw suspicion. The waiter noticed Walt's look and moved to the closest table. A brief address to its occupants, and they left for a complimentary round of margaritas at the bar. Then the waiter moved to the next table and made the same offer. Again the patrons moved inside. Walt's prearranged signal had cleared the sidewalk portion of the café. There was no hiding the move from Alexov or Naht Synn. Both watched then looked over at Walt in satisfaction. It was a nice and subtle move that impressed them both. Now, Alexov could make his offer in the privacy of the empty street.

"Ming, I've got a pretty good idea of what your people are doing over in that southeastern corner of your continent. I've also been able to find out from some friends over in our intelligence section exactly what Uncle Sam and his friends have been doing. A few weeks ago, an opportunity came my way, and I got in touch with Walt. He said that maybe, if your friends have a setback like the one in Vietnam the other day, that you might be interested."

Alexov reached into his back pocket and pulled out seven pieces of paper stapled together and folded four ways. They were crumpled and tattered from their journey. It was his offer. He leaned back in his chair, unfolded the packet and tried in vain to smooth it out so that it might look something like a professional proposal. There was no hope, so he passed it to his general friend and watched him read it.

The papers offered to transfer 35 Soviet-built SS-20 intermediate range ballistic missiles, their mobile launch trailers, 11 command and control vehicles, 8 fuel trucks, two truckloads of spare parts, and 105, 150-kiloton warheads (3 per missile). The cost: 63 tons of gold or about $1 trillion dollars. It also offered the services of the Star City Laser Research Foundation to assist in blinding the American DSP satellites. Last but not least, there was an agreement to allow all missiles launched from the People's Republic of China to fly over the Russian Federation during a one-hour window. The agreement required that there was a full day's notice to Alexov, and an additional $10 million dollars was provided for incidental fees; i.e. bribes.

Naht Synn read the document. When he was finished, he sat back and looked at Alexov. Seconds passed before he even blinked. The general couldn't tell what was going on. The defeat in the Black River Valley was embarrassing, but it was hardly reason to prepare for the worst. The Chinese were a long way from being defeated. Yet, Naht Synn also knew that if their fortunes didn't change, and they continued to lose in the jungles, then there were no realistic contingencies on the drawing board.

There was also the problem of pushing the Americans too far. Since the first day of the war, the Chinese had worried that the Americans might overreact to a defeat in battle and begin using nuclear weapons. The Chinese arsenal would take years to build up. Until then, they had a pitiful number of weapons that really couldn't be counted upon as a deterrent force. If he agreed to deal, they would finally be able to initiate a serious counterattack to any American nuclear strike.

He also knew that Alexov was probably the only man in the world who could make this deal happen. Others might have access to the missiles, or the lasers, or the Russian nuclear security officials, but no one had connections with all three. No one had the experience to put it all together except Alexov.

Still, there were questions. The lasers were privately operated, so both they and the Russian nuclear security people would have to be bribed. Where would he get the missiles, and most importantly, why would these two risk nuclear war? They weren't bloodthirsty people, at least Alexov didn't used to be one. Walt was a purebred capitalist, so he was definitely in it for the money. There was no doubt that he'd get hundreds of billions in commission on the deal. Naht Synn had to ask.

"Why would you make such an offer, Alexov? You know what this can lead to. I know you do. Is it the money? I'm sure that's what Walt's got in it. Are you in some sort of trouble? I can help you if you are. What's going on here?"

Alexov filled a glass to the top with vodka, flipped it back and down his throat, then threw the glass at a building across the street. He took a coffee mug from an empty place sitting and poured another glassful.

"Ming, do you remember back in February when that Russian Task Force got caught in a cruise missile crossfire? They were out there, in the Sea of Japan, on a routine patrol, and those North Korean idiots fired over them at the Americans. Then the damned Yankees decided to fire blindly down the direction that the missiles had come from. Do you remember that? Well, let me tell you something. My son was on a Krivak class destroyer at the time. He's at the bottom of the sea now thanks to those bastards. I thought about signing up with the Russian volunteers who have joined the Eastern Alliance, but that won't give me enough satisfaction."

Alexov was getting bitter, and he took another drag from his coffee mug.

"A few months ago, I met an old buddy that worked with me when I was on duty with the Dead Hand system back in '91. He told me that the old SS-20 replacement program was finally getting funding again. The new SS-28's are already on line, but they haven't decided the cheapest way of getting rid of the old missiles and all their associated hardware. A couple days later, I found out a little more about all the hardware. Around that same time, I got the usual investment opportunity call from Walt. That's when I thought up this whole crazy scheme. I think it's about time someone takes those damned Yankees down a bit."

Naht Synn looked at Walt who was still and poker-faced.

"Alexov, this is a lot of money. I don't think I can just pull this out of my own budget. I'd have to go through channels, and then everybody would find out about it."

Walt finally spoke up.

"Actually general, I had a thought on that. If you could simply arrange for a few signatures from some specific people, I can arrange for some documents to be drawn up that would authorize the transfer of gold from other nations into private banks. You see, the gold doesn't even have to be transported or anything. I just give you a list of names, the appropriate forms, and you convince these people to sign portions of their liberated nation's gold over to a list of accounts that I provide. I imagine that a Special General in charge of Internal Security and Counter Espionage might have a gift for encouraging people to do things like sign things they ordinarily wouldn't want to sign."

Naht Synn thought about it and shared a serious glance with each man. Time stopped as they waited for his answer.

Finally, the waiter brought more beverages. When he left, the general looked around to see if anyone was listening. Then came his answer.

"Walt, I'll need to review the documents, and the list of names before I can commit to anything, but I think this is a pretty good safeguard. Have the papers in my office first thing next week. Alexov, I'm sorry to hear about your son, but you need to understand that if I do this, it will be so that we do NOT have to use them. If we do have to use them, then…I can't even think it. We're too close to that already. Just get me the papers and I'll consider this whole issue. I'll consider it all very, very seriously. Right now, it looks like we'll have a deal."

All three sat still. There was nothing except the sounds of the bar inside, and the Gulf of Mexico down the hill. The rush of the waves equaled the white noise in their minds.

Saturday, August 6, 2011

1st Lt. John Chamberlin -Personal Log

Somewhere in the South China Sea

Today was the anniversary of the first atomic bomb attack on Hiroshima back in 1945. The Chinese reminded us by trying to take out our Task Force. A well-coordinated submarine, missile frigate, and air attack brought about our usual morning general quarters alarm. Everyone in the well deck went to their stations like we had in the past, but we all could tell by the voice on the 1MC that this was more than the usual air raid. When the ASW (Anti Submarine Warfare) helicopters left the ship, we knew the sub threat was close.

The sea was relatively calm. The water was unusually clear, and the sky was hazy, but from our station, I could see all the way to the horizon. Through my binoculars, I could barely make out our distant escort ships miles away launching SAM's at unseen missiles and planes. At the same time, four Harriers from our ship took off to tangle with either missiles or Chinese planes. We heard the familiar thunderclap sounds of missiles being shot down somewhere in beyond the horizon.

I watched the Harriers fly towards the west, off our starboard side. They were still clearly visible when they let go of a few air-to-air missiles. Their targets, apparently some anti-ship missiles, closed on us at hundreds of miles an hour. Even the slowest missile travels at three or four hundred miles an hour. I watched the Harrier's missiles leave long white strings of smoke. One of the strings ended in a black puff. A few seconds later I heard a double boom announcing its destruction. The Harriers fired two more pairs of air-to-air missiles, but the long white strings lead to nowhere.

Then the SAM launchers on the Wasp went to work putting thin white lines of smoke into the sky. Our missiles, four of them, arced and went to starboard. They went all the way to the horizon. Another set of four went up and over. Two ended in orange flashes and black puffs, this time much closer to the ship; maybe five or six miles away.

I still didn't see any targets, but the 20mm phalanx turrets began firing. Their rate of fire left a ripping, tearing sound, and a flood of brass casings poured out of the guns. It continued until there was a huge explosion and splashes of debris kicked their way toward the ship from maybe a mile or two away. The 1MC sounded with warning, "BRACE! BRACE! BRA-!" An explosion amidships interrupted.

We took a hit dead center on the starboard side of the island. Its concussion moved the ship a few feet to port, and everyone fell to the deck no matter how hard they were braced for the hit. I looked to make sure no one was knocked overboard, but they were all still there. When I looked up and across the deck to see how badly we were hit, some sailors were already fighting small fires, and thick black smoke was making its way high into the sky. We all hoped it was over, but two more Chinese missiles hit the starboard side in rapid succession. There was no warning to brace, and everyone was thrown to the deck again. The sailors on the deck managed to hold onto their firehoses, and when I got back up to survey the scene, they were already on their feet fighting the fire, like we did the Chinese at the Black River. This was their hand-to-hand combat. Lucky for us all, there were no more missiles headed for our ship.

The Wasp continued moving at speed, so I knew we weren't too badly hit. I wanted to send the platoon up to help out the sailors, but soldiers from below deck came up with more hoses, and I didn't want to put too many people on the job. Too many cooks might ruin the meal, and the same type of effect might be felt fighting the fire. Bruce did the same, and Hector didn't give us an order to help out, so I'm pretty sure it was the right thing to do. The fires on the island took about half an hour to contain.

Below decks, Colonel Ghetty's rag tag makeshift damage control and fire parties really went to work. The Army and the new Marines on the ship swarmed over fires. They beat them out with blankets, mattresses, ponchos, fire extinguishers, and the few hoses that weren't already assigned. There were a lot of 1st and 2nd degree burns among the thousands of people in the well deck, but it was nothing that field dressings couldn't handle. All total, the ship somehow managed to only lose 15 people. There were some serious injury victims, but they'll be well treated in our first-class med center. Everyone on the ship has gone through basic medical training, and there were plenty of medics to give first aid too.

The civilian welders and electricians were running new wires and cutting away damaged steel before the fires were even contained. It was something to see. A few months ago, solid hits like that almost put this ship underwater. Today, they didn't even slow us down.

The attack continued, but the remaining missiles went after other ships in the Task Force. I watched eleven plow into a little frigate about ten miles off the port side. It just took hit after hit. The first one or two must have blown it apart. I couldn't see through the black smoke and orange fireballs. Then another hit, and another hit. It was awful to watch. When the smoke finally cleared a minute or two later, there was nothing-not even debris left on the surface. If I had to guess, I'd say that the wreckage from the first two hits probably looked like a huge ship in the Chinese missile's radar. They must have locked onto that, then hit, made another explosion of debris, drawn in more missiles, and so on until there were no more missiles to attract. If those missiles would have gone past the frigate and hit on a carrier, or an amphibious ship like this one, thousands of lives could have been at lost.

No sooner had the air attack slowed down than another frigate, this one a few miles behind us, blew up. I happened to be looking at it when it happened. I wanted to remember what the other frigate had looked like, then this one just lifted right out of the water 10 or 20 feet! The middle of the ship went up, and then came right back down, snapping the ship in two under its own falling weight. It's stern rolled over on its port side and slipped below the surface in seconds. The bow slowly rose and pointed toward the sky at about 45 degrees. Then another explosion underneath it, and the bow was blown off. Helicopters headed in from all directions.

Even with the sky as clear as it was, I hadn't seen them a few minutes before. They all started dropping sonobouys, and I actually saw one drop a torpedo with a parachute on its tail end. Then another torpedo was dropped a few miles farther behind us. The missile attack happened so fast, and then this submarine attack seemed to take forever. I never saw if they had gotten the sub or not.

A few hours after the fires were out, and the 1MC was brought back on line, we were allowed to stand down from general quarters. I think I might just bring a sleeping bag and some tarps and just re-station the platoon up here. It's a lot more comfortable, and since we don't have a full complement of aircraft anymore, the flight deck isn't anywhere near as noisy as the well deck.

Sunday, August 7, 2011

Correspondence

From Corporal Matthew McKenna,

U.S. Army, 1st Infantry Division, 12th CHEM To Jonathon McKenna (*Grandfather*) Yo Gramps!

I called mom again yesterday, but she wasn't home. I'm not sure what time it was back home. It might have been the middle of the day, or the middle of the night. It gets confusing because of my location, but I can't say much more. I'm sure everything's okay with everybody back home. I just wanted to let her know that I was doing well. Please let her know for me.

I got your letter about the FBI roundup. It sounds like everything is finally going to be a big step closer to normal back there now that the Chinese spies are being arrested. I still can't believe that the FBI took so long to figure out that all those attacks were Chinese commandos and not wacked-out terrorists. At least they finally got 'em, right?

Thanks for the encouragement in your letter. You're right. There are a lot of similarities between this war and your experience in Vietnam. Most of those similarities are probably related to the geography. From what I've read and what I've seen on TV, no one wanted to be involved in Vietnam. The same thing is true here. None of us want to be here, but we have to be here. I don't mean we have to be here because we were drafted (remember: I enlisted!). We have to be here because this is where the enemy is, and someone has to fight them. Someone has to make them pay for what they did us in the Yellow Sea, in Korea, in the Korea Strait, and now in Taiwan. No one wants to be here, but we're here because we all want to be the ones who stop them.

Vietnam was a war that didn't need to be fought. This is a war that cannot be avoided. I mean, what're we supposed to do? Should the rest of the world just sit around and let the communists do whatever they want to whoever they want? Should we just let them bomb the shit out of us with ICBM's while we disarm and go to the movies?!

I know you guys are all proud of us back home. Everyone hears it when they call home. We all see it on the news, and we read it in the letters that come from home. You should know that we are just as proud to be here. We don't wanna be here, but we're proud to be here.

Next week, Colonel Saunders is going to have us start putting in groundwater monitoring wells in a new area. It's still miles from the front, but we'll be close enough and high enough to see it. Everyone's a bit excited because we know that we did our jobs and made a difference the other day. Now, we wanna do it again. There's a lot of pride in this unit right now. I'll be careful.

Gramps, you're right about the similarities between this place and Vietnam. Both Indochina wars are hot, wet, muddy, ugly, and fought in places so beautiful that they should never have been scarred by battle. There is a major difference, though. This time, in this war, we're fighting for the right reasons. This time, we really are over here to protect America. It's not like before when people said support the troops and not the war. Until someone can find a better way to convince the Chinese and the North Koreans to stop lobbing ICBM's at Los Angeles and to stop invading our allies, this is the only way to go, and both the troops and the war need to be supported. This is a DIFFERENT war. Vietnam is over, and Flower Power doesn't seem to work anymore.

-Matt

Tuesday, August 9, 2011

1st Lt. John Chamberlin -Personal Log

Somewhere in the South China Sea

Well, Colonel Ghetty met with Bruce, Hector, and I, and the reason we're not getting any replacements is because they're breaking up the company. Bruce's platoon, including the remnants of Hector's platoon, are all going to be our battalion's contribution to a new division being formed in Australia. The details are limited, but it sounds like they're putting together a Coalition riverine assault division with lots of river patrol boats and Marines from around the world. About 2/3's of the force seems to be either R.O.K. or

Taiwanese expatriates. The rest are going to come from U.S., British, Australian, Canadian...the list goes on.

My platoon, what's left of it, is going to work directly for Colonel Ghetty now. He's been put in charge of Task Force Stingray's special assault teams. We're not Special Forces. You have to volunteer to be in a Special Forces team. You also have to have a lot of training. Instead, Colonel Ghetty's being assigned a lot of other units that have been as shot up as we've been. I think we're going to do a lot of TRAP work again. With the daily dogfights going on around the world, I would think such teams would be in high demand.

Bruce didn't get a lot of time to hang around after the announcement. He was told in the morning, and they were gone before dinner. We barely had enough time to make sure that we had each other's e-mail address. I hope he fares well. I hope they all do.

Hector's being sent over to the H.M.S. Invincible (a ship similar to the Wasp). He's going to get a fresh company of U.S. Marines. They were on the U.S.S. Iwo Jima when it was hit. The Brits picked them up, and they've been waiting to be reassigned ever since. He's going over tomorrow morning to meet with them, check out their condition, maybe train them too. Then they'll all be brought back over here to be reattached to other Marine Corps units.

Wednesday, August 10, 2011

1st Lt. John Chamberlin -Personal Log

Somewhere in the South China Sea

Colonel Ghetty and I saw Hector off this morning. I get the feeling I'll never see him again, not that I think he's in some sort of grave danger. It just feels like we've gone our separate ways for good. On the other hand, Bruce and I are bound to get together again. I don't know why, I just had that feeling.

After Hector left, Ghetty (Dennis, as I'm supposed to call him privately now) told me to go ahead and get the platoon ready to leave. We're going over to the U.S.S. Ronald Reagan tomorrow. The ship took a lot of damage from some terrorists when the war first started, and its catapults are out of action until it can get into a dock that handles carrier rebuilds and modifications. The higher-ups didn't want the ship to be out of action at all, so they sent it to sea and now it only operates Harriers, helos and Osprey. The Marines don't have enough of the three types to really make use of the ship, so Dennis got together with some Colonel he met from the 82nd Airborne Division, and they proposed an idea for the ship. Back in the late '90's, the Army operated a brigade from a carrier to invade Haiti. Ex-President Jimmy Carter leaked it to the Haitian government, and they suddenly went from threatening all U.S. forces to asking for our presence. The guy that thought up that idea went on to become Secretary of the Navy!

The 82nd was one of the first U.S. reinforcement forces into South Korea after the North Koreans and Chinese crossed the DMZ. They're choppers never made it to the peninsula, and the air transports, C-130's, C-141's, and C-17's mostly went down to communist fighters within the first few days. The few that remained had to be used to cycle supplies in from the states, and wounded back out. The 82nd had to fight like a light infantry unit against heavy Chinese armored columns and strong North Korean divisions. They lost as many people as our division, maybe more. When their helicopters did arrive, it was the Reagan that brought them. Now all they needed were people who knew how to use them. That's where we came in. Dennis volunteered his shot-up units. Tomorrow, we're going over.

Thursday, August 11, 2011

1st Lt. John Chamberlin -Personal Log

Somewhere in the South China Sea

WOW. All I can say is wow. This ship is huge. The crew is about 2500 people, and not counting ourselves or the other Marines that came over with us from the Wasp, there are about 6300 soldiers and Marines on board. Normally, in peacetime, the ship is supposedly cramped with 2500 crew and 2500 airwing personnel. Compared to the Wasp, this is heaven.

There's even room for more aircraft than usual. It was designed to hold about 90 fighters and support aircraft, but since we're using helos, a few Osprey, and small Harriers, we've got a lot more than 90. There's three squadrons of Harriers at about ten planes each, give or take a few. There's also at least ten Cobra gunships, a quartet of Apaches, about twenty Osprey, and well over a hundred UH-60M Blackhawks.

The sailors on board complain a lot about how crowded it is, but I think it's only because the corridors aren't as wide as they are on a ship designed to have fully-armed and loaded Marines moving about. We've actually got our own racks, also not as big as on the Wasp, but still much more comfortable than the aft port catwalk on the Wasp. I think we'll all get lost a few times before we really get to know the ship, but I found the head, the mess, and at least one way onto the deck, the rest is just extra. Home ship home.

Friday, August 12, 2011

1st Lt. John Chamberlin -Personal Log

Somewhere in the South China Sea

We continued steaming around in the South China Sea today. More Army and Marine helicopters came on board throughout the day. I spent a few hours walking around the ship, trying to learn my way around. It looked to me like the hangar deck was packed full with aircraft, but more keep landing. I know there's not a lot more room for Marines and troops. The corridors are getting turned into berthing areas by default. It's starting to look as crowded as the Wasp was, but at least we don't have to eat MRE's on the ship. This ship's mess hall was built to feed five thousand people. It's being asked to do more than that, but I didn't have to wait in line for any of my meals.

The only downside to the ship is its condition. Everything still smells like smoke from the fires that were started when it was attacked in Pearl Harbor back in mid-February. Some Chinese commandos tried to sink America's newest carrier with some scuba gear and mines. When they went off, the results weren't as spectacular as the attack on the U.S.S. Ranger in San Diego, but the ship did catch fire. There was only a skeleton portside crew on board, so the fire spread quickly.

I was talking with a sailor while waiting in line at the mess for breakfast. It's that fire that put Marines and soldiers on the Navy's prize capital ship, it's last supercarrier. The Reagan, unlike its predecessors, was supposed to bridge the gap between the supercarriers that had formed the backbone of U.S. fleets since the 1960's, and a newer, smaller, high-tech, more efficient carrier. Some of the high-tech improvements that are being built into new carriers are electromagnetic catapults instead of steam-powered, better electrical systems throughout the ship, a smaller and more digital-controlled island, more elevators, and endless small improvements like a new valve here, or a modified compartment there.

When the fires on the ship started getting out of control, the crew shut the huge fire doors that can divide the hangar into four sections. That stopped the fire's progression and probably saved the ship, but the small crew, even with the aid of additional fire crews from the docks and other ships, just couldn't get the fires out in time. The heat wound up getting so hot that the deck warped, the new catapults were ruined, and most of the new electronics were melted. The Navy couldn't launch planes on the Reagan with the catapults destroyed, and it couldn't replace the entire warped deck in any reasonable time frame either.

A few surviving senior officers from some of the amphibious ships that were lost in the Battle of Yellow Sea and the Battle of Korea Strait liked Dennis' idea to put a Marine helicopter assault unit on the Reagan. The higherups in the Navy balked at first, but then had to give in since the ship wasn't good for much else. Somehow, the Army got wind of the idea, and, since we Marines have taken some serious losses, the Pentagon said why not. Of course, it's one of those situations that's so messy that no one of any real authority wants to take charge.

Yet, on the other hand, it's novel enough that all the commanders involved want to at least keep their current status and presence. That's why more and more helicopters keep coming on board, and no one has taken the time to unite all of these different units yet. It seems like the type of situation that Colonel Ghetty looks for; General Shapiro too. I'm sure all the higher-ups are getting together, and I have equal faith that the cream of the crop will rise to the occasion, that being a Marine like Shapiro or Ghetty.

Monday, August 15, 2011

1st Lt. John Chamberlin -Personal Log

Somewhere in the South China Sea

It looks like our little combined seaborne helicopter assault group is finally taking shape. The senior commanders have worked out a chain of command, and we've all been given a packet of papers that outlines all of the details. We're to be called the 1st Composite Air Assault Regiment with General Shapiro in command. The ship will still going to be run by its Captain, and Admiral Pender is still in charge of the Task Force. Under General Shapiro, there are four separate assault battalions. One Marine battalion is made up of units like ours that have been shot up and sent here to make a jigsaw puzzle Marine special assault battalion. The Army has sent the remnants of six different brigades here, and they've been put together into two helicopter assault battalions and an aerial artillery battalion. The aerial artillery battalion uses all of the Army's helicopter gunships. The Marine Harriers are under the direct control of the ship's Captain, since they'll primarily be used as air defense against missile and bomber attacks.

Tomorrow we're going to begin a series of training maneuvers to help us all get aquainted with each other and to put the entire unit's capabilities to the test. No one has ever operated this many ground units or this many helicopters from a carrier before. There was a time when three Army brigades operated from two carriers off the coast of Haiti, but that was a temporary operation, and there was no where near as many helos as there are here on this one ship. That's the difference between "low intensity" conflicts during peacetime, and full-blown all-out war. During a peacetime deployment, officers have to worry about their careers if an operation doesn't go perfectly. During wartime operations, survival rules first in an officer's list of things to do; worrying about "doing things by the book" doesn't even take a close second. Here on the burnt-out hulk of the U.S.S. Ronald Reagan, with a hangar deck and flight deck devoid of high-tech fighter planes, but packed solid with Marine and Army helicopters, there is no better evidence of that idea.

Tuesday, August 16, 2011

1st Lt. John Chamberlin -Personal Log

Somewhere in the South China Sea

The morning started off with large groups of people gathering into huge meetings around the ship; a thousand here, two thousand there. It's incredible how many people are on this ship. When I look at all the equipment and people, I wonder if it's actually big enough! We had our meeting on the forward flight deck between what used to be Catapults 1 and 2. All of the Marines in our new battalion got to listen to a speech from Dennis. I'm not sure if he jumped at the opportunity or if he was volunteered, but he's definitely serious about what we as Marines are being expected to contribute to this Regiment.

He's breaking us down into helicopter and Osprey teams instead of platoons and companies. We've been assigned one of the few V-22 Osprey, but we've been designated as Helo Troop 513. Actually, all of the Marines have been divided into Helo Troops.

There's just enough of us left to make the Osprey seem crowded when we all cram in. I'd rather have the uncomfortable load than more seating due to losses. Since the war's a long way from ending, I can't kid myself about that reality. I looked around and wondered whose seat we would wind up sharing in the future, and when. A lot of the Helo troops are made up of people from all over the Corps. We're lucky enough to still be a single unit, so we're still going to call ourselves 2nd platoon. Most of the other Troops have begun choosing flamboyant names like "Death From Above," "The Heartbreakers," or "The Airborne Cowboys." It seems kind of stupid to us, so we decided to just keep being called 2nd platoon. After enough lead's been thrown at you, you don't need a flashy name to remind you of who you are, or what you're capable of doing. We know who we are, and we know what we can do.

After our morning briefing with Colonel Ghetty, we all listened to some of the flight deck personnel tell us how and what we're supposed to do during flight ops. We'd all been through it before back on the Wasp, but some of the new people needed to hear it. Actually, it was good for everyone, ourselves included. It reminded us of the dangers on a flight deck, especially one that's launching twenty helos every minute. That's a lot of blades to look out for!

Our meeting ended a little earlier than the ones for the Army battalions, and since our flight deck liaison officer who had given the briefing was more than aware that the Marines have always been part of the Navy, he let us go to take over the mess hall before the Army had a chance. The ship is so crowded now that we're only served two mess hall meals a day now. The third one has to be an MRE. There are 1100 of us in the Marine battalion, and we all really appreciated the few minutes head start toward the mess. Sometimes it's good to be part of the Navy.

After our stint in the mess hall, we had to go back up to the flight deck for the first round of maneuvers. There aren't enough flight deck personnel to move around the aircraft, so each Troop is responsible for getting their helo or

Osprey into one of the ten take off and landing spots. We've been assigned landing spot 4, just forward of the #1 elevator; the one forward of the carrier's island. We're very lucky to be right next to an elevator (right next to it). Some of the other troops have to push their helos as much as a quarter of the distance of the flight deck, just to get to the elevator. Then it's a few hundred more feet to their landing spot!

There are only four HUMMV's being used to tow the helos around. When the carrier's air wing was reassigned, they took a lot of the ship's aircraft equipment. A lot of the deck and maintenance crews went too, so in addition to moving our own birds around, we're learning to repair and maintain them too. There are a lot of Army mechanics on board, a few hundred we're told, and I hope they can keep up with the job. I think it's probably a good idea since it will help bond the Army and Marines together with a new common enemy; breakdowns.

After we got our Osprey into position, we followed the instructions of one of the few deck hands left on board. He had us line up on the elevator. I suggested that the catwalk would be safer, and he liked the idea. Now, all of the Troops are to line up on the catwalk until their birds are warmed up and ready. We waited, along with the rest of our Marine battalion, around the perimeter of the flightdeck. When all of the helos and Osprey were ready, the loudspeaker on the ship's island told us all to begin boarding, single file. It was really exciting to watch 1000+ Marines climb out of the catwalk onto the deck and run as fast as they could, single file, into their rides.

It only took a few seconds. We had enough time to get settled, then we waited. Not long after, although it seems like it took forever, the speaker sounded again, loud enough to hear over the rotor blades from the helicopters and the twin rotors from the Osprey. The speaker gave the order to lift off, and the real excitement began.

In groups of 10 and 12, (56) UH-60 Blackhawks and (20) V-22 Osprey lifted off and joined a growing airborne train that was circling the carrier at about 1000 feet. It moved between 50 and 100 miles and hour, and created a very dizzy effect. As more and more aircraft joined the circle, it expanded until it was a few miles wide. They all flew in a single file with about 500 feet between them. It didn't even take five minutes to get the entire battalion in the air.

While we were circling, we watched the Army's first battalion get their birds up from the hangar deck. Each one of the four elevators brought up another four helos. I have to give it to those Army grunts, they pushed a lot of metal a long distance, and in a short amount of time. We hadn't been in the air fifteen minutes when they lined up along the catwalks while their helos fired up. A few minutes after that, they we taking off and joining our flight train as it doubled in diameter to about ten miles wide.

The cycle continued. We rode in circle after circle for almost two hours. We watched both Army battalions and the aerial artillery battalion all take-off and join our train. It was an impressive sight. We watched over two hundred and fifty helicopters and Osprey circle the small task force. Once we were all up, the two Army battalions broke formation and climbed. One battalion, about a hundred Blackhawks, leveled out in their own train formation at about 2000 feet. The other Army battalion continued rising until it formed into its own train at about 1000 feet. Then the rings closed and all three rings drew tighter until we were about two miles wide and circling the carrier. It was really something to see, and it was even more impressive to watch on video when we finally landed.

Landing was the tricky part of the operation. It took almost two hours to get us all airborne, and none of us carried four hours of fuel. There was no margin for error. Since we were the first in the air, we landed first. The Army followed and landed in the same order that they had lifted off.

The Reagan, as big as it is, doesn't have a lot of extra space with three hundred birds on board. When they're all on board, each one has to have its rotors turned and tucked in, and they're parked in alternating directions to put them all as close together as possible. When we take off, it'll be in the opposite order that we landed because there's not a lot of space to move all the birds around. Since we took off first, we land first, and tomorrow, we'll take off and land last because of the way we're parked.

For now, we were all just looking forward to end the dizzying course we'd been flying. We were also looking forward to our second mess call. The only downside was that, once we hit the deck, we had to get out and push our Osprey back over to the elevator and into a parking spot below deck in the hangar.

The Osprey holds about twice as many people as a Blackhawk, but it's also twice as big. I'm not sure if it seems heavier because it's bigger, or if it's any easier to push around than the smaller Blackhawks because we have twice as many people. It's probably a trade off, but it's still not a walk in the park.

By what seems like a stroke of incredible luck, all the aircraft wound up landing without incident, and everything went smoothly when it came to parking them. I'm sure most of these pilots are well seasoned from the past few months of combat, but it still amazes me. Tomorrow, we're supposed to do it again. I guess Shapiro wants to see how fast and precise we can get an operation going. Either that, or he wants to watch the flying three-ring circus until he feels safer giving the take off order.

Wednesday, August 17, 2011

Correspondence

From Corporal Matthew McKenna,

U.S. Army, 1st Infantry Division, 12th CHEM To Jonathon McKenna (*Grandfather*) Yo Gramps!

We moved to a new position closer to the frontlines a few days ago. Our Command Post is still in a small town. There are hills on three sides with the ones to the south making a ridgeline toward the mountains in the north. When we go up on top of a nearby hill, we can see part of the frontlines.

It's quite a sight! When you look to the west, you can still see the huge city of XXXX XXX XXXXXX. All of its buildings are still reflective glass, and there are still golden temple spires poking through the dark green jungle around the city. The highways are still busy, and everything looks a lot like it would back home in PA. When you turn around, you can see a few miles of the frontlines-they call them The Sewer. The jungle is even thicker all the way up to the lines, then there's a red-dirt scar from all of the B-52's. It goes on for miles, and it's like a red strip of the Moon or Mars. I haven't seen any shelling or any activity at all, but sometimes we can hear the automatic weapons fire. We also hear artillery, mortars, and bombs, but I haven't been able to see any of them hitting. The red, no-man's land is probably about ¼ mile wide, and I have no idea how long. It snakes through a valley and almost everything between the two crests of ridges has been blasted down by bombs. It kinda looks like a very sloppy logging operation.

Colonel Saunders is having us put in a picket line of groundwater monitoring wells all along the reverse slope of the ridge where the frontline runs along the crest. We'll be getting with a few hundred yards, but we'll be behind some very tall hills, and the Chinese will never know we're there. Even if they did, I doubt they would see us a much of a threat, and it's even more doubtful that they could do anything to us from the other side of the next valley over. I'll be careful.

Oh yeah! I forgot! The FBI called Colonel Saunders the other day. It turns out that one of the kids I went to high school with was connected with the Chinese commandos that were rounded up! His name is Choi Paek. He moved to California right after school. I'm not sure what he did, or any of that, but I thought you might want to let some of the folks in the neighborhood know. I guess Saunders convinced the FBI that I wasn't a Chinese commando, because right after I met them, they smiled, shook my hand, and dismissed me. Me! A Chinese commando! I DON'T THINK SOOOOOO! If it wasn't so stupid, it wouldn't be so funny. Of course, Choi seemed like a normal guy in high school, so I can't blame them for looking into it.

Well, I've got a lot of work to do. Saunders wants almost 300 wells put in right away! When it's all divided up between the different field teams, I'll probably have to help put in at least 25-30 wells! We've been getting 5-7 done per day, but Saunders wants them all in by the weekend! Gotta go!

Take Care,

Matt

Thursday, August 18, 2011

Correspondence

From Corporal Matthew McKenna,

U.S. Army, 1st Infantry Division, 12th CHEM

To Jonathon McKenna (*Grandfather*)

Yo Gramps,

I guess we're not as safe as we thought. I'm okay, but came under some sniper fire today. Lieutenant Satterfield was hit in the chest, and two of the people in his field team had some small cuts from ricochet fragments. Satterfield's gonna be okay.

We all wear some really nice body armor. It's a new version of the old Kevlar vests. This version has all kinds of layers and different shock-absorbing materials. It also has pockets for putting in special Boron/ceramic/aluminum plates. The nice thing is, it all weighs less than 15-pounds! It's still hot to wear in this area, but a little extra sweat is a lot better than the alternative!

We also have to wear special sunglasses. They're nice since the sun is mega-bright in the afternoon. They also keep rocks from getting in our eyes during drilling operations. The guys in the trenches use them too since they're even hard enough to stop shrapnel! We think the real reason that the higherups have everyone wearing them is because both sides are using lasers to blind each other's troops. They're like sniper rifles that don't have to worry about range, trajectory, wind, or powder grain. So far, the glasses seem to work. None of us are blind, and the rocks from the drilling haven't injured anyone yet either. The only real bad thing about them is that everyone has this raccoonlike tan from wearing them in the sun!

Snipers are a rare occurrence behind our lines. We have a special way of dealing with them; there's a specially modified Bradley M-2 fighting vehicle that goes around and takes them out. The Bradley looks like any other, but it has a small video camera mounted on the turret. When a sniper incident is reported, they bring the Bradley over and park it with the turret pointed in the general direction of the sniper. If the sniper fires again, it can actually detect digital changes on all of the digital images as the bullet flies! It can't see the bullet, but it can highlight the differences between the different images. As the bullet flies, it distorts the air like hot sun on asphalt. The images are then combined, and you can trace back the origin of the bullet by looking at a series of brackets that highlight the differences on the images. The Bradley then turns its turret toward the bullet's origin and cranks out a few 25mm cannon shells into the area. Problem solved.

Satterfield was running a different drilling operation almost a mile from me. I wasn't even close to the sniper fire. I could hear the Bradley when it opened up with that 25mm, though! There was no way anyone made it through that. Some grunts up on the line are still searching for the remains.

This is a much more dangerous area than I thought, but it's not at all as bad as it could be. I'm still really safe, and I'm not going to take any chances. We all want to be here, but we all want to go home in one piece too!

I've got a lotta work to do, so I'll leave it there for now. Say hi to everyone, and tell them all that I love them very much.

-Matt

Saturday, August 20, 2011

Correspondence

From Corporal Matthew McKenna,

U.S. Army, 1ˢᵗ Infantry Division, 12ᵗʰ CHEM

To Jonathon McKenna (*Grandfather*)

Yo Gramps,

A few of us went up to the top of the ridgeline and into The Sewer this morning. What a sight. The trenches look a lot like pictures of World War I. The grunts there made sure we kept our heads down and didn't poke up above the parapets. It's mostly a sniper and small unit war from the frontlines, and I was told that there are snipers on both sides who shot anything that moves, then worry about it later. We kept our heads down.

It wasn't too hard to stay away from the parapets. The section of sewers that we were in is almost 15 feet deep. The walls and floors are thick bamboo (4"-8" thick). All the bamboo runs horizontally. It's filled with mud, then tied together with palms and staked in place with more bamboo. They tell us it's better than sandbags (The grunts prefer to use those for portable, disposable latrines.).

The trenches don't have a lot of organization to their shape. They sort of twist along the backside of the ridgeline's crest, but they also work down the front of the ridge to some underground bunkers that the grunts have set up. We went down into one and looked through a firing port at The Zone (That's what they call no-man's land).

From the firing port, I saw bodies for the first time. We're told they're all Chinese, but I couldn't tell. They were shaped like people, but they were all covered with mud, bloated, and still in the same positions as they had fallen. Some were still standing-held up by Chinese barbed wire. Others were holding each other in bomb craters. It was too far away to see them in detail, but they faced all different directions, and they were evenly spread across The Zone. I couldn't see much detail-thankfully.

There were times when we could smell the bodies, and that was as gross as I had expected. It's windy at the top of the ridgeline, so the air carries a lot of smells-bodies, gunpowder, napalm, jungle flowers, rain, and more than anything else-mildew. The place stinks of it.

The reason they call the trenches The Sewer is because it's really a series of firebases and bunkers connected by 8'-wide corrugated plastic storm sewer pipes. The pipes are an off-white with a glow in the dark additive that makes them kinda green at night and puke-colored, pea-green during the day. The pipes run between the all the firebases and bunkers along the ridgeline. They also run down the backside of the ridge to a road that runs the length of the valley. This way, we can get reinforcements to any firebase right away. All someone has to do is call for help. Trucks drive reinforcements to the nearest rear entrance to The Sewer, and the position is reinforced. It must be a good system because the Chinese haven't been able to crack it yet.

There was no fighting while we were in The Sewer. We heard a few sniper shots and bursts from automatic weapons in the distance, but we were safe. If there had been trouble, we would have just left through the green pipeline to the rear. I was safe.

We finished installing all the monitor wells that Colonel Saunders wanted. Now, we have to go around and measure each one twice a day. Each well takes about 5-10 minutes to find, and another 5 minutes to gauge. We have a water probe that looks like a giant tape measure. When you lower it down into the well, it beeps when it touches water. Each of the wells has been surveyed, so we type the survey elevation, the monitoring well name/number, and the depth-to-water reading into a palm PC. Our field officer for the day downloads all the palm PC data into a laptop, and uses a computer program to create a digital model of the water table. Walking from well to well, through the jungle, along roads, up the sides of the ridges, and even in the middle of roads all sucks, but it's better than sitting in The Sewer and waiting for a Chinese attack.

Speaking of work, I'd better get back to it. Say hi to everyone, and take care of yourself!

-Matt

Sunday, August 21, 2011

1st Lt. John Chamberlin -Personal Log

Somewhere in the South China Sea

We've been flying formation ops for several days now, and it still doesn't get old. Up until today, we'd only been flying circles. Today, we went out on a mock operation. Our battalion took off first. We didn't wait for the Army's. As soon as we were all up, all our Osprey left the single file formation and headed west. The wings rotated into a fully horizontal position, and we flew like regular airplanes at about 500 feet.

Our target, an abandoned oilrig, came into view. About half an hour later all twenty of the Osprey rotated their wings into the vertical. We circled the oilrig and slowed down until all twenty were completely stopped and hovering, each about 500 meters from the rig. The lead Osprey fired a green flare, and we all headed back to the Reagan. We rode with our wings horizontal for most of the trip, so we were able to pass the rest of the Marine battalion before it had even reached the oilrig. Once we made it back to the carrier, we had to circle at 5000 feet until the two Army battalions lifted off and headed out to the rig. Only then were we allowed to land. By the time all twenty Osprey were on the deck, the rest of the Marine force was beginning to show up. It was another hurry up and move the bird operation, but we're getting a lot more comfortable pushing around 'ole 513. I think the rest of the grunts on board are feeling the same way about their rides too.

Monday, August 22, 2011

1st Lt. John Chamberlin -Personal Log

Ko Tao Island, Indonesia in the Gulf of Thailand

Today we made our first real assault. Yesterday was apparently a test to see how well we could fly to a point and back. Today, we tried actually flying to a point and landing. Since we made it back first yesterday, we took off last today. The Army's battalions broke into formations of ten choppers each. Our Blackhawks followed in single file. The Osprey took off last, but we rotated the wings forward and passed almost everyone. Only the first ten Army Blackhawks touched down before us. We landed on this near-deserted island out in the middle of nowhere. The LZ is a series of empty grass fields along the coast at the base of two hills. As soon as 513 came close to the ground, we started running and jumping off the rear ramp. Bob was out first, and he directed the rest of the platoon towards a small pile of partially buried rocks about a hundred feet behind where we had landed. The other Marine Troops did the same and spread out around the same area that we had. The Army was landing about a quarter mile closer to the hills, and they were already moving up both. All the Osprey and Blackhawks headed back to the ship as soon as we were dropped off.

When we landed, we had been told that the island was abandoned, but that it was strategic, and it had to be held at all costs. Colonel Ghetty gave us a briefing before we left the Reagan, and he made it sound like this was a real operation. If it is, it's a cakewalk. The other two battalions came in after we did and they landed wherever they could find space. Our battalion set up to cover the beach and we left about a hundred Marines to cover the LZ and act as a reserve. The Army moved a battalion up to the top of each hill, and we all dug in for the night.

It was a very quiet and peaceful night. It was so quiet and so clear that even though the Army was almost a mile away in some cases, we could hear them talking, and watch their lips through binoculars. Above us, a cloudless night sky glittered with stars, the Milky Way's cloud, and a beautiful near-half moon. The only sound we heard was the soft surf gently rolling up onto the wide white beach. It was beautiful, peaceful, and extremely relaxing. Everyone took full advantage of the moment, and I had Bob rotate the watch so everyone got to enjoy a quiet night's sleep; something non-existent on the Reagan. For a few hours we were all able to steal a few moments away from the war. Dennis is pretty busy working with the two battalion commanders from the Army, so I'll have to thank him when we get back to the Reagan.

Tuesday, August 23, 2011

1st Lt. John Chamberlin -Personal Log

Somewhere in the Gulf of Thailand

The sun rose this morning. There were no Chinese, North Korean, or other Eastern Alliance troops on the island. They're weren't any twisting white contrail lines from dogfights and missiles in the sky. There were no patrol boats off the shore. The sun rose this morning, and the war was nowhere to be seen.

Some locals came over this morning with as much food as they could scrounge up. There were only a few hundred (maybe not even that many), so there wasn't enough for everyone in all three battalions, and it wasn't even close to being as decent as the food on the Reagan. It was a lot better than MRE's, though! I had some sort of fried eggs with some sort of fried beef. I don't want to know what it really was. I'll just trust my imagination. Since there wasn't enough food for everyone, most people had to eat at least something from their MRE's, but the gesture was well appreciated.

At a location as remote as this, I would have imagined that these people wouldn't even know that a war was going on. Yet they obviously do, and they seemed so happy to see us. A good night sleep, no sign of the war, a beautiful tropical island, happy and grateful locals with home cooked food, it all makes me wonder. Anytime something this good happens, it seems that there should be a price to pay. I wonder what it will be? When will we have to pay that price? I hope we already have, but my instinct tells me something big is in the works. This just seems like it was too nice to be a training mission.

It wasn't until late afternoon, almost sunset, that the Osprey and Blackhawks began to arrive; our rides back to the war. The crew of 513 remembered where they had dropped us off, and in less than a minute, we were all back on board. We lifted off and climbed to about 1000 feet where the rest of the Osprey joined us. As soon as the last one was almost at the rendezvous altitude, they all began rotating their wings forward for conventional flight. The Blackhawks would have to hover their way home. On our way back to the Reagan, we passed the rest of the 270 Blackhawks that were on their way to the island. It was quite a sight. Instead of feeling pride or awe, all I could think about was getting to the mess hall first. I guess the big formations have finally become routine.

Wednesday, August 24, 2011

Correspondence

From Corporal Matthew McKenna,

U.S. Army, 1st Infantry Division, 12th CHEM To Jonathon McKenna (*Grandfather*) Yo Gramps!

Well, we've paid the price for our little visit to The Sewer. Colonel Saunders found out about it, and he sent me back up there to do some surveying and to tie-in the monitoring well elevations. I've only surveyed once before, but Saunders said that was enough. It's not like it was 40 years ago. All I needed to do was bring the 1-meter, benchmark-stadia reflector to the top of the hills along the rigde. I set up the laser range-finder (It looks like a large can of soup sitting on a tripod) close to the command post where it had a direct line-of-sight to all the wells. After I turned it on, I just had to walk around and put the stadia reflector at the top of each well for a few seconds. The laser spins around looking for the stadia reflector. When its beam is reflected back, the laser stores the distance and angle of elevation to the reflector in a small memory chip. After I had done that for each of the wells, I went to the tops of the hills along the ridgeline and held up the reflector. The highest points on the hills are all on our maps, so even if one changes, the computer could take an average from all of them and find a good benchmark. When it was done, I just plugged in my palm PC to the laser near the CP, and the data was ready for Colonel Saunders. The whole thing took about 9 hours, but I took my time while I was up in The Sewer.

I might have painted a bleak image of The Sewer in my last letter. It stinks of mildew, and the view of The Zone is unreal, but the trenches and bunkers in The Sewer are in excellent form. Most of the dugouts-places where the grunts sleep and wait out Chinese barrages-are very nice. Their walls are usually 105mm howitzer ammo boxes that have been filled with dirt. They have floors and tent ceilings. Moisture runs off the tents and into 55-gallon drums or buckets. These dugouts range in size from 8'x8' to 20'x20'. They usually have interlocking rooms with plastic storm sewer pipes leading to even more rooms, emergency exits on the backside of the ridgeline, and-of courseto the trenches and forward bunkers.

The dugouts have a lot of personality too. Almost all of them are lit by strings of battery-operated Christmas lights. Solar battery chargers-about the size of 9, CD-Roms, can be found every few feet in the forward trenches. A lot of them have portable CD and MP3 players, and noise discipline is mostly ignored since the Chinese know exactly where we are. The forward CP's all have laptops, countless radios, and even old-fashioned telephones with wires running between them. Everyone has their own personal space where they can store their gear and get some rest. The latest trend in the dugouts is to have small, portable air-conditioners shipped in from camping supply stores back home (Yes, American Parcel delivers right to your dugout almost every day!).

It is a bit odd to see a small, portable fire extinguisher in every bunker and every open trenchline, but it's just as weird seeing the obligatory 12-gauge shotgun next to each and every one. There are a lot of unusual sights and combinations around here. Still, it makes you wonder…things that common must be there for a reason!

On the backside of the ridge, there are some open bunkers with tent covers where platoons seem to have an undeclared contest going to see who can set up the best outdoor grilling facility. The smell is phat! Thank God for gas grills! There are even dugouts devoted to retail shopping and online shopping via laptop terminals connected to the division telephone wires that lead out from The Sewer! It's very rough, but the grunts do what they can to make it bearable.

I hung out for an extra hour or so and talked with some of the grunts. There's a guy name Chip-Sgt. Chad Chambers, 21. He and I were talking about our different MOS' (Military Occupational Specialty). Chip's job is a lot like mine. The real difference is that instead of groundwater monitoring wells, he leads a small squad into The Zone and places electronic sensors to detect Chinese advances. They also plant devices that can prematurely explode incoming "stuff" (They call all artillery, mortars, and air attacks "stuff."). It was interesting talking with him. He gave me a couple of beers in exchange for one of the Navy sonobouys that the Lt. Commander's been using. I think he can spare one, or one can conveniently break down on him. Chip and his guys wanna see if it's more sensitve than the microphones that they've been placing in The Zone.

It's getting late, and I've got a big day ahead tomorrow. Any news about the Chinese commando attacks back home? How about that stint you had put in? Is it still working well? Say hi to everyone, and tell mom I'm being careful.

Later,

Matt

Thursday, August 25, 2011

Correspondence

From Corporal Matthew McKenna,

U.S. Army, 1ˢᵗ Infantry Division, 12ᵗʰ CHEM To Jonathon McKenna (*Grandfather*) Yo Gramps!

Colonel Saunders says that we might have found two more tunnels. They're both probably still underneath The Zone, so it's hard to detect them. The Squid officer has had us put some more sonobouys into nine more wells, and he agrees with Saunders. The plan is to put in another 30 wells and 10 more sonobouys. That should help us narrow it down. We're also coordinating our ops with the grunts in The Sewer. Hopefully, they'll find some similarities between our computer model and some of their observations.

I tried to call mom again last night, but there was no answer. I hope everything is okay back there. I'm thinking about ya'll a lot. Take care of yourselves. I've gotta get back to work now.

Take care of yourselves!

-Matt

P.S. Sorry about the condition of this letter. It rains for at least a few hours every day here, and today was a very wet one. I don't think you can get this wet while taking a shower! When it's this bad, everything seems to get soaked, and this was the driest piece of paper that I could find. Sorry.

Friday, August 26, 2011

Correspondence

From Corporal Matthew McKenna,

U.S. Army, 1ˢᵗ Infantry Division, 12ᵗʰ CHEM To Jonathon McKenna (*Grandfather*) Yo Gramps!

I got your another one of your letters today. Sorry to hear about your heart medication. I hear that stuff's always been expensive. If it's any consolation, we've gotta take a fistful of different pills every day over here too. There's all kinds of tropical diseases, so there's seven different pills we have to take for those (things like malaria, beri beri, dysentery, etc.). We also have to take some pills to help build up our immunity to any chemical or biological weapons that the Chinese might start to use. So, there's another three pills for things like anthrax, blistering agents, etc. For whatever reason, it seems the Army thinks we don't get enough vitamins in our MRE's (Meals-Ready-to-Eat; i.e.: rations). So, we get a healthy dose of multi-purpose vitamin pills too (three more pills). Finally, there are the big heavy ones that we have to take for classified purposes. I wouldn't bother to try and tell you what's in them even if any of us knew. The only thing that's clear is that we have to take a pair of big, heavy pills twice a day, and it's like trying to swallow steel weights! You'd think those drug companies were making enough off of their sales to the U.S. Army, but if they need to charge you an arm and a leg too…well, I guess we have to trust that they're doing the best with what they've got. At the very least, every soldier in this area can sympathize with you and your calendar box of meds. I do know this: very few people get sick over here, so something must be working!

There's not too much new over here right now. The Chinese have stepped up their shelling against some of our division's positions a few miles to the north. Saunders is still convinced that a pair of tunnels are in the area. He's ruffled a few feathers among the officers in some of the units being shelled. It seems that Saunders borrowed four seismographs that were being used to triangulate the location of the local Chinese artillery. He had a few of us learn how to use them, and then set them up in the low areas of The Sewer. Normally, they work just like the larger ones used for detecting earthquakes, but these are smaller (about the size of a large trashcan), and they detect the vibrations caused by Chinese artillery firing. By matching up the graphs, we can determine the distance that the artillery is from each seismograph, then we can overlay all those range circles and pinpoint the artillery. Saunders is trying to determine the exact position of the tunnels by determining the distance between each seismograph and from the sounds of Chinese engineers digging. The tunnels are big (like the highway ones that go through mountains back home), and the equipment used to dig them is just as big and noisy. We just set them up this afternoon, so we'll have to wait a day or two before we can pinpoint the tunnel(s) any better. Until then, those guys getting pounded up north will just have to dig deeper, I guess.

It's been a long day, so I'm gonna bail. Keep taking your meds, and I will too. Let mom know that I'm still playing it safe, and say hi to everyone else too.

Thanks,

Matt

Saturday, August 27, 2011

Correspondence

From Corporal Matthew McKenna,

U.S. Army, 1st Infantry Division, 12th CHEM To Jonathon McKenna (*Grandfather*) Yo Gramps!

I got another letter from you today. I didn't know you were in the Iron Triangle back in Vietnam! Saunders says that's where he first got involved in tunnel hunting. I guess he was a tunnel rat. Saunders says that the tunnels back there were so small that at one point the Army only took people under 5' for the tunnel rat service. He says to tell you that he thinks he remembers you now. So you know, the tunnels over here are a lot bigger than the ones back in Cu Chi. Saunders says those were a maze of fortifications, barracks, and bunkers. It's more like an underground city. These are just like large mines or train tunnels. The North Koreans are the ones that started digging them back in the 1970's. The ROK Army found eleven over the years, and everyone suspects that's how so many of their units made it through (under) the DMZ earlier this year.

We're still trying to narrow down the location of the tunnels that we know are out there. Apparently there's a geologic layer of granite that's messing up all of our data. It completely screws with the hydrogeology of the area, and our monitoring wells are almost useless (None of us are too happy about that after all the effort we've spent putting in the wells and monitoring them all everyday!). The seismographs are coming up useless since there's a constant shelling going on up north, and it's got them all shaking so bad that we can't even hope to pick up a feint digging sound. The Navy Commander is having similar problems with the sonobouys, and he's really pissed at everyone about the ones that "someone" seems to have "lost." Tomorrow, I'm gonna go back up into The Sewer and see if I can get them back from Chip.

I'm still playing it safe. There's a constant stream of flares every night, so we all know that we're within Chinese artillery range. No one's really scared about it, though. We've got Shortstops all over the place. Shortstops are small, shoebox-sized transmitters. Each one has a cat food can-sized antenna that detects incoming artillery shells and broadcasts a signals that prematurely setsoff their fuses. Since they broadcast radio signals, they might be detected, and we have to keep them behind the ridgeline to block the signals' strength. They're great for us, but the poor grunts in The Sewer can't use them. If they did, the Chinese would triangulate the signals from the Shortstops and pound the hell out of our guys with armor-piercing shells that don't have proximity fuses. Until someone comes up with a better idea, the only safe places are deep in The Sewer's system of tunnels, or behind the ridges-where we are.

Saunders wanted me to tell you, "Hi," so there's that. I'm gonna get going now. I think if I get my data from today's monitoring well gauging turned in early,

I might be able to get up to The Sewer and get the sonobouys and stuff back from Chip.

Take care of yourselves. Tell mom how safe I am, and don't work too hard.

Later,

Matt

Sunday, August 28, 2011

Correspondence

From Corporal Matthew McKenna,

U.S. Army, 1ˢᵗ Infantry Division, 12ᵗʰ CHEM To Jonathon McKenna (*Grandfather*) Yo Gramps!

I spent the night up in The Sewer last night. I was trying to get the sonobouys back that I had loaned to Chip and his team. They had already taken it out and were getting data from it, but they were kind enough to go out and bring it in for me. I watched from the firing port of a bunker at the bottom of the valley-almost right in the middle of The Zone. Those guys crawled out of the same tiny firing port and slipped away into the darkness. Illumination flares were lighting everything up, but I lost sight of the entire team when they were only a few meters out. The last time that I looked out into The Zone, it was all mud, craters, and mangled jungle debris. Last night, the flares were so bright that I could actually see the green bamboo and grass that's coming up through the twisted foliage. Chip and his team were gone for about an hour, and they finally came back. They climbed in through a different bunker and carefully covered up their trail as best they could.

After I got the sonobouys back, they asked if I wanted to hang out with them for a bit. In their bunker, they've got a keg of beer and a small wet bar all set up. The beer wasn't just warm, it was hot and stuffy-like the bunker and the entire Sewer-but it tasted great. Their flashing Christmas tree lights really added to the atmosphere. It's really four bunkers connected to one larger one that leads up to the bunkers at the top of the ridge-overlooking The Zone, and down to the Sewer entrance at the back of the ridge. I could have gone either way, but instead, we all hung out, drank beer, and even watched some XXX DVD's on Chip's laptop. It was great-like R&R!

The only thing that really sucked was the rats. They grow 'em big down here, and they…are….everywhere! The guys tell me that they have to sleep with their boots on, a towel wrapped around their heads, and their cover over your faces. I've never seen so many ugly critters. Most of 'em are about the size of a guinea pig, but some are like small dogs! It's incredible-and disgusting! Some people have brought cats up here and bred them, but they just run away when the rats come. Rat poison is a favorite, and there never seems to be enough. The guys say that the rats usually prefer to feed on the corpses on the other side of The Zone.

They also tell me that if I can find a good supply of rat poison-one that they'll actually eat, they'll take me out on a microphone-planting op. Those are their easiest and safest missions. They call 'em milk runs, so I'm not too worried. I'm not sure how Saunders will take it though, so he might not find out about it right away. With any luck, I'll plant some microphones that help pinpoint those tunnels! It'd be nice to do some good around here. Until then, I'll find the rat poison anyway-just to thank them for the nice time last night.

How're things back home? Are there still commando attacks going on? How's everyone doing? Oh yeah, if you get a chance, could someone send me some spices, sauces, or seasonings? The MRE's could use some help-ANY HELP!

Take Care,

Matt

Monday, August 29, 2011

Correspondence

From Corporal Matthew McKenna,

U.S. Army, 1ˢᵗ Infantry Division, 12ᵗʰ CHEM To Jonathon McKenna (*Grandfather*) Yo Gramps!

Thanks to the U.S. Air Force, we have to re-do almost all of our work in this area. Someone decided that the Chinese needed to have another B-52 raid. So, The Zone was hit with another 84, 2000-pound bombs today. We haven't been able to check all of our monitoring wells, but every one that we've checked out so far has collapsed under the concussion of the blasts on the other side of the ridgeline. The seismographs are designed to detect tiny vibrations from a single cannon miles away. They were never designed to survive the concussion of 2000-pound bombs hitting ¼-mile away, and they all jammed after pegging-out the strongest readings possible. Saunders called it "electronic suicide." We call it a pain in the ass-and that's the diluted terminology!

No one knows if the Chinese took any losses. The level of stuff that they're throwing into The Zone and The Sewer seems about the same. It's still infrequent, inaccurate, and inconvenient. There haven't been any more infantry pushes, but those are rare anyway. This is a war of sniper and small unit action. There's always a sniper shot ringing out, or a burst of machinegun fire, but Chip and the team tell me that on the few occasions when the Chinese made large-scale assaults, they were cut to pieces by our stuff. Apparently, there were a few occasions when they attacked along a long stretch of frontline, and I'm told that it sort of diluted our firepower. The guys say that it was only really rough on one or two occasions.

The vast majority of the time is spent in The Sewer or sneaking around in The Zone where individuals and small teams avoid each other as much as possible.

Saunders is screaming up the chain to get the Air Force to air-drop some sensors into the area as a stop-gap measure until we can get some new wells dug and some new sensors placed. His concern is that the Chinese know we're hunting their tunnels after we took out the one last month. They might figure out that our sensors are down and use the time to open their tunnel and attack in force.

Well, I don't have a lot of time to write today. There's a lotta work to be done! Say hi to everyone, and take care of yourself. I got a letter from mom saying that you're stressin' out about some sort of stock market fund you have that's taken a loss. Take it easy. It'll come back. Even if it doesn't, at least you'll still have your health, right? How is that stint and your medication going?

Later,

Matt

Tuesday, August 30, 2011

Correspondence

From Corporal Matthew McKenna,

U.S. Army, 1ˢᵗ Infantry Division, 12ᵗʰ CHEM To Jonathon McKenna (*Grandfather*) Yo Gramps!

The Air Force told Saunders that they'd drop a load of BFS 109's (airdelivered motion and infra-red detection sensors) in the area, but it'll be a few days. Until then, the Chinese are sure to be advancing their tunnels and shifting their positions. Most of our crew is either digging new groundwater monitoring wells, inspecting the old ones, or collecting data from the handful that remain useful.

I owned up to the Navy commander about how I loaned his sonobouy to some grunts. He wasn't pleased, but it wasn't much more than an obligatory dressing down. Afterward, he told Saunders that he wanted me and "my grunts" to go out and place microphones in The Zone. I had seen it coming, so it wasn't a surprise to me. When I brought the box of "mikes" up to Chip and his team, they didn't seem too surprised either. Of course, that squid commander didn't bother to go out into The Zone with us.

We headed out early this morning-before sunrise. The usual layer of fog was nice visual cover. We couldn't see anything through our night vision, and we had to use a thermal sight that the team had "procured" from a Dragon missile launcher. There were no Chinese anywhere. So, after about an hour's worth of watching and listening from a forward bunker deep in the valley, we started to make our way through the firing port.

Piotre Kowalski, 19, is the team sniper, and he led the way. Chip says that Piotre is the worst marksman that the Army's Sniper School has ever created. The School's motto is "One Shot, One Kill," but Piotre's never hit anything or anyone with the first shot. He was issued an M24 bolt-action sniper rifle, but traded it in for a German, semi-auto, Heckler & Koch, 7.62mm NATO, PSG-1 rifle. Piotre is very quick to acknowledge his relatively poor marksmanship, but he's just as quick to point out that he always gets his target in the end-it just takes more than one shot. The PSG1 is advertised as capable of putting 50+ rounds into an 80mm circle at a quarter mile+ range. Piotre uses a 20-round magazine and standard, match-grade ammunition. When we crawled out into The Zone, he was carrying as much ammo as the next guy. He doesn't wear one of those ghillie suits that snipers usually don, either. He just looks like a regular grunt with an unusual rifle. Piotre's a nice guy, kinda quiet, a bit indecisive, not very opinionated, but always seemed relaxed and non-chalant.

A minute or two after Piotre crawled out into the fog, "Bad Luck" Harry followed. Private Harry Meadows, 18, is a really nice guy who just always finds some sort of bad luck. I guess he's grown used to it because nothing seems to get him down in the least. Chip and Piotre tell me that Harry's the best point man on that the world has ever seen because there isn't a trap or an ambush that he doesn't spring himself. I think if anyone's that prone to problems, and he's still alive, then he's actually very lucky!

"Tattoo" Joe and "Fart Boy" Pepe headed out a few minutes after Harry. They went out together to carry all the team's extra gear. Tattoo Joe has a bag of claymore mines that he wants to put out, and Pepe's M249 SAW (squad automatic weapon) is the team's firepower, so there was a lot of extra ammo to bring along.

Corporal Joe Toland, 19, is the team demolitions freak. It's not his MOS (military occupational specialty), but I'm told he knows everything there is to know about mines and boobytraps. All I know for sure is that he acts like a hardass, drinks a LOT whenever I've visited the team, and the rest of the guys tease him about his family being trailer park trash. He has four tattoos on each arm with different girls' names. The problem is one of the tattoos says "MOM." I'm sure it's taken out of context, but they still treat him like white trash because of what it implies. I didn't say anything about our family's little scandals since I figure every family has their share.

I'm not sure how Pepe Kelly, 18, got his nickname Fart Boy. He doesn't fart any more than the rest of the team. He's actually a nice guy, and he readily admits that-like Piotre-his marksmanship is very poor. He claims that carrying the SAW lets him throw enough lead to let statistics and odds replace his aim. His only problem is that he's from New York City, and he NEVER stops talking about it. After hangin' with the team for only a few days, it's already worn on me, and I don't know how the rest of the guys put up with it!

Chip and I headed out last. Since this little mission was being conducted at my expense, I had to drag the bag with all the mikes in it. There were 97 mikes (4 oz. each), 100 batteries (12 oz each), 9600 feet of poly-wire, the bag, some MRE's for each of us, a spare canteen for each of us, and my own gear. I felt like a mule! Dragging the bundle in the mud was like trying to bulldoze, so I had to put it over my back and swim my way through the muck and debris to stay up with the guys. Chip was nice and never let them all get too far ahead.

The Zone in our area is actually two parallel valleys. The Chinese hold the northern valley that's filled with jungle debris. We hold the southern valley that's barren mud, and has odd clumps of fast-growing bamboo shooting up in spots. Our Sewer system of bunkers, tunnels, trenches, and firebases is extremely well-camouflaged, and I had a tough time recognizing any of the places where I had looked out into The Zone. The ridgeline in the middle is where small units and snipers like to work. It took us a few hours to get up there even though it's barely 2 clicks (kilometers) away. From there, we could see the Chinese trenchline. It's an imposing and infinite zipper-pattern of trenches. They're all lined with sandbags and supported by bamboo or logs. I didn't see any of the Chinese moving around, but we all knew that they were there.

At the top of the Ridgeline, we split up. Piotre headed east-toward the right side of our lines. Harry led Tattoo Joe and Fart Boy to the west to get a covering position. Chip and I set up a central location, and then we both started setting up microphones at about 10 meter intervals. Each one had a separate wire that we both brought back to the bag. We also numbered them all and jotted their locations down in my notebook. About an hour before dawn, we connected all the wires together in a data ribbon and Chip went out to recall the team.

I was alone for a few minutes, and it was only then that I started to get a little nervous. The rest of the time, this was just a job, and it had to be done. Until Chip came back, I had a chance to look out and really take in my surroundings. The constant yellow and white flickering of Chinese flares, the distant sounds of machineguns and "stuff," the mangled jungle, the smells, the occasional sights of dead Chinese, it all made me very edgy. All of a sudden, Harry's face appeared over my left shoulder. I swung my weapon around to shoot him before I recognized him. His luck held, and I didn't have time to pull the trigger. Chip followed him with Tattoo Joe and Fart Boy.

We headed back without Piotre. Chip said that he always comes in alone since he works alone. There was even a chance that he'd already headed back! We all pictured Piotre sitting in the relatively dry bunker, surrounded by gentle Christmas tree lights and drinking the keg dry. It was a good motivator, and we made it back to The Sewer System just after sunrise, but before the fog burned off.

Tattoo Joe was the last guy in. While he shoved his gear through the firing port, we heard a sniper shot from near where our central wire location was. Then we heard a lot of Chinese firing from the other side of the middle ridge. There was a quick succession of sniper shots, and then we saw Piotre running down our side of the middle ridge as fast as gravity could bring him down. He hurdled over stumps and craters, tripped more than once, and finally dove into the ground headfirst about 50 meters to our left. He'd found another bunker to enter The Sewer.

The guys were tired, so Harry and Fart Boy passed out. Chip helped me wire the microphones into a small receiver that's connected to my laptop. Tattoo Joe was nursing the spigot to the keg like it was his mother's nipple. At about 0930, I finally finished connected the mikes and began to correlate the sounds smooth enough to determine each one's location within a foot or so. At about the same time, Piotre finally came back into to the bunker. He had a tray with 6 plates of bacon, eggs, sausage, and toast. He also had a dozen sealed cups of cold orange juice in his buttpack.

It's true that The Zone is very eerie. It's true that The Sewer is a dirty, smelly, rat-infested hole in the ground. It's also true that these Americans are making the very best of the situation, and whenever possible, every effort is made to make it more like home. The bacon and sausage were even freshgrilled from the company grill bunker on the protected backside of The Sewer!

Take Care!

-Matt

Wednesday, August 31, 2011

Correspondence

From Corporal Matthew McKenna,

U.S. Army, 1ˢᵗ Infantry Division, 12ᵗʰ CHEM To Jonathon McKenna (*Grandfather*) Yo Gramps!

The mikes seem to be working out. I uplinked my data to Saunders and ran the ribbon down through The Sewer all the way to our CP. Some of my locations are off a bit, but while Saunders was sorting them out, we were able to listen to a Chinese patrol crawling around right where we were yesterday! No one wanted to call in an artillery strike on them unless they started some trouble. A strike would take out our mikes, and we'd have to go out there and put in a new set. It's a lotta work just dragging them up there!

I'm sure you guys are getting worried about my recent experiences in The Zone. I'm not in a hurry to get hurt, so know that I'm being as careful as possible, and that the guys I've gone out with are the best. They know exactly what to do, what not to do, and how to do things in The Zone. It's my job to go out there and help detect the bad guys, and I'll admit that it can be a bit scary, but this is what I trained to do. We've all got to do our jobs-however unpleasant, and besides....I can't exactly say, "No! I don't feel like going!"

I got another letter from you today. It's the one where you talk about the mud in Vietnam. It still rains everyday here, and I've gotten used to most of the different types of mud. There's the chunky, sticky mud. There's the weightless, foamy mud. There's smelly mud, there's mud in red, orange, a thousand shades of brown, and even green, algae-laden mud in some areas. Thanks for the tip about walking in the mud, but while you guys may have found it easier to go barefoot, we don't dare try it here. There's shards of razorsharp shrapnel in every square foot of the muck. We do step down with our toes first like you suggested. I'm not sure if we were ever taught to walk that way, or if it's just a natural instinct. Either way, that was a good tip. Thanks!

All that shrapnel really plays havoc with the poor engineers trying to find mines. This place is loaded with them! We've got claymores all over The Zone. Hell, Tattoo Joe put some out just the other night while I was setting up the mikes. On the Chinese side of The Zone, our artillery guys have been kind enough to fire FASCAM rounds right in front of their trenches and on their side of the middle ridge. The FASCAM artillery rounds explode a few hundred feet in the air and scatter small, plastic, anti-personnel mines all over.

They're easy to detect, but motion sensors activate when they hit the ground, and it's a real job to get close enough to disarm them. The Chinese have fired a few into our area of The Zone, but theirs are pressure-activated, and our engineers just walk over and disarm them. Sometimes, small patrols from both sides will set up mines like Tattoo Joe did last night, so we're careful about where we step and what we do. Like anything else over here or anywhere in the World, if that bus has your number on its front grille, you're not gonna escape it.

A bigger problem than the mines-and adding to the metal detection difficulties-is all of the unexploded ordnance. A single Air Force F-16 drops thousands of cluster munitions into an area all at once. I've heard that more than 5% of them never explode because of poor detonators or impacts in soft areas like sand or mud. This entire area is soggy, so it's a lot higher than 5%. The Air Force has found a way around that: they drop more cluster munitions. That means that there are millions of these yellow, softball-sized bombs all over. Most are easy to see, but I don't think 5-minutes goes by where we don't hear one go off under some poor sniper somewhere out in The Zone. Bigger bombs have the same problem, and it's not uncommon to have a relatively quiet moment destroyed by a 2000-pound bomb exploding under some sniper or grunt on patrol. At least they never knew what hit them!

Well, I'm gonna get going. Rest easy, I'm scheduled to install more wells over the next week. It's cake duty-probably as a reward for going out and setting up those mikes. No complaints here! Say hi to everyone, and take care of yourselves. Thanks again for the tips about the mud. If you think of any more combat tidbits from your days in the field, write 'em down! We might be able to use them here too!

Thanks,

-Matt

Thursday, September 1, 2011

Associated Press International

~~Old Dogs Teach New Tricks~~

In a battle lasting less than one day, four World War II Iowa class battleships, having been reactivated and operating with a minimal escort, single-handedly destroyed every last ship and plane in the Bangladesh arsenal. Then, Air reinforcements from the small British aircraft carrier Invincible damaged most of the airfields in and around the former Bangladesh.

The Chinese presence was missing as the P.R.C. thought it was more likely that the U.N. would try to retake South Korea, and its valuable industrial potential, before they move on with an operation as large as the reinforcement of Southeast Asia. They were wrong.

Mid-Sunday afternoon eight American warships met with two supertankers and over a dozen cargo ships. They made their way towards Malaya-south of Myanmar-towards the port of Singapore. A patrolling British Nimrod flying ahead of the course detected a possible hostile submarine contact. However, but the previous day's rough weather, coupled with the shallow water and mechanical wear on the planes equipment made maintaining the contact and destroying it difficult; actually impossible. For safety's sake, a course change was made, reluctantly, to head north, and closer to the coast of Myanmar; formerly known as Burma.

At 7:04 p.m., a lurking Chinese submarine detected the convoy. The submarine crawled at 5 knots to intercept the convoy. At 8:34 p.m., a captured Taiwanese Frigate fired 8 anti-ship missiles at the surface radar contacts it had detected. The American ships turned on their radars (They had been turned off to avoid their own detection.), and fired surface-to-air missiles in self-defense. The fight had begun. The Chinese submarine rose to periscope depth, having detected the American radars, and then possessing specific target locations, it fired missiles and torpedoes at the Americans.

The convoy called in air cover from all over. The Nimrod was able to detect the SSGN Chakra submarine almost immediately and dropped a torpedo less than one mile from its stern. Two French Rafale fighters en route to Sidney turned to try and intercept the missiles from the Frigate. An American KC-135 mid-air refueling plane took off from Singapore to make sure that the Rafales didn't run out of fuel. Antisubmarine helicopters took off from the convoy's escorts to aid in the attack against the hostile submarine.

The Rafales were too far away to stop the current attack. Since there was little time to ready their defenses, the convoy was unable to stop most of the missiles, and it took losses.

The Spruance class destroyer U.S.S. Cushing fired off several Sea Sparrow SAM'S, and while maneuvering to bring its various weapons to bear, launched radar-confusing chaff rockets. As three more missiles homed in on the Cushing, the fast anti-submarine destroyer began firing all of its different guns. There was to be no salvation. All three SS-19 Shipwreck missiles slammed directly into the ship. What was left of the destroyer after the ensuing explosions sank in less than a minute. There were no survivors.

Two more missiles hit the Tanker Yoko Maru. The ship exploded instantly leaving nothing but a black, 10,000foot high column of smoke where it had once been.

The concussion from the blast damaged a Kidd class guided-missile destroyer, the U.S.S. Scott, over three miles away. All of its radar and radio masts were blown off. A series of small fires also ignited on the stern of the ship, and it was defenseless against the next series of missiles.

Another missile hit the transport ship Lopez. That ship was able to minimize its radar silhouette to the incoming missile by turning directly away from it. However, the missile wasn't completely confused, and it hit square in the center of the stern only a few feet above the waterline. Hundreds of troops from the Democratic Indian Army were on board and many were killed. Another 200+ were seriously wounded.

As the ship lay dead in the water, the troops helped to fight the fires and flooding, but dense smoke and language difficulties made the situation confusing. In one of the war's unknown miracles, the ship was saved, and it was able to make 7 knots less than 24 hours later.

The first missile attack was over, having lasted less than 5 minutes. The Communists rallied more forces to the scene, but the U.N. was unable to send more units in a timely manner. Another submarine, this one an old diesel, and several heavily-armed patrol boats headed toward the action from the ports in Bangladesh to the north.

While the Chinese-controlled Taiwanese frigate launched another volley at the small convoy, the SSGN Chakra took its first hit from a torpedo. Most submarines only take a single hit from modern anti-submarine weapons, but Soviet-built ones have stronger hulls, and the Chakra was still able to fight. The torpedo had hit between the conning tower and the stern. Flooding was slow and difficult to control. Another line of torpedoes came out from the Chakra in response. The Chakra would take three more torpedo hits from the British Nimrod in the skies above. After the last torpedo hit, the pressure of the water crushed it. It went on one last dive with no lives left on board.

The U.S.S. Scott was defenseless, but not incapable of offensive action. All eight of its Harpoon missiles were launched in a tight spread down the direction from which the enemy missiles had come from. The missiles used their own built-in radars and their own computers, so they needed no help from the ship except to launch them. The Frigate and patrol boats stood no chance against the 'smart' missiles, and all were later considered 'classic kills'.

A helicopter from the Scott flew over what its crew believed to be a possible submarine contact, and the Captain of the Scott ordered them to drop a torpedo just to be sure. A few minutes later, the diesel submarine Kasura exploded under the waves, and a small mountain of water leapt from the green sea a few miles from where the helicopter had dropped the torpedo.

Only a few weeks earlier, officers from all four of the services had come together with a plan to reinforce Myanmar and deter further Chinese infiltrations into Indochina. To support the brainstorm of an operation, the

U.N. sent its nearest naval task force into the Bay of Bengal to sweep out any of the small combatants that could have been lurking.

Task Force Almighty, consisting of the last 4 battleships afloat in the world, a Spruance class antisubmarine destroyer, the guided-missile cruiser U.S.S. Sterett, and an oil replenishment ship, moved at 16 knots with its radars on. They had been sent to pick a fight with the enemy and win. Their radars scanned the sky for hours, watching the communist air patrols lazily fly along the coast. Finally, the enemy attack began. Vice Admiral John Strase commanded:

"We watched 'em take off, circle their airfields, and join in formation. They knew we were looking for'em, but it was actually kind of funny to watch'em drop altitude in the hopes that we wouldn't see'em. That's when I gave the order to prepare the group for air attacks. We went to general quarters air, and the boys was a waitin'."

"It was really a pretty good attack plan on their part. They broke into three groups, stayed well out of our SAM range, and then waited till their commander gave the order. We had their frequencies monitored, so we heard it too."

"Each group broke into 5 smaller ones, consisting of 2 & 3-plane formations. They were at about 500 feet, and

then I got the only surprise in the hole thing."

"We'd also been watching their naval assets, if you will. They had a considerable number of missile patrol boats, and 5 frigates along the coast. All of their ships began launching missiles at us!"

"It really pissed me off! We had their entire air force cold as ice in our sights, then they pulled this crap on us. Only two of my ships had any real SAM capability, and no one knew if a battleship could defend itself with gunfire, let alone if it could take a hit?! Anyway, we changed course to bring every gun we had to bear on

these guys. It was a real sight to see all 4 battleships in a line for battle. The Sterett, Spruance, and Willamette pulled into a line behind the battleships."

"We put up a wall of flak so thick, I thought for a second that some of our rounds are probably hitting each other and opening gaps in the wall!"

"About then, their timing went to hell. If their missiles would have been hitting us at the same time as their planes, then we'd a been in a bad way. Lucky for us, somebody didn't do their math right, and the missiles came a few minutes ahead."

"Most of 'em got blown away, but a handful made it through, and that's really all ya need after all. We lost the Spruance right away, then the Willamette and Sterett got hit. Sterett was able to control its damage by turning into the wind to help blow out the flames. It made port a few days later under its own power. The New Jersey took a hit right in the bridge, but most everyone came out ok."

"I had everyone get their act together, reload, stow the wounded and take care of any accidents or damage. Then we actually had about a minute or so to wait for the planes."

"The air attacks were exciting, terrifying, and sad. In a way, I thought how terrible it must have been to be in one of those planes. We put a lotta rounds into the sky. Flak was really, really, really intense. Only six survived out of the 463 that came at us. All the planes were 1950's vintage, and we blew 'em away with 21st-century radar-guided anti-aircraft cannons mounted on battleships that had been designed in the 1930's. Strange."

"The New Jersey took 7, 500-pound bomb hits. It looked like hell afterwards, but it still worked."

"The strangest part about the whole thing wasn't how long it took, but rather how fast it ended. Just like that (snapping his fingers), the entire communist air threat in the Bay of Bengal was gone. I launched a remote-controlled targeting spy plane to fly over the country as we moved closer to the coast. The airfields were almost empty. The few planes that were there were cold on the infrared scan so we knew they weren't operable. I suppose it was like the smoke had lifted, and there was no one left to fight."

Friday, September 2, 2011

1st Lt. John Chamberlin -Personal Log

Somewhere in the Bay of Bengal

This morning's briefing with Colonel Ghetty included a personal news update from General Shapiro. Task Force Almighty, commanded by Admiral John Strasse, took on two-thirds of the military forces in Bangladesh yesterday. He wasn't looking for a fight, but he wound up kicking their asses! Bangladesh and Myanmar have squabbled over some border territory since Bangladesh was separated from India decades ago. Now, with Myanmar's defense forces busy trying to keep out Chinese infiltrators, it seems that somebody decided it was a good time to walk into Myanmar and take the disputed territory. The only problem was that Task Force Almighty was on patrol in the area, and it's well-known by now that the Coalition is staking a claim in Indochina. We held in Vietnam, but to make that victory last, we've got to protect all of Indochina, Indonesia, and the bulk of the South Pacific.

So, Shapiro volunteered our little assault force for a possible interdiction mission out in the Bay of Bengal. We're supposed to rendezvous with Task Force Almighty as soon as possible. We should be there in about a week. Admiral Strasse seems to have taken care of most of the threats in the area, so we're only taking two destroyers and a Taiwanese frigate along. We'll meet up with a small task force from Myanmar as soon as we get around the Malaya Peninsula. We'll stay close to the coast the whole way so as to keep friendly ground-based air cover close at hand.

It was nice to see General Shapiro paying such close attention to what was happening. I know he did in the past, but it was nice to be kept so informed. I think he wants to tell us as much as he can to make us all feel a little special.

Saturday, September 3, 2011

1st Lt. John Chamberlin -Personal Log

Somewhere in the Bay of Bengal

The latest news from Myanmar is that Bangladesh is getting a lot of support from its new Eastern Alliance neighbor to the north; China. I watched part of a news program that was being fed throughout the ship on our closedcircuit TV system. There were brand new Chinese-built Mig 29's flying over some troops marching through recently occupied Sittwe, Myanmar. I hope they were Chinese instead of ones from Russia. We've got a hell of an outfit put together here, but making an amphibious and air assault against the Russians is a tall order, no matter how good we all think we are! Just a few more days and we'll find out regardless of who's in the jungle or at the beach.

Sunday, September 4, 2011

Correspondence

From Corporal Matthew McKenna,

U.S. Army, 1st Infantry Division, 12th CHEM To Jonathon McKenna (*Grandfather*) Yo Gramps!

I've been gauging wells for the past week now, and Saunders is finally letting me do something a bit more interesting. I'm supposed to go out with Chip and his team again. This time we're going to set up some remote microcameras, like the one on your computer monitor or those video phones you see in conference rooms. The tricky part is: we have to run a wire back from The Zone to The Sewer. If the Chinese ever found the wire, it would lead them right to our location. So, Chip's going to have Tattoo Joe booby trap each of the cameras and microphones-including the ones we planted last time.

We're also going to put out some sticky foam cans. These are small cans-about the size of two soup cans or a small fire extinguisher. The jungle's too torn up in The Zone to make very good use of barbed wire. The Chinese have tried, but only close to their positions, and our effort has been half-assed at best. These sticky foam cans-the guys call 'em "Booger Bottles"-can be set off like a claymore; electronically or by trip wire. When they go off, they squirt a beige liquid all over the place. As soon as the liquid hits the air, it expands and sticks to EVERYTHING! It's a lot like packing foam, but the Marines dreamed the stuff up in their Clinton-era "non-lethal" weapons days. They actually first used it when they were pulling out of Somalia (there's a note of pride-NOT!).

When Chip heard that we were getting some booger bottles, he laughed out loud. I guess the Chinese hate the stuff. He said that the first time that they put some cans out in The Zone, the Chinese thought it was funny-until they came under fire. Then they tried to run away but stuck to every twig, branch, and rock that they touched. Chip says he heard grunts laughing all up and down the line. They're a rare find now, and I'm kinda excited about getting to use 'em. It's a real privilege to get to do all the things that I'm doing right now.

I found out why they call Harry "Bad Luck Harry." While we were dividing up the Booger Bottles, he accidentally set one off in the bunker he shares with Tattoo Joe. The stuff is everywhere! Joe was a lot of fun to watch. He was so pissed off, but there was nothing he could do about it besides bitch. His arms were stuck to his body. His feet were stuck together. When he went to block the squirt from the can with his hands, they were glued together. It was the funniest thing. Poor Harry turned to take cover, and he got goo all over his back.

On his way out of the bunker-away from Tattoo Joe-he turned so that he wouldn't have to look him in the eye. It was the worst thing he could do. Joe squirmed until some of the foam on his left arm touched the foam on Harry's back. The two were stuck together! Joe tried desperately to hit him, but he couldn't move. Harry was kicking, pulling, and trying to drag himself away, but he just became more entangled. Finally, Harry started to laugh uncontrollably, and that really set off Joe. Chip, Piotre, and I had to leave the bunker because we were laughing so hard!

I think the worst part of it all was that Pepe stayed to laugh directly at them. He said they looked like underside of a seat on a New York Subway car (everything is somehow related to New York for Pepe.). Those poor guys had to sit there and listen to him go on and on about New York, and there was nothing they could do about it. Pepe yapped. Joe bitched, and Harry laughed and laughed. What a scene!

Chip and Piotre are on the hunt for some diesel fuel. They tell me that it turns the sticky booger bottle goo into a slime that's a little easier to get off. The stuff is everywhere in the bunker. When they do find some diesel, it'll take hours for all of them to clean that place, and when they do, it'll stink of rotten fuel and half-melted plastic foam. What a mess!

I'll let you know how things go tomorrow. Say hi to everyone, and take good care of yourselves!

-Matt

Monday, September 5, 2011

Correspondence

From Corporal Matthew McKenna,

U.S. Army, 1st Infantry Division, 12th CHEM

To Jonathon McKenna (*Grandfather*)

Gramps,

We didn't get to leave for The Zone until almost sunrise this morning.

Harry and Joe were still getting the Booger Bottle goo out of their gear. Piotre, Chip, and I headed out first, and they followed about a half hour laterminutes before sunrise. We all slipped out the same firing port from the same bunker in the same way that we did last time. Chip and I both dragged the bags of cameras, mikes, Booger Bottles, and mines, while Piotre crawled ahead and found an overwatch position.

It was still dark when we reached our rally point on the middle ridgeline. Our night vision goggles are thermal, and I could look through some of the thicker debris at the Chinese positions on the next ridge to the north. They looked like little white blobs going back and forth through the zig-zag trenches. I looked through the low light enhancement sight on my M16 to see if there was any difference. That sight's a lot older, and all I could see was the shiny green reflection from each set of Chinese eyes. They looked like exactly like swarms of green lightning bugs moving in a regular pattern. It was a bit scary seeing so many people, so close and ready to kill. It was also very sobering. I set up a camera, a mike, and a trip-flare. This way we could keep an eye on the Chinese, listen for any patrols that we didn't visually detect, triangulate their heavy weapons, and illuminate them if they found my little electronic observation post (EOP). Chip waited for Tattoo Joe and Bad Luck Harry to come up, but he was nice enough to separate the gear so that each buttpack carried enough equipment for two locations (It's a lot easier than constantly going back and forth to the rally point.).

While we were up there, the usual sniper and machinegun fire increased as the sun was coming up. This made things a lot more dangerous. I was extracareful, but that also meant that I had to move slower and take more time. Around noon, Piotre spotted a Chinese patrol coming out of their trenchline. I heard him fire the first time, then the second, and then the Chinese cut loose with everything they had along the line! Piotre kept firing-faster and faster-as the Chinese patrol tried to move in on him under cover from their trenchline. Chip, Harry, and Joe crawled through the tangled brush to help him out. Fart Boy stayed at the rally point. I figured that since the Chinese had their attention stuck on Piotre, I could move a little faster and get things done before anyone came up to give us any trouble.

Piotre must have gotten all the Chinese on the patrol because after I setup four more EOP's, they were all waiting for me at the rally point. Chip said that we needed to start heading back before the Chinese started to send us some stuff. So, I packed up the rest of the gear while Joe started to put out the rest of the mines, Booger Bottles, and other nasty little devices. As soon as the cameras and mikes were packed, Pepe and Harry started to drag them back down toward The Sewer-less than a click away.

I was entering the positions of the different EOP's into my laptop when the first of the Chinese stuff arrived. We all heard the mortars thump away from behind the northern ridge. The sun was up, and a quick glance into the sky gave quick glimmers of the mortar rounds as they reached their apex and started down on where Piotre had been when he picked off the Chinese patrol.

They fired at least twenty rounds, and then the artillery stuff started to come down. We heard it thunder in the distance, and knew right away what was coming. Harry, Joe, Pepe, and Piotre ran like hell to get back to The Sewer. Chip grabbed me and my laptop and we started to join them. The rounds were falling about 200 meters to our east, but three of them fell short and landed within 50 meters of Chip and I. I could hear the shrapnel cutting through the dead trees and snapping through all the fallen brush. The concussion from the closest one knocked me off balance, and I fell forwarddown the ridge toward The Sewer.

Chip fell too, but not because he was thrown off balance. He was dead. When I stopped rolling, I turned to see where he was, and his body just slid down the hill past me like a pile of laundry down the clothes chute. Harry and Pepe saw him fall, and they had turned around to help out. They caught his body at the bottom of the valley. Joe and I joined them and we all carried him back in through a bunker's firing port while Piotre watched for any Chinese to come up behind their barrage. It was still going on, but after the close call that killed Chip, it slipped from my mind's focus. I knew it was there only like I know that the sun feels warm and the wind is blowing.

Chip had taken a racquetball-sized piece of Chinese steel between his shoulders. It definitely cut his spine, but it must have gone through his heart too. There was blood everywhere. I had no idea a human being carried so much blood. His body armor was so filled that the layers of Kevlar, cloth, and padding separated and swelled like a sponge.

617

The bunker that we all climbed into couldn't have been more than 5'x5'. There were already four grunts in there manning a Mk19 (a fully automatic, belt-fed, 40mm grenade launcher). The four of us joined them, and we all found ways to step back and give Chip's body some space. It was like we were waiting for him to catch his breath, cough, and sit up as though he had drowned. The grunts who were in the bunker stepped back into The Sewer and gave us some more space.

We needed it. Harry (who's always fun and happy regardless of the situation or his luck) cried openly and loudly. Joe held Chip's hand, kept shaking his head in obvious denial, and I could feel his frustration evolving into total rage. Pepe kept his hand behind Chip's back in an unconscious attempt to apply pressure to the already fatal wound. Piotre stayed outside, and I just sat there like an idiot.

Chip's dead, and it's because I waited to dick-around with the laptop instead of leaving when everyone else was heading back. I know it's the oldest story in the world, but I know that this is really my fault. He was a nice guy, a good guy, a good leader, and maybe the last connection that any of us had with normal people. Over here, everything seems out of place, weird, or surreal. People take on the characteristics and egocentrics of their surroundings. Chip didn't. He was the last normal guy here, and now he's dead.

Saunders had some of the other guys in the 12th CHEM take care of his body. Chip's CO is writing the family, and so is Saunders. I want to, but I don't know how to tell them that he died because I screwed up and waited. I'm embarrassed to have let this happen, and I don't even know how I'm going to face the rest of the team. Worst of all, Saunders says that we have to go back out there tomorrow and finish the job!

The guys say it's not a problem, and everyone tells me that it's not my fault, but I'll be damned if I screw up like that again and get someone else killed. Harry's got the worst luck in the history of the world. Joe's usually an ass. Pepe is as annoying as a yipping kick dog on caffeine, and Piotre's the worst sniper in the Army. Still, I think they're the finest group of grunts we've got, and I'll do anything to earn their respect back.

-Matt

Tuesday, September 6, 2011

Correspondence

From Corporal Matthew McKenna,

U.S. Army, 1st Infantry Division, 12th CHEM

To Jonathon McKenna (*Grandfather*)

Gramps,

We headed back out into The Zone last night. It's a good thing we left early, too. A full day of rain left the valley filled with soft, foamy mud, and the middle ridgeline was covered with greasy mud. We had to climb up using the debris for handholds. The rain also made it real foggy. A steady stream of yellow Chinese artillery flares in the sky made the whole scene seem like a piss-colored heaven.

No one's blamed me for Chip's death. They've all gone out of their way to make me feel better, actually (These guys are so forgiving.). Tattoo Joe and I set up another 19, EOP's. As soon as I had a camera and mike wired, Joe moved in and booby-trapped it. It was quiet. The artillery stuff, the machinegun bursts, and the sniper duels are all just background noise now. We're well aware of it, but it only draws our attention when it's incoming stuff, or the rounds are starting to snap through nearby brush.

While I was typing in the locations of all the EOP's, Joe placed the Booger Bottles all along the data ribbon from the top of the middle ridgeline down to the valley at the bottom of The Sewer. When I was done, I was ready to go back, but the guys had other ideas. They wanted to get a little payback.

I didn't even bother to mention our mission. It's not that I didn't care about our orders. They wouldn't have listened. Besides, I wanted to get some revenge for Chip too. I wanted to remind the Chinese that no matter how hard they try, we're not giving up. We're not going away. We're not going to roll over and let them walk all over Asia or the World. So, we crawled away from the EOP's to keep them from taking damage from any possible return fire.

We found a nice little spot where a large bomb had blown a hole-about 100' across-in a pile of torn jungle. It was like a circular bunker complete with circular brush for cover. Piotre climbed back up to the top of the middle ridgeline behind us, and he disappeared into his own little nook. No one wanted to be near Harry if there was going to be any sort of shooting, so we had him go back up to the top of the middle ridgeline and find his own nook too. Joe cleared out a firing port from the bottom of the brush a few meters to my left, and 'Fart Boy' Pepe did the same on my right, but a few feet closer. I set up a Shortstop artillery buster in the back of our crater, and then I checked to make sure all my gear and magazines were easily accessible.

When everyone was comfortable-and we all had escape paths picked outeveryone motioned to each other that we were ready. The Chinese were invisible in the night vision sights on our weapons, but our thermal, helmetmounted pieces cut right through the fog. Piotre fired once, then a few times in succession. His first shot never found its mark, but he swears that the rest always do.

After the second or third shot from Piotre, the Chinese trenchline on the ridge to the north opened fire on us. Tracer rounds came from everywhere on that ridge. Normally, we'd see a lot of tracer rounds (about one out of every five Chinese bullets is a glowing, yellow tracer bullet). Since bullets come out of guns almost red-hot, we saw every single one through our thermal sights. It was terrifying. I tried to crawl into the ground, but I was afraid to go deeper into the crater and lose sight of the Chinese.

Piotre stopped firing for a little bit and crawled down to tell us what was going on. He had spotted an officer in the Chinese trenches, and actually managed to hit him with the first shot. It was the first time in months! He was embarrassed and proud at the same time. I think his reputation as a poor "One shot-one kill" sniper was bothering him, and now he's obviously glad it's over. Piotre thought the Chinese officer was doing an inspection or checking The Zone for something. It didn't matter. He had a group of officers and security troops with him, and when Piotre popped him, they all zeroed in on our sniper.

Tattoo Joe told Piotre to go back up and keep an eye peeled for patrols. He didn't need to. Pepe cut loose with a few short bursts from his M249 SAW. Then he let out a long burst-maybe 50 rounds-and yelled over to us. Piotre crawled back to his position while Harry, Joe, and I turned around to see what was happening. Joe started firing next, and then Harry. I still didn't see anyone to shoot at.

Then the stuff started raining in on us. Just as before, the Chinese started off with some small infantry mortars-60mm and 82mm stuff. The Shortstop that I brought along must have been working. We could hear the mortars fire, and we could hear the explosions above us, but there was nothing more than the sound of steel raining into the brush piles a few hundred feet behind us. The larger howitzers followed; 8-inch stuff. Bullets were still slapping through the air from all along the Chinese trenchline and from the patrol that I still didn't see. When we heard their artillery thumping away in the distance, all of us got down and braced for the inevitable. It never came. The Shortstop must have set off their fuses high in the air because there were these incredibly loud flashes of thunder high in the sky. Noise from the guns firing, the bullets snapping through the sound barrier as they passed by, the mortars raining steel behind us, and the heavy artillery exploding in the sky all combined to leave our ears ringing. I didn't hear anything after a while, but I could always feel the concussion from the mortar, artillery, and then grenade explosions.

Tattoo Joe, Harry, and Piotre must have had good views of the Chinese patrol. I didn't see anything. My thermal night vision piece was too confusing to make out anything through the bullets, shrapnel, and anything else. When I took them off, all I could see was the flickering yellow fog with glowing yellow tracers coming through it like a swarm of lightning bugs or hornets on acid.

It was time to go. I told Joe and Harry that we had to get out of there and run the EOP ribbon back to The Sewer. I grabbed the Shortstop and planted a pair of Booger Bottles. Harry fired off a burst into through the brush and headed back along where we had buried the ribbon. Joe was pissed, and it wasn't until he was down to his last magazine of ammunition that he was ready to go. Pepe finished his belt, reloaded, and let out a few long bursts for cover fire (all the while swearing at the Chinese and saying something about Brooklyn) as Joe, Piotre and

I followed Harry home. The rest of us ran, crawled, ducked, and hurdled our way through the maze of debris on that middle ridgeline.

A few minutes later, we stopped hearing Pepe firing, and so we stopped and waited for him to catch up. The Chinese never let up, and their stuff was starting to land all around the crater. Finally, Pepe came jumping over a pile of brush. He landed right next to Harry-surprising both. They shot at each other, but both dodged the others bursts and recognized one another right away. Pepe bitched that the place was as crowded as Times Square, then he pushed Harry out of the way and headed back to The Sewer with the rest of us. It was a close call, but we all understood how it could happen, and we were all grateful that no one was killed!

We made it back into The Sewer about an hour and a half later. Piotre picked off three Chinese (probably from the patrol that was sent after him) as they came out of the debris at the bottom of the middle ridgeline. All three only took five shots, so he must be getting better. I noticed that the spent casings from his rifle really eject a long distance-almost 10-feet! He says it's normal, but I'm gonna look around on the internet tonight and see if H&K has a brass catcher that we can order and have delivered (APS delivers right to The Sewer- overnight).

I've spent most of the day trying to locate in all the different EOP's that we set up last night. So far, all but two of them work very well, and one of those is still sending microphone data, so I just have to figure out what's wrong with it specifically. The cameras are 1990's-era technology, and their resolution is cheap, but it's fun to watch. I can call up any EOP on my laptop and both watch and listen to what it's picking up. It's like having security cameras monitoring the Chinese lines. As soon as they step out, we can call in the artillery right from the comfort of our air-conditioned bunker (while the rest of the guys cook internet-ordered steaks on the grills out back). Until then, division artillery refuses to fire (and reveal their positions) unless they've got a serious attack coming. Now, Saunders thinks we might be able to use them a bit more effectively even without large-scale Chinese attacks bearing down on us. We'll see.

I've got a lot more work to do, so I'm gonna go. Take care, and say hi to everyone.

-Matt

Wednesday, September 7, 2011 Correspondence

From Corporal Matthew McKenna,

U.S. Army, 1st Infantry Division, 12th CHEM

To Jonathon McKenna (*Grandfather*)

Gramps,

We had a big cookout today to celebrate the EOP network. Last night, Saunders and a few of the guys got together and connected all our laptops. They were up all night, but they were finally able to get all the different programs to compile their data and send it to Saunders' program in the correct format. It enables Saunders to send data to Division, Corps, theatre command, and even the Pentagon. He's getting combined data from all the groundwater monitoring wells, all of the sonobouys, all of the mikes, all of the cameras, and even from some ESM (radio signal detectors) that another team has been putting out into The Zone. He figures that in a few days, he'll have an accurate map of the Chinese positions, their strength, AND what they're doing. It started off as a simple tunnel hunt, and now, we've got the opportunity to blow a hole in the Chinese lines and end this trench-warfare crap!

I wasn't sure how much I would want to celebrate so soon after Chip's death. None of us were. Once we got there, the cheeseburgers, fries, steaks, and barbecued chicken took hold. It's only been a few days, but somehow, it seems like years since Chip died. We don't talk about it, but we haven't forgotten him. Maybe we're all just too busy, or maybe we've just seen so much since then? I don't know.

I got another one of your letters today. Thanks for trying to cheer me up about Chip. It was very hard at first. You said that you still have memories of Vietnam like they happened yesterday, but I have a tough time remembering everything that happened when Chip was killed-even though it's only been a few days. Maybe the memories will come back. Thanks anyway.

I'm sorry to hear that you have to go back to work. It sounds like you really got ripped off on that IRA. I guess you can't trust anyone-banks included-to be honest with your money for decades and decades. Sure, the economy's supposedly kicking ass right now, but I figure if one guy's making money, he's gotta be getting it from somebody who's losing it, right? If it's any consolation, I'd love to come back and work so that you don't have to! This place is wonderous, exciting, and not exactly too uncomfortable, but sometimes the dangers-and realities like Chip's death-can really get to us. No one's afraid of the pain from getting hit, we just don't wanna go home in a chair or a bag (They call it "Rollin' an floppin'" over here.).

Don't work too hard. Do you have any idea what kind of job you're gonna look for? Whatever it is, try and get one of those medical plans that covers "physical maintenance." I hear that those plans actually cover health clubs and sometimes even spas! One thing we do not have over here is a spa, but there is a rumor that the Ukranian division to our east has a few bunkers decked-out for hookers. It's probably not the first time for an army in the field, but I have a hard time believing it.

Take care,

-Matt

Thursday, September 8, 2011

Correspondence

From Corporal Matthew McKenna,

U.S. Army, 1st Infantry Division, 12th CHEM

To Jonathon McKenna (*Grandfather*)

Yo Gramps,

We've gotten some spastic intel on the Chinese trenchline from our EOP's and all of our monitoring well adventures. Saunders says that there's probably a tunnel coming diagonally through The Zone. It looks like it starts up north about two miles from our CP. According to the results from the groundwater monitoring wells, they're digging it under a layer of rock that splits the aquifer into two distinctly different layers. Saunders figured that it must be a leaky nightmare, so he looked around for signs of pumps or dewatering devices. He never saw any, but he was able to use one of the cameras from one of my EOP's to spot a nice little stream flowing from a low point in the Chinese trenches. Some of us figured it was just to de-water their trenches (most of their line doesn't have any sort of roofing at all!), but this water is extremely muddy, and very dark colored-different from the soil in the area near the stream's exit from their trenchline. Our ESM sensors have also indicated that there might be a Chinese forward CP in the same area where Saunders thinks the tunnel starts. Best of all, it's not too far from where Piotre popped the inspecting Chinese officer the other day. Taken individually, these clues don't mean a thing, but when you see the data on the map, it's very clear that there's a good, hard, and valuable target not too far from here.

The question was, what do we do about it? Saunders went up to Division to see if we can get an artillery strike on the area, but General Simms doesn't think that even our big guns are strong enough to do much to the tunnel and CP bunkers. He said he'd try to get III Corps to send in a heavy airstrike tomorrow-something like a few of those 2000-pound Bunker-Busters, or maybe even a 15,000 pound Daisy Cutter. We'll have to wait and see!

Most of my time has been spent monitoring my EOP's, collecting data, and working with Saunders to come up with some results. I'm only getting a few hours of sleep anymore, but it's more than most of the grunts up in The Sewer. They have to rotate their post duties every 4 hours. Since most of the firing pits and forward bunkers only hold two people, that means no one gets more than 4 hours of sleep. Back here in the rear, we're still getting 6-7.

I couldn't complain if I wanted to. It's true that we live in dirt homes, surrounded by mud and debris. It does rain everyday, and there are always the threats of some random stuff coming over from the Chinese side and popping people when they least expect it. Still, we have all the comforts of home-even satellite TV, retail stores in some bunkers, and the ever-important backside of the ridge grill pits. We've got things fairly good around here.

You guys back home sound like you're having it rough. You have to worry about your mortgages, medical bills, even your food. Traffic probably hasn't gotten much better than the hell it was a few months ago, and we hear that the communist commandos are keeping the crime rate on a steady climb into the stratosphere. We actually worry more about y'all than we do ourselves over here!

It's time for me to get some sack time. Take care of yourselves-I mean it. Don't work too hard. Don't sweat the small stuff, and as long as you look out for stray buses racing down the road, I'll watch out for that random Chinese stuff coming over the ridge. Say "Hi" to everyone for me.

-Matt

Friday, September 9, 2011

Correspondence

From Corporal Matthew McKenna,

U.S. Army, 1ˢᵗ Infantry Division, 12ᵗʰ CHEM

To Jonathon McKenna (*Grandfather*)

Yo Gramps,

Today was an Air Force day-for both sides! This morning, the Chinese had a pair of fighters come haulin' ass through The Zone. They dropped a string of bombs on the top of the middle ridgeline. It wasn't too out of the norm. The do it every few days. This time was different though. This time, they dropped E-bombs.

E-bombs look like regular 500-pounders. Inside, they're just a small stick of high-explosives. The explosives are surrounded by tightly packed coils. Those coils are charged by capacitors in the back of the bomb that activate when the bombs are released from their planes. As soon as the bomb hits the ground, its fuse sets-off the stick of explosives in the middle. In the instant that the explosion goes from the nose of the bomb to the tail, it blows apart the coils as the explosion goes back to the tail of the bomb. Fewer coils means more charge to the ones near the back. In less than a second, the explosion breaks all the coils, and releases the charge-creating a huge magnetic field around the bomb. They don't do too much physical damage, but just one can knock out everything electrical in a two-mile radius. These are real easy to make, and both sides have been using them throughout the war. Of course, ours are a little more powerful, and we can tune them to only block out certain frequencies, but the effect is still generally the same.

They dropped eight on us today. We were so pissed. After all that effort to put out all the EOP's, after Chip died to get the job done, after all those hours collecting data, the Chinese wiped it all out in an instant. On second we're watching them on TV and listening to their jokes through the mikes in our EOP's. The next second, we're back to World War I and carrier pigeons for communications. We didn't have any pigeons, so Saunders sent a runner back to Division to let them know that the Chinese must have found one of our EOP's. Since they know that we know where they are, he expected that they might attack at any time.

Division thought the same thing. They had scheduled an air strike on the suspected tunnel entrance for this morning. A pair of F-15E Strike Eagles was going to put down 8, 2000-pound Bunker Busters. The new attack threat meant that additional firepower was called for. The runner came back and talked with Saunders for a few seconds. Then he turned and told everyone around to pass the word, "Get down and get ready for the biggest bomb in the world!" Even people who were outside passing stopped. We all thought the nukes were flying. In less than a minute, the entire Sewer was buttoned up and bracing for miles around.

The Strike Eagles came in first. They came down the same flight path that the Chinese planes had used when they dropped their E-bombs. At 1000mph, we all saw them dive out of the overcast, level out, drop their bombs, and disappear back into the clouds-all before anyone heard their engines. They had out-raced the sound! When we finally did hear them, the bombs were already cooking off deep underground.

A 2000-pound bomb is a wonderful thing. It leaves a crater well over a hundred feet wide, depending on the type of soil, rock, or mud that it hits. The concussion is so strong that it easily knocks everyone off their feet and onto the ground within a mile+ radius. Anyone within a few hundred feet of the actual crater is instantly deaf-I mean their eardrums actually burst from the overpressure. It's no good hiding in a foxhole either, because the concussion will actually crush your body against the dirt walls. The only way to survive is to watch from a distance-like we did.

Ground-penetrating "Bunker Busters" are even better than the average 2 Grand bomb. They're steered to their targets by Global Positioning System satellites. Someone back at a CP or an HQ enters the longitude and latitude where the target is, and its computer homes in on that location by triangulating the direction of different signals from all of the military GPS satellites. The computer just flies the bomb to that location like a simple flight sim game. When the bomb hits, it can be programmed not to explode until a certain time has passed since it impacted with the ground. It can also be set to explode after penetrating one, two, three, up to seven separate layers of hardened bunker. When they do go off, the explosion is tamped or compressed by the surrounding soil-sometimes even trapped inside a deep bunker or tunnel. Normally the energy of the explosion spreads out 360x360 degrees. When it's trapped, all of the energy bounces off the surroundings and finds the nearest, easiest path to open air. As powerful and devastating as a 2 Grand bomb can be, the underground ones are at least 100 times more potent, and the sound of their blast is actually painful to hear-even miles away.

The first Strike Eagle dropped 2000-pounders, but the second dropped a pair of 5000-pound Bunker-Busters. The tunnel that Saunders gave coordinates for was dug underneath a natural layer of rock. We think that the 2000-pounders probably created some micro-fissures or cracks in that layer of rock. The first 5 Grand bomb was huge. We were miles away, deep in the Sewer bunkers, and it knocked everyone down. When I got back up and looked out our firing port, a wave of dust rolled up and into The Sewer. The second 5 Grander hit just a second later. About half an hour later the dust finally cleared, and we all saw a deep groove winding through The Zone. The entire Chinese tunnel collapsed and all the earth above it looked like a new ravine. In some spots, small streams even changed course and flooded into the crack. It just doesn't get more complete than that!

You would think that everyone would be thrilled to see that all of the effort spent finding that tunnel finally paid off. No one talks about Chip or any of the grunts who have died in the past few days, but those guys are still on our minds. When I looked out that narrow, camouflaged firing port and saw that groove carved in the landscape from the tunnel's collapse, I didn't care about the work. I still don't give a rat's ass about all the pinko troops buried inside. All I can think about is that now we have to go out and plant a new network of sensors because we just blew all the old EOP's and wells to crap. There's a ton of work to be done now, and with that work, we all know that there will be some more lives lost.

Everyone else seemed to have that same thought-at least in the bunker I was in at the time. I know because when I turned away from the firing port so that the next guy could look out at the collapsed tunnel, there was no one in line to go up to the firing port. There were six of us in that small hole in the ground, and we were all looking at each other with the same look. It's that look that says, "I wonder which one of you is gonna buy it now?"

I'm gonna go now. No one's given any specific orders yet, but we all know the work orders are coming. I'm gonna try and get some rest while I can Take care. Stay safe, and say hi to everyone for me.

-Matt

P.S. Saunders just called a meeting a few minutes ago. It was short and sweet. Tattoo Joe, Pepe, Harry, and Piotre have been reassigned as my own personal bodyguard team. It sounded cool, but I remembered one of the things that you told me about the military, "There's a reason for everything they do." I knew that if I was given a team of bodyguards, then my body was going to need guarding. I mentioned it to Saunders, and he tells me that Division wants us to setup a line of

EOP's a few miles behind the Chinese trenchline. Orders are orders. I'm not too thrilled, but I didn't come over here to sit in the rear with the gear. Besides, I've got a good group of guys looking out for me, right?

Saturday, September 10, 2011

1st Lt. John Chamberlin -Personal Log

Off the coast of occupied Sittwe, Myanmar

Today it's our turn to attack. The U.S.S. Saipan has made some small helicopter and amphibious landings. The four WWII Iowa class battleships are supporting with 27-round volleys of 16-inch shells. Our carrier, the Ronald Reagan, will be close enough to launch a helo assault before noon. General Shapiro briefed everyone on the hangar deck last night after dinner was served. When I say everyone, I mean every single Marine and soldier on the ship was there.

A few days ago, a rapid reaction armored ground force from Myanmar broke through a thin frontline of troops from Bangladesh. The armored force encircled occupied Sittwe and began squeezing the perimeter. Dense jungle and steep terrain are keeping the Eastern Allies from breaking through to the city from the north. Instead, they're trying to build up for an offensive from the south. The U.S.S. Roosevelt pulled close to the coast a few days after Sittwe was surrounded and has been launching airstrikes against Bangladesh strongpoints ever since. Our four battleships moved even closer and have been pounding any bad guys moving within 30 miles of the coast.

The platoon is scheduled to take off after dinner so we'll get one more hot meal than most people will. Our destination is the Sittwe telephone office, right between Sittwe University and the city police station. Those two targets are assigned to some of the U.S. Army airborne troops that were attached to our unit.

We don't have a building layout, but the aerial photos give a pretty good idea of the place. The biggest problem is that we have to land in the street. Choppers and powerlines don't get along very well, so we'll have to rope down. I'd prefer to rope in on the rooftop, but no one knows how strong it is. Besides, the chopper pilots are telling us that they'll get us within 20 or 30 feet of the ground. It's not really the fall, or even the landing that concerns everyone. The problem is the amount of time that we're sitting in the open, a few feet from people who'll be shooting at us, while we have to untie ourselves from the ropes. Normally, we'd rappel down, but I'm having our platoon rope in untied. If someone falls and breaks a foot or a leg, at least we'll all have time to get down on the ground and get in a building for some cover.

Sunday, September 11, 2011

1st Lt. John Chamberlin -Personal Log

Occupied Sittwe, Myanmar

Telephone office building,

The 1st Composite Air Assault Regiment has made its mark. We lifted off from the Reagan late last night. We were supposed to leave after dinner, but air traffic management had some problems, and we didn't make it into town until after midnight. When we got there, most of the fighting was done. There was no opposition. In fact, the city police were waiting outside with eleven Bangladesh officers in chains. Scuttlebutt is that the first assault wave came in so quick and hard that they took their targets in only a few minutes. After that, the second wave came in unopposed and pushed north until they linked up with the Myanmar armored troops who were only two miles away. By the time we came in, all that could be heard was the choppers and the occasional sniper.

Sittwe is like another planet! The city is barely a town by American standards. The university is like an old high school, and the rest of the town seems like one big bazaar. It reminds me a lot of Vietnam, but even in the midst of battle, with bullet holes on every building and craters at every intersection, the place has a peaceful, supernatural feel. It might have something to do with the temples, shrines, and religious icons that fill every viewpoint. I can see other Marines' eyes wandering and checking out the scenery too, so I know that I'm not the only one who's somewhat enamored.

All in all, our debut went better than expected. From the news and intelligence reports that we were getting, we all expected to be landing in a hornet's nest. Instead, it almost looks like we missed the fight. More accurately, I think we swatted a fly with a sledgehammer. We did accomplish our mission, and I don't expect that the regiment took too many casualties. We took the city, we shattered the enemy, and we made one hell of an impression.

Even after all these weeks of maneuvers, it was spectacular watching all our Blackhawks and Osprey spinning about the city and back and forth from the carrier. Anyone watching from an opposing side must have been stunned with awe. If there were any Chinese or Russians in the city, they must have hightailed it out without anyone seeing. I've no idea where the troops and planes that we saw on the ship's TV disappeared to.

Monday, September 12, 2011

1st Lt. John Chamberlin -Personal Log

Occupied Sittwe, Myanmar,

Telephone office building

The Chinese are still around. They hit the city with an artillery and mortar barrage this morning. A few minutes after each one, we would watch a couple of F-18 Hornets race over in formation. They'd fly overhead at about 500 feet. After they disappeared over the jungle, we'd hear their engines tearing through the air, then there was a rumble in the distance. After two barrages and two air strikes, it was over.

Our resupply Osprey dropped off some MRE's and ammo after the second attack. The crew chief said that she heard from one of the deck hands back on the Reagan that the scuttlebutt on the ship is the Chinese ran off into the jungle as soon as they saw the huge helo assault the other night. I'm not looking forward to jungle fighting again, especially if we have to do the attacking. Maybe they won't stop running until they get to some open terrain. Then our planes and artillery can pound the hell out of'em. Of course, maybe we can find out where they're headed and the 1st Composite Air Regiment can leapfrog ahead and ambush them sort of like we did in Vietnam!

Tuesday, September 13, 2011

Correspondence

From Corporal Matthew McKenna,

U.S. Army, 1ˢᵗ Infantry Division, 12ᵗʰ CHEM

To Jonathon McKenna (*Grandfather*)

Yo Gramps,

S'up, mon?! Greetings from behind enemy lines! How many emails start off like that?! It's taken a few days, but we're finally here. I won't say specifically, but we're over a hundred miles north of The Zone-far enough that we shouldn't have had to worry about running into a frontline Chinese unit. I won't say exactly how we arrived here or how we made it across The Zone, but I can tell you that we took a regular airline turboprop flight to another city here in Indochina. Then we took a ferry upriver, and finally, we trudged through helluva jungle/swamp for the last three rainy days! We're somewhere near Popa Hill *(Spirit Mountain)*, south of Mandalay-near the Irrawaddy River. That only narrows it to a few hundred square miles, but you can at least find us on a map now (sorta).

It's been a VERY strange journey. Sitting in The Sewer the other day, as we looked across The Zone at the Chinese trenchline, it seemed impossible to get behind enemy lines. We've since found out that there are huge areas that are

629

completely open for attack. The Shan State here in Eastern Myanmar (Burma) is about the size of West Virginia. It's all rough mountains, jungles, swamps, flooded rivers, and combat areas.

Thailand, Laos, and China border the State, but there are a lot more people fighting in it! Besides a Coalition Corps made of troops from Myanmar, Laos, and Thailand, there are all kinds of local opposition groups who seem to want their individual towns or even the entire Shan State to be under their control-or their warlord's control. My briefing data file has details for groups like: The Karenni Army (KA) about 1000-2000 fighters, about 2000-3000 fighters for The All-Burma Students Democratic Front, at least 4000 fighters for the Karen National Liberation Army (KNLA), the Mong Tai with at least 8500 fighters, the Mong Tai Army (MTA) with about 3000-4000 fighters, The Shan State Army with at least 3000 fighters, the Myanmar National Democratic Alliance Army (MNDAA) with at least 2000 fighters, the National Democratic Alliance Army (NDAA) with another 1000 fighters, and the Democratic Karen Buddhist Organization (DKBO) with almost 1000 fighters. Even the smallest villages and encampments all have their own militia, and we're told that alliances change every day-sometimes several times a day! The people of Thailand have also seen fit to send almost 10,000 of their Hunter Soldiers (Yes, that's 10,000 well-trained snipers and assassins) into the Shan as a token of their appreciation for American support in the region. There are also tribes of honest-to-God headhunters! Best of all, there are at least six Chinese divisions reported in the Shan State, and at the heart of all our thoughts we know that we've been sent here to set up EOP's because SOMEONE back at Division Intel thinks there might be a lot more Chinese than we already know about!

There's a lot of conventional fighting going on around the cities in this area. Myitkyina and Mandalay are battlefields in the most traditional sense. The Chinese trenchline and our Sewer seem to go around this entire region. It's just too dangerous to move large forces through it. Between the cities, are thousands of miles of the most twisting roads in the world. The roads are all mined daily, and we're told it's like driving through one long ambush that goes on for days. The paths through the jungle and the mountains are all controlled by local militias, and their loyalties to the different combatants are constantly changing.

While most of our trip was little more than a few routine airline tickets, the last few miles were equally unusual. With all the trillions of dollars spent on fighters, bombers, destroyers, aircraft carriers, Abrams', Bradleys, and helicopters, you'd think the U.S. government could put people just about anywhere in no time. Well, our flight was on a civilian plane (There were even flight attendants!), and after we landed…WE HAD TO WALK! Yes, that's right, it all comes down to a $29.95 pair of boots. We got off the plane, grabbed our gear, and walked off into the mountains. It was actually that simple-sorta.

On the way to the mountains, we had to negotiate swamps, rice paddies, and villages with unknown affiliations and loyalties. It was a bit dicey, but it was a learning experience. Do you remember that letter you sent me about the mud in

Vietnam? You talked about how you guys found it easier to walk barefoot sometimes. Well, I told the guys, we tried it, and you're right! It's a lot easier to walk through the swamps and rice paddies with nothing on. I guess it makes sense, since the locals have been doing it that way for a long long time. It's easier, It's faster, and it's a LOT quieter. Thanks! Keep those tips comin'!

Well, it's been a long, hot, and wet day (as usual), and we're gonna get going soon. I wanna make it to the top of the mountain tomorrow or the next day. Saunders sent me an email saying that Division's getting restless and wants to see some video soon. They'll get it when they get it, but I don't want to get him into any trouble. He's been good to us so far (although Chip's family might argue that.). I'll send you another email when we get there. Right now, it's getting dark and that's when it's easiest for us to avoid the all the shooters. I'll take care if you do. Say "Hi" to everyone for me.

-Matt

Thursday, September 15, 2011

1st Lt. John Chamberlin -Personal Log

Sittwe Airport,

Occupied Sittwe, Myanmar

There's a monsoon coming in sometime in the next few days. The Myanmar armored forces have a pretty good hold on the situation now, so most of us are going back to the Reagan to ride out the storm. Seems odd that it would be better to ride out the storm on a ship rather than on land, but since Sittwe's a flat, wet, coastal town already, who knows what it would be like during a monsoon?

Sometime tonight, our troop is supposed to leave from the airport about a kilometer and a half to the South. We handed over the building to the policeman that greeted us on our arrival, and then we marched out. The Osprey will be giving rides back to the ship from 2:00pm until the weather gets too rough. I hope we're not on the last flight.

There's no sign of the Chinese or the Bangladeshi troops anymore. Even the Navy's close air support strikes are winding down. The battleships are still pounding away, though. They're too far to the South and too far out to see to watch, but last night, their muzzles flickered on the horizon. I thought they would eventually run out of ammunition, but they didn't stop until well into the morning. Somewhere, across the river and bay, deep in the jungle, they've got targets; very hammered targets.

Friday, September 16, 2011

1st Lt. John Chamberlin -Personal Log

U.S.S. Ronald Reagan

Approximately 30nm SSW of Sittwe, Myanmar

Murphy's Law hit again. The day before yesterday, the sky was clear blue, and there was hardly any breeze to fight the tropical heat. Yesterday we were told there was a monsoon coming even though there were hardly any clouds throughout the whole day. Still, we had our orders to bug out, so we waited at the airport for a ride back to the ship. As the wind started blowing, and the night's stars faded away behind the clouds, we waited. We didn't catch the last ride, but they stopped the airborne ferry service less than half an hour after ours lifted off from Sittwe.

The ride was awful. The wind picked up in minutes. At first I thought it was just prop wash from the rotors, but it got worse and worse as we headed out to sea. About the same time that I lost sight of the coast, the seas started to get bigger and bigger. I thought the pilot was just getting lower until we came in on our approach to the Reagan. I never imagined that a ship that big could twist and turn so badly. It's as big as the Empire State building lying flat, but it was bobbing like a kid's toy in a bathtub. It took our pilot three tries to get us on the deck. Finally, he timed the ship's rise and fall just right, and we slammed down on the deck.

When I got below, our berthing area had been taken over by some of the former Army Airborne troops. We tried to kick them out, but they weren't budging. We'll spend the night in the central corridor and steal it back when they all head out for 1st mess in the morning. We might be on the same team, but that's our berth, our home. We'll get it back, even if there has to be a "disagreement" tomorrow.

Friday, September 16, 2011

Correspondence

From Corporal Matthew McKenna,

U.S. Army, 1st Infantry Division, 12th CHEM

To Jonathon McKenna (*Grandfather*)

Gramps,

Well, just when you thought things couldn't get any stranger…. Harry was on point last night, and he fell in an honest-to-God Tiger Trap. Oh wait, it gets better! There was a tiger in the damn thing! It was dark. We were off the path to avoid booby-traps, ambushes, and snipers, and things were going well. We only had to stop a few times to do security checks, and each time, it turned out to just be jungle noise (Why it's louder at night than in the day I'll never understand!).

So, it's triple-canopy stuff, and we have to keep kind of close to still see each other-even with the thermal night vision gear. Harry was about 50' in front of Pepe, and I was in the back with Piotre. All I heard was this CRASH-SPLASH-ROAR-"SHIT!" We all stopped and looked around for an ambush while he splashed about in the tiger trap.

Finally-just as Harry was crawling out-Tattoo Joe ran up and saw what had happened. The trap (roughly 20'x20') was filled with water. A huge, muddy, and dried blood-covered tiger was still alive, but it must have been stuck on one of the bamboo spikes still under the water. Harry-with his typical luck-fell in, but somehow it didn't land on a spike. As soon as he splashed down (the water was about 6' below the trap's roof), the tiger must have woken up. It scared the shit out of all of us-EXCEPT Harry! He seems used to his luck. We're all wondering if his luck is good or bad?! It was definitely bad luck for the rare tiger, but whaddya say about Harry?

Now, normally, Harry's really good about his luck. If he gets a crummy MRE, or trips, or almost gets whacked on patrol he usually just gets back up, shrugs it off, smiles in disbelief, and then walks away like nothing's happened. This time, Pepe couldn't control his mouth, and he made some sort of wisecrack about how the tiger didn't seem as big as a New York City Subway rat. Harry popped him so hard in the jaw that his helmet fell over his eyes. THEN, Harry smiled and walked on. Pepe was pissed, but none of us heard a word he said (as usual). He bitched and babbled about this and that, but since everything he said had a New York connection, and since we've taken to ignoring his repetition, we didn't hear a word. After about 15-minutes, Tattoo Joe finally made him shut up (He's a big guy, and when Tattoo Joe gives makes a suggestion, people listen!).

I thought we were getting closer to the mountain, but it seems to be taking forever to get there. I'm gonna get goin' so we can get a few more miles under our belt by sunrise. I'll zap ya another email as soon as I can, but I'll be checking it often. So, if you remember anymore bush tricks from your days in 'Nam, SEND 'EM. That barefoot in the mud one is really good!

Thanks, Matt **Saturday, September 17, 2011**

1st Lt. John Chamberlin -Personal Log

U.S.S. Ronald Reagan

Approximately 250nm west of Sittwe, Myanmar

Last night was something to remember, but most people won't. The Airborne troops that had taken our berthing area knew we were going to try and take it back as soon as they left, so they never left it unoccupied. They went to third mess in shifts and always left 10-15 people in the compartment as squatters sentries. Dennis got wind of our situation when I was turning in our after-action reports. He wanted to now where he could find me if he had any questions, so I had to explain it to him. I think he wanted to see how delicately I could handle it. He didn't seem too happy about the prospect of any sort of brawling on the ship, so Bob, Wess, and I decided to handle it a little more diplomatically.

Before we went into Sittwe, Wess met another Airborne sergeant from an entirely different troop. This other troop had found an empty compartment on the carrier, and they made a still using a steam line that passed through the compartment. Every night, these guys would "help" out after third mess and they'd take care of the trash. Some went into the disposal. Some went to the incinerator. Some went to different recycling areas. Some went to the compactor, and some went back to their secret compartment.

The secret starboard side compartment was actually a ballast room that was built into the ship to help counter flood against a port side list. It had been welded shut except for a fill port for the flooding, and a vent pipe that ran into the ship's main vent lines. Back when the Reagan was attacked at Pearl, the compartment had been flooded to keep the ship level, but the drain was plugged, so the repair crews had to cut their way in to let the ballast water out. After the ship put to sea, they never sealed the compartment, and the Army was able to find a use for it.

Wess and Bob hooked up with this troop and convinced them all that a victory celebration should be held to honor our Sittwe assault. Wess really played up the valuable role that the former airborne troops had played, and he suggested that a certain group of former airborne troops should be invited; specifically, the troops that were squatting our berthing compartment. They agreed. Bob asked if they had enough of their special beverage, and they all just silently looked over at three 55-gallon drums that had been stuffed into the ballast compartment.

It wasn't long after the idea was discussed that the party began. People started filling their canteens in the drums, and the corridor/barracks became a line to the bar. Wess and Bob came back to the corridor by our occupied berth. Bob conveniently leaked the information about the party to our squatters, and Wess organized our troop.

First, everyone in our troop went off to the party, loaded up their canteens and then disappeared. When the squatters thought they were safe because we had all gone to the party, they made a break for the drums of homemade jungle juice. As soon as Wess saw that the squatters had arrived at the party, he went and found the rest of our troop. They all came back to our berth, and switched our gear with the squatters, then sent waves of spies back to the growing party to watch for when the squatters were heading back.

It didn't take an hour for the drums to be emptied. Hundreds of troopers, Marines, and sailors showed up. Everyone loaded their canteens and disappeared. A few officers found out about what was going on, myself included. I can only guess that they either approved, or assumed that since no one was breaking it up that it had been deemed acceptable by higher authorities. The drums were emptied, our berthing area was reclaimed, though protested very vocally, and hundreds of people spent the night in celebration of our victory over the hardly impressive Bangladeshi armed forces.

Sunday, September 18, 2011

1st Lt. John Chamberlin -Personal Log

U.S.S. Ronald Reagan

Somewhere in the Bay of Bengal

The monsoon is getting rougher. There's a rumor on the ship that it's actually a hurricane, but they only call them hurricanes in the Atlantic. Out here, a hurricane is either a cyclone or a typhoon. In any event, the deck access is being limited, and we're all getting a little claustrophobic. The seas are huge and the rumor mill has them ranging from anywhere between 20' and 100' high. I have no idea how big they are, but this ship is almost 100,000 tons of steel and when I look down the central corridor that runs the length of the ship on this deck, I can see the ship twist, rise, fall, and bend like it was a noodle. Sometimes, you can't even see the other end because it's distorted so badly. The sailors tell us it's normal, and that it's supposed to be flexible, but I for one find it very unnerving.

Monday, September 19, 2011

1st Lt. John Chamberlin -Personal Log

U.S.S. Ronald Reagan

Somewhere in the Bay of Bengal

The heavy seas continued today. With all these people crammed into the ship, it had a pretty bad smell to begin with. Add to that the effects of seasickness...well, I could use some fresh air. Someone else had the idea this afternoon, and they opened up the door to the forward port elevator. The elevator was up to prevent it from being damaged by rogue waves, and as soon as the door opened, water started splashing in. Every wave must have poured in thousands of gallons. No sooner had the door been fully opened than someone had the good sense to close it. It went right up and came right down, but there was just enough time to catch a glimpse of the conditions outside. I saw it at the last second, and it was tough to gage the scale of the waves without something to reference them next to; especially since the ship keeps rolling, rising, and falling. Since the waves were splashing in, I'd guess that they were about 50'. Some of the sailors claim that they're in the seventies with rogue waves over 100'. Usually I'd write that off to inter-service boasting, but looking outside, it might be the truth!

Normally a carrier does about 30 knots so when it launches planes that need 200 knots of airspeed to take off, they'd already have 30 knots under their wings. All of our aircraft can take off vertically, and since the Reagan's reactor was so heavily damaged back at Pearl, so we usually only do about 20 knots. A lot of ships would be more difficult for a submarine to detect if they went slower and made less noise, but there's really no hiding the sound of an aircraft carrier; except in the noise from a storm's waves. In seas like this, I'm told it's nearly impossible for a submarine to detect any specific man-made noise. All they would hear is static from the waves. If they tried to locate us by periscope, they'd come right out of the water between swells.

Someone on the bridge understood this, realized our condition, and changed the deteriorating attitude on the ship with one act of intelligence. No one knows who did it, but somebody started broadcasting tunes through the ship's 1mc. While I was watching the hangar door close with water flooding in, I heard old 1960's songs over all of the ship's speakers. It was weird at first, then it got kind of distracting. It took our minds off of the rough weather, even the war in general.

As I walked back to our berth, I passed by other compartments where people were dancing. In the corridors, small groups of people were singing along. It was almost surreal. I noticed that the ship's bow came down off the crest of the waves in almost perfect company to the beat of "Give Me Shelter". It was great to say the least.

When I got back to my troop, they were much more relaxed. Some were singing like the people in the corridors. No one was dancing, but most were swaying or bobbing their heads. When The Satisfaction Song came on, I think everyone on the ship, 10000+ people, sang along at the tops of their lungs. It was great.

For hours we listened to the 60's tunes, rap music, country-western, some heavy metal, and even some techno-thrash-rap from a group called C3, Crusading Commies for Christ. Around 2200hrs, they put on a lot of 40's music. I went up to the hangar deck, and there was an old-style swing dance going on.

After that, the music only lasted about another hour. Then it ended without notice or explanation. Outside, I think the storm might even be getting worse, but inside the U.S.S. Ronald Reagan our spirits remain high. Our thoughts of seasickness and war have been put away for the time being. Today, we found shelter and lived one more day.

Tuesday, September 20, 2011

1st Lt. John Chamberlin -Personal Log

U.S.S. Ronald Reagan

Somewhere in the Bay of Bengal

The wind and sea are still rough, but the rain stopped, and the sun came out today. The ship's making a steady 20 knots to wherever our next assignment is going to be. We're heading into a 20-knot wind with gusts occasionally going over 50 knots. I went up to the deck when none of the sailors were looking this morning. It was like standing on the hood of a car while it was in Southern California rush hour traffic on the Pacific Coast Highway. I was surrounded by steel on three sides, and I thought I was going to be sucked right into the air and tossed overboard. There's so many people on this ship that if someone did go over the side, I can see where no one would even notice for hours, maybe days.

When I was back down in the hangar deck, I ran into Dennis. He pulled me aside and asked me how the troop was doing. I told him that the music really helped yesterday, and now everybody is coming back into a ready to fight type of spirit. He was pleased. He told me that he knew about the way we handled the squatting thing. In fact, he was told about it from Gen. Shapiro himself.

The word from above was that there would be no point in any punitive action since all the jungle juice was gone. Dennis said no one was really upset, but some of the navy officers-not including the captain-were a little ruffled over the party.

Ever since the war began, there's a new unwritten rule: the more combat a person has been in-the more that they can get away with; regardless of age, sex, religion, race, etc. There's just no room for those kind of things in the military right now. We've had our backs against the wall since the beginning, and only now are we being able to hold our own. If some combat vets who are locked in a ship during a typhoon want to have a few drinks, it seems that most everyone thinks they might as well. On the Reagan, we all fit that criteria.

Dennis also told me about some of the options for our next operations. The most likely one is the most ambitious. General Shapiro was really surprised by how easy our Sittwe operation had gone, and now he was talking about something bigger. The powers that be are working on a plan to fly us into the capital of Bangladesh, Dhaka, to take over the city.

Aside from the obvious shock, Dhaka is almost 150 miles inland. Sittwe was right on the coast. We would have to either get the Reagan right up to the beach, or make it a one-way trip! It's right at the limit of our helicopters' range. I couldn't believe it when Dennis told me about the plan. They did have a lot of the details worked out, and theoretically we could land right on top of all our targets at the exact same time, but it's still an ambitious plan. I don't know if we could defend the city even if we could take it. At least this storm is going to give everybody time to really make sure that our next plan is as completely thought out as the Sittwe operation.

Wednesday, September 21, 2011 Correspondence

From Corporal Matthew McKenna,

U.S. Army, 1st Infantry Division, 12th CHEM

To Jonathon McKenna (*Grandfather*) Gramps,

We're starting to wonder if we'll ever make it to that f- - king mountain! Up until today, we've only been moving at night. After today, we'll probably go back to dark-only ops. Harry-point man for life-was working his way through the tallest elephant grass we'd ever seen. Except for a few brief security stops, we had walked all night. After sunrise, we all decided to try and get another mile or two under our boots, so we kept moving. Not even an hour later, Harry was knocked down. A second after that we heard a crack from a sniper rifle. Harry took the hit square in the middle of his body armor. It even hit the ceramic/metal plate insert (The question remains: is this good or bad luck?)! Pepe dragged him back toward us. It wasn't any safer, but Harry was in a lot of pain, and we wanted to try and calm him down. Tattoo Joe and I looked for the sniper but didn't see anything. Piotre disappeared the instant that Harry was hit. While we debated where the sniper was firing from, we kept dragging Harry back toward a ditch inside the closest jungle treeline.

Piotre and the sniper were playing a cat-and-mouse game the whole time. It took almost half an hour to crawl into the jungle from the elephant grass. During that time, we kept hearing their little duel. The sniper would fire once, then a few minutes later, Piotre would pop off a dozen rounds or so. It's the sniper equivalent of spray-and-pray shooting, but it has gotten the job done for Piotre in the past.

Once we were in the jungle, Tattoo Joe and I tried to camouflage our position while Pepe worked on Harry. We all knew he was going to be fine, but he's probably got a few broken ribs. More than anything, he completely had the wind knocked out of him, and his entire chest is one big ugly bruise. It looks like a tie-die t-shirt until you get up close. Then the rings of white, black, grey, red, and purple all blend together. He's in a lot of pain, but typical of Harry, he just grins, shrugs his shoulders at his luck, and sometimes even laughed while he was still trying to catch his breath.

The four of us curled up in that small jungle hole all day long. Occasionally we heard the sniper or Piotre, but most of the time, there was just the quiet noise of a daytime jungle. I wanted to call for help, but there was no way to tell where the sniper was. Only Piotre seemed to know. It would have been useless anyway. We're so far from any supporting friendly forces that calling for help would just give away our position to the bad guys all around. Even the smallest villages know how to detect and triangulate a radio signal. They may look like a Stone Age exhibit at the Natural History Museum, but every village has a hooch with a CD-player, a solar battery recharger, and some sort of radio. Stone Age, my ass!

The sun set a little bit ago, and we still can't move on without Piotre. I got an email from Saunders asking why we're not on Popa Hill yet. I'll respond tomorrow. Maybe we'll get there then. Popa Hill's actually in Bagan-the next region/state over. Either way, I'll let him know about our progress and the trouble with the sniper today.

In the meantime, Piotre has finally come back. He's kinda shaken up. It was a Thai Hunter Soldier who shot Harry. They're supposed to be our allies, but right now, over here, the only people you can COUNT on are those grunts in the hole with you. Piotre took some splinters in the face from a very close call with the Hunter Soldier, but otherwise he's physically fine. Mentally, he needs to regain his discipline. Both snipers spent all day popping off rounds at each other. Piotre says that he almost bought it at least three or four times. The only thing that carried the day for him was the darkness. As soon as the sunset, his thermal night vision picked up the Thai sniper easily. Piotre told us that the poor guy didn't even wear any camo. He was wearing a loincloth, a bandolier of ammo, a bag of rice, and a map of the area. That was it. Camo would have helped mask his thermal image, but without it, his sunburned chest, arms, head, and legs stood out like a lighthouse on the sea.

The map that the Hunter Soldier had showed the areas where four other snipers were assigned. We don't know exactly where they are, but we have a good idea. I'd send Piotre off to clear the path tonight, but I think he's fried. Harry's still too banged up to take point, but Pepe volunteered. He's got some skills, and we should be able to avoid the sniper-ridden areas a bit better now.

I've gotta get going again. It's been a long day. The sun has been down for a few hours, and we're mega-behind schedule. Saunders is gonna be pissed when I zap him an email tomorrow, and the more ground we can cover tonight the better. That damn hill just never seems to get closer!

Take care,

-Matt

Thursday, September 22, 2011

The New Cleveland Press Newspaper

Trouble in the Jungle

For decades, the mountainous jungles of Southeast Asia have been home to insurgency, counter-insurgencies, secret wars, and guerrilla attacks. Today, the usual sniping and odd mortar barrages have grown in frequency and intensity until finally full battalions are engaging each other. Chinese forces have been supplying multiple rebel factions in several countries since the fall of Nationalist China half a century ago. From time to time, they've clashed openly with some of the regional governments, most recently in Vietnam.

Throughout the day, all along the Chinese border from The Bay of Bengal to the Gulf of Tonkin, large numbers of Chinese troops have engaged border patrols and regional troops from Myanmar, Laos, and Vietnam. Casualty reports are difficult to gather from any of the parties involved, particularly the Chinese who are denying that there is any P.R.C. presence south of their borders. Unofficial reports and Red Cross estimates put the killed, wounded, and missing reports in the hundreds.

The question of Chinese involvement has yet to be proven and documented, but the filling hospitals, roads of refugees, and smoke on the northern horizon indicates that deep in the jungle, high up in the hills, people are fighting.

Friday, September 23, 2011

1st Lt. John Chamberlin -Personal Log

U.S.S. Ronald Reagan

Somewhere in the Bay of Bengal

Tomorrow morning, we're going to town. Shapiro held an officers meeting in the forward hangar deck this morning. Everyone was given a 3-ring binder with the entire operation plan detailed. We can all see what each and every troop is supposed to be doing, where they're going, and when they'll get there. I've got until 3rd mess tonight to learn as much as I can about our mission-and everyone else's. The notebooks are going to be collected after we eat. Shapiro was very clear that unless all the books were collected, the mission would be scrubbed. It's time to hit the books!

Saturday, September 24, 2011

1st Lt. John Chamberlin -Personal Log

Dhaka, Bangladesh

Like clockwork. Everything is running like clockwork. The operation plan was easy to read yesterday, and by the time 3rd mess was done, I was ready to turn it in. Everyone else did too, and the mission was a go.

Just like all of the maneuvers that we did before Sittwe, the entire regiment took off and circled the Reagan until almost everyone was up. The Osprey, like the 513 that we've been assigned to again, took off last, but we passed the formation on its way to Dhaka, and we landed before them.

Our troop's mission was to land in Old Dhaka and set up a defensive perimeter around one of the countless bazaars in the area. Once the perimeter was secured, reinforcements would rappel in from a pair of UH-60 Blackhawks. Our only fire support goes through the regimental forward air controller who will determine how badly we need an airstrike, and what planes to direct in, etc.

Just after sunrise, when the market tables were being setup, our Osprey came in low, pulled up its nose, rotated the propellers skyward, and dropped us right in on the market. Tables were blown away, fruit and all kinds of goods flew in all directions like a bomb had gone off. The people ran away just as fast. Bob had the squads deployed in less than five minutes, and the reinforcement Blackhawks were already circling. I called them on the radio and gave them the okay to land. Over a hundred Army airborne troops slid down their ropes to back us up. There was no shooting at all in our landing.

Around the city, it was hard to hear any shooting over all the rotors from the helicopters and Osprey. I did hear some, but it was sporadic at best. The radio net told a different story. Some of the Marines and troopers had landed right into a hornet's nest. The "prized" Bangladeshi 9th infantry division was in town for some sort of event, and they were quick to react to our landings.

All total, our operation plan called for over 100 different landing sites and mission targets. No matter how "prized" or renown any infantry division is, it could never react to that many attacks all at once. I assume that's why Shapiro had the entire regiment go in all at once. That way, anyone trying to defend the city would have no choice but to split their forces and try to fight us all off. I think it also makes the size of our force seem a lot bigger too.

Around lunchtime, I heard the first airstrikes. They were sporadic at first, but it didn't take long for them to become more frequent. They seem to average one or two every five minutes. That's a lot of bombs. I haven't seen any of the planes, bombs, blasts, or resulting fires, but the sky is one big gray and black cloud from smoke. I talked it over with the officers from the other troop, and we all agreed that if the city did catch fire, the bazaar that we're holding is basically a firetrap. We'll try and fight it if it comes, but we also plan on calling for an evac flight out of here. All of the buildings were built with wood scraps God knows how many years ago. It could have been 5, 10, 20, 50, 100, maybe more. There's no point defending ashes.

By midafternoon, I could hear more and more small arms fire from around the city. It increased until sunset, and then seemed to end almost as soon as the sun dropped below the smoke-filled horizon. During the night, it sounded like there was a lot of sniper fire, and return fire. Occasionally-about twice an hour-I could hear heavy firing from different places around the city. I think it was probably some of our people taking advantage of our night vision equipment and rousting some of the Bangladeshis.

We still haven't seen any fighting near our position, and it's giving us time to dig in. Even if we get rushed and overwhelmed, it'll take some effort for somebody to climb over the barricades that we've put up. We've taken all the tables, crates, pallets, whatever from around the bazaar and build a nice wall around the entire place. It's probably about 200 yards across at its widest point. There's a bunch of small buildings inside and built into it. That's where we have our M249 SAW's placed for fire support. The M60's are spread around and setup to be quickly moved. They'll serve as our fire brigade against any attack except an actual fire. The civilians disappeared before we even hit the ground, so we're all sitting around, piling stuff up, and watching the silent surroundings of this ancient supermarket. Every where you turn, you can hear fighting, but around us there's nothing but silence-and waiting.

Sunday, September 25, 2011

0908hrs

40000 feet, 415 nm west of Hanoi Vietnam

Near the Vietnam/Laos/Myanmar/China border

Number 500. As Al Meyers lead a loose and spreadout formation of fighter planes over the mountains, his attention drifted with the light turbulence, and he reflected on the day. Today he joined an elite few. Today was flying his 500th combat mission. Only a few other Coalition pilots had survived 500 missions so far in the war. There were now almost as many new 500 pilots as there were in all of aviation history before the war.

Al's composite squadrons, made up of pilot-refugees from numerous countries, had staked their claim and found a home at a small airfield in northwest Vietnam. The people there appreciated the efforts of the refugee pilots. In fact, they were often treated as heroes. How odd it was that the local peasants, all proud communists, should be so revering towards pilots who saw themselves as defending their democratic rights. Dedication and heroism seem to have drawn a respect that could cross-political boundaries. After all, this war, like so many of its predecessors, was being fought more for the rights of people to be self-determining in their form of government, rather than have the will of other nations imposed on them from some far away capital; like London, Washington, or Beijing.

The F-16A that Al had been given from the fleeing South Korean Air Force was showing its age. It wasn't just the age, and it wasn't the miles. It was the hours. The flight time that Al had accumulated in the F-16 as part of his 500 missions was well beyond the plane's mandatory overhaul point. Sometime soon, he was going to have to get some more technicians into his unit.

There were a lot of other planes that needed work too. Al looked around outside his canopy. 5 Miles below him, Al's squadron of Taiwanese pilots were skimming the mountaintops and searching for Chinese fighters. About the same distance to his left and right, he could see a squadron of Vietnamese Mig-21's left over from their war with the United Sates. In an V-shaped formation behind his plane, Al's squadron of South Korean pilots kept their eyes open for trouble. In just the quick shake of his head, Al had glanced at over fifty planes, all in need of some serious maintenance.

There was no time to dwell on it. The group radio frequency suddenly went to static. They were being jammed. Everyone motioned to each other with hand signals that knew no language barrier. Eyes opened a bit wider, and heads swiveled faster. Somewhere close by, the Chinese were up to something.

After a few tense seconds, the Vietnamese squadron on Al's left began a wide turn to the left. Equally aware of each others actions as they were of any potential enemy activity, all of Al's other pilots made a slow turn in the same direction. When Al's squadron and the low squadron leveled their wings along the same path as the Southern Vietnamese squadron, the Chinese came into view. They were in a river valley almost 15 miles ahead.

The river was filled with camouflaged barges for miles. They fired no AAA or SAM's at the oncoming aircraft. Instead, they had brought along their own air cover. An entire group of sophisticated Russian built Su-35 fighter planes circled as though they were driving on an eighteen-mile oval racetrack in the sky. There were over forty Russian planes with Chinese markings.

The last Vietnamese squadron finally lined up with the other three flights at about the same time that Al's low squadron went to work. They popped up over the top of the river valley's Eastern ridgeline. White streaks from heatseeking and radar-guided air-to-air missiles drew lines over the deep green jungle canopy. The fighters from the low squadron seemed to ride the white contrails like trains on rails. At the other end of the white lines, yellow and orange explosions surrounded themselves with black puffs of smoke and twinkling pieces of debris that were only moments before Chinese fighters and pilots.

The high cover Vietnamese squadron got off the second volley of missiles just before the low squadron closed into a dogfight with the Chinese. Their missiles found fewer marks than those fired in the initial shots from the low squadron. Still, the Chinese numbers were dwindling fast. They had been outnumbered to begin with, but now it was their turn.

After the Vietnamese missiles passed through their formation, the Chinese planes used their 21st-century thrust-vectoring controls to great advantage. The nozzles at the end of the engines on each plane had been modified to redirect the engines thrust up, down, or to either side of the plane. This allowed them to do incredible maneuvers in dogfights; maneuvers that 25-50 year old planes could never keep up with.

The low squadron passed over the line of barges at an average speed of over 700 miles per hour. In response, the Chinese simply pulled their planes' noses straight up, stopped their forward movement, turned around like ballerinas, and gave chase. They were in excellent firing position against Al's low squadron, but this provided an equally excellent position for the other Vietnamese squadron. Al and another Vietnamese squadron circled above watching for Chinese reinforcements. They also waited to serve as a reserve force for their brothers in arms below.

The Chinese pilots spit out a barrage of heat-seeking missiles toward the hot exhaust coming from the tails of Al's low squadron. Some found their mark, and more of Al's friends died. They were heroes to the Vietnamese, devils to the

Chinese, defenders of democracy for their individual nations, and friends to Al. Not a single one of those thoughts entered Al's mind. All he knew was that some of his coworkers wouldn't be at lunch tomorrow...or the next day, or the next day, etc. If only the Chinese would get some reinforcements, Al could get to work. He could get some payback.

In a twenty-mile radius, hopping in and out of the river valley, almost a hundred fighter planes were spinning in a giant furball of jet exhaust contrails. Each line twisted and turned to get into a better position to attack the tail end of the one in front. They were weaving in the sky. When the Vietnamese squadron that had initially spotted the Chinese formation finally came down and joined the fray, Chinese reinforcements started coming in from the north.

Al's high cover Vietnamese squadron saw them first. They were only five miles closer than Al's squadron, but that made all the difference. The glittering specks on the horizon were Chinese copies of the Soviet-built Mig21; the exact same plane and color scheme as the Vietnamese planes.

They came in groups of four. The Vietnamese engaged each group with their entire high cover squadron. The first four were met with the typical missile barrage, of which only one Chinese plane survived to join the dogfight in the valley. The next four only lost a single plane, and the remaining three made it into the valley for the furball fight. Another flight of four got into its own dogfight with the high cover Vietnamese. While they were scrapping, two more Chinese flights raced into the scene.

By now the high level dogfight had fallen into the valley and joined the crowded treetops. Al could wait no longer. There was no point anymore. He wagged his wings to signal the other members of his squadron, and then they dropped into the dogfight. Al noticed that the spinning combat seemed to be moving counter clockwise almost without exception. He decided that it was time to take back some of the initiative that had been lost to the Russian built SU-35's with their high-tech engines. Al lead his squadron into the dogfight backwards, clockwise.

Once inside the man-made hurricane of combat, they passed friend and flow alike with closing speeds well in excess of 1500 miles an hour. At such speeds it was near suicide to try and engage any targets, but this was just as true for the Chinese. They couldn't chase Al's low squadron or the Vietnamese as long as they had Al's squadron passing them so fast that radar and even human vision couldn't recognize them as targets. All that anyone was seeing were blurs streaking past. It was a game of chicken between almost 150 jet fighters. All this with autocannons blazing, tracers streaming, missiles chasing, and a primordial triple canopy jungle below filled with thousands of Chinese troops watching on camouflaged barges.

It was too much for most. No one side pulled out first. Chinese and Coalition planes alike started to pop out of the smoky knot in the sky. The Coalition planes headed east toward the relative safety of Vietnam. The Chinese climbed for altitude in an attempt to regroup. Al rode the roller coaster. Sooner or later, something was going to either hit him or offer itself as a target to his 20mm cannon.

At 1500+mph, he didn't have to wait long. An SU-35 came right at him, and he fired a burst without even thinking of aiming. Some of the 20mm rounds went down the Chinese plane's left engine intake. It flashed with flame as it passed within 50 feet of Al's canopy. Through his peripheral vision, he watched black smoke flow out from the enemy planes high-tech engine nozzle. It spun into the jungle and disappeared without so much as a spark of flame or recognition. The trees just swallowed it up.

After only a few laps on the racetrack in the sky, Al had had enough. He pulled up out of the dogfight and turned to the outside until he was facing the other direction. Then he went back in. Immediately, a pair of SU-35's came up from ahead and below his plane. He had stumbled into a perfect firing position, though a little too close. The two planes, obviously wingmen, crisscrossed their flight paths to confuse their prey and anyone like Al who might get behind them. The concentration of the dogfight kept them from noticing him until it was too late. Al waited until they crossed for the second time in front of him, then he fired a long burst of cannon fire at the spot in the sky where they appeared as one. The closer plane took the brunt of his fire, but it went into a roll and collided with the lower SU-35. Both went down and splashed in the river, just missing a barge.

It was time to get out. More Chinese flights kept arriving, and it was getting too confusing to tell who was Vietnamese, and who was Chinese. The remaining SU-35's were leaving to rally at higher altitude for another attack. Al's remaining planes were going back to Vietnam with as much speed as their engines could muster. Al "whipped his pony" and aimed his F-16 toward home.

It looked like the skirmish was over, but the Chinese had other thoughts. Looking in his rearview mirrors, Al watched as the rest of his low squadron and what appeared to be many of the Vietnamese planes all left the dogfight and began following him back to the base. Above and behind him, the Chinese that had regrouped were doing the same. Al counted seven of the Chinese SU35's diving on the planes behind him. With the altitude edge, they would soon overtake his planes and make mincemeat of them. He had to turn and engage. As soon as he did, the Vietnamese and the remains of the low squadron saw why. One by one and in pairs, they lit their afterburners and tried to make a break for it.

Al headed straight at them and pulled up as they passed underneath. He aimed his plane at the diving Chinese who all opened fire on him with cannons and air-to-air missiles. Al popped off some chaff bundles and flares to confuse the missiles, but they all continued on in a straight path. The game of chicken had worked in the dogfight, and he tried it again. Neither the Chinese nor Al changed course. He passed through their formation again at over 1000 miles an hour. No one could possibly tell how close they had come to colliding.

Al chopped his throttle back, opened his airbrake, and pulled the nose over backwards until the plane stalled and dropped. He was falling on the tail of the

Chinese in an upside down dive, so he rolled the plane level and took aim. Missiles were getting scarce for the Coalition. It was even worse for his semi-outlaw unit operating from a jungle airfield in the backcountry of Vietnam. Al had only brought two. He lined up an SU-35 and listened for the growling sound in his helmet telling him that the missile had locked on its target. When he heard it, he fired and tried to lock on another Chinese plane. Before his first missile found its mark, Al launched a second and lined up his cannon on a third. His cannon rounds and the second sidewinder heat-seeking missile both met their targets within milliseconds. The Chinese broke off their attack. They scattered and headed back to the river with their mission accomplished. They had successfully protected the barges and their troops.

Sunday, September 25, 2011

1st Lt. John Chamberlin -Personal Log

Old Bazaar area of Dhaka, Bangladesh

We got a resupply chopper in this morning. An old Marine CH-46 Seaknight came in with a huge pallet in a net hanging below it. They didn't even stop over the LZ. The pilot just flew over, pulled up, and swung the net down into the LZ. The pallet kissed the ground, and as soon as there was some slack in the net, the pilot dropped it. He never once stopped. It was an excellent display of experience and skill. The pallet was packed with enough ammo to double everyone's supply and then leave some left over! It also had a bunch of those new heat-and-eat MRE's.

They're really nice. All you have to do is open the plastic outer bag, pull out the pouch of food that you want to heat, and stick it in the thick heating pad. The heating pad has a vial inside with a chemical that reacts with another one that fills the remainder of the bag. When the two chemicals meet, they generate heat and warm up whatever pouch is in the heating bag. This way, when you open the pouch of food, it's nice and warm, sometimes even hot if you leave it in long enough; say half an hour or so.

We ate in shifts so that we could always keep a wary eye out for trouble. My troop was just about to pull out their nice heated food when our friends in the "prized" Bangladesh 9th Infantry Division decided to pay us a visit. The Army's airborne people were nice and well fed. They might have been ready for a little after lunch siesta, though, because it was someone in our hungry Marine troop that noticed the first infantry coming down the alley.

I'll give the 9th this much credit: they did approach us very professionally. They came in from three different sides, all at the same time, and they moved in pairs while other pairs covered them. It seemed like they knew exactly where we were. They also had pretty good fire control discipline right up until we opened fire. I think they were hoping to get the jump on us, but it didn't work.

Our fire discipline wasn't as good. One of my troopers with an M249 SAW decided on her own that it was time to open fire regardless of any consequences to the rest of the troop. She was in the first floor of a building, and the "Banger's" as we've come to call them did get really close to her, maybe even less than 10 meters before she cut the first one in half with a long burst of full auto. PFC Christina Boyle had taken over a SAW when her fire team leader, Corporal Lee Burke, was killed back in Taiwan at the Choo Beachhead evacuation. Christina, like all Marines, had a little training with the weapon, but she's never been one to exhibit a great deal of self control.

That lack of discipline showed itself very well in her new role. When Christina began firing, everyone else on both sides opened up too. We all fired in 3 round bursts, maybe more for the M60's, but Christina kept her finger squeezed until her entire belt of ammo was gone. Her barrel was white hot, and she had to change it before she could even think about firing off another 200 rounds. The Bangers didn't wait, and they poured lead into the slot that she was firing from. She had to leave her weapon and crawl out before the statistics caught up with her. Hundreds of bullets were going through the wood walls, and all it takes is one.

I called in the contact report back to the regiment HQ. They acknowledged, and so did the forward air controller. He happened to have just flown over the area a few minutes earlier. Since he didn't spot the Bangers, he must have felt a little guilty and he offered us a direct air strike. I had never heard of that before, so I took it. We traded shots with the Bangers for a minute or two, then all hell broke loose.

We never heard the planes, missiles, or whatever it was that came in. All I know is that there was huge explosion. It knocked everyone down, backwards or both. There were more explosions right after the first one, but they were muffled from the lack of hearing we all had after the first one.

The next thing I heard, after my hearing came back, was the regiment radio on the ground, next to the barricade, about 20' away. I crawled over in case there were any Bangers still around. The forward air controller wanted a report to see if we needed a follow-up. Again he was offering to send in a strike. I looked into the dust that was surrounding our position. As it settled, there was nothing there. The buildings, the power lines, the Bangers, even the tiles on the streets were all gone. There was nothing but a big dirt crater in front of me. I told the controller that we appreciated it, but he should pass on the strike to someone else who really needed it. One was all it took.

I never saw or heard the planes, but I recognized the handiwork of a 2000lb bomb very well. After the dust settled, I looked around the perimeter with the other troop's officer. Wess took a squad out into the crater zone to recon around, and a few minutes later, he came back and told us the story.

It's obvious that the entire bazaar is gone-not just destroyed or left in ruin. It's gone! The Bangers? They're gone. Wess found some weapons, gear, and some remains on the far side of the craters. There's no way to tell how many there were, or how many escaped. There are some certainties, though. While we were crowded into an urban shooting gallery before, now we've got a big field of fire all around. Another certainty is that sooner or later, someone is going to come looking for that unit that was here.

I called in the battle damage assessment report. Dennis came on the net and asked if we could hold out. I told him that it should be a lot easier now, and he was glad to hear it. Some of the other troops have run into serious trouble. Dennis said that we've even lost contact with some entire units.

They probably won't come again tonight, not with our night vision advantage. Maybe that's when someone will come looking for the missing Bangers, but we won't see a good-sized attack. The Airborne troop said they'd take the night watch so we could serve as a ready to go reserve. Bob made sure a few people stayed awake though, just to make sure.

Monday, September 26, 2011

1st Lt. John Chamberlin -Personal Log

Old Bazaar area of Dhaka, Bangladesh

A fresh group of Bangers came at us this morning. They started off as a few snipers, but that just kept us alert. No one caught a round-at least not in my troop. More and more snipers started harassing us throughout the morning. Eventually, somebody had to take a hit. The other troop's officer, my equal, didn't hide his rank and position from the snipers well enough. He took a single shot right between the eyes, just under his helmet. He died instantly. I'd like to think that the Bangers don't have that kind of skill, so I wrote it off to luck and started meeting with all the NCO's one by one. Everybody was as dug in as they could reasonably be.

Eventually, the Bangers tried another full-fledged attack on us. It started around 1500 with a mortar barrage that lasted almost half an hour. I had the impression that they were using every round they had on us. It felt like a lastditch effort for whoever was in charge. When the mortars lifted and the dust settled in the craters, the Bangers started coming out of the rubble and buildings on the far side of our firing zone. Most were walking, about 5-10 yards apart, but I knew that for every man walking, there were probably two waiting to provide cover fire.

Everyone in both troops showed excellent fire control discipline this time. I don't think anyone so much as blinked until I gave the word. I wanted to get as many as possible in the first opening shots, so I waited and waited. In fact, I waited until some actually climbed over our barricades and into the perimeter, unknown to them. Then I yelled out "Open fire!" I think by the time the "O" was out of my mouth, and before the "P" could sound, every gun cut loose. The Bangers in the compound took the most hits, but the ones in the free-fire zone went down fast too. I thought they were just scrambling to eat dirt and take cover, but after a few seconds without moving, I realized that they were all hit and either dead or dying. Their covering fire came into the perimeter very well, but our positions are very good, and getting bullets through the firing ports from the other side of the craters was nearly impossible.

I passed the word to cease-fire and keep under cover. If they had used up all their mortar rounds, then they'd have to come into the open again, at least to get their wounded. The Banger commander must have realized this too, so he kept his men in their positions and didn't let anyone help their injured.

We listened to the moaning, crying, and indiscernible pleas for aid. It lasted till sunset. I hoped that the Banger CO would send someone after dark. I'd even resolved to pass the word not to shoot them if they would just help make the sounds of the dying cease. No one was sent. I couldn't send one of our people, and I'm still convinced that no one would have volunteered. Wess said he'd take care of it. I knew perfectly well what he meant, and I let him do it.

Wess has a night vision scope for his M-16. He attached it, fired two shots at something, hopefully someone, on the other side of the craters, and then he began picking off the Banger wounded. He was careful to make each shot count. We all hated watching it, and I felt guilty at first for letting in go on, but there was nothing else we could do.

I know how immoral it sounds to give the order to shoot wounded soldiers, but in a way, it was also the right thing to do morally. I can't justify ordering someone to die in the hopes that they might be able to save the life of an enemy who only hours earlier was trying their best to kill us all. I can't sit by and let men suffer. A hunter doesn't do that to a deer or a squirrel, so why should we be so cruel to men?

Wess went from one position to another. Each time, he'd fire two or three shots, then wait a minute or two and move on slowly and silently. Tonight, he is the Angel of Death, and I think even some of the wounded might thank him for taking their pain.

Monday, September 26, 2011

The New Cleveland Press Newspaper Chinese Breakthrough!

A veteran Myanmar armored brigade suffered heavy casualties today. It was driven back from a defensive position called Phase Line Gilbert this morning by a concentrated Chinese infantry assault. Officials report that a great number of their vehicles were destroyed or damaged during the attack's preemptive artillery barrage. After the barrage, P.R.C. aircraft were reported over the area, and the acting commander of the armored unit was forced to give the withdrawal order. It's rumored that an entire Chinese division has already occupied their former position while additional units are moving further south. An untold number of infantry battalions have been ordered to the area to plug the hole in the line that runs from just South of Hikabo Rizi Mountain (19296 ft) down to the Irrawaddy

River. Officials are confident that their presence will be enough to halt the Chinese advance.

Tuesday, September 27, 2011

Correspondence

From Corporal Matthew McKenna,

U.S. Army, 1ˢᵗ Infantry Division, 12ᵗʰ CHEM FWD Team ASD3

To Jonathon McKenna (*Grandfather*)

Gramps,

Well, I'm pretty sure that things can NEVER be as rough as they were today! We were supposed to set up a network of EOP's on Popa Hill, but the Chinese had a few surprises for us. At about 10 or 11 this morning, I was setting up yet another camera/mike-style EOP somewhere on Popa Hill (It's really a 5000'+ mountain with a huge, bombed-out, temple complex on top). Tattoo Joe was helping me with the wiring. Harry and Pepe were carrying all the gear, and Piotre was off someplace watching out for the Chinese. The rain was letting up, and it looked like it was going to be a nice day-typically hot as hell-but otherwise nice. Never let those thoughts go through your head. A few hours before sunset, Piotre came haulin' ass through down our small path from higher up the mountain. He was seriously frazzled. I thought he was coming apart (As you know from your days back in Vietnam, the jungle can mess with your head).

Piotre said that the Chinese were sending an entire army down the Irrawaddy in broad daylight. It seemed too screwed up to be true. Who would dare move hundreds of thousands of troops down a river in daylight?! It was like begging for an airstrike. We stopped what we were doing, and went the rest of the way up the hill with Piotre (There are over 700 steps to get to the top from the base of the mountain's vertical face, but that's just the last few hundred feet!). As sure as it rains every morning, there were river barges packed with Chinese troops coming down that fat, brown river. The river's a long way away, but with Piotre's rifle sights, we could clearly see 'em comin'. I don't think it was an entire Army, but there were at least a few battalions. One of your letters said that back in Vietnam, whenever you ran into a company of VC, you knew there was a battalion. I figured that if there were a few battalions, there were probably a few brigades!

Tattoo Joe started putting out claymores in a circle-about 100 meters around us and lower on the mountain. Pepe and Harry found some nice cover behind some of the broken statues in the temple ruins. Piotre stayed with me for a while I tried to call in some air support (thankfully). The river is a mile below us and a few miles away, but there were a lot of troops, and we wanted to be as safe as possible when the air support came in. We all knew that as soon as the air cover came in, those barges would head for the river banks. They'd drop off troops, and as fast as they could run those miles, we'd be up to our asses in bad guys.

After a few minutes of trying to raise some air support, we were about ready to give up. I tried all our pre-arranged frequencies. I even tried a few frequencies that I remembered from when I first arrived here. Hell, I even sent Saunders an email! No one was answering.

That's when things went from bad to worse. All my work on the radio must have drawn the attention of the Chinese. The third barge turned toward us and began to dump troops on our side of the river. There were at least a couple hundred, and that would have been bad enough, but then we spotted the aircraft. I counted10, but Pepe and Tattoo Joe said they saw 19.

Harry wanted to take a shot at one that was coming toward us, but Joe convinced him otherwise. He said that they were Soviet-built Su-25 Frogfootsground attack planes. They're supposedly loaded down with armor and even hard to drop with an air-to-air missile (let alone one of our puny short-range anti-tank jobs). Joe says he read that their pilots get to sit in a layered titaniumarmor bathtub. We stood still to prevent the pilot from spotting our movement and dropping a load on us. It worked-for a while.

It had taken us days to get to this hill, and a long time to climb up to our position. Sure, we had done it at night to prevent being detected, and the Chinese would have to do it during the day, but that didn't calm our nerves very well. The Frogfoot buzzed us, and we watched as the Chinese troops bailed out of their barge and disappeared into the jungle. Everyone dug in while I set up an EOP on our own position.

When I was done, I gave the radio another try. This time, I actually contacted someone. There was a group of friendly aircraft somewhere in the area. They didn't know the day word, or the challenge password, so I wasn't sure if they were for real or if it was a Chinese trick. The guy sounded for real, and he asked me to turn on our infrared IFF (identification friend or foe) beacon so he could get a better bearing on our position. If it was a Chinese trick, it was the perfect way to let them know where we were. Harry, Joe, and Pepe were getting a little edgy (I think the broken statues were starting to wear on Pepe's nerves) so I agreed to turn on the beacon. It was against all the rules in the book, but I guess when you're outnumbered a few thousand to one…well, I guess in that kind of a situation you just gotta go on gut feel, and this felt like the right thing to do.

The flight leader I was talking with-some guy calling himself The Wolfman-said that he could see our beacon. They were far from our location, but he said he'd join us in about 10-minutes. It was going to take a lot longer than that for the Chinese troops to get up to us, but we were already within their heavier weapons' range. I was more concerned with the Frogfoots than anything else, and friendly aircraft were exactly what we needed to deal with those!

I told the guys that help was coming. They seemed relieved-especially when this Wolfman guy radioed back "…hold on buddy. We're comin', and we're bringin' a little bit o' hell for them Chinese, too. Wolfman One One out." It was a re-assuring feeling, but right then, a pair of Frogfoots circled around and lined up on our position. We all froze-just like when a fresh flare pops into the sky. Even when their small rockets started to explode in the jungle around us, we still had a lot of hope.

The rocket attacks from the Frogfoots were sporadic and scattered. They knew we were on the hill, but it's a big hill, and they were hitting everywhere but our little nest. Finally, this Wolfman guy called back on the radio and asked us to light up our IFF beacon one more time. I didn't bother to answer back, since I knew the Chinese would home in on the signal. So, I turned on the beacon, and Tattoo Joe fired a purple smoke grenade toward the riverbank where the Chinese had landed. The Frogfoots all seemed to turn at once and head right for us, but all of a sudden, this F-16 dropped out of a small cloud and blew the shit out of a Frogfoot. Two more F-16's followed and did the same. In just a few seconds, the sky was filled with all kinds of Coalition fighters. I even saw some U.S. Navy F-18 Hornets! The Frogfoots tried to beat feet and bug out, but our flyboys had 'em ambushed and outnumbered.

I was busy watching the aerial display when that Wolfman guy called back. He asked us to pop a flare to identify our position. We all knew that it would draw all kinds of trouble, but I did it anyway. The beacons are great for night or from far away, but during daylight, you really need to have something more precise to pinpoint your own location. Without the flare, our own flyboys would have shot the hell out of us. After all, that is EXACTLY what they did to the entire hill!

No one seemed to bring any air-to-ground ordnance. It only took a few planes to hunt down all the Frogfoots, and the remaining 30-40 friendlies took turns strafing around us. They all seemed to make runs on the river barges too. It would have been something to watch, but as soon as we popped that flare, we started getting ready to bug out. I made sure that the EOP was up and ready, and Joe put together a nice grenade and Booger Bottle boobytrap for anyone who messed with it. The Wolfman flyboy told us that they were running low on ammo, so I thanked him and told him that we would bug out to the southeast. Instead of answering, he brought his F-16 right down past us and wagged his wings. We all looked, and I think I even saw the guy waving!

We ran like hell down the path. It had taken us days to get up to the top of Popa. Within an hour, we passed one of our earlier encampments. By the end of the day, we were halfway down to the bottom.

We stopped a few times to do a security check, and Tattoo Joe always took the time to put out a frag-in-a-can boobytrap. I'm sure you remember it from Vietnam. It's just an empty MRE bag or can with a frag grenade inside. You tie a trip-wire from a tree to the grenade and pull the pin when you leave the security stop. As soon as someone trips the wire, the grenade is pulled out and wammo. It's simple, lightweight, and effective. I think he's really just doing it to lighten the load in his rucksack.

Whenever we leave a security stop, Piotre is still behind. Somehow, whenever we stop again he pops right up. It's nice having him watch our backs, but he must not have been following too far behind.

Harry never seems to stop. Despite his luck, he's great on point. He does spot a lot of potential ambush areas. He keeps us out of places where snipers might have good fields of fire. He occasionally spots a boobytrap, and he always keeps us off the beaten path. I'm not sure you can ask anything more of a point-man besides keeping you out of trouble, and he does that very well.

Pepe continues to ramble on about New York City despite Joe's warning. We're all a bit homesick, and civic pride is one thing, but Pepe is obnoxious! The guy never stops babbling. We all talk about home from now and then, but when we're running from thousands of Chinese; I think it's time to shut up. What kind of a time is that to be babbling about the Yankees, Times Square, Broadway, The Empire State building, the Brooklyn Bridge, or his family's GD taco stand?! We've put a good deal of distance between ourselves and the Chinese, but if he doesn't shut up soon, I think one of us might just pop him with more than a fist!

In all the excitement of the day, I never really had a chance to get scared. We've stopped for a 3 hour rest/security break. Harry's been edgy since the tiger trap incident. Pepe's mouth probably flutters faster as a way of venting his stress. Tattoo Joe keeps getting more and more angry despite the fact that we're likely to make it out of what was once a VERY bad situation. Piotre has moved to the next level of alertness after his sniper duel the other day. I'm trying not to show it, but it feels like we're drowning in jungle now. Up until our hasty retreat from Popa Hill, I felt safer in the jungle. Now it feels like it's wrapping up around me-like a tight, wet blanket or a python. I'm holding out okay for the time being, and I'm sure I'll make it till we get moving again. Once we get back to The Sewer, everything'll be fine again. Until then, I think today's memories will stay vivid for a long time to come. This must be the kind of thing you were talking about that caused you guys to get DSS. Well, I HAVE to get some rest. We're moving out in a bit. There's a long way to go, and it's all through Indian Country.

We're all confident that we'll make it, but the trip in was rough, and the one out promises to be rougher. I'll drop you another email as soon as I can. You watch your back, I'll watch mine. Say hi to everyone for me.

-Matt

Tuesday, September 27, 2011

1st Lt. John Chamberlin -Personal Log

Old Bazaar area of Dhaka, Bangladesh

Last night, Wess capped almost 40 wounded Bangers. When the sun rose, and the morning mist finally lifted, the dirt from the craters was littered with corpses. Sometimes, the sweet smell makes your cheeks pinch like sour candy. I saw a lot of people grimacing from it. The airborne people have been through almost as much as we have. Some of them may have been through more, but everyone gets the death smile sometimes.

The Bangers kept us all in our positions most of the day. I'm sure they were pissed about Wess' work last night, but if they just would have helped their own guys, I would have made sure that no one capped their medics. It's almost as much their fault as it is mine that someone had to go out and murder the wounded. Still, their fire was a little more accurate today, and it was higher in volume than the usual sporadic sniper fire. Generally speaking, my two units controlled their fire fairly well. The few retaliatory bursts were well directed, and they were few and far between.

All in all, the day was relatively non-eventful. No one was scratched under my command, and the Bangers didn't make another push on us. The only really down side was the heat. It had to be somewhere in the 90's, and that really doesn't help a breeze that's rolling over the hundred-plus corpses.

Tuesday, September 27, 2011

1535hrs

40000 feet above the Myanmar, Thailand, Laos border

150nm east of Mandalay, Myanmar

The remains of three squadrons that had been pulled together months ago whisped through the thin tropical air. The 1st Composite Fighter Group had suffered grievous casualties in this area only a few days earlier, and now they were looking for another fight. In the earlier fight, a pair of Vietnamese fighter squadrons had accompanied the mixed group of fighter planes. After the heavy air combat action in this area, they were ordered by Hanoi to remain within Vietnamese airspace.

The composite group's commander, Al Meyers, wasn't exactly reporting to any higher authority. He had been lost in the paperwork shuffle by the U.S.A.F, and most of his pilots didn't even have working governments to take orders from anymore. By consensus, they wanted some revenge for the other day's dogfight. Al was more than willing to lead them back into the area, but he was driven more out of a sense of duty than anything else. There was no point keeping fighter pilots away from dogfights. Their mission briefing was succinct and ended with a simple speech from Al.

Through their translators, he told his pilots, "Dogfights are where fighter pilots live, and where they eventually wanna die-if they have to die at all!" The irony was that Al was a reconnaissance pilot by trade who had cut his teeth as a bomber pilot. It was only by accident and necessity that he wound up driving an F-16 in the darkest days of the second Korean war.

In the thick jungle-covered mountains almost eight miles below them, the Chinese were moving entire armies into Indochina. Al had no way of knowing that even though he was still over "friendly" Thailand and Laos, there were a lot of unfriendlies waiting for him to come lower.

Communication in the air remained tricky. Even if someone had tried to call him on the radio, he never would have gotten the message. His plane had been without maintenance for so long that almost every time he took it into the sky, some new system stopped working. Most of his pilots didn't speak English, so Al had to communicate with hand signals. He done this for so long, and listened to the sporadic silence from his radio for weeks that when he finally did hear someone speaking in English, it almost stunned him stiff.

Through heavy static Al heard a feint and frantic voice.

"Bravo Mike Three One, this is Three Three, come in. Repeat, Bravo

Mike Three One, this is Three Three, come in."

There was silence, but that was an American somewhere out there.

"Bravo Mike Three One, this is Three Three, come in. We are under heavy air and infantry attack. Repeat, heavy air and infantry attack. Bravo Mike Three One, do you copy?" There was no answer.

"This is Bravo Mike Three Three, I am sending but not receiving. We are lighting our IR transponder beacon. Request immediate fire mission danger close on the north side of our position. They're coming down the Irrawaddy in barges, and they've got fighters all over the place."

The voice was getting more and more urgent as it repeated again and again. Al looked through his Forward Looking Infrared (FLIR) Camera. It rarely worked anymore, but the infrared beacons that ground forces used to show their position should have been easy to detect. It would simply be a bright black line on his screen. All he needed was the radio signal code to broadcast and activate it. He wasn't even remotely in touch with anyone in the region outside his squadron, so he had no way of knowing what the secret code of the day was.

There was an even chance that it would work, so he tried his radio.

"Bravo Mike Three Three, this is Wolfman One One. We're not from around here, and we can't toggle your beacon. Can you turn it on for a few seconds? Maybe we can help you out."

There was silence. Al was about to try his radio again when the voice crackled through the static again.

"Wolfman One One, this is Bravo Mike Three Three. We are under heavy air and infantry attack. We cannot reveal our position to just anybody. Do you know the day word, or the challenge?"

Al was relieved to know that his radio was working, but frustrated that he couldn't help the young man somewhere on the ground.

"Bravo Mike Three Three, this is Wolfman One One. Understand your situation. If it's that bad, advise that you turn on that beacon anyway. It sounds like they already know where you are. It's your call. We operate pretty far out to the east usually, and it just happens that we're in the area. No idea what the day code, the challenge, or the beacon codes might be."

Again there was silence. This time, Al figured that it was the people on the ground trying to decide. When the voice came back, Al could barely understand it through the static.

"Wolfman One One, this is Bravo Mike Three Three. We'll give you a ten second beacon starting....now."

Al looked at the cockpit monitor that showed the FLIR's display. He rotated its ball shaped head along the horizon and saw nothing. He tried a second time, this time panning slower, and spotted a small thin line way off on the edge of the horizon. When the line disappeared, Al noted its bearing, and called back on the radio.

"Bravo Mike Three Three, this is Wolfman One One. I've got a bearing on your position, but you are way out there. We're on our way, but it'll be at least ten mikes (minutes) before we get to your pos. I'll call again for another beacon spot when we get closer. Till then, hold on buddy. We're comin', and we're bringin' a little bit'o hell for them Chinese too. Wolfman One One out."

Al motioned to some of the other planes in his formation. There were only 21 left after the other day, and he couldn't communicate to them all so he had to rely on his wingmen to pass on the message. With his hands alone, Al told them that he was going to attack a ground target far away to the westsouthwest. He let the message pass through the formation for a few seconds. As each pilot received it, they wagged their wings and passed it on. When the last plane wagged, Al headed towards where he had seen the beacon. All the while, he listened to the sporadic voice on the radio calling for other to come to his aid. No one else answered.

Ten minutes and almost a hundred miles further from home, Al asked the voice on the radio to turn on the beacon again. This time, there was no response. He tried a second time, and still there was nothing. Al watched the FLIR monitor and panned the horizon again and again waiting for a response on his radio. There was no response, but then again, maybe it was his own radio that wasn't working again. Finally, Al saw a line appear on his monitor. It was still a far distance away, but much closer than it had been. He guessed that it was between 25 and 50 miles straight ahead. That meant that somewhere, close by air combat standards, there were enemy aircraft. Al motioned to his wingmen for them to pass the word, stay alert. Watch for fighters.

"Bravo Mike Three Three this is Wolfman One One. We're still a few more minutes away. Have seen your beacon. If possible, mark the target with colored smoke."

As Al was sending his message, one of the Taiwanese F-5 fighters broke formation and pulled up on the right side of Al's plane. His wingman saw it coming and climbed out of the way. The Taiwanese pilot pointed to the north and down, then indicated that he had seen enemy aircraft in that direction. Al wagged his wings and panned his FLIR around. His radar had given out weeks ago, and he had never really known how to use it.

The FLIR was identical to the one used on the F-111 bomber that he had flown for years so he found it relatively easy to use, when it worked. It was nice because the enemy couldn't tell that they were being watched by a FLIR, where as a radar could be detected, jammed, or evaded by enemy planes.

It took a few pans, but Al finally found the Chinese formation. It looked like a squadron of Soviet-built SU-25 Frogfoot ground attack planes. They were heading South towards where Al had last seen the ground unit's beacon. He looked around and saw a small smudge of purple smoke coming from deep in the jungle a few miles ahead. The Su-25's were indeed heading right at the friendly ground unit.

Al rolled his F-16 on its back and pulled back on the stick forcing the plane to dive steeply. He aimed the nose toward the SU-25's, knowing that the rest of his formation was following his lead. They were right behind him, accelerating, falling, and taking aim.

The Chinese were coming in low, winding through mountain passes and using the Irrawaddy River as a highway. Al's planes had to drop almost 8 miles to get to the same altitude. He lined up one of the SU-25's and locked on it with one of the two Sidewinder missiles that he had on his plane. When the missile growled in his headset to signify that it had locked on, he fired it off and locked on to another SU-25. Three more missiles zipped past his plane from behind. The other planes in his group were finding their targets.

Just before Al's missile was about to hit its target, the SU-25's spotted the ambush. They all popped out decoy flares to confuse the heat-seeking missiles. Most of the missiles were confused and went after the decoys instead of the hot engines of the SU-25's. Al's first missile was among those that missed, his second found its mark as the Chinese formation broke and spread out for combat. An SU-25 exploded, but kept flying. More Sidewinders found their marks, but the planes weren't going down. They had been designed for ground attack and were built with more armor plating than some light tanks. The small fragmentation warheads were only damaging the Chinese engines, not exploding their fuel tanks, killing pilots, or blowing off wings.

Al got on the tail of one of the Chinese that were still heading for the purple smoke in the jungle. He imagined the pilot to be a die hard, perhaps the squadron leader. He was definitely someone who was determined to complete his attack run. He might have made it if it was anyone other than a fellow bomber pilot on his tail.

The SU-25 got down low over the river. He was so low that every time there was a bend in the river, he had to pull up so that his wing didn't kiss the muddy Irrawaddy River. Most of the fighter pilots in Al's group were more comfortable in a dogfight, but Al liked flying as fast and low as possible. Upon later reflection, he would determine that the Chinese pilot wasn't discouraging his pursuer, he was making the chase irresistible.

It was so irresistible that even when Al had the plane in his sights, he didn't always fire his cannon. From time to time, he'd squeeze out a few hundred rounds, either passed the SU-25, or into the water ahead of it. Al was unconsciously toying with the low-tech bomber. It might have gone on endlessly, but when the fighter made a sudden turn to the left, Al temporarily lost sight of it. He expected to see it again as soon as he made the same turn, but when he did, he saw much more than the one plane.

Al saw barges, lots of them. He was travelling at over 500 miles an hour, so everything that wasn't almost directly in front of him was blurry. Still, Al counted at least 15. Each one was loaded with troops, just like the ones that he had seen a few days earlier. Distracted by the barges for only a split second, Al lined up the SU-25 in his gun sight and fired loud tearing stream of 20mm cannon rounds into its engines. Both exploded, and the plane slammed down into one of the barges only a few feet below. Immediately, Al had to pull up to avoid the debris filled explosion and subsequent splash. He put the F-16 into a steep 45-degree climb and saw that he was only a few miles from the purple smoke.

Al tried his radio again. "Bravo Mike Three Three this is Wolfman One One. I've a visual on your smoke, can you pop a flare to indicate your position?"

His answer was mostly static, but a white signal flare did pop out of the jungle. The smoke was directly between the Irrawaddy and a small stream. Al could see the barge that had landed on the Irrawaddy and most likely dropped off the infantry that was causing so much trouble.

He continued his climb and looked around for his fighter group. They were scattered as far as he could see, chasing down runaway SU-25's. Al brought his plane back up to 6500 feet and circled for his attack run. He called the ground unit and lined up the purple smoke in his sight.

"Bravo Mike Three Three from Wolfman One One, get your people down, I'm coming in guns hot."

There was no response from the ground. They were too busy taking cover. When the smoke was in the middle of Al's cockpit heads up display, he opened fire with his cannon. Al let out a full second and a half's worth of 20mm rounds, then pulled up just before his F-16 was about to hit the tops of the jungles highest canopy. As he climbed and circled for another attack run, Al spotted three Taiwanese F-5 Freedom Fighters diving on the smoke and firing their twin 20mm cannon mounts. Al made another dive, this time on the barge, when he was pulling up, the Taiwanese F-5's were diving on the purple smoke again. They had no idea what they were shooting at, and they hadn't heard a word on the radio, but Al, their squadron leader was attacking purple marker smoke, so they did too.

After Al's second pass, he was down to just over 150 rounds left in his Vulcan cannon. The F-5's carried even less, so they broke off the attack and circled with Al. Four South Korean F-16's arrived and strafed the area until they too were low on ammo. The small flight grew in number as planes from the mixed group returned to the area from chasing down the SU-25's. Every plane that Al had left Vietnam with eventually came back and joined the formation. After a few more strafing runs on the smoke and all of the barges in the river, Al called back to the ground unit.

"Bravo Mike Three Three from Wolfman One One, the clouds are dry so the thunderstorm is over. Hope we did some good down there."

His answer was rewarding. *"Wolfman One One, who the hell are you guys?! That was incredible! We're buggin out, but thanks a million. Hope we get to call on you guys again sometime. Bravo Mike Three Three out."*

Al smiled. It was the first time an American had complimented like that in a long time. He gave a hand signal to his wingmen and headed for home, almost 1000miles to the east in Vietnam.

Wednesday, September 28, 2011

Victory in Bangladesh-Score one for the Good Guys!!!

Just as quickly as the assault began on Bangladesh began, the fighting is over. While American Special Forces teams swarmed in and above the Bangladesh capital city of Dhaka, it's leaders were video conferencing with Swedish diplomats in an effort to broker a peace agreement. In the pre-dawn hours this morning, the word went out from those same leaders to military commanders around the country; cease-fire.

The war with Bangladesh is over. Per specified arrangements with the Coalition, Bangladesh signs the Geneva Peace Agreement of 2011, and declares itself a neutral country. It will be occupied by Coalition troops, and new elections will be held on a staggered basis beginning at the local level in 1 year, and moving up 1 level of government each year thereafter. A formal signing of the agreement will take place sometime in the next few weeks, but until then, the fighting has ended via the first ever digital transmission of a peace agreement.

Wednesday, September 28, 2011

Myanmar Air Cavalry Unit Relieves Besieged U.S. Troops Near Myitkina, Myanmar.

The cavalry came to the rescue again today when a small American garrison deep in the heart of Myanmar heard the closing thunder of helicopter rotor blades. They had been sent to the city to reinforce the local authorities against Eastern Alliance backed guerrilla attacks. With little or no warning, an untold number of Chinese troops began flooding down from the mountains through the Irrawaddy river valley. The 509th airborne infantry battalion was sent in to help stem the flow, but they didn't even have time to get out of the city.

For two days, Chinese artillery, mortar and infantry wrapped a noose around the entire city and just as they were building up for a direct assault, three helicopterborne air cavalry infantry battalions landed up river. The Chinese immediately began pulling back and concentrated their efforts on the new threat. More Coalition reinforcements are expected to arrive over the next few days, bolstering forces in the area.

Wednesday, September 28, 2011

1st Lt. John Chamberlin -Personal Log

U.S.S. Ronald Reagan

22nm South of the Ganges River Delta, Bangladesh

It's over! Last night I got the word on the regimental net. Bangladesh's government has resigned and an interim government has proclaimed their neutrality. They requested an immediate armistice, and their troops were told to turn in their weapons at noon today. At exactly 1200 Hrs, I saw a white flag from across our little no man's land, and out came the Banger CO.

He spoke English very well, and we discussed the situation. I asked him to have his men come across the craters with their weapons gripped at the barrel and held above their heads. He agreed, and out they came. By 1345, we had disarmed the remnants of an entire battalion, about 600 men. Our troops shared what food we had with them. I had a squad escort one of their platoons down to the river where they could gather water for everyone, and all told, things went incredibly well.

They were upset about Wess' actions, especially after I told them that we would have let them get their wounded. I think part of the reason that they were so upset was that they understood. Had it been our people out there, most of the prisoners that I spoke to said that they would have done the same.

At 1700 hrs, a column of trucks came through the cratered area. They were Australian Infantry that had been flown in to the airport after we had secured it on the first day. I met with the Australian officer and introduced her to the Banger

CO. He and I then briefed her on what had transpired over the past few days. She thanked us, relieved me, and we hitched a ride on the trucks back to the airport. Two hours later, we were choppered back to the Reagan. My two teams bunked down, and I began writing up my formal report for Dennis.

Thursday, September 29, 2011

1st Lt. John Chamberlin -Personal Log

U.S.S. Ronald Reagan

Somewhere in the Bay of Bengal

It is soooo good to be back on the ship. We might be packed to the gills with people and equipment, but air conditioning and cafeteria/galley food never felt this good before. Most importantly, there's no smell. In Dhaka, the air was filled with so many different scents that you couldn't tell good from bad after a while. Cinnamon and spices from the market start to make you sick. The smell of death, garbage, and "fertilizer" never smell good, but every once and a while they were a welcome change to the spices.

After chow this morning, I dropped of my report with Dennis. He was happy to see we made it back ok. Even though the operation was an overwhelming success, there were a few of our units that took some serious casualties. One was almost wiped out entirely!

Dennis pulled me aside into an old pilot's berthing area to talk to me for a while. He wanted to find out more about Wess' shooting of the Banger wounded. I basically told him the same story that was in my report, but I think he wanted to me to give it to him in my own voice, not just my own words. Dennis tried to give me a lecture about how terrible it is to shoot wounded, or prisoners as some others were doing, but he couldn't fool me, and he knew it. The issue was simple. We're in a war. The higher up you go in the chain of command, the more responsibility you have to win that war. The lower you are on the totem pole, the more your responsibility is simply to follow orders, and survive. We talked for almost and hour, then he let me go. I don't think he likes the idea any more than I did, but I'm certain he understands it. Maybe he even has a little empathy for those of out in the field?

Friday, September 30, 2011

Correspondence

From Corporal Matthew McKenna,

U.S. Army, 1st Infantry Division, 12th CHEM FWD Team ASD3 To Jonathon McKenna (*Grandfather*) Su'p Gramps?!

We made it through the rough spot in just a few days. It was a hard run, but we did it. The Pinkos never really caught up with us, and we only had one tense moment.

Harry tripped a homemade Bouncing Betty mine. When it popped out of the ground, it ricocheted off the bottom of his ruck and rolled over to Joe before it went off. They're supposed to pop out of the ground and blow up at waist height, but this one took just a little longer to go off. We've seen others go off in The Zone, and the explosives are usually just powerful enough to wound one man. The idea is to wound one man and make two other carry him from the field (eliminating 3 grunts from the action). This mine seems like it was a re-build, and the explosives must have gotten wet (Gee? I wonder how that could have happened in a place where it rains at least once a day and fills up to 12' a year?!). Well, the mine rolled back to Joe, and went off on the ground-right in front of him. When it bounced off Harry's Ruck, Joe stopped in case he was on a mine too. He wasn't, but he did have to watch Harry's come rolling over to him and then explode.

The explosion was a lot bigger too. With wet or damp powder, the burning process goes slower. That meant that as the explosion started, it tossed the surrounding powder out into the air. It mixed with the O2, and slowly ignited. That also meant that the blast wasn't fast enough to make the shrapnel (nails in this case) hit with enough force to do much damage. Joe's got some nice cuts on his face, hands, and legs, but otherwise, he was just knocked back a little-maybe as much as 20'. Afterward, Joe rolled around on the ground, and all the nails that were sticking in him kept getting pushed farther in. It was like having a cactus blow up in your face. It was nothing that a little liquid bandage sealant couldn't handle, but he was soooo pissed! Piotre disappeared as soon as the mine went off (looking for a sniper ambush that never came). Harry fell to his knees laughing. So it was up to Pepe and I to calm him down and start pulling out the nails.

Pepe made some sort of New York reference (somehow!), and Joe punched him. I stepped out of the way. Pepe rolled out of the way and stepped much farther out of his way, and we left Joe to his misery. This all happened two days ago, and the guy is still so pissed that Pepe won't come anywhere near him. He even ducks down and slips out of the bunker when Joe comes in now.

We had a while to walk yet, and when Joe was finally calmed down, we headed out again. As soon as Piotre was back, and Harry had moved forward, Joe started kicking and bitching again. Everyone froze and looked around to make sure we weren't being ambushed. We were-at least Joe was. The poor SOB hadn't even taken 10 paces when the biggest porcupine we had ever seen ran right across his path. It was the size of a large dog, and the long white quills looked just like punji sticks. It didn't even look like an animal, so we all thought it was another boobytrap at first. Well Joe was still pissed about the mine, so he shot the hell out of the critter. It took a full burst of 5.56mm and still rolled around on the ground like a snake or a chicken with its head cut off. Joe fired another 3-round burst and stepped back, but he was too slow. Sure as shit, it got him. The big ball of quills rolled right into his leg, and you could see Joe fighting back the scream. Harry, Piotre, and I grabbed him and got him moving out of the area since we knew his fire would attract bad guys. After he hobbled for 100-meters or so, he was able to limp on his own. Piotre and Harry moved forward to keep us going, and Pepe was happy to trail far behind Joe. It wasn't his fault that the porcupine was startled, but we all knew Joe was going to blame him for it. There just seems to be a simmering anger between the two, and it's all coming from Joe. Talk about your bad luck?

We're back in The Sewer now. Saunders was pissed that we didn't keep in touch with him a little better, but he cooled down when I uplinked the image and audio feeds from the different EOP's. You could tell he was about to come down on me, but when he saw the footage from the EOP's-especially the last one that I left on Popa Hill, he huffed and left the bunker. A few later, he came back and had me send the feed to division so they could start coordinating air strikes. I told him that I had already called in that Wolfman guy, and the rest of the guys told the story. Saunders was cool about it. He nodded his head, told us that we had done "a decent job," and then left us alone. We must have done a bit more than just a decent job since we're all off the duty roster until Monday-two whole days to sit back, burn meat, sleep, and recover from "The Big Run," as Harry calls it.

I hope all is well back home. We've heard some news that the U.S. Navy and some composite air assault unit kicked ass against Bangladesh and prevented them from breaking through on our western flank. Everybody's real happy about that. It seems like the tide is finally turning. At least we're not getting our butts kicked all over Asia anymore!

Piotre, Joe, and Harry tell me that there's a bit of a gathering going on over at Alpha Company's grill pit. I'm gonna head over there now. Our bunkers are all air-conditioned, but it's still 105-degrees outside with so much humidity that the ground is steaming. Beer sounds VERY good right now!

Take care,

-Matt

Saturday, October 1, 2011

Correspondence

From Corporal Matthew McKenna,

U.S. Army, 1st Infantry Division, 12th CHEM FWD Team ASD3 To Jonathon McKenna (*Grandfather*) Su'p Gramps?!

We're having a blast on our unofficial R&R. The Alpha Company gathering at the grill pit was a lot of fun. Some females from Bravo showed up and it was like a feeding frenzy as everyone tried to hook up. It's funny how on the line everyone does their job, and we all seem like professional soldiers instead of the

80+% draftees that we are. Behind the lines is a different story. That's when we all cut loose a little and try to live it up while we can. Some people go a little too far sometimes, but hell, this time last year most of us were making faces at each other in High School Study Halls. I guess it's understandable. Well, the burgers and beer were in good supply, and we had a great time.

We got back late last night, and Joe finally got his revenge on Pepe. It's either been one New York City story too many, or just plain bottled-up aggression from the past couple days. For whatever reason, Tattoo Joe decided to play a little joke on Pepe. It was a joke that might have gone a bit too far.

We were all having a good time at the gathering, and one of the guys from Alpha showed us one of his Nat statues. There are about a million different religions in this country, and each one seems to have a hundred temples. Out in the Bagan area-where Popa Hill is located-there's a big group of people who worship Nats-their word for spirits. Well! When we were back on top of Popa, Joe remembered that the broken Nat statues sorta freaked out Pepe. So, he bought this 5' statue from the guy in Alpha Company.

The statue has that typical Buddha-style face, but it's eyes are open, and no matter where you are when you look at it, they seem to stare right at you.

The arms aren't wrapped around the belly either. Instead, they rest on the knees like it's waiting to do something-or it's waiting for you to do something. It has a ton of jewelry carved all around it, so it looks a lot like one of those Aztec or Egyptian statues-very unusual for this area. It's just a very freaky statue.

Well, at about 0330, we dragged it through The Sewer (This was NO easy task!). Pepe was passed out in his dugout (a coffin-shaped area cut into the walls of his bunker and used for sleeping). Joe stood the statue up next to Pepe's face, and then crept out of the bunker. He was smart enough to warn everyone around otherwise he would have gotten in a lot of trouble. Saunders thought it would be a nice diversion, so instead of putting a cap on the fun, he gave Joe a Screamer.

A Screamer is a noise grenade. It's another one of those "non-lethal weapons" that were developed back in the 90's. We mostly use 'em for fun, but they're also good for ambushes. When these things go off, they're just about as loud as a jet engine, but only for about 15-seconds. Then, your ears are left with the same ringing that you get after a barrage.

The Sewer was packed with people, and we all covered our ears to see what was going to happen. Joe tossed in the Screamer. It was so loud that most of the crowd headed back to their bunkers to get away from the temporary sound. It was a lot of fun at the time. Joe and I were the first ones into the bunker when it stopped screaming, and that's when I lost the fun. Pepe was curled up in his dugout and shaking. Joe was still laughing, but he toned it down a lot when he saw the total terror in Pepe's eyes. He was also smart enough to prevent anyone else from coming in and seeing Pepe like that. I came up to him and helped him laugh it off real quick before the crowd managed to get past Joe. I don't think anyone saw Pepe in bad shape. If they did, no one acknowledged it, and everyone tried to highlight the fact that it was nice to joke around instead of just sitting and waiting for the Chinese to hit us again. The strangest part about it is that Pepe is still walking on eggshells around Joe, but there are far fewer New York City references coming from him, and Joe is a lot more tolerant all of a sudden. That Screamer brought them to some sort of unspoken understanding, I think.

It's time to get going again. Rumor has it that Alpha Company has secured more beer and burgers. We're gonna go over and take advantage of it before whoever they're stealing from catches on and ruins a good thing! Hope all is well there. Say hi to everyone, and tell mom I'm playing it safe. She might not believe it, but maybe hearing it will help.

Thanks,

-Matt

Sunday, October 2, 2011

Correspondence

From Corporal Matthew McKenna,

U.S. Army, 1st Infantry Division, 12th CHEM FWD Team ASD3

To Jonathon McKenna (*Grandfather*) Su'p Gramps?!

Ya know, a lot's changed around here since our little hike to Popa Hill the other day. The Chinese made a coordinated series of small unit and sniper pushes into The Zone right after we left. A guy from Alpha Company told us that Division called in every available air asset in the area to deal with the problem, but nothing seemed to stop the small attacks. Finally, someone rerouted a C-130 Hercules transport that was carrying a Daisy Cutter (A 15,000pound, Fuel Air Explosive bomb).

They flew over The Zone and dropped it on the other side of the middle ridgeline right in front of Alpha. The truck-sized bomb rolled out and parachuted down with everyone on both sides watching. It probably caught the attention of the Chinese AAA people too since only a few shot at the slow C130. Well, those bombs are designed to explode above the ground. I guess the altitude of that center ridgeline must have set off the fuse to the bomb's dispersal charge a bit early. This guy from Alpha said they had all been warned to strengthen their bunkers, but it really just made everyone look. The dispersal charge made an opaque cloud rain down on the other side of the middle ridge. The detonating charge went off at the correct altitude, and this guy from Alpha told us that it looked like a nuclear mushroom cloud rising from the other side of The Zone.

Everyone watching from The Sewer ducked to get away from the heat (We just thought it had been unusually sunny and everyone had gotten bad sunburns).

Then the mud and debris from the other side of The Zone was blown over the crest of the middle ridge, and the guy from Alpha says it was raining trees and branches for almost 5 full minutes! Everyone was very impressed-except the Chinese on the receiving end. Most of the EOP's that we set up are gone-AGAIN! So, battalion had every company send out some patrols.

They took digital pics of the other side of The Zone. We thought it looked like the moon before. That Daisy Cutter wiped the place clean. There's nothing left. It's like one gigantic pit was scraped out of the northern valley. All the other craters around where it went off are gone-filled in with mud and debris from the big blast. The craters that remain on the Chinese side of the middle ridgeline all seem very small-even the ones from the big guns and bombs from our planes. In comparison, the 2000-pound bombs can leave a crater almost 100-feet wide in soft soil. The Daisy Cutter leaves a crater about the size of a football field and scours the area around it for almost 2-miles. Since this one went off in a valley, its shock wave disappeared on the crests of the two ridgelines (one in the middle of The Zone and the other along the edge of the Chinese trenchline). The shock waves that went up and down the valley were just as concentrated as the ones on the ridges were, and they wrecked everything in the zone for over 5-miles (2-miles to the east, and 2-miles to the west). They dropped the Daisy a few days ago, and normally grass or bamboo will sprout in any new crater within 48 hours, but the digital pics that we saw were from today, and there still isn't a blade of green anywhere within the bomb's burst area. A lot of firepower was spent on just a few snipers and squads of Chinese.

The mud, debris, and general trash (The Zone seems more like a landfill than a battlefield sometimes) was tossed over the middle ridgeline. We've got snipers and patrols up there, but only to serve as early warnings against any more Chinese attacks. Saunders has a few other teams putting in fresh EOP's all along the crest of that middle ridge. We're supposed to hang out and rest up for the next few days. Then, he says he'll cut us loose.

I got a little bored and tried to tap into my old EOP's to see if any were still sending signals. Out of 126 webcams and 319 mikes, I was only able to get signals from 3 cameras and 9 mikes. I'm sure the mikes have been blown all over the place, and there's no hope of triangulating any sounds from them until I survey them again. All 3 of the cameras are covered in mud, knocked over, and resting in front of debris. Only 1 is able to give any real imagery, but I wish it hadn't.

I've got a camera/mike combo left that's still kind of operable. The screen is very dusty, and I thought the image was a rock or a helmet at first, but it's not. It's a head-a Chinese head. The poor bastard must have been blown into some soft, foamy mud and then sank. I would have ignored it completely since the camera has too much dust on it to see more than a few feet, but I checked the mike's record, and occasionally, it picks up a moan or some garble. The head-the Chinese soldier who's buried up to his chin-is still alive! I have some idea of where the camera/mike EOP is located, but it could be anywhere within 100' of where I left it. We'd probably send someone to grab the guy-for intelligence purposes of course, but the Daisy Cutter fell days ago. The guy's clearly wacko now. He just stares motionless and mutters, "Ni hao ma? Wo xiang-yao yi-bei bing-shui." My computer translating program says that means, "How are you? I'd like a cup of iced water." The mud, the sun, the rain, the shock from the bomb, lack of food, and dehydration all probably sent him over the edge a long time ago. Now, it's just a human animal dying a horrible death.

That's not real news around here, but I wonder what would happen if the other Chinese troops could see this video of their friend dying? It's too bad I can't email it to them. I could, but then they'd know more about our EOP's, and I don't think it'd be worth it. Piotre asked where the EOP was last located, and I showed him. He says that the next time he goes out, he'll go over and pop him to put him out of his misery. The guy's head is already gone, his body just doesn't know it yet.

I watched the video for over an hour. It's very hard. The guys came and saw it too, but they all walked away-ignoring it as much as possible. I showed it to Saunders who just said, "I guess that EOP's wasted. Make sure to disconnect it to open up more (computer) memory for the replacement ones, and do it ASAP." I disconnected it, but the imagery still plays in my head. I still see that scared face, and I still hear the recordings of his babble. This all happens while I'm still awake, and I definitely don't want to go to sleep tonight.

There's a gathering over at battalion HQ tonight. The word is that some honeys from division intel will be there. Gatherings at higher levels tend to have less beer and more honeys, but we heard about it from Alpha Company, and they're always finding new sources of beverage, so it should be wet enough.

I'm off to get some grub before we head out.

-Matt

Monday, October 3, 2011

Stow Herald

Myitkina, Myanmar:

A joint American and Myanmar Rapid Reaction ground force is on its way to reinforce the city's garrison after several days of guerrilla-style attacks. These attacks began as typical terrorist bombings and sniping incidents, but over the past few nights the local authoritiesincluding the Myanmar garrison of almost 3000 men-has come under mortar and even light artillery strikes. Rumors abound that this might be the forefront of yet another Chinese offensive into Indochina.

Prior to the terrorist attacks, at least two battalion-sized Chinese infantry units had attacked the city, and even occupied parts of it. Only when a Myanmar air cavalry brigade was redeployed from further west were the Chinese driven back into the jungle. It was widely accepted that the Chinese had been completely driven out of the area, but the recent guerrilla attacks are proving otherwise.

Wednesday, October 5, 2011

Miami Daily

Reinforcements Make it to Myitkina

After a two-day trek through jungles, mountains, and guerrilla ambushes, a relief column has finally arrived only to find the city besieged by local communist insurgents and Chinese irregulars. American Army officials claim that there were several indescribable incidents along the way. When asked to comment further on the incidents, officials would only say that an international war crimes investigative team was going to be needed. It would appear that the traditional stench of jungle warfare, atrocities, has finally reared its ugly head.

Thursday, October 6, 2011

Correspondence

From Corporal Matthew McKenna,

U.S. Army, 1st Infantry Division, 12th CHEM FWD Team ASD3 To Jonathon McKenna (*Grandfather*) Su'p Gramps?!

It seems that the Chinese have a new toy! We have some small electronic surveillance teams that use special remote-control airplanes to look over The Zone. Our guys have been working on making smaller, grenade-sized planes that every soldier could carry, but we're told that the ones they finally came up with were too expensive. Well, the Chinese aren't morons. They're very sharp, and they seem to have found a way to make the small remote-controlled planes affordable.

Officially, their called Micro-Air-Vehicles or MAV's. Each one is about the size of a dove. They're controlled by individual soldiers. They're very hard to hear until they're right on top of you, and they're so fast (maybe 20-30 miles an hour), that by the time you hear 'em, they're already past you. Each one has a tiny video camera and transmitter built in, and we assume that they can send back real-time imagery of our positions.

So far, the Chinese just seem to be learning to use them. We started seeing them yesterday morning. At first, it was just one or two buzzing their way over The Zone. There have been times when the sky was filled with them, and now hardly a minute goes by when you can't hear that annoying buzz sound. Everyone thinks that when the Chinese figure out how to use them well enough, they'll map out our positions and call in some accurate artillery on us. There hasn't really been an increase in stuff coming over from the other side, but we're all careful to stay under cover-most of the time.

The Sewer is loaded with 12-guage shotguns. They're as common as fire extinguishers. There must be thousands of the things in our division's area alone! We have them all over just in case the Chinese ever managed to infiltrate inside or come right up on us. American ingenuity has taken over, and a new hobby has been created. Every single bunker seems to have at least one or two people watching for the new Busy Bees (our nickname for the MAV's). When a Busy Bee flies within range of a bunker, it's like duck season, and everyone with a 12-guage opens fire. The buckshot works well, and we've downed lots of 'em so far. It's a lot of fun, and as long as the Chinese don't pair up the Bees with their artillery or other stuff, we're having a good time. If/when that happens, I think things will be a lot safer in The Sewer than they are in the rear with the gear!

Murphy's Law has come into play now. After our days off, Saunders has us putting in new groundwater monitoring wells. It used to be the safest job in the area. It still is, but if the Chinese use their Bees to direct artillery and other stuff on the rear areas, we'll be out in the open. I almost feel like it was safer back near Popa Hill-ALMOST! It's just for a few days though, and so far the Chinese seem to be taking their own sweet time to learn how to use their Bees. I should be fine.

After the monitoring wells are in place, I'll have to go around and gauge their depth-to-water levels. It should only take a few days if I haul ass. Saunders wants at least three full rounds of data from each well. If I pack MRE's and only sleep for 3-5 hours between sets, I should be able to get it all done by Monday. That would be great. Then I might be able to take a few days off and hang in The Sewer before the Chinese start matching their MAV's and stuff together.

As usual, I've gotta run. I'm already loaded and set to head off for the well installs. I haven't been getting many emails from y'ins back home. Wassup? DROP ME A LINE!!!!! Say "Hi" to everyone, and take care of yourselves.

-Matt

Monday, October 10, 2011

Correspondence

From Corporal Matthew McKenna,

U.S. Army, 1st Infantry Division, 12th CHEM FWD Team ASD3

To Jonathon McKenna (*Grandfather*)

Gramps,

Things are getting a bit trickier over here. We put in all our monitoring wells, and I was able to get three full rounds of data, but Saunders had everyone out on EOP missions today-even if we had our well data complete. We got a new truckload of seismographs and foot traffic sensors last night, and he's hot to have them all in place. The guys were off duty today, but Piotre was nice enough to go out there with me.

He wanted to pop that Pinko we saw stuck in the mud last week. The guy should have been dead when we saw him on the EOP monitor, and we didn't expect to find him alive. It was awful. I lead Piotre to the area where the EOP had been. We both walked around for a minute or so, and then I found himSTILL ALIVE! He'd been there so long that the daily rain had flowed mud around him like a river sand bar. That same rain probably kept him alive. Instead of looking at the pitiful sight, I went and found Piotre right away. He came over and saw the guy. Under that mud-covered face his eyes were spinning, and he was mumbling continuously. It was VERY clear that his mind was totally gone. Even if we were able to dig him up, he was useless as an intelligence source, and neither of us could see someone that mental ever coming back. Piotre looked at me in disbelief, and I'm sure I gave him the same look in return. Then he looked around for any other Chinese and saw how visible we were to their trenchline. As soon as he looked back, he took the butt of his rifle and smashed in the buried Chinese guy's skull. It would have been a lot faster and easier to watch if he'd have shot him, but we both knew that it would rain lead from their trenchline the second that they heard the shot.

I found the old EOP, disconnected it, and set up a foot traffic sensor. As soon as everything was in place (and well-camouflaged) we headed back. Neither of us spoke. Killing that Chinese grunt was an ugly thing to do and to watch, but it had to be done-for everyone's sake. We crawled past plenty of dead bodies and other grotesque scenes on the way back to The Sewer, and we had definitely seen more than our share in the past. Still, that one grunt has remained in our minds.

Piotre had killed plenty of Chinese and looked into their eyes every time, but he never seemed so cold as he does now. I see him differently. He still seems normal enough, but I can see that emotional black hole inside of him. Maybe it's because I feel a touch of that too, now?

Saunders has more EOP's that he wants me to set up, and then he wants another round of monitoring well readings. He's convinced that the Chinese are up to something big. I've gotta cruise now if I wanna get any sleep.

Take care,

-Matt

Wednesday, October 12, 2011

Correspondence

From Corporal Matthew McKenna,

U.S. Army, 1st Infantry Division, 12th CHEM FWD Team ASD3

THE X-MAS WAR

To Jonathon McKenna (*Grandfather*)

Gramps,

The Zone just keeps getting stranger every time I look out at it! Last night, the Air Force (love 'em or hate 'em) decided to make a big show. Some sort of jets flew low over the middle ridgeline and dropped huge illumination flares on the Chinese valley to the north. It silhouetted the middle ridge, and made the northern horizon look like it was a yellow-white inferno. A few seconds later, we heard the sound of a turboprop plane, and then we watched as thousands and thousands of tracer rounds streamed down on the Chinese trenches. It was probably an AC-130 gunship. While that was going on, a few more turboprop planes came down our side of The Zone. They were silhouetted against the illumination flares to the north, and we could see that they were C-130's also.

When the flares burned out, dropped, or simply died down, the gunship stopped firing and disappeared to the west. I think everyone woke up to watch the show. Those planes carry 2, 20mm cannons that fire up to 1200 rounds a minute each. They also have 2, 40mm cannons and even a 105mm howitzer! Joe says that their 20mm cannons alone can put a round in every square inch of a football field in one minute! I believe it!

The illumination and tracer round display were wild enough to watch. As darkness came back to the area, and everyone's night vision came back, the completely strange stuff started. Small, fuzzy patches of glowing green started to appear all over our valley-right up to the crest of the middle ridgeline. We didn't know what it was, so Pepe went back to battalion to find out. We all prepped for an attack, but he came back about 10-minutes later and told us to relax.

He found out that the C-130's that flew low over our valley had sprayed bacteria all over our side of The Zone. The bacteria are harmless, so we didn't have to put on our CBW (Chemical Biological Warfare gear-gas masks etc.). Turns out that the Air Force has been using the stuff since we went into Bosnia back in the 90's! This bacteria is attracted to explosives, and when it finds even a trace amount of any explosive, it feeds on it and glows like a microscopic lightning bug! The small, fuzzy patches of glowing green showed us the location of all the Chinese mines. Saunders knew about the C-130 Chemstrike, but couldn't tell anyone-obviously. I'd been briefed on the stuff back during training, but seeing it was another thing entirely! Pepe says that every engineer unit in our corps was in our area and heading out into The Zone. Those poor guys…now they can find the mines, but when they remove 'em, they're gonna glow from all the residue!

Like I said, this place gets stranger every day (or night)!

Take care,

-Matt

Friday, October 14, 2011

Correspondence

From Corporal Matthew McKenna,

U.S. Army, 1ˢᵗ Infantry Division, 12ᵗʰ CHEM FWD Team ASD3

To Jonathon McKenna (*Grandfather*)

Yo Gramps,

Saunders is giving us a few days off. It's been raining for two days straight now, and according to the online weather radar, it looks like it's going to rain for some time to come. All this water is throwing the data from our groundwater monitoring wells outta whack. If we went into The Zone and planted any EOP's or sensors, they're sure to get washed away with any camouflaging we might do. So, we get a couple of days off.

It sounds great, but the constant rain is a lot worse than the morning showers we've been getting. It's always washing away the camo to the firing ports and trenches in The Sewer. Every night we have to crawl out into the muck and try to re-cover our positions. It's tough. The mud here is mostly that red, greasy kind, and people are always sliding down the hill-sometimes all the way to the bottom of the valley.

The heavy rain is also driving the jungle critters into The Sewer. The rats take care of most of the larger ones, but the bugs are insane. Beetles (the size of golfballs) latch on to you in the middle of the night using the hooks on the ends of their legs. Lice, mosquitos, gnats, and all kinds of flyers make you want to hold your breath forever.

Everyone agrees that the leeches are definitely the worst! They slide in through the mud, and they're so small and slippery that you never notice until they're already attached. Most are black with brown and red speckles. When they bite, they have a sedative in their mouth, so even then, you still don't feel it. We've used up all the salt from the cooks, and now (even though smoking is strictly prohibited), everyone carries a lighter to burn them off. Their mouths also have some sort of natural anti-coagulant. So, even after they're off, the holes bleed and bleed until you put some minor wound liquid sealant on 'em. That sealant stuff helps a lot, but it's clear and shiny, and we all look like we've got glossy measles all over. And YES, the leeches attach everywhere, just like I'm sure they did to you when you were in Vietnam.

The rain, bugs, critters, rats, and choking mildew all suck, but add the Chinese seem to be up to something too. Our engineers were able to clear away most of the mines a few days ago, and yesterday, the Pinkos started raining in FASCAM rounds (artillery shells that burst open and scatter small anti-personnel mines). They're also making mine-laying runs down The Zone with those huge H-5 bombers. I don't know how those planes survive because I think everyone in the entire corps pops off at least a few bursts at each one as it comes down the middle ridgeline. Division is keeping track of where the Chinese are putting down mines, and I'm sure they're pre-planning artillery strikes on those clear areas in case the Pinkos try to attack through them (If they wanna attack through their own minefields, that's fine with all of us!).

When it rains, the drops are usually big, fat, and heavy. We all have a tough time keeping our eye protection (sunglasses) clean, and lots of people have taken to going without them-until today. A fresh batch of replacements was getting a tour of The Sewer this morning. When they were moving through a trenchline at the top of the ridge, the rain must have become inconvenient to them. A bunch of them-almost an entire platoon-took off their eye gear and zap: those non-lethal lasers that we were warned about came into play. Somewhere in The Zone (probably not too far away), a Chinese sniper used a colorless laser on the replacements. First one guy started complaining about how his eyes hurt. While he tried to blink them clear, the next guy in line noticed it too. Then the next guy was blinded…and so on, and so on, and so on. They never saw the sniper with the laser. It made no sound. It somehow made it through the rain, and the pinko just pointed it at one grunt's face, then another, and then another, until they finally realized what was happening. 28 people were blinded, and the rumor is that most will never see again. Sad. It could have been so easily prevented. Now, everyone in the division is wearing their eye gear ALL the time. You can even see people sleeping with it on their faces. I never stopped, and I certainly won't now! What a way to get sent home?!

I'm gonna get going and check out Alpha Company's grill pit. Rumor has it that they bought an entire ox, and they're BQ'in it tonight. Alpha always seems to have beef and beer somehow. Their scrounger has got some serious skill!

Take care,

-Matt

Thursday, October 20, 2011

Correspondence

From Corporal Matthew McKenna,

U.S. Army, 1ˢᵗ Infantry Division, 12ᵗʰ CHEM FWD Team ASD3

To Jonathon McKenna (*Grandfather*) Gramps,

The entire line went on alert last night. I was at the Alpha grill pit when I got the word. The place emptied in 2-seconds flat. Everyone was at their bunkers and ready in just a few minutes. Since we're officially still a rearechelon unit, we don't have our own bunkers. We're supposed to reinforce the trenchline at the top of the ridge. Since I've spent so much time with Joe, Pepe, Harry, and Piotre, Saunders let me join them in their bunker (those trenches at the top of the ridge don't have a roof, and the rain comes right in!).

About an hour after everyone was in place, some supply guys stopped by and handed out drum mags of ammo. Normally, we use the new 50-round straight magazines. These are a lot shorter, easier to move around, less prone to jamming, and hold 200-rounds! We love 'em, but-once again I'm reminded of your message about the military: "They do everything for a reason." I figure someone back at division thinks we're gonna need a lot more ammo to fight a lot of Pinkos. We're all a bit scared, but we're also confident in our training. It's a strange mixture, and in the end, all any of us are feeling is a huge sense of anticipation.

I noticed that Piotre has a new rifle that he only uses in the bunker. He's back to using a bolt action sniper rifle. He's also wearing more and more ghillie suit-style camo. I think that as he's regaining his aim, he's regaining his pride and feeling more like the sniper that he is. Still, Piotre's Piotre. He always has to do things a bit differently, and he's taken to using small, armorpiercing, discarding, sabot rounds. It's still a 7.62mm round, but instead of a steel-jacketed piece of lead, the operating end of the bullet is a long, thin, heavy, hardened steel pin that's only .22cal. The space between the pin and the barrel is taken up by a plastic bushing that comes apart as soon as it leaves the barrel. What this does is allow the total mass of the bullet to hit a smaller area. Getting hit with a regular bullet might be like trying to catch a falling refrigerator, but the sabot rounds are like trying to catch a refrigerator that's 50' long and only ½ inch x ½ inch, and you'd have to catch it by the small end-not the length! Piotre says that they go through anything; all body armor, sandbags, huge trees, vehicle engines, even the thinner armor on a tank. The only downside is that they're tricky to load because of their length, and each one has to be individually hand-chambered. For that reason, Piotre still has PSG1 handy at all times-just in case.

It's been raining for over a week now, and everything seems soaked. Along with the 100+degree heat, the humidity is choking. It's also made a warm weather fog. It's really just the moisture steaming out of the mud in The Zone, but anyway you look at it…it just sucks! And of course, both of the portable, camping air-conditioners in this bunker have died. We think that it's just that the batteries need to be recharged, but without sunlight, the solar charger doesn't do very much to the batteries.

The Sewer was filled with that mildew smell before. Now it's almost unbearable. There's also a lot of new people in The Sewer now. With the alert, division has called everybody to The Sewer. There's almost no one behind the ridge anymore. All the reserves have been moved in so that if the Chinese use the intel from their MAV's to hit the rear areas, they won't hit very much, and if the Pinkos try to break through The Sewer, they'll face every available weapon that the Big Red One can bring to bear on 'em. All these bodies are making it even tougher to breathe in here, but if the Chinese attack, and division seems to be confident that they will, then I can't complain about the added firepower. Then again, how much value does a supply clerk have when it comes to frontline duty? Dunno.

It's getting dark now. The night vision sights on our weapons are having a tough time seeing through the fog/steam. The thermal imaging gear mounted on our covers (helmets) cuts right through the steam, but mist from the rain tends to screw them up. There's also some sort of fungus that's prone to growing inside them (the only thing more common than mosquitos and mildew is fungus over here), and sometimes it's tough to keep them clean and operable. Anyway, we have to be a little more alert at night for a million reasons, so I'm gonna bail and help keep watch for a few hours. The more eyes on The Zone, the better off we are.

Hope you're staying drier than we are.

-Matt

Monday, October 24, 2011

Correspondence

From Corporal Matthew McKenna,

U.S. Army, 1st Infantry Division, 12th CHEM FWD Team ASD3

To Jonathon McKenna (*Grandfather*)

Gramps,

I don't understand this war anymore. I never really understood it in the first place, but now-after today-nothing makes sense. Why do the Chinese even bother to try and attack? It seems like such a pitiful waste of human life.

The big push that we've been expecting them to make finally started this morning. Everyone's been on alert for days now, and even though we sleep in shifts, we were all still half-awake and waiting for the call that the Chinese were coming. It rained all night (It hasn't let up in weeks now), and the resulting steam/fog never seems to go away. At night, both sides put up a steady flow of flares, and it makes the fog flicker and glow.

The normal shelling increased a few days ago. They seem to fire in 1015 round salvos, and then-not even a minute later-they move on to a different target area-probably using different batteries for each barrage to confuse our counter-fire detectors. Inside The Sewer, we've been relatively safe. Except for a few close hits around our bunker's firing port, we haven't even been muddied. Up on top of our ridgeline, Saunders and the rest of my unit are being hit kind of hard. The ground is too soft to dig into, and the tunnels in the Sewer are packed with gear and people trying to stay dry and safe from Pinko stuff coming down on top of our ridge.

675

When the sun came up, we were all on watch. The Chinese have some night vision, but nowhere near as much as we do, so we knew they'd probably attack during the day, and that meant an early morning start. So, every day we all stood by the firing port to the bunker and waited for the sun to rise. Today, the flares were still making the fog glow a piss-yellow color. A few days ago, before we went on alert, some replacement grunts were blinded by a Chinese sniper with a laser. Since then, everyone always wears their eye protection, but that makes it even harder to see into the glowing fog, so most of us were already looking through our thermal night vision gear (It's becoming second nature now). At the top of the middle ridgeline in The Zone, everyone started seeing the little white blobs of movement on our 1"x1" monocle monitors. Each white blob was the thermal image of a person's body heat-Chinese body heat. It wasn't any of our people. Even our snipers had been pulled in when the alert was put in effect.

No one had to say a thing, but in the background, deep in the pale-green glow of The Sewer's tunnels, we could hear people passing the word that the Chinese were on their way. The Sewer's always echoed with the low drone of activity. When the word was passed, everything seemed to go silent. I still think I could hear all 11,000 people in the division clicking off the safety switches on their weapons. No one spoke. We all knew what was coming, and what had to be done.

With every second, the number of white blobs seemed to double-even triple. It was eerie seeing the featureless yellow fog through the firing port, and looking right through it with that 1"x1" monitor. I felt like I had X-Ray vision. After all, it was close!

The Chinese had almost a mile to advance between their lines and the crest of the middle ridgeline in The Zone. Before they could get within unassisted visual range of our position, there was probably another mile of twisted debris, greasy mud, and fast-growing bamboo to cross. We could all see them slipping and sliding down the hill toward us.

I don't know about everyone else, but I started to get nervous. It was one thing when I could look at that tiny monitor and see a lot of bad guys coming down the hill. It was another thing when their numbers made it clear that this was a serious attack. When it was clear that there were lots more Chinese than I had bullets, that's when I became nervous-though I don't think I showed it. It was an unusual change from anticipation to….I guess fear. My stomach shriveled and felt like a rock. My legs were heavy, wobbly, and hard to move even a little bit. I couldn't consciously make myself do anything other than remain at my position and wait to open fire. That's what I was trained to do, and just like a dog that's trained to sit or stay, I didn't budge. No one else did either.

After almost half an hour, the Chinese were already at the bottom of our valley in The Zone, and no one in the division had fired a round. They flowed over the crest of the middle ridge like a river. It seemed like the flow never stopped or even slowed. When the first Chinese troops were within 100-yards of our most forward positions, the division opened fire.

There was no glorious command of "FIRE!" from any of the officers. I think what happened was that the guys in the closest hole just finally decided that it was time, and they cut loose. As soon as the other 11,000+ people in The Big Red One heard that, we all began to pop our targets. The Chinese in the front stopped and took cover. As more followed them, they began to bunch up and I could see them squabbling for the little cover that there is. But, there was no time to be a witness. It was killing time, and there was a lot of work to do.

I don't know how many I shot. I fired off 3, 200-round drums from my M-16A3. It was set on semi-auto, so every shot was aimed, and I hit most of the digital, white blobs on the first or second try. I'm not particularly skilled, but they were 500' away, and it doesn't take a lot of aim at that range. Since I was only shooting white blobs, I never felt like I was killing anyone. It was more like target practice, or some sort of game.

The rest of the guys were firing too. Joe was using an 8-round, 37mm grenade launcher, but as soon as he fired off his grenades, he spent just as much time reloading. Harry was working his M-16 just as well as I was, but he seemed a lot calmer than I felt. Piotre ignored his bolt-action rifle, went right to the PSG1, and I think his shots fired/shots hit ratio improved radically. Pepe had connected five 200-round belts of 5.56mm ammo together, and he was intent on firing all 1000 rounds as fast as his barrel could handle it (He did have to replace it three times, though). We could hear the rest of the division firing too, and through the thermal sights we could see every bullet flying. It looked like the Chinese were trying to crawl through a horizontal hail of lead. It was incredible.

They never really had a chance. Our small arms fire could have done the job, but someone must have thought that's just not the American way. As expected, the Chinese had run down the middle ridge through large gaps in their minefields. Division artillery had pre-planned attacks on those areas. When the Chinese went to ground looking for cover, they bunched up, and their attack seemed to hit a wall. Fresh troops coming over the ridge ran down and joined the crowd. That's when someone at division called in the artillery.

We hit 'em with Willy Pete (WP-White Phosphorous; i.e.: incendiary/smoke /fragmentation warheads). Our thermal vision let us see each fragment from the blossoming rounds. It was a massacre. I don't know how many died at the bottom of our valley, but it had to be hundreds-maybe thousands. Out of my left eye, all I saw was a shapeless white steam/fog/smoke combination. In my right eye, I could see the negative digital image of white, blob-shaped people being blown apart. I saw millions of white dots-bullets and WP fragments-going in every direction. It became so confusing that I had to blink my eyes to try and figure out what I was seeing on my monitor.

The artillery lasted for almost half an hour. Finally, with troops still advancing over the crest of that middle ridgeline, the Chinese survivors turned and headed back. Only a few kept coming at us, and they were easy to pop. The rest saw their own troops heading back to their lines, turned around, and joined them.

We didn't lose a single person in the entire division. There were more wounds from hot shell casings ejecting into each other than there were from the Pinkos. When it was over, our adrenaline was pumping away at max speed. We were all filled with a feeling of pride and ego that was like nothing else I've ever felt. The fog never really lifted, but as our adrenaline wore off, and fatigue set in, we could still look out with our thermal vision and see the piles of white blobs-Chinese bodies, and we could see the small white blobs scattered all around them-limbs? Right now, I think every single muscle in the entire division aches. During the attack, we all tensed up. The adrenaline kept us from feeling it at the time, but now…sleep is my best friend!

I guess this is victory.

Right?

Take care,

-Matt

Tuesday, October 25, 2011

Correspondence

From Corporal Matthew McKenna,

U.S. Army, 1ˢᵗ Infantry Division, 12ᵗʰ CHEM FWD Team ASD3

To Jonathon McKenna (*Grandfather*)

Hi Gramps,

Well, it's still raining, and the Chinese have tried to hit us again. I can't believe that any half-competent leader would send people out into such a killing ground! The rain hasn't let up in over three weeks, and The Zone is a mess. High heat makes movement in The Sewer tiring, and I can't imagine what it's like to have to run across The Zone. Steamy fog has normal visibility down to (at the best of times) 100-yards. Our thermal sights and our night sights are all having fungus-related glitches, but we can still see all the way to the top of the middle ridgeline in The Zone with enough clarity to aim and shoot the Pinkos. It's insane for them to keep rushing us like this!

We've been told that our division-The Big Red One-is being attacked by the Chinese 513ᵗʰ Red Banner Infantry Division, the 617ᵗʰ Infantry Division, and the 612ᵗʰ Mechanized Infantry Division (dismounted). We've got a little over 11,000 people in our division holding a 6-mile section of The Sewer. The 29ᵗʰ Infantry Division is holding our western/left flank, and the 19ᵗʰ Ukrainian Infantry Division is holding our eastern/right flank. They've got their own problems to deal with, but the three Chinese units that have been coming across The Zone to try and break us began with almost 40,000 troops. Division tells us that we might have already taken down as many as 8500 of them. It seems like a lot more to me, but who knows about military intelligence?!

As usual, the Chinese swarmed over the middle ridgeline just after sunrise. We were waiting for them again, and they were met with a curtain of lead from all of our trenches, bunkers, and firing pits. This time, they made it past our forward-most bunkers, and some of them even made it into The Sewer (They were dealt with right away), but the bulk of their advance was pinned down about 100-yards from our bunker. Without the thermal sights, we never would have seen them, but with them, it was a turkey shoot.

The U.S.A.F. decided to get in on the action today. We had two pairs of A-10 Warthogs hit The Zone. They came in from behind our lines and must have turned around somewhere behind the Chinese trenches. After they dropped their loads, we never saw them again. The first A-10 dropped 12, tanks of napalm all over the bottom of our valley where the Chinese were bunching up. The heat from the blasts temporarily blinded almost everyone in the division, and I could hear our small arms fire stop for a few seconds. The second A-10 followed right behind the first and dropped a load of 12, Snakeye 500-pound fragmentation bombs on the top of the middle ridge. As our thermal sights came back from the blinding napalm, all we saw was a rain of hot fragments bursting off the top of the hill. The next 2, A-10's dropped 12, cluster bombs each. Each cluster bomb holds about 5000, golfball-sized bomblets with the power of 4 or 5 grenades each. The bombs were dropped over the valley, burst open about 100' above the middle ridgeline, and rained bomblets on the Chinese side of The Zone. It sounded like a million fireworks echoing through The Zone.

We're miles away from where they landed, and the ground vibrated so badly from all the small explosions that part of our bunker collapsed. Some of the 105mm ammo crates that are stacked to make the rear wall shook their way loose and mud pushed them away from the wall. Joe and Harry ran off for a few moments. They came right back with more crates of mud and built a new wall between us and the collapsing one. Then we all went back to work shooting the digital, white blobs on our little monitors.

Pepe used two more barrels today, but he shot off 6, 200-round belts. Harry and I only used 1, 200-round drum each. Joe shot off an entire 100round case of 37mm, high-explosive grenades, and now he's started on a fresh case of WP's. Piotre only shot off a single magazine, and we all gave him congratulatory nods for his improved marksmanship. Ammunition is still in infinite supply. Water…well, it rains more than we can ever drink, so there's plenty of that. We're all throwing down at least 4 MRE's a day, but sometimes Pepe throws at least one back up. Morale is through the roof since the Chinese are getting their butts kicked worse and worse everyday. Privately, I think we're all getting a little disgusted by the incredible waste of humanity, but they're still just digital, white blobs to us, and everyone's grateful that we don't have clear visual images to go with the new stink in the air!

Once more, sleep is everyone's friend. The Chinese attacks last for hours, and when they're done, we're all exhausted. It's a miracle anyone's still awake to watch out for a renewed attack! I'm gonna get some sack time.

-Matt out!

Tuesday, October 25, 2011

New York Post

Myitkyina Defense is Strong

Myanmar ground forces consolidated their positions as they finished moving into the city of Myitkyina. The local garrison commander vows that the city will be defended to the last. A proud man who lost his brother in a sniper attack only weeks ago, he made no bones about telling anyone that would listen exactly how he would do it. The mountains to the northwest along the Irrawaddy river will need to be retaken from the Chinese in order to create an L-shaped perimeter and protect the small city. This will also create a massive free-fire zone into which it is hoped that Chinese will force either the withdrawal or the destruction of any new Chinese offensives in the area. A series of hilltop firebases, linked by trenches and other earthworks, will maintain the perimeter's strength.

Thursday, October 27, 2011

1st Lt. John Chamberlin -Personal Log

U.S.S. Ronald Reagan

Singapore, Malaya

After a month of getting tossed around in the typhoon-stirred Bay of Bengal, we've been given a chance to get off the boat. Under cover of darkness, the Reagan pulled into the port of Singapore last night. All of the higher-ups kept it such a secret that no one even knew we had left the coast of Myanmar until we were pulling into port. Even then, we had no idea it was Singapore!

Shapiro's letting us all have a four-day pass with the understanding that no one goes any farther than 50 miles from the ship; in case there's an emergency or an attack. The word went out through the ship during morning mess, and everyone immediately sucked down their food to get back and get ready for the port call. My troop doesn't have any dress uniforms. All we've got are the same beat-up BDU's that we were issued back at Cam Rahn Bay. Oh, well! I, for one, am getting off this ship! I know everyone is looking to get some terra firma time too.

At noon, the entire regiment formed up on the flight deck in formation. We got a speech from Shapiro about how to behave, but it was brief; very brief. Everybody knows that we are going to leave our mark everywhere we go, and there's little I or any officer can do to curb it. Hell, I haven't met an officer yet who wasn't going to go out and cut loose, myself included. When Shapiro finished, the 1MC announced that the is unit dismissed until Halloween. Everyone took a step back, turned around to their right, took another step, then bolted for the gangways; almost 10,000 people running as fast as they could to get down three 5' wide bridges between the ship and the dock.

It was fun, but I completely lost my entire troop in the shuffle. I tried to get everybody together when I was on the dock, but nobody found me. I was on my own. Oh well!

Thursday, October 27, 2011

0850 hours

6000 feet above the Vietnamese/Laotian border

Back at Phuc Ya airfield, there was rumor that the Chinese were going to try another river assault, this time down the Nan Hou River between the Mekong and the now infamous Black River. Al Meyers got the word and scrambled his ad hoc international fighter force. They were airborne and in formation within fifteen minutes. His Vietnamese liaison officer had told him that the terrain was so mountainous and remote that there probably wasn't a sizable Coalition force within a hundred miles. That would give the Chinese just too much of a head start for Al.

Hundreds of combat missions had warped him from a cool levelheaded professional pilot into an aggressive and gung ho hero. Only his trained responses kept him alive now. The professionalism had been replaced with a fire born out of frustration and loss, fueled by victories and well-witnessed kills. The 1st Composite Air Group was on its way. The Chinese didn't know it, but the Wolfman was a comin'.

Below the formation a thick layer of overcast from 2000 to 5000 feet shrouded the terrain. Only mountaintops poked through. High above, a layer of stratocumulus dimmed the sun. Everything was white. Most of the pilots in the group spoke different languages, but they all had the same thoughts. How would they attack a ground force through that overcast? Maybe they'd get lucky and there would be a break in the clouds. Even if there wasn't, they had to try.

The flight from the airfield to the rumored target area only took 7 minutes. Still, the overcast was unrelenting. Al led all three of his squadrons in a 10-mile wide circle, then he motioned for his wingman to go down and take a look. It was an incredibly risky move. Most pilots would have seen it as a suicidal task, diving down into thick overcast in an enemy area where mountains and jungle could pop up from nowhere. The South Korean pilot that Al motioned to never hesitated. He rolled his F-4 Phantom on its side and spun into the clouds. A second later he was gone.

It didn't take Anyone in the group very long to realize that the rumors were probably true. Tracer rounds and airbursts of shrapnel started coming through the clouds. Then the South Korean pilot shot straight up through the overcast like a missile. His plane was trailing a plume of fuel behind. There was a hole through his port wing from the AAA. He leveled out and pulled up next to Al's plane. A simple nod between their two canopies was all that was needed. Al responded with a salute, and the Phantom headed back to the airfield for repairs. The game was afoot!

As soon as the Phantom was clear of Al's plane, the three squadron commanders moved into position next to him. Al motioned for them to form the entire group into a line astern formation and follow him in on his attack. A bomber pilot at heart, Al would take the lead himself.

He twisted his head and watched as the squadron leaders drifted back to their formations and passed on his silent orders with hand signals. When the group began to form the long spinning chain, Al corkscrewed into the overcast. Everything disappeared into the thick white mist. All of a sudden, he was under the clouds. There were green jungle-covered mountains on both sides of him. Below, a murky brown river snaked its way through the valley. He had just enough time to look around and get his bearings, then the AAA opened fire on him. It was a bomber pilot's dream while an unblemished nightmare to most. There were tracer rounds and explosions all around him. He was alone, at low level, walls on both sides, clouds only a few hundred feet above, and nothing but enemy in front. The tracers and explosions added to his adrenaline.

It wasn't hard to spot the barges, they were the source of all the tracers. There were too many to count. It seemed like one long water-borne train going an unknown length up the river. On top of that, Al thought, how many were behind him?!

He looked at the closest one and instantly realized he was too close to line up an attack on it, so he aimed his worn out F-16 at the next one. As he looked through his heads-up display, Al could see what looked like hundreds of troops on the barge. Not a one tried to run or jump. They were all aiming their rifles at him and shooting. As the first of the tracers from the small arms started to surround his plane, Al fired off a pair of bursts from his 20mm cannon. The rounds exploded in the crowd of troops, and their fire ended abruptly.

There was no time to dwell on the matter. There were more barges ahead. Al pulled out of his shallow dive, banked to the right, then left, and leveled out. Two more barges zipped underneath his plane. He lined up on another and repeated his attack. More troops fired at him. More bursts from his cannon. More deaths. He pulled up again and followed the river for almost five miles. When he finally pulled into the clouds to go around again, he had left hundreds of Chinese killed, wounded, or swimming from seven sinking barges.

Once on top of the clouds, He looked around. There was a Vietnamese squadron of MIG 21's orbiting the area to protect his group from any enemy fighters, but his three squadrons were all below the clouds messing things up for the Chinese. Al joined with the Vietnamese. One by one, his planes came up from beneath the overcast and joined in the long circling formation near 10,000 feet.

They circled until they were close to the point where Al had originally dove into the clouds, then Al went back down. His group followed. Once more Al broke through the clouds between the valley. This time, he had come in too close to the Southern mountains, and he had to veer to the right to avoid crashing. A quick correction, and he was back on the river's path. There was far less AAA this time, and the reason was obvious. Most of the barges were sinking, burning, or both. He followed the river for five minutes but only found one barge to attack. Al knew the battle had been won, but it lacked satisfaction. He fired one long streaming burst from his 20mm cannon into a barge-almost cutting it in half from bow to stern. As he pulled up and into the clouds, Al glanced over his shoulder just in time to see it filling with water. He also saw a few people swimming, but then the white overcast wrapped itself around him again.

Al looked around, found the Vietnamese squadron, and began reforming his group. When everyone was back in line, he pulled out to the side and wagged his wings; the signal for his squadron leaders to move in close for a little hand signal communication. Three planes left the long circling chain and pulled next to Al's right wing. He checked their losses. There were none. It was time to head back to Phuc Ya Field Vietnam; home.

Thursday, October 27, 2011

Correspondence

From Corporal Matthew McKenna,

U.S. Army, 1st Infantry Division, 12th CHEM FWD Team ASD3

To Jonathon McKenna (*Grandfather*)

Gramps,

Well, I thought I'd seen it all by now, but I was wrong-waaaaaaay wrong! The Pinkos came at us again this morning. As usual, they came in large numbers, right after sunrise, through the fog and rain (It hasn't stopped raining in weeks!). They stayed to the areas between the minefields that they put down a couple days ago, and once more, we shot the hell out of 'em. They just kept coming and coming! Ammo was still as plentiful as the targets, so we weren't that worried (Everyone's getting used to their rushes, and few are as nervous as before their first attack). The Pinkos bunched up at the bottom of our valley again, and some almost made it into The Sewer via our farthest bunkers.

Instead of artillery or air support, division brought in a trio of AH-64 Apache helicopter gunships. I was up in the trenches at the top of our ridge talking with Saunders when they arrived. The Apaches hovered behind our hill for a while, then one lifted up and took a radar scan of The Zone. It came back down, and must have sent the data to the other Apaches because they were firing missiles blindly over the hill-never exposing themselves to the Chinese. Each Apache carried 16, Hellfire MkIV missiles. Through my thermal vision, I watched the missiles strike in a grid pattern all along one of the Chinese routes down the middle ridge. There was no time to count their casualties, but it was ugly! There had to be a few hundred dropped. The Apaches could have turned for home at this point, but instead, they all rose from the back of our hill and hovered a few feet above our trenches. Then they cut loose with the 30mm chain guns mounted under their noses. Hot casings rained down on us from the nearest Apache, and we had to clear our trench for a little bit. They were working the Pinkos over something serious.

I had just enough time to look out from The Sewer and see the helicopter that was right above us get hit. The Pinkos fired a shoulder launched SAM (surface-to-air missile). It was so close that the pilot didn't have enough time to react. They were hit in the engine, and thick black smoke instantly started to spew from underneath the main rotor. The Apache started to spin, but the pilot maintained altitude and somehow managed to move it away from our trenches. It disappeared in the fog as it headed away from The Zone.

The other Apaches scattered and headed back to the reverse slope of our hill, but another one was hit. It took a missile square in the cockpit and fell from the sky like a brick. The last Apache made it to the back of our hill. Then it accelerated toward the east and suddenly popped back over the hill toward The Zone. The pilot brought the helicopter right down to the ground-not more than 5' above the mud, and the gunner began to chop away at anything he saw. All of our people were under cover, so it was a free-fire zone. The pilot also started popping out flares to decoy any more heat seeking missiles, and the fog in The Zone glowed like the sun. Without thermal imaging, it was terrifying, but through the thermal imaging system, it was like an incredible video game with full surround sound; just incredible!

That's when I heard more helicopters. The Pinkos had sent two of theirs up to deal with us, but they arrived late for the attack. They looked like Mi-17 Hip's, but they carried rocket pods on their sides. It's was tough to tell through the rain, the fungus-ridden thermal vision, and the chaos of the scene. Besides, most of the Pinko birds look the same to me. They all look like Soviet leftovers from the 1960's. Anyway, the Hip's started to fire at our last Apache, and missiles were landing all around it-throwing decoy flares everywhere. The Apache pilot should have bugged out, but he climbed and headed right for them. The gunner on the Apache was out of missiles, and there couldn't have been too much left in that 30mm, but I watched him pound away at the closest Hip, and then the one farther to the west. In order, both Pinko birds fell from the sky like someone had cut the strings that were holding them up. They fell on their own troops, and HUGE explosions climbed above the fog. The last Apache banked hard as it passed between them, and then it headed for homepopping flares and shooting at Pinko infantry all the way out of The Zone.

Still, the infantry kept coming. Our guys down in those low bunkers really must have nerves of steel, because it looks like it's face to face fighting down at the bottom of our valley. I'm sure it helps having 11,000+ grunts backing you up with cover fire from above and behind, and we definitely poured out the lead!

The Pinkos also brought some other new tricks this time. Joe pointed out that there was a steady series of explosions around one of our lowest bunkers. It looked odd, and blasts rarely come in such a long a steady rotation, so we all kept one eye peeled for anything out of the norm while the other eye kept working on the Pinkos. There were three more explosions after Joe told us to watch for the unusual. Then, Piotre stopped firing and stepped away from the firing port. He grabbed his bolt action rifle and told us that he found "it." Pepe was swept away in the moment, and his SAW was making him deaf, so he yelled at Piotre who wasn't even 3' feet away "WHAT'S 'IT'?!"

Piotre told us to look at the western edge of the crest on the middle ridge. Everyone looked, and we all saw it at the same time. The Pinkos were setting up ATGW (anti-tank guided weapons/missiles). It was too far away to tell the type. They looked a lot like the French-made MILAN's, but they were probably the Soviet-made copies, AT-4 Spigot's or Faggot's. The missiles are just a long fat tube, about 1'x4'. They sit on a huge tripod with a huge electronic thermal sight. When fired, the missile has a thin wire connected to the launcher, and it spews out this wire as it travels. The Pinko who fired it keeps his target in the crosshairs of his thermal sight, and the missile is steered to target by a computer built into the thermal sight. We use a similar system in our TOW II missiles, but all of our stuff is supposed to have much more reliable electronics. The Pinkos were using the anti-tank missiles to take out our farthest bunkers and gain access to The Sewer, but they kept missing. Maybe it was poor electronics, maybe it was poor marksmanship, maybe it was the long range, but whatever it was, they weren't hitting our guys in those far holes! Piotre was going to make sure of that.

He crawled through the firing port and spread out in the mud in front of us. With the barrel of his rifle resting on a piece of log, he loaded one of his sabot rounds. We stopped firing to let him concentrate for a few seconds. With a loud crack, he fired the rifle, and a few seconds later, we could see one of the Pinkos on the crest of the middle ridge fall over! While the Pinko was still squirming, Piotre had already reloaded and fired off another round. Once again, we all watched in amazement. The Pinkos were almost a mile away, and Piotre-of all people-was picking them off one by one! He squeezed off three more shots and dropped two more of the ATGW team members. Then the rest grabbed their gear and ran back over the ridge-out of sight. It was an incredible display of marksmanship for any sniper-let alone Piotre!

The Pinkos that were closest to us-about 150 yards away, had heard Piotre's distinct rifle, and they were spraying lead toward us, so we pulled him back in and patted him on the back. He was grinning up a storm, but there wasn't a lot of time for congratulations. He got a "Nice job!" or a "Way to go!" from each of us, then we all went back to work on the Pinkos. By noon, they had given up, and they headed back up that middle ridge toward their own lines. It was another disgusting waste of humanity, but we didn't focus on that. Instead, we all kind of feel like a team that's won the Superbowl for 5-years in a row. We feel almost invincible.

Tomorrow's another day. In the meantime, we've got some minor repairs to make to the bunker. There's also water to be bailed out (After weeks of contstant, heavy rain, we've developed a leak somewhere in our tent ceiling), there's food to be thrown down (or up if you're Pepe), and we all wanna make sure we have a fresh supply of ammo. There's just a lot to do, and I'm gonna get to it.

Take care of yourself,

-Matt

Friday, October 28, 2011

New York Post

Give and Take with the Chinese

Indochina continues to be both a major hot spot and stumbling block for the Chinese on their road to Asian domination. Yesterday, Chinese forces that had been advancing in the area of Myitkyina were stopped cold in their tracks by Coalition forces from Germany, France, Poland, and the Ukraine. The Chinese were able to wrest control of the city of Man Hpang, Myanmar, but that seems to be the extent of their advance into western Indochina. Further to the East, acting only upon rumors coming out of the jungle, the 1st Composite Air Group found a Chinese amphibious group coming down the Nan Hou River somewhere between the Mekong and Black River in Vietnam. After inflicting grievous losses on the Chinese, further Coalition air strikes were called in on the area. As many as 12 B-52 bombers left Clark Field in the Philippines and carpet-bombed the area. Bomb damage assessment has been difficult because of low overcast cloud cover and the dense debris left from the jungle. U.S. Air Force officials pointed out that bomb damage assessment is often very difficult in regards to B-52 strikes because they are so devastating that they rarely leave even a trace of their targets.

Saturday, October 29, 2011

Air Force Weekly Down The River Again!

The People's Liberation Army has tried yet one more time to move its troops deep into Indochina via river assault. Once more, they appear to have failed. This time an estimated corps-sized force tried to come down the Salween River east of the city of Mandalay, Myanmar. Regional Myanmar forces, supported by American and British fighter-bombers operating from Mandalay; itself engaged the Chinese on the Shan Plateau. The Chinese were able to escape the normal slaughter on the barges, but are currently scattered throughout the area. While not deemed as mopping-up operations because of the large numbers of Chinese involved, a Myanmar military official has made the comparison noting only that things could become very dangerous if the Chinese are able to regroup. Polish and German armored brigades are en route to the area to prevent the Chinese from collecting themselves and renewing their attacks.

Monday, October 31, 2011

Hudson Springs Tribune

Easter Alliance brings new terror to Halloween

Most have said that it was long overdue, but chemical warfare has finally been introduced into the Third World War. A North Korean division north of Myitkyina, Myanmar attempted to break through the city's perimeter by using a hallucinogenic artillery-delivered gas attack. The effect was devastating, horrific, and not exactly what the North Koreans had expected.

The gas, identified as GSXL15 by a forward-deployed Coalition chemical warning team, is colorless and odorless. When mixed into a normal artillery barrage the only way that the frontline troops could have known they were being gassed was the taste. Most of them reported tasting something super sweet like watermelon candy just before they began feeling its effects.

The chemical affects the nervous system after it enters the bloodstream through inhalation. The result is supposed to be mass panic attacks in the troops exposed, but the North Koreans failed to concentrate it enough. Some of the Coalition troops in and around Myitkyina did panic and run from their positions until the gas wore off, but most remained. Those that did not run describe seeing horrific nightmare-like visions. Some saw the attacking North Koreans as figures from horror movies, others saw them as demons and mythical villains. One man said every time that he looked at a North Korean, he saw his 2nd grade elementary school teacher!

What no one had expected was the bottom line of the gas' effect. Those that didn't run said that they felt they had fought even tougher because it was a chance to kill their own personal demons. Some are calling that it group therapy!

After the hallucinations passed, the troops that were exposed suffered from headaches and nausea, but thanks to the lack of concentration, the physical effects of the gas wore off within a few hours. Because of the attack, chemical warfare detection equipment is being calibrated to detect the gas' presence. In the future, fewer troops will have to be exposed without immediate warning.

In addition to the usual official condemnation of the incident, a Coalition spokesman has not ruled out the retaliatory use of chemical weapons. Officials did say that, to date, none has been used on any of the Eastern Alliance forces, and the future use of such weapons would be done only after serious discussion.

There was no word on whether or not there are any Coalition chemical weapons already deployed for action in the Indochina theatre of operations.

Monday, October 31, 2011

1st Lt. John Chamberlin -Personal Log

U.S.S. Ronald Reagan

Singapore, Malaya

Everybody survived their 4-day pass, but I'm not sure survived is the right word for it. Bob looks like he's a walking wounded case, along with most of the troop. Even the legendary Wess was throwing down aspirin. Of course, I do have to take some serious pride in the fact that every one of my people made it back to the ship. There are hundreds of Marines and airborne troopers still missing. I suspect most are either in jail, or simply passed out somewhere.

Except for when the seas get really, really rough, morale has always been fair. When it comes to fighting spirit, it's exceptional. I don't think there's a single man or woman on board who doesn't thirst for the chance to let loose some rounds at hostiles.

Sometimes it gives me a rush. I feel like I'm doing something, not necessarily something that's going to save the world, but at least I'm trying. If my aim is good enough, and if I lead my troop skillfully, I might make at least a little difference in this screwball war. Sitting around, working out, training, reading, maintaining equipment, that's all worthwhile, but it's a far cry from actually taking aim on some bastard who's just shot who knows how many of your fellow troops-especially when their next shots are directed straight at you!

Tuesday, November 1, 2011

Correspondence

From Corporal Matthew McKenna,

U.S. Army, 1ˢᵗ Infantry Division, 12ᵗʰ CHEM FWD Team ASD3

To Jonathon McKenna (*Grandfather*)

Gramps,

I'm sure you've heard by now that I've been wounded. Don't worry, I still have all my arms, legs, fingers, toes, etc. Hell, I think I'm fine, but the docs told me that writing to someone would help with the healing process. It actually makes good sense, so I'll tell you what happened, and then we'll see if it helps. At least you'll understand what's wrong with me.

After their attack on the 27th, the Pinkos shelled the shit out of us. They hit us all day and every night for almost a week. Every kind of stuff imaginable came across The Zone. We were hit with 100mm, 120mm, 130mm, 152mm, and those huge 8-inch artillery rounds. They mixed in some mortars-mostly 82mm and 100mm, but there was plenty of 120mm and some ungodly 160mm stuff. We took in a lot of rockets too. Truckloads of 120mm, 130mm, and these giant 273mm rockets all screamed on to our hill. It was a continuous rain of big stuff, and if we hadn't had those plastic tunnels connecting everything in The Sewer, I don't know if anyone would have made it.

Our trenches were all hit, and after the first day they were unrecognizable. Behind The Sewer, everything was fair game (Looks like they figured out how to use the intel from those MAV's after all!). They even leveled all of our grill tents! Nothing was safe, and our supplies had to be moved back 8-miles to the South. Most of our bunkers were okay-including ours, but we all took damage. Almost a month's worth of rain made repairs difficult, but we coped. Even though it was always-ALWAYS-raining, and even though the Pinkos threw over everything they had at us, our morale remained fair to good. After all, we had totally kicked their asses repeatedly, and everyone's confidence was very high.

On the night of the 30th, the rain finally let up. We've been through every kind of wet weather you can imagine (except snow), and only a light drizzle/mist remained. The ground steamed and a fog seemed to lift out of the ground. By midnight, the fog had gathered and lifted over a hundred feet in the air. We could see all the way to the middle ridge for the first time in weeks.

Smoke from all the Pinko stuff was rising with the fog. It had masked the sight of the valley in front of us, but it had also masked the smell. I remember looking through my thermal night vision goggles during the attack on the 27th, and I remember wondering what all those bodies looked like. We had killed thousands in a very small area, and yet we'd rarely seen anyone through the fog and smoke. As Halloween approached, under the low clouds of smoke/fog/steam (glimmering with the yellow and white flares high above them), we finally saw the carnage.

The Chinese bodies were all bloated, battered, and tossed in piles. I had assumed that we had killed a few hundred in the earlier attacks. Division told us that thousands had been capped. When we finally saw them, it was hard to pick out what was a leg, an arm, a body, or a tree stump. The debris and the disorder made it look like there were millions of them.

At first glance, they just looked like inflated piles of dirty laundry. After a more determined look, we could see their equipment, then their weapons, and finally features like arms, legs, feet, and even hands. All of their skin was black from rot, but the artillery, mortars, and rockets had speckled them with mud, and they looked more like giant leeches instead of dead human beings. It was disgusting, but no one got sick. We all just looked at them with this blank, matter-of-fact stare.

A few minutes after midnight, the Pinkos stopped their shelling. There was no need for an alarm. We all knew what was coming. Everyone tripledchecked their weapons and ammo. Then we all took our places in The Seweren in the torn apart trenches on top of our hill. Everyone in the entire division waited-all 10,000+ of us! I was just about to turn to Harry on my right and ask him if he thought this might be a feint, but I stopped. On the other side of The Zone, a bugle wailed. Then three more sounded. I don't know how many we heard after that, but the entire Zone was echoing.

With the fog/smoke/steam lifted, it was going to be a lot easier to aim. The night vision sights on everyone's weapons are the older low-light amplification type. The previous attacks were during the day, but we all expected to have an easier time with this attack since we could see the Pinkos with our thermal imaging goggles, AND we'd be able to really use the sights on our weapons for the first time (The fog/steam had prevented that in the earlier attack.). Hell, the low cloud layer was glowing so much from all the flares above it that it felt like we could just use our eyes!

We all waited for the Pinkos to come running down the slope of that middle ridge, but they never came. The bugles stopped after a minute or so, and then there was another long and quiet moment of nothing. We didn't even hear shelling from the divisions on our flanks! The entire Zone was quiet.

Instead of infantry coming out of the haze as they ran down the middle ridge, the Pinkos hit us with some more stuff. We had to get back into The Sewer to avoid the small pieces of shrapnel that were coming through the firing port in our bunker. It was a quick attack, and it ended as fast as it had started. Everyone went back to their posts and waited again.

There was fresh steam coming from the mud again. It mixed with the cooler, moist air and made another fog. There was a lot of smoke too. It seemed like the Pinkos might have thrown in some smoke rounds to help mask their attack. This time, the haze sank and formed a 3' thick blanket over the ground. It flowed like water between the debris and bodies. Some of it even flowed into our bunker through the firing port. It stank of mildew, rain, burnt explosive, and the sweet smell of the rotting corpses was back.

The sights on our weapons were useless again. That made us a bit edgy. We'd been on alert and at our posts for weeks, and that made us kind of jumpy. The Chinese were coming at night, and that bit of irregularity was unnerving. Everything suddenly seemed wrong-or rather like things weren't as right as they had seemed when we first expected the attack. Something just didn't feel right. Maybe it was the long periods of silence, the constant tension, or the incredibly strange visuals in The Zone? I didn't know what was wrong all of a sudden, but when I looked at everyone else, I could see that they felt something was out of place too.

We were all watching with our thermal sights when all of a sudden it seemed like the entire Chinese Army appeared at the top of the middle ridge. There were a few white blobs, then some more, and finally, the entire crest of the ridge was white with people. There were a lot of Pinkos in the earlier attacks, but nowhere near this many. Joe suggested that the earlier attacks must have only been with one of the divisions facing us, and that this one must be all three Chinese divisions.

After a while, the Pinkos started up with those damned bugles again. They use them to communicate between units. It's no different than the sound of a cavalry charge in the old west or a bayonet charge in the Civil War. I guess it's a good idea. It's hard to knock out a bugler, but all of our electronics get freaked out by a simple electro-magnetic pulse from their cheap E-bombs! They're also great at wearing on people's nerves!

When the bugles wailed, the digital, white blobs started running down the side of the middle ridge. This time, they ignored their own minefields and ran right into them. If Joe was right, the newer divisions must not have been informed of where the mines were located. It was good and bad. The Pinkos who went into their own minefields were getting blown away, but they were also clearing the minefield, and the troops who were following the first ones had a clear run into the valley.

Our artillery was prepared to hit the usual paths between the minefields. So, when the Pinkos cleared them with their own troops, the remainder of the troops came down the hill almost unopposed. Of course, our arty put down a load of Willy Pete rounds on the guys who came down the clear paths for the third time!

While there was a flashing and glowing cloud layer above us and a flowing mist below waist-height, the Willy Pete and its orange-glowing fragments seemed to fountain smoke and embers all over the middle ridge. Visibility from the bottom of the valley forward was zero, and we all had to watch the spectacle through our thermal vision goggles. It seemed like the Pinkos were being wasted in huge numbers, but more kept coming over the top of the ridge.

The smell was getting very bad too. By this time, everyone in the division was firing, and the mildew smell of our bunker was replaced by the familiar choking scent of spent bullets. Occasionally, we caught a whiff of the Willy Pete outside or the bodies. They had gone from a putrid rotten smell to a sweet, almost candy-like smell. It reminded me of watermelon gum or candy, and whenever we smelled it, it was so overwhelmingly sweet that we all started to gag. Those bodies stank something awful!

We were all to busy to really take in everything. Piotre was already on his second magazine. Pepe was already on his third replacement barrel for the SAW. Harry's M-16 had already jammed twice, and he was getting ready to pitch it in favor of the bunker's shotgun. Tattoo Joe had already fired off three loads of grenades (almost 20), and his brand new box was soon to be empty. I was picking and choosing my targets a little more than in the previous attacks, and I was still on my first 200-round drum, but things were happening fast.

The Pinkos made it to the bottom of the valley again. Just like before, they started to bunch up right in front of our farthest bunkers. It looked like our artillery was going to have a good night again, but the troops who had followed their late brothers-in-arms through their own minefield were a new problem. They were able to get between two of our farthest bunkers, and then the hell began.

The Pinkos have never really had a great regard for human life. We knew that. They sent thousands of troops on a frontal attack against an enemy (us) with far superior firepower and deep fortifications. They did this in broad daylight. They usually left their wounded behind (sometimes stuck in the mud), and probably sent their own guys into those minefields on purpose! When those Pinkos got between our forward bunkers, they stepped down another level. They used flamethrowers on our guys.

I wasn't sure what I was seeing when I looked through my thermal goggles. There was just this long, white line, and then a white blob all over one of the forward bunkers. The white line seemed to fade in and out. Then a new line went from the same origin to the other forward bunker. Piotre was the first to figure it out, and he let everyone know it! Over the deafening artillery and machinegun fire, he yelled out, "They're using flamethrowers on us!" We all dove for the firing port and tried to climb out. I think Piotre and I were the last ones out. It was dangerous to be in the open, but it was better than getting torched inside!

We found a log above and behind the firing port to our bunker, and we used that as our new position. While we were bailing out of The Sewer, the

Pinkos began to move up our side of the hill. All of us were firing away, and Piotre claimed to have bagged at least one of the flamethrower guys. Harry tried to crawl forward and get back in the bunker, but the Pinkos seemed to have found us, and they were pouring out the lead. It reminded me of when Chip bought it, and I was happy to stay behind the log. So was everyone else. The thermal sights are great, but the sight of every single red-hot bullet coming at you can really get your blood pumping!

When the Pinkos made it past our forward bunkers, they also made it past our pre-planned artillery areas. They were also clear of the smoke from the Willy Pete, and we could see them with the older light amplification sights on our weapons. At first they looked like a field of green lightning bugs flying around in pairs. Then the shadows of their silhouettes filled in around the glowing green eyes. Since we could finally use the sights on our weapons, we started to drop a lot more of them, and as the dead or wounded Pinkos fell into the waist-high mist, the rest dove into it for cover.

I switched back to my thermals, and it seemed like every single Pinko was shooting at us. Tiny white blobs came at us from every direction. There were so many that it seemed like static, or a winter white-out back home. Pepe started to freak out, and he began to yell as he fired. Harry's M-16 jammed for a third time, and I could see by the way he was trying to clear it that he was unnerved. Joe emptied his grenade launcher, tossed it back into the firing port of our bunker, and began to fire with his 9mm Beretta pistol. Piotre slipped away to hunt and be a sniper. The sight of all those rounds coming at us was too much to take. It was too distracting, so I gave up on the thermals and used both my eyes and the sight on my M-16.

The Pinkos were a lot closer at this point. In the past, they never came within 150-yards. They were already about 50-yards away when I gave up on my thermals. Around that same time, I had to switch drums. While I was messing around with the fresh drum, I tried to focus on something other than the entire Chinese Army on the other side of the log, or the curtain of lead that was raining above my head. The only thing I could think about was that damn watermelon candy smell. It was so thick I could taste it! Even thinking about it now is making my stomach fill with bile from that sugar-sweet rotting corpse smell! I'm sure I reloaded in record time-just to get my mind back on the fight instead of that smell!

When I was up and firing again, Piotre came running back without his rifle. He had left with his PSG1, and his bolt-action was still in the bunker. It was clear that something had rattled his cage. He dove down and hid behind our log. Pepe asked him where his rifle was, but all Piotre could say was, "I've gotta pay the man in the loincloth! I've gotta pay him! I've gotta pay him!" Pepe and Joe stayed focused on the Pinkos while Harry and I tried to calm him down. It was no use. Piotre was completely freaked out. He jumped over the log, into the hail of Chinese bullets, through the firing port, and into our bunker. Harry and I went back to work, and a little bit later we heard Piotre firing away with his rifle from inside the bunker.

Piotre's panic added to the stress of the situation. Harry was firing his weapon faster and faster. Pepe became silent, and I knew his mind was probably drifting instead of focusing on putting out the lead. Joe was yelling so hard that he seemed to even laugh occasionally. Once, I definitely saw him giggling while he was reloading his pistol. I tried to put everything out of my mind, but whenever I tried to think of something else, that gagging watermelon candy taste filled my mouth, and my stomach started pumping like I was about to puke.

The smoke along the ground started to thin out, and in a lot of places, we could see all the way to the ground. The Pinkos were more than silhouettes with glowing green eyes now. I could see camouflage-twigs and mud mostlystuck into the netting on their helmets. I could see their web gear, their rifles, and their hands.

When I looked away from the green digital display of my weapon's sight, I saw something entirely different. Coming out of the flowing, waist-high smoke, I saw naked Pinkos clawing their way up our hill. They weren't completely naked. Each one still had on a helmet, ammo pouch, some web gear, and they all had their weapons, but their BDU's were all gone. No onenot a single Pinko-was wearing anything more than a rag around his privates. I couldn't figure out what had happened to their uniforms. Sure, it's always 100+degrees with 100% humidity, and sometimes we've gone without a blouse or even a t-shirt, but this was extreme! Who in their right mind would attack wearing nothing but camo underwear?!

It was clear that these were the Pinkos who had freaked out Piotreprobably because they looked like the Thai Hunter Soldier who tormented him for an entire day back when we moved on Popa Hill. He was shaken up that day, and I guess all these near-naked Pinkos must have sent him over the edge. It didn't really matter too badly. Piotre was back in the bunker and firing away at them again. He was probably using his long-range, armor-piercing rounds on them, but at least he was still in the fight. That's all that really mattered.

The naked Pinkos didn't have much of an effect on the rest of us. Joe and Harry were busy clearing another jam in Harry's M-16. Pepe was still working his SAW like he was in a trance. The strange sight coming up the hill confused me for a little bit, and that might have kept some of them alive for a few more seconds, but I was back at 'em right away.

While I was picking and choosing which half-naked Pinkos to pop off, they took the opportunity to throw some more stuff and some smoke on our hill. We were out in the open, and it was rough for a little bit. Most of their rounds fell on top of the hill-around our top trenchline. We only had one close call.

A gigantic 8-inch shell met the stump of the log that we were hiding behind. The log was about 5'-10' thick, and it was stopping all the shrapnel and small arms fire that came our way. It was also tall enough to allow us to reload or repair our weapons in some safety. Well, the stump had to be 20' across. When that 8-inch round hit it, the shell didn't explode; it was a smoke round, and the vent was stuck in the stump. Since the entire tree was completely rotten, a beige smoke vented out the top of it like a chimney. We could still see the giant shell. Some Pinko had written on it, "Ching Ai bough bai" (Honey Baby!). Harry was closest to the stump, so I think everyone subconsciously wrote it off as his kind of luck.

We were trying to squeeze under our log when the stuff first started coming in, and when that 8-inch smoke round hit, there was just enough time to look at it, read it, and shake our heads in disbelief. When I looked around to see if anyone else had seen the close call, everyone else was still shaking their heads and looking around in confusion too! Harry's luck seemed to be spreading. The sight of that shell in that tree stump somehow made everything more serious, but less realistic. It was like we were in a dream. Nothing was the way it was supposed to be-or the way it normally was.

The Pinkos who were coming up the hill never stopped advancing. They came up and into their own artillery fire. It was a great example of fool's courage-or idiot's courage! There was beige smoke all around us, and that's when I figured out that the watermelon candy smell wasn't from all the dead bodies, it was from their smoke shells. It was hard to stomach the stench before, but when that smoke poured out the top of that stump and fell back down on us…oh….my ….God! Did it ever stink?!! I know it wasn't just bothering me because Pepe puked all over his weapon. He never stopped firing though! He just sat there and tossed out another MRE while he fired in a daze.

I could feel the sweetness bringing a strong bile taste to the back of my tongue, so I told Joe that we needed to get back in the bunker. He just looked at me with an anger that I had never seen before. I thought he was pull out his knife and kill me just for suggesting it! Joe was almost over the edge. He stood up to look over the log and down the hill at the Pinkos. Right then, it sounded like a grenade hit the other side of our log. We were fine, but Joe threw himself backwards and started screaming that he'd been hit. There was no way. The log totally protected him, but Harry went over to check him out. Joe rolled on his back yelling at Pepe and calling him all kinds of names like it was his fault, but Pepe just kept firing away and puking on his weapon.

Things were out of control. Pepe was robotic. Joe was having some sort of flashback to his close call with that homemade Bouncing Betty mine from a few weeks ago. Piotre was in the bunker below us screaming about how he had to pay the loincloth man. I was thinking about watermelon candy. The sky was hidden by a blanket of low clouds that constantly strobed from yellow and white illumination flares. It seemed like a million half-naked Pinkos were coming up the hill through their own artillery, and Harry's luck seemed to have spread to all of us. Then it got weird.

Harry and Joe were about 30' behind the log and a little higher on the hill. While Joe was having his little flashback, Harry looked back at me and panicked too! He grabbed his M-16 and tried to fire it in my direction. It jammed (AGAIN!), and he threw it at me! I had been looking over the log before he panicked, and when he threw his weapon at me, I blocked it with my left arm. As soon as it bounced to my left, a Pinko came over the top and tackled me. I slipped on the mud, and we rolled back into the log. The Pinko was pinned between the mud and the log, and then he went limp. I turned around, and Joe squeezed out three more rounds from his Beretta. The Pinko flinched with every hit, but then he was down for good. I nodded in thanks to Joe, and then I looked for Harry to give him a quick nod too.

Harry was right behind me. He grabbed the Pinko's Type 95 assault rifle and was about to try and scrounge some more ammo from its owner's body. As soon as I was out of his way, Harry lost it. He jumped back and emptied the Type 95's magazine into the corpse. Then he fell back on his ass, rolled over on his stomach, and started crawling frantically up the hill. The whole way, he kept yelling, "Tiger! Tiger! Get me outta here!!"

This Pinko wasn't half-naked like the others coming up the hill. He was wearing a mud-covered set of tiger-stripe BDU's. With all the debris and mud stuck to him, it must have reminded Harry of the tiger in the trap that he fell into during our Popa Hill trip. He was crawling up the hill just as fast and importantly as he had when he climbed out of that trap. The more I looked at the Pinko, the more I saw the resemblance. After a while, even his face seemed cat-like, and his BDU's started to look more like mud-caked fur. I had to blink my eyes to wash out the image because I actually saw the Pinko change from a grunt into some sort of tiger-man. It was like a computer morph image, but it was happening right before my eyes! I knew it was stress at the time, so I shook it off, tried to ignore it, and moved on.

Harry was on his way to the top of the hill. Joe went after him-still bitching about Pepe for some reason. Meanwhile, Pepe had melted his last barrel. He told me that he was going to go get another one. Then he calmly climbed over the log-through the Pinko lead that was flying, and started to slide into the bunker. Piotre stopped firing to let him in, but as soon as I lost sight of Pepe's feet, I heard both guys screaming inside. I tried to get over the log, but there were too many Pinko bullets in the air, and I just knew that if I stuck my head up again, I'd get it blown off. I heard some shotgun blasts inside, and then their screaming stopped after a few seconds.

I was all alone. Since I couldn't go forward or around the log, and I couldn't get back into the bunker, I headed up the hill to find Joe and Harry. Another tiger-striped Pinko came over the log and tried to jump on me, but I rolled over and took care of him. They were getting closer, so I rolled back over and started to crawl my way back up the hill.

A few more Pinkos (half-naked ones) appeared on the log and tried to draw a bead on me. There was just enough time to see them raise their weapons when all of a sudden I heard that wonderful sound of a .50 cal machinegun. It barked away from the top of the hill, and the tree started to come apart. The Pinkos went right to pieces-for real! Those huge, .50 cal rounds blew through them, and the shock waves liquefied all of the flesh surrounding the exit wounds. I used the time to get crawling, and that .50 cal kept hammering the log.

By this time, the sun was starting to come up, but a sunrise under all that smoke, fog, steam, and overcast was more like a dimmer switch in a house. The Pinkos must have had orders to break our line during the night because while everything was starting to get brighter, they put out more lead. My M-16 jammed with mud, and the only thing keeping the Pinkos from tagging me was that .50 cal on top of the hill.

I was having a rough time getting up the hill. Pinko rounds were snapping past me. There was almost no cover, and the .50 cal above me was busy barking away, but its bullets were going just above my head too. As I inched my way toward the top of the hill, there was less and less cover. Days of Pinko shelling had left plenty of shell holes, and I wound up going from one to another. About 100' from the top, I was crawling into a small crater that was next to a large, rain-filled crater. A big Pinko shell landed in the middle of that rain-filled crater, and all the water splashed out.

Right after it hit, another round exploded in the bottom of my crater. I wasn't wounded, but it threw me all the way into the large crater. That crater was very slippery. I slid to the bottom and sank past my knees. Rainwater that had been blown out and up the walls of the crater flowed back down and it started to fill up again. I tried to get out, but the more that I squirmed, the more I sank. I tried calling for help, but with all the artillery and gunfire around, no one heard me. The crater filled up quickly, and in no time, I was up to my armpits.

Instead of Pepe, Joe, or some other grunt coming to my rescue, the Pinkos showed up! One either jumped in or was blown into my crater. I grabbed his arms and held him under the water until he stopped resisting. Then I used stood on his body and used it as a firm foothold to get closer to the edge of the crater (The crater was at least 50' wide). Two more Pinkos crawled up to the bottom lip of the crater and tried to pop me, but that .50 cal took care of them. It also blew some holes in the lip of the crater, and it started to drain like a broken dam. I tried to get out, but it was still too slippery. Another Pinko came over the bottom lip, but he couldn't get a grip, and as he slipped back down the hill, he opened the lip of the crater deeper-letting out a lot of water! I knew that my luck was running out, and that the Pinkos were getting close, so I searched around under the water for the first Pinko's Type 95 rifle. When I found it, I cocked it to make sure that it could cycle (It did!).

Then I tried to get out again. The mud seemed to be sucking me under. Another pair of Pinkos came up to the bottom lip of the crater. I took down one, and someone else got the other. By this point I was sliding back toward the bottom of the crater, and the mud/water was up to my neck. It reminded me of the one EOP that showed the nutball Pinko who was left behind-buried up to his neck in mud for days. I remembered how Piotre cracked open his skull in cold blood rather than risk digging him out or shooting him and giving away our position. It gave me a splitting, sympathetic headache.

Everyone's heart rate jumps when the shooting starts. If they say it doesn't, then they're either lying or psychopathic. Mine had been beating hard all night. Everything was wrong about the whole attack. The scenery was straight from Hell. The smell made my stomach contract so much that my ribs were hurting. I could feel myself panicking-losing control, and there was nothing I could do to stop it.

I screamed for help. I yelled so loud that my ears hurt, and they hurt so bad that I couldn't even hear myself anymore. It sounded like someone else was screaming. After a bit, I wasn't sure if it was someone else, or what was going on, so I tried to yell louder. By the time that the sun was all the way up, I had lost hope. I was trapped. There was no freedom of movement. There was no freedom to figure out how to escape. Without freedom of any kind, there's just no hope, and when someone's hope fades, there's nothing left. I wasn't going to give up and drowned in mud, but I did lean on the wall of the crater and wait for the next Pinko to come over the low lip and pop me. It was the only thing that I could do-that and suck in the filling mud/water mix!

My thoughts kept going back to that crazed Pinko that I saw in The Zone. I remembered that he just kept muttering something about how he wanted water or iced tea or something. There I was ready to die, and the biggest thought in my mind was a glass of ultra-cold iced tea. I totally ignored all of the devastation that was going on around me.

Around 0900, someone came over the high side of the crater. I don't know who they were, but they were grunts like me. They made a human chain and pulled me out. Then we all went the rest of the way to the top.

I noticed that there was hardly any firing going on. The artillery had piddled down to the usual trickle instead of the downpour that it had been for the past few days. The attack was over.

And it was raining again. This time, it was a soft and steady rain. Mud flowed down the hill. Camo around bunkers slid away, and everything was washed clean again. I held out my hands to catch some fresh water to drink, and they filled up with a grayish beige rain. The rain was even cleaning the sky of all the smoke, soot, and ash in the air. When it came down, it almost looked black…a black rain.

At the top of the hill, I saw the .50 cal that had been protecting me. The mud-filled trench around it had been blown apart by the days of artillery, but the two guys standing behind the big gun were up to their knees in spent brass casings. They looked at me like zombies, or like they were looking through me. It was Harry and Joe. Both we okay, but their faces were numb.

We stood in the rain and talked for a while. They were both shaken up, but they were alive. Neither knew that it was me in the rain-filled crater. They seemed glad to have been such a help, but Harry told me that they were really just taking down Pinkos as they saw them, and there really wasn't a special effort put forth to protect me. It doesn't matter. I don't care. However it happened, they watched my back while I was in a bad way! THAT'S all that matters.

I wasn't embarrassed to tell them how scared I was in that crater. Those guys are my bros. I knew they'd understand, and I was there when they were scared. All three of us were still shaking, and hours had passed since the Pinkos went home. Joe-normally a kind of hard ass-said that "…when that grenade went off, it felt exactly-" like when the mine bounced off Harry's pack and almost got him. He said it "startled him" so badly that his body just replayed the mine incident. Harry <u>still</u> thinks he saw a tiger-man come over the log and not some Pinko, and the thought of it makes him nervous. He looked around when he was talking about it like some giant tiger was gonna hear him and attack. All three of us agreed that we'd never been in a worse situation, and we all hoped that this was as rough as the Pinkos could make it for us.

Time seemed to pass quickly, and we decided to try and dig up some lunch. On our way into The Sewer, Piotre and Pepe climbed out into the trench with us! They had made it! Joe made the mistake of trying to be sarcastic, andacting like a hard ass again-he blurted out, "Are you guys done hiding and holding each other?" Piotre froze and gazed right through him, but Pepe was PISSED! He stepped out from behind Piotre with this bowling ball-sized rock and threw it at Joe! A few choice words were exchanged for a few minutes, but they calmed down. Harry picked up the rock and saw that it was actually the head of the Nat statue that Joe had stuck in front of Pepe's dugout a few nights ago-the one that scared the piss out of him.

Well, Piotre said that one of the Pinkos almost zapped him, and his mind "…clicked" back to his duel with the Thai Hunter Soldier near Popa Hill (like I thought). That made him panic, and he didn't start to get his head together until he got back into the familiar bunker again. Then he just went to work with his bolt-action gig until Pepe jumped in. When Pepe squeezed through the firing port, Piotre had to step back into the bunker to let him through. As soon as Pepe was all the way in, he saw movement in The Sewer. It was probably just Piotre, but Pepe grabbed the shotgun and started shooting at the silhouette. Piotre though Pepe was trying to shoot someone behind him, so he stepped out of the way and saw a silhouette too. Then they both started to yell as they shot the hell out of the statue that Joe had stuck next to Pepe's dugout. The statue came apart, but ricochets and fragments cut the two up, and they thought they were being shot at. So they ran into The Sewer and all the way over to Alpha Company! Both of them found an empty bunker and used the shotguns in there to take down Pinkos. When it was over, they helped Alpha get their gear back together until someone told them that we were up on top of the hill.

So it's over now. We made it through the big push. The Pinkos almost made it to the top of our hill, but we held. We held, and this morning the entire division has been pulled from the line. Why? Because we're all officially wounded now?

The beige smoke that smelled and tasted like watermelon candy was something called GSXL15. It's a hallucinogenic gas similar to LSD. It's supposed to make people paranoid, and attacking a position that's defended by paranoid troops should have been a lot easier. No one's sure why we didn't panic and run. Some people say that the heavy rain and high humidity dissipated the drug. There's a rumor that one of the pills we have to take everyday has built up an immunity or a block to the drug.

My theory is actually yours! You once told me that "Heroes aren't made. They're cornered." I believe that now. I certainly was no hero, but I think that plenty of people were cornered by the worst fears that they could have imagined. The people with the strongest fears were the ones with the toughest corners, and those are the people who didn't see a way out of the situation. Those are the people who fought the worst things that they could imagine just to stay alive.

Hell, a lot of people might have actually run away if they thought they could. Pepe was in that bunker, and the only thing he could do was throw everything he had at that statue. After that, the Pinkos were easy for him. Piotre was terrorized by a million of the best snipers that he had ever faced, but when he managed to get into a familiar surrounding, they didn't stand a chance against him. Joe thought he had been blown up-again, but when he realized that it was nothing, he worked that .50cal with a passion! Poor old "Bad Luck" Harry had to face the trapped tiger again, but once he escaped from the imaginary tiger-man, the Pinkos were just targets! It was the same all over the division. After people survived being attacked by the worst things that they could think of, the Pinkos seemed tame. I think the GSXL15 made everyone use up their fear in an instant, and then the Pinkos faced people with who had no more fear to give.

The doctors have classified the entire Big Red One as WIA because of the high exposure to the GSXL15. According to their research, the stuff can stay in your system for years. Besides being the largest unit ever WIA in U.S. Army history, we're also the largest group to be pulled from the line for what used to be called Shell Shock. They also guaranteed each of us a lifetime of DSS (Delayed Stress Syndrome). Oh, joy.

There's a rumor that we're to be shipped home ASAP. They want to get us into familiar environments right away since that seems to begin the mental healing process. So, tell everyone that I'll be home soon. Tell them that I'm fine, but there's gonna be a lot of support group meetings, and I might do some funny things for a while.

Love, -Matt

Wednesday, November 2, 2011

Washington Post

Eastern Alliance Buildup in Indochina

After days of speculation and rumor, Pentagon officials confirmed today that the Eastern Alliance is massing troops all over northern Indochina. Since the retaking of the Korean Peninsula months ago, most of the combat in Asia has been with primarily the Chinese forces. Along with confirmation of the buildup, a Pentagon spokesman has revealed that frontline troops from several other Eastern Alliance nations have been moved into the area. Estimates of the force's size have ranged widely from 50,000 to 250,000.

However, a few sources inside the Pentagon have unofficially put that number at close to 2,000,000 troops. This includes Iraqi, Iranian, Saudi, and Syrian divisions. Smaller contingents from a long list of other countries have also been detected in the area including an undetermined number of Cuban and Russian volunteers.

Numbers aside, military planners spoke privately of the real threat; material. Up until now, most of the Chinese incursions have been failed river assaults via barge traffic, and small-scale raids coming out of the dense jungle. Satellite intelligence has revealed that large numbers of mechanized forces, including tanks and heavy artillery, have been spotted among the new Alliance arrivals.

When questioned further about the growing number of Alliance forces in Southeast Asia, a White House spokesman was quick to downplay the reports. She is quoted as saying that, "There are almost a million Coalition troops lined up and waiting for just such an attack."

Friday, November 11, 2011 Correspondence

From Colonel Dick Saunders,

U.S. Army, 1ˢᵗ Infantry Division, CO: 12ᵗʰ CHEM FWD

To Jonathon McKenna

(Grandfather of Corporal Matthew McKenna)

Dear Jon,

It is with a great deal of pride that this letter is sent to you. I'm sure you have been informed about your grandson Matthew's wounds and about his forthcoming return home. By the time that this letter finds you, he may very well be in your company once again. Matthew has served his country with honor, pride, and deep distinction. That is the primary purpose for this communication.

By act of Congress, an investigation has been ordered to determine the specific events of 27, September 2011 in the vicinity of Popa Hill, Myanmar. In question are the actions of Sgt. Matthew McKenna U.S. Army, 1ˢᵗ Infantry Division, CO: 12ᵗʰ CHEM FWD, and his team; ASD3. The findings of this investigation will be used to determine the merit of presenting Sgt. McKenna and his team with The Congressional Medal of Honor (CMH). If the award is to be presented as expected, each member of the team will be awarded the CMH individually, as it is has never been awarded to an entire unit.

Jon, Matt has always proven to be an exemplary individual. I remember him telling me that you once advised him to "Never Volunteer." He tried hard to follow that advice, but it was not in his marrow. In the civilian world, young men and women Matt's age are often warily given the keys to the family car. Here they are entrusted with the lives of thousands of their peers and equipment that can cost millions of dollars. In the civilian world, parents fear that their children might "fall in with the wrong crowd," and either wind up in trouble or in physical danger. In this place, Matt fought the urge to volunteer, but he saw a duty and chose it. He chose to work with people who were basically "the right crowd." His team was not trained for elite duty. Their files are not immaculate, but they did their jobs in the face of indeterminate danger, and they never wavered. Too often-in a war zone-that kind of a crowd can get you killed. This was the case with that team's initial leader, Sgt. Chad Chambers. Despite the loss, Matt and his team continued their operations unabated.

On 27, September 2011, Matt and his team were on a mission to set up electronic surveillance sensors on and around a tall rocky feature at the top of Popa Hill. The location was almost 200-miles behind Chinese lines. They had to march through the most difficult and dangerous terrain on the planet to get there; the Shan State. On the way, they evaded almost 110,000 enemy troops and other hostile forces. At Popa Hill, they accomplished their mission. Before leaving, they spotted a massive force of river barges coming down the Irrawaddy River. The barges carried an entire Corps of Chinese troops estimated at over 55,000 troops in strength. Communications in the area were limited at the time due to an electronic warfare operation that was being conducted by other Coalition forces far to the south. Despite this, Matt was able to notify elements of the 1st Composite Air Intercept Group and direct a series of air strikes on the river barges. The attack was a complete success.

During the attack, Chinese forces landed and attempted to stop Matt from communicating with the 1st Comp Air Int. Grp. Hundreds, perhaps thousands of PLA troops were deployed west and north of his team's position. Under air cover, they escaped to the southeast-back into the Shan State. Once again, they marched through thousands of hostile forces. Early on 30, September 2011, they returned to Coalition lines.

Matt and his team survived the mission despite insurmountable odds. They also orchestrated an adhoc ambush that single-handedly prevented a major Chinese attack. Subsequent to their detection and destruction of the Chinese Corp-sized infiltration, Coalition air assets stepped up their monitoring of all southbound river traffic from the People's Republic of China into the Indochina Theatre of Operations. These reconnaissance efforts yielded fantastic results. On 28, September 2011, two more Corps of PLA infantry were spotted and attacked as they attempted to move down the northern Mekong River. On 25, October 2011, another PLA Corp was spotted and attacked as it attempted to infiltrate into Indochina via the northern Salween River. On 27, October 2011, two more PLA Corps were spotted and attacked as they attempted to infiltrate into Indochina via the Nan Hou River. All of these attacks on massive PLA units are the result of Matt's detection and air strike coordination efforts on 27, September 2011.

I am convinced that on that day in September, Matt probably thought that he was just doing his job. He was. He was also looking out for the interests of the people in his team when he directed the air strikes to clear a path for their escape from Popa Hill. Matt was looking out for the other soldiers in his unit, his division, even his Army.

I joined the Army back in 1969. Since then, I've seen lots of strange and remarkable things. More than one person has suggested that I retire, but every few years, I find a way to stay in. When I first enlisted, I was full of piss and vinegar. I volunteered for Vietnam when I was still only 17 years old.

During investigations such as this, people often and reasonably ponder the question, "Where do such men come from?" During a convoy escort operation in the Iron Triangle, our unit came under fire. Our Lieutenant was killed, and my platoon Staff Sergeant took over. The Lieutenant knew how to manage the platoon, but our sergeant knew how to lead people, and he found a way to keep us alive. That night, I told him that he was my hero. He shook his head and told me that he was just trying to keep us alive, and that in the end,"…we aren't fighting for flags, Presidents, or even a great Saturday night. We fight for each other." Matt has always done his job, and he's always done it for the people around him-not some Hollywood image of mom, the flag, and apple pie. I've never forgotten that moment in the Iron Triangle or that message, and if my memory serves me correctly, you were that sergeant. Thank you.

Over the next few weeks, you and everyone you know can expect to be contacted by a long list of government agencies and representatives. They'll be looking into Matt's CMH. I expect he'll get it. He deserves it. Whether he likes it or not, Matt is a hero, and this war could use all the heroes it can get.

Take care of Matt. He'll need your understanding and support for years now. Give him my warmest regards, thanks, and congratulations. It has been an honor to have served with him. Your grandson is an example for us all.

Sincerely,

Colonel Dick Saunders

Saturday, November 12, 2011

Stow Herald

Dragons from the Jungle!

In yesterday's early morning fog, American and other Coalition forces protecting the citadel of Myitkyina, Myanmar awoke to one of the largest artillery barrages in history. Thousands of Chinese, North Korean, Arab, and Russian artillery pieces opened fire on the city and its protectors. The barrage lasted over an hour. Before the damage could be assessed, armored vehicles, including heavy main battle tanks, came roaring out of the jungle, down the dusty dirt roads, and into the city. Behind the armor, hundreds of thousands of Eastern Alliance troops from around the globe attacked and effectively split the Coalition's 13th and 11th Armies.

By noon, any seriously-organized resistance from Coalition and American troops was decimated, and the Eastern Alliance armor was rolling South into rear areas unchecked. There are still rumors of friendly troops holding out in the city itself, but officials report that there is little real hope of rescuing them in such a dense concentration of enemy forces.

An adhoc relief force has been sent to the area composed of new arrivals to Indochina, reserve units, and rear echelon cadres. From Southeast of the city, the 1st Free Republic of Korea Mechanized Infantry Division, supported by a pair of Infantry brigades from Thailand-is attempting to stem the tide. West of Myitkyina, the U.S. 29th infantry division, 4th Armored Division, and remnants of the 2nd Marine Regiment are trying to hold the line. Fresh to the area, the Polish 1st Armored Division, Ukrainian 3rd Mechanized Division, British 24th Light Armored Brigade, French 18th Infantry Division, and German 1st Panzer Brigade are all racing to the scene from the small port of Sittwe over 100 miles away. Every available air asset has also been called to the area, including the majority of the Australian, New Zealand, and Philippine Air Forces. The United States is contributing by operating in a support role with the carriers Enterprise and Carl

Vinson operating in the Bay of Bengal, and from the Nimitz and Independence operating in the Gulf of Thailand.

Saturday, November 12, 2011

1st Lt. John Chamberlin -Personal Log

U.S.S. Ronald Reagan

Somewhere in the Gulf of Siam

There's trouble in the jungle once again, and they want us to put an end to it. It sounds like the Chinese have really thrown their hat in the ring on this one. They've been trying to sneak forces into Indochina for months now. We stopped 'em when they tried to come in through Vietnam, but they never really gave up. Since then, they've tried to send smaller units down jungle paths and rivers. Now, they've just decided to try a good old-fashioned armored thrust. You'd never guess that Indochina would be a good place for tanks, but I guess the French thought the same thing about their Ardennes forest before Rommel drove his panzers through back in 1940. I dunno. I've never been a tanker, just a plain old mud Marine. All that I really know about is hitting a beach from sea or now from the air.

Anyway, the Chinese have really come out in force this time. When they came into Korea last January, they only sent a few hundred thousand troops (only!). Dennis tells me that they used about 500,000+ in Taiwan. They've also tried sending a couple thousand here and there through the jungles and rivers, but local forces in Myanmar, Laos, and Thailand have been pretty good at bushwhacking them in the thickets. Coming down rivers was a good way to move larger numbers, but it seems to always draw air attacks from our guys in the sky. Too bad for them, and hats off to our flyboys!

Now, Dennis is telling me that the Chinese have come south with well over a million men, lots of armored vehicles, some serious artillery, and enough aircraft to keep us from owning the sky. They've also brought their friends. Some of our guys have already captured troops from a half dozen different Arab nations, Korea, and even some Russian volunteers. I passed all this information on to my troop, and their only concern was that we'd have enough ammo. I guess it's not really such a cavalier thought. After all, we did run out the last time we met the Chinese in the jungle; back at the Black River Valley in Vietnam. That was not a fun moment. I think I'll make sure that we all bring a few extra magazines when we do board the Osprey.

Tuesday, November 15, 2011

Miami Herald EXTRA

Coalition Wins Battle of Myitkyina!

In an early morning clash, a makeshift group of Coalition reinforcements on their way to Myitkyina slammed into elements of the Chinese 23rd Army Group. The Eastern Alliance had pushed its way over 75 miles south of the city when the lead elements of the Jordanian 13th Armored Brigade found themselves entering a valley where the German 1st Panzer Brigade had chosen to dig in and stop their advance.

For the first time in over 50 years, German troops opened fire on an enemy force, the Jordanians. Within 15 minutes, all 46 of the Jordanian's British-built Khalid/Chieftain main battle tanks were torn to pieces. The Germans, with their brand new Leopard 2A5 tanks shrugged off the 1970's era Arabs and immediately counterattacked down the valley.

The German attack was launched as soon as the Jordanian destruction was reported to Coalition command. With only 23 tanks of their own, the Germans only hoped to temporarily halt the Eastern Alliance attack. To their north, a small valley opened up into a large jungle and paddy-strewn plain with hundreds of thousands of enemy troops and vehicles. Losses were quick and severe, but the Germans did manage to halt the enemy attack for almost an hour at the cost of less than half of their Brigade.

Ground attack aircraft were already on their way to the scene, and a heavy toll was further imposed on Alliance units while more and more Coalition reinforcements arrived and formed into a defensive line in the hills south of the main Eastern Alliance attack.

As night fell, the ebb and flow of the frontline ended. A storm system moving in from the Bay of Bengal has seriously hindered low altitude air operations and is expected to for the next few days. More and more Coalition ground units are arriving every hour from around the world. The battle continues, but the attack beyond Myitkyina has been halted. As soon as the weather clears, military officials are promising to move back through the Eastern Alliance forces in the plains to the north and retake Myitkyina. Most are already describing the battle as won-save the mopping up.

Tuesday, November 15, 2011

1st Lt. John Chamberlin -Personal Log

U.S.S. Ronald Reagan

South China Sea-off the Southern Coast of Vietnam

We joined up with two other carriers and their escorts last night. Later today, we should join with Task Force Zebra. Zebra's the largest task force since World War II. We'll have a total of 7 super carriers, (including ours), five amphibious carriers, and three small carriers from other Coalition countries. We'll also have almost 150 escort ships ranging from stealthy new light frigates to those four old World War II Iowa class battleships. We'll also have a few hundred fighters and bombers flying cover over us as they cycle back and forth between Clark Field in the Philippines and airfields all over Vietnam. Even with all this, I think it's safe to assume that the Chinese won't just let us cruise around in the China Sea at our leisure. They'll be coming for us. I'm sure of that, and so is everybody else on the ship.

Wednesday, November 23, 2011

Associated Press International

<u>Thanksgiving in the Trenches</u>

After weeks of fighting the Eastern Alliance attack into Indochina continues. Myitkyina remains in enemy hands, and while the attack into Coalition rear areas south of the city has been halted, Chinese forces are deploying further to the east and west of the area where the attack was originally halted. In that area, armored forces have been stretched thin, and infantry units have been forced to dig in.

From the Bay of Bengal in the west, through the swampy plains, the steep mountains, twisted jungles, and all the way to the Black River Valley in Vietnam, infantry are settling into an old style of combat: trench warfare. Full-scale infantry attacks have been nothing but fodder in the terrain that seems designed for ambush. Small-scale squad and platoon-level probing actions are the mainstay. For 1500 miles, a contiguous line of slit trenches and bunkers stretches on both sides; between them i.e. snipers, mines, and pre-planned fields of fire. It's a kind of no-man's land called "the zone".

Wherever possible, the old rule of holding the high ground has been abandoned. Coalition and Alliance troops have both found it easier to survive in the thick lowland areas, with mountains in the middle of the zone. Anyone foolish enough to climb up for a view is finding themselves silhouetted against a low overcast and often rainy skyline; a perfect shot for even a novice sniper.

As the war continues and as hundreds of thousands of troops rush to the area from the north or South, the bullet and shrapnel-shredded jungles promise to evaporate and further resemble the nightmarish hell of places like Verdun, Stalingrad, Okinawa, Chosin, Hue, or more recently, Seoul and Taipei.

Friday, November 25, 2011

1st Lt. John Chamberlin -Personal Log

THE X-MAS WAR

U.S.S. Ronald Reagan

Singapore, Malaya

We're going in tomorrow. There's some trouble smack dab in the middle of the area where China, Myanmar, and Laos intersect. It seems that the Chinese have been smuggling troops into there through the jungle for a few weeks now. It's also one of the areas where our flyboys chewed up a river assault that was trying to get a big force in.

The Chinese have really hit the western portion of Indochina, predominately Myanmar, very hard. I guess they even took Myitkyina. That city's been the linchpin to the entire region for centuries. It took almost the entire duration of World War II to take it back from the Japanese. Now that the Chinese have it, we're pouring all of our reinforcements into the area.

In the Eastern portion of Indochina, the Vietnamese have really stuck it to the Chinese ever since we stopped them back in the Black River Valley. Since then, the Chinese have really only tried to infiltrate in smaller units. It didn't work. I don' t know what they were thinking trying to go toe to toe in a jungle guerrilla war with the Vietnamese? Our Vietnamese allies are sweeping them up by the hundreds. It's nice to have them on our side now!

With the east and west tied up, there aren't a lot of forces to send into the central part of Indochina. The Laotians have thrown 5 infantry divisions into the 1st United Nations Expeditionary Force, the main unit between the Black River Valley and the Myitkyina area. There's also 5 Thai infantry divisions and 2 Thai Special Forces divisions. We'll be the only Americans around there. All total we'll have about 100,000 troops, including our unit. It sounds like a lot, but we have to hold the line in an area that's the same length as the one to our east or the one to our west. Both of those are held by almost a million troops each! This promises to be a little on the hairy side.

SECTION 7

"Courage can win a fight, but logistics will win a war."

Saturday, November 26, 2011

1st Lt. John Chamberlin -Personal Log

Approximately 50 miles south of Keng Tung, Myanmar

W e left the Reagan this morning about 1000 hours. The entire Regiment began leaving in one long airborne train at about 0800. I guess they wanted to let us get one more good meal in before we had to start working over the MRE's again.

Within a few minutes, we were over Vietnam again. We passed a bit to the South of the Black River Valley, but I could recognize it pretty clearly. Not much has changed. I guess that makes sense. We were only there a few months ago, but it feels like decades.

When we left the Reagan, we had lots of U.S. Navy F-18s and F-14's flying escort, but when we got over Vietnam, they turned around and these Vietnamese planes pulled alongside. They looked old as dirt, but some help was a hell of a lot better than none. I can't imagine what it would be like to have some Chinese Migs come down and see a few hundred low and slow helos and Osprey just putzing their way into deepest Indochina. We'd make a fighter pilot's fantasy come true.

After we flew over Vietnam, it took us about another hour and a half to get to our landing zone, LZ Bravo65. This place is truly in the middle of nowhere. I think someone must have looked at a map and said, "Yeah, that looks like it's far enough from civilization." It also wasn't big enough to land all our aircraft at once, so every chopper and Osprey had to make a touch, drop, and go style landing. It doesn't matter, we all made it in. By 1400 hours, we had all landed, gathered our gear, met our local U.S. Army Liaison officer, and we were marching toward the front-somewhere in the jungle.

At 1745 we finally stopped marching. We're still in the jungle. I can't see anything around us, or above us. The trees are at least a hundred feet tall, and the underbrush is so thick with bamboo that I feel like we're fleas on a dog's back.

Dennis came over and told us to dig in, then he showed me where we are on the map. There's supposedly a mountain range running north to south about 5 miles to our east, and another one about 20 miles to the west. Somewhere in between is one of the rivers that Chinese tried to come down and got massacred. All I can see is jungle. All I can hear is bugs. All I can feel is tired. It's November, but it's still in the 90's. When the sunset and the evening mist came up from the dank ground, the fog was so thick, it felt like a drizzle.

I met with Bob and told him our situation, then I walked up and down the line of our troop to check out everyone's foxhole or trench. Our frontal field of fire is only a few yards, but they do look good. We're also blessed in having a troop on our left and another troop on our right. They're both close enough to where we can see them about as well as we can see each other.

The most serious problem we have, aside from the mosquitoes and exotic diseases that they carry, is the lack of any sort of fire support. Almost all of the Coalition aircraft are either flying ground support around Myitkyina to the west, or they're flying combat air patrols over our Task Force in the northern South China Sea. We didn't bring any artillery or mortars either. There was no place to set them up in this jungle. If the Chinese hit us, we could be in a serious fight.

Thursday, December 1, 2011

1st Lt. John Chamberlin -Personal Log

Approximately 50 miles south of Keng Tung, Myanmar

Well, we've been here for a few days and, as usual, there's no sign of Chinese or any other Alliance troops in the area. I've been running an around the clock patrol rotation that keeps everybody on their toes. Everyone in the troop is either digging in deeper, resting, or patrolling.

Our position is really shaping up. We're using machetes to widen the field of fire in front of us. The brush is mostly small saplings, thick ferns, and monstrous bamboo. We're using the cut-down pieces of bamboo to line the bottoms of our trenches, dugouts, and foxholes. Then we use the saplings and ferns as camouflage. We've also been working on three trenches, each about a foot wide, in the middle of the fire zone. Inside the three small trenches, we used the smaller pieces of bamboo to make punji sticks, then, again, we cover it with the saplings and ferns. I noticed that some of my people are filling the thick pieces of bamboo with mud and lying them down in front of their positions like a fence or a wall. It's a pretty good idea, so while I'm digging my little rifle pit deeper, I'm doing it too.

The more we open up that field of fire, the more solid our positions become. When we got here, there was absolutely zero open area in front of us. Today I measured the area in front of our forward most foxholes. It was almost 100 paces! The best thing is that the jungle is so tall here, no one could possibly fly over and see our new killing field or us.

The other troops to our left and right are digging in and running patrols too. So far, we've been able to coordinate pretty well, and we haven't ambushed each other yet. I'm afraid that might only be a matter of time. I tried working out a schedule with the two other commanders, but they rarely stick to it. As a safety measure of sorts, I have all my patrols go forward at least a kilometer before they move parallel to our position in any way, shape, or form. It seems to have kept us out of trouble with the other two troops, but it guarantees that if and when the Chinese ever hit us, my people will get hit first. When that happens, they'll be too far away for us to cover them in this thick jungle. I can see moving that field of fire out another 100, maybe 150 paces, but no way 500.

Saturday, December 3, 2011

1st Lt. John Chamberlin -Personal Log

Approximately 50 miles south of Keng Tung, Myanmar

I got word from the troop on our left's lieutenant that one of his patrols found a Chinese can of rations just on the other side of our killing field. He showed it to me, and it didn't look like something that had been sitting in the jungle for very long. I warned our patrols to be extra careful. Dennis comes down the line on his every other day or so visit, and we told him about the ration can. He says some of the other troops in the area have found footprints and the like.

With no air, artillery, or mortar support, I'm starting to feel a little like Custer in Indian country; especially since the only sounds I can hear are still bugs, birds, machetes, Marines whispering, and the near-daily rain. The jungle's volume is always very VERY loud, even at night. Somehow, even with all the wild racket going on, it feels silent. I honestly think if I dropped a pin in the mud, I could hear it drop.

Sunday, December 4, 2011

1st Lt. John Chamberlin -Personal Log

Approximately 50 miles south of Keng Tung, Myanmar

Last night, a patrol from the troop on our right had a Chinese patrol walk within a few feet of where they were hiding. They choose not to engage the Chinese because they didn't know how many there were, and their job was only to detect any enemy presence. The Chinese passed them by and still don't seem to have noticed that we're here! I got together with the other troop lieutenants and we've decided to put claymore mines at the far edge of our killing field. Each of us marked a passage through the mines, but we're also going to temporarily halt the patrols. If we get pounced on, we'll need everybody in their holes and ready to fight. I for one don't want anyone caught out in the open when we do open fire.

I've restricted the reconditioning of the bunkers, etc. to limit the amount of noise we make, but I also let everyone know that they need to keep their camouflage up to date with fresh ferns and saplings. I can just see some Chinese point man walking into this clearly man-made open area, in the middle of the world's most dense jungle, and he stops when he notices a eleven piles of dead plants. I wonder what would happen?!

Most everyone's already doing a fine job. The bunkers are deep and well protected with bamboo that we've filled with mud. There's plenty of both around. They've also all got bamboo floors and roofs. The roofs really help camouflage them. It also helps keep the hot sun and never-ending rain out. Hopefully they'll keep the grenades out too.

Tuesday, December 6, 2011

1st Lt. John Chamberlin -Personal Log

Approximately 50 miles south of Keng Tung, Myanmar

We listened to some rumbling off to our left all day today. The rain only lasted for about an hour this morning, and I'm pretty sure that it's artillery, probably on the other side of that mountain range to our left. I can't be sure. If it is artillery, it's gotta be some big stuff to be heard all the way over here.

Everyday I try make my way around from bunker to bunker. Everybody seemed to get tired of climbing in and out of their positions, so driven by laziness or sun-induced fatigue, we've all connected our positions together by trenches. It's not a very good trench and not very deep in most spots, but there are places where bamboo walls have been added to make it higher so you don't have to duck as low when visiting. There are even a few places where bamboo roofs and/or floors have been added.

Morale isn't as good as it was back on the Reagan. It's not bad, but after sitting in the jungle and getting eaten by mosquitoes that are big enough to show up on radar, we're all getting a little anxious for a fight. The constant rain and mud don't help.

The latest headache is the news from Dennis that the Reagan's under attack. In fact it's been under air, missile, and submarine attack since the day that they dropped us off. As a result, we haven't been getting any supply drops. Water is <u>absolutely</u> not a problem with all this rain, but food's starting to get a little tight. Dennis is having everyone dropped to half rations until our next drop. No one really loves an MRE, but food is food, and we're all getting a little hungry.

The unspoken word that Dennis passed on when he gave us the news about he Reagan, was that if they can't get supplies in, they can't get us out either. No air, no artillery, no mortars, spotty communications, no good intelligence, and no way out except by foot...all that, and the only thing we want is a fight. Everybody does. There's a feeling that if we can just start doing our job, fighting Chinese, everything else will fall into place. It's not practical at all, but it is understandable. I feel it too.

Wednesday, December 7, 2011

1st Lt. John Chamberlin -Personal Log

Approximately 50 miles south of Keng Tung, Myanmar

How's that saying go..."Be careful what you wish for. You might just get it." It rained all night, then just after we would have eaten breakfast, the jungle came alive. There was no preemptive artillery or mortar attack. We heard a bugle and then every Chinese soldier on the face of the planet came out of nowhere, ran across our killing field in less than 10 seconds, and went right on past us into the jungle to our rear. They caught us by surprise so badly that no one, myself included had time to do anything but look around and watch them jump over our positions. The truly amazing, truly terrifying part about it is that they never even saw us! While every person on all three troops was searching around in shock, ducking for cover, or trying to find a weapon, the Chinese passed right over us. We never even got a single shot off. For that matter, we never even got our heads up to take aim! I have no idea how many jumped right over my bunker;5, 10, 100, 1000. It's like it all happened in the blink of an eye.

When they finally all went by, everyone started crawling to check out each other's positions. We whispered through all the jungle's noise, through all the Chinese rustling in the brush behind us. We whispered as though making any sound at all would get us all killed. Everyone was ok, shaken very badly, but ok.

The entire event couldn't have taken a full minute. Maybe it did, but it felt like the blink of an eye. The rest of the day, no one really made a sound. No one's getting any real sleep tonight. I don't think anyone's even eaten despite how hungry we all are.

No one wants to let their guard down even the slightest now. When we prepared for a Chinese attack from the north, we cleared about 150 meters of brush out of the way and made a nice killing field. Behind us there can't be more than a 10 meters, maybe 15 in some spots. I think that the Chinese commander probably lined up his troops in front of the killing field and then had them charge across in case someone, like us, was waiting for them. We were waiting, but not very well. In fact, we screwed up; bad! Luckily, they did too. Someone's sure to stop them, maybe even drive them back. When they do, we'll be here, and this time, even if we don't sleep a wink until they do, we <u>will</u> be ready. We can't afford not to be.

Saturday, December 10, 2011

1st Lt. John Chamberlin -Personal Log

Approximately 50 miles south of Keng Tung, Myanmar

Out of my original platoon, my first platoon, my first assignment in the Marine Corps, there are only a handful of people beside myself still here; that was until today. I lost two more today.

Late last night, it was raining pretty bad, but we're getting pretty good at telling the difference between high explosive thunder and the real thing. It's particularly easy when you hear them both at the same time-like last night. There was thunder and lightning all around us, but high explosive about a mile or two to our rear.

I didn't have to tell anyone, we all knew what was going to happen. The Chinese had overrun us, now, well into our rear area, someone was trying to stop them. If they did stop them, there was a good chance the Chinese would fall back over our positions again. If our guys in the rear couldn't stop them, we're going to be stuck even farther behind the lines.

The rain, thunder and artillery continued all night. Around the time that the sun should have been rising through the smoke, fog, and drizzle, I heard a rifle fire behind us. Everyone in all three troops locked their attention to our south.

During the night, Wess, Steve Burns, and Steve Hauser (the three best marksmen in my troop) had all crept out of their bunkers and taken up positions as snipers further in the jungle. None of the three realized that the other two had left, and no one told me. It wasn't hard to figure out. Our M-16's sound entirely different from any Chinese weapon.

I also knew that if one of our people was out there playing sniper, it wasn't hard to guess that it would at least be one of those three. They're Marines, grunts, and air assault troopers, but in their hearts Wess, Steve, and Steve love the idea of 'one shot=one kill'. They've always been among the coolest under fire, the most accurate, and the most satisfied with their work. I knew who it was out there as soon as I heard the shot.

Of course, after the one shot, I heard all kinds of rustling and Chinese orders being shouted. That was followed a few seconds later by lots and lots of small arms fire probably directed at my sniper. Through it all, I heard several more individual M-16 shots coming from two other directions. That's when I knew that all three snipers were out in the jungle.

After we were overrun the other night, I had Bob take some people out to the edge of our killing field and move half of our claymore mines to a line about 50 meters to our rear. It's half of the firepower, but something's better than nothing. I knew that I had at least three people out in the brush behind us, so I made sure everyone else knew to check their targets before firing off their claymores. I didn't want to accidentally blow away our three best marksmen.

The Chinese small arms fire increased with every shot M-16 shot we heard in the brush. Finally, I saw both Steves come crawling back into our tunnel/trench system. Wess was still out there someplace causing trouble for the Chinese. After nearly an hour's worth of sniper hunting, the Chinese lost out. Wess slithered out of the thick brush and back into his hole. The bad guys just kept on firing blindly for a few more minutes, but when they realized that there were no more M-16 shots ringing out, they stopped and began their search for what they obviously thought was a dead sniper or two.

I'm not sure how long we waited, but we finally saw some of the Chinese coming through the brush. We were all watching so intently that there was no way anyone was going to look away to check their watch.

The first Chinese to come through stepped back in when they saw the killing ground we had made. Then more and more gathered along the brush line. Some even hid among the camouflage palms and saplings that were hiding our covered trenches. It was clear that they were going to form up and charge across the open ground just like they did when they overran us before. Either these were different troops than the ones that ran over us the other night, or they had forgotten about the killing field. Either way, it was only going to be a matter of time before they took a careful look at what was literally underneath them, and we'd be spotted.

I don't know who fired first, but in one single instant, everything exploded. First the claymores, then, almost at the same moment, grenades and machineguns from both sides lit up. I joined in. After all, there was a commie on both sides of my bunker and three looking right at the barrel of my gun.

I took out all three in two bursts. The other two that were next to my bunker disappeared in the smoke. I tossed a pair of grenades into the brush, one about 20 feet, another about 50. There was no chance of aiming since visibility was near zero.

The Chinese answered by firing in all directions. I don't think they had any idea where we were even after we opened fire. The ones that were right on top of us must have been killed outright, and the rest must have thought that they were being overrun or even surrounded. Without being able to see more than a few feet, it was hard to tell what was going on, especially with all the noise.

I stopped firing and tried to see. My night vision had stopped working in all the days of rain, and I really missed it now. It would have let me see right through all the smoke and mist. During my rounds over the past few days, one by one, everyone else was having the same problem. I prayed that the Chinese were having it too.

The rest of my troop stopped at looked for targets too, but the two troops next to mine just faced the jungle, then sprayed and prayed. They were using up their ammo, and I'm certain that they'll regret it in the next few days.

Both sides kept up their fire for nearly three hours. My troop stayed silent. Bob crawled through the trench and delicately whispered the casualty report to me. I lost two.

Barbra "Barbie" Warner was dead. Someone rolled a grenade in her covered foxhole, and she didn't make it out. Barb was one of the prettiest women in the unit-blonde hair, blue eyes, and pale skin, athletic build. She had made her rounds on the dating scene with several of the guys in the platoon, but never seriously. Instead, she became a friend to everyone. She was intelligent and fun-loving, but very conservative in her nature. I think she was one of the few of us that could still live a normal life after the war. Now, her body's spread all over a 4' x 4' x 4' foxhole with a bamboo roof as a grave marker. She'll definitely be going home in a bag.

It's very sad. I think whenever any of us talked with her, we felt closer to home, closer to a reality that seems so far away now. It feels like a different dimension. Along with my night vision, my palm PC has stopped working so I can't even send the usual email to her family.

Missy "Bitchy" Jansen was our other casualty. It looks like she was one of the few that the Chinese could actually see. I'm not sure if she got hit by one bullet through the forearm and neck or if it was two. It doesn't matter. She's dead. She's not wasting away in the middle of Indian country like the rest of us.

In contrast to Barbie, no one in the troop liked Missy. She had a real tough guy hardass attitude that she just couldn't back up. In fact, she was quite the wimp when challenged. I remember back in Taiwan she got a small piece of shrapnel in her hand, and we all had to listen to her whine about it for weeks. She never really fit the Corps. I'm not glad to see her dead (there are plenty of others who feel that way), but I will miss her rifle on the line more than I will her.

There were a few other people who took either some shrapnel or rounds. Most everybody has some sort of new scratch or bruise, but our body armor is really proving its weight. In this tropical heat, that weight is a lot more than it normally would be.

Tonight, we wait again for the Chinese to overrun us. No one, not even our snipers, is going to venture out of the trenches tonight.

Sunday, December 11, 2011

1st Lt. John Chamberlin -Personal Log

Approximately 50 miles south of Keng Tung, Myanmar NO RAIN TODAY!

I slept on my rifle facing out the back of my bunker last night. The Chinese never came. Most of the troop is starting to get pretty worn-out from fatigue. Another problem is that we'll use up the last of our MRE's today. Nobody's too concerned about it either. We all feel like we could be overrun at any second. 24-hours a day, we wait through each and every second for the bullets to start flying again. So far nothing.

Yesterday's grenades and claymores opened up a little bit of a killing field in the brush, about 20 paces. The fallen brush is helping to refresh the aging palms and saplings that's our camouflage. It's not enough, but it's better than nothing.

There was a quick mortar barrage around noon, but it was directed at the far end of our original killing field on the north side of our position about 150 meters away. I don't know who called it in or if it was our mortars or theirs. I haven't heard from either of the two other troop commanders since yesterday's point blank shootout. I don't even know if they're still there at all.

Tuesday, December 13, 2011

1st Lt. John Chamberlin -Personal Log

Approximately 50 miles south of Keng Tung, Myanmar

It's been several days since our last little firefight. An ungodly heat has replaced the rain. It's amazing how even though we're underground, under the shade of trees the size of small city buildings, I still feel like the sun is burning right down on me. The heat's also making all the corpses from the other day's shootout really stink. It's very demoralizing when you know that the awful smell that's making it hard to breathe is coming not only from the dead Chinese soldiers all over the place, but from the two women we lost the other day. No one even wants to look at them now let alone try and think about them. I'd send someone over to bury them or fill in their foxholes, but it might give us away if the Chinese have some spotters in the brush trying to see if we're still here.

My three snipers have been begging me to let them out into the brush, so I'm letting them go out on 1-man patrols for 4 hours at a stint. It's nerve wracking, but so far it's worked. Steve Burns made it over to the two other troops next to ours and was able to contact them. They're still alive, but they both thought we had been wiped out.

Dennis crawled into our trench system this afternoon. He said that the Reagan's been hit pretty badly. They lost a lot of our aircraft so it might be a few more days before we get re-supplied. It was not the kind of news I needed. Combat wears people out ten times as fast as even the hardest labor. In this kind of environment, my people are near complete exhaustion. The weaker we all get, the more susceptible to all those weird jungle diseases we become.

Only one person's gotten sick so far. Yesterday Louie "Looeye" Madill came down with some sort of measles-like thing. It happened overnight. He's got black and blue splotches all over his body. I think he's running a fever too. I told everyone to stay away from him unless they want to put on a CBW suit. I knew no one would want to do that in this heat, so he's effectively quarantined. With no food, no fresh water in almost a week, and no medical treatment, I think he's as good as gone. It's a shame. He's a really nice, funny, cool-headed Marine, and a very good fighter too. When it comes to a fistfight or anything hand-to-hand, he's one of those guys I would really like to have nearby, until now. Now, I look at his black and blue face, and I think to myself, that could be all of us in a few days; maybe tomorrow.

Wednesday, December 14, 2011

1st Lt. John Chamberlin -Personal Log

Approximately 50 miles south of Keng Tung, Myanmar

We got pounded today. Someone decided our little hidden spot in the jungle would be a good place to put some artillery-or maybe all the artillery! For nearly three hours, some really really heavy stuff blasted my troop and the two adjacent ones. Even when it ended, they waited for a few minutes than hit us about twenty or thirty more times again. When the dust settled, the sun was coming through the jungle. Our top level of the triple canopy jungle is gone. So is the second one. The bottom level of the jungle, the brush, is just a big pile of fallen trees, broken branches, and smoldering leaves. The jungle is gone. It's so thick and torn up that I don't know if we can even walk out of here now!

Our trenches and bunkers were all at ankle height, now, we've got to come up with new positions that can see and fire over the debris. In some places that debris is at least 20 feet tall. The only good thing is that no one, not infantry, and certainly not tanks will be able to overrun us now; not without getting hand to hand. There is no field of fire. There's no way to clear one, and there's no real way to cut a path. The only way anyone can move around is in our trench system. That might as well be a tunnel network now.

Dennis came over when it was done. He brought some of the other Troop commanders over too to show them our trench setup. I guess ours is a lot better than theirs is, and he wants to use it as an example. They were impressed. I told them that all it takes is hard work, and constant maintenance. Every person in my troop is responsible for their own foxhole or bunker, but we all work on the interlinking trenches everyday-cleaning out mud, putting up fresh camouflage, reinforcing the walls, floor, and ceiling with mud-filled bamboo. We work a little at it everyday, and it's paid off-until now. Now we'll have to build it taller so our bunkers and foxholes don't just look forward to a big pile of debris 3 feet away.

Saturday, December 17, 2011

1st Lt. John Chamberlin -Personal Log

Approximately 50 miles south of Keng Tung, Myanmar

Louie "Looeye" died sometime last night. When Dwaine found his body, still in his foxhole, he looked awful. The next time someone comes down with the black and blue splotches, I might have Wess take care of them. It was like he decayed while he was still alive. I didn't know what to do with his body because it was clearly diseased. We had to do something, but we don't have any body bags, and I didn't want anyone exposed if it was contagious. I had three people put on their old CBW suits and cover him with dirt and leaves. I wouldn't let anyone take out his personal belongings, even though he had some sort of crumpled picture in his hands. That's how he died, and that's how he'll spend eternity, in a foxhole, in the middle of nowhere, with someone's picture in his black and blue hand.

Louie's death had a bad effect on the sinking morale of the troop. I called off the sniper patrols and had everyone build up their bunkers in the debris. At least it kept them busy instead of just staring at a pile of debris wondering how they'll die-Chinese, disease, starvation, or worse?

I went around to everyone and spent sometime doing a bit of 'rah-rah-rah' moral reinforcement. It worked as well as I could have hoped. We're really in a pinch, and I'm starting to wonder if we've been forgotten, or what good we could possibly be doing in the middle of nowhere. I would never pass that on to the troops, but they feel the same way and sometime they'll be able to see through me; if they can't already.

Some of the bunkers are really getting elaborate now. Most are at least two story; the original bunker or foxhole as a basement, and the new aboveground bunker built up in the jungle's debris. I can't believe how industrious everyone still is. With no food, less and less water, hot sun, bugs everywhere, the smell of death around us, and deep in enemy territory, my Marines are making some very, very good positions. When you walk through the old trench/tunnel and into one of the bunkers, it's almost like you're inside a small log cabin, but instead of logs, it's a mixture of woven branches, fat mud-filled bamboo, and a few fallen tree trunks. Some even have steps to reach the firing ports like they were parapets. It's very impressive. I can't wait to see what Dennis and the other troop commanders think when they see ours again.

Monday, December 19, 2011

1st Lt. John Chamberlin -Personal Log

Approximately 50 miles south of Keng Tung, Myanmar

What a day. We woke up to a heavy rain, and we liked it. Fresh water was running low, and down it came, by the bucket. It also washed away a lot of that stuffy dead body smell.

It turns out that our tunnel network has it's own drainage system. Somehow, completely by chance, all the water that does make it through the bamboo and palm roof collects on the bamboo floor and runs east to a shellhole. It collects in the shellhole, but that's fine with us as long as it's not in our bunkers. I think the other troop commanders are really going to be impressed now. We'll make sure to tell them that we planned it that way.

After we all drank our fill and topped off our canteens, everybody settled back in to their daily routine of waiting and watching. The rain continued most of the day, until around 1500hrs. When it stopped, the moist ground steamed as soon as the clouds broke, and the sun beat down. A few hours later the sun set and a thick fog formed.

Just after dark, flares went up from someplace in the south. I heard my sniper pop off a few rounds and knew that the Chinese were coming again. After the first M-16 shots there was the usual hail of Chinese small arms fire trying to find the sniper. Steve Burns didn't hangout long. He made it back into our system (now nicknamed the sewer by most of the troop), and we ran through it until he made it into my bunker. Steve said that there were hundreds, maybe thousands of them not even 200 meters out to the south. I wasn't surprised, and I don't think anyone else was when I had Steve pass the word.

More man-made stars were tossed into the air. Yellow flares hung from small parachutes and the entire world flickered. I was frustrated by the night attack until

I realized that the heavy use of flares could well be a sign that the Chinese night vision equipment wasn't working either. At least we'd all be on the same playing field; there would just be a lot more of them than us, but that wasn't anything new at all.

I didn't have to tell anyone to get ready. They had been for days. Everyone knew where to look. Everyone knew to hold their fire as long as possible. It had been days since they came at us, but the attack was anything but a surprise. There was no doubt it would come, and most of us welcomed the chance to break the monotony. It was also a chance to vent our frustrations; frustrations from casualties, from the heat, the rain, and from the hunger. It can actually be said that we were hungry for blood, not as food, but as a way of quenching our thirst for payback. The Chinese would have to pay dearly for our hardships, and pay, they did.

The shooting started somewhere to my right. I hadn't even seen a single enemy soldier until then. As soon as everyone in the area heard that M-16, I think people went into overdrive. I tossed out a grenade just in case there were some bad guys in front of my bunker. There were. It only went about 10 meters, but when it went off, I heard that familiar sound of someone getting hit.

Chinese yellow tracers flashed through the fog and smoke, over the mangled jungle debris, and into the mud-filled bamboo firing port of my bunker. They had my number, so I jumped down into my original bunker and dove into the tunnel. When I turned around, there were so many tracers hitting the firing port and going into the bunker that it was actually bright inside. I scrambled further into the tunnel and then there was a loud explosion. Someone did have my position spotted, and they had tossed in a grenade. I got up and ran down the tunnel a few more feet, but two more grenades went off in my bunker, and I was knocked down. Outside, there was all kinds of racket.

A few seconds passed and there were no more explosions in my bunker, so I figured that they had given me up for dead. I turned and watched down the dark tunnel to see if anyone came in, but no one did.

Since the Chinese had given up on me, I went back to my bunker to see if I could surprise them. The roof was gone and the floor was filled with debris. I could see fog slowly pouring in like a thick soup. I saw two soldiers jump over the hole, and a third fell in. I pulled out my bayonet and took care of him as quickly and silently as a could. Then I slipped back into the tunnel waiting for another to fall in.

I didn't have to wait long. One, then two, then another fell in. They all had no idea I was in the tunnel. It must have been well-hidden by the roof's debris. I knew it was too risky to use my bayonet against four people. It would have been risky to use my rifle. I tossed in my last grenade and ran down the tunnel again. When it went off, I fell down and turned to watch for survivors. None came.

I crept back down to the bunker again. My tactic was working well for me, so I thought I'd try it again. When I got close enough for the flare's flicker to light the inside, I peeked in and saw an ugly scene. There was no time to get sick or anything so I focused on what needed to be done, and I watched the open roof for signs of more intruders. There were none. I surveyed its entire perimeter twice and started a third time, but the Chinese blew their bugle again, and the retreat was on. A few soldiers jumped or ran by the bunker, but I held my fire, and no one else fell in.

The battle was over. They only had one attack in them this time. I don't think they're going home, and even if they wanted, they'd have to go through us now. We'll be meeting again.

Tuesday, December 20, 2011

1st Lt. John Chamberlin -Personal Log

Approximately 50 miles south of Keng Tung, Myanmar

We lost another one last night. Ted Knox left his bunker and got out into the open during last night's fight. No one knows why, but it looks like they bayoneted him-a lot. We didn't count. Anything more than 5 times is beyond brutality. It's outright mutilation. He was really mutilated.

Ted was a tall husky guy, not too quiet, not too vocal. He was kind of argumentative, but generally a good Marine. In fact, he was a really good shot, so, we'll miss him sorely when we get hit again. I was never too close to him. He was always the professional; not a loner, but not the kind of guy who ever developed a relationship beyond the fellow teammate.

No rain today. Instead, we get more of that clear sky and scorching sun. Everything really stinks now. We've made it a policy not to bury the Chinese if it's going to give away our position, so there are some that have been rotting in the sun for almost two weeks now. My bunker is pretty gross right now, and it's too big to fill in with dirt. We did close off the entrance to the sewer though.

Lucky me, I get Ted's old hole. My new bunker is kind of sloppy. Ted had an organizational system that was his alone. Instead of trying to build it up, I just put up new a new bamboo wall and supports. It hides the old, makes it my hole, and probably makes it stronger.

Most everyone thought I had bought it last night, but that was easily rectified. Dennis came by around noon, saw the mess in my bunker and was visibly moved. I was flattered, and a bit embarrassed when I had to walk over to him and let him know I was still around. We both wanted to laugh, but hunger has a way of making everything very serious. The good news is that Dennis heard from the ship, and the told him they'd try and get us some supplies tonight. A daylight drop is too risky and would definitely give us away.

When the sun finally set, I heard my snipers go to work. The Chinese didn't answer except with flares and single shots of their own. It looks like they've sent in their own snipers to take mine out. Tomorrow will be a completely new day-if we get those supplies.

Wednesday, December 21, 2011

1st Lt. John Chamberlin -Personal Log

Approximately 50 miles south of Keng Tung, Myanmar

Rear echelon bastards! I waited all night and didn't even hear a single aircraft. There wasn't even much of a mist last night. The flares were sporadic. I waited and watched but saw nothing but stars, planets, and satellites. This morning, we took a few mortar rounds and thought maybe the Chinese were coming again, but nothing else happened. No attack, no food, no supplies, no reinforcements, no pickup-NOTHING.

I keep trying to hide my frustration, but it's hard to tell my people they haven't been forgotten when all I did yesterday was get their hopes up. Hunger has a strong influence on how well we maintain our morale around here. I'm getting the feeling that some of the people are starting to think about leaving. I can't blame them. After all, is it desertion if they just wandered off in search of food?

Napoleon once said that an Army fights on its stomach. Ours are empty, but still we fight. I wonder if he was right, if we're special, if we're losing it, or if we're too worn out to leave?

Thursday, December 22, 2011

1st Lt. John Chamberlin -Personal Log

Approximately 50 miles south of Keng Tung, Myanmar

Today I didn't wake up until well into the morning. The fatigue is starting to get to me. Everyone else is sleeping in later and later. The sun gave way to yet another full day of rain not long after I got up.

When I made my daily rounds throughout the sewer, I picked up the unmistakable smell of marijuana. If I could smell it, so could the Chinese. I moved faster and faster, eventually running from bunker to bunker looking for the idiot. A lot of the bunkers were empty. Finally, I found them.

Wess, Tom "Tattoo" Terry, Dan "Hoser" Farnum, Kurt "Stoner" MacDougall, Amy "Party Girl" Broady, Portia, and Christina were all in Wess' bunker getting stoned. I yelled as much as I could, but they laughed right at me! I was furious. Wess got up and stumbled over to me. I didn't know if he was going to try something or not, so I reached for the first weapon that I could, my machete.

He put his arm around me and motioned me back into the sewer. Wess' theory was that anything that could help morale right now was worth it, and the worst that could happen was that the Chinese might try something while they were "loosening up". I had made my position about drugs pretty clear a long time ago so I started back to the bunker to give them another dressing down. This was no light matter.

In fact, it turned out to be a life and death matter. No sooner had I taken a step back toward the bunker than I heard it. First there was this deafening whoosh or roar, then the screams. I dove back to Wess just in time to realize what had happened. The Chinese had smelled the pot and used it as a beacon for a flame-thrower team. All 6 people still in the bunker getting high were covered with burning napalm. The liquid flames poured into their bunker and then down into the sewer. All Wess and I could do was run to get out of the way. They screamed and no one will ever forget it, even Wess.

At first he was furious that he had been proven so terribly wrong. In classic Wess form, he ran over me, down the tunnel, into the nearest bunker, and went to work on the flame-thrower team. They of course turned on him with the fiery hose and he had to take cover with me again.

Once they had found the bunker's firing port, they moved in closer and closer to get more of the liquid flame into our tunnel system. I ran back towards his bunker to get away, but Wess ran straight into the flames to get to the other side of the tunnel. As he passed the bunker he had just been in, he blindly threw a grenade in. When it exploded, the blast tossed a good deal of the liquid back out of the bunker and onto the closing flame-thrower team. When they were distracted, he ran into the burning bunker, tossed another grenade out through the firing port and bailed back into the tunnel.

It was a real medal of honor moment, except for the fact that it was his fault the flame-thrower even found us. I'm still mad. I'll report all of his actions as soon as I can, and the higher-ups can decide how to deal with him. No one will be able to adequately punish him. He's carrying their deaths pretty heavily on his back. Then there's the matter of the burns he got. His entire head, hands, and parts of his neck and chest are all badly burned; red and blistered. I don't see how he can fight off any infection or diseases now. He probably won't last the night anyway.

Friday, December 23, 2011

1st Lt. John Chamberlin -Personal Log

Approximately 50 miles south of Keng Tung, Myanmar

The supply Osprey came in last night. Wess is on his way to safety, and we got some well needed supplies.

We got all kinds of good things last night. Food, lots of ammo, claymores, new night vision gear that can switch between low light level detection, thermal imaging, or ultraviolet (our old stuff was just low light level). We got new palm PC's that actually work (for how long I don't know). We got a little remote-controlled airplane to play with (It's got a camera on board, and we're supposed to use it as a frontline spyplane.). We also got a few medkits and a chemical, nuclear, biological detection kit. I'm not sure if I like the message that the higher-ups are sending by putting this in our little care package.

It rained hard all day, but all of the new electronic toys we got seem to be at least water-resistant. I think it's actually the mildew that kills them. We'll see.

I had four MRE's today: spaghetti and garlic bread, a lasagna and some sort of pastry, a spicy meat pack with toast and peach cobbler (worth a million dollars in the field), and a chicken Alfredo with crackers. The rain's starting to be kind of relaxing. Coupled with the food, ammo, and lack of Chinese attacks, this place really doesn't seem anywhere near as bad as it did a few days ago. I think the rest of the troop is feeling the same way.

The news on my new palm PC is definitely good. It seems that the Chinese broke through our stop gap line south of Myitkyina a few days ago, but the British I Corps and the French II Corps, both from the 2nd Army, have counterattacked with 2 armored divisions, 4 mechanized infantry divisions, and 6 infantry divisions. They haven't gone very far, but the reports are that they are kicking ass and taking names.

They must have been fed.

Saturday, December 24, 2011

1st Lt. John Chamberlin -Personal Log

Approximately 50 miles south of Keng Tung, Myanmar

They finally did it. They began hitting us with mortars and artillery just after sun-up, and it went on until well after lunch. There was no rain today, so visibility was very good except for the smoke and dust. Everyone in my troop, myself included, has come to the realization that when your number is up, there's nothing anyone can do about it. Ducking and feeling all afraid is completely useless.

By the time it ended I had already eaten 3 MRE's: a supreme pizza, barbecued beef with cornbread, pepper steak with rice and a fortune cookie. My fortune was: "Now is the time to finish that real estate deal." How clairvoyant?

They came at us hard this time. It's tough to tell if there really were more of them than usual, but it felt that way. It might be that there were so few of us thanks to Wess' attempt at raising morale. Either way, it was hard.

After the barrage, we heard their bugles. This time they used whistles too. I wish they would just use radios like the rest of the world. We all waited till the last second to open fire. It was nice to see all those people trying so hard to climb up and over all that jungle debris. Again, someone else fired the first bursts, then the Chinese answered with everything they had. The noise made all communication impossible, and it was everyone for themselves again.

I took out at least 3, maybe as many as 10. Then I saw someone toss a backpack sized satchel charge over the closest pile of twigs. I heard it thump onto the roof of my bunker, Ted's old one. It took a few seconds for its fuse to burn down, so I had enough time to get into the tunnel and make it to the next bunker, Steve Hauser's.

Steve was happy to see me even though it made things very crowded in his little 4'x 5' x 6' covered trench. He was picking and choosing his targets and firing semi-auto single shots in true sniper form. I was going through magazines like there was no tomorrow. At the time, I didn't know if there was going to be one.

The Chinese kept coming through the piles of twisted and torn brush. Normally they would have retreated by now. This time, they were using those satchel charges well. Another one came at Steve's bunker. It landed just in front of his firing port and we both went running into the sewer.

When it went off, it collapsed Steve's bunker and closed off one entire portion of our sewer. We headed for the next bunker, Rick and Bob's.

Rick and Bob had actually setup a small rifle pit under a huge fallen tree. The tree had to be at least 6 feet thick. The roof was a row of thick logs, with small branches, then bamboo, and finally palms on top. It was the biggest and most waterproof position we had. It was also the closest to ground level. The field of fire in front of the pentagonal hole in the ground was no greater than 20 feet at its farthest. When we got there, there were already enough Chinese bodies right in front of the firing ports to hide the jungle debris behind them. It looked like they were attacking on the bodies of their comrades. I guess they were.

After we were in the rifle pit for a few minutes, we heard another satchel charge go off somewhere. Not long after, Barry Gallagher came running in. They blew up his position. It was obvious that they were being methodical. The more bunker they blew up, the easier it would be to find the remaining ones. When it got down to one, it would be real easy. I looked at Bob, Rick, and Steve. They were thinking the same thing. We could either hide in the tunnel or hope that they ran over us, or we could get out and make a fight of it.

We all looked at each other and heard another charge go off someplace. Barry "Jesus Joe" Gallagher closed his eyes and said "'Lo, though I walk through the Valley of the Shadow of Death I-" Keith Brabble interrupted and finished with the ole Marine Corps boot camp version "-shall fear no evil because I am the biggest, meanest mother in the valley!" It was melodramatic to the extreme, but the mental ups and downs of the past few weeks were making everything seem unreal, almost like it wasn't happening. Life had a joke-like feel to it now.

We climbed out of the rifle pit and into the debris, towards the Chinese. Barry hosed down the left flank with his M249 SAW. Rick took care of the right flank. Bob popped off a few bad guys trying to come straight at us. Steve picked off anyone behind us, and I felt like I was spinning in circles.

We crawled over the pile of bodies, under a wall of twigs, between two big trees that were perpendicular to our old positions, and that was where we were going to die. No one had to say it. We knew it. We accepted it. We almost welcomed it.

As soon as we got settled in, they blew the bugles and whistles. Game over. The visiting team was going home; again. Everything suddenly felt fine. We had made it.

Some of the Chinese had made it over our former position and they had to come past us to get back to their line. At least twenty came out from under the wall of twigs, all at once. Jesus Joe emptied his belt into the pile. Steve actually fired in bursts instead of single shots. Bob and Rick were shooting in all directions too.

The distance was just too close, and there were just too many of them. Before we knew it, we were fighting hand to hand. I don't know about the other guys, but I clubbed one. Then I shot another in the face, chest, and groin with a 3 round burst in each location, stabbed another in the chest with my bayonet (it got stuck), and then started hacking away with my machete.

They had machetes too (who wouldn't in this part of the world!), and before you knew it, we were sword fighting. I took out one more. Another one slashed my body armor's chest plate before I could get him in the face. Two more came at me, but Jesus Joe was right there to stop them. He nearly cut them both in two with a fresh belt in his gun. We turned around to help Bob, Rick, and Steve. Steve was already down. He'd taken a machete in the groin, and he was bleeding very badly. Barry helped him control it, and we dragged him back into the sewer.

I have no idea how many Chinese we killed today or any other day, but this place has so many bodies lying around now that you'd never be able to tell the difference between any of Dante's levels of hell and this little corner of central Southeast Asia.

Flares and sniper fire went on through the night with the occasional three round mortar barrage. I don't think I actually sleep anymore except for maybe five or ten minutes here and there. Does anyone anymore? Will anyone ever again?

Saturday, December 24, 2011

1314hrs

Approximately 100 nm north of Mandalay, Myanmar

45,000 feet over the border of Myanmar and the People's Republic of China

THE X-MAS WAR

Three dozen brown-and-green camouflaged, ancient and war-weary, Vietnamese Mig 21 Fishbed fighter planes whisped their way west. Ten miles behind them, the Coalition's 1st Composite Fighter Group, 77 planes from half a dozen countries bobbled in a loose formation. Pilots whose units and sometimes countries had been decimated seemed to find their way to their sleepy home airfield in the Vietnamese jungle. In command of the group, was the now legendary Wolfman, Al Meyers.

Four miles below them, a thick layer of overcast kept the ground from view. Poking out of the white blanket, round thunderstorms rose and twisted their way upward, some as high as 50,000'; pillars of soft white cotton that glittered with lightning even in the bright mid day sun. Inside the docile looking fluffy structures, violent turbulence and heavy rain waited for the wary, the ignorant, and the bold.

The Wolfman's original gift from the fleeing South Korean Air Force, an American-built F-16A Falcon, had finally worn out. In its place, he had commandeered an old Australian F-111 Aardvark bomber on a supply visit to Clark Field. The F-111 had been sitting idle and in good condition near the field's north tower. The crew was nowhere to be seen, and so he left his shot up and worn out F-16 parked next to where the forty year old F-111 was parked with a note saying, "-Traded your plane for this one-1st Coalition Composite Air Group/ thanks and enjoy!"

Al had flown F-111's for years before his assignment to special flight operations in the Nevada Desert; special flight operations like the one that he was conducting when the war really started back in January.

Combat had take its toll on everything, and even his new plane was tired. So was Al. He'd flown over a thousand missions in less than 9 months. He had over 150 confirmed kills. He also couldn't even remember what a good night's sleep was. His average was only 4 hours. Methamphetamines washed down with coffee and drops for his eyes had become his life's new staples.

The big F-111 normally had a second crewman sitting next to Al, but the weapons officer he normally flew with was wounded on that fateful night in January, so Al was flying alone in the cockpit. It didn't really matter. His wingman could navigate for the patrol, and he wasn't carrying any air-to ground ordnance on the bomber's movable wings. Instead, he had all 22 of the plane's weapon mounts loaded with short-range heat-seeking air-to-air missiles.

They were almost a thousand miles from their home field, and fuel was becoming an issue. If someone ran low or ran into trouble at this point, they might have to try and land at Coalition-controlled Mandalay. With the semi typhoon weather below, no one wanted to do that, and they were all thinking the same thing in many different languages. It was time to turn back.

Al was about to make the call, but his feel for combat switched on. Without knowing which sense had spotted the trouble, Al snap rolled his plane upside down and pulled back on the stick heading for the clouds below. Everyone else followed except the Vietnamese squadron. No one had seen what happened to them. They were gone. Al had missed it too, but he knew something was wrong, and so he began the defensive maneuver.

When fighter planes get in trouble, they need speed. Speed kills, but it can also save lives in air-to-air combat. There were two ways to get speed: light up the afterburners and burn fuel at twelve times the normal rate, or dive hard. Al was flying a bomber, so as soon as the dogfight would begin, he knew that he'd be outclassed by even an antiquated fighter. He planned on using speed and lots of missiles to make up the difference.

As his airspeed passed 1000 knots and the overcast layer came closer and closer into view, Al rolled upright, leveled out from his dive, and began heading to where he had last seen the Vietnamese. He spotted them in an instant; what was left of them. Most of the planes were missing. A few gray puffs and trails leading down into the clouds showed that at least some had probably been shot down. Those that remained were clearly visible. Their green and brown bodies silhouetted with great contrast against the bright blue sky and blinding white overcast. Where was the enemy?

He watched one of the green specks and tried to see if he could spot anything behind or in front of it. Finally, a little white speck came through the clouds and turned in unison with a doomed Vietnamese pilot. A white line popped out from the speck and drew its way into the tailpipe of a friendly Mig-21. Al knew it must have been a heat-seeking missile drawn to the engine's exhaust. While he was figuring out the weapon's nature, it quietly exploded in a puff of black around an orange fireball and invisible shrapnel. The Mig ingloriously disintegrated.

At 800 knots, Al closed in on the white speck quickly, and he recognized it as a Russian Su-27 Flanker; one of the biggest and best fighters in the world. The only thing worse than an Su-27 was its younger brother, the Su-37, with an extra set of wings near the nose and engine nozzles that could steer the thrust in different directions.

In his helmet, the Wolfman heard the familiar locked-on alarm of a heatseeking missile at the tip of his wing. The alarm, more of a growl, told him that the missile had a target and was ready to be let loose. He fired it without caring what it was locked on to.

If it hit a Vietnamese plane, that was too bad. More and more white painted Flankers were coming out of the clouds, and with as many planes in the air as there were, it was anything goes. He was out to stay alive at this point.

The missile disappeared from his wingtip and drew a white line to its target; one of the Russian planes. More missiles passed him from behind as the other 76 members of his ad hoc unit began their attack. The Russians answered with more missiles, and the white scribbles of air-to-air missiles in the blue sky passed each other at Mach 10+.

He saw the missile he'd fired only a second earlier pop into a black puff as it showered a Flanker with shrapnel. The Flanker's fuel tank must have been punctured and leaked because it immediately trailed a brilliant orange flame and a long black cloud of smoke. Behind him, the other members of his group were dodging missiles or taking hits. He couldn't tell, but he saw tracers from a friendly cannon stream past and into the wounded Flanker. Another kill shared with a friendly pilot. They'd figure out who gave the assist after the mission, but he knew it was at least half a credit for him.

Al popped off a few flares to confuse the Russian heat-seeking missiles, then he started jinking the plane from left to right to confuse the pilots. At 800 knots he blasted through the swarm of white Russian Flankers and the few remaining green Vietnamese planes in less than half a second. His F-111 couldn't even hope to turn in a dogfight with the fighter planes, so he continued on for a few miles; about 2 seconds later, Al pulled its nose up until he was at almost 60,000 feet, over 11 miles in the sky.

The Wolfman was in his element and in his favorite type of tool, the F111. He snap-rolled his plane onto its back, chopped the throttles until it stalled and dropped, then pulled back on the stick to aim the plane at the middle of the giant dogfight. He popped another pair of flares in case someone was trying to get a missile lock on him. Then he picked out a white Russian speck and bored in for the kill. His speed picked up as the heavy bomber dropped from above. Soon he was back in the 700-800 knot range and accelerating.

Al saw a missile coming out of the twisting white contrails of the dogfight. It was headed his way, so he popped out some more flares. As he was doing so, he heard one of his Sidewinder missiles growl in his ear to let him know that it had found a target. While the fingers on his left hand popped out the flares, his right hand fingers danced through around his flight stick and let loose the hungry air-to-air missile.

The Russian missile missed, but Al's found its mark on the other side of the dogfight, about 8 miles ahead. It was too far to tell if the missile had killed its target, or even if it was in fact a Russian and not one of his own! Again, a missile growled in his ear, and Al let it loose. Out of the corner of his eye, he noticed bright yellow tracers coming past his plane from behind and to the right. Al banked left, then right, popped out more flares, and headed for the clouds miles below. After a few seconds he was still alive, and he had passed through the furball of white smoky lines in the wild blue yonder. The fact that he was still alive told him that his adversary to his rear must have found better prey or had to take evasive action of his own.

Again, Al headed towards his 60,000-foot limit and away from the dogfight. This time, he was almost 25 miles away when he pointed the plane up at the sun, rolled the plane over on its back, and dove back into the fight upside down as usual. His helmet was already growling, but he was too far away to see any planes, just specks and white contrails. It didn't matter, he fired. One of his missiles went ahead for only a mile, then it dove straight down, and Al flew over it. He lost sight of it as it passed beneath him on someone else's tail.

The dogfight came into clear view again. A Flanker chasing one of his Taiwanese F-5 Freedom Fighters crossed his path, and Al tried to turn in behind it. He had too much speed, and not enough maneuverability to stay in the chase, but one of his sidewinders was up for it; a growl in his helmet let him know that much. He let it fly, and watched it go after the pretty white Russian plane. The Russian popped out flares, but it was too late, Al's Sidewinder had gotten close enough to set off its internal proximity fuse. It exploded, and the Flanker came apart as shrapnel from the little black puff shredded the plane.

Al leveled out and tried his run-through tactic again. He was definitely meeting with great success. About 10 miles beyond the dogfight, Al pulled the nose of his big plane back toward the sun. This time, he chopped the throttles and let the plane loop back into the fight. It might be more fuel efficient, and he was already thinking about alternative landing areas.

As his plane reached the top of the loop, with the view through the top of the cockpit showing the entire fight, Al saw an Su-27 coming up after him with its 30mm cannon cranking out rounds of yellow tracer and high explosive armor piercing rounds. There was nothing he could do except hope that he didn't get hit too badly. At the top of the loop, his airspeed was under 150 knots. It was like he was a sitting target.

Most of the rounds passed on his left side, then a few on his right. Al knew that he was being lined up for the kill. The last chance he had was to get out of the loop and regain some airspeed. Al lit the afterburners. Fuel wouldn't matter if he was shot to pieces.

Valves shot the jet fuel into the hot exhaust of the plane's two huge engines. The fuel exploded and turned into giant blowtorches, each 5 feet wide and 150 feet long (twice the length of the plane itself!). Instantly, Al was kicked back into his seat. The swinging wings of the plane swept back and formed the shape of an arrow for better high-speed aerodynamics. In almost no time at all, it kicked over the top of the loop and dove into the dogfight at over 1500 knots. With the plane's speed still increasing with the dive, Al shut off the burners and tried to slow down before he tore the plane apart.

On the way over the top, he heard one of his Sidewinders lock onto the Flanker that was climbing toward him, so he fired it. Just as the two planes were about to pass each other at almost 3000 miles an hour, his missile impacted on the Flanker's cockpit and blew it to pieces. He could hear some of them hitting his plane.

By this time, the dogfight had descended to almost 30,000 feet, just above the overcast. It didn't take long for Al to line up another Flanker and fire at it from above. Again, the Russian tried to pop out flares to confuse the Sidewinder's heat-seeking guidance system, but again, it failed, and another Russian came apart in a fireball masked by black and gray smoke.

Al leveled out and tried to break through the dogfight again. This time, a Flanker that was about to cross his path from the lower right turned parallel to his flight path, and he was right away in firing position for Al. The Flanker pilot planned it that way. As soon as the Russian's wings were level, it yanked it nose up so fast that its back end kept going forward! The plane flipped itself end over end and now it was facing Al! The Russian and Al fired at the same time. Both missiles missed, and Al went under the inverted Russian with only 100 feet to spare at over 1000 miles an hour.

Now the Russian righted his plane and headed back toward Al. As soon as his wings were level, he was in a perfect firing position. Al knew what was coming. He chopped his throttles to almost nothing, opened his airbrake, dropped his flaps, and pulled back on the stick. The huge F-111 seemed to stop in mid air, with its nose 90 degrees toward the sky. The Russian pilot couldn't believe that the historic fighter-bomber was actually pulling off the maneuver! Of course, Al's engines were starving for oxygen, and he couldn't hold his position for more than a full second. As the white Flanker passed underneath, Al fired his afterburners to keep the plane from falling backwards and blowing out his engines like candles on a birthday cake. The last thing that the Russian pilot saw was a pair of glowing white plasma plumes coming from the Aardvark's afterburners. The Flanker hit the white-hot exhaust and exploded into a million pieces of ash and molten metal.

Al's effort to keep his green and brown monster aloft failed. The afterburners helped, but it wasn't enough to keep the plane vertical with no airspeed, and the plane's nose pitched over violently as it lost all lift. Everything in the world turned red as blood raced to his head with the force of 4 negative g's. Blood filled his eyes. His head ached with pressure from the inside. Finally, the plane's nose aimed toward the ground and it began moving again. The negative g's ended.

More Flankers shot past him as he had appeared to be a sitting duck again, but this time, no one fired. His nose had dropped so fast, and the plane's position so different, that no one knew where it was going next-except Al. He'd done this before many times over the years.

The Aardvark's wings were swept all the way outward again to maximize its lift. Now, the heavier end of the plane, the rear section with the huge engines, sank and the nose began to lift again. Al fired the burners in two short bursts to right the F-111 and get some forward air speed. It was time to get out of Dodge!

Another Russian tried to pull the same trick as the last one by flying in front of Al and then pulling up quickly. This time, Al was still going so slow that he fired off a Sidewinder and broke the white plane's back. It was the first plane he had seen all day where the pilot was actually able to eject.

Now his airspeed was building. He was up to 500 knots and accelerating his way out of the dogfight. Suddenly, a Russian missile flew by his right side from behind. Al popped out ten flares, aimed down toward the clouds, and toggled the afterburners.

Through the top of the canopy, he saw the Russian overfly him and explode as a friendly plane shot it down from its right side. In front of him three more Su-27's were coming at him from 11 o'clock low; almost dead ahead. All three were already firing their cannons, and he knew their missiles were locking on. Just then , the afterburners cut out, and the engines began to shut down. He was out of fuel, soon to be without electricity, and in a very bad situation. At least his helmet was growling again.

Al fired off the remainder of his original 22-missile load. The Flankers tried to escape by scattering into different directions, but there was no hope. The more they spread out, the easier it was for the missiles to lock on to more than one plane. They popped flares like it was the Fourth of July, but the missiles hit home. One Flanker was hit in the engines and trailed heavy black smoke. Another lost a wing and cartwheeled towards Al. The last one lost one of its two tails, but it flew on; damaged and dying.

Al tried to avoid the cartwheeling Su-27, but it was too late. The white plane surrounded by orange flames clipped Al's right rear stabilizer. He was out of gas, damaged, and out of ammunition. The impact sent his plane into a spin, and he fell another 5000 feet until he could recover. By then, he was in the clouds.

The heavy rain sounded like machinegun fire on his plexiglass canopy. The wind was like an artillery barrage buffeting the wounded and falling Aardvark. He had no way of telling whether there were mountains, valleys, or plains below him, and ejecting in a heavy thunderstorm was just asking for a lightning strike. Al decided to ride it in. As the white blanket of overcast turned farther and farther from gray and then into black, he thought to himself, maybe-just maybe-he might make it....

Sunday, December 25, 2011

1st Lt. John Chamberlin -Personal Log

Approximately 50 miles south of Keng Tung, Myanmar

Last night I tried to sleep, but it takes hours for the adrenaline to wear off. It took that long for the wounded Chinese to shut up anyway. Until then, I just sat and looked out over the debris; ignoring the bodies. The sounds of the jungle left weeks ago. Now there's near silence. Sometimes you can hear some dogs, frogs, or crickets, but the jungle noise went as soon as the artillery knocked everything down and the green rotted away. The only natural beauty left was the stars and the planets.

When it's not raining here, you can see thousands more stars than usual. Every once and a while, I could see a satellite cruise overhead, and I had to wonder if anyone could see us? Whose satellite was it? Is it even a military one, or is it some cable network funneling valuable infomercials to folks back home? Just as soon as I'd get interested in some constellation, the Chinese would send off a flare and ruin the view for 15 minutes. It was like counting sheep, but as soon as you got the last one, someone shakes you and tells you you're snoring.

When the sun came up this morning, I gathered everybody together in Rick and Bob's rifle pit. We're going to completely close off the sewer entrances to the bunkers that were blown yesterday. That will leave us with one bunker to the right of the rifle pit, and one to the left. The rifle pit is well concealed, but we'll have to move the bodies from in front of it or else the sight and smell will drive us all nuts. Bob and Barry are going to take care of that.

I think Barry will enjoy the chance to say a few words over them. Prayer helps him. That said, it's true that there are no atheists in foxholes. I know we've all mumbled a few words to the big ref in the sky in the hopes that we don't get kicked out of the game. So far it's worked, at least for those of us still alive.

Anyway, Steve Burns and Steve Hauser will rotate 12-hour shifts as snipers to give us some early warning of the next Chinese attack. The rest of us will move to a 20 hours on, 4 hours off sleep schedule. There's so little left to protect, I think I might as well get some rest.

The rest of the day was quiet. I think the ole Eastern Alliance was licking its wounds. We sort of did too. Everyone's into the regular schedule of things now: try to sleep, wake up, eat at least one MRE, dig, rebuild damage to the tunnel or bunkers, eat, clean weapons, stand watch, eat, and finally try and sleep again. I had a Sloppy Joe with cornbread, a broccoli rice casserole dinner for lunch, and burritos with soft nachos for dinner.

I must have been out here too long. We all must have been. No one minds the smell. The scenery isn't depressing anymore. Morale is neither good nor bad, and worst of all, I think everyone is starting to like the MRE's-a LOT!

SECTION 8

"A weapon not used is a useless weapon."

Monday, December 26, 2011, 1303 hrs. local

Task Force 63, U.S.S. Enterprise CVN

Northern Gulf of Tonkin

T he old carrier's Combat Information Center (CIC) was busy. On the three large blue monitors every ship, plane, submarine, missile, and decoy was visible. There were hundreds of Coalition ships in the area serving no strategic purpose. Officially, they were there to support the Vietnamese against their most recent Chinese invasion.

After months of trying to infiltrate their way into Indochina, the Chinese finally opted for a straightforward advance from their border into Vietnam. The Vietnamese had resisted fiercely, but the attack came in the northeast corner of Vietnam, where there are no natural boundaries to bolster a defensive perimeter. They were unabated until they reached Highway 106 from Haiphong harbor to Hanoi. There, the raised road was turned into a wall; a wall that the Chinese had to advance upon through nothing taller than the paths between muddy rice paddies. It was the perfect killing ground.

Task Force 63 was sent to help out with the slaughter, but once Marines and naval infantry units were landed, there was nothing else to do. The Chinese solved that problem by hitting the tightly packed Task Force with near round the clock missile attacks, air raids, and even a few commando attacks. All were repulsed, but inside the small room where the entire task force was managed, the blue monitors were packed with symbols. Even the computers were having trouble managing all the data.

Radios, telephones, and small conferences with officers were all kept at a low volume, but the sound was still the equivalent of a roaring fan, a jet engine, or a stormy sea. Over all the rumbling mumble, one tired monitor watcher cried out with conviction.

"Vampire, Vampire, Vampire! I have multiple missile launches, sources unknown."

The young technician knew that once he grabbed everyone's attention, he'd better explain his cryptic description of the missile's origins. He also knew that he'd have to do it quickly! His mouse-controlled cursor moved with lightning speed on the monitor he was charged with watching. He clicked the left mouse button and dragged the cursor to the lower right, highlighting an area on the map with a box. Then he clicked a few times to have his display shown on one of the three big monitors in the room. Everyone instantly saw what was going on.

All over Indochina, missiles were being launched. The white lines on the big blue screen moved south from several indiscernible locations on the map. None were being launched from any population center or known Eastern Alliance threat areas. All but one of them were coming from hundreds of miles behind the no-man's land of a front line.

Admiral Sam Pender stood silently in the back of the room. Arms crossed, khaki uniform untucked and unbuttoned, his face went icy and cold under his untrimmed grey beard. He knew what it was. Sam had recently returned to field duty after months of Pentagon desk duty. There was little doubt in anyone's mind that his desk duty was a punishment for the Battle of Yellow Sea that had started the war. A Congressional investigative committee looking into the matter had tried to use him as a scapegoat, but Sam turned it right back around at the hypocritical congressmen and women. He was a temporary national hero among the people, military brass, and even the President, but Congress was not amused.

In Washington, such action usually ended with the firebrand being put out to pasture. This was especially true when upsetting a member of congress, let alone several important ones on a very public inquisition. Sam spent months doing duty assignment after assignment until finally he was sent back to the fleet; most likely to get his less than politically correct and highly vocal nature away from the politicians and his superiors.

Sam was aware of his situation. Now, the crotchety old sea salt had been sent back out to sea. He was back in his element, and filled with the frustrations of Washington, he was even more explosive than ever.

When the missile tracks began sprouting from the jungles of Indochina, Sam knew that the moment of truth was at hand. He'd been through more missile combat than any other American commander, but this was something knew. It was also something that everyone had dreamt of; everyone had their nightmares. Given that the missiles were pouring out of unseen locations deep in the jungle, and given the speed and spread of their courses, they had to be short-range ballistic missiles.

If there were that many in the air, all launched simultaneously, they were either conventional warheads acting in disguise of a nuclear attack, or it was the real deal. There was no point in feigning a fake nuclear attack for the Chinese. It would serve no purpose except to terrify the Coalition, and that might trigger a nuclear first strike on China and the Eastern Alliance. It had to be the real deal.

When the missile tracks came on the big monitor, everyone looked up and recognized it too. This was it. Silence hit the room.

Sam stepped out of the shadow under the red EXIT sign at one of the room's two entrances. Even though his sneakers let him move silently, everyone turned at looked at him. Radios and phones were still mumbling, but he had the room's full attention.

"Alright people. Let's take care of business. I'm sure the folks back at NORAD already have this on their screen too. Pass the word to all aircraft and ships in the area to engage these targets at all expense. That means every available weapon. Get those THEAD (Theatre Air Defense) missiles and Patriots in the sky.

NavComm, I want the entire Task Force to execute evasion maneuver pattern one, max speed; repeat max speed. At thirty knots, were won't outrun those suckers, but we might have a chance with any follow-up launches."

Sam paused and glared that the screen. Everyone else did too. They were just in time to see the first missile separate from its booster. It was the one that had been launched from right near the Chinese frontline. Everyone rushed to relay his orders. One technician glanced at his display and confirmed what was painfully obvious.

"Sir, the computer's classifying the exhaust pattern of these targets as Soviet-built SS-20 mobile IRBM's."

Sam had suspected the Chinese version of the SCUD missile. The SS-20 was similar in that it was a big rocket carried and launched from a truck, but it was far more sophisticated than your average SCUD or Al Hussein missile. It also could carry as many (2) 150-kiloton warheads. That's what intelligence had reported, but Sam had become very skeptical of intelligence reports.

When the missile separated, another white line appeared on the screen as the booster section. Then more lines appeared. The warheads, three of them, had been fired from the missile. Now they were tracking 6 targets. Finally, a Patriot battery almost 30 miles southwest of the targets fired off three missiles. It was too late.

If there was anyone in the room that was still clinging to the hope that this was not a nuclear attack, it disappeared in an instant. A large white circle appeared on the screen, then another, and another. The Chinese were firing their intermediate ballistic missiles at the Coalition front line. The barren, debris-strewn no-mans-land that had been carved across Indochina from the Bay of Bengal in the west to the Gulf of Tonkin in the east was all about to be stripped clean of any Coalition presence.

The room fell silent yet again, except for Sam. He was vocal; very vocal, very loud, very distinct, and very angry. This war started with him getting ambushed and overwhelmed by Chinese and North Korean missiles. Now, it stood a good chance of ending the same way. He was as profane and poignant as only a man who had been at sea his entire life could be. It conveyed his anger very clearly, and focused everyone in his vicinity on their duties. Some imagined that he might explode before being nuked.

More white lines separated on the big blue screen. 48 missile launches were detected and tracked, all within less than a minute. As the missiles entered the lower atmosphere, they expended and discarded their booster packs, then they fired their warheads at the frontline. A few missiles were stopped by Patriots and ship launched THEAD anti ballistic missiles, but there were too many targets to track and kill now. The 48 had become almost 300.

Then, a new threat emerged from the deep. In the northern end of the Yellow Sea, not far from where Admiral Sam and Task Force Stingray had been ambushed last winter, two more locations were sprouting white lines from the middle of nowhere. At sea, that meant that the two Chinese ballistic missile submarines were getting in on the action. They were a few minutes late, but it didn't matter. Not many weapon systems could stop their missiles once launched and on their way to their targets at 15,000 miles an hour.

The war was getting bigger. Sam knew that the Chinese nuclear force was small and closely guarded by the political elite in their military. To have fired off so much of it almost at once was similar to a boxer using everything he had except his knockout punch. No one would fire off as much ordinance as the Chinese were doing without having the intent of going for a knockout. Sam waited to hear that their ICBM's, their hammer, their knockout punch was on its way to the U.S.

He watched the screen as some of the SS-20 IRBM's were picked off by Patriots, THEAD's, and even a laser-equipped 747 that was patrolling the area over the Gulf of Thailand far to the south. Still, the ominous white balls appeared all over Indochina on the monitors. Soon the room that had been dimly lit by the blue monitors became white. Outside, missiles flew faster than bullets trying to hit other missiles falling at even faster speeds. Sam also knew that 1000 miles to his west, millions of people were dying in flames, torn apart by airborne refuse, and poisoned by radiation.

Then the one phone in the room that had been silent finally rang.

Sunday, December 25, 2011, 0004 hours local

NORAD Command facility

Cheyenne Mountain, Colorado

United States Air Force General John McDonald had served for thirty years. His position as Commander-In-Chief of North American Aerospace Defense/United States Commander-In-Chief of Space Command (CINCNORAD/USCINCSPACECOM) was the pinnacle of his topsy-turvy career. Like Admiral Sam Pender, McDonald was old school. He was one of the last of the Cold Warriors. Many saw him as too aggressive, but he had always been lucky and found himself on the side of the victorious whenever he scrapped.

In Vietnam, he had been one of the first U.S. B-52 pilots to bomb Hanoi during Nixon's famous Christmas bombings. On one mission, his plane had taken two direct hits from Vietnamese SA-2 SAM's, and another four near misses. Fuel leaked from the big bomber so fast that it stripped the paint. He brought the plane back and ran out of gas on final approach. Still, with two wounded crewman on board, he glided the BUFF ("Big, Ugly, Fat, Fuck"nickname for the B-52 Stratofortress) into a successful landing. John won a Silver Star for bravery on the mission, and a Distinguished Flying Cross for the landing. Since then, he was almost untouchable.

Just about anything he did was acceptable, within reason. During the Cold War, he served as Air Force liaison to the Army Chief of Staff, and the two men never got along. He was quickly reassigned and never saw another rewarding command until the late 1990's. Then, as had happened with Admiral Pender, McDonald found himself on the career fast track, but it was mainly because he was one of the few high-ranking Air Force generals to have fought in Vietnam and throughout the Cold War. It wasn't nostalgia. It was more of a 'last man standing' situation.

Now here he was, facing the unthinkable. Short and intermediate-range ballistic missiles were raining down nuclear warheads all over Indochina. The Third World War had raged for almost a year, and now, with the bulk of the American and Coalition forces being wiped out, the scope of the Chinese nuclear attack was still unknown. Already, at least 5 minutes had passed since the first missile launches were detected. John knew that in a nuclear war, 5 minutes could be enough time to decide the fate of the world-twice!

Technically, General McDonald was the commander of both NORAD and United States Space Command. The primary charge of both units was to detect and track aerospace threats to North America, yet neither had actually detected anything yet. The Chinese short and intermediate range launches were detected by local Coalition radar centers in southern Indochina, and in Task Force 63 of the northern coast of Vietnam. John had already alerted the President, and various operations were already under way to protect American command and control assets. Still, he could not even confirm that the missiles had been launched or even had detonated! In order to do so, policy required that he be able to provide "Dual Phenomenology."

For years, it had been decided that two methods be used to confirm that a hostile airborne attack was underway against North America. The purpose was to prevent a counterattack from being launched by some sort of technical error, trick, or general system fluke. Hundreds of billions of dollars had been spent by the United States and Canada building immense radar stations all along the Arctic Circle (the Distant Early Warning radar network-DEW Line). More radar nets were built on both coasts of the continental United States (the PAVE PAWS radar sites), and on a Satellite Early Warning System (SEWS) . Hundreds more billions of dollars were spent by Space Command to put four Dual Support Program (DSP) satellites in overlapping, geo-synchronous orbits.

The DSP satellites, loaded with incredibly sensitive infrared detection equipment were supposed to warn of a ballistic missile launch by spotting the hot exhaust plume of a missile's booster. The radar satellites and ground stations would also detect the launches. In the worst case, local military units close to the launches would detect the launches and relay the information to a central command center in Cheyenne Mountain, Wyoming. After that, John was supposed to inform the National Military Command Center in Washington where the President and the Secretary of Defense would decide how to respond to the attack.

So far, only Admiral Sam Pender and the other field commanders could confirm that any missiles were in the sky. Some of the commanders confirmed their detonations with the radio silence left by their own deaths. Still, policy was policy, and he could not advise the President on how to respond without Dual Phenomenology.

John pulled his closest advisors into the war room at NORAD. They ran in and took their seats. Everyone could see the big screen on the back wall through the room's glass walls. There was no reason to hold a briefing, but they needed that second method of confirmation before they could confirm the attack. Without it, the President could not order a counterattack. John pressed a button on the speakerphone and was immediately connected with Admiral Pender on the other side of the globe.

"Admiral Pender, this is General John McDonald CINCNORAD/USCINCSPACECOM, I need you to report the situation to me in your own words, sir."

The phone paused a second, then the static broke with a feint and raspy voice.

"The situation is simple General, they're unloading on us. Your intelligence reports said that they might have about 60 intermediate range missiles, and we've had over a hundred come our way. Your reports also said they had about...hold on, lemme see here. I've got the list...here it is. It says, '...about 150, CSS-6...' short range ballistic missiles. Given your margin of error, we should have something like 3 or 400 more to come. I'd like to meet the guy that dug up all this crap. There're probably a few hundred thousand of our guys out there in the jungle thinking the same thing. Now, I'm looking on one of the screens in my CIC and it shows a missile coming right at us. I've got about a minute or two left unless somebody gets lucky. How's that for your report?!"

John had had nightmares about this moment. Every commander who was ever part of the possible nuclear war had also. He swallowed hard and explained the situation to Admiral Pender.

"Sam, the Russians have blinded the infrared sensors on our DSP satellite with a pair of ground-based lasers left over from their Star Wars programs back in the 80's. China's too far south to detect launches with our ground stations, and something's going on with our other satellites. The President's not going to order any counterattacks until he gets his Dual Phenomenology. Until they get something within range of those ground radar stations in Canada, you'll be the only one counterattacking. I'm sure you've realized that your rules of engagement allow you to return fire on the sources of nuclear attacks with any weapons of mass destruction that you deem fit. That includes nuclear."

There was another second's worth of static as the communication, unable to go through satellites for some reason, passed from city to city around the globe. Then Admiral Pender's response came back.

"Well, you're luck is changing there, John. One of my picket destroyers was able to stop that inbound. Looks like I get to live a little longer. That'll really piss off all the right people! Listen, I've already got planes in the air and on their way to the coordinates where the Intermediate missiles came from. I'd use the Tomahawks, but the launchers might drive away by then. You Air Force boys'll be happy to here that there's at least one occasion where a pilot will be better than a drone. Besides, I wanna hold onto the Tomahawks in case they decide to sortie some surface units on us, too. I've also got a strike prepping to take out their known forward command and control areas. I doubt they even knew the attack was coming, but just to be sure, we're gonna take'em out. I've also had an S-3 Viking on ready reserve with nuclear depth charges just in case. It's already on its way to the first of those two subs that were throwing out the SLBM's (Submarine Launched Ballistic Missiles). My screen shows those are on their way to the Philippines and unknown points in your direction. Better have somebody try and scramble out of Clark Field and Subic Bay; hell, maybe Wake Island and Pearl Harbor too, for that matter."

As Admiral Pender spoke, General McDonald's staff was taking his advice. One by one, people ran in and out of the room to make things happen. Soon alarms were going off all the Philippines. Clark Field Air Force Base and Subic Bay Naval Station immediately put began putting planes in the air, and ships on their way out to sea; most with skeleton crews. There was no time to wait for everyone to come back from their shore leave.

Sam continued.

"Something else struck me, John. You said the Rooskies zapped our DSP's with lasers. That's why you didn't spot the heat from the missile's boosters, but I think the important thing here is what it tells us. That is, the Russians knew about this before the birds went up. They're in on it with them Chinese. If that's the case, you could be staring down the barrel of a full scale ambush, and you wouldn't even know it."

Time was passing far too quickly for such debate. General McDonald finally heard an alarm go off in the room on the other side of the glass. The Distant Early Warning line of radar stations along the Arctic Circle had detected multiple targets closing in on North America at over 25,000 miles an hour. They were being picked up over Russia; confirming Admiral Pender's suspicions of Russian collaboration, but if the Russians wanted to toss their hat in the ring with the Chinese, why weren't they launching their missiles too?. At least the President would finally have his Dual Phenomenology!

McDonald picked up a phone to the National Military Command Center in Washington and immediately reported the news. Early indications were that the incoming targets' radar signature was consistent with Chinese DF-5 and DF-6 ICBM delivery vehicles. Intelligence reports had estimated that no more than 10-15 of these missiles existed, but there were 20 on the screen. That number almost instantly doubled when each of the ICBM's used the last of the fuel in their booster stages and discarded them. The boosters continued on course with the secondary stages of the ICBMs, but at a dwindling speed, and surrounded by debris from the separation. Total radar targets: 47.

Ever since the North Koreans fired an ICBM at Los Angeles early in the war, the 1980's-era ballistic missile defense program was immediately revived with a surge of new funding. Time was not friendly to the various facets of the program, and only a few operational platforms were available for use when the Chinese decided to turn the conventional Third World War into the second nuclear war.

Short range Anti-Ballistic Missiles (ABM's) were deployed on ships around the world, including Admiral Pender's Task Force 63. Loaded with brand new 'BOB' guidance software, one such missile, a Theatre High-Energy Air Defense type, had actually saved Pender's Task Force already. A modified 747 with a sophisticated tracking system and a 10' wide laser mounted in its nose had also stopped other Chinese submarine launched missiles. This stopped a handful of the missiles, but far from all of them. All of these defenses were enabled and sent into action as soon as the first Chinese missiles started launching from the jungles.

Now that a full-scale Chinese nuclear attack was under way, and it had been confirmed through two separate methods of detection, more strategic defenses were enabled automatically. Some American military communications satellites had also been equipped to focus microwave energy at targets, not in an effort to make them explode or disintegrate, but rather to short out the sensitive solid state electronics inside that manipulate their controls. A pair of these was close enough to the incoming targets to track and confuse the Chinese ICBM's, but only one was thrown off course. That missile never achieved orbit and simply continued onward into outer space-forever. Total radar targets: 46.

When the missiles left the Russian Confederation and flew over the arctic ice cap, they separated once more by discarding their secondary booster stages. Again, the radar targets increased. Without the infrared signature from a booster's ignition, there was no longer an accurate way to discern the boosters, debris caused by separations, or the increasingly important warhead delivery vehicles. Total radar targets: 71.

Over the ice cap, the much-vaunted American Star Wars missile defense system finally was able to get to work. Two brand new satellites were waiting. They sat over the North Pole in eternal sunlight using huge solar panels to maintain an ever-ready charge in their batteries. Each was bigger than a semitractor trailer truck, and armed with the same hydrogen/fluorine laser as found on several modified 747's that had been strategically placed around the globe.

As big as they were, there was no room left to attach an adequate targeting and tracking system. Instead, they relied on a data downlink with U.S. Space Command in Cheyenne Mountain, Wyoming. The data was sent. Both fired, and both missed.

A small group of technicians huddled around a single 22-year-old 1st Lieutenant and his small console in the back right corner of the command center's main room. He typed in the data again and fired. There were no results. With just enough power remaining in the batteries for one more shot, the data was adjusted and sent again. This time, the lasers both found their targets.

Traveling at over 25,000 miles and hour, the missiles drew closer and closer to the invisible beams of energy. As they did, the beams became more and more intense, finally, the targets exploded. When they went off, both laser satellites were knocked off line, and some of the other incoming targets changed course radically. The young Lieutenant didn't know it, but he had just saved millions of people, specifically those living in Jacksonville, Tampa Bay, and Miami, Florida. Total radar targets: 65.

While the third and last rounds of lasers were fired, the next line of defense was lining up its targets. A small automated space station of three satellites, all three even bigger than the laser defense satellites, was taking aim on its first prey. The space-based particle accelerator also gained power from the sun, but the two smaller satellites docked with it served as backup batteries to augment its own solar-panel-charged system on board. It also had its own targeting system. Though limited in range, it could tell the difference between a wrench and a warhead; something very few other defense systems were able to do after the missiles finished their boost phases.

When the first radar targets came within 300 miles of the satellite/space station, it detected and took aim on them. A silent and invisible beam of electrons reached out, but was pulled down and away from the target by the gravitational force of the Earth. After a second and a half, it stopped, realigned itself, and tried two more times. All three attempts failed. Targets began passing by. Finally, the beam connected with a target only 70 miles away. The invisible rows of electrons eroded their way through the thick re-entry heat shield of the warhead setting off its 150-kiloton nuclear payload. The blast put the satellite out of commission, but, again, several other targets disappeared or bounced off course. Total radar targets: 61.

While the laser and particle beam satellites were working, controllers at U.S. Space Command facilities around the globe began overriding the control codes on communication, navigation, research, and weather satellites. Once control their controls taken over, the Space Command technicians plotted out collision courses and sent the satellites on their way. They had become Kamikaze killer satellites. All over the northern sky stars flickered, planets shined, and satellites rammed. Total radar targets: 43.

It was time for the more experimental ballistic missile defense satellites to have a try. A space-based X-ray laser aimed at 3 separate delivery vehicles and was able to damage the guidance systems on 2 of them. An electromagnetic rail gun, almost 150 yards long and powered by the sun, fired 100 acrylic pellets at speeds in excess of 30,000 miles an hour. When they hit their the delivery vehicles, they went right on through the outer shells, through the tough heat shields on the warheads, and shattered the metal components inside; not by mass, but by the sheer velocity of objects colliding at almost 50,000 miles an hour (13 miles/73,000 feet every second).

Time had almost run out. The noses of the ICBM's were popping off and their warheads were being aimed. When the delivery vehicles were within range of their targets, they repositioned themselves and began firing the warheads. Total number of radar targets: 121.

It was time for the last ditch defenses to go into action. Patriot and THEAD missile batteries positioned all over the United States began launching their missiles. Even the fastest computers and the most reliable radars could never accurately steer the ABM's as they closed in on their targets at over mach 7. With the missile batteries placed in target areas, the crews were inherently committed to hitting their targets. Failure meant certain death.

As the first wave of ABM's took their toll on the incoming warheads, Chicago, Cleveland, Detroit, and Pittsburgh were saved. There was barely enough time to get off another shot, so every missile battery's crew opted for a shotgun effect. They launched all their missiles hoping one would come close enough to save their lives. New Orleans and Charleston were saved.

(Detonation)

Seattle was hit first. A Soviet-built 150-kiloton warhead surrounded by the thick black cone-shaped heat shield came toward the city at over 15,000 miles an hour. No one saw it in the cloudy Christmas night sky. When it was only 1348 feet above the Bremerton Naval Yard, a ball of plastic explosives received electrical pulses from an onboard battery. The explosives went off.

The energy of each was tamped by a hardened steel shell, and the bulk of their energy was reflected to the inside of the plastic ball where a plutonium core, surrounded by thin layers of deuterium and tritium, was hypercompressed. In a billionth of a second, the atoms of the fissionable materials were crushed to the point of breaking. As the first one broke, all 235 protons and neutrons of a single uranium 235 atom came apart. Their bond had been broken. The energy released from the tiny atom was virtually nothing, just enough to send the protons and neutrons racing outward. The nearly immeasurable mass of each proton and neutron meant that even the smallest amount of energy sent them hurling at speeds close to that of light. Around the broken uranium atom, other atoms were reaching states of hyper-compression.

When each proton and each neutron of the first broken atom collided with another compressed uranium atom, the atom exploded and released another 235 protons and neutrons. The chain reaction continued unabated in such a short period of time that the warhead, traveling at over 22,000 feet per second, had barely moved a 1/16 of an inch before it was over. Every atom had been crushed, broken, and disintegrated.

Protons, neutrons, flew outward and began breaking atoms and molecules from less radioactive materials. Nitrogen, oxygen, water, carbon dioxide, and even air pollutants continued the reaction. With every break of an atom, a little more energy was released. The softball-sized core of the nuclear warhead was turned into a fireball of energy, protons, neutrons, and radiation.

While the incredibly small atomic particles and radiation weren't visible to the people who saw the explosion, the light was. In fact, the light was so intense that anyone in the city who was looking in the direction of the blast 10 miles west of the city was instantly sunburned and blinded. For most, the blindness was temporary, but the closer they were, the more permanent was their damage.

Of course, the closer they were to the blast, the less they really had to worry about blindness. The light was so intense that it turned the night into day for almost 2 full seconds. It came in through windows and burned witnesses wherever clothing didn't cover their skin. It set flammable materials in houses and offices within 3 miles on fire. Everything that wasn't in the shadows was burned. Then, without the proton and neutron-rich radioactive material to fuel it, the explosion ended. The fireball remained as oxygen in the atmosphere burned from the heat of the initial detonation.

It took no time at all for the fireball to expand to over 1580 feet in diameter.

When the shock wave from the airburst reached the shipyard, it crushed ships, buildings, piers, and even the ground itself, then it bounced straight up; back into the sky. It passed through the fireball and sent it spinning in on itself like a fiery donut in the night. The donut turned itself inside and out from the bottom up and rose almost 2 miles before it was swallowed by the dust and debris from its blast.

For a brief moment, there was a 530' wide crater between the piers. No one ever saw it. There was no one left alive, and the water from the bay quickly filled it in. Everything in a two-mile radius of the water filled crater was crushed, leveled, shattered, or simply blown away by the 250-mile-anhour, 1200-degree wind. Even asphalt on the roads had been pulled up and thrown away. Save sinking and smoldering piles of twisted steel, the area was clean. 1.8 seconds earlier those same piles of crumpled steel had been naval warships and submarines. The fleet was gone.

Downtown Seattle was nearly untouched. Almost every single western facing window was blown out, there were innumerable fires ranging in scale for someone's apartment draperies to oil refineries and chemical plants. The city's fire department tried to tackle some, but it was hopeless. All they could do was try and prevent them from spreading. Within an hour, Seattle's neardaily rain rinsed the sky of debris, dust, and ash. As it hit the ground, the drops splashed black.

Not even 30 seconds after Seattle, San Francisco took a direct hit. Another 150-kiloton warhead detonated a quarter mile above Yerba Buena Island. Most of downtown San Francisco and Oakland were completely destroyed. Hundreds of thousands died instantly from the heat and blast. All of the port facilities were swept away. All of the bridges tumbled into the bay, including the Golden Gate and Bay Bridges. Even their towers toppled. It was like the city never existed.

St. Louis was hit almost at the exact same instant. This time the warhead exploded nearly half a mile in the night sky-almost directly above the famous arch. The blast crushed all the buildings in downtown St. Louis and East St. Louis, but the warhead had gone off too high, and the damage was half as severe as San Francisco. The city was toppled, but not swept clean. For a while the Mississippi's flow stopped as the shock wave actually pushed the water north and south of the city. This caused the Chain of Rocks bridge several miles north of the city to be lifted off its base and knocked over by the river a few minutes later as it returned to its normal flow.

In Seattle and San Francisco, fires popped up all over the city. Here, heavy winter snow kept things cool enough to slow their spread. While the orange mushroom cloud disappeared in dust, the city's nighttime skyline was gone. All that remained was the silhouette of the blackened arch lit by the fire in the dusty sky above.

Philadelphia, the nation's original capital, suffered next. This warhead, another 150-kiloton Soviet-built black cone filled with radioactive material, plastic explosives, and sensitive electronics, exploded over Camden Petty Island in the middle of the city. The flash of light energy from the fission-style atomic chain reaction set the old city ablaze. Downtown, piers were washed away, skyscrapers shattered and fell.

Prior to the Christmas night nuclear massacre, a seven day indian summer with temperatures in the 50's melted the snow that had fallen earlier in the winter. The ground was clear, cool, and dry. The fires would spread unstopped for hours. It didn't take long for a firestorm to form. As the flames from different fires joined together, they formed a single large fire around the pile of dust and debris that was once one of America's most historic places. Flames pulled in fresh air from the surrounding countryside. More and more air was pulled in, the flames grew more intense until finally they arced almost a mile into the sky and sucking in air that was more like an inescapable 60 mile an hour wind. Everything burned. The west side of the city was in ashes, the east side in dust. A little over a hundred miles away, the First Lady watched the horizon glow and flicker to her northeast. She wondered, how much longer would she be spared?

In New York, luck was taking a strange shape. The warhead that had managed to impact the city carried no nuclear warhead. During the Soviet Union's Cold War buildup, there wasn't always enough nuclear material available. Rather than leave a missile empty or launch less than fully loaded, many warheads were loaded with other munitions; some conventional, some chemical, and some biological. Through a combination of luck and beaurcratic snafu, New York's harbinger of death was mislabeled during manufacture. When the Chinese bought their missiles, no one had any idea that there was nothing nuclear inside one of the warheads in one of the missile's nose cone delivery vehicles. All they knew was that the serial number A06171 indicating a nuclear payload. No one followed-up on the smudged last digit. That 1 was actually a smudged I indicating inert or non-nuclear payload.

New York's warhead didn't detonate over the city. Instead it impacted 2 blocks north of Carnegie Hall, at the southern edge of Central Park. The crater was only 15 feet wide, but the bottom was almost 70' deep. Dirt, asphalt, cement, wires, and pipe fragments surrounded the surface of the area. Down below, a single gallon of DSKVD34 concentrated nerve agent dripped out of the cracked black cone. The clear drops soaked the soil and mixed with the winter's salty road slush that was seeping into the hole. A deadly mixture then spread down through the soil and into the groundwater. It slipped its way into the local waterlines and sewer lines. Before dawn, hospitals would fill with over 250,000 people who were dying from exposure. Hours later, the city's water had to be cut off. 8 Million people had nothing to drink, nowhere to bathe, and sewage was almost instantly unbearable. Most would, however, live to tell about it.

Shortly after New York, Los Angeles took a near miss. High in the mountains to the northeast, between Strawberry Peak, Mount Gleason, and Pacifico Mountain, over the small town of Singing Springs, another 150kiloton warhead came down. The warhead was programmed to hit between Los Angeles and Long Beach, and explode at 1735 feet above sea level. The high terrain where it actually hit meant that the warhead impacted the earth, plowed 100 feet into the top layer of sandy loam, and exploded just above the hard granite of the Great Southern California Batholith geological region. With nothing but solid rock below and around it, the explosion was reflected in three directions rather than the full 360 degrees. It also sent twice as much force straight up into the sky. Unlike the other mushroom clouds, this one had carried so much of the sandy loam that was above it high into the sky that there was very little flame; just a big, tall, thin mushroom cloud with a stem twice as high as any other that had hit the U.S.; almost 25,000 feet high.

The mountains also protected the city from all but the comparably minor effects of minor burns, thousands of broken windows, and temporary blindness of anyone who happened to be looking to the northeast. Also unlike the other explosions, the heavy amounts of sandy load that had been tossed into the air by the below-ground detonation dumped thousands of tons of radioactive dust and fallout in a plume stretching all the way across the state and into Arizona. Desert populations rather than urban ones suffered mostly from this blast and predominately from the radiation. Still, if the missile had been launched 1/10000 of a second earlier, its one surviving warhead would have hit its target and the casualties would have been untold.

Command and Control was temporarily lost when Cheyenne Mountain took a hit. The facility had been built to take direct hits from warheads much bigger than 150 kilotons, so the people inside were able to survive. Outside, all communications with the facility were destroyed, along with the entire city of Cheyenne. The city's small fire department was unable to quench the post detonation flames, and the resulting firestorm sucked in all the flammable debris and oxygen it could get. The city was completely burned to the ground. Even the roads melted from the 2000-degree heat.

The last to die were also the last saved. San Diego was just too far for an accurate guidance system, and it suffered terrible damage from the resulting near miss. The last of the 150-kiloton warheads to hit the U.S. overshot its target area in San Diego by over a mile and a half. Instead of exploding in the air to create a larger blast area, the black cone splashed down south of North Island. There was typical damage from the searing flash of light-lots of fires and burns on everything not protected in the shadows or by clothing. All the southwestern facing windows were knocked out in the city. However, the explosion was far enough away from the city to cause very little in the way of serious blast damage. The worst damage came from the surf.

When the warhead splashed down, it exploded underwater. Millions of gallons of radioactive water were tossed a mile into the sky. The shock wave also sent a 30' high circular wave in all directions. When the wave hit North Island, it swept across and washed everything even remotely loose into the bay on the north Side. Meanwhile, a small portion of the wave entered San Diego Bay past Point Loma. Here, it was compressed and amplified by the tight geographical confines. It didn't take long for the water to breach the banks and wash all over the area north of the Bay. The Airport was immediately swamped, and airliners were tossed about like toys in a tub. The wave also flooded downtown and completely washed away the famous convention center. Another part of the wave flushed over the jetty between Imperial Beach to the south and North Island. The Pacific Fleet was damaged as ships were tossed about while still moored to their piers. Finally, the radioactive water came back down from the mushroom cloud and irradiated all of North Island, all of the Pacific Fleet, and a good portion of the downtown area.

Meanwhile, the President had decided that he would wait to see how much damage was inflicted before deciding on an appropriate level of retaliation. If none of the missiles had gotten through, and he had ordered a total strike, he would be the first billion-person murderer even though he was the head of a nation that had been brutally attacked. There was no doubt that the vast American nuclear arsenal would survive for the retaliation. Whatever method of counterattack he decided upon would serve only as a signal to the Chinese as to how much damage they could expect in return for their actions in the present and future.

With the damaged assessed, he read the different War Plans that had been prepared decades earlier. Each plan was pulled from a computer that took the location of all American nuclear forces around the globe, their stockpiles of weapons, their capabilities, and the location of likely targets all organized in strike packages. If the President wanted to respond in a missile for missile, warhead for warhead manner, there was a plan. If he wanted to inflict the same number of casualties that had been inflicted on the U.S., there was a plan. There were even plans to inflict twice as many casualties, three times as much industrial destruction, and ten times the command and control damage. Whatever he wanted to target, and to whatever degree he wanted, there was a plan.

The President reviewed the options once more. He had been flipping through the files since the first explosions in Indochina were reported. Now it was time. In fact, he may have already waited too long. If the intelligence reports as to how many missiles the Chinese had were grossly inaccurate and understated, then there very well might be untold numbers still over there, ready to launch a second strike. There was no telling what might lie ahead. The Chinese had taken the initiative, they threw the first punch. It had landed, and now it was time to swing back. He couldn't hope to turn the other cheek and pray that the Chinese had emptied all their silos. If he took that risk and was proven wrong, how many millions of Americans would die?

He had to launch a counter strike. The President decided not to hit the Chinese with everything in the arsenal. That would leave nothing to deter the Russians from attacking. Without the threat of retaliation, they would have no fear, no reason not to let loose their destruction on their old foe. Clearly the Russians had given aid to the Chinese. Because of their apparent permission to let Chinese nuclear missiles fly over their country, the Russians could not be trusted explicitly. Besides, there was no need to use the entire arsenal.

Spread out in the South China Sea, the Yellow Sea, and the Philippine Sea, 8 American Ohio Class Ballistic Missile Submarines (SSBN's; i.e.: "Boomers") had crept into firing position. Each submarine was loaded with 24 Trident D-5 Submarine Launched Ballistic Missiles (SLBM's). Each missile carried 10 independently targeted warheads with enough range to reach all the way to the Russian border. All total, the subs were loaded with 1920 warheads. The closer their targets were to the subs, the less reaction time the Chinese would have if they still had missiles for a second strike.

The President reviewed the plans. He knew that he had to send a clear message to the Chinese that they could not fight and win a nuclear war, he would have to inflict at least twice the casualties, twice the damage. Given the 4 to 1 population ratio of Chinese to Americans, he would have to kill 8 Chinese for every American. Computer models indicated that 25-50 millions Americans were already dead or dying, and another 15-25 million were at risk from all kinds of other damage including radiation, lack of heat in the northeast, food and medical shortages, etc. Total casualties estimated 75 million of 280 million. The Chinese would pay with the loss of at least 600 million people; nearly half their population.

The President wondered if that would be enough to stop them. Would it be enough to end the war? Would the other members of the Eastern Alliance end the war if he killed 600 million? The numbers were incredible, but there was no time to consider them. He had to end the war, now, and by whatever means possible.

The President flipped through the War Plans and strike packages again. The only way he could be sure that the Chinese would not attack again and the rest of the Eastern Alliance would capitulate was to inflict devastating damage. He remembered watching a Chinese general give a TV interview at Taipei Airport in Taiwan even while it was under attack. The general was asked about the huge numbers of casualties that they had taken in the war, and his response was blank, stoic, and genuine. "...So we lose a million or two." The President saw that single interview as a testament to Chinese will, to the philosophy, and as a window into a different cultural perspective on human life. In the same way that the United States would never surrender, the President felt that the Chinese would be equally stubborn. It would take more than 600 million Chinese deaths to end the war. He was sure of that. In fact, they might fight even harder; Americans would.

Still, the order did not come easy. He kept telling himself that there had to be a better way. Even if he could make contact with the Chinese government, why would they negotiate or surrender unless he had already demonstrated his resolve to use the nuclear hammer. Simply by attacking first, they had already proven that the threat of a nuclear counterattack was considered acceptable.

The National Military Command Center was full of military and civilian leaders. Even the people operating the myriad of communications consoles were no lower in rank than Captain. While the warheads were landing in the U.S., and millions of Americans were dying, the room was silent, waiting for the go ahead to initiate a counterattack.

The President's eyes were clinched shut. There had to be a better way than killing half the Chinese in the world, but he couldn't find it, and none of the plans he had been paging through gave him any good options either. If he waited much longer, the Russians might publicly offer to shield the Chinese by joining the Eastern Alliance. Then he would have to either face a Russian attack, or fight the war indefinitely, and all the while, the Chinese could fire missile after missile at the United States without threat of retaliation. He looked around and saw the professional faces in the crowded room. There was vengeance behind their eyes. The President asked openly if there was any other way, but none of the experts had another solution.

With the continued silence, and the drooped shaking of heads, he pulled out a war plan and issued the command.

"I want history to record that this decision has been taken with the utmost of consideration. I believe that the only course of action remaining is to begin a counter strike immediately. This strike will have to be devastating enough to deter our enemies from attacking the United States ever again. It will also have to be done immediately and without warning to have such an effect. I bear the responsibility of being the sole person ordering this attack, and I want that conveyed to everyone involved as the order is passed down to the men and women on those submarines who will carry it out."

No one moved or spoke. He let the words sink in, and then he continued.

"I also want it to known that we have endeavored not to use weapons of mass destruction of any kind in this terrible war until now. None of this will matter to the hundreds of millions of people who will die in just a few minutes. They will never hear my claims. It won't matter to the friends and families of those people. It certainly won't matter to the millions of people who have already suffered and or perished. They say that history is written by the victors, and that might be. It might be that we are even considered heroes, but we should all bear in mind that if we do not make this attack, if the attack fails, if it isn't conducted with the utmost urgency and ferocity, then there may not be any victors to remember history. Once again, this is my call."

The room was still silent. Many heads bowed in personal prayer for the dead, dying, and soon to be incinerated. The President let his speech set in then he continued.

"Execute War Plan Operation Tango Tango Oscar. I want the four Boomers that are closest to the targets to be the primary attack units. They are to expend all, repeat all ordnance. The Boomer nearest to Task Force 63 in the Gulf of Tonkin is to be released to that Task Force for tactical use by Admiral Pender against targets of his discretion. Admiral Pender's Task Force is to change all current operational objectives to secondary objectives. Effective immediately, Task Force 63 is to do whatever is necessary to remove the Eastern Alliance nuclear threat in Indochina. I want them to take out those mobile missile launchers in the jungles first, and then get to work on all their units in the field."

He paused for a moment and studied the map of Eastern Asia.

"I want Pender to take the two Boomers in the Bay of Bengal and use them too, but let him know that I prefer they be used only as a reserve, and I want to be notified if he intends to put them into the fight too. I think he'll only need the one sub that's right by his Task Force right now."

No one in the room was moving or speaking. Everyone was a zombie. They all had the notorious "thousand yard stare," but in this case it was more of a ten thousand mile one. The President was satisfied that his orders were complete enough, so he nodded and told them to make it happen. In a flash, every communication device in the Center went to work. On the other side of the globe, low frequency radio waves went out to the Ohio Class subs bearing Emergency Action Messages (EAM's) from the U.S. east coast. It was payback time-the likes of which mankind had never even accurately conceived.

It only took a few minutes for all of the codes and targeting data to be understood, confirmed, and sent to the missile's computer systems. As soon as the EAM's were received, the submarines had begun their quiet rise to periscope depth. The countdown began. All six of the submarines would begin launching their missiles at the same time; only 4 minutes after the President gave the order. Once the data had been entered, and the subs were in firing position, everyone waited for the seconds to tick by. Hardly a soul even breathed.

In the National Military Command Center, the clock on the wall finally snapped to 12:34am local time. On the other side of the world, the submarines went to work. Enormous Trident II D-5 SLBM's (Submarine Launched Ballistic Missiles) were flushed from their tubes with compressed steam from each submarine's nuclear reactor. The steam catapulted the missiles, one by one, through 60 feet of water and another 40 feet into the air. When each missile escaped its blanket of steam and began falling back to the sea, small sensors detected the change in g-forces and ignited the solid rocket boosters.

At various points in the Philippine Sea, the Yellow Sea, and the South China Sea, in the middle of the afternoon, white, gray, and black waterspouts were spiraling out of nowhere. At their crests, 44 foot long, 6-foot wide black and white rockets trailed a glowing white and orange blossom of fire. 144 Missiles were launched in only 3 1/2 minutes. Their internal computers, fresh with new target data, immediately took over and arched the SLBM's toward their targets. A minute later, the nose cones were popping off, and warheads were being fired at military and industrial targets all over China; predominately on cities and bases nearest to the coast. 1440-man-sized black cones sprinted to their coordinates at over 20,000 feet per second.

In the National Military Command Center, there was stillness once again. They watched a data downlink from Task Force 63's radar coverage of the area, and that of several U.S. bases in Japan. There was no time for a Chinese or Russian response. Some of the missiles were still being launched when the first of the black warheads rained down and detonated. The monitor's yellow outline indicating the coast of China was spotted, then speckled, then blanketed with over 1400 white circles. Each circle represented a single 475kiloton explosion from a single warhead. The black cone-shaped warheads were detonating. Millions were dying. In less time than it took to issue the nuclear attack order, it was over.

On six Ohio class ballistic missile submarines, missile tubes that once carried China's fate were now filled with seawater. Their hatches were closed, and the subs snaked their way back to the west coast of the U.S.

On one other submarine, the strike was taking longer. Task Force 63 had sent multiple reconnaissance aircraft into northern Indochina in an effort to find the mobile missile launchers that had carved a 5-mile wide no-man's land across the peninsula. Each time one of the large 12-wheeled trucks was spotted, the coordinates were radioed back to the Task Force. Then the 560 foot long U.S.S. Alabama, a boomer, opened a hatch while still 60 feet underwater. Soon after there was the familiar blast of steam, a long white plume. A Trident II D-5 missile spewed white-orange flame from at its peak. Less than 3 minutes after launch, the missile's 10 warheads, fired in a shotgunlike pattern, erased the jungle from where the Chinese mobile missile launcher was reported. 49 Minutes after the first Chinese missiles had come from the jungle; the last of their launchers was destroyed.

The American strike had been aimed at destroying China's military bases, ports, communication centers, and industrial areas. Many were in the cities along the coast, a handful were farther inland. Most were also imbedded in highly populated areas. While the U.S. President had hoped to keep the counter strike balanced to twice the effect that was impaired on American citizens, given the population concentration in China's targeted areas, 11 died for every American that was either killed or dying. Including the troops killed in Indochina by the U.S.S. Alabama, 829,000,000 Chinese were dead or dying. Unknown numbers of Vietnamese, Laotian, Burmese, and Thai civilians were never heard from again. Anywhere between 5-20 million people is the best estimate ever determined. 3,119,213 Coalition troops were killed from the initial Chinese mobile missile attack.

The American response to the Chinese attack was based on early U.S. casualty estimates, but those were grossly overestimated. The outset of the war caused Congress to learn a new fiscal responsibility immediately. The bulk of new Congressional budgeting refurbished and restructured the American health care system. As a result, it was able to save millions of people who would have died elsewhere. The initial estimates of American casualties (upwards of 75 million), were grossly reduced. 23,008,743 Americans died. 57,061,682 would suffer from wounds for the rest of their lives.

There never was a military response from Russia. Instead, it condemned both nations for their actions and offered aid. Still, the condemnations were weak and never followed up by real action. They were more a matter of political requirement and historical positioning. American resolve to defend itself and return fire, even nuclear fire, was very VERY clear.

The Chinese government was silent. Reports coming out of the barren nation were scattered and inconsistent. There was no surrender. There simply was no one left to either fight in her defense or to sign the document of capitulation. Instead, the Coalition signed an armistice with the remaining Eastern Alliance nations 3 days after the nuclear phase of the war.

North Korea didn't surrender, but instead agreed to have multi-tiered free elections beginning with village mayoral contests, and followed by higher and higher levels of government every six months. Middle-Eastern members of the Alliance agreed to pay restitution for the war to a Coalition slush fund, but the restitution was payable in oil based on pre-war values. Given that oil was more prevalent than water in most of those nations, they hardly paid the true price at all. It was fair enough since their participation was minimal beyond logistical. In the end the armistice was fair, and created little animosity, or food for another global catastrophe.

America survived, but liberty would forever need to be defended. As such, unlike the post-war period in other American wars and conflicts, she finally learned her lesson. America would forever fund her strong military, not to conquer or destroy, but to serve as an instrument used to defend the peace and liberty that had cost so much in blood.

The war was over.

Monday, December 26, 2011

1st Lt. John Chamberlin -Personal Log

Approximately 50 miles south of Keng Tung, Myanmar Merry Christmas.

Sometime after 1300 last night, the war finally took that last step towards oblivion. I was drifting somewhere between sleep and consciousness, just like everyone here has since we first arrived. We were lying on the bottom of what remains of the sewer. Steve Hauser's leg was still pretty bad from the machete that tagged him the other day, so he was keeping his 12 hour watch from Rick and Bob's rifle pit. The rest of us were scattered around, resting, thinking, eating, meditating, etc. It was actually pretty quiet. The artillery and mortars were had been as sporadic as usual, but no more and no less.

I had just finished making my middle of the night rounds; checking on everyone. As I turned and stepped into the sewer, I saw this glow light up the tunnels and the inside of the rifle pit. Almost immediately, the glow became a bright light. I thought for a moment that there was a spotlight on our position, but when Steve started screaming, I knew what it was. We all did. Everyone curled up into balls and leaned into the walls of the sewer. Steve was still stumbling around and screaming in pain. His face was burning right in front of my eyes, but I had to cover up. There was no time to help him. The bright light was already coming in through the tiny gaps in our roof. Before I closed my eyes, I noticed the dried out bamboo, saplings, and palms that made up our roof igniting.

No sooner did I get my headgear on and my body on the floor next to the sewer's wall, then there was the blast. It was like being in a hurricane. The burning roof cover was ripped away, and I could hear the shock wave pushing all the debris that was around our position over towards the Chinese lines. The wind was so strong that it pulled up all the loosed MRE wrappers, empty magazines, spent shell casings, and pretty much anything that was loose in the sewer right up and out. The light went away, then the wind died down, and the air choked with dust. Bob and Rick took care of Steve. His face was badly burned, mostly 2nd degree. He was also blind from the light. We'll see if his sight comes back.

Where we used to have a tunnel, we now had an open trench. Above and probably three or four miles away, I watched the orange and red fireball twist into a mushroom and roll high into the sky. In less than 2 seconds, we had gone from no-man's land quiet to nuclear numbness.

The Chinese never followed up the attack though. I guess our lines were close enough that they probably nuked their own guys just to get us. It wouldn't be the first time generals wasted their own guys for better results. I think the correct term is acceptable losses or collateral casualties.

I had the guys quickly scrounge up anything they could and put another roof over the rifle pit. We layered it with ponchos and we'll use that as a place to get away from the fallout for as long as we can.

SECTION 9

"Another New World Order?"

Tuesday, December 27, 2011

1st Lt. John Chamberlin -Personal Log

U.S.S. Enterprise

Somewhere in the Gulf of Tonkin

Dennis came over this morning, just before sunrise. We're pulling out. Our position is too close to the irradiated area, and besides, the Chinese have had it. They're not surrendering, and they're not retreating. He says that most reports are more like desertion en masse.

I told the guys, but there wasn't any cheering. There was a little disbelief. Everyone, myself included, is wondering why the hell we were here anyway?! Of course, there couldn't be a lot of cheering anyway. There're only a few of us left; myself, Bob, Rick, Steve and Steve, Barry, and Keith. Everyone else is either dead, missing from last night, or I lost them along the way.

Having lost people in six major engagements, I don't think any of them died for any cause bigger than each other. I can't remember any of them whispering that their death was for mom, apple pie, the American way, national interests, or the politics of Saturday night. They were all professional Marines; each and everyone. They died trying to keep each other alive in dangerous situations.

An Osprey came in and picked us up a few hours later. The new guys that got dropped off a few day ago are all gone. I can't even remember their lieutenant's name. I can't even picture his face for that matter. We put the remnants of three other troops on the same Osprey as mine. It should have been 150 men and women, but there was still room for more. Even with the crew, there were only 21 people on the dirty gray bird.

The Reagan's getting towed back to Subic Bay for repairs, so they dropped us off on the Enterprise. It's a lot different than the Reagan. There are hundreds, maybe thousands of other bombed-out troops, Marines, soldiers, and even sailors from other ships on board. I guess the Navy's had a rough couple of days too. We went below and found some showers. I think all seven of us just stood there for at least an hour. I have no concept of time anymore. The showers weren't just for winding down. It was the best way to decontaminate ourselves. There were others there too. I don't remember how many. I rarely opened my eyes. We all just stood there and scrubbed in slow motion.

Bob stole us some uniforms from a berthing area. Now we're all dressed as sailors. I threw the old stuff overboard. I figure if someone tries to give any of us any trouble, I just give them a choice: they can either get us some Marine uniforms, or we can walk around naked. It's pretty warm, and none of us really care. So far no one's given us any trouble.

The ship is a combination of strange and wonderful. I sort of miss the soft dirt, the clean rain, and the jungle's symphony; with Chinese mortars for bass, of course! The showers seemed to rinse away the entire jungle experience; a little bit of memory with every drop that went down the drain.

Then there was the head; the toilets. We all feel weird using them now after going in sandbags for a month. It felt unnatural at first, but I think we'll adapt. Rick just sits there flushing and flushing in amazement. It's like he's never seen one before.

Finally, the food. The squids all complain, but even after the decontamination and the new uniforms, you can still tell who's been in the bush for a while. All of us grunts chow down in silent ecstasy on what the sailors call paper pizza. Half the mess is quiet, the other half is ripe with complaint. I think it's safe to say that we weren't the only ones who went without food for long time. There's so many grunts feasting that they setup an Army mobile kitchen unit in the forward hangar bay.

We couldn't find a place to bunk, so we're just sleeping in the hangar for now. No blankets, no sleeping bags, no hammocks, no cots, no bed, no pajamas, just a hard steel deck in a loud echo filled room, and NO Chinese artillery.

Tuesday, December 29, 2011

1st Lt. John Chamberlin-Personal Log

U.S.S. Enterprise

Singapore

We pulled into Singapore late in the day. Admiral Pender announced over the ships 1MC that there would most likely not be any more hostilities with the Eastern Alliance. He says that the Chinese not only carved a nuclear no-man's-land all the way across Indochina, but they also threw a couple at our Task Force, and even the continental U.S.! Some of our cities took some hits, so our brilliant President actually stepped up to bat and hit the ball out of the park. Pender says most of China's a parking lot now. The rest of the Eastern Alliance is sending delegates to sign some sort of armistice. I guess the war's as over as it's going to be. It sounds like we won, but I don't know what we got out of it. Most of my people are gone, and it's readily apparent that we lost a lot of others too.

Then there's the billion or so Chinese that are dead or dying to think about. I can't say I feel too bad about them. I kind of hate to see so many people taken out over a fight that might not have been every Chinaman's, but then again, what were we supposed to do, sit around and let them pound us into the dirt? I don't think so.

Back in the States, it sounds like we came out of everything better than expected; depending on where you lived. I guess Philadelphia and San Francisco are pretty much gone. New York is being evacuated because of some sort of contaminated water. Los Angeles, Seattle, and San Diego took near misses. There were some more terrorist attacks, but we didn't get a lot of news about those. Generally speaking, we're told life is pretty normal for the rest of the country except that the weatherman on the evening news has to give warnings about clouds loaded with black radioactive soot and fallout along with his usual thunderstorms and cold fronts.

I thought about sending some emails to the families of the people I lost, but it'll have to wait. I'm not exactly sure what I would say, and my form letter was lost when my old palm PC died back in the jungle. I guess all those microprocessors and stuff on all our new toys just weren't designed to hold up against that kind of moisture, fungus, mud, and abuse.

Most everyone is walking around the ship in a daze. We're all completely worn out. Even the crew doesn't seem to be moving with a great deal of purpose right now. Sometimes I wonder if anyone's even driving.

During one of those moments, I wandered towards the bridge-just to snoop around and fight the boredom. While I was coming up into the conning tower, I literally bumped into Admiral Pender himself on his way in from the flight deck. I don't know why, but I introduced myself and let him know that I was the one who was in charge of the TRAP mission back at LZ Delta in Korea. It just sort of popped out. I guess deep down inside, I must have still had some issues about my responsibility for that whole thing, maybe even for this who war.

Pender's always been known for being a hardass, but when I told him, he reached out to shake my hand, and when I reached for his, he wrapped his other arm around me like I was a long lost son. Then he told me to follow him to the flag bridge so we could talk.

When we were up on vulture's row, Admiral Pender sat down in his barber chair behind the thick plexiglass overlooking the flight deck, and we began talking about Korea. He said he felt terrible about putting us into that awful situation, and that it took him a long time to come to grips with it. He said that after the Battle of Yellow Sea, when his lifeboat was finally picked up, he gave up on the entire responsibility thing.

Admiral Pender had been called to Washington to testify in front of a Senate subcommittee that was investigating Delta and the Battle of the Yellow Sea. One of the Senators tried to put the blame on him, but he remembered that the same exact Senator was the one who had pushed the Secretary of the Navy into saving money by combining Carrier Battlegroups and Marine Amphibious Units. If it hadn't been for that one Senator, he would have had enough escort ships to win the battle. That's when he snapped. I didn't see it, but ole Pender must have really let into her, because a few weeks later, he was sent back out here to command one of our Task Forces.

I told him about my experience throughout the war, and we drew some comparisons, but nothing specific. Basically, we both lost people, and we both saw a lot of people die for virtually no reason. We talked about it for at least an hour or two. There's not much else to do right now.

I told him that I once believed in fate as though history had already been written and there was nothing we could do but go along for the ride. He said he used to think that way too, but now he thinks that one event just naturally leads to another. When you squeeze a trigger, a pin hits a bullet's primer, and lead comes out the barrel. Of course, we all have the choice as to whether or not to pull that trigger. Events will happen, and we can change them, but if we don't then history seems to ricochet off anything and everything. I think we both accepted the fact that we were just caught in some sort of historical storm that seems to come around every generation or two. When it does, there's nothing anyone can do except hope for the best, do their job, and ride the whirlwind.

Pender says Subic Bay and Clark Field both got taken out by some of the first Chinese nukes. Japan is being paranoid about more nukes, so none of the Coalition forces are allowed to use their ports even though the war's over. So, we'll have to stop in Singapore for a few days to pick up more supplies.

None of the grunts are allowed off the boat, but the some of the squids are getting shore leave for the first time in months. I guess they deserve it. Besides, I'd rather steal someone's rack and try sleeping on a mattress again. Right now, that's pretty much all I care about in the world.

Sunday, January 1, 2006, 8:21am

Isle of Grenada, T.K.'s Tequila Kafe'

Walt and Alexov sat at their usual table with their usual beverages. They had arrived on the island just before noon the previous day and their drinking began almost immediately. When they met they were glad to see each other. They had planted the seeds for a nuclear war together just to escape the grasp of the Chinese, and now it was comforting to them to know that their desperate act of survival had worked. As the night progressed, and the beverages continued to flow, they became more melancholy as they reflected on their actions. Neither man had the mettle to confide even the slightest guilt to the other. They were, after all, co-conspirators; not confidants.

As the previous year ended, nearly a billion people had died because of their dealings; their international meddling. Old acquaintances like Walt's assistant and Alexov's son, had not been forgotten. Eventually, the sun's early morning glow heated the New Year, and the two stoic men stared at their drinks.

The sun had risen slowly. It was hidden by ocean mist and a high-level microscopic fallout that was still suspended in the stratosphere. The sun had crawled from the sea's horizon nearly two hours ago, but the brilliant orange and pink clouds were still getting brighter instead of fading. Alexov knew the after effects of a nuclear war better than most, so while Walt reflected on the sun's beauty, it only reminded Alexov of the nuclear horror that had masked their disappearance. The nuclear dust would enhance the sun's rise and set for years.

Just before dawn began, both men had switched to coffee. After watching the sunrise for a while, Alexov stood, walked over to the bar, and helped himself to another bottle of vodka. Walt believed that the Russian was feeling at least some of the same guilt that he had.

While Alexov stood at the bar pouring a healthy glass of spirit, he looked toward Walt. The two men stared into each other's eyes. Alexov saw some of the guilt, but it was buried under Walt's poker face and an overwhelming selfish satisfaction; the satisfaction of still being alive.

Walt saw nothing in Alexov's face. He was a soldier at heart, and emotion didn't bode well for such people. Walt knew that there was more inside the man. He had lost a son and avenged his death nearly a billion-fold. Maybe that was what he was hiding?

Behind Walt, coming up the street from the seaside, Alexov saw someone walking alone. As the man came closer, he recognized the familiar features of his face. It was his old college roommate. Not a particularly fearful man, Alexov calmly drank his vodka, looked past Walt, and waited to see what was about to happen. All that he knew for certain was that neither he nor Walt was armed, and his old roommate, Chinese General Naht Synn, had something palm-sized and heavy inside the right front pocket of his sport coat.